PRICING DERIVATIVE SECURITIES

SECOND EDITION

T W E P P S
University of Virginia, USA

PRICING DERIVATIVE SECURITIES

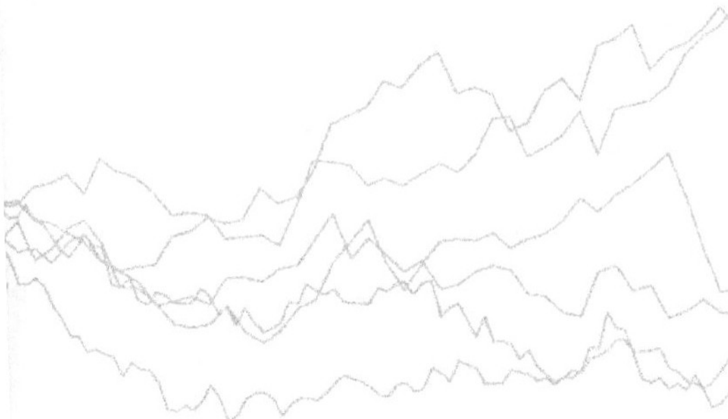

SECOND EDITION

World Scientific

NEW JERSEY · LONDON · SINGAPORE · BEIJING · SHANGHAI · HONG KONG · TAIPEI · CHENNAI

Published by

World Scientific Publishing Co. Pte. Ltd.
5 Toh Tuck Link, Singapore 596224
USA office: 27 Warren Street, Suite 401-402, Hackensack, NJ 07601
UK office: 57 Shelton Street, Covent Garden, London WC2H 9HE

Library of Congress Cataloging-in-Publication Data
Epps, T. W.
Pricing derivative securities / by Thomas W. Epps. --2nd ed.
p. cm.
ISBN-13 978-981-270-033-9 -- ISBN-10 981-270-033-1
ISBN-13 978-981-283-397-6 (pbk) -- ISBN-10 981-283-397-8 (pbk)
1. Derivative securities--Prices--Mathematical models. I. Title.
HG6024.A3E66 2007
332.64'57--dc22
2007006660

British Library Cataloguing-in-Publication Data
A catalogue record for this book is available from the British Library.

Typeset by Stallion Press
Email: enquiries@stallionpress.com

Contents

PART I

PRELIMINARIES

Preface

This book was developed during many years of teaching derivatives to doctoral students in financial economics. The aim in writing it was to help fill the gap between books that offer a theoretical treatment without much application and those that simply present the pricing formulas without deriving them. The project was guided by the beliefs that understanding is not complete without practice at application and that applying results one doesn't understand is risky and unsatisfying. This book presents the theory but directs it toward the goals of producing practical pricing formulas for derivative assets and of implementing them empirically.

Teaching the theory of derivatives to Ph.D. students in economics or to doctoral finance students in business requires teaching some math along the way. I have tried to make the book self-contained in this regard by presenting the required basics of analysis, general probability theory, and stochastic processes. This development starts with chapter 2, which presents (and is devoted entirely to) the general mathematical background needed throughout the book. It continues in chapter 3 with the specific tools required for continuous-time finance, including the special theory needed to handle processes with discontinuous sample paths. Since teaching math is not the book's main purpose, the treatment is necessarily brief and emphasizes examples rather than proofs. Because appendices (like prefaces) tend to be ignored by students, I have chosen to present this background in the main text. Paging through chapters 2 and 3 will serve as a refresher to those who have had the required math. For those who have not, the material is there for more careful thought, as a guide to the mathematical literature, and as a reference to be consulted in the later chapters. The math-prep chapters and the general overview in chapter 1 comprise Part I: Preliminaries.

The book has two distinctive features besides the math lessons. First, the chapters in Part II: Pricing Theory, are organized around the assumptions made about the dynamics of underlying assets on which the derivatives depend. This allows us to progress from the relatively simple models that require little mathematical sophistication to the more complex ones that require a great deal. Thus, chapter 4 begins with pricing bounds and relations such as European put-call parity that apply very generally and are derived from simple static-replication arguments. Chapter 5 progresses to Bernoulli dynamics and the associated binomial approach to pricing in discrete time. Modeling the evolution of prices in continuous time begins in chapter 6 with the basic Black-Scholes theory for pricing European-style derivatives under geometric Brownian motion. Chapter 7 applies the same dynamic framework to American options and some of the many "exotic" varieties of contingent claims. Chapters 8 and 9 deal with more elaborate models based on diffusions with *ex ante* uncertain volatility and with discontinuous processes, respectively. Much of this material comes from very recent research literature. Finally, chapter 10 brings in stochastic models for interest rates and introduces readers to the literature on pricing interest-sensitive instruments.

The book's second distinctive feature is its detailed treatment of empirical and numerical methods for implementing the pricing procedures. This material is presented in Part III: Computational Methods, which comprises chapters on simulation, on the numerical solution of partial differential equations, and on computation. The last chapter is a summary list of FORTRAN, C++, and VBA programs on the accompanying CDROM that implement many of the basic pricing procedures discussed in the text: binomial pricing, Black-Scholes, the constant-elasticity-of-variance model, options on assets following mixed jump-diffusions, and others. There are also basic routines for generating pseudorandom numbers, for testing for randomness and adherence to specific distributional forms, for numerical integration and Fourier inversion, and for solving p.d.e.s numerically. The FORTRAN and C++ programs are presented in source form in order to guide readers in producing their own applications. Code for many of the Visual Basic for Applications routines can be viewed as macros within the spreadsheets that facilitate the input and output.

I have tried to supply enough detail in the derivations of pricing formulas to enable readers to follow the development without having to puzzle over each line. The amount of such detail declines as we move along, in the expectation that readers are acquiring a mastery of technique as they progress.

The second edition is a complete revision of the first and adds several topics from recent research in the field. The chapter on fixed-income derivatives has been significantly expanded to include sections on the LIBOR market model and default risk. The chapter on discontinuous processes now includes two new models, one that allows jumps in volatility as well as price and another that accommodates stochastic variation in the mean frequency of price breaks. A new section on regime-change models opens up a number of flexible and computationally attractive strategies for parsimonious modeling of noisy processes. The chapter on stochastic volatility models presents a new and very fast technique for pricing European-style derivatives off the underlying characteristic function. Finally, the chapter on simulation now includes a detailed treatment of its application to American-style derivatives. Both to save space and to facilitate their use, the computer routines included in the first edition have been placed on an accompanying CDROM. An outline and general description still appears as chapter 13.

Besides graduate students in economics and doctoral students in business, others who should find this book useful include masters students or upper-level undergraduates in applied math courses, students in masters-level computational finance and financial engineering programs, those with prior math background who are being trained as practitioners, and non-specialist researchers who are trying to acquire some familiarity with the field of derivatives. Whether this book does help to fill the gap between the too-theoretical and the too-applied, I leave to the market to decide!

I am grateful to the many students who have commented on and corrected the course notes from which this book evolved; to Hua Fang and, especially, to Todd Williams, who proofed and gave detailed comments on several chapters in the first edition; and to Michael Nahas, who painstakingly translated my FORTRAN code into C++. The second edition owes much to Ubbo Wiersema, who suggested several additions, provided detailed comments on one of the chapters, and assisted in developing the VBA programs. For his capable work in producing these, I thank Ben Koulibali. Finally, I am indebted to my wife Mary Lee for her bounteous support, advice, and editorial assistance.

Queries about the book or reports of errors in the text or in the computer codes should be directed to twe@virginia.edu. Please include the initials "PDS" in the subject field. A current list of errata will be maintained at http://www.worldscibooks.com/economics/6243.html.

PART I

PRELIMINARIES

1

Introduction and Overview

Derivative securities are financial instruments whose values are tied contractually to values of other assets, called underlying assets, at one or more points in time. One of the simplest examples is a forward contract. This is an agreement between two parties to exchange some specified quantity of a commodity or asset for a certain sum of money at a stated future time and location. For example, a forward buyer may agree to accept delivery of $1,000,000 face value of U.S. Treasury bills of six months' maturity from the forward seller on a specific date and to pay $968,000—the forward price—at that time. Delivery would likely be to an account with a bank or a broker. Once the bills are delivered and paid for, the agreement expires. The forward contract is thus a derivative on the underlying Treasury bills. To the buyer, the per-unit value of the contract at expiration is the difference between the underlying asset's spot-market price and the forward price of $968,000; to the seller, it is the negative of this. The spot-market price is the price at which one could buy 6-month Treasury bills on the open market for immediate delivery.

Derivatives are distinguished from primary assets, such as stocks, bonds, currencies, and gold. Values of these are not tied contractually to other prices, although they and other prices are surely interdependent as a consequence of the connections between markets in general economic equilibrium. We make no distinction between primary financial assets, such as stocks, bonds, and currencies, and primary "real" assets or commodities, such as gold. It is sometimes necessary to distinguish commodities that are held primarily for investment purposes—gold being a prime example—from other commodities like wheat and petroleum that are often held in inventory for use in production. Forward contracts and certain other derivatives on such underlying assets commonly come into being as firms and individuals do business and attempt to limit risk.

Forward contracts, and derivatives generally, fall into the broader class of contingent claims—contracts that mandate payments that are contingent on uncertain events. As regards the forward contract, it is the commodity's future spot price that is uncertain. Casualty and life insurance contracts, in which damage and death are the contingencies, are members of this broader class that are not normally considered derivative assets. However, the distinction begins to blur as the field of financial engineering becomes more creative. For example, one can now buy and sell bets on outcomes of temperature and snowfall in specific localities, and these are often referred to loosely as "derivative" products. Our use of the term *derivative* and our concern in this book will, however, be limited to those financial instruments described in the first sentence of this chapter.

It is precisely because derivatives' values are tied contractually to other assets that "pricing" them is a distinct subfield within economic science. While economics helps us understand the determinants of prices of goods, services, and primary assets—for example, Coca Cola and the price of Coke stock—economists might disagree profoundly were they asked to specify the prices at which these commodities "should" sell. Indeed, since no good answer is apt to be forthcoming, the question is not often asked. On the other hand, financial economists who agreed about the dynamic behavior of the price of an underlying asset would likely supply very similar answers when asked to value, or "price", a derivative. Before we can begin to see why this is so, we will need to know more about the kinds of derivatives that are available, how they are used, and how they are bought and sold.

1.1 A Tour of Derivatives and Markets

1.1.1 *Forward Contracts*

Forward agreements to deliver and accept delivery of commodities or financial instruments at a future date are alternatives to simply waiting and trading cash for commodity at the spot price that prevails at that date. Such agreements are made because at least one party finds the risk of an adverse price fluctuation undesirable. For example, the parties to the exchange of Treasury bills discussed in the introduction might be a bank and a dealer in government securities. The bank's purpose in contracting to accept forward delivery—that is, to take a "long" forward position—would be to lock in the prices of these instruments until the funds to purchase them became available. The dealer's purpose would be to earn a fee, in the form of a markup, for taking the opposite or "short" side of the transaction. Similarly, a U.S. importing firm that had contracted for a shipment of goods from a Japanese supplier might enter into a forward agreement with a bank for the purchase of Japanese yen, thus protecting against an increase in the $/¥ exchange rate. In the same way, a producer of cotton yarn might contract to deliver a shipment of such goods to a textile manufacturer at a specified price, allowing both firms to avoid the risk of adverse price fluctuation. As these examples illustrate, forward agreements are typically made between financial institutions, between productive enterprises and financial institutions, and between enterprises that produce commodities and those that use them as inputs. Because the delivery terms—date, place, and quantity—are tailored to suit the demands of the originating parties, positions in forward agreements are rarely resold to others. In particular, they are not traded on any organized exchange.

Given the delivery terms, the forward price is set so that neither party requires compensation for entering the agreement. That is, the forward price is such that the initial value of the contract is zero to both parties.[1] However, once the agreement is concluded and the spot price of the underlying commodity (or other relevant factors) begins to change, the agreement starts to acquire positive or negative value. Letting T be the time to delivery for a contract initiated at time $t = 0$, we will represent the forward price itself as $\mathsf{f}(0,T)$ (or simply as f if the times are understood) and the time-t value of the long position in the contract as $\mathfrak{F}(t,T;\mathsf{f})$, $t \in [0,T]$. Thus, $\mathfrak{F}(0,T;\mathsf{f}) = 0$ and $\mathfrak{F}(T,T;\mathsf{f}) = S_T - \mathsf{f}$, where S_T is the spot price at T. In chapter 4 we will see how to price a forward contract at any $t \in [0,T]$ and to find the forward price, $\mathsf{f}(0,T)$, that satisfies $\mathfrak{F}(0,T;\mathsf{f}) = 0$. We will also see that the special feature that makes these tasks very easy is that the contract's terminal value is linear in the spot price.

[1] But why would they enter the agreement if it had no value to them? The statement in the text, like many others to follow in this book, pertains to conditions that would apply in markets free of impediments to borrowing and transacting. We shall see in chapter 4 that in such frictionless markets either party could replicate the agreement by taking appropriate positions in the underlying commodity and riskless bonds. It is the common value of this replicating portfolio that would be zero at the agreed-upon forward price.

1.1.2 *Futures*

Futures contracts involve the same general commitments that forward contracts entail; namely, to exchange a commodity for cash at a future date. Unlike forward agreements, which are made through direct negotiation between private parties, futures commitments are bought and sold in active markets conducted on organized exchanges. Examples of these are the Chicago Board of Trade (CBOT), the Chicago Mercantile Exchange, the London International Financial Futures Exchange, the New York Futures Exchange, the Tokyo International Financial Futures Exchange, and the Toronto Futures Exchange. Most of the distinguishing features of futures contracts can be traced to the need to assure that markets for them will be active and liquid, and therefore cheap to trade in.

One such feature that distinguishes futures from forwards is that the terms of futures contracts are standardized as to quantity and quality of the commodity, as to delivery dates, and as to delivery locations.[2] For example, wheat contracts on the CBOT call for delivery of 5,000 bushels of wheat of specified types and grades to approved warehouses in Chicago (and certain other specific locations) during a particular delivery month. Besides wheat contracts, one can buy and sell futures commitments for many other agricultural products, for certain industrial commodities, and for various financial instruments. Among the last are U.S. Treasury bonds and bills, certain stock indexes, and major currencies. Trading of futures is typically done through a broker, who transmits clients' orders to the exchange. The transactions arising from the continual flow of buy and sell orders determine the futures price, which in turn determines the net cash outlay that the buyer must provide at delivery. This is explained further below.

Besides standardizing the contracts, an important step to promote market liquidity is assuring buyers and sellers that counterparties will comply with their obligations to deliver or take delivery. To this end, the exchanges operate clearing houses that serve as intermediaries to the parties and back the compliance ("performance") of buyers and sellers with their own capital and that of member brokers. Thus, one who buys (takes a long position in) wheat futures on the CBOT deals not directly with the seller of the

[2]As do some forward agreements, futures contracts typically allow the delivering party some flexibility in certain of these terms; for example, some contracts allow for delivery on any business day of the expiration month.

contract but with the CBOT's clearinghouse. Since the house simultaneously commits to deliver to the buyer and to take delivery from the seller, it has no net exposure to price risk; but it *is* exposed to risk of default by parties to the trade. It is to limit this risk that exchanges follow a practice known as marking to market, which from our perspective constitutes the major difference between futures contracts and forward contracts.

Here is how marking to market works. An individual who places an order to buy or sell futures must post a surety deposit, called the initial margin, with the broker who handles the trade. This deposit usually amounts to just a small proportion of the total value of the commodity for which delivery is contracted, because, like forward contracts when they are first arranged, futures positions at first have no net value to either party. This changes, however, as the futures price fluctuates through time and one of the parties acquires a liability—the obligation to sell at a price below the spot or to buy at a price above. To help keep unsecured losses from getting out of hand, futures contracts are, in effect, renegotiated at the end of each trading day to provide for delivery at the current futures price. If this is lower than the original price, then the long position acquires negative value equal to the product of quantity and price difference, while the short position acquires a positive value of equal magnitude. These losses and gains are then deducted from and added to the respective traders' accounts, which are thereby "marked" to market.

For example, suppose one buys a contract for 5,000 bushels of July wheat at futures price \$3.00, posting initial margin of \$500. At the end of the first day, with the futures price settling at \$2.97, the loss of (\$3.00-\$2.97) $\times$ 5,000 $=$ \$150 is deducted from the margin balance, the original commitment to buy at \$3.00 is terminated, and a new contract is opened for delivery at the futures price \$2.97. When losses reduce the balance below some threshold, another deposit is required. If this deposit is not forthcoming, the position is closed out by the broker and the balance, if any, returned to the client. A large intraday price fluctuation that exhausted the margin balance would have to be made up by the broker were the client unable to do so.[3]

The net effect of marking to market can be seen as follows. Let $\{S_t\}_{t\geq 0}$ be the spot-price process and $\{\mathsf{F}(t,T)\}_{t\in[0,T]}$ be the futures price for a

[3]To reduce the chance of such an occurrence, exchanges limit daily price fluctuations by halting trading for the day when prices make "limit moves" from the previous day's settlement value.

contract expiring at T. Of course, at $t = T$ the futures price and spot price coincide, so that $\mathsf{F}(T,T) = S_T$. Taking the trading day as the unit of time, a long position in a contract initiated at price $\mathsf{F}(0,T)$ is worth $\mathsf{F}(1,T) - \mathsf{F}(0,T)$ per commodity unit at the end of the first trading day, $\mathsf{F}(2,T) - \mathsf{F}(0,T)$ per unit at the end of the second day, and so on. On the delivery day itself the value of the position will be $S_T - \mathsf{F}(0,T)$, just as for a forward contract initiated at $\mathsf{F}(0,T)$. Upon paying $\mathsf{F}(T,T) = S_T$ at delivery, the net cost to the long party is $S_T - [S_T - \mathsf{F}(0,T)] = \mathsf{F}(0,T)$ per unit, which was the futures price when the contract was first created. One can see, then, that the effect of marking to market is merely to distribute the net gain or loss over time, rather than allowing it all to come due at T. We consider in chapters 4 and 10 under what conditions and how this difference in timing affects the relative values of futures and forward positions and the relation between forward and futures prices.

The example just given pertained to futures positions that were held open to delivery, but in fact only a small proportion of positions are held this long. Instead, parties usually cancel their commitments to the clearing house by taking offsetting positions before expiration; that is, going long to cancel a short, and *vice versa.* For those using futures to hedge the risk of adverse price fluctuations, this eliminates the requirement to meet the exact quantity, quality, location, and time requirements specified in the contract. The futures position still provides a partial hedge for spot purchases and sales in local markets, to the extent that local spot prices in the commodity are correlated with those on which futures contracts are written.[4]

One should bear in mind that futures prices, like forward prices, are not themselves prices of traded assets. Instead, it is the futures position that is the asset, the futures price being merely the quantifiable variable that fluctuates so as to keep the values of new futures positions equal to zero.

1.1.3 *"Vanilla" Options*

Put and call options give those who hold them the rights to transact in underlying assets at specified prices during specified periods of time. There are many variations on this theme, but we consider in this section just the most basic instruments, often called "vanilla" options. One who owns (is long in) a vanilla put has the right to sell the underlying asset at a fixed

---

[4]Kolb and Overdahl (2006) describes in detail the operation of futures markets and the use of futures in hedging.

price, called the "strike price" or "exercise price", during or at the end of a specified period. If this option is exercised, the counterparty who is short the put has the obligation to take the other side of the trade; that is, to buy the asset at the strike price. Calls work in just the opposite way: one who is long a call has the option to buy at the strike price during or at the end of a specified period, while the counterparty must sell at this price if the holder chooses to exercise. Options not exercised by the end of the specified period are said to expire. Even vanilla puts and calls come in two flavors, having different conventions as to when they can be exercised. American options can be exercised at the discretion of the holder at any time on or before the expiration date, whereas European options can be exercised only at that specific time. The names have nothing to do with where these options are traded today.

The seller of an option is often referred to as the "writer", a term that originated before the advent of exchange trading. In those days all option contracts were negotiated between buyers and sellers or purchased from dealers, known as put and call brokers. Since opportunities for resale were limited, options were almost always held until they either expired or were exercised. Exchange trading of options was begun in 1973 by the Chicago Board Options Exchange (CBOE). Currently, exchanges in major financial centers around the world list European- and American-style options on a variety of underlying assets, including common stocks, stock indexes, and currencies. In the U.S. exchange-traded stock options are almost always American,[5] but there are European-style options on various stock indexes, such as the Dow Jones Industrials and the S&P 500. Index options, like index futures contracts, are settled in cash rather than through delivery of the underlying.[6] Exchange-traded options on spot currencies are also available, in both American and European style. Besides these spot options there are also exchange-traded futures options. A futures option gives the right to assume a position in a futures contract that expires at some later date than the option itself. One who exercises a futures put gets a short position in a futures contract plus cash equal to the difference between the strike price and the futures price. Exercising a futures call conveys a long position in the futures plus the difference between the futures price and

[5] "Flex" options available through the CBOE can be obtained in European form.

[6] One who exercises a call on an index receives from the seller in cash an amount proportional to the difference between the index value and the strike price. For puts, the exercise value is proportional to the difference between the strike price and the index.

the strike.[7] There are exchange-traded futures options on many underlying assets, including stock indexes, currencies, short- and long-term bonds, and various agricultural commodities.

As for futures contracts, the key features of most exchange-traded options—time to expiration, strike price, and quantity—are standardized in order to promote a liquid market. For example, stock options are typically for 100 shares and expire in six months or less.[8] Buying or selling an exchange-traded option (of any sort) works much like buying or selling an exchange-listed stock: an order placed through a broker is transmitted to the exchange, where it is either matched with another individual's order or accepted by the market maker.

Sales of options either reduce or close out existing long positions arising from previous purchases, or else they create short positions. Short positions may be covered or uncovered. A covered short is one that is backed by a position in the underlying (or in another derivative) that automatically enables one to meet the obligation imposed by the option holder's decision to exercise. For example, a short position in calls would be covered if backed by a long position in the underlying, which could be delivered if the call were exercised. By contrast, to meet the terms of an uncovered or "naked" short position in calls one must have the resources to purchase the underlying at the prevailing spot price. To reduce the chance of nonperformance, brokers demand deposits of cash or other securities as surety for naked short positions, just as they do for short positions in stocks. However, whether they are covered or not, short sales of options are not bound by the tick rules that restrict short sales of stocks on certain exchanges.[9] As with futures, short positions in options can be eliminated simply by buying options of the same type. As for futures also, option exchanges act as intermediaries to all transactions in order to reduce counterparty risk to traders.

---

[7] Note that, unlike the payoffs of spot options, payoffs of futures options are linked to futures prices and therefore not to prices of traded assets. We will see that this difference has important implications for pricing futures options.

[8] The CBOE's flex options, designed to appeal to institutional investors, do offer flexible features, but they must be traded in large quantities. There are also long-term options, called "LEAPS" with lives of up to 39 months.

[9] Securities and Exchange Commission rules in the U.S. prohibit the short sale of a stock at a price below that of the last preceding trade. A short sale at that price is allowed only if that price is itself above the last preceding different price.

Introducing the general notation used throughout for prices and contract features of standard options, we let T represent the expiration date; X, the strike price; and S_t, the value at t of the market-determined price to which the option's payoff is tied at expiration—that is, the underlying price. We use more specific notation for the underlying price in certain cases; for example, F_t for the underlying price of a futures option. Time-t prices of generic calls and puts are represented as $C(S_t, T-t)$ and $P(S_t, T-t)$, where $T-t$ is the remaining time until expiration. Other arguments are added when it is necessary to specify contract features or other parameters affecting value; for example, $C(S_t, T-t; X_1)$, $C(S_t, T-t; X_2)$ for calls with different strikes. Superscripts are used to identify specific types of options, as C^A, C^E for the American and European varieties. Taking $t = 0$ as the initiation date of an option contract, the initial values of puts and calls are thus $P(S_0, T)$ and $C(S_0, T)$, and terminal values—values at expiration if not previously exercised—are $P(S_T, 0)$ and $C(S_T, 0)$. As for futures contracts, our convention is that these represent values to one who has a long position in the option.

Terminal values of vanilla options can be stated explicitly in terms of the underlying price at T and the strike alone. Since exercising an option is not obligatory, a call (which gives the right to buy at X) would not (rationally) be exercised at expiration unless $S_T > X$. Otherwise, the option conveys nothing of value, and is said to be "out of the money". When $S_T > X$ an expiring call is "in the money" and, abstracting from transaction costs, has net value to the holder equal to $S_T - X$. Likewise, a put would not be exercised at expiration unless $S_T < X$ and would then have value $X - S_T$. Terminal values of puts and calls can therefore be expressed as

$$P(S_T, 0) = \max(X - S_T, 0) \equiv (X - S_T) \vee 0 \equiv (X - S_T)^+$$
$$C(S_T, 0) = \max(S_T - X, 0) \equiv (S_T - X) \vee 0 \equiv (S_T - X)^+.$$

Notice that when the underlying price is bounded below by zero and unbounded above (the usual case), then the terminal value of the put is bounded above by the strike price, whereas the call's value has no upper bound. One who is short a put can therefore lose no more than X, whereas one who is short an uncovered call has unlimited potential loss.

Figure 1.1 represents these elementary payoff functions at expiration. Unlike payoffs of forwards, they are clearly not linear functions of S_T. We will see in chapter 4 that this fact has far-reaching implications for pricing the options at dates before expiration; that is, for finding $P(S_t, T-t)$ and

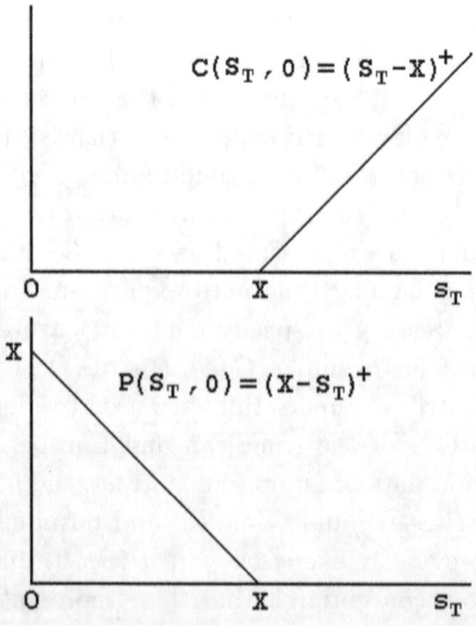

Fig. 1.1. Payoff functions of call and put options.

$C(S_t, T-t)$ for $t < T$. Pricing American options faces the additional complication that exercise can occur at any time before T. We begin in chapter 5 to develop specific techniques for pricing both American and European options, extending and developing these procedures in later chapters.

1.1.4 *Other Derivative Products*

Besides the many exchange-traded derivatives there are available over the counter—that is, through negotiated transactions with dealers and other private parties—an immense and rapidly growing array of structured derivative products. Serving institutional investors, firms involved in manufacturing, finance, and commerce, public enterprises, and governmental units, these products make it possible to hedge risks of price movements that would adversely affect cash flow and/or the values of assets. Of course, as several well-publicized incidents have demonstrated, derivatives also support purely speculative activity that can have disastrous financial consequences if there is insufficient internal oversight.

One form of specialized derivatives sold over the counter are the so-called "exotic" options, which can be grouped according to how they modify the features of vanilla puts and calls. Here is a sampling. We treat most of these in chapter 7.

1. Variations on the terminal payoff function. There are "digital" calls that pay a fixed sum when the underlying price at expiration is above the strike, and "threshold" calls that pay $S_T - X$, but only when $S_T > K > X$ for some threshold K. There are various path-dependent options whose payoffs depend on the entire path of price up to expiration; for examples, (i) "down-and-out" puts that pay $X - S_T$ provided S_t remains above some level $K < X$ for $0 \leq t \leq T$; (ii) "lookback" options whose payoffs depend on the extrema of price over $[0, T]$; and (iii) "Asian" options whose payoffs depend on the average price.
2. Variations on the underlying. There are options on other options. There are "basket" options with payoffs depending on the value of some portfolio; options on the minimum or maximum of two or more underlying prices; and "quanto" derivatives, whose payoffs in domestic currency are in fixed proportion to prices of assets denominated in a foreign currency, irrespective of changes in exchange rates.
3. Variations on what right the option confers. There are "chooser" options that let one choose at some future date whether the option is to be a put or a call.
4. Variations on the option's term. There are "forward-start" options that come to life at some future date, t^*, as at-the-money puts or calls (that is, with $X = S_{t^*}$) that expire at some $T > t^*$. There are "extendable" options that must be exercised if in the money at one or more discrete dates but are otherwise extended one or more times. There are "Bermudan" options that can be exercised at certain discrete times but do not have to be.

There is also a huge over-the-counter market in interest-rate derivatives, whose values depend primarily on prices of fixed-income assets. These include: (i) options on bonds; (ii) interest-rate "swaps", which are exchanges of payments on floating-rate loans for payments on fixed-rate loans; (iii) currency swaps, which are exchanges of loans in one currency for loans in another; (iv) options on swaps, called "swaptions"; (v) "caps" and "floors", which provide upper and lower limits to variable-rate loans; and (vi) options on caps and floors. Chapter 10 serves as an introduction to the large literature on these instruments.

1.2 An Overview of Derivatives Pricing

The theory of derivatives pricing is founded on two observations about the prices we find in markets for economic goods:

- Two identical commodities, offered for sale at the same time and place, usually sell for the same price.
- One cannot ordinarily obtain for free something that people value.

The first observation, minus the qualification *usually*, is often referred to in economics as the law of one price. The second observation—again minus the qualification—corresponds to the trite expressions, "You can't get something for nothing" and "There's no such thing as a free lunch". Of course, we recognize that both qualifiers are needed, since we have all seen exceptions to the unqualified statements. For example, the first condition can fail when not everyone is aware of what goods are available in the marketplace or not well informed about their characteristics; and the second can fail when one who owns some commodity either has some charitable motive or is simply unaware that the commodity has value to others. Of course, both conditions can fail when some coercive authority controls prices by fiat or limits individuals' ability to transact. However, few would disagree that the unqualified versions of these statements characterize situations to which things tend over time in free, competitive markets. As individuals learn through experience or from others, their actions to obtain more of the things they value will cause prices of goods, services, and assets to attain levels consistent with their marginal social values.

The failure of either of these conditions to hold in a freely functioning and frictionless market for assets presents an opportunity for arbitrage. Specifically, there is an arbitrage opportunity if one can either:

- Obtain a sure, immediate return in cash (or other good) by trading assets, or
- Get for free a claim on cash (or other good) that has a positive chance of paying off in the future.

In either case something of value—cash, good, or valuable claim—can be obtained for sure at no cost. Thus, when transacting is costless a failure of the law of one price confers the opportunity for a sure, immediate cash return. To earn it, one simply sells the more expensive asset and buys the cheaper one. Likewise, selling one asset and buying another that costs the same but adds potential cash payments conveys a valuable claim for no net outlay.

1.2.1 *Replication: Static and Dynamic*

In its simplest form pricing some financial asset by arbitrage works by (i) finding a way to replicate its payoffs exactly by assembling a portfolio of other traded assets whose prices are known, then (ii) invoking the law of one price. Arbitrage pricing would thus allow a financial firm or its client to judge the value of a derivative asset that is not already traded but that the firm might be asked to sell. The method's usefulness relies on two key conditions:

1. That markets function well enough to eliminate significant opportunities for arbitrage, and
2. That the introduction of the new asset does not appreciably change the value of the replicating portfolio.

If these conditions hold, then the law of one price equates the value of the asset to that of the replicating portfolio. What distinguishes derivatives from primary assets is that it *is* sometimes possible to replicate their payoffs and price them by arbitrage. In chapter 4 we will state formally the sufficient conditions on markets, but—depending on the derivative—we shall see that other conditions may be needed on the dynamics of the underlying price as well.

The use of arbitrage arguments for pricing financial assets can be traced at least to John B. Williams' classic 1938 monograph, *The Theory of Investment Value*. Applying what he called the "principle of conservation of investment value", Williams argued that the true worth of a firm is invariant under changes in its capital structure—that is, changes in its mix of debt and equity. Modigliani and Miller (1958) developed a rigorous proof of this proposition for firms in a frictionless economy without taxes and free of transaction costs and the legal costs associated with contracting and bankruptcy. In Miller and Modigliani (1961) they argued on similar grounds that a firm's dividend policy is also irrelevant to its total market value. All of the arguments rely on the law of one price to show that the total value of claims on a stream of earnings should not depend on how the receipts are allocated between dividends and interest. For example, individuals could in effect create their own desired debt/equity mix by combining the firm's stock with borrowing and lending on their own account, thereby replicating any preferred allocation.

Modigliani-Miller's recipe for combining stock and bonds to attain a particular degree of financial leverage is an example of "static" replication. It is static in the sense that no further trades are required once the replicating portfolio is put in place. We shall see in chapter 4 that static replication is also possible for certain derivatives. For example, one can mimic a long position in a forward agreement by buying the underlying asset now, borrowing the present value of the forward payment, and repaying the loan at delivery. Although just one transaction is involved, the portfolio's market value nevertheless equals that of the forward position at any time until expiration. However, static replication is not usually possible for more complicated derivatives whose payoffs are not linear functions of the underlying price. It remained for Black and Scholes (1973) and Merton (1973) to show how nonlinear derivatives such as European options can be valued by "dynamic" replication. Dynamic replication works by forming a self-financing portfolio of traded assets and trading back and forth between them in such a way that the portfolio is sure to have the same value as the derivative when the derivative expires. The term *self-financing* means that purchases of one asset are always financed by sales of others, so that new funds need never be added nor withdrawn. For example, we shall see in chapters 5 and 6 that the terminal payoff of a European stock option can be replicated with a self-financing portfolio of stock and bonds, assuming that the prices of these assets behave in certain ways. Again, by invoking the law of one price we can conclude that the arbitrage-free value of the derivative in a frictionless market must be that of the replicating portfolio.

1.2.2 *Approaches to Valuation when Replication is Possible*

This is all well and good, but how exactly does one find a dynamic replicating portfolio? Paradoxically, once the existence of a self-financing, replicating portfolio can be verified, it happens that the value of the derivative can be worked out directly without first finding the portfolio weights. Indeed, the solution will tell us what those weights must be. There are two general ways to work out the derivative's arbitrage-free value.

Pricing by Solving P.D.E.s

The first approach, following the path laid out by Black and Scholes (1973) and Merton (1973), is to infer from the replicating argument that the price of the derivative asset must follow a certain fundamental equation of motion. This fundamental equation is either a partial difference equation or a partial differential equation, depending on whether the underlying price is modeled in discrete time or in continuous time. In either case it describes how the arbitrage-free price of the derivative changes with respect to time and the various state variables, such as the prices of assets that comprise the replicating portfolio. The solution gives price as a weighted sum of the prices of the primary assets and thereby indicates precisely what the replicating portfolio is. Moreover, the solution specifies the portfolio weights as functions of time and the state variables, thus prescribing how the replicating portfolio is to be adjusted as the state variables evolve.

Risk-Neutral/Martingale Pricing

The second way to find a derivative's arbitrage-free price and replicating portfolio relies on the no-free-lunch property of arbitrage-free markets. Adopting Arrow's (1964) and Debreu's (1959) characterization of assets as bundles of state-contingent receipts, we can view the market value of an asset in an arbitrage-free market as simply the sum of the values of the contingent receipts it offers. In other words, it is the sum of products of the cash amounts received in various states and times multiplied by the market prices associated with those time-and-state-dependent payoffs. Each time-state price is, of course, the value of a unit cash receipt. It reflects both the likelihood of the state and the subjective trade-off between having cash to spend now versus having a unit payoff at the particular time and in the particular set of circumstances that the state represents.

These state prices have two important properties. First, state prices are positive if and only if state probabilities are positive. This is because the right to a positive chance of receiving a receipt at some date T always commands a positive price in an arbitrage-free market, whereas there can be no value to a payoff that is tied to a contingency that has no chance of occurring. Second, in an arbitrage-free market the sum of all the state prices for contingent payoffs at T must equal the price of a default-free, T-maturing, discount "unit" bond that returns one unit of currency at T in every state—i.e., a *riskless* bond. If we divide the state prices by the price of such a riskless bond, the resulting normalized state prices then have all the properties of probabilities; that is, they are nonnegative and sum to unity. Indeed, the normalized prices represent a new probability measure over states that is "equivalent" to the real one, in the technical sense that both measures assign positive values to the same sets of contingencies.

This equivalent measure affords another way to characterize an asset's arbitrage-free price. Summing products of payoffs and state prices is equivalent to summing products of payoffs and pseudo-probabilities and then multiplying the result by the price of the riskless bond. Of course, multiplying by the price of the bond is the same as discounting at the average rate of interest that the bond pays over its lifetime. Therefore, the asset's price can be thought of as the mathematical expectation of its payoffs in the pseudo-probability measure, discounted to the present at the riskless rate of interest. Since assets would actually be valued at their discounted expected payoffs in the true measure if people were risk neutral, it makes sense to refer to this pseudo-measure as the "risk-neutral" measure.

Once one has found the equivalent risk-neutral measure, there is a simple recipe for pricing a derivative security whose value at expiration (date T) is a known function of an underlying price: (i) find the mathematical expectation of the derivative's value at T in the risk-neutral measure and (ii) discount it to the present at the riskless rate. Derivatives with payoffs at more than one date can be decomposed into the dated claims, which can be priced individually in this way and then added together. Thus, we can calculate an arbitrage-free price for any derivative if we can find a new probability measure in which other assets, including its underlying, are priced as though they were riskless. The one concern is whether the risk-neutral measure and the price obtained from it are unique. As we shall see, the ability to replicate the derivative's payoff with traded assets is precisely what is needed for uniqueness.

This risk-neutral approach to pricing followed from the insights of Cox and Ross (1976), who noted that the Black-Scholes (1973) and Merton (1973) formulas for values of European puts and calls can be interpreted as discounted expected values of their terminal payoffs—$(X - S_T)^+$ or $(S_T - X)^+$—in a different measure. Later, Harrison and Kreps (1979) showed that suitably normalized prices of assets behave as martingales in the equivalent risk-neutral measure; that is, they are stochastic processes whose current values equal the conditional expectations of their values at any future date. The usual normalizing factor or numeraire is the current value of a savings account that grows continuously at the riskless rate of interest. From this perspective risk-neutral pricing of a derivative asset becomes "equivalent-martingale" pricing, and the recipe becomes:

- Choose a numeraire—a strictly positive price process by which to normalize all other relevant prices;
- Find a measure in which normalized prices of assets (and derivatives) are martingales;
- Calculate the derivative's current normalized price as the conditional expectation of its future value; and then
- Multiply by the current value of the numeraire to obtain the derivative's arbitrage-free price in currency units.

We shall fill in the details of this procedure in later chapters and apply the recipe repeatedly.

1.2.3 *Markets: Complete and Otherwise*

In the abstract, at least, martingale pricing seems simple enough, assuming that one can find the mysterious martingale measure. But even if we take for granted that it can be found, an obvious question is whether the measure and the prices derived from it are unique. If not, then a derivative could have more than one price that was consistent with the absence of arbitrage opportunities—hardly a satisfactory situation. As it turns out, uniqueness depends ultimately on the dynamics followed by the underlying assets. When they are such that the payoff of any contingent claim can be replicated with a portfolio of traded assets (dynamically or otherwise), then the market for these claims is said to be "complete". As shown by Harrison and Pliska (1981) and more rigorously by Delbaen and Schachermayer (1994), there is only one equivalent-martingale measure in a complete market—and one set of prices that denies opportunities for arbitrage.

Our first applications of martingale pricing are to models for underlying prices that do allow replication. Chapter 5 treats discrete models, in which the underlying price evolves in discrete time and has discrete outcome states. The binomial model considered there both develops our intuitive understanding of martingale pricing and provides a practical numerical method for pricing many complicated derivatives. We take up in chapter 6 the continuous-time/continuous-state dynamics that Black and Scholes used to price European options. Chapter 7, where we price various exotic options under Black-Scholes dynamics, highlights the complementarity between their p.d.e. approach to valuation and the modern martingale approach. Other complete-markets situations are encountered in chapter 10, where we take up the modeling of stochastic interest rates and the pricing of interest-sensitive derivatives.

1.2.4 *Derivatives Pricing in Incomplete Markets*

The hard fact is that certain models that allow for more realistic behavior of the dynamics of underlying assets do not allow all contingent claims to be replicated with traded assets—even dynamically. In the incomplete markets that such models generate, infinitely many measures exist that could potentially price the claims. Of these measures the market actually uses just one, but we cannot discover which one it uses without more information. What this means in practice is that derivatives cannot be priced by arbitrage arguments alone when markets are incomplete. Instead, prices will depend on free parameters that are usually interpreted as reflecting peoples' tastes for risk bearing. While these parameters are in principle discoverable, they are not readily observable.

Since the risk parameters are in fact embedded in the observed market prices of traded derivatives, one way of discovering them is to see what values of the parameters make predictions of the pricing model accord with observation. That is, if we can somehow find the equivalent-martingale prices that correspond to any given set of parameters—either by solving the p.d.e.s or by calculating the conditional expectations—then we can grope around in the parameter space, repricing at each point until we land on one at which there is a good fit with the prices quoted in the market. Chapters 8 and 9 provide the first look at these incomplete-markets models, for which the relevant parameters for pricing must be filtered in this way from the observed prices of traded derivatives.

The program ahead of us clearly involves some math. In the next two chapters we review the principal concepts of analysis, probability, and stochastic processes that will be applied in financial modeling, and we learn to use the special tools of stochastic calculus for treating processes that evolve in continuous time.

2

Mathematical Preparation

This chapter summarizes the basic tools of analysis and probability theory that are needed in the applications to financial derivatives. It also introduces the continuous-time stochastic processes that will be the subject of the next chapter. The following summarizes notation that will be used throughout the book:

$\{a\}, \{a,b\}, \{a,b,\ldots\}$	sets of discrete points in a space
$[a,b], (a,b), [a,b), (a,b]$	closed, open, half-open intervals of reals
$\aleph$	the natural numbers, $\{1, 2, \ldots\}$
$\aleph_0$	$\aleph$ with 0 added, $\{0, 1, 2, \ldots\}$
$\Re$	the real numbers, $(-\infty, \infty)$
$\Re^+$	the nonnegative reals, $[0, \infty)$
$\Re_n$	n-dimensional space, $n \in \aleph$
$\mathcal{Q}$	the rational numbers
$\mathfrak{C}, \mathfrak{C}_n$	the complex numbers, n-vectors
$\subseteq, \supseteq; \subset, \supset$	subset, superset; proper subset, superset
$\cap, \cup, \backslash$	intersect, union, and difference of sets
$\emptyset$	empty set
$A^c = \Omega \backslash A$	complement of set A relative to space Ω
$\{A_n\}_{n=1}^{\infty} \uparrow, \{A_n\}_{n=1}^{\infty} \downarrow$	nondecreasing, nonincreasing sequences
$\mathbf{a}, \mathbf{b}$	column vectors or matrices
$\mathbf{a}'\mathbf{b}, \mathbf{a}\mathbf{b}'$	inner product, outer product
$\mathbf{1}_A(x)$	indicator function $= \begin{cases} 1, & x \in A \\ 0, & x \in A^c \end{cases}$
$f(x\pm)$	right/left-hand limits: $\lim_{n\to\infty} f(x \pm 1/n)$
$a \vee b, a \wedge b$	greater of a or b, lesser of a or b
$\rightsquigarrow$	convergence in distribution

2.1 Analytical Tools

We begin with some basic results from analysis and measure theory that will be essential in the study of derivatives.[1]

2.1.1 *Order Notation*

It is often necessary to approximate functions and to specify how the approximation error varies with an argument of the function. *Order notation* is useful for this purpose. Consider, for example, a sequence of real numbers $\{a_n\}_{n=1}^{\infty}$, which is a mapping from the positive integers, $\aleph$, to the reals, $\Re$. One often wants to approximate a_n to some specified degree of accuracy by a function $\hat{a}_n$ that has simpler properties. Thus, since the function $a_n = \frac{n}{n-1} = \frac{1}{1-1/n}$ can be expanded as

$$\begin{aligned} a_n &= 1 + 1/n + 1/n^2 + \cdots + 1/n^{m-1} + 1/n^m + 1/n^{m+1} + \cdots \\ &= \sum_{j=0}^{m-1} n^{-j} + 1/n^m + 1/n^{m+1} + \cdots \\ &= \sum_{j=0}^{m-1} n^{-j} + n^{-m} \frac{n}{n-1}, \end{aligned}$$

one could write $a_n = \sum_{j=0}^{m-1} n^{-j} + O(n^{-m})$. This both represents a_n in terms of an approximating polynomial of degree $m-1$ in n^{-1} and indicates how the maximum discrepancy between a_n and the polynomial behaves as n increases. In this example we would say that discrepancy is of order n^{-m} as $n \to \infty$. A quantity that is $O(n^{-m})$ satisfies

$$\lim_{n\to\infty} n^m O(n^{-m}) = c,$$

where c (equal to unity in the example) is a constant not equal to zero. Thus, the statement $a_n - \hat{a}_n = O(n^{-m})$ specifies the slowest rate at which the difference vanishes or, if $m < 0$, the fastest rate at which it increases with n. The definition implies that a quantity is $O(n^0)$ or, equivalently, $O(1)$ if it neither vanishes nor diverges as $n \to \infty$.

In many cases it is necessary to know only that an approximation error vanishes faster, or increases more slowly, than a specified rate.

[1] References that may be consulted for systematic treatment and proofs of the theorems include Ash (1972), Billingsley (1986), Royden (1968), Rudin (1976), and Taylor (1966).

The small-o notation is used for this. We would write $a_n - \hat{a}_n = o(n^{-m})$ if $\lim_{n\to\infty} n^m(a_n - \hat{a}_n) = 0$ and would then say that the approximation error is of order *less than* n^{-m}. Thus, by adding one more term to the summation in the example above, we could write

$$a_n = \sum_{j=0}^{m} n^{-j} + o(n^{-m}).$$

A quantity that is $o(n^0)$ or, equivalently, $o(1)$ simply vanishes (at an unspecified rate) as $n \to \infty$.

The following arithmetic of order notation is obvious from the definitions. For $k, m \in \{0, \pm 1, \pm 2, \ldots\}$

$$\begin{aligned}
O(n^k) \pm O(n^m) &= O(n^{\max(k,m)}) \\
o(n^k) \pm o(n^m) &= o(n^{\max(k,m)}) \\
O(n^k) \pm o(n^m) &= \begin{cases} o(n^m), & m > k \\ O(n^k), & m \le k \end{cases} \\
O(n^k) \times O(n^m) &= O(n^{k+m}) \\
o(n^k) \times o(n^m) &= o(n^{k+m}) \\
O(n^k) \times o(n^m) &= o(n^{k+m}).
\end{aligned}$$

In approximating functions of real variables a variant of this notation is often used in which a real variable h, corresponding to n^{-1}, approaches zero. For example, $f(h) = \hat{f}(h) + O(h^k)$ as $h \to 0$ means that

$$\lim_{h\to 0} h^{-k}[f(h) - \hat{f}(h)] = c \neq 0,$$

while $f(h) = \hat{f}(h) + o(h^k)$ as $h \to 0$ means that

$$\lim_{h\to 0} h^{-k}[f(h) - \hat{f}(h)] = 0.$$

2.1.2 *Series Expansions and Finite Sums*

Taylor series gives polynomial approximations for differentiable functions, as

$$\begin{aligned}
f(x) &= f(a) + f'(a)(x-a) + \frac{f''(a)}{2}(x-a)^2 + \cdots + \frac{f^{(k)}(a)}{k!}(x-a)^k + R_{k+1}, \\
&= \sum_{j=0}^{k} \frac{f^{(j)}(a)}{j!}(x-a)^j + R_{k+1},
\end{aligned}$$

where

$$R_{k+1} = \frac{f^{(k+1)}(x_a)}{(k+1)!}(x-a)^{k+1}$$

and $|x - x_a| < |x - a|$. Thus, x_a is a point between x and a, its precise value not specified. Two important applications (both with $a = 0$) are

$$e^x = \sum_{j=0}^{k} \frac{x^j}{j!} + o(x^k) \text{ as } x \to 0 \tag{2.1}$$

$$\ln(1+x) = \sum_{j=1}^{k} \frac{(-1)^{j-1}}{j} x^j + o(x^k) \text{ as } x \to 0. \tag{2.2}$$

Formula (2.1) is valid for $k = \infty$ and any complex number x, while (2.2) holds for $k = \infty$ only when $|x| < 1$. We will also need the following equalities, which can be verified by comparing Taylor expansions of each side about $\zeta = 0$:

$$\sin\zeta = \frac{e^{i\zeta} - e^{-i\zeta}}{2i}, \quad \cos\zeta = \frac{e^{i\zeta} + e^{-i\zeta}}{2}, \quad i = \sqrt{-1}. \tag{2.3}$$

From these follows Euler's formula,

$$e^{i\zeta} = \cos\zeta + i\sin\zeta. \tag{2.4}$$

Setting $f(x) = (1+x)^s$ for $s \in \Re$ and $|x| < 1$ produces the binomial series,

$$(1+x)^s = \sum_{j=0}^{\infty} \binom{s}{j} x^j. \tag{2.5}$$

The binomial coefficients are given by

$$\binom{s}{j} \equiv \begin{cases} s(s-1)\cdots(s-j+1)/j!, & j \in \aleph \\ 1, & j = 0 \\ 0, & \text{else} \end{cases} \tag{2.6}$$

When $s \in \aleph$ the right side of (2.5) comprises just $s+1$ terms, all terms in powers higher than s vanishing by virtue of (2.6). In this case the expansion holds without restriction on x. Note that definition (2.6) implies the identity

$$\binom{s}{j-1} + \binom{s}{j} = \binom{s+1}{j}. \tag{2.7}$$

The following elementary expressions for finite sums will often be useful:

$$\sum_{j=1}^{n} j = n(n+1)/2 \tag{2.8}$$

$$\sum_{j=1}^{n} j^2 = n(n+1)(2n+1)/6 \tag{2.9}$$

$$\sum_{j=0}^{n} a^j = \begin{cases} \dfrac{1-a^{n+1}}{1-a}, & a \neq 1 \\ n+1, & a = 1 \end{cases} \tag{2.10}$$

$$(a+b)^n = \sum_{j=0}^{n} \binom{n}{j} a^j b^{n-j}. \tag{2.11}$$

2.1.3 *Measures*

A measure is simply a special type of set function; that is, a function that assigns numbers to sets. The explanation of the concept begins with an overall space, Ω, of which everything we would want to measure is a subset. For example, Ω would be $\Re$ if we wanted to measure various sets of real numbers. The next ingredient is a class $\mathcal{M}$ of measurable sets, the collection of sets that actually can be measured, given the requirements we impose for how a measure should behave. Measures with the right properties can be defined on classes of sets known as σ-fields or σ-algebras. These are collections that contain all the sets that can be built up by countably many of the ordinary set operations (union, intersection, complementation) on some fundamental constituent class. For example, the Borel sets, $\mathcal{B}$, are a σ -field of sets of the real line that contain all countable unions and intersections of the open intervals (x, y) (or closed intervals $[x, y]$, or half-open intervals $(x, y]$, etc.), together with their complements.[2] The Borel sets of the line comprise all those sets (intervals, points, unions of intervals, ...) to which a meaningful notion of length could be assigned. Indeed, they are the smallest such collection that can be built up from the intervals. Another way to put this is to say that $\mathcal{B}$ is the σ-field "generated" by the intervals in $\Re$.

[2]For examples, (i) an open interval (x, y) can be built from the class of half-open intervals $\{(x, z] : x \in \Re, z \in \Re\}$ as $(x, y) = \cup_{n=1}^{\infty}(x, y - 1/n]$; (ii) the closed interval $[x, y]$ can be built from open intervals $\{(w, z) : w \in \Re, z \in \Re\}$ as $[x, y] = \cap_{n=1}^{\infty}(x - 1/n, y + 1/n)$; and (iii) the singleton set (point) $\{x\}$ can be constructed from intervals in any of the following ways: $\cap_{n=1}^{\infty}(x - 1/n, x]$, $\cap_{n=1}^{\infty}[x - 1/n, x]$, $\cap_{n=1}^{\infty}[x, x + 1/n)$, $\cap_{n=1}^{\infty}[x, x + 1/n]$.

A measure μ on $(\Omega, \mathcal{M})$ is then one of the class of set functions that map $\mathcal{M}$ into the reals, but it has two special properties that set it apart from other functions in the class that are not measures. The required special properties are (i) $\mu(\cdot) \geq 0$ (non-negativity), and (ii) $\mu(\cup_{j=1}^{\infty} A_j) = \sum_{j=1}^{\infty} \mu(A_j)$, when $A_j \cap A_{j'} = \emptyset$ (countable additivity). In short, μ is a countably additive mapping from $\mathcal{M}$ into $\Re^+ \cup \{+\infty\} \equiv [0, +\infty]$. Property (i) just rules out having negative lengths, masses, probabilities, etc.; while property (ii) assures that the measure of the whole is the sum of the measures of its (disjoint) parts. The triple $(\Omega, \mathcal{M}, \mu)$ forms a measure space, all three ingredients being needed in order to specify exactly what the measure is. Typically, measures are defined in some intuitive way on a relatively simple class of sets that generate $\mathcal{M}$, and then it is proved that they can be extended—uniquely—to $\mathcal{M}$ itself. With $\Omega = \Re$ and $\mathcal{M} = \mathcal{B}$, Lebesgue measure is defined for a simple class of sets, the intervals, to conform with our common notion of length; e.g., $\mu((a, b]) = b - a$. The extension of this measure to the class $\mathcal{B}$ (some sets of which are extremely complicated!) can then be made. There are, however, subsets of $\Re$ that are not in $\mathcal{B}$ and cannot be measured if we insist that measures have properties (i) and (ii).[3]

Our main interest will be in probability measures, which are introduced in the next section, but we will need to know a few facts and some terminology pertaining to measures generally and some properties of Lebesgue measure in particular. First, we shall require a fundamental result about the measures of monotone sets. A sequence of sets $\{A_n\}_{n=1}^{\infty}$ is said to be monotone if either $A_n \subseteq A_{n+1}$ or $A_n \supseteq A_{n+1}$ for each n. In the first case the sequence is said to be weakly increasing or, equivalently, nondecreasing; and in the second, weakly decreasing or nonincreasing. These features of the sequence are also expressed symbolically as $\{A_n\} \uparrow$ and $\{A_n\} \downarrow$, respectively. If $\{A_n\} \uparrow$, then we define $\lim_{n \to \infty} A_n$ as $\cup_{n=1}^{\infty} A_n$; and if $\{A_n\} \downarrow$, we define $\lim_{n \to \infty} A_n$ as $\cap_{n=1}^{\infty} A_n$. For the fundamental result about measures of such sets let $\{A_n\}_{n=1}^{\infty}$ be a monotone sequence of sets in $\mathcal{M}$, and let μ be a measure on $(\Omega, \mathcal{M})$. Then if either (i) $\{A_n\} \uparrow$ or (ii) $\{A_n\} \downarrow$ *and* $\mu(A_1) < \infty$, we have

$$\mu(\lim_{n \to \infty} A_n) = \lim_{n \to \infty} \mu(A_n). \tag{2.12}$$

[3]While all sets in $\mathcal{B}$ are Lebesgue-measurable, there are some measurable sets that are not in $\mathcal{B}$ (Taylor, 1966, p. 94). Examples of nonmeasurable sets are (happily) not easily constructed. For one such, see Billingsley (1986, pp. 41–42).

In words, this says that (under the stated condition) measures of limits of monotone sequences of sets equal limits of the sequences of their measures. This is referred to as the "monotone property" of μ. As an example of why $\mu(A_1) < \infty$ is needed in the decreasing case, take $(\Re, \mathcal{B})$ as measure space, $A_n = [0, n^{-1}] \cap \mathcal{Q}$ (the rational numbers on $[0, n^{-1}]$), and let $\#$ be counting measure ($\#(A)$ equals the number of elements of A). Thus, $\#(A_n) = \infty$ for each n but $\#(\lim_{n\to\infty} A_n) = \#(\{0\}) = 1$.

Now, some terminology. A measure μ is "finite" if $\mu(\Omega) < \infty$. It is "σ-finite" if there is a countable decomposition, $\Omega = \cup_{n=1}^{\infty} A_n$, such that $\mu(A_n) < \infty$ for each n. Thus, Lebesgue measure on $(\Re, \mathcal{B})$ is not finite, but it *is* σ-finite. Next, suppose some condition C holds for all members of Ω except those comprising a measurable set A_0 such that $\mu(A_0) = 0$. Then we say that C holds "almost everywhere" (a.e.) with respect to μ. For example, a function g such that $g > 0$ except on such a set would be said to be positive almost everywhere with respect to μ, or, for short, to be positive a.e. μ.

Finally, consider some aspects of Lebesgue measure in particular. Define the sequence of sets $\{A_n\}$ as $A_n = (a - 1/n, a]$ for $n \in \aleph$. Then with μ as Lebesgue measure we have $\mu(A_n) = 1/n$, conforming to the ordinary notion of length; and, by the monotone property, $\mu(\{a\}) = \mu(\cap_{n=1}^{\infty} A_n) = \lim_{n\to\infty} \mu(A_n) = \lim_{n\to\infty} 1/n = 0$. Thus, the Lebesgue measure of any point in $\Re$ (any singleton set) is zero. Now let B be any countable set of points in $\Re$, so that $B = \cup_{n=1}^{\infty}\{x_n\}$ for real numbers $x_1, x_2, \ldots$ Then by countable additivity of measures, $\mu(B) = \sum_{n=1}^{\infty} \mu(\{x_n\}) = 0$. Countable sets in $\Re$ are thus sets of Lebesgue measure zero. Applying the "a.e." terminology, we would say that a condition that holds except on a countable set in $\Re$ holds a.e. with respect to Lebesgue measure.

2.1.4 *Measurable Functions*

A mapping f from a measurable space $(\Omega, \mathcal{M})$ to a measurable space $(\Psi, \mathcal{N})$ is said to be a "measurable" mapping if the inverse image under f of each set in $\mathcal{N}$ is a set in $\mathcal{M}$; that is, if $f^{-1}(D) \equiv \{\omega : f(\omega) \in D\} \in \mathcal{M}$ for each $D \in \mathcal{N}$. Similarly, a real-valued function $f : \Omega \to \Re$ is measurable (that is, $\mathcal{M}$-measurable) if for each Borel set $B \in \mathcal{B}$ the inverse image, $f^{-1}(B) = \{\omega : f(\omega) \in B\}$, is in $\mathcal{M}$.[4] Knowing that f is a measurable

[4]Notice that f here is not a set function, which is a mapping from $\mathcal{M}$ to $\Re$. Instead, it is a "point" function that maps Ω into $\Re$ and therefore assigns real numbers to elements ω of Ω.

mapping means that we can associate with a statement like "f takes a value in D" a probability, a length, or a value of whatever measure it is that has been defined on $(\Omega, \mathcal{M})$. That a function f is $\mathcal{M}$-measurable is often expressed as $f \in \mathcal{M}$ for brevity.

For example, working with $(\Re, \mathcal{B}, \mu)$, where μ is Lebesgue measure, let $f(x) = e^x$, so that f maps $\Re$ onto $\Re^+ \backslash \{0\} = (0, \infty)$. Then the Borel sets $(0, 1]$ and $[1, e]$ have inverse images $(-\infty, 0]$ and $[0, 1]$ under f, and these are themselves Borel sets with $\mu((-\infty, 0]) = +\infty$ and $\mu([0, 1]) = 1$. For functions of real variables it can be shown that the condition $f^{-1}((-\infty, x]) \in \mathcal{M}$ for each $x \in \Re$ is sufficient for measurability. (Of course, the necessity of this is obvious.) This simplifies considerably the task of checking that a function is measurable. The functions of real variables that arise in ordinary applications of analysis are, in fact, Borel measurable. Thus, right- and/or left-continuous real functions are measurable, as are compositions, sums, products, maxima and minima, suprema and infima, inverses of (non-vanishing) functions, and limits of sequences of measurable functions (on the sets where the limits exist).

2.1.5 *Variation and Absolute Continuity of Functions*

A collection $\mathcal{S}$ of subsets of a set S constitutes a partition of S if each element of S belongs to one and only one member of $\mathcal{S}$. Thus, if $\mathcal{S}$ is a partition that comprises the sets $s_1, s_2, \ldots$, then $s_j \cap s_{j'} = \emptyset$ whenever $j \neq j'$ and $\cup_j s_j = S$. Considering now subsets of $\Re$, let $\mathcal{S} = \{(x_{j-1}, x_j]\}_{j=1}^n$ be a partition of the interval $(a, b]$ into a finite number of subintervals, where we take $x_0 = a$ and $x_n = b$. The "variation" of a measurable function $F : \Re \to \Re$ over $(a, b]$ is defined as

$$\mathbb{V}^{(a,b]}F \equiv \sup \sum_{j=1}^{n} |F(x_j) - F(x_{j-1})|,$$

where the supremum is over all such finite partitions. F is said to be of "bounded variation" over $(a, b]$ if $\mathbb{V}^{(a,b]}F < \infty$. Monotone functions on $(a, b]$, whether nondecreasing or nonincreasing, necessarily have bounded variation, equal to $|F(b) - F(a)|$. Likewise, F has bounded variation if it is a sum or a difference of monotone functions, in which case it can always be represented as the difference between two nondecreasing functions. Indeed, it is true that a function is of bounded variation on a finite interval if and only if it can be represented as the difference in nondecreasing functions.[5]

We know from elementary calculus that a function $F : \Re \to \Re$ is continuous if $|F(x') - F(x)|$ can be made arbitrarily small (less than an arbitrary positive ε) by taking $|x' - x|$ sufficiently small (less than some $\delta(\varepsilon)$). Absolute continuity imposes the following more stringent restriction: for each $\varepsilon > 0$ there is a $\delta > 0$ such that $\sum_{k=1}^{n} |F(x'_k) - F(x_k)| < \varepsilon$ for each finite collection of nonoverlapping intervals such that $\sum_{k=1}^{n} |x'_k - x_k| < \delta$. Unlike simple continuity, absolute continuity imposes extra smoothness requirements that ensure differentiability a.e. with respect to Lebesgue measure. Clearly, no function of unbounded variation can be absolutely continuous, although simply continuous functions can have unbounded variation. Chapter 3 will present a prominent example of the latter—realizations of sample paths of Brownian motions.

An important feature of absolutely continuous functions of real variables is that they (and only they) have representations as indefinite integrals. That is, $F : \Re \to \Re$ is absolutely continuous if and only if there exists a "density" f such that

$$F(x) - F(a) = \int_a^x f(t) \cdot dt, \tag{2.13}$$

where $f = F'$ a.e. with respect to Lebesgue measure and where the integral is interpreted in the Lebesgue sense.[6] Our next task is to define such integrals.

2.1.6 *Integration*

This section defines several constructions of integrals of real-valued, measurable functions that are encountered in the applications to financial derivatives. Following a quick review of the ordinary Riemann integral from calculus, we extend to Riemann-Stieltjes integrals of functions of real variables and then develop the more general concept of integrals on abstract measure spaces, of which the Lebesgue and Lebesgue-Stieltjes integrals are special cases.

---

[5]For this see Billingsley (1986, pp. 435–436).

[6]For a proof, see Ash (1972, theorem 2.3.4). We shall see in chapter 3 that the only-if restriction of this theorem makes the Lebesgue construction unsuitable as an interpretation of stochastic integrals with respect to processes of unbounded variation on finite intervals.

Riemann Integral

Defining the definite Riemann integral, $\int_a^b g(x) \cdot dx$, for a suitable function $g : \Re \to \Re$ begins by partitioning $[a, b]$ as

$$a = x_0 < x_1 < \cdots < x_{n-1} < x_n = b$$

in such a way that $\delta_n \equiv \max_{1 \le j \le n} |x_j - x_{j-1}| \to 0$ as $n \to \infty$ and setting

$$\bar{g}_j \equiv \sup_{x \in (x_{j-1}, x_j]} g(x)$$
$$\underline{g}_j \equiv \inf_{x \in (x_{j-1}, x_j]} g(x).$$

It does not matter whether the intervals from x_{j-1} to x_j are open or closed. Next, construct upper and lower approximating sums, $\bar{S}_n = \sum_{j=1}^n \bar{g}_j (x_j - x_{j-1})$ and $\underline{S}_n = \sum_{j=1}^n \underline{g}_j (x_j - x_{j-1})$. If $\lim_{n \to \infty} \bar{S}_n$ and $\lim_{n \to \infty} \underline{S}_n$ both exist and are equal, then g is said to be Riemann integrable on $[a, b]$, and $\int_a^b g(x) \cdot dx$ is defined as the common value of the two limits. Clearly, the limits will be the same if g is continuous, for then each $\bar{g}_j$ and $\underline{g}_j$ can be made arbitrarily close by taking δ_n sufficiently small. Bounded functions having a countable number of jump discontinuities are also Riemann integrable, as are finite sums, differences, products, and compositions of integrable functions. Since $\lim_{m \to \infty} |\int_{a-1/m}^a g(x) \cdot dx| \le \lim_{m \to \infty} m^{-1} \sup_{x \in [a-1/m, a]} |g(x)| = 0$ if g is Riemann integrable, there is no distinction between integrals over closed intervals, half-open intervals, and open intervals. (The vague notation $\int_a^b g(x) \cdot dx$ admits any of these interpretations.) The improper integral $\int_a^\infty g(x) \cdot dx$ exists if $\lim_{b \to \infty} \int_a^b g(x) \cdot dx$ exists; and $\int_{-\infty}^b g(x) \cdot dx$ and $\int_{-\infty}^\infty g(x) \cdot dx$ are defined in a like manner.

The definition of the Riemann integral grants substantial flexibility in its construction (when it exists). Since $\underline{g}_j \le g(x_j^*) \le \bar{g}_j$ for any $x_j^* \in (x_{j-1}, x_j]$, the condition $\lim_{n \longrightarrow \infty} \bar{S}_n = \lim_{n \longrightarrow \infty} \underline{S}_n$ guarantees that

$$\lim_{n \longrightarrow \infty} \sum_{j=1}^n g(x_j^*)(x_j - x_{j-1}) = \int_a^b g(x) \cdot dx.$$

Example 1 *To illustrate the process of evaluating a Riemann integral from the definition, take* $g(x) = x$, $a = 0$, $b = 1$, *and* $x_j = j/n$. *Using* (2.8) *to* *express* $\bar{S}_n$ *and* $\underline{S}_n$, *we have*

$$\begin{aligned}\lim_{n\to\infty} \bar{S}_n &= \lim_{n\to\infty} \sum_{j=1}^{n} \frac{j}{n}\left(\frac{1}{n}\right) \\ &= \lim_{n\to\infty} \frac{n(n+1)}{2n^2} \\ &= \frac{1}{2} \\ &= \lim_{n\to\infty} \frac{n(n-1)}{2n^2} \\ &= \lim_{n\to\infty} \sum_{j=1}^{n} \frac{j-1}{n}\left(\frac{1}{n}\right) = \lim_{n\to\infty} \underline{S}_n.\end{aligned}$$

The Riemann-Stieltjes Integral

Picking an arbitrary origin c and representing the Lebesgue measure of $[c, x]$ as $\mu([c, x]) = x - c$ for $x \geq c$, the Riemann integral can be expressed as

$$\int_a^b g(x) \cdot d\mu([c, x]) = \lim_{n\to\infty} \sum_{j=1}^{n} g(x_j^*) \cdot \Delta\mu([c, x_j]),$$

where (i) $c \leq a$, (ii) $x_j^* \in (x_{j-1}, x_j]$, with $\{x_j\}_{j=0}^n$ partitioning $[a, b]$ as before, and (iii) $\Delta\mu([c, x_j]) = \mu([c, x_j]) - \mu([c, x_{j-1}]) = x_j - x_{j-1}$. The message is that in the Riemann sum the representative value of g in each interval is being weighted by the Lebesgue measure of that interval. The Stieltjes integral extends the concept of integration by allowing other measures to be used as weights.

Suppose F is a nondecreasing function on $\Re$. Then a measure μ_F on $(\Omega, \mathcal{B})$ can be defined by taking $\mu_F((a, b]) = F(b) - F(a)$ for intervals and then extending to the entire class $\mathcal{B}$, just as Lebesgue measure is extended from the same fundamental class. A function g is Riemann-Stieltjes integrable with respect to F on $(a, b]$ if the limits

$$\lim_{n\to\infty} \sum_{j=1}^{n} \bar{g}_j \left[F(x_j) - F(x_{j-1})\right], \quad \lim_{n\to\infty} \sum_{j=1}^{n} \underline{g}_j \left[F(x_j) - F(x_{j-1})\right]$$

exist and are equal, in which case the common value is written as

$$\int_{(a,b]} g(x) \cdot dF(x).$$

Again, if the integral exists then g could as well be evaluated at any $x_j^* \in (x_{j-1}, x_j]$. Unlike the Riemann integral, however, Stieltjes integrals over open and closed sets may differ, because $\mu_F(\{a\}) = F(a) - \lim_{m\to\infty} F(a-1/m) \equiv F(a)-F(a-)$ is not necessarily zero. Thus, the notation for Stieltjes integrals must indicate explicitly whether intervals include the end points, as $\int_{(a,b]} g(x) \cdot dF(x)$, $\int_{[a,b]} g(x) \cdot dF(x)$, $\int_{[a,b)} g(x) \cdot dF(x)$, etc. One should note also that $\int_{\{a\}} g(x) \cdot dF(a) = g(a)\,[F(a) - F(a-)]$ and $\int_{[a,b]} g(x) \cdot dF(x) = \int_{\{a\}} g(x) \cdot dF(a) + \int_{(a,b)} g(x) \cdot dF(x) + \int_{\{b\}} g(x) \cdot dF(a)$.

Two common special cases that allow other representations of the Stieltjes integral are when (i) F is absolutely continuous and (ii) F is a right-continuous step function. In case (i) there exists a density f that allows a representation for F itself as a Riemann-integral:

$$F(b) - F(a) = \int_a^b f(x) \cdot dx.$$

Since absolute continuity implies differentiability with respect to Lebesgue measure almost everywhere, we have $f(x) = F'(x)$ except possibly at isolated points where F' fails to exist. On that (countable) set of zero Lebesgue measure f can be defined arbitrarily. In the limiting sum that defines the integral the mean value theorem can be used to approximate $F(x_j) - F(x_{j-1})$ on $(x_{j-1}, x_j]$ as $f(x_j^*)(x_j - x_{j-1})$ for some $x_j^* \in (x_{j-1}, x_j)$. Passing to the limit shows that the Stieltjes integral with respect to F can be evaluated as an ordinary Riemann integral of the product $g(x)f(x)$; that is,

$$\lim_{n\to\infty} \sum_{j=1}^{n} g(x_j^*) f(x_j^*)(x_j - x_{j-1}) = \int_a^b g(x) f(x) \cdot dx.$$

The other common situation allowing an elementary representation of the Stieltjes integral on $(a,b]$ is that F is a right-continuous step function with jump discontinuities at a countable number of points $x_1', x_2', \ldots$. At any such point x_k' we have $\mu_F(\{x_k'\}) = F(x_k') - F(x_k'-)$, so that the definition of the integral in this case gives

$$\begin{aligned}\int_{(a,b]} g(x) \cdot dF(x) &= \lim_{n\to\infty} \sum_{j=1}^{n} g(x_j^*)\,[F(x_j) - F(x_{j-1})] \\ &= \sum_k g(x_k')\,[F(x_k') - F(x_k'-)]\,.\end{aligned}$$

The Stieltjes integral of g is thus evaluated as the sum of its values at the points where F is discontinuous, weighted by the jumps in F at these points.

Mixed cases in which F is absolutely continuous on some set C and has jump discontinuities at points in a countable set D are handled by decomposition:

$$\begin{aligned}\int_{(a,b]} g(x) \cdot dF &= \int_{(a,b]\cap C} g(x) \cdot dF + \int_{(a,b]\cap D} g(x) \cdot dF \\ &= \int_{(a,b]\cap C} g(x) F'(x) \cdot dx + \sum_{x \in (a,b]\cap D} g(x) \left[F(x) - F(x-)\right].\end{aligned}$$

A list of the important common properties of Riemann and abstract integrals appears on page 35.

Integrals on Abstract Measure Spaces

A more general concept of the integral is sometimes needed than that provided by the Riemann construction. A limitation of the Riemann construction is that a set A over which the integral is evaluated (e.g., $A = [a, b]$) must have a natural ordering, such as the real numbers have. This is needed because the set itself has to be partitioned according to the values of its members. A more general construction is possible if the range of the integrand g over the domain A is subdivided, rather than A itself. As it turns out, the more general construction achieved in this way also has properties that are more convenient and robust in mathematical analysis.

To develop the abstract integral, consider a measure space $(\Omega, \mathcal{M}, \mu)$ and a measurable function $g : \Omega \to \Re$. Here Ω could be $\Re$ (or $\Re_k$, $k > 1$), but it could as well be a collection of physical objects or of abstract entities, such as the outcomes of a chance experiment. The function g assigns a real number to any such element ω of Ω. The abstract integral, written either as $\int g(\omega) \cdot d\mu(\omega)$ or as $\int g(\omega) \cdot \mu(d\omega)$, is defined in a series of steps in which the allowed class of integrands becomes increasingly more general. Step 1 pertains to the class $\mathbb{S}^+$ of nonnegative simple functions, which are functions that take nonnegative, constant values on each of a finite number of subsets that partition Ω. Thus, we suppose initially that Ω is the union of disjoint sets $\{A_j\}_{j=1}^n$ such that $g(\omega) = g_j \geq 0$ for each $\omega \in A_j$. Using indicator functions that take the value unity on the designated sets and zero elsewhere, simple functions can be written compactly as $g(\omega) = \sum_{j=1}^{n} g_j \mathbf{1}_{A_j}(\omega)$. For $g \in \mathbb{S}^+$ the integral is defined in a very intuitive way:

$$\int g(\omega) \cdot d\mu(\omega) = \sum_{j=1}^{n} g_j \mu(A_j).$$

Here, the constant value of the function g on each set in the partition is simply weighted by the measure of the set, and the integral is just the sum of these weighted values.

Step two extends the definition to arbitrary nonnegative measurable functions $g \in \mathbb{M}^+$. This is done by showing that any member of $\mathbb{M}^+$ can be attained as the limit of a sequence of functions $\{g_n\}$ in $\mathbb{S}^+$. A common construction is to define sets

$$A_{j,n} = \begin{cases} \left\{\omega : \frac{j-1}{2^n} \le g(\omega) < \frac{j}{2^n}\right\}, & j = 1, 2, \ldots, 2^{2n} \\ \{\omega : g(\omega) \ge 2^n\}, & j = 0 \end{cases}$$

and take

$$g_n(\omega) = \begin{cases} \frac{j-1}{2^n}, & \omega \in A_{j,n},\ j > 0 \\ 2^n, & \omega \in A_{0,n} \end{cases}.$$

Notice that the sets $\{A_{j,n}\}$ are indeed based on a subdivision of the range of g, namely $\Re^+$. It is obvious that the sequence of such simple functions $\{g_n\}$ has the first two of the following properties, and the third is not hard to show: (i) $g_n(\omega) \le g(\omega)$, (ii) $g_n(\omega) \le g_{n+1}(\omega)$, and (iii) $g_n(\omega) \uparrow g(\omega)$ for each $\omega \in \Omega$ and each $g \in \mathbb{M}^+$. Since $g_n \in \mathbb{S}^+$, the definition of the integral extends naturally to this broader class of nonnegative measurable functions as the limit of the sequence of integrals of the simple functions that converge to g:

$$\int g(\omega) \cdot d\mu(\omega) = \lim_{n \to \infty} \int g_n(\omega) \cdot d\mu(\omega).$$

It can be shown that the limit is unique even though there are many ways of constructing such a sequence of approximating simple functions.[7] The limit may well be infinite, however. In that case the integral is said not to exist, and g is said to be not μ-integrable. This can happen if $g(\omega) \ge \varepsilon > 0$ on some set A_ε such that $\mu(A_\varepsilon) = +\infty$. It can happen also if g is positive

[7]For a proof of uniqueness see Taylor (1966, pp. 112–113).

and unbounded on a set of finite measure in such a way that the sequence of integrals diverges for all approximating simple functions.

Finally, step three extends the definition to the entire class of measurable functions $\mathbb{M}$ without the positivity restriction. This is done very simply. Setting $g^+(\omega) \equiv g(\omega) \vee 0$ and $g^-(\omega) \equiv g(\omega) \wedge 0$, we have $g^+ \in \mathbb{M}^+$, $-g^- \in \mathbb{M}^+$, and $g = g^+ - (-g^-)$. The natural definition of the integral for $g \in \mathbb{M}$ is therefore

$$\int g(\omega) \cdot d\mu(\omega) = \int g^+(\omega) \cdot d\mu(\omega) - \int [-g^-(\omega)] \cdot d\mu(\omega).$$

As thus defined, the integral is said to exist if and only if the integrals of g^+ and $-g^-$ are both finite—a point of some importance that is discussed below.

Many familiar properties of the Riemann integral are preserved in this construction. In the following list it is assumed that $\{A_j\}_{j=0}^{\infty} \in \mathcal{M}$ (the σ-field of sets on which the measure μ is defined), that $A_j \cap A_{j'} = \emptyset$ for $j \neq j'$, and that $\mu(A_0) = 0$. As well, a and b are arbitrary real numbers, and g and h are measurable functions.[8]

1. $\int_{A_1} g(\omega) \cdot d\mu(\omega) = \int g(\omega)\mathbf{1}_{A_1}(\mu) \cdot d\mu(\omega)$
2. $\int_{\cup_{j=1}^{\infty} A_j} g(\omega) \cdot d\mu(\omega) = \sum_{j=1}^{\infty} \int_{A_j} g(\omega) \cdot d\mu(\omega)$
3. $\int a \cdot d\mu(\omega) = a \int d\mu(\omega) = a\mu(\Omega)$
4. $\int bg(\omega) \cdot d\mu(\omega) = b \int g(\omega) \cdot d\mu(\omega)$
5. $\int [a + bg(\omega) + ch(\omega)] \cdot d\mu(\omega) = a + b \int g(\omega) \cdot d\mu(\omega) + c \int h(\omega) \cdot d\mu(\omega)$
6. $g(\omega) \leq h(\omega)$ a.e. $\mu \Rightarrow \int g(\omega) \cdot d\mu(\omega) \leq \int h(\omega) \cdot d\mu(\omega)$
7. $g(\omega) = h(\omega)$ a.e. $\mu \Rightarrow \int g(\omega) \cdot d\mu(\omega) = \int h(\omega) \cdot d\mu(\omega)$
8. $\left|\int g(\omega) \cdot d\mu(\omega)\right| \leq \int |g(\omega)| \cdot d\mu(\omega)$
9. $\mu(A_0) = 0 \Rightarrow \int_{A_0} g(\omega) \cdot d\mu(\omega) = 0.$

While these properties are familiar because they are shared by the Riemann integral, the abstract integral has other important properties that

---

[8]To apply these to the Riemann contruction (assuming that g and h are Riemann integrable), take $\omega \in \Re$, $A_j \in \mathcal{B}$, $\mu(A_j)$ as either Lebesgue measure or as the measure induced by a nondecreasing function F on $\Re$, and interpret $d\mu(\omega)$ as either $d\omega$ or $dF(\omega)$.

Note that property 1 implies that $\int g(\omega) \cdot d\mu(\omega)$ has the same meaning as $\int_\Omega g(\omega) \cdot d\mu(\omega)$. Property 7 just states that if $g(\omega) = h(\omega)$ outside a set of measure zero, then the integrals also are equal.

the Riemann construction does not necessarily confer. One of these is the implication that $\int g(\omega) \cdot d\mu(\omega)$ exists if and only if $\int |g(\omega)| \cdot d\mu(\omega)$ exists. Since $|g| = g^+ - g^-$, this property follows from the way the definition is extended from nonnegative to arbitrary measurable functions. Another very important implication of the definition that does not hold in general for Riemann integrals is the dominated convergence theorem. This essential tool in analysis gives a simple condition under which the operations of integrating and taking limits can be interchanged.[9]

Theorem 1 *Suppose g_n is a sequence of $\mathcal{M}$-measurable functions converging (pointwise) to a function g, with $|g_n| \le h$ for some integrable h. Then g itself is integrable, and*

$$\lim_{n\to\infty} \int g_n(\omega) \cdot d\mu(\omega) = \int \lim_{n\to\infty} g_n(\omega) \cdot d\mu(\omega) = \int g(\omega) \cdot d\mu(\omega).$$

Thus, the interchange of limits and integration is justified if there is a dominating, integrable function h.

The Lebesgue Integral

The integral having been defined on an abstract measure space $(\Omega, \mathcal{M}, \mu)$, the Lebesgue integral can now be viewed just as the special case in which $\Omega = \Re$, $\mathcal{M} = \mathcal{B}$, and μ is Lebesgue measure. Since then $\mu((c, x]) = x - c$ for $c \le x$, the integral can be expressed simply as $\int g(x) \cdot dx$. Also, as for the Riemann integral, $\mu(\{a\}) = \mu(\{b\}) = 0$ implies that $\int_{[a,b]} g(x) \cdot dx = \int_{(a,b)} g(x) \cdot dx = \int_{[a,b)} g(x) \cdot dx$, etc., so that the ambiguous representation $\int_a^b g(x) \cdot dx$ still works perfectly well. Finally, letting $F : \Re \to \Re$ be a nondecreasing function and defining the measure $\mu_F((c, x])$ as $F(x) - F(c)$ gives us the Lebesgue-Stieltjes integral: $\int_\Re g(x) \cdot d\mu_F(x) = \int_\Re g(x) \cdot dF(x)$. As in the Riemann-Stieltjes construction the fact that $\mu_F(\{a\})$ and $\mu_F(\{b\})$ are not necessarily zero requires specifying whether sets are open or closed, as $\int_{[a,b]} g(x) \cdot dF(x)$, $\int_{(a,b]} g(x) \cdot dF(x)$, etc.

When g is Riemann integrable, the Riemann and Lebesgue constructions give the same values for $\int_a^b g(x) \cdot dx$. Likewise, when g is Riemann-Stieltjes integrable with respect to F, then $\int_{[a,b]} g(x) \cdot dF$ is the same in

---

[9]The proof can be found in standard texts on real analysis and measure; e.g., Ash (1972), Billingsley (1986), Royden (1968), and Taylor (1966).

either construction. Thus, when Riemann integrability holds, the ordinary integration formulas from calculus still apply to evaluate the Lebesgue integral. The greater generality of the latter comes into play for integrals of functions that are highly oscillatory or discontinuous.

Example 2 *Set $g(x) = 1$ for $x \in \mathcal{Q}$ (the rational numbers) and $g(x) = 0$ elsewhere, and partition (a, b) as $a = x_0 < x_1 < \cdots < x_n = b$ so that $\max_{j \in \{1,2,\ldots,n\}} x_j - x_{j-1} \to 0$ as $n \to \infty$. Then g is not Riemann integrable, since $\bar{g}_j \equiv \sup_{x \in (x_{j-1}, x_j]} g(x) = 1$ and $\underline{g}_j \equiv \inf_{x \in (x_{j-1}, x_j]} g(x) = 0$ for each j and n imply that*

$$\lim_{n\to\infty} \sum_{j=1}^{n} \bar{g}_j (x_j - x_{j-1}) = b - a \neq \lim_{n\to\infty} \sum_{j=1}^{n} \underline{g}_j (x_j - x_{j-1}) = 0.$$

However, g is Lebesgue integrable because the splitting of (a, b) into subsets is based on values of g rather than of x:

$$\begin{aligned}\int_a^b g(x) \cdot dx &= \int_{(a,b)\cap\mathcal{Q}} g(x) \cdot dx + \int_{(a,b)\cap(\Re\backslash\mathcal{Q})} g(x) \cdot dx \\ &= 1 \cdot \mu\left[(a, b) \cap \mathcal{Q}\right] + 0 \cdot \mu\left[(a, b) \cap (\Re\backslash\mathcal{Q})\right] \\ &= 1 \cdot 0 + 0 \cdot (b - a) \\ &= 0.\end{aligned}$$

This greater generality of the Lebesgue construction applies also when integrating over the entire real line. However, in cases where integrals over the positive and negative axes are not both finite, it is possible to come up with finite values for the Riemann integral when Lebesgue integrability fails, depending on how the limits are taken. For example, evaluating $\int_{-\infty}^{\infty} x(1+x^2)^{-1} \cdot dx$ as $\lim_{a\to\infty} \int_{-a}^{a} x(1+x^2)^{-1} \cdot dx$ (which is Cauchy's interpretation of the improper Riemann integral from complex analysis) gives 0 as the unambiguous result, even though the integrals over the positive and negative axes individually are $+\infty$ and $-\infty$, respectively. That is, in the Riemann construction the positive and negative parts simply cancel out if the limits are taken symmetrically, whereas $\int_{-\infty}^{\infty} |x|(1 + x^2)^{-1} \cdot dx$ fails to exist no matter how the limits are taken. However, if a function g is absolutely integrable in the Riemann sense, meaning that $\int_{-\infty}^{\infty} |g(x)| \cdot dx < \infty$, then $\int_{-\infty}^{\infty} g(x) \cdot dx$ has the same value in the Lebesgue and Riemann constructions.

Uniform Integrability

Noting that property 2 of the integral implies that

$$\int_{\Re} g(x) \cdot dx = \int_{\{x:|g(x)|>n\}} g(x) \cdot dx + \int_{\{x:|g(x)|\le n\}} g(x) \cdot dx$$

for each n, and that

$$\lim_{n\to\infty} \int_{\{x:|g(x)|\le n\}} |g(x)| \cdot dx = \int_{\Re} |g(x)| \cdot dx,$$

it is clear that we must have $\lim_{n\to\infty} \int_{\{x:|g(x)|>n\}} |g(x)| \cdot dx = 0$ if g is to be integrable in the Lebesgue sense. For a collection $\{g_\alpha(x)\}_{\alpha\in\mathcal{A}}$ of integrable functions, it is therefore true that $\lim_{n\to\infty} \int_{\{x:|g_\alpha(x)|>n\}} |g_\alpha(x)| \cdot dx = 0$ for each member of the collection. This means that each integral can be made smaller than an arbitrary $\varepsilon > 0$ by taking $n > N(\alpha, \varepsilon)$, where N may depend on α as well as ε. However, it may be that there is no single $N(\varepsilon)$ that simultaneously bounds $\int_{\{x:|g_\alpha(x)|>n\}} |g_\alpha(x)| \cdot dx$ for *all* members of the collection. On the other hand, if

$$\lim_{n\to\infty} \sup_{\alpha\in\mathcal{A}} \int_{\{x:|g_\alpha(x)|>n\}} |g_\alpha(x)| \cdot dx = 0 \tag{2.14}$$

then this simultaneous control is in fact possible, and the family $\{g_\alpha(x)\}_{\alpha\in\mathcal{A}}$ is said to be "uniformly integrable". (The set $\mathcal{A}$ here is unrestricted; it may be uncountable.)

Example 3 *Functions* $\{g_\alpha(x) = x^{-1}\mathbf{1}_{[1/\alpha,1]}(x)\}_{1\le\alpha<\infty}$ *are integrable, with* $\int g_\alpha(x) \cdot dx = \ln\alpha$, *but are not uniformly so because* $\sup_\alpha \int_{\{x:|g_\alpha(x)|>n\}} |g_\alpha(x)| \cdot dx = +\infty$ *for each* n.

Multiple Integrals and Fubini's Theorem

Taking $\Omega = \Re_2$ (the plane) and $\mathcal{B}_2$ the σ-field generated by the product sets $(-\infty, x_1] \times (-\infty, x_2]$ gives a measurable space $(\Re_2, \mathcal{B}_2)$. On this space there exists a *product* measure, $\mu_2((-\infty, x_1] \times (-\infty, x_2]) = \mu((-\infty, x_1]) \cdot \mu((-\infty, x_2])$, that, with μ as Lebesgue measure, corresponds to areas of the indicated rectangular sets. Given $g : \Re_2 \to \Re$, construction of the integral $\int g(x_1, x_2) \cdot d\mu_2(x_1, x_2) \equiv \int\int g(x_1, x_2) \cdot dx_1\, dx_2$ proceeds in much the same way as in the one-dimensional case. Fubini's theorem states conditions under which the double integral can be evaluated as an iterated integral and in either order; that is, conditions under which

$$\int\int g(x_1, x_2) \cdot dx_1\, dx_2 = \int \left[\int g(x_1, x_2) \cdot dx_1\right] \cdot dx_2$$
$$= \int \left[\int g(x_1, x_2) \cdot dx_2\right] \cdot dx_1. \qquad (2.15)$$

Fortunately, the conditions for this are very general. Equalities (2.15) hold under any one of the following conditions: (i) g is a nonnegative function (in which case all the integrals may be infinite); (ii) $\int\int g(x_1, x_2) \cdot dx_1\, dx_2 < \infty$; (iii) $\int \left(\int |g(x_1, x_2)| \cdot dx_1\right) \cdot dx_2 < \infty$.

Changes of Variable in the Integral

Given a definite integral $\int_A g(x) \cdot dx$, where $A \in \mathcal{B}$ (the Borel sets of $\Re$), suppose we convert to a new variable y as $y = h(x)$, where h is a (Borel) measurable function. Let $h(A)$ be the image of A under h; that is, the set onto which h maps A. Suppose further that h is differentiable and that $0 < |h'(x)|$ on A. This ensures that the transformation is one-to-one, so that a unique inverse, $h^{-1}(y)$, exists. It also ensures that the derivative of $h^{-1}(y)$ is finite; that is, $|dy/dx| > 0 \Rightarrow |dx/dy| < \infty$. Making the substitution $x = h^{-1}(y)$ then gives the change-of-variable formula,

$$\int_A g(x) \cdot dx = \int_{h(A)} g\left[h^{-1}(y)\right] \cdot \left|dh^{-1}(y)\right|, \qquad (2.16)$$

where $|dh^{-1}(y)| \equiv |dh^{-1}(y)/dy| \cdot dy$.

Example 4 *In $\int_0^a e^{-x^2} \cdot dx$ change variables as $y \equiv h(x) = x^2$. Since $h'(x) > 0$ on $(0, a)$, there is a unique inverse, $x = h^{-1}(y) = +\sqrt{y}$, with $dh^{-1}(y) = \frac{1}{2}y^{-1/2} \cdot dy$. Since $h : (0, a) \to (0, a^2)$, we have*

$$\int_0^a e^{-x^2} \cdot dx = \frac{1}{2}\int_0^{a^2} y^{-1/2} e^{-y} \cdot dy.$$

Now suppose that h is not one to one and has no unique inverse on A but that A can be decomposed into disjoint subsets as $A = \cup_{j=0}^k A_j$ such that (i) h has a unique inverse on each of $A_1, A_2, \ldots, A_k$ and (ii) $\mu(A_0) = 0$, where μ is Lebesgue measure. Then since $\int_{A_0} g(x) \cdot dx = 0$, we have $\int_A g(x) \cdot dx = \sum_{j=1}^k \int_{A_j} g(x) \cdot dx$ by properties **9** and **2** of the integral, so that

$$\int_A g(x) \cdot dx = \sum_{j=1}^{k} \int_{h(A_j)} g\left[h^{-1}(y)\right] \cdot d\left|h(y)^{-1}\right|.$$

Example 5 *In* $\int_{(-\infty,\infty)} e^{-x^2} \cdot dx$ *change variables as* $y \equiv h(x) = x^2$. *Decomposing* $(-\infty,\infty)$ *as* $(-\infty,0) \cup \{0\} \cup (0,\infty)$, *we note that (i)* $x = -\sqrt{y}$ *is one-to-one on* $(-\infty,0)$ *with* $dh^{-1}(y) = -\frac{1}{2}y^{-1/2} \cdot dy$; (ii) $x = +\sqrt{y}$ *is one-to-one on* $(0,\infty)$ *with* $dh^{-1}(y) = \frac{1}{2}y^{-1/2} \cdot dy$; *and* (iii) $\mu(\{0\}) = 0$. *Since* $h : (-\infty,0) \to (0,\infty)$ *and* $h : (0,\infty) \to (0,\infty)$ *we then have*

$$\begin{aligned}\int_{(-\infty,\infty)} e^{-x^2} \cdot dx &= \int_0^\infty y^{-1/2} \left| -\frac{1}{2}e^{-y} \right| \cdot dy + \int_0^\infty y^{-1/2} \left| \frac{1}{2}e^{-y} \right| \cdot dy \\ &= \int_0^\infty y^{-1/2} e^{-y} \cdot dy.\end{aligned}$$

Now let us extend the change-of-variable formula to the multivariate case. Given the integral $\int_A g(\mathbf{x}) \cdot d\mathbf{x}$, where $\mathbf{x} = (x_1, x_2, \ldots, x_n)'$, $A \in \mathcal{B}_n$ (the Borel sets of $\Re_n$), and $g : \Re_n \to \Re$, introduce new variables as $\mathbf{y} = h(\mathbf{x})$, $h : \Re_n \to \Re_n$. Let $h(A)$ represent the image of A under h, which is the set $\{\mathbf{y} = h(\mathbf{x}) : \mathbf{x} \in A\}$. Suppose that the partial derivatives $\partial y_j / \partial x_k$ exist and that J, the determinant of the $n \times n$ matrix $\{\partial y_j / \partial x_k\}$, is not zero on A. J is called the "Jacobian" of the transformation from $\mathbf{x}$ to $\mathbf{y}$. That J is not zero signifies that the transformation is one to one and guarantees that the determinant of the inverse transformation from $\mathbf{y}$ to $\mathbf{x}$, denoted J^{-1}, is a finite number. Conveniently, it happens that $J^{-1} = 1/J$. Then the multivariate change-of-variable formula in this one-to-one case becomes

$$\int_A g(\mathbf{x}) \cdot d\mathbf{x} = \int_{h(A)} g\left[h^{-1}(\mathbf{y})\right] \left|J^{-1}\right| \cdot d\mathbf{y}. \tag{2.17}$$

The standard integration formulas for shifts to polar coordinates are just special cases of (2.17).

Example 6 *In* $\int_0^\infty \int_0^\infty (x_1 x_2)^{-1/2} e^{-x_1 - x_2} \cdot dx_2\ dx_1$ *change variables as* $x_1 = r^2 \cos^2\theta$ *and* $x_2 = r^2 \sin^2\theta$, *thus defining directly the inverse transformation from new variables* r, θ *to* x_1, x_2. *Then* $A = (0,\infty) \times (0,\infty)$, $h(A) = (0,\infty) \times (0,\pi/2)$, *and*

$$J^{-1} = \begin{vmatrix} 2r\cos^2\theta & 2r\sin^2\theta \\ -2r^2\cos\theta\sin\theta & 2r^2\cos\theta\sin\theta \end{vmatrix} = 4r^3\cos\theta\sin\theta$$

is finite on $h(A)$. Therefore

$$\begin{aligned}\int_0^\infty \int_0^\infty (x_1 x_2)^{-1/2} e^{-x_1 - x_2} \cdot dx_2\, dx_1 &= 4\int_0^{\pi/2} \left[\int_0^\infty re^{-r^2} \cdot dr\right] \cdot d\theta \\ &= 2\int_0^{\pi/2} d\theta \\ &= \pi.\end{aligned}$$

As in the univariate case, situations in which h is not one-to-one on A can sometimes be handled by decomposition.

2.1.7 *Change of Measure: Radon-Nikodym Theorem*

Suppose there are two different measures, μ and ν, on the same measurable space $(\Omega, \mathcal{M})$. For example, $\nu = 100\mu$ would be one obvious possibility, corresponding to a simple change of units. Suppose, as in this simple case, it happens that $\nu(A) = 0$ for any set A for which $\mu(A) = 0$. That is, sets that are "null" under μ are also null under ν. Then we say that ν is "absolutely continuous" with respect to μ and express this symbolically as $\nu << \mu$. The term *continuity* is apt here since the condition $\nu << \mu$ implies (assuming that ν is finite or σ-finite) that $\nu(A)$ can be made arbitrarily small by choosing A so that $\mu(A)$ is sufficiently small. Two measures are said to be "equivalent" if each is absolutely continuous with respect to the other. Equivalent measures thus have the same null sets.

With this terminology in mind and a measure space $(\Omega, \mathcal{M}, \mu)$ at hand, let us construct a new measure ν on $(\Omega, \mathcal{M})$ in a particular way. Introducing a nonnegative integrable function $Q : \Omega \to \Re^+$, define a set function ν as

$$\nu(A) = \int_A Q(\omega) \cdot d\mu(\omega), \quad A \in \mathcal{M}. \tag{2.18}$$

Property **6** of the integral (page 35) implies that $\nu \geq 0$; property 2 implies that ν is countably additive; and the integrability of Q implies that ν is finite. Therefore, ν thus constructed is a finite measure on $(\Omega, \mathcal{M})$. Moreover, we have $\nu \ll \mu$ by property **9** of the integral. Thus, given a measure and an integrable function, it is easy to create a new measure that is absolutely continuous with respect to the first. What is not so easily seen to be possible, but what is often desirable to do, is to go the other way; that is, to know whether (and how) to convert one measure to another. The following theorem tells us that under appropriate conditions on the measures such a conversion is in fact possible.

Theorem 2 (*Radon-Nikodym*) *If μ and ν are σ-finite measures on a measurable space $(\Omega, \mathcal{M})$, and if $\nu \ll \mu$, then there exists an essentially unique function $Q : \Omega \to \Re^+$ such that* (2.18) *holds for each $A \in \mathcal{M}$.*

Here, *essentially unique* means that if there is another function Q' that satisfies (2.18), then $Q = Q'$ a.e. μ. Clearly, absolute continuity is necessary here, for (2.18) could not possibly hold if $\nu(A) > 0$ for a set for which $\mu(A) = 0$. By analogy with the integral representation for absolutely continuous functions of real variables, $F(b) - F(a) = \int_a^b F'(t) \cdot dt = \int_a^b (dF/dt) \cdot dt$, the function Q is sometimes written as $d\nu/d\mu$ in applications of the Radon-Nikodym theorem. Q is called the "Radon-Nikodym derivative". The implication of the theorem is that if one wants to change from one measure μ to a given measure $\nu \ll \mu$, then there is a function that can be integrated with respect to μ that does this.[10] In the case of equivalent measures, the function Q that takes us from μ to ν must clearly be strictly positive on sets of positive measure. In this case $d\mu/d\nu = Q^{-1}$. We will find that such changes from one *probability* measure to another are a critically important tool in the pricing of derivative assets. Application of Radon-Nikodym to probability measures specifically is treated in section 2.2.4 below.

2.1.8 *Special Functions and Integral Transforms*

Many useful functions in analysis are simply defined as definite or indefinite integrals of other functions or as solutions to differential equations. For our purposes the most important of these types are the gamma function, the error function, and the modified Bessel functions. We will also encounter Laplace transforms and will have *frequent* contact with Fourier transforms.

Gamma and Incomplete Gamma Functions

The gamma function is defined as

$$\Gamma(t) \equiv \int_0^\infty x^{t-1} e^{-x} \cdot dx, \quad t > 0.$$

Integration by parts establishes the recursive property

$$\Gamma(t+1) = t\Gamma(t), \quad t > 0. \tag{2.19}$$

[10]The Radon-Nikodym derivative that works in the introductory example, $\nu = 100\mu$, is obviously $dv/d\mu = Q(\omega) = 100$. Less trivial examples will be given in section 2.2.4 in connection with probability measures.

This then gives the explicit values $\Gamma(n) = (n-1)!$ for $n \in \aleph$ and, since $\int_0^\infty e^{-x} \cdot dx = 1$, explains the convention that $0! = 1$. Taking $t = 1/2$ and making use of example 6 on page 40 yield another explicit form:

$$\Gamma\left(\frac{1}{2}\right) = \sqrt{\left(\int_0^\infty x^{-1/2} e^{-x} \cdot dx\right)^2}$$

$$= \sqrt{\int_0^\infty \int_0^\infty x^{-1/2} y^{-1/2} e^{-x-y} \cdot dy\, dx}$$

$$= \sqrt{\pi}.$$

Using this and (2.19) gives explicitly for all odd multiples of $1/2$

$$\Gamma\left(\frac{2n-1}{2}\right) = \left(\frac{2n-3}{2}\right) \cdot \left(\frac{2n-5}{2}\right) \cdot \cdots \cdot \frac{1}{2} \cdot \sqrt{\pi}, \quad n = 1, 2, \ldots .$$

The "incomplete" gamma function is defined as

$$\Gamma(t; y) = \int_0^y x^{t-1} e^{-x} \cdot dx, \quad t > 0, \ y \geq 0. \tag{2.20}$$

Obviously,

$$\Gamma(t; \infty) \equiv \lim_{y \to \infty} \Gamma(t; y) = \Gamma(t). \tag{2.21}$$

Error Function

The error function of a positive real variable x, usually denoted "erf(x)", is defined as

$$\text{erf}(x) = \frac{2}{\sqrt{\pi}} \int_0^x e^{-t^2} \cdot dt, \quad x \geq 0.$$

By example 4 on page 39 the change of variable to $y = t^2$ relates this to the incomplete gamma function, as

$$\text{erf}(x) = \frac{\Gamma(1/2; x^2)}{\Gamma(1/2)},$$

and this with (2.21) shows that $\text{erf}(+\infty) = 1$.

Bessel Functions

Bessel functions are solutions of the differential equation

$$z^2\frac{d^2f}{dz^2} + z\frac{df}{dz} + (z^2 - \nu^2)f = 0, \tag{2.22}$$

which is known as "Bessel's equation of order ν". Both z and ν may be complex, although our applications will involve only real values of ν. The two linearly independent solutions to this linear homogeneous second-order equation are the Bessel functions of the first kind of order ν and second kind of order ν, denoted $J_\nu(z)$ and $Y_\nu(z)$, respectively.[11] If z is replaced by iz, where $i = \sqrt{-1}$, and if $d^k f/d(iz)^k$ is interpreted as $i^{-k} \cdot d^k f/dz^k$, then (2.22) becomes

$$z^2\frac{d^2f}{dz^2} + z\frac{df}{dz} - (z^2 + \nu^2)f = 0. \tag{2.23}$$

Applying (2.4) to write iz as $e^{i\pi/2}z$ and normalizing the solution $J_v(iz)$, one gets the "modified" Bessel function of the first kind of order ν, as

$$I_\nu(z) = e^{-i\nu\pi/2}J_\nu(e^{i\pi/2}z).$$

Among several representations is the ascending series

$$I_\nu(z) = (z/2)^\nu \sum_{k=0}^{\infty} \frac{(z/2)^{2k}}{k!\Gamma(\nu + k + 1)}. \tag{2.24}$$

The second solution to (2.23), a modified Bessel function of the second kind, is usually written as

$$K_\nu(z) = \frac{\pi}{2}\frac{I_{-\nu}(z) - I_\nu(z)}{\sin(\nu\pi)} \tag{2.25}$$

when ν is not an integer. When $\nu = n$, an integer, this is defined by continuity and has the representation[12]

$$K_n(z) = (-1)^{n+1}\left[\gamma + \ln(z/2)\right]I_n(z) + \frac{1}{2}\sum_{k=0}^{n-1}(-1)^k\frac{(n-k-1)!}{k!}(z/2)^{2k-n}$$

$$+ \frac{(-1)^n}{2}\sum_{k=0}^{\infty}\frac{(z/2)^{n+k}}{k!(n+k)!}\left(\sum_{j=1}^{k}j^{-1} + \sum_{j=1}^{n+k}j^{-1}\right), \tag{2.26}$$

---

[11] For various representations of these as infinite series or as integrals, see Abramowitz and Stegun (1970) and Gradshteyn and Ryzhik (1980). General references for Bessel functions are Kotz *et al.* (1982), McLachlan (1955), Relton (1946), and (for the definitive treatment) Watson (1944).

[12] See McLachlan (1955, p. 115).

where γ is Euler's constant: $\gamma = \lim_{n\to\infty}\left(\sum_{j=1}^{n} j^{-1} - \ln n\right) \doteq 0.577216$. Useful differentiation/recurrence formulas for $K_\nu(\cdot)$ are

$$K'_\nu(z) = \nu z^{-1}K_\nu(z) - K_{\nu+1}(z) \tag{2.27}$$

$$K'_\nu(z) = -\nu z^{-1}K_\nu(z) - K_{\nu-1}(z). \tag{2.28}$$

Laplace Transforms

The two-sided Laplace transform of a function $g(x)$ is defined as

$$\mathcal{L}_2 g = \int_{-\infty}^{\infty} e^{\zeta x} g(x) \cdot dx \equiv \lambda_2(\zeta), \quad \zeta \in \Re,$$

when the integral exists, as it necessarily does at $\zeta = 0$ if g itself is (Lebesgue) integrable. If g is defined just on the positive reals and ζ is restricted to be negative, then the transform often exists even when g itself is not integrable. Restricting to the one-sided case, the definition of the Laplace transform is commonly presented as

$$\mathcal{L}g = \int_0^{\infty} e^{-\zeta x} g(x) \cdot dx \equiv \lambda(\zeta), \quad \zeta > 0. \tag{2.29}$$

Three examples of one-sided transforms are given in table 2.1.[13]

More general is the one-sided Laplace-Stieltjes transform,

$$\mathcal{L}G = \int_0^{\infty} e^{-\zeta x} \cdot dG(x),$$

which specializes to (2.29) when G is absolutely continuous with density g.

Table 2.1. Laplace transforms of certain functions on $\Re^+$.

$g : \Re^+ \to \Re$	$\mathcal{L}g : \Re \to \Re$
$x^\alpha, \alpha > -1$	$\Gamma(\alpha+1)/\zeta^{\alpha+1}$
$x^{\alpha-1}e^{-x}, \alpha > 0$	$\Gamma(\alpha)(1+\zeta)^{-\alpha}, \zeta > -1$
$e^{-\alpha x^2}, \alpha > 0$	$e^{\zeta^2/\alpha}\sqrt{\frac{\pi}{4\alpha}}$

[13]Extensive tables are presented in Abramowitz and Stegun (1970) and Gradshteyn and Ryzhik (1980).

Fourier Transforms

Related to the two-sided Laplace transform, but applicable to a much broader class of functions, is the Fourier transform. This is defined as

$$\mathcal{F}g = \int_{-\infty}^{\infty} e^{i\zeta x} g(x) \cdot dx \equiv f(\zeta), \tag{2.30}$$

where $\zeta \in \Re$ and $i = \sqrt{-1}$. The identity (2.4) shows that $|e^{i\zeta x}| = 1$, so that by property **8** of the integral (page 35), $\mathcal{F}g$ exists for all ζ so long as g itself is integrable. In the case $\int_{-\infty}^{\infty} |f(\zeta)| \cdot d\zeta < \infty$ the inversion formula

$$g(x) = \mathcal{F}^{-1} f(\zeta) \equiv \frac{1}{2\pi} \int_{-\infty}^{\infty} e^{-i\zeta x} f(\zeta) \cdot d\zeta \tag{2.31}$$

allows g to be recovered from f. Some other important properties are (where primes denote derivatives)

1. $\mathcal{F}(bg + ch) = b\mathcal{F}g + c\mathcal{F}h$, where b, c are constants
2. $\mathcal{F}g' = -i\zeta \cdot f(\zeta)$ if $\lim_{|x|\to\infty} g(x) = 0$
3. $\mathcal{F}g'' = -\zeta^2 \cdot f(\zeta)$ if $\lim_{|x|\to\infty} g'(x) = 0$.

We also have the important convolution property:

$$\mathcal{F}(g * h) = \mathcal{F}g \cdot \mathcal{F}h, \tag{2.32}$$

where $(g * h)(x) \equiv \int g(x - y)h(y) \cdot dy$ represents the convolution of the two real functions g and h. The fact that this rather complicated process of convolution is reduced to the multiplication of transforms accounts for much of the usefulness of the Fourier transform in analysis and probability. We will encounter many applications. Three examples of Fourier transform pairs are in table 2.2, where $\mathfrak{C}$ represents the complex numbers.[14]

Table 2.2. Fourier transforms of certain functions on $\Re^+$.

$g : \Re^+ \to \Re$	$\mathcal{F}g : \Re \to \mathfrak{C}$
$\mathbf{1}_{[\alpha,\beta]}(x)$	$(e^{i\zeta\beta} - e^{i\zeta\alpha})/(i\zeta)$
$x^{\alpha-1}e^{-x}\mathbf{1}_{[0,\infty)}(x), \alpha > 0$	$\Gamma(\alpha)(1 - i\zeta)^{-\nu}$
$e^{-\alpha x^2}, \alpha > 0$	$e^{-\zeta^2/\alpha}\sqrt{\frac{\pi}{4\alpha}}$

[14]Abramowitz and Stegun (1970) and Gradshteyn and Ryzhik (1980) give extensive tabulations.

More general than (2.30) is the Fourier-Stieltjes transform,

$$\mathcal{F}G = \int_{\Re} e^{i\varsigma x} \cdot dG(x).$$

We encounter applications of Fourier-Stieltjes and Laplace-Stieltjes transforms in the next section.

2.2 Probability

This section presents the basic tools of probability from a measure-theoretic perspective and serves to introduce the continuous-time stochastic processes that will be the main subject of chapter 3.[15]

2.2.1 *Probability Spaces*

We begin with a collection Ω of outcomes of some chance experiment and a σ-field, $\mathcal{F}$, containing those subsets of Ω whose probabilities we will be able to determine. Although probabilities are just measures with particular properties, some new terminology is applied. Sets in $\mathcal{F}$ are now "events", and we say that event A "occurs" when the chance experiment has an outcome ω that is an element of set A. Disjoint sets are now "exclusive events". The σ-field $\mathcal{F}$ contains all countable unions and intersections of some basic class of events of interest, together with their complements. This implies, among other things, that $\Omega \in \mathcal{F}$. We take for granted that all events in the discussion that follows do belong to $\mathcal{F}$.

A probability measure $\mathbb{P}$ is a mapping from $\mathcal{F}$ into $[0, 1]$. $\mathbb{P}$ is thus a finite measure. The triple $(\Omega, \mathcal{F}, \mathbb{P})$ represents a probability space. The usual (frequentist) interpretation is that $\mathbb{P}(A)$ represents the proportion of replications of the experiment in which an event $A \in \mathcal{F}$ would occur if the chance experiment were repeated indefinitely while holding constant the conditions under one's control. If $\mathbb{P}(A) = 1$, then we say that A occurs "almost surely" (a.s.). The meaning is the same as "A occurs a.e. $\mathbb{P}$". If $\mathbb{P}(A) = 0$, then A is said to be a "null" set or, more explicitly, a "$\mathbb{P}$-null" set.

---

[15] Good general references, roughly in order of prerequisite knowledge, are Bartoszynski and Niewiadomska-Bugaj (1996), Williams (1991), Billingsley (1986), Chung (1974), and Laha and Rohatgi (1979).

The following fundamental properties of $\mathbb{P}$ are just assumed:

1. $\mathbb{P}(\Omega) = 1$
2. $\mathbb{P}(A) \geq 0$
3. $\mathbb{P}(\cup_{j=1}^{\infty} A_j) = \sum_{j=1}^{\infty} \mathbb{P}(A_j)$ whenever $A_j \cap A_{j'} = \emptyset$.

Of course, the first property is the only one that distinguishes $\mathbb{P}$ from other measures on $(\Omega, \mathcal{F})$ that are not also *probability* measures. From these three properties we then deduce the familiar complement rule, addition rule, etc. A familiarity with these basic ideas is assumed. Recall, also, that events comprising a finite collection $\{A_j\}_{j=1}^{n}$ are independent if $\mathbb{P}(A_{j_1} \cap \ldots \cap A_{j_m}) = \prod_{i=1}^{m} \mathbb{P}(A_{j_i})$ for $m \in \{2, 3, \ldots, n\}$ and each such subcollection $\{A_{j_i}\}_{i=1}^{m}$ of m distinct sets. Events in the infinite collection $\{A_n\}_{n=1}^{\infty}$ are independent if those in each finite collection are independent. The practical implication of independence is that the assessment of the probability of an event is not altered by knowledge that other, independent events have occurred. Other essential properties of $\mathbb{P}$ are

1. The monotone property. If $\{A_n\}_{n=1}^{\infty}$ is a monotone sequence of events that converges to a limit A, with either $A = \cup_{n=1}^{\infty} A_n$ (in the case that $A_n \uparrow$) or $A = \cap_{n=1}^{\infty} A_n$ (in the case that $A_n \downarrow$), then $\lim_{n\to\infty} \mathbb{P}(A_n) = \mathbb{P}(A)$. This follows from the monotone property of measures generally, (2.12).
2. The law of total probability. If $\Omega = \cup_{j=1}^{\infty} A_j$ with $A_j \cap A_{j'} = \emptyset$, then $\mathbb{P}(B) = \sum_{j=1}^{\infty} \mathbb{P}(B \cap A_j)$.
3. Borel-Cantelli lemma (convergence part). $\sum_{n=1}^{\infty} \mathbb{P}(A_n) < \infty$ implies $\mathbb{P}$(infinitely many of the events $\{A_n\}$ occur) $= 0$.
4. Borel-Cantelli lemma (divergence part). If $\{A_n\}_{n=1}^{\infty}$ are independent, $\sum_{n=1}^{\infty} \mathbb{P}(A_n) = \infty$ implies $\mathbb{P}$(infinitely many of the events $\{A_n\}$ occur) $= 1$.

The event that infinitely many of the $\{A_n\}$ occur is expressed symbolically as $\cap_{n=1}^{\infty} \cup_{m=n}^{\infty} A_m$ or as $\lim_{n\to\infty} \sup A_n$. The complement of this event is $(\lim_{n\to\infty} \sup A_n)^c = \lim_{n\to\infty} \inf A_n^c \equiv \cup_{n=1}^{\infty} \cap_{m=n}^{\infty} A_n^c$. This represents the event that no events in the sequence occur from *some* point forward.

Example 7 *A sequence of independent, random draws is made from an urn containing, at the* n*th stage of the sequence, one red ball and* $n-1$ *white balls. At stage* n *the probability of event red is thus* $1/n$*. A draw is made and then another white ball is added before proceeding to stage* $n+1$*. What is the probability that red occurs infinitely many times? Since* $\sum_{n=1}^{\infty} \mathbb{P}($*red* *at stage n)* $= \sum_{n=1}^{\infty} n^{-1} = \infty$*, and since events at different stages are independent, the divergence part of Borel-Cantelli implies that a red ball is certain to be drawn infinitely many times in an infinite sequence of draws. Put differently, this tells us that there is* never *a stage beyond which it is certain that only white balls will be drawn.*

Example 8 *In an otherwise identical urn experiment, suppose the number of white balls at stage* n *is* $2^{n-1} - 1$*, so that the probability of red at stage* n *is* 2^{-n+1}*. A draw is made and then* 2^{n-1} *more white balls are added before proceeding to stage* $n + 1$*. Since* $\sum_{n=1}^{\infty} 2^{-n+1} = 2 < \infty$*, the convergence part of Borel-Cantelli tells us that a red ball is certain* not *to be drawn infinitely many times. Put differently, there is some (undetermined) stage beyond which we can be sure that only white will be drawn.*

2.2.2 *Random Variables and Their Distributions*

A (scalar-valued) random variable on a probability space $(\Omega, \mathcal{F}, \mathbb{P})$ is an $\mathcal{F}$-measurable mapping from Ω to $\Re$. (Vector-valued random variables are treated below.) That is, as a mapping from Ω to $\Re$, a random variable just assigns a number to an outcome; for example, the number 4 to the outcome $\{::\}$ in the experiment of rolling a 6-sided die, or the number 52 to the outcome $\{A\spadesuit\}$ in drawing a card from an ordinary deck. Such functions are written as $X(\omega), Y(\omega), Z(\omega), \ldots$ etc., or simply $X, Y, Z, \ldots$ for short. Focusing on some particular random variable X, recall that $\mathcal{F}$-measurability requires that the inverse image under X of any Borel set B be contained in $\mathcal{F}$; that is, that $X^{-1}(B) \equiv \{\omega : X(\omega) \in B\} \in \mathcal{F}$ for each $B \in \mathcal{B}$. Recall, too, that this is guaranteed if sets of the specific form $X^{-1}((-\infty, x]) \equiv \{\omega : X(\omega) \leq x\}$ belong to $\mathcal{F}$ for each $x \in \Re$. The shorthand notation $X \in \mathcal{F}$ is often used to state that X is $\mathcal{F}$-measurable. Examples of measurable and nonmeasurable functions are given below.

Since $\mathcal{F}$ is the collection of events to which probabilities can be assigned, the measurability of X implies that $\mathbb{P}\left(X^{-1}(B)\right)$ exists for each Borel set B and defines a new mapping, $\mathbb{P}_X$ say, from $\mathcal{B}$ into $[0, 1]$. Since the single-valuedness of functions implies that $X^{-1}(B_j) \cap X^{-1}(B_{j'}) = \emptyset$ whenever $B_j \cap B_{j'} = \emptyset$, and since $\mathbb{P}$ is countably additive, it follows that $\mathbb{P}_X\left(\cup_{j=1}^{\infty} B_j\right) = \sum_{j=1}^{\infty} \mathbb{P}_X(B_j)$ for a sequence of such exclusive events. Thus, $\mathbb{P}_X$ is itself a probability measure and is referred to as the measure "induced" by the random variable X. Thus, introducing a random variable on a probability space $(\Omega, \mathcal{F}, \mathbb{P})$ gives rise to an entirely new, induced probability space,

$(\Re, \mathcal{B}, \mathbb{P}_X)$. Stated formally in words, $\mathbb{P}_X(B)$ represents the probability of the event that an outcome ω occurs to which X assigns a value in B. Stated more simply, it is the probability that X takes a value in B. Often, we will express this in an abbreviated form as $\mathbb{P}(X \in B)$.

The fineness of the structure of the σ-field $\mathcal{F}$ governs the amount of detail we can have about the results of a chance experiment. How fine the structure needs to be in modeling a given experiment is determined by our interest.

For example, suppose we are to roll an ordinary six-sided die. If what matters is only whether an even number or an odd turns up, then it is enough to know the probabilities of just $\{\cdot, \cdots, :\cdot:\}$ and $\{:, ::, :::\}$—plus the events that can be built up from these by intersection and union. We can then take as the overall space the set Ω_2 that contains just these two elementary outcomes and can work with the σ-field $\mathcal{F}_2$ (which in this case is just a field[16]) that is generated from these two events. The collection $\mathcal{F}_2$ comprises the two events themselves together with their intersection and union, $\emptyset$ and Ω_2. A random variable $Y(\omega)$ with, say, $Y(\{:, ::, :::\}) = 0$ and $Y(\{\cdot, \cdots, :\cdot:\}) = 1$, is then measurable with respect to $\mathcal{F}_2$, since we can tell whether Y takes on a value in any arbitrary Borel subset of $\Re$ if we know whether each event in $\mathcal{F}_2$ occurs.

On the other hand, the very limited information structure $\mathcal{F}_2$ will not suffice if we need to know more about what happens. For example, suppose we want to determine the probability associated with each possible number of dots that can turn up on the roll of the die. To define the random variable $X(\omega)$ = number of dots on side ω, we will have to work with an overall space that is at least as detailed as the set Ω_6 that comprises the elementary outcomes $\{\cdot, :, \cdots, ::, :\cdot:, :::\}$. We will then need a field $\mathcal{F}_6$ that contains all 2^6 subsets of Ω_6, including $\emptyset$ and Ω_6 itself. Now $X \in \mathcal{F}_6$ (that is, X is measurable with respect to $\mathcal{F}_6$), whereas $X \notin \mathcal{F}_2$ since one cannot tell whether events like $X = 1$ and $3 \le X \le 5$ occur just from the events in $\mathcal{F}_2$. On the other hand, the random variable Y defined in the previous paragraph is $\mathcal{F}_6$-measurable as well as $\mathcal{F}_2$-measurable, since $\mathcal{F}_6$ contains all the sets of $\mathcal{F}_2$. This makes $\mathcal{F}_2$ a subfield of $\mathcal{F}_6$. Being the

[16] "Fields" of subsets of some space Ω are collections closed under *finitely* many set operations (unions, intersections, complementations), whereas σ-fields remain closed under *countably* many such operations. When Ω comprises a finite number of elementary outcomes, the field of sets built up from these is also a σ-field, since any set can be represented both as the union and as the intersection of countably many copies of itself.

coarsest information structure that makes X measurable, $\mathcal{F}_6$ is said to be the σ-field "generated" by X; likewise, $\mathcal{F}_2$ is the σ-field generated by Y.

More detail about the die experiment might be required than even $\mathcal{F}_6$ provides. For instance, one might also want to assess probabilities of events describing the die's orientation after it comes to rest or its position in some coordinate system. Depending on how precisely these are to be measured, Ω might then have to comprise an infinite number of elementary outcomes.[17]

Representations of Distributions

A "support" of a random variable X is a Borel set $\mathcal{X}$ such that $\mathbb{P}(X \in \mathcal{X}) \equiv \mathbb{P}_X(\mathcal{X}) = \mathbb{P}\left(X^{-1}(\mathcal{X})\right) = \mathbb{P}(\{\omega : X(\omega) \in \mathcal{X}\}) = 1$. Clearly, $\Re$ itself is always a support since $\mathbb{P}_X(\Re) = \mathbb{P}\left(X^{-1}(\Re)\right) = \mathbb{P}(\Omega) = 1$, but $\mathcal{X}$ may also be a proper subset of $\Re$. Thus, $\mathcal{X} = \{1, 2, \ldots, 6\}$ supports X in the die example. The induced probabilities associated with events of the form $X^{-1}((-\infty, x]) \equiv \{\omega : X(\omega) \leq x\}$ show how the unit mass of probability is distributed over the real line. Because it is often more convenient to work with point functions than with measures, we can make use of an equivalent representation in terms of the cumulative distribution function (c.d.f.) of the random variable X. The c.d.f. of X at some $x \in \Re$ is defined as

$$F(x) \equiv \mathbb{P}_X((-\infty, x]) \equiv \mathbb{P}(X \leq x).$$

This mapping from $\Re$ into $[0, 1]$ has the following properties:

1. $F(-\infty) = 0$, since (by the monotone property)

$$\lim_{n \to \infty} \mathbb{P}_X((-\infty, -n]) = \mathbb{P}_X(\cap_{n=1}^{\infty}(-\infty, -n]) = \mathbb{P}_X(\emptyset).$$

2. $F(+\infty) = 1$, since (again by the monotone property)

$$\lim_{n \to \infty} \mathbb{P}_X((-\infty, n]) = \mathbb{P}_X(\cup_{n=1}^{\infty}(-\infty, n]) = \mathbb{P}_X(\Re).$$

3. $F(y) - F(x) \geq 0$ when $x < y$, since

$$\begin{aligned} \mathbb{P}_X((-\infty, y]) - \mathbb{P}_X((-\infty, x]) &= \mathbb{P}_X((-\infty, x] \cup (x, y]) - \mathbb{P}_X((-\infty, x]) \\ &= \mathbb{P}_X((x, y]). \end{aligned}$$

[17] In that case the distinction between finite and countable additivity—between fields and σ-fields of subsets—would indeed arise.

4. $F(y) - F(y-) \geq 0$, since

$$\begin{aligned} F(y) - \lim_{n\to\infty} F(y - 1/n) &= \lim_{n\to\infty} \mathbb{P}_X((y - 1/n, y]) \\ &= \mathbb{P}_X(\cap_{n=1}^{\infty}(y - 1/n, y]) \\ &= \mathbb{P}_X(\{y\}). \end{aligned}$$

5. $F(y+) - F(y) = 0$, since

$$\lim_{n\to\infty} F(y + 1/n) - F(y) = \mathbb{P}_X(\cap_{n=1}^{\infty}(y, y + 1/n]) = \mathbb{P}_X(\emptyset).$$

The nature of F is the basis for a taxonomy of random variables. When there is a countable collection of discrete points $\{x_j\}_{j=1}^{\infty}$, called "atoms", such that $\sum_{j=1}^{\infty} F(x_j - x_j-) \equiv \sum_{j=1}^{\infty} \mathbb{P}_X(\{x_j\}) = 1$, then X is called a "discrete" random variable. In other words X is a discrete random variable when an atomic set is a support. In this case F is a step function that increases only by discontinuous jumps at these discrete points. Since F is simply flat between the atoms, the probability distribution of X can be represented more compactly by just identifying the atoms themselves and specifying how much probability mass is concentrated at each one. Defining $f(x) = F(x) - F(x-)$ as the "probability mass function" (p.m.f.) of the discrete random variable X, we see that $f(x_j) > 0$ for each x_j in the atomic set and $\sum_{j=1}^{\infty} f(x_j) = 1$. Models for the various discrete distributions we encounter are described in section 2.2.7 (table 2.3). Most common are the lattice distributions, in which the atoms are equally spaced, typically comprising some subset of the nonnegative integers, $\aleph_0 = \{0, 1, 2, \ldots.\}$.

If the distribution of X contains no such atoms, then F is continuous, and X is a "continuous" random variable. If, in addition, F is absolutely continuous, then there exists a density f such that $\mathbb{P}_X((a, b]) = F(b) - F(a) = \int_a^b f(x) \cdot dx$ for arbitrary $a \leq b$. In that case X is an "absolutely continuous" random variable, and f is called the "probability density function" (p.d.f.) of X.[18] We will not encounter the other type of continuous random variable, "continuous singular". Since absolute continuity of F implies differentiability a.e. with respect to Lebesgue measure, $f(x)$ can be identified with $F'(x)$ except on the $\mathbb{P}_X$-null set of exceptional points

[18] C.d.f. F is absolutely continuous if and only if induced measure $\mathbb{P}_X$ is absolutely continuous with respect to Lebesgue measure μ. Thus, sets that are μ-null are also $\mathbb{P}_X$-null, and p.d.f. f corresponds to Radon-Nikodym derivative $d\mathbb{P}_X/d\mu$. If X is a discrete random variable on a finite space, then $\mathbb{P}_X$ is absolutely continuous with respect to counting measure $\#$, which maps subsets of countable spaces into $\aleph_0 \cup \{\infty\}$. P.m.f. f then corresponds to Radon-Nikodym derivative $d\mathbb{P}_X/d\#$.

where F' fails to exist. At these exceptional points f can be defined to have any value that makes the representation convenient.[19] Clearly, properties **1** and **2** of the c.d.f. imply that $\int_{-\infty}^{\infty} f(x) \cdot dx = 1$. Models for the various continuous distributions we encounter are summarized in section 2.2.7 (table 2.3).

"Mixed" random variables that combine the two pure types are also possible, and indeed are encountered often in the study of financial derivatives. In these cases X has a countable set D of atoms such that $0 < \sum_{x \in D} [F(x) - F(x-)] < 1$, but there is also a density f such that

$$\mathbb{P}_X(B) = \int_B f(x) \cdot dx + \sum_{x \in D \cap B} [F(x) - F(x-)] \tag{2.33}$$

for each $B \in \mathcal{B}$.

Example 9 *A fair pointer on a circular scale of unit circumference is spun and eventually comes to rest at some physical point ω. Taking an arbitrary ω_0 as the origin, let $X(\omega)$ be the clockwise distance from ω_0 to ω, as depicted in figure 2.1, and define $Y(\omega) = \max[X(\omega), 1/2]$. Then if we model the c.d.f. of X as*

$$F_X(x) = \left\{ \begin{array}{ll} 0, & x < 0 \\ x, & 0 \leq x < 1 \\ 1, & x \geq 1 \end{array} \right\} = x\mathbf{1}_{[0,1)}(x) + \mathbf{1}_{[1,\infty)}(x)$$

the implied model for the c.d.f. of Y is

$$F_Y(y) = \left\{ \begin{array}{ll} 0, & y < 1/2 \\ y, & 1/2 \leq y < 1 \\ 1, & y \geq 1. \end{array} \right\} = y\mathbf{1}_{[.5,1)}(y) + \mathbf{1}_{[1,\infty)}(y).$$

The model for F_X implies that X is absolutely continuous, with p.d.f. $f(x) = \mathbf{1}_{[0,1]}(x)$ (where we arbitrarily define $f(x) = 1$ at $x = 0$ and $x = 1$, where F' does not exist). However, the implied model for F_Y is such that Y is neither purely discrete nor purely continuous.

[19] Since F is nondecreasing by property **3** above, it follows that $f \geq 0$ at points where F' does exist, and for consistency it is conventional to choose nonnegative values also at the exceptional points.

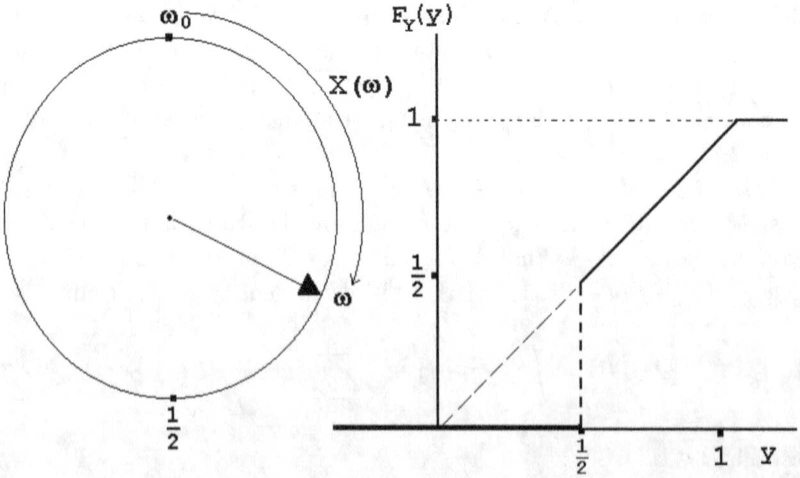

Fig. 2.1. Experiment and c.d.f for mixed random variable Y.

The Multidimensional Case

The concepts of random variables and the representations of their distributions extend to higher dimension. Thus, on the space $(\Omega, \mathcal{F}, \mathbb{P})$ the $\mathcal{F}$-measurable mapping $\mathbf{X} = (X_1, X_2, \ldots, X_k)' : \Omega \to \Re_k$ defines a vector-valued random variable. For examples, an experiment could involve drawing one individual at random from a population and recording k quantitative characteristics (age, years of schooling, annual income, etc.), or it could involve sampling k individuals at random from the same population and recording one quantifiable characteristic for each. Induced probability measure $\mathbb{P}_{\mathbf{X}}$ and joint c.d.f. $F(\mathbf{x})$ and joint p.m.f. or p.d.f. $f(\mathbf{x})$ are developed in ways that parallel the one-dimensional case. Thus, if $B \in \mathcal{B}_k$ (the Borel sets of $\Re_k$) then $\mathbb{P}_{\mathbf{X}}(B) = \mathbb{P}(\mathbf{X}^{-1}(B))$, and $F(\mathbf{x}) = F(x_1, \ldots, x_k)$ is the $\mathbb{P}_{\mathbf{X}}$ measure of the product set $(-\infty, x_1] \times \ldots \times (-\infty, x_k]$. From the induced measure $\mathbb{P}_{\mathbf{X}}$ one obtains the induced "marginal" probability measure corresponding to a subset $X_1, \ldots, X_j$ as

$$\mathbb{P}_{X_1,\ldots,X_j}(B) = \mathbb{P}_{X_1,\ldots,X_j,X_{j+1},\ldots,X_k}(B \times \Re_{k-j}),$$

where $B \in \mathcal{B}_j$ and $j < k$. Likewise, the marginal c.d.f. is

$$F_{X_1,\ldots,X_j}(x_1, \ldots, x_j) = F_{X_1,\ldots,X_j,X_{j+1},\ldots,X_k}(x_1, \ldots, x_j, +\infty, \ldots, +\infty).$$

In the absolutely continuous case the joint p.d.f. is developed as the derivative of the c.d.f. where this exists, as it does a.e. $\mathbb{P}_X$:

$$f(x_1, \ldots, x_k) = \partial^k F(x_1, \ldots, x_k) / \partial x_1 \cdot \ldots \cdot \partial x_k.$$

Random variables $X_1, X_2, \ldots, X_k$ are independent if and only if events $X_1^{-1}(B_1), X_2^{-1}(B_2), \ldots, X_k^{-1}(B_k)$ are independent for each finite collection of sets $\{B_j\}_{j=1}^k$ with $B_j \in \mathcal{B}$. This is the case if and only if the joint induced probability measure of a product set factors as the product of marginal measures, as

$$\mathbb{P}_{\mathbf{X}}(B_1 \times B_2 \times \ldots \times B_k) = \mathbb{P}_{X_1}(B_1)\mathbb{P}_{X_2}(B_2) \cdot \ldots \cdot \mathbb{P}_{X_k}(B_k).$$

Equivalently, independence holds if and only if the joint c.d.f. factors into marginal c.d.f.s; that is,

$$F_{\mathbf{X}}(\mathbf{x}) = F_1(x_1)F_2(x_2) \cdot \ldots \cdot F_k(x_k),$$

for each $\mathbf{x} \in \Re_k$. Likewise, random variables that are all purely discrete or purely absolutely continuous are independent if and only if the joint p.m.f. or p.d.f., respectively, factors as the product of the marginals:

$$f_{\mathbf{X}}(\mathbf{x}) = f_1(x_1)f_2(x_2) \cdot \ldots \cdot f_k(x_k).$$

The implication of independence is that assessments of probabilities of events defined in terms of any subset of $X_1, \ldots, X_k$ are not altered by information about the realizations of the remaining variables.

2.2.3 *Mathematical Expectation*

Integrals of functions on abstract measure spaces were defined in section 2.1.6, and the concept can be carried over directly to random variables. As usual, however, some new jargon is attached. The "expected value" or "mathematical expectation" of the random variable X, denoted EX, is defined as

$$EX \equiv \int X(\omega) \cdot d\mathbb{P}(\omega). \tag{2.34}$$

Recalling the properties of the abstract integral, we say that EX exists when X is absolutely integrable; that is, when $\int |X(\omega)| \cdot d\mathbb{P}(\omega) < \infty$. For some $g : \Re \to \Re$ we can likewise define the expected value of the composition as

$$Eg(X) \equiv \int g\left[X(\omega)\right] \cdot d\mathbb{P}(\omega) \tag{2.35}$$

whenever $E|g(X)|$ is finite. Of course, (2.35) encompasses (2.34) on taking $g(X) = X$. $Eg(X)$ can be expressed in other ways. Translating to the induced probability space $(\Re, \mathcal{B}, \mathbb{P}_X)$, we have

$$Eg(X) = \int_{\Re} g(x) \cdot d\mathbb{P}_X(x),$$

or, in terms of the c.d.f.,

$$Eg(X) = \int_{\Re} g(x) \cdot dF(x), \tag{2.36}$$

where the latter is interpreted as a Lebesgue-Stieltjes integral.

In the case that X is a purely discrete random variable with atomic set D expression (2.36) can be written in terms of the p.m.f. as

$$Eg(X) = \sum_{x \in D} g(x)f(x);$$

while if X is absolutely continuous it can be written in terms of the p.d.f.:

$$Eg(X) = \int_{-\infty}^{\infty} g(x)f(x) \cdot dx.$$

In the mixed case the expectation can be expressed as

$$Eg(X) = \sum_{x \in D} g(x)\left[F(x) - F(x-)\right] + \int_{-\infty}^{\infty} g(x)f(x) \cdot dx,$$

where D is the countable set of atoms and f is the density that appears in (2.33).

Example 10 *With F_Y as in example 9 and $g(Y) = e^Y$ we have $D = \{0.5\}$ and $F(.5) - F(.5-) = 0.5$. Taking $f(y) = \mathbf{1}_{(.5,1)}(y)$ gives*

$$Ee^Y = .5e^{.5} + \int_{.5}^{1} e^y \cdot dy = e - \sqrt{e}.$$

The Stieltjes representation, as in (2.36), is especially convenient because it encompasses all of these cases.

The nine properties of the integral listed on page 35 imply the following properties for expectation, where a, b, c are constants, $\{B_j\}_{j=1}^{\infty}$ are disjoint Borel sets, and $\mathbb{P}_X(B_0) = 0$:

1. $E\mathbf{1}_{B_1}(X)\,g(X) = \int_{B_1} g(x) \cdot d\mathbb{P}_X(x)$
2. $E\left[\mathbf{1}_{\cup_{j=1}^{\infty} B_j}(X)\,g(X)\right] = \sum_{j=1}^{\infty} E\mathbf{1}_{B_j}(X)\,g(X)$
3. $Ea \equiv E[a\mathbf{1}_{\Re}(X)] = a\mathbb{P}_X(\Re) = a$
4. $E\left[bg(X)\right] = bEg(X)$
5. $E[a + bg(X) + ch(X)] = a + bEg(X) + cEh(X)$
6. $g(X) \leq h(X)$ a.s. $\Rightarrow Eg(X) \leq Eh(X)$
7. $g(X) = h(X)$ a.s. $\Rightarrow Eg(X) = Eh(X)$

8. $|Eg(X)| \leq E|g(X)|$

9. $\mathbb{P}_X(B_0) = 0 \Rightarrow E\mathbf{1}_{B_0}(X)\, g(X) = 0.$

Property 1 implies that $\mathbb{P}_X(B_1) \equiv \mathbb{P}(X \in B_1)$ can be represented as an expectation through the use of indicator functions: $\mathbb{P}_X(B_1) = E\mathbf{1}_{B_1}(X) = \int_{B_1} d\mathbb{P}_X(x) = \int_{B_1} dF(x)$.

Expectation in the multivariate case is treated by regarding the variables as vectors in the expressions above and treating sums and integrals as multiple. The new result in the case of independent random variables is that the expectation of the product of functions of the individual variables equals the product of the expectations, given that all the expectations exist. That is,

$$\begin{aligned} &Eg_1(X_1) \cdot \ldots \cdot g_k(X_k) \\ &\quad = \int_{\Re_k} g_1(x_1) \cdot \ldots \cdot g_k(x_k) \cdot dF_{\mathbf{X}}(\mathbf{x}) \\ &\quad = \int \int \cdots \int g_1(x_1) \cdot \ldots \cdot g_k(x_k) \cdot dF_{X_1}(x_1) \ldots dF_{X_k}(x_k) \\ &\quad = Eg_1(X_1) \cdot \ldots \cdot Eg_k(X_k), \end{aligned} \tag{2.37}$$

where the last line follows from Fubini's theorem.

Uniformly Integrable Random Variables

The concept of uniform integrability of functions, defined in (2.14), carries over to families of random variables. Random variables $\{X_\alpha\}_{\alpha \in \mathcal{A}}$ defined on $(\Omega, \mathcal{F}, \mathbb{P})$ are said to be uniformly integrable if

$$\lim_{n \to \infty} \sup_{\alpha \in \mathcal{A}} E|X_\alpha| \mathbf{1}_{\{|X_\alpha| > n\}} = 0. \tag{2.38}$$

Since $\sup_{\alpha \in \mathcal{A}} E|X_\alpha| = \sup_{\alpha \in \mathcal{A}} E|X_\alpha|\mathbf{1}_{\{|X_\alpha| > n\}} + \sup_{\alpha \in \mathcal{A}} E|X_\alpha|\mathbf{1}_{\{|X_\alpha| \leq n\}}$ it follows that

$$\lim_{n \to \infty} \sup_{\alpha \in \mathcal{A}} \left[E|X_\alpha| - E|X_\alpha|\mathbf{1}_{\{|X_\alpha| \leq n\}} \right] = 0$$

when (2.38) holds and therefore that $\sup_{\alpha \in \mathcal{A}} E|X_\alpha| < \infty$. Thus, uniform integrability implies that the expectations $E|X_\alpha|$ are bounded; that is, that there exists $b < \infty$ such that $E|X_\alpha| \leq b$ for each $\alpha \in \mathcal{A}$.[20]

[20]On the other hand, $\sup_{\alpha \in \mathcal{A}} E|X_\alpha| < \infty$ does not imply that the $\{X_\alpha\}$ are uniformly integrable. For a counterexample see Williams (1991, p. 127). However, the condition $\sup_{\alpha \in \mathcal{A}} E|X_\alpha|^p < \infty$ for some $p > 1$ does suffice. Obviously, it is also true that random variables that are a.s. uniformly *bounded* are uniformly integrable.

Moments

The kth "moment" about the origin of a scalar-valued random variable X is defined as[21] $EX^k = \int x^k \cdot dF(x)$ and denoted μ'_k. The kth moment exists if the corresponding absolute moment, $\nu_k \equiv \int |x|^k \cdot dF(x)$, is finite. Of course, $\mu'_k \leq \nu_k$, with equality guaranteed both when k is even and when $\Re^+$ is a support for X. While we sometimes encounter moments of negative or fractional order, the positive-integer moments ($k \in \aleph$) are usually the ones of interest. It is easy to see that the existence of μ'_k implies the existence of μ'_j for $j = 1, 2, \ldots, k$. Of course $\mu'_0 = 1$ invariably. The first moment about the origin, EX, is represented simply as μ and is called the "mean" of X. This corresponds to the center of mass of the probability distribution of X.

The following representation of the mean is often useful and is easily proved using integration by parts:

$$EX = -\int_{-\infty}^{0} F(x) \cdot dx + \int_{0}^{\infty} [1 - F(x)] \cdot dx. \tag{2.39}$$

When X is supported on $\Re^+$, this becomes simply

$$EX = \int_{0}^{\infty} [1 - F(x)] \cdot dx = \int_{0}^{\infty} \mathbb{P}(X > x) \cdot dx. \tag{2.40}$$

In dealing with discrete random variables having integer support (that is, variables with lattice distributions) we will sometimes encounter the "descending factorial moments", defined as

$$\mu_{(k)} = EX^{(k)} \equiv E\left[X(X-1) \cdot \ldots \cdot (X-k+1)\right].$$

"Central" moments, or moments about the mean, are defined as $E(X - \mu)^k$ and denoted μ_k without the prime. The binomial expansion

$$(X - \mu)^k = \sum_{j=0}^{k} \binom{k}{j} X^{k-j}(-\mu)^j$$

shows that the kth central moment (and, for that matter, central moments of all integer orders less than k) exists if μ'_k exists. The second central moment, μ_2, is the "variance" of X and is usually denoted specially as

[21] Throughout, expressions like EX^k or $Eg(X)$ always have the same meaning as $E(X^k)$ or $E[g(X)]$, parentheses or brackets being used only when there is ambiguity about what the operation E applies to. Parentheses are invariably used when expectations appear as arguments of functions, as $(EX)^k$, $g(EX)$, $(EX^j)^k$, etc.

σ^2 or as VX. The standard deviation, σ, is defined as $+\sqrt{\sigma^2} = +\sqrt{VX}$. The transformation $X' = (X - \mu)/\sigma$ puts random variable X in "standard form", with $EX' = 0$ and $VX' = 1$. The third and fourth moments of standardized random variables are often used as summary indicators of asymmetry (skewness) and tail thickness (kurtosis). Specifically, the coefficients of skewness and kurtosis are

$$\alpha_3 \equiv \mu_3/\sigma^3$$
$$\alpha_4 \equiv \mu_4/\sigma^4.$$

Distributions with long right (left) tails are said to be right (left) skewed. Typically in such cases the sign of α_3 indicates the direction of skewness—positive for right- and negative for left-skewed. Obviously, symmetric distributions possessing third moments have $\alpha_3 = 0$. The coefficient of kurtosis is a common (though often unreliable) indicator of tail thickness relative to the normal distribution, for which $\alpha_4 = 3.0$. The quantity $\alpha_4 - 3$ is sometimes called the "excess" kurtosis. One typically thinks of symmetric distributions with $\alpha_4 > 3.0$ as having thicker tails than a normal distribution with the same μ and σ^2, and higher density near the mean. Such distributions are said to be "leptokurtic". Empirical marginal distributions of price changes of financial assets are often of this sort.

The only additional concept needed in the multivariate case, $\mathbf{X} : \Omega \longrightarrow \Re_k$, is that of "product moments". These are expectations of the form $E(X_{j_1} - a_{j_1})^{m_1} \cdot \ldots \cdot (X_{j_\ell} - a_{j_\ell})^{m_\ell}$, where the a's are constants—usually equal either to zero or to the respective means. By far the most important case is the bivariate concept of "covariance". The covariance between random variables X_1 and X_2 is defined as $E(X_1 - \mu_1)(X_2 - \mu_2)$ and denoted σ_{12} or $Cov(X_1, X_2)$. The coefficient of correlation between X_1 and X_2, denoted ρ_{12} or $Corr(X_1, X_2)$, is the ratio of the covariance to the product of the standard deviations: $\rho_{12} \equiv \sigma_{12}/(\sigma_1\sigma_2)$.

Inequalities

The following inequalities for random variables are frequently used:

1. Jensen: If $g(\cdot)$ is a convex function on a support $\mathcal{X}$ of X and $E|X| < \infty$, then $Eg(X) \geq g(EX)$. [Proof: let a and b be real numbers such that $g(x) \geq a + bx$ for $x \in \mathcal{X}$ and $g(EX) = a + bEX$. Convexity implies that such a and b exist, not necessarily uniquely. Then $Eg(X) \geq E(a + bX) = a + bEX = g(EX)$.]

2. Markov: If $g(\cdot) \geq 0$ on a support $\mathcal{X}$ of X and if $\varepsilon > 0$, then $\mathbb{P}(g(X) \geq \varepsilon) \leq Eg(X)/\varepsilon$. [Proof: let $\mathcal{G}_\varepsilon \equiv \{x : g(x) \geq \varepsilon\}$. Then $Eg(X) = \int_{\mathcal{G}_\varepsilon} g(x) \cdot dF(x) + \int_{\mathcal{X}\backslash\mathcal{G}_\varepsilon} g(x) \cdot dF(x) \geq \int_{\mathcal{G}_\varepsilon} g(x) \cdot dF(x) \geq \varepsilon \mathbb{P}_X(\mathcal{G}_\varepsilon)$.]

3. Schwarz: If $EX^2 < \infty$ and $EY^2 < \infty$, $(EXY)^2 \leq EX^2 \cdot EY^2$. [Proof: For some $n \in \aleph$ put $X_n = |X| \wedge n$, $Y_n = |Y| \wedge n$. EX_nY_n exists since $X_nY_n \leq n^2$, so $\forall t \in \Re$ we have

$$\begin{aligned} 0 &\leq E(X_n + tY_n)^2 \\ &= (1, t)\begin{pmatrix} EX_n^2 & EX_nY_n \\ EX_nY_n & EY_n^2 \end{pmatrix}\begin{pmatrix} 1 \\ t \end{pmatrix}. \end{aligned}$$

The matrix is thus positive semidefinite, which implies $(EX_nY_n)^2 \leq EX_n^2EY_n^2$. Now let $n \longrightarrow \infty$ and apply dominated convergence to conclude that $EX_n^2 \uparrow EX^2$, $EY_n^2 \uparrow EY^2$, and $(EX_nY_n)^2 \uparrow (E|XY|)^2 \leq EX^2 \cdot EY^2$.] The special case $\mathbb{P}(X = a_j, Y = b_j) = 1/n$ for $j \in \{1, 2, \ldots, n\}$ and $\mathbb{P}(X = x, Y = y) = 0$ elsewhere gives the deterministic version:

$$\left(\sum_{j=1}^n a_j b_j\right)^2 \leq \sum_{j=1}^n a_j^2 \sum_{j=1}^n b_j^2. \tag{2.41}$$

The special case that X and Y are in standard form gives $-1 \leq \rho_{XY} \leq 1$ as bounds for the coefficient of correlation.

Generating Functions and Characteristic Functions

The moment generating function (m.g.f.) of a random variable X is equivalent to the two-sided Laplace-Stieltjes transform of its c.d.f.:

$$\mathfrak{M}(\zeta) = Ee^{\zeta X} = \int e^{\zeta x} \cdot dF(x).$$

The m.g.f. exists when the integral is finite for all real ζ in some open interval about $\zeta = 0$. Of course, $\mathfrak{M}(0) = 1$ always. Differentiating $\mathfrak{M}(\cdot)$ k times with respect to ζ and interchanging the operations of integration and derivation, as is justified by dominated convergence, shows that $\mathfrak{M}^{(k)}(\zeta) = \int x^k e^{\zeta x} \cdot dF(x)$ and hence $\mathfrak{M}^{(k)}(0) = EX^k$. This is the sense in which $\mathfrak{M}(\cdot)$ "generates" the moments. Existence of $\mathfrak{M}(\cdot)$ implies the existence of *all* positive integer moments of X, but the converse does not hold.[22]

[22]The lognormal distribution (section 2.2.7) has finite moments of all orders but no m.g.f.

The natural log of the m.g.f., $\mathfrak{L}(\zeta) \equiv \ln \mathfrak{M}(\zeta)$, is the "cumulant generating function" (c.g.f.) of X. It can be expanded in Taylor series about $\zeta = 0$ to give

$$\mathfrak{L}(\zeta) = \sum_{j=1}^{\infty} \kappa_j \zeta^j / j!,$$

where the $\{\kappa_j\}$, which are defined from this relation, are the "cumulants" of X. The first four cumulants are

$$\begin{aligned}
\kappa_1 &= \mu \\
\kappa_2 &= \mu_2' - \mu^2 = \mu_2 = \sigma^2 \\
\kappa_3 &= \mu_3' - 3\mu\mu_2' + 2\mu^2 = \mu_3 \\
\kappa_4 &= \mu_4' - 4\mu_3'\mu - 3(\mu_2')^2 + 12\mu_2'\mu - 6\mu^4 = \mu_4 - 3\sigma^4.
\end{aligned}$$

It follows that the first three derivatives of $\mathfrak{L}(\zeta)$ evaluated at $\zeta = 0$ are the mean, the variance, and the third central moment, respectively, while $\mathfrak{L}^{(4)}(0)$ is the fourth central moment minus three times the square of the variance. The coefficients of skewness and excess kurtosis can be calculated from the c.g.f. as

$$\alpha_3 = \mathfrak{L}^{(3)}(0)/[\mathfrak{L}''(0)]^{3/2} = \kappa_3/\kappa_2^{3/2} \tag{2.42}$$

$$\alpha_4 - 3 = \mathfrak{L}^{(4)}(0)/[\mathfrak{L}''(0)]^2 = \kappa_4/\kappa_2^2. \tag{2.43}$$

The related "probability generating function" (p.g.f.) is useful in characterizing distributions supported on the nonnegative integers. This is defined as

$$\Pi(\zeta) \equiv E\zeta^X = \sum_{x \in \mathcal{X}} \zeta^x f(x),$$

for $\zeta \in \mathfrak{C}$ (the complex numbers) and restricted to some region of convergence. Clearly, there is always convergence for $|\zeta| \leq 1$; in particular, $\Pi(1) = 1$. The obvious identity $\Pi(\zeta) = \mathfrak{M}(\ln \zeta)$ for real, positive ζ shows that convergence in a neighborhood of $\zeta = 1$ implies existence of the m.g.f. and hence of all the moments. In this case the derivatives of $\Pi(\cdot)$ evaluated at $\zeta = 1$ equal the descending factorial moments, as

$$\Pi^{(k)}(1) = \sum_{x \in \mathcal{X}} x(x-1) \cdot \ldots \cdot (x-k+1) f(x) = \mu_{(k)}.$$

In any case the derivatives evaluated at $\zeta = 0$ generate the probabilities, as $\Pi^{(k)}(0)/k! = f(k) = \mathbb{P}_X(\{k\})$.

The most important of the generating functions for our purposes is the "characteristic function" (c.f.). The c.f. of a random variable X is the Fourier-Stieltjes transform of the c.d.f.:

$$\Psi(\zeta) = Ee^{i\zeta x} = \int e^{i\zeta x} \cdot dF(x), \zeta \in \Re, i = \sqrt{-1}.$$

The joint c.f. of a vector-valued random variable $\mathbf{X}$ with joint c.d.f. F is given by

$$\Psi(\boldsymbol{\zeta}) = Ee^{i\boldsymbol{\zeta}'\mathbf{X}} = \int_{\Re_k} e^{i\boldsymbol{\zeta}'\mathbf{x}} \cdot dF(\mathbf{x}), \boldsymbol{\zeta} \in \Re_k,$$

where $\boldsymbol{\zeta}'\mathbf{x}$ is the inner product. Since $|e^{i\boldsymbol{\zeta}'\mathbf{x}}| = 1$, we have $\left|\int e^{i\boldsymbol{\zeta}'\mathbf{x}} \cdot dF(\mathbf{x})\right| \leq 1$, so existence of $\Psi(\boldsymbol{\zeta})$ is guaranteed for each real $\boldsymbol{\zeta}$. Focusing on the univariate case, here are the properties we will use:[23]

1. $\Psi(0) = 1$

2. $|\Psi(\zeta)| = [\Psi(\zeta)\Psi(-\zeta)]^{1/2} \leq 1$

3. $\Psi(\zeta)$ is uniformly continuous on $\Re$.
[Proof: for $B > 0$

$$\begin{aligned}\frac{|\Psi(\zeta+h) - \Psi(\zeta-h)|}{2} &= \left|\int_{\Re} e^{i\zeta x}\sin(hx) \cdot dF(x)\right| \\ &\leq \int_{\Re} |\sin(hx)| \cdot dF(x) \\ &\leq F(-B) + 1 - F(B) + \int_{-B}^{B} |\sin(hx)| \cdot dF(x) \\ &\leq F(-B) + 1 - F(B) + h\int_{(-B,B]} |x| \cdot dF(x),\end{aligned}$$

which can be made arbitrarily small by taking B sufficiently large and h sufficiently small, independently of ζ.]

4. C.f.s of absolutely continuous random variables approach $(0, 0)$ as $|\zeta| \to \infty$, whereas c.f.s of random variables supported on a lattice are periodic and never damp to the origin.

5. If kth derivative $\Psi^{(k)}(\zeta)$ exists at $\zeta = 0$, then moments to order k exist if k is even, and moments to order $k-1$ exist if k is odd. By the contrapositive, when k is even the nonexistence of $\Psi^{(k)}(0)$ implies that the kth moment does not exist. If the kth moment about the origin does

[23] See Lukacs (1970) for proofs of those not given and not obvious.

exist, it can be generated as $\mu'_k = i^{-k} d^k \Psi(\zeta) d\zeta^k \mid_{\zeta=0}$. Similarly, the logarithm of the c.f. serves to generate the cumulants:

$$\kappa_k = i^{-k} d^k \ln \Psi(\zeta)/d\zeta^k \mid_{\zeta=0} . \tag{2.44}$$

6. If kth derivative $\Psi^{(k)}(\zeta)$ exists at $\zeta = 0$, then Ψ can be expanded as

$$\Psi(\zeta) = 1 + \Psi'(0)\zeta + \Psi''(0)\zeta^2/2 + \ldots + \Psi^{(k)}(0)\zeta^k/k! + o\left(\zeta^k\right),$$

where $\zeta^{-k} o\left(\zeta^k\right) \to 0$ as $\zeta \to 0$—the crucial point being that the remainder vanishes faster than ζ^{-k} even if the $(k+1)$st derivative at $\zeta = 0$ does not exist.

Example 11 *If X is a continuous random variable with p.d.f. $f(x) = (2\pi)^{-1/2} e^{-x^2/2}$ for $x \in \Re$, the c.f. is*

$$\begin{aligned} \Psi(\zeta) &= \int_{-\infty}^{\infty} (2\pi)^{-1/2} e^{i\zeta x - x^2/2} \cdot dx \\ &= e^{-\zeta^2/2} \int_{-\infty}^{\infty} (2\pi)^{-1/2} e^{-(x-i\zeta)^2/2} \cdot dx \\ &= e^{-\zeta^2/2} \operatorname{erf}(+\infty) \\ &= e^{-\zeta^2/2}, \end{aligned}$$

where the next-to-last step follows by setting $z = (x - i\zeta)/\sqrt{2}$. Note that $\Psi(\zeta)$ is real-valued and positive and that $\Psi(\zeta) \to 0$ as $|\zeta| \to \infty$.

Example 12 *If X has a lattice distribution with p.m.f. $f(x) = \theta^{1-x}(1-\theta)^x$ for $x \in \{0,1\}$ and $\theta \in (0,1)$, then $\Psi(\zeta) = 1 + \theta(e^{i\zeta} - 1)$. Applying (2.3) shows that $|\Psi(\zeta)|^2 = (1-\theta)^2 + 2\theta(1-\theta)\cos\zeta$, which oscillates indefinitely as $|\zeta|$ increases.*

A crucial property of c.f.s is expressed in the uniqueness theorem. This states that if c.f.s of two random variables X and Y agree for every $\zeta \in \Re$ then the distributions of X and Y are also the same, in the sense that $\mathbb{P}_X(B) = \mathbb{P}_Y(B)$ for all $B \in \mathcal{B}$. In other words, there is a one-to-one correspondence between c.f.s and probability distributions. Combined with the convolution property of c.f.s and the continuity theorem for c.f.s this will make it possible to deduce the distributions of sums of independent variables and the limiting distributions of certain sequences of random variables. We derive the convolution formula now and present the continuity theorem in section 2.2.6.

Let us consider how one would find directly—that is, without the benefit of c.f.s—the distribution of the sum of two independent random variables

X and Y with marginal c.d.f.s F_X and F_Y. Letting $Z = X + Y$, the c.d.f. of Z can be found by integrating over the half plane as

$$\begin{aligned} F_Z(z) &= \mathbb{P}(X + Y \le z) \\ &= \int_{x+y\le z} dF_Y(y)\, dF_X(x) \\ &= \int_{\Re} \left[\int_{(-\infty, z-x]} dF_Y(y) \right] \cdot dF_X(x). \end{aligned}$$

This gives $F_Z(z) = \int_{\Re} F_Y(z-x) \cdot dF_X(x)$ or, switching the roles of X and Y, $F_Z(z) = \int_{\Re} F_X(z-y) \cdot dF_Y(y)$. Either of these integrals represents the convolution of F_X and F_Y, denoted $F_X * F_Y$. From this it follows that $F_Z(z)$ is absolutely continuous if either X or Y is a continuous random variable; for if Y is continuous, differentiating F_Z gives the density as $f_Z(z) = \int_{\Re} f_Y(z - x) \cdot dF_X(x)$, while if F_X is absolutely continuous it gives the density as $f_Z(z) = \int_{\Re} f_X(z-y) \cdot dF_Y(y)$. If both F_X and F_Y are continuous, either expression is equivalent to

$$f_Z(z) = \int_{\Re} f_Y(z-x) f_X(x) \cdot dx, \tag{2.45}$$

which is the convolution of the p.d.f.s: $f_X * f_Y$. These are the operations involved in order to work out directly the distributions of sums of independent random variables. Sums of more than two have to be handled in this way, one step at a time.

Working instead through the c.f. and using (2.37), we have

$$\Psi_Z(\zeta) = Ee^{i\zeta Z} = E\left(e^{i\zeta X} \cdot e^{i\zeta Y}\right) = Ee^{i\zeta X} Ee^{i\zeta Y} = \Psi_X(\zeta)\Psi_Y(\zeta).$$

Therefore, the c.f. of the sum of independent random variables is just the product of their c.f.s. This extends to any finite number of independent random variables; for if $Z_n = \sum_{j=1}^n X_j$ and the $\{X_j\}$ are independent, then $\Psi_{Z_n}(\zeta) = \prod_{j=1}^n \Psi_{X_j}(\zeta)$. In particular, if $\{X_j\}_{j=1}^n$ are independent and identically distributed (i.i.d.) then $\Psi_{Z_n}(\zeta) = \Psi_{X_1}(\zeta)^n$.

Notice that if X_1 and X_2 are i.i.d. random variables with c.f. $\Psi_X(\zeta)$, then $|\Psi_X(\zeta)|^2 = \Psi_X(\zeta)\Psi_X(-\zeta)$ is the c.f. of $X_1 - X_2$. In general, as in this case, random variables distributed symmetrically about the origin have real-valued c.f.s.

Example 13 *In the pointer experiment of example 9 the c.f.s of X and Y are*

$$\Psi_X(\zeta) = \frac{1}{i\zeta}(e^{i\zeta} - 1)$$

$$\Psi_Y(\zeta) = \frac{e^{i\zeta/2}}{2} + \frac{e^{i\zeta} - e^{i\zeta/2}}{i\zeta}.$$

If the pointer is spun k times in such a way as to generate independent random variables $\{X_j\}_{j=1}^k$ and $\{Y_j\}_{j=1}^k$, then the c.f.s of $S_X \equiv \sum_{j=1}^k X_j$ and S_Y are $\Psi_{S_X}(\zeta) = \Psi_X(\zeta)^k$ and $\Psi_{S_Y}(\zeta) = \Psi_Y(\zeta)^k$.

Deriving the full power of the convolution property requires a way to retrieve the c.d.f. (or p.d.f. or p.m.f.) from the c.f. There are a number of inversion formulas that enable this to be done under appropriate conditions. The easiest version, which corresponds to Fourier inversion formula (2.31), requires the c.f. to be absolutely integrable, meaning that $\int_{-\infty}^{\infty} |\Psi(\zeta)| \cdot d\zeta < \infty$. In this case one obtains the p.d.f. as

$$f(x) = \frac{1}{2\pi}\int_{-\infty}^{\infty} e^{-i\zeta x}\Psi(\zeta) \cdot d\zeta. \tag{2.46}$$

Example 14 *Applying* (2.46) *to the c.f. of the standard normal distribution in example* 11 *gives*

$$f(x) = \frac{1}{2\pi}\int_{-\infty}^{\infty} e^{-i\zeta x}e^{-\zeta^2/2} \cdot d\zeta$$

and (changing variables as $z = (\zeta + ix)/\sqrt{2}$)

$$\begin{aligned} f(x) &= \frac{e^{-x^2/2}}{\sqrt{2\pi}}\int_0^{\infty} \frac{2}{\sqrt{\pi}}e^{-z^2} \cdot dz \\ &= (2\pi)^{-1/2}e^{-x^2/2} \cdot \text{erf}(+\infty) \\ &= (2\pi)^{-1/2}e^{-x^2/2}. \end{aligned}$$

The requirement that $\Psi(\zeta)$ be absolutely integrable clearly restricts formula (2.46) to continuous random variables, since only these satisfy the necessary condition $\lim_{|\zeta|\to\infty} |\Psi(\zeta)| \to 0$. However, not all continuous variables have absolutely integrable c.f.s. For example, X in example 13 does not. The following more general results retrieve from the c.f. the probability associated with a finite or infinite interval under the condition that no end point of the interval is an atom.

Theorem 3

(i) If $\mathbb{P}_X(\{a\}) = \mathbb{P}_X(\{b\}) = 0$,

$$F(b) - F(a) = \lim_{c\to\infty} \int_{-c}^{c} \frac{e^{-i\zeta a} - e^{-i\zeta b}}{2\pi i\zeta} \Psi(\zeta) \cdot d\zeta.$$

(ii) If $\mathbb{P}_X(\{x\}) = 0$,

$$F(x) = \frac{1}{2} - \lim_{c\to\infty} \int_{-c}^{c} \frac{e^{-i\zeta x}}{2\pi i\zeta} \Psi(\zeta) \cdot d\zeta. \tag{2.47}$$

Notice that the integrals correspond to the Cauchy interpretation of the improper integral, $\int_{-\infty}^{\infty}$, which need not exist in the Lebesgue sense when Ψ is not absolutely integrable.

Example 15 *With* $\Psi_X(\zeta) = \frac{1}{i\zeta}(e^{i\zeta} - 1)$ *as in example* 13 *and* $a = 0, b = 1$

$$\begin{aligned} F(1) - F(0) &= \lim_{c\to\infty} \frac{1}{2\pi} \int_{-c}^{c} \frac{2 - (e^{i\zeta} + e^{-i\zeta})}{\zeta^2} \cdot d\zeta \\ &= \frac{1}{\pi} \int_{0}^{\infty} \frac{\sin^2(\zeta/2)}{(\zeta/2)^2} \cdot d\zeta \\ &= 1. \end{aligned}$$

2.2.4 *Radon-Nikodym for Probability Measures*

Given a σ-finite measure μ on a measurable space $(\Omega, \mathcal{M})$, we have seen that an equivalent σ-finite measure ν can be constructed as

$$\nu(A) = \int_A Q(\omega) \cdot d\mu(\omega), A \in \mathcal{M},$$

where Q is a μ-integrable function that is strictly positive on sets of positive μ measure. From the Radon-Nikodym theorem we know also that an essentially unique such Q exists that connects any pair of equivalent measures, with $Q = d\nu/d\mu$ as the Radon-Nikodym derivative. Let us see how these statements carry over to probability measures in particular and also to the probability measures induced by a random variable X. Of course, σ-finiteness always holds in this context because probability measures are finite, but there are much more interesting things to see.

Given a probability space $(\Omega, \mathcal{F}, \mathbb{P})$, it is clear that any function $Q > 0$ such that $\int Q(\omega) \cdot d\mathbb{P}(\omega) = 1$ defines a new probability measure,

$$\hat{\mathbb{P}}(A) = \int_A Q(\omega) \cdot d\mathbb{P}(\omega), A \in \mathcal{F},$$

that is equivalent to (has the same null sets as) $\mathbb{P}$. In this context the mapping $Q \equiv d\hat{\mathbb{P}}/d\mathbb{P}$ from Ω to $(0, \infty)$ is just a positive random variable with unitary expected value—the Radon-Nikodym derivative. Indeed, any positive, integrable random variable can be used to create such a new measure if the random variable is normalized by its expected value. Conversely, Radon-Nikodym tells us that for two such equivalent probability measures there always does exist a positive random variable Q that connects them in this way, and that it is essentially unique.[24]

Now introduce a random variable X and consider the probability measure $\mathbb{P}_X$ on $(\Re, \mathcal{B})$ that it induces, together with its c.d.f., $F(x) = \mathbb{P}_X((-\infty, x])$. Let $q(\cdot)$ be a positive function on a support of X such that $Eq(X) < \infty$, and set $Q(X) = q(X)/Eq(X)$. Then from $\mathbb{P}_X$ and F the new measure $\hat{\mathbb{P}}_X$ and c.d.f. $\hat{F}$ can be generated as

$$\begin{aligned} \hat{\mathbb{P}}_X(B) &= \int_B Q(x) \cdot d\mathbb{P}_X(x) = \int_B Q(x) \cdot dF(x) \\ \hat{F}(x) &= \hat{\mathbb{P}}_X((-\infty, x]). \end{aligned} \tag{2.48}$$

Conversely, Radon-Nikodym tells us that given two induced measures or two distribution functions there is a nonnegative and integrable $Q = d\hat{F}/dF$ that connects them. Finally, if g is such that $E|g(X)| = \int |g(x)| \cdot d\mathbb{P}_X(x) < \infty$, then the expected value with respect to $\hat{\mathbb{P}}_X$ is

$$\begin{aligned} \hat{E}g(X) &= Eg(X)Q(X) \\ &= \int g(x)Q(x) \cdot dF(x) \\ &= \int g(x)Q(x) \cdot d\mathbb{P}_X(x). \end{aligned} \tag{2.49}$$

Example 16 *Consider the c.d.f.s*

$$\begin{aligned} F(x) &= \left(1 - e^{-x}\right) \mathbf{1}_{(0,\infty)}(x) \\ \hat{F}(x) &= \left[1 - (1+x)e^{-x}\right] \mathbf{1}_{(0,\infty)}(x). \end{aligned}$$

---

[24]That is, if there are two such functions, Q and Q', then they agree almost everywhere with respect to $\mathbb{P}$ and (by equivalence) with respect to $\hat{\mathbb{P}}$.

The supports being the same, the corresponding induced measures $\mathbb{P}_X$ and $\hat{\mathbb{P}}_X$ are equivalent, so there must be a $\mathbb{P}_X$-almost-surely positive Q such that (2.48) holds. Since F and $\hat{F}$ are absolutely continuous with densities f and $\hat{f}$, we have

$$Q(x) = \frac{d\hat{F}(x)}{dF(x)} = \frac{\hat{f}(x)}{f(x)} = \frac{xe^{-x}}{e^{-x}} = x$$

for $x > 0$.

Example 17 *Consider the c.d.f.s*

$$F(x) = (1-\pi)\,\mathbf{1}_{[0,1)}(x) + \mathbf{1}_{[1,\infty)}(x)$$
$$\hat{F}(x) = (1-\hat{\pi})\,\mathbf{1}_{[0,1)}(x) + \mathbf{1}_{[1,\infty)}(x)$$

where $\pi, \hat{\pi} \in (0,1)$. Both being supported on $\{0,1\}$, the induced measures are equivalent. Since F and $\hat{F}$ are step functions with corresponding p.m.f.s f and $\hat{f}$, we have

$$Q(x) = \frac{d\hat{F}(x)}{dF(x)} = \frac{\hat{f}(x)}{f(x)} = \begin{cases} \dfrac{1-\hat{\pi}}{1-\pi}, & x = 0 \\ \dfrac{\hat{\pi}}{\pi}, & x = 1 \end{cases}.$$

2.2.5 *Conditional Probability and Expectation*

Given a probability space $(\Omega, \mathcal{F}, \mathbb{P})$ and two events $A, A' \in \mathcal{F}$, the conditional probability of A given A', written $\mathbb{P}(A \,|\, A')$, is defined as $\mathbb{P}(A \cap A')/\mathbb{P}(A')$ whenever $\mathbb{P}(A') > 0$. Intuitively, $\mathbb{P}(A \,|\, A')$ is the probability assigned to A when we have partial information about the experiment, to the extent of knowing that A' has occurred.

Introducing the bivariate random variable $\mathbf{Z} = (X,Y)'$ on $(\Omega, \mathcal{F}, \mathbb{P})$, the familiar concepts of conditional p.m.f.s and conditional p.d.f.s can be developed directly from the definition of conditional probability. Thus, in the discrete case that $\mathbf{Z}$ is supported on a countable set of points $\mathcal{Z} = \mathcal{X} \times \mathcal{Y} \subset \Re_2$ we can take $A = \{\omega : X(\omega) = x\}$, $A' = \{\omega : Y(\omega) = y\}$, and define the conditional p.m.f. for $(x,y) \in \mathcal{Z}$ as

$$f_{X|Y}(x \mid y) = \frac{f_{\mathbf{Z}}(x,y)}{f_Y(y)} = \frac{\mathbb{P}(A \cap A')}{\mathbb{P}(A')} = \mathbb{P}(A \mid A').$$

For the continuous case, putting $A = \{\omega : X(\omega) \in (-\infty, x]\}$ and $A' = \{\omega : Y(\omega) \in (y - h', y]\}$ for $h' > 0$, and considering just the values of y

such that $\mathbb{P}(A') > 0$ for all $h' > 0$, we get the conditional c.d.f. of X given the *zero-probability* event $Y = y$ as

$$F_{X|Y}(x \mid y) = \lim_{h' \longrightarrow 0} \frac{\mathbb{P}(A \cap A')}{\mathbb{P}(A')} = \lim_{h' \longrightarrow 0} \mathbb{P}(A \mid A').$$

The conditional p.d.f. is then found as $f_{X|Y}(x \mid y) = dF_{X|Y}(x \mid y)/dx$. Alternatively, putting $A = \{\omega : X(\omega) \in (x - h, x]\}$ for $h > 0$ and leaving A' as before, the p.d.f. can be obtained as

$$f_{X|Y}(x \mid y) = \frac{f_{\mathbf{Z}}(x, y)}{f_Y(y)} \equiv \lim_{h, h' \longrightarrow 0} \frac{\mathbb{P}(A \cap A')/(hh')}{\mathbb{P}(A')/h'} = \lim_{h, h' \longrightarrow 0} h^{-1}\mathbb{P}(A \mid A').$$

From the conditional p.m.f.s and p.d.f.s conditional expectations can be developed in the usual way as weighted sums or integrals; e.g.,

$$E[g(X, Y) \mid Y = y] = \begin{cases} \sum_{x \in \mathcal{X}} g(x, y) f_{X|Y}(x \mid y) \\ \int g(x, y) f_{X|Y}(x \mid y) \cdot dx \end{cases}.$$

Of course, all this should be familiar from elementary probability theory. For the finance applications that lie ahead, we now need to extend the concepts to probabilities and expectations conditional on collections of sets that comprise σ-fields.

Probabilities Conditional on σ-fields

We have explained the sense in which σ-field $\mathcal{F}$ in a probability space $(\Omega, \mathcal{F}, \mathbb{P})$ represents an information structure and have discussed sub-σ-fields that contain coarser (that is, less) information than $\mathcal{F}$ about the results of the chance experiment. Letting $\mathcal{F}' \subseteq \mathcal{F}$ be a sub-σ-field of $\mathcal{F}$ and exhibiting some event $A \in \mathcal{F}$, let us consider the symbol $\mathbb{P}(A \mid \mathcal{F}')$ to represent the probability of A given that we know the information represented by $\mathcal{F}'$. "Knowing $\mathcal{F}'$" means that we know whether each (nonempty) event $A' \in \mathcal{F}'$ has occurred. This is equivalent to knowing which of the elementary events has occurred of which events in $\mathcal{F}'$ are composed.[25] Thus, we can think of $\mathbb{P}(A \mid \mathcal{F}')$ as an $\mathcal{F}'$-measurable random variable—a mapping from the events that generate $\mathcal{F}'$ to the interval $[0, 1]$. $\mathbb{P}(A \mid \mathcal{F}')$ is $\mathcal{F}'$-measurable in the sense that we will know the actual value of $\mathbb{P}(A \mid \mathcal{F}')$ once we obtain the $\mathcal{F}'$ information. Note that $\mathbb{P}(A)$ itself can be thought of as conditional

[25] For example, if σ-field $\mathcal{F}'$ were generated by a countable collection of disjoint sets that partitioned Ω, then these would be the "elementary events" of $\mathcal{F}'$.

on the trivial σ-field $\mathcal{T} = \{\emptyset, \Omega\} \subseteq \mathcal{F}'$, which represents the knowledge just of what things are possible; thus, $\mathbb{P}(A) \equiv \mathbb{P}(A \mid \mathcal{T})$. At the other extreme, $\mathbb{P}(A \mid \mathcal{F})$ is a Bernoulli random variable that takes one of two values—unity or zero, since $\mathcal{F}$ includes the knowledge of whether $\omega \in A$ or not.

Example 18 *Referring to the die example on page 50, take $\mathcal{F} = \mathcal{F}_6$ to be the σ-field generated by the random variable $X(\omega)$ = number of dots on side ω when a fair six-sided die is thrown, and let $\mathcal{G} = \mathcal{F}_2$ be the sub-σ-field generated by the random variable Y, where $Y(\omega) = 0$ or $Y(\omega) = 1$ as the number of dots is even or odd. Knowing $\mathcal{F}_2$ means that we know which of the odd-even outcomes, $\{\cdot, \cdots, :\cdot:\}$ or $\{:, ::, :::\}$, has occurred. Since $\{:\} \in \mathcal{F}_6$, we interpret $\mathbb{P}(\{:\} \mid \mathcal{F}_2)$ as a random variable taking the value $\mathbb{P}(\{:\} \mid \{:, ::, :::\}) = 1/3$ if $\{:, ::, :::\}$ occurs or $\mathbb{P}(\{:\} \mid \{\cdot, \cdots, :\cdot:\}) = 0$ if $\{\cdot, \cdots, :\cdot:\}$ occurs.*

The same statements naturally apply to measures induced by random variables. Thus, if X is a random variable on $(\Omega, \mathcal{F}, \mathbb{P})$ and $(\Re, \mathcal{B}, \mathbb{P}_X)$ is the induced space, then for $\mathcal{B}' \subset \mathcal{B}$ and $B \in \mathcal{B}$ we think of $\mathbb{P}_X(B \mid \mathcal{B}')$ as a $\mathcal{B}'$-measurable random variable. For each of the collection of events B' that generate $\mathcal{B}'$ the variable $\mathbb{P}_X(B \mid \mathcal{B}')$ expresses the probability that $X \in B$ given that $X \in B'$—i.e., $\mathbb{P}(X \in B \mid X \in B')$. As before, we can think of $\mathbb{P}_X(B)$ as conditioned on the trivial field $\mathcal{T}' = \{\emptyset, \Re\}$ and can recognize $\mathbb{P}_X(B \mid \mathcal{B})$ as a Bernoulli random variable.

Expectations Conditional on σ-fields

With these concepts in mind, we can now develop the concept of conditional expectation given a σ-field of events. For each fixed sub-σ-field $\mathcal{F}' \subseteq \mathcal{F}$, $\mathbb{P}(\cdot \mid \mathcal{F}')$ is a mapping from $\mathcal{F}$ to $[0, 1]$, and so $(\Omega, \mathcal{F}, \mathbb{P}(\cdot \mid \mathcal{F}'))$ is a probability space. Defining a random variable X on Ω, statements such as $\mathbb{P}(X \le x \mid \mathcal{F}')$ can be interpreted as the probability that will be assigned to the $\mathcal{F}$-measurable event $\{\omega : X(\omega) \le x\}$ once the information $\mathcal{F}'$ is known. Since $\mathbb{P}(\cdot \mid \mathcal{F}')$ is a probability measure, the conditional expectation of X given $\mathcal{F}'$ can then be defined in the usual way as

$$E(X \mid \mathcal{F}') = \int X(\omega) \cdot d\mathbb{P}(\omega \mid \mathcal{F}').$$

This conditional expectation can be interpreted as the $\mathcal{F}'$-measurable random variable that represents what the expected value of X will be once the information $\mathcal{F}'$ becomes available. In particular, $E(X \mid \mathcal{T}) \equiv EX$ when $\mathcal{T}$

is trivial, while $E(X \mid \mathcal{F}) = X$. Defined in this way, $E(\cdot \mid \mathcal{F}')$ is automatically endowed with the nine properties of mathematical expectation given on pages 56–57. In particular, corresponding to $E[bg(X)] = bEg(X)$ for some constant b we have $E[Yg(X) \mid \mathcal{F}'] = YE[g(X) \mid \mathcal{F}']$ if $Y \in \mathcal{F}'$ (Y is an $\mathcal{F}'$-measurable random variable).

There is also the parallel interpretation in terms of induced measures. Thus, $E(X \mid \mathcal{B}') = \int x \cdot d\mathbb{P}_X(x \mid \mathcal{B}')$ corresponds to the expectation of X given the partial information that $X \in \mathcal{B}' \subseteq \mathcal{B}$.

There is one other very essential property, known as the "tower" property of conditional expectation:

$$E[E(X \mid \mathcal{F}')\mathbf{1}_{A'}] = E(X\mathbf{1}_{A'}), \tag{2.50}$$

where A' is any set in $\mathcal{F}'$. In particular, with $A' = \Omega$ we have $E[E(X \mid \mathcal{F}')] = EX$. A way to understand the tower property intuitively is from the identity

$$E(X\mathbf{1}_{A'}) = E[E(X \mid \mathcal{F}')\mathbf{1}_{A'}] + E\{[X - E(X \mid \mathcal{F}')]\mathbf{1}_{A'}\}.$$

Thinking of the conditional expectation $E(X \mid \mathcal{F}')$ as the best prediction[26] of X given the information $\mathcal{F}'$, it is clear that the second term should be zero, since prediction error $X - E(X \mid \mathcal{F}')$ should be orthogonal to anything in $\mathcal{F}'$. Likewise, writing EX as $E[E(X \mid \mathcal{F}')] + E[X - E(X \mid \mathcal{F}')]$, we can understand that the second term should be zero because it represents the expectation given *less* information ($\mathcal{T}$) of the prediction error based on *more* information ($\mathcal{F}' \supseteq \mathcal{T}$).[27]

While this development of conditional expectation is very natural, the usual treatment, traced to Kolmogorov (1950), is simply to define $E(X \mid \mathcal{F}')$ as an $\mathcal{F}'$-measurable random variable having the property (2.50). More precisely, $E(X \mid \mathcal{F}')$ is regarded as an "equivalence class" of random variables satisfying (2.50). This allows different "versions" to exist, members of any countable collection of which are equal with probability one. With conditional expectation thus defined, conditional probabilities can be deduced as conditional expectations of indicator functions—e.g., $\mathbb{P}(A \mid \mathcal{F}') = E(\mathbf{1}_A \mid \mathcal{F}')$ for $A \in \mathcal{F}$ and $\mathbb{P}_X(B \mid \mathcal{B}') = E(\mathbf{1}_B \mid \mathcal{B}')$ for any

---

[26]It is in fact the best prediction in the mean-square sense, in that $E[(X - m)^2 \mid \mathcal{F}']$ attains a unique minimum (among all $\mathcal{F}'$-measurable m) at $m = E(X \mid \mathcal{F}')$.

[27]Williams (1991) gives a formal derivation of the tower property from conditional probabilities in the case that Ω is a finite set.

$B \in \mathcal{B}$ (the Borel sets) and any $\mathcal{B}' \subseteq \mathcal{B}$. The fact that conditional expectations have the usual properties of expectation can then be shown to follow from the definition. While less insightful, this standard treatment has the advantage of making it easy to verify that some $\mathcal{F}'$-measurable random variable Y is in fact a version of the conditional expectation $E(X \mid \mathcal{F}')$: simply show that $E(Y\mathbf{1}_{A'}) = E(X\mathbf{1}_{A'})$ for any $\mathcal{F}'$-measurable A'.

Conditional Expectations under Changes of Measure

Now, given the measure $\mathbb{P}$ on $(\Omega, \mathcal{F})$ and the sub-σ-field $\mathcal{F}'$, let us see how to express conditional expectations in a new measure $\hat{\mathbb{P}}$. As in section 2.2.4, let $\mathbb{P}$ and $\hat{\mathbb{P}}$ be equivalent probability measures on $(\Omega, \mathcal{F})$, with random variable $Q = d\hat{\mathbb{P}}/d\mathbb{P}$ as the Radon-Nikodym derivative. Thus, $\mathbb{P}(Q > 0) = \hat{\mathbb{P}}(Q > 0) = 1$ and $\int Q \cdot d\mathbb{P} = 1$. Let X be a random variable with $\hat{E}|X| \equiv \int |X| \cdot d\hat{\mathbb{P}} = \int |X| Q \cdot d\mathbb{P} < \infty$. Then if $A' \in \mathcal{F}' \subseteq \mathcal{F}$,

$$EXQ\mathbf{1}_{A'} = \hat{E}X\mathbf{1}_{A'} = \hat{E}\left[\hat{E}(X\mathbf{1}_{A'} \mid \mathcal{F}')\right] = E\left[\hat{E}(X\mathbf{1}_{A'} \mid \mathcal{F}')Q\right],$$

where the second equality follows from the tower property and the first and third equalities follow from (2.49). But, again by the tower property,

$$\begin{aligned} E\left[\hat{E}(X\mathbf{1}_{A'} \mid \mathcal{F}')Q\right] &= E\left\{E[\hat{E}(X\mathbf{1}_{A'} \mid \mathcal{F}')Q] \mid \mathcal{F}'\right\} \\ &= E\left[\hat{E}(X\mathbf{1}_{A'} \mid \mathcal{F}')E(Q \mid \mathcal{F}')\right], \end{aligned}$$

since $\hat{E}(X\mathbf{1}_{A'} \mid \mathcal{F}')$ is $\mathcal{F}'$-measurable. That this holds for each $A' \in \mathcal{F}'$ proves that the expression in brackets is in fact (a version of) the conditional expectation of XQ given $\mathcal{F}'$; i.e., $E(XQ \mid \mathcal{F}') = \hat{E}(X \mid \mathcal{F}')E(Q \mid \mathcal{F}')$ a.s. Finally, since $Q > 0$ we have

$$\hat{E}(X \mid \mathcal{F}') = \frac{E(XQ \mid \mathcal{F}')}{E(Q \mid \mathcal{F}')}, \tag{2.51}$$

which is sometimes called "Bayes' rule for conditional expectations".

Example 19 *Let $X = Y + Z$, where Y and Z are independent random variables with p.d.f.s $f_Y(y) = (2\pi)^{-1/2} e^{-(y-1)^2/2}$ and $f_Z(z) = (2\pi)^{-1/2} e^{-z^2/2}$. Take $\mathcal{F}$ to be the σ-field generated by Y and Z (so that X is $\mathcal{F}$-measurable) and $\mathcal{F}'$ to be the sub-σ-field generated by Y alone. Convolution formula (2.45) shows the p.d.f. of X to be $f_X(x) = (4\pi)^{-1/2} e^{-(x-1)^2/4}$. Thus, for any Borel set B we have $\mathbb{P}_X(B) =$*

$\int_B (4\pi)^{-1/2} e^{-(x-1)^2/4} \cdot dx$. *To define a new measure* $\hat{\mathbb{P}}_X$, *put* $Q = d\hat{\mathbb{P}}_X / d\mathbb{P}_X = d\hat{F}_X / dF_X = e^{-X/2+1/4}$. *Then* $Q > 0$ *and*

$$EQ = \int_{-\infty}^{\infty} \frac{1}{\sqrt{4\pi}} e^{-x/2+1/4-(x-1)^2/4} \cdot dx = \int_{-\infty}^{\infty} \frac{1}{\sqrt{4\pi}} e^{-x^2/4} = 1,$$

as required. Thus, for integrable $g = g(X)$ *we have* $\hat{E}g = E(Qg)$. *In particular,* $\hat{E}X = 0$ *and* $\hat{E}X^2 = 1$. *Now let us use (2.51) to find* $\hat{E}(X \mid \mathcal{F}')$ *and* $\hat{E}(X^2 \mid \mathcal{F}')$. *Since knowledge of* $\mathcal{F}'$ *confers knowledge of realization* $Y(\omega) = y$, *we can write these alternatively as* $\hat{E}(X \mid y)$ *and* $\hat{E}(X^2 \mid y)$. *Under original measure* $\mathbb{P}_X$ *the conditional p.d.f. of* X *given* $Y = y$ *is*

$$f_{X|Y}(x \mid y) = \frac{d}{dx} F_{X|Y}(x \mid y) = \frac{d}{dx} F_Z(x - y) = \frac{1}{\sqrt{2\pi}} e^{-(x-y)^2/2},$$

and so

$$\begin{aligned} E(Q \mid y) &= \int \frac{1}{\sqrt{2\pi}} e^{-x/2+1/4-(x-y)^2/2} \cdot dx \\ &= e^{-y/2+3/8} \int \frac{1}{\sqrt{2\pi}} e^{-[x-(y-1/2)]^2/2} \cdot dx \\ &= e^{-y/2+3/8}. \end{aligned}$$

Applying (2.51) for arbitrary, integrable g, *we have*

$$\begin{aligned} \hat{E}[g(X) \mid y] &= \frac{E[g(X)Q \mid y]}{E(Q \mid y)} \\ &= \frac{e^{-y/2+3/8} \int \frac{1}{\sqrt{2\pi}} g(x) e^{-[x-(y-1/2)]^2/2} \cdot dx}{e^{-y/2+3/8}} \\ &= \int \frac{1}{\sqrt{2\pi}} g(w + y - 1/2) e^{-w^2/2} \cdot dw \\ &= \begin{cases} y - 1/2, & g(X) = X \\ 1 + (y - 1/2)^2, & g(X) = X^2 \end{cases}. \end{aligned}$$

2.2.6 *Stochastic Convergence*

This section deals with sequences of random variables $\{X_n\}_{n=1}^{\infty}$ that, in some sense, settle down or converge as they progress. There are various senses in which this can occur. We begin with the most basic one, in which the distributions of the $\{X_n\}$ approach some particular form.

Convergence in Distribution

Definition 1 *Let F be a c.d.f. The sequence $\{X_n\}$ converges in distribution to F if the sequence of c.d.f.s $\{F_n\}$ converges (pointwise) to F for each $x \in \Re$ that is a continuity point of F.*

The term "weak convergence" is also used. We write $X_n \rightsquigarrow F$ to indicate convergence in distribution. Since there is a one-to-one correspondence between F_n, the c.d.f. at stage n, and its induced probability measure, $\mathbb{P}_{X_n}$, it is clear that the measures are approaching some common form that corresponds to the c.d.f. F that represents the limiting distribution. The weak requirement that the c.d.f.s converge to F just where F is continuous handles the awkward transitions from sequences of continuous distributions to limiting forms with one or more atoms.

Example 20 *Let X_1 have c.d.f.*

$$F_1(x) = x\mathbf{1}_{[0,1)}(x) + \mathbf{1}_{[1,\infty)}(x),$$

corresponding to the random variable X in example 9, and define $X_n = X_1^n$ for $n \in \aleph$. Then X_n has c.d.f.

$$F_n(x) = \mathbb{P}(X_1^n \le x) = \mathbb{P}(X_1 \le x^{1/n}) = x^{1/n}\mathbf{1}_{[0,1)}(x) + \mathbf{1}_{[1,\infty)}(x).$$

Plainly, the probability mass for X_n becomes more and more concentrated near the origin the larger is n. Taking limits,

$$F_n(x) \to \mathbf{1}_{(0,\infty)}(x) \equiv F_\infty(x).$$

F_∞ is not a c.d.f. since it is not right-continuous at $x = 0$, but defining $F(x) = F_\infty(x)$ for $x \neq 0$ and $F(0) = 1$ does produce a c.d.f. to which $\{F_n\}$ converges except at discontinuity point 0. The continuous distribution of X_n can therefore be approximated arbitrarily closely for large enough n by the degenerate distribution F that puts unitary probability mass at the origin.

In many applications it is difficult to express the c.d.f.s of the $\{X_n\}$ and therefore hard to ascertain the limiting distribution from the definition. For example, when each X_n is a sum of independent variables, finding F_n directly requires calculating a sequence of convolution integrals. In such cases the following "continuity" theorem for characteristic functions tells us that the limiting distributions can be deduced from the limits of the c.f.s.

Theorem 4 *X_n converges in distribution to F if and only if the sequence of characteristic functions $\{\Psi_n\}_{n=1}^{\infty}$ converges (pointwise) to a function Ψ_∞*

that is continuous at $\zeta = 0$. In this case Ψ_∞ is a c.f., and the one that corresponds to F.

Example 21 *In example 12 on page 63 take $\theta = 1/2$ and let $\{X_j\}_{j=1}^{\infty}$ be independent random variables, each having c.f. $\Psi(\zeta) = 1 + (e^{i\zeta} - 1)/2$. The linear transformation $2X_j - 1$ sets the mean to zero and the variance to unity. Let Z_n be $n^{-1/2}$ times the sum of the first n of the scaled, centered variates: $Z_n = n^{-1/2}\sum_{j=1}^{n}(2X_j - 1)$. The limiting distribution of $\{Z_n\}$ can be found via the continuity theorem, as follows:*

$$\Psi_n(\zeta) = Ee^{i\zeta Z_n} = \prod_{j=1}^{n} Ee^{i(2X_j-1)\zeta/\sqrt{n}} = e^{-i\zeta\sqrt{n}}\left[1 + (e^{2i\zeta/\sqrt{n}} - 1)/2\right]^n.$$

Expanding $e^{2i\zeta/\sqrt{n}}$ to order n^{-1} as $1 + 2i\zeta/\sqrt{n} - 2\zeta^2/n + o\left(n^{-1}\right)$ and taking logs give

$$\begin{aligned}
\ln \Psi_n(\zeta) &= -i\zeta\sqrt{n} + n\ln\left[1 + i\zeta/\sqrt{n} - \zeta^2/n + o\left(n^{-1}\right)\right] \\
&= -i\zeta\sqrt{n} + n\left[i\zeta/\sqrt{n} - \zeta^2/n + \zeta^2/(2n) + o\left(n^{-1}\right)\right] \\
&= -\zeta^2/2 + o(1),
\end{aligned}$$

so that $\Psi_n(\zeta) \to e^{-\zeta^2/2}$. One can either recognize this from example 11 as the c.f. of the standard normal distribution, or else invert it as in example 14 to see that the limiting distribution of Z_n is represented by p.d.f. $f(z) = (2\pi)^{-1/2}e^{-z^2/2}$.

Central limit theorems give conditions under which averages or sums of random variables, appropriately centered and scaled, converge in distribution to the standard normal. In the case of independent and identically distributed random variables the Lindeberg-Lévy theorem states the weakest possible conditions for this result. Example 21 was just a special case, but the proof of the general result is even easier.

Theorem 5 *If $\{X_j\}_{j=1}^{\infty}$ are independent and identically distributed (i.i.d.) with mean μ and finite variance σ^2, then $\left(\sum_{j=1}^{n} X_j - n\mu\right)/(\sigma\sqrt{n})$ converges in distribution to the standard normal as $n \to \infty$.*

Proof: Put $Z_n = \left(\sum_{j=1}^{n} X_j - n\mu\right)/(\sigma\sqrt{n}) = n^{-1/2}\sum_{j=1}^{n} Y_j$, where $\{Y_j \equiv (X_j - \mu)/\sigma\}$ are i.i.d. with $EY_1 = 0, EY_1^2 = 1$ and c.f. Ψ. Then

Z_n has c.f.

$$\begin{aligned}\Psi_n(\zeta) &= \Psi\left(n^{-1/2}\zeta\right)^n \\ &= \left[1 + \frac{i\zeta}{\sqrt{n}}EY_1 - \frac{\zeta^2}{2n}EY_1^2 + o\left(n^{-1}\right)\right]^n \\ &= \left[1 - \frac{\zeta^2}{2n} + o\left(n^{-1}\right)\right]^n,\end{aligned}$$

and $\lim_{n\to\infty} \Psi_n(\zeta) = e^{-\zeta^2/2}$. □

Stochastic Convergence to a Constant

In example 20 the limiting distribution of the sequence $\{X_n\}$ was degenerate, in the sense that all probability mass was concentrated at one point. In such cases we say that $\{X_n\}$ "converges stochastically to a constant". Again, there are several modes of such convergence.

Definition 2 *The sequence of random variables $\{X_n\}$ converges stochastically to the constant c*

1. *In probability, if the limiting distribution of X_n is $F(x) = \mathbf{1}_{[c,\infty)}(x)$.*
2. *In mean square, if $\lim_{n\longrightarrow\infty} E(X_n - c)^2 = 0$.*
3. *Almost surely, if for each $\varepsilon > 0$*

$$\mathbb{P}(|X_n - c| > \varepsilon \text{ for infinitely many } n) = 0.$$

The usual notations are $X_n \to^{\mathbb{P}} c$, $X_n \to^{m.s.} c$, and $X_n \to^{a.s.} c$, respectively. An equivalent condition for convergence in probability is that $\mathbb{P}(|X_n - c| > \varepsilon) \to 0$ for each $\varepsilon > 0$. In other words, the requirement is that the probability of a significant deviation at stage n be made arbitrarily small by taking n sufficiently large. Almost-sure (a.s.). convergence is a stronger form, since it requires there to be some point beyond which there will never be a significant deviation. The convergence part of the Borel-Cantelli lemma provides a sufficient condition:

$$\sum_{n=1}^{\infty} \mathbb{P}(|X_n - c| > \varepsilon) < \infty \Rightarrow \mathbb{P}(|X_n - c| > \varepsilon \text{ for infinitely many } n) = 0.$$

Taking $g(X_n) = |X_n - c|^2$ in the Markov inequality shows that convergence in mean square implies convergence in probability. On the other hand, mean-square convergence is not implied by either of the other two forms, since neither a.s. convergence nor convergence in probability requires the existence of moments.

Laws of large numbers give conditions under which averages of a collection of random variables converge stochastically to some constant μ. For i.i.d. random variables with mean μ Kolmogorov's strong law of large numbers gives the best possible result:

Theorem 6 *If* $\{X_j\}_{j=1}^{\infty}$ *are i.i.d. with* $EX_j = \mu$, *then* $n^{-1}\sum_{j=1}^{n} X_j \to^{a.s.} \mu$ *as* $n \to \infty$.

2.2.7 *Models for Distributions*

Important properties of most of the models for distributions we will encounter in applications to financial derivatives are summarized in table 2.3. For the moments the table gives whichever has the simplest form—moments about the origin, μ_k'; moments about the mean, μ_k; or descending factorial moments, $\mu_{(k)}$. Any of these suffices to determine the others. "$\boldsymbol{\Sigma} >> \mathbf{0}$" for the multinormal signifies positive definiteness of the matrix. Moment generating functions exist for all distributions in the table except the lognormal and Cauchy. These can be obtained from the expressions for the c.f.s by replacing ζ by ζ/i. The lognormal c.f. has no closed form, and the Cauchy law has no positive-integer moments.

Normal and Lognormal Families

The normal and lognormal distributions are of special importance in the study of derivatives. If random variable X is distributed as (univariate) normal with parameters θ_1 and θ_2^2, we write $X \sim N(\theta_1, \theta_2^2)$ for short. Since X then has mean θ_1 and variance θ_2^2, we typically represent the distribution as $N(\mu, \sigma^2)$. The case $N(0,1)$ is the "standard" normal. The normal distribution is often called the "Gaussian" law. One can see from the expressions for the central moments in table 2.3 that the coefficients of skewness and kurtosis are $\alpha_3 = 0$ and $\alpha_4 = 3.0$, respectively. Special symbols for the standard normal p.d.f. and c.d.f. are $\phi(x) = (2\pi)^{-1/2}e^{-x^2/2}$ and $\Phi(x) = \int_{-\infty}^{x} \phi(z) \cdot dz$. A change of variables in the integral shows that $\operatorname{erf}(x/\sqrt{2}) = \Phi(x) - \Phi(-x)$ and therefore

$$\Phi(x) = \frac{1}{2}\left[1 + \operatorname{erf}\left(x/\sqrt{2}\right)\right].$$

Table 2.3. Common distributions and their properties.

Name Abbrev.	P.d.f./ P.m.f.	Support $\mathcal{X}$	Parameter Space	Moments $\mu_k', \mu_k, \mu_{(k)}$	Characteristic Function, $\Psi(\zeta)$
Bernoulli	$\theta^{1-x}\bar{\theta}^x$	$\{0,1\}$	$\theta \in (0,1)$	μ_k' :	$1-\theta+\theta e^{i\zeta}$
$B(1,\theta)$	$\bar{\theta} \equiv 1-\theta$			θ	
Binomial	$\binom{n}{x}\theta^{n-x}\bar{\theta}^x$	$\{0,1,\ldots,n\}$	$\theta \in (0,1)$	$\mu_{(k)}$:	$\left(1-\theta+\theta e^{i\zeta}\right)^n$
$B(n,\theta)$	$\bar{\theta} \equiv 1-\theta$		$n \in \aleph$	$\frac{n!\theta^k}{(n-k)!}$	
Poisson	$\theta^x e^{-\theta}/x!$	$\aleph_0$	$\theta > 0$	$\mu_{(k)}$:	$\exp\left(\theta e^{i\zeta}-\theta\right)$
$P(\theta)$				θ^k	
Neg. Bin.	$\frac{\Gamma(\varphi+x)}{x!\Gamma(\varphi)}\theta^\varphi\bar{\theta}^x$	$\aleph_0$	$\theta \in (0,1)$	$\mu_{(k)}$:	$\left(\frac{\theta}{1-\bar{\theta}e^{i\zeta}}\right)^\varphi$
$NB(\theta,\varphi)$	$\bar{\theta} \equiv 1-\theta$		$\varphi \geq 1$	$\frac{\Gamma(\varphi+k)}{\Gamma(\varphi)}\frac{\bar{\theta}^k}{\theta^k}$	
Uniform	$(\varphi-\theta)^{-1}$	(θ,φ)	$\theta < \varphi$	μ_k' :	$\frac{e^{i\zeta\delta}}{\delta}\sin\left[\frac{\zeta(\varphi+\theta)}{2}\right]$
$U(\theta,\varphi)$				$\frac{\varphi^{k+1}-\theta^{k+1}}{(k+1)(\varphi-\theta)}$	$\delta \equiv \frac{\varphi-\theta}{2}$
Gamma	$\frac{\varphi^{-\theta}x^{\theta-1}e^{-x/\varphi}}{\Gamma(\theta)}$	$\Re^+$	$\theta > 0$	μ_k' :	$(1-i\zeta\varphi)^{-\theta}$
$\Gamma(\theta,\varphi)$			$\varphi > 0$	$\varphi^k\frac{\Gamma(\theta+k)}{\Gamma(\theta)}$	
Exponential	$\varphi^{-1}e^{-x/\varphi}$	$\Re^+$	$\varphi > 0$	μ_k' :	$(1-i\zeta\varphi)^{-1}$
$\Gamma(1,\varphi)$				$\varphi^k k!$	
Normal	$\frac{e^{-(x-\mu)^2/2\sigma^2}}{\sigma\sqrt{2\pi}}$	$\Re$	$\mu \in \Re$	μ_k :	$e^{i\zeta\mu-\zeta^2\sigma^2/2}$
$N(\mu,\sigma^2)$			$\sigma > 0$	$0, k$ odd	
				$\frac{\sigma^k k!}{2^{k/2}[k/2]!}$, else	
Multinormal	$\frac{e^{-(\mathbf{x}-\boldsymbol{\mu})'\boldsymbol{\Sigma}^{-1}(\mathbf{x}-\boldsymbol{\mu})/2}}{\sqrt{(2\pi)^k\lvert\boldsymbol{\Sigma}\rvert}}$	$\Re_k$	$\mu \in \Re_k$	$\boldsymbol{\mu}_1' = \boldsymbol{\mu}$	$e^{i\boldsymbol{\zeta}'\boldsymbol{\mu}-\boldsymbol{\zeta}'\boldsymbol{\Sigma}\boldsymbol{\zeta}/2}$
$N(\mu,\boldsymbol{\Sigma})$			$\boldsymbol{\Sigma} >> 0$	$\boldsymbol{\mu}_2 = \boldsymbol{\Sigma}$	
Lognormal	$\frac{e^{-(\ln x-\mu)^2/2\sigma^2}}{x\sigma\sqrt{2\pi}}$	$(0,\infty)$	$\mu \in \Re$	μ_k' :	no closed form
$LN(\mu,\sigma^2)$			$\sigma > 0$	$e^{k\mu+k^2\sigma^2/2}$	
Cauchy	$\frac{1}{\pi\left[1+\left(\frac{x-\theta}{\varphi}\right)^2\right]}$	$\Re$	$\theta \in \Re$	$\nexists$	$e^{i\zeta\theta-\lvert\zeta\rvert\varphi}$
$C(\theta,\varphi)$			$\varphi > 0$		

Two useful inequalities for bounding probabilities of large deviations are

$$\mathbb{P}(Z > c) = 1 - \Phi(c) \leq \int_c^\infty (2\pi)^{-1/2} z e^{-z^2/2} \cdot dz = \phi(c) \qquad (2.52)$$

$$\mathbb{P}(|Z| > c) = 2\mathbb{P}(Z > c) < e^{-c^2/2} \qquad (2.53)$$

for $c \geq 1$.

For the k-variate version of the normal we write $\mathbf{X} \sim N(\boldsymbol{\mu}, \boldsymbol{\Sigma})$, where $\boldsymbol{\mu} \equiv E\mathbf{X}$ is the k-vector of means and $\boldsymbol{\Sigma} \equiv E(\mathbf{X} - \boldsymbol{\mu})(\mathbf{X} - \boldsymbol{\mu})'$ is the $k \times k$ matrix of variances and covariances. Thus, $\boldsymbol{\Sigma}$ is a symmetric, positive-semidefinite matrix whose $j\ell$th element, $\sigma_{j\ell}$, is the covariance between X_j and X_ℓ when $j \neq \ell$ and is the variance of X_j when $j = \ell$. The c.f. of the affine function $a_0 + \mathbf{a}'\mathbf{X} = a_0 + \sum_{j=1}^{k} a_j X_j$, where $\{a_j\}_{j=0}^{k}$ are constants, is

$$\Psi_{a_0+\mathbf{a}'\mathbf{X}}(\zeta) = e^{i\zeta a_0}\Psi_{\mathbf{X}}(\zeta\mathbf{a}) = \exp\left[i\zeta(a_0 + \mathbf{a}'\boldsymbol{\mu}) - \zeta^2\mathbf{a}'\boldsymbol{\Sigma}\mathbf{a}/2\right],$$

which shows that $a_0 + \mathbf{a}'\mathbf{X} \sim N(a_0 + \mathbf{a}'\boldsymbol{\mu}, \mathbf{a}'\boldsymbol{\Sigma}\mathbf{a})$.

When the covariances are all zero, $\boldsymbol{\Sigma}$ then being a diagonal matrix, the multivariate normal p.d.f. factors as the product of the marginal p.d.f.s of $X_1, \ldots, X_k$; that is,

$$\begin{aligned} f(\mathbf{x}) &= (2\pi)^{-k/2}|\boldsymbol{\Sigma}|^{-1/2}\exp\left[-\frac{1}{2}(\mathbf{x} - \boldsymbol{\mu})'\boldsymbol{\Sigma}^{-1}(\mathbf{x} - \boldsymbol{\mu})\right] \\ &= (2\pi)^{-k/2}\prod_{j=1}^{k}\sigma_j^{-1}\exp\left[-\frac{1}{2}\sum_{j=1}^{k}\left(\frac{x_j - \mu_j}{\sigma_j}\right)^2\right] \\ &= \prod_{j=1}^{k}\frac{1}{\sigma_j\sqrt{2\pi}}\exp\left[-\frac{1}{2}\left(\frac{x_j - \mu_j}{\sigma_j}\right)^2\right]. \end{aligned} \tag{2.54}$$

Uncorrelated normals are therefore independent. Also, since independent random variables with normal marginal distributions have (2.54) as joint p.d.f., any collection of independent normals is necessarily multivariate normal. A sum, S, of such independent normals is distributed as normal with mean $\sum_{j=1}^{k}\mu_j$ and variance $\sum_{j=1}^{k}\sigma_j^2$. If the $\{X_j\}$ are also identically distributed, then $S \sim N(k\mu, k\sigma^2)$. The normal family is therefore closed under convolution.

In the bivariate case, $k = 2$, the p.d.f. is

$$f(x_1, x_2) = \frac{\exp\left\{-\frac{1}{2(1-\rho^2)}\left[\left(\frac{x_1-\mu_1}{\sigma_1}\right)^2 - 2\rho\left(\frac{x_1-\mu_1}{\sigma_1}\right)\left(\frac{x_2-\mu_2}{\sigma_2}\right) + \left(\frac{x_2-\mu_2}{\sigma_2}\right)^2\right]\right\}}{2\pi\sigma_1\sigma_2\sqrt{1-\rho^2}},$$

where $\rho \in (-1, 1)$ is the coefficient of correlation. The bivariate standard normal form for random variables Z_1, Z_2 with zero means and unit variances is

$$\phi(z_1, z_2) = \frac{1}{2\pi\sqrt{1-\rho^2}}\exp\left[-\frac{z_1^2 - 2\rho z_1 z_2 + z_2^2}{2(1-\rho^2)}\right].$$

A little manipulation presents this as

$$\phi(z_1, z_2) = \left\{ \frac{1}{\sqrt{2\pi}\sqrt{1-\rho^2}} \exp\left[-\frac{(z_1 - \rho z_2)^2}{2(1-\rho^2)} \right] \right\} \cdot \left(\frac{1}{\sqrt{2\pi}} e^{-z_2^2/2} \right),$$

which shows that the conditional density of Z_1 given Z_2 is

$$\phi(z_1 \mid z_2) = \frac{1}{\sqrt{2\pi}\sqrt{1-\rho^2}} \exp\left[-\frac{(z_1 - \rho z_2)^2}{2(1-\rho^2)} \right], \quad (z_1, z_2) \in \Re_2.$$

Thus, for a bivariate standard normal pair Z_1, Z_2 the conditional distribution of Z_1 given $Z_2 = z_2$ is itself normal, with mean proportional to z_2. Changing variables as $X_1 = \mu_1 + \sigma_1 Z_1$ and $X_2 = \mu_2 + \sigma_2 Z_2$ gives a corresponding result for the general case:

$$f(x_1 \mid x_2) = \frac{1}{\sqrt{2\pi}\delta} \exp\left[-\frac{(x_1 - \alpha - \beta x_2)^2}{2\delta^2} \right],$$

where $\beta \equiv \rho\sigma_1/\sigma_2$, $\alpha \equiv \mu_1 - \beta\mu_2$, and $\delta \equiv \sigma_1\sqrt{1-\rho^2}$. Thus, the conditional expectation of X_1 given X_2 is a linear function of X_2 when (X_1, X_2) is bivariate normal, whereas the conditional variance does not depend on X_2 at all.

Introducing a standard normal variate, $U \sim N(0,1)$, that is independent of X_2, the c.f. of $\alpha + \beta X_2 + \delta U$ is

$$\begin{aligned} e^{i\zeta\alpha} E e^{i\zeta\beta X_2 + i\zeta\delta U} &= e^{i\zeta\alpha} \exp\left(i\zeta\beta\mu_2 - \zeta^2\beta^2\sigma_2^2/2\right) \exp(-\zeta^2\delta^2/2) \\ &= e^{i\zeta(\mu_1 - \beta\mu_2)} \exp\left[i\zeta\beta\mu_2 - \zeta^2\rho^2\sigma_1^2/2 - \zeta^2(1-\rho^2)\sigma_1^2/2\right] \\ &= e^{i\zeta\mu_1 - \zeta^2\sigma_1^2/2} \\ &= \Psi_{X_1}(\zeta). \end{aligned}$$

This shows that when X_1, X_2 are bivariate normal X_1 is distributed as an affine function of X_2 plus an independent normal error. The affine function of X_2 is the "regression function" of X_1 on X_2.

We will often use the following relation between normal and bivariate normal c.d.f.s:

$$\int_{-\infty}^{\gamma} \Phi(\alpha - \beta y) \cdot d\Phi(y) = \Phi\left(\gamma, \frac{\alpha}{\sqrt{1+\beta^2}}; \frac{\beta}{\sqrt{1+\beta^2}}\right), \tag{2.55}$$

where α, β, γ are arbitrary real numbers and $\Phi(\cdot, \cdot; \rho)$ represents the bivariate normal c.d.f. To see that this holds, write the left side of (2.55) as

$$\int_{-\infty}^{\gamma} \int_{-\infty}^{\alpha - \beta y} \frac{1}{2\pi} e^{-(w^2+y^2)/2} \cdot dw\, dy$$

and change variables as $x = \beta y/\sqrt{1+\beta^2} + w/\sqrt{1+\beta^2}$. Defining $\rho = \rho(\beta) \equiv \beta/\sqrt{1+\beta^2}$ then gives

$$\int_{-\infty}^{\gamma} \int_{-\infty}^{\alpha/\sqrt{1+\beta^2}} \frac{1}{2\pi\sqrt{1-\rho^2}} \exp\left[-\frac{y^2 - 2\rho xy + x^2}{2(1-\rho^2)}\right] \cdot dx\, dy,$$

which corresponds to the right-hand side of (2.55).

Noting that $\Phi(+\infty, \cdot; \rho)$ is the marginal c.d.f. of the second variate, sending $\gamma \to +\infty$ in (2.55) gives another useful relation:

$$\int_{-\infty}^{\infty} \Phi(\alpha \pm \beta y) \cdot d\Phi(y) = \Phi\left(\alpha/\sqrt{1+\beta^2}\right). \tag{2.56}$$

If $X \sim N(\mu, \sigma^2)$ then $Y \equiv e^X$ is distributed as lognormal with the same parameters: $Y \sim LN(\mu, \sigma^2)$. Lognormal variates are clearly strictly positive. Their moments can be deduced from the m.g.f. of the normal, as

$$EY^k = Ee^{kX} = \mathfrak{M}_X(k) = e^{k\mu + k^2\sigma^2/2}.$$

In particular,

$$\begin{aligned} EY &= e^{\mu+\sigma^2/2} \\ VY &= e^{2\mu+\sigma^2}(e^{\sigma^2} - 1). \end{aligned}$$

While all the moments exist—even fractional and negative moments, the m.g.f. does not, since $E\exp(\zeta e^X) < \infty$ only if $\zeta = 0$.

Since the normal family is closed under convolution (i.e., sums of independent normals are normal), it is immediate that products of independent lognormals are lognormal.

Mixtures of Distributions

Distributions that serve as good models for financial data can sometimes be constructed by mixing standard models probabilistically. The resulting forms are often referred to as "hierarchical" models. As a simple example, suppose realizations of a random variable X are generated by drawing with probability p from the distribution $N(0, \sigma_1^2)$ and with probability $1-p$ from the distribution $N(0, \sigma_2^2)$, where $\sigma_1^2 \neq \sigma_2^2$. Here the variance represents the mixing parameter. The resulting mixture of normals that this generates can be thought of in the following way. Regard the mixing parameter as a discrete random variable V, with $\mathbb{P}(V = \sigma_j^2) = p^{2-j}(1-p)^{j-1}$ for $j \in \{1, 2\}$, and think of X as having the distribution $N(0, V)$ conditional on V. With this interpretation the marginal c.d.f. of X is

$$F(x) = E\mathbb{P}(X \leq x|V) = p\Phi(x/\sigma_1) + (1-p)\Phi(x/\sigma_2).$$

Mixtures of this particular type are symmetric about the origin and bell shaped, but they have thicker tails than the normal and kurtosis exceeding 3.0.

More elaborate hierarchical schemes can be generated by choosing other distributions for V, including continuous distributions that are supported on $(0, \infty)$. Of course, the mixture class is not limited to mixtures of normals. One has the freedom to choose among various models for the conditional distributions, among the mixing parameters, and among models for the distributions of those parameters.

Infinitely Divisible Distributions

A distribution F is infinitely divisible if for each positive integer n it can be expressed as the convolution of n identical distributions, F_n. In other words, X has an infinitely divisible distribution if for each n there are i.i.d. random variables $\{X_{jn}\}_{j=1}^n$ such that $X = \sum_{j=1}^n X_{jn}$. The c.f. of an infinitely divisible distribution can therefore be written as $\Psi(\zeta) = \Psi_n(\zeta)^n$ for some c.f. Ψ_n. Taking $\Psi_n(\zeta) = \exp\left[n^{-1}(i\zeta\mu - \zeta^2\sigma^2/2)\right]$ shows that the normal distribution is infinitely divisible, as are the Poisson, gamma, and Cauchy laws. We will make use of the following properties of infinitely divisible distributions.

1. An infinitely divisible c.f. has no zeroes; i.e., $\Psi(\zeta) > 0\ \forall \zeta \in \Re$ if $\Psi(\zeta)$ is infinitely divisible. [Proof: By infinite divisibility $\Psi(\zeta)^{1/n}$ is a c.f. for each n, and therefore $\Upsilon_n(\zeta) \equiv \Psi(\zeta)^{1/n}\Psi(-\zeta)^{1/n} = |\Psi(\zeta)|^{2/n}$ is a real-valued c.f. Since $|\Psi(\zeta)| \leq 1$, it follows that $\Upsilon_\infty(\zeta) \equiv \lim_{n\to\infty}\Upsilon_n(\zeta) = 0$ or $\Upsilon_\infty(\zeta) = 1$ according as $\Psi(\zeta) = 0$ or $\Psi(\zeta) \neq 0$. But since $\Psi(\zeta)$ is continuous and $\Psi(0) = 1$ there is an $\varepsilon > 0$ such that $\Psi(\zeta) \neq 0$ for $|\zeta| < \varepsilon$. Thus, $\Upsilon_\infty(\zeta) = 1$ for $|\zeta| < \varepsilon$ and is therefore continuous there. But by the continuity theorem $\Upsilon_\infty(\zeta)$ is a c.f. and therefore continuous everywhere. Accordingly, $\Upsilon_\infty(\zeta) = 1$ and $\Psi(\zeta) \neq 0$ for all $\zeta \in \Re$.]
2. If $\Psi(\zeta)$ is the c.f. of an infinitely divisible distribution with mean zero, then there exists a σ-finite measure M such that

$$\ln \Psi(\zeta) = \int_\Re \frac{e^{i\zeta x} - 1 - i\zeta x}{x^2} M(dx). \tag{2.57}$$

Property 2, the Lévy-Khintchine theorem, gives one of several canonical representations of an infinitely divisible c.f.[28] Here are two examples of infinitely divisible distributions in the canonical form:

Example 22 *Defining the integrand of* (2.57) *at zero by continuity and setting* $M(\Re) = M(\{0\}) = \sigma^2$ *give* $\Psi(\zeta) = \exp(-\zeta^2\sigma^2/2)$, *the c.f. of* $N(0, \sigma^2)$.

Example 23 *Setting* $M(\Re) = M(\{1\}) = \theta$ *gives* $\Psi(\zeta) = e^{-i\zeta\theta} \exp\left[\theta(e^{i\zeta} - 1)\right]$, *the c.f. of a centered Poisson distribution.*

Recall from (2.44) that the variance (second cumulant) of a random variable X with c.f. $\Psi(\zeta)$ is $d^2 \ln \Psi(\zeta)/d(i\zeta)^2 \mid_{\zeta=0}$. Differentiating the right side of (2.57) gives $VX = \int_{\Re} M(dx)$, which shows that VX exists if and only if M is a finite measure, as it was in the two examples.

2.2.8 *Introduction to Stochastic Processes*

A stochastic process is a family of random variables with some natural ordering. To represent the family requires an index and an index set that indicates what the ordering is, such as $\{X_j\}_{j \in \mathcal{J}}$ or $\{X_t\}_{t \in \mathcal{T}}$. Most applications involve orderings in space or in time. For example, $\{X_d\}_{d \in \mathcal{D}}$ might represent a feature of a topographic map that shows the heights above sea level at points on some curvilinear track on the earth's surface; and $\{X_t\}_{t \in \mathcal{T}}$ might represent the evolution through time of the concentration of a chemical reactant, of the incidence of solar radiation on a panel of receptors, or of the price of an asset. Since our interest is just in the last of these, we consider only processes that evolve in time. Among such there are discrete-time processes, for which index set $\mathcal{T}$ is countable (usually with equally spaced elements), and continuous-time processes, for which $\mathcal{T}$ is a continuum (usually a single bounded or unbounded interval). For example, a continuous record of bid prices for a common stock would represent a continuous-time process, while a daily record of settlement prices for commodity futures would be a process in discrete time.

Regardless of how time is recorded, any individual X_t in the family can be either a discrete, a continuous, or a mixed random variable. For example, bid prices for stocks evolve in continuous time but could be modeled either as continuous or as discrete variables supported on a lattice of

[28] For proofs see Feller (1971, chapter 17) and Lukacs (1970, chapter 5).

integer multiples of the minimum price increment. Thus, "discrete-time" and "continuous-time" refer to how the process is indexed, while the state space of the process defines the random variables' support. Of course, everything in the real world is discretely measured, so how a particular process is regarded usually hinges on the predictive power and convenience of the probability models from which we can choose to represent it.

Filtrations and Adapted Processes

The modeling of a stochastic process proceeds, as usual, with reference to a probability space $(\Omega, \mathcal{F}, \mathbb{P})$ that defines events and their measures. However, there are some subtleties. First, a random draw ω from Ω now produces one realization of the entire process, $\{X_t(\omega)\}_{t\in\mathcal{T}}$. Second, as observers of the process evolve along with it, they acquire new information that includes, at the least, the current and past values of X_t. Thinking of $\mathcal{F}$ and various sub-σ-fields of $\mathcal{F}$ as information structures gives a natural way to characterize this evolution as a progression of σ-fields—a "filtration". Represented as $\{\mathcal{F}_t\}_{t\in\mathcal{T}}$, a filtration is an increasing family of sub-σ-fields of $\mathcal{F}$, meaning that $\mathcal{F}_s \subseteq \mathcal{F}_t \subseteq \mathcal{F}$ for $s < t$. The process $\{X_t\}$ is then required to be "adapted" to $\{\mathcal{F}_t\}$, meaning that all the history of the process up to and including t is $\mathcal{F}_t$-measurable. A discrete-time example will help to show what all this means.

Example 24 *Consider the experiment of flipping a coin three times in succession, and let Ω comprise the sequences of heads or tails that could be obtained, as*

$$\Omega \equiv \{HHH, HHT, HTH, HTT, THH, THT, TTH, TTT\}.$$

Next, take as the general information structure $\mathcal{F}$ the collection of all 2^8 subsets of these eight elementary outcomes, and define the probability measure $\mathbb{P}$ as

$$\mathbb{P}(A) = \#(A)/\#(\Omega) = \#(A)/8, \ A \in \mathcal{F}.$$

Here "#" is counting measure that enumerates the elementary outcomes in sets. Having fixed the overall probability space, we can now describe a filtration that represents how information evolves as one learns, successively, what happens on each flip. We can also define an adapted process $\{X_t\}_{t\in\{0,1,2,3\}}$ that represents the total number of heads obtained after t flips.

At stage $t = 0$, before the first flip is made, we know only that one of the outcomes in Ω will occur and that an outcome not in Ω will not *occur. In this case our initial information set is the trivial field $\mathcal{F}_0 = \{\emptyset, \Omega\}$. Since we start off with $X_0 = 0$, a known constant that does not depend on the outcome of the experiment, it follows $X_0 \in \mathcal{F}_0$ (is $\mathcal{F}_0$-measurable). Now partition Ω into the exclusive sets*

$$\underbrace{\{HHH, HHT, HTH, HTT\}}_{H}, \quad \underbrace{\{THH, THT, TTH, TTT\}}_{T}.$$

Once the first flip is made and we see whether the coin came up heads or tails, we acquire the information $\mathcal{F}_1 \equiv \{\emptyset, H, T, \Omega\}$. In other words, we then know whether each of the 2^2 events generated by the partition has occurred. In particular, we know the value assumed by X_1, which means that random variable X_1 is measurable with respect to $\mathcal{F}_1$. Of course, we retain the trivial knowledge we had at the outset, so $\mathcal{F}_0 \subset \mathcal{F}_1$, and thus X_0 is also $\mathcal{F}_1$-measurable. However, $\mathcal{F}_1$ clearly does not pin down the value of X_2, the total number of heads on the first two flips. That value will be known once we get the information $\mathcal{F}_2$ that represents the history to that stage. This comprises the 2^4 events generated by the finer partition

$$\underbrace{\{HHH, HHT\}}_{HH}, \quad \underbrace{\{HTH, HTT\}}_{HT}, \quad \underbrace{\{THH, THT\}}_{TH}, \quad \underbrace{\{TTH, TTT\}}_{TT}.$$

But each set in $\mathcal{F}_1$ is also in $\mathcal{F}_2$, so we still know what happened on the first flip. For example, $H = HH \cup HT \in \mathcal{F}_2$. Therefore, all of X_0, X_1, and X_2 are $\mathcal{F}_2$-measurable. Finally, making the last flip gives us $\mathcal{F}_3 = \mathcal{F}$, the information structure of the 2^8 events generated by the finest possible partition of Ω (given its original description) into the eight elementary outcomes. $\mathcal{F}_3$ tells the full story of the experiment and determines the values of all of X_0, X_1, X_2, and X_3. Our discrete-time stochastic process $\{X_t\}_{t \in \{0,1,2,3\}}$ is then adapted to the filtration $\{\mathcal{F}_t\}_{t \in \{0,1,2,3\}}$, in the sense that the information structure at each t reveals the entire history of the process to that time.

Notice in the example that $\mathcal{F}_2$ is not the smallest information set that determines X_2, in the sense that it is not built from the coarsest possible partition that tells the number of heads after two flips. This minimal σ-field, the field $\sigma(X_2)$ that is generated by X_2, comprises just the 2^3 sets built up from

$$\{TTH, TTT\}, \quad \{THH, THT, HTH, HTT\}, \quad \{HHH, HHT\}.$$

However, X_1 is not measurable with respect to $\sigma(X_2)$ and is therefore not a random variable on the probability space $(\Omega, \sigma(X_2), \mathbb{P}_2)$ (where $\mathbb{P}_2$ is the appropriate probability measure on $(\Omega, \sigma(X_2))$). In fact, $\mathcal{F}_2$ is the smallest information set that determines the entire history of X_t through $t = 2$; that is, $\mathcal{F}_2 = \sigma(X_0, X_1, X_2)$. Similarly, $\sigma(X_3)$ is based on a coarser partition than $\mathcal{F}_3$, namely

$$\{TTT\},\ \{TTH, HTT, THT\},\ \{HHT, HTH, THH\},\ \{HHH\},$$

but $\mathcal{F}_3 = \sigma(X_0, X_1, X_2, X_3)$.

It is worth emphasizing that $\mathcal{F}_t$ tells us the history up through t but does not support prophecy!

Example 25 *Suppose we start with some fixed initial capital C_0 and place bets on the outcomes of the coin flips in example 24, winning one unit of currency each time a head turns up and losing one unit each time there is a tail. Then the process $\{C_t = C_0 + X_t\}$ is adapted to the filtration $\{\mathcal{F}_t\}$, since we know the value of C_t and its past at each t; however, the process $\{U_t = C_3 - C_t\}$ (the total amount yet to be won) is not adapted.*

Does it make any sense to call $\{\mathcal{F}_t\}$ a "filtration"? Indeed, the sequence of information sets really is like a series of filters. $\mathcal{F}_0$ catches just the coarsest possible information that lets us distinguish between certainties and impossibilities but allows all the finer detail to pass through. The knowledge of what happened on the first flip is caught by $\mathcal{F}_1$, yet $\mathcal{F}_1$ allows the still finer information about the second flip to pass, and so on. The fineness of information that resides in each $\mathcal{F}_t$ corresponds precisely to the fineness of the partition of Ω from which $\mathcal{F}_t$ is constructed. It does makes sense, then, to think of progressing through information sets as a filtering operation and to describe the process $\{X_t\}_{t \in \mathcal{T}}$ as being adapted to a filtration. Notice that the overall σ-field $\mathcal{F}$ in the underlying probability space $(\Omega, \mathcal{F}, \mathbb{P})$ contains information that is at least as fine as the finest in $\{\mathcal{F}_t\}$.

We often refer to a probability space endowed with a filtration as a "filtered probability space" and represent it as $(\Omega, \mathcal{F}, \{\mathcal{F}_t\}, \mathbb{P})$.

Probabilities and expectations conditioned on sub-σ-fields, as $\mathbb{P}(\cdot \mid \mathcal{F}_t)$ and $E(\cdot \mid \mathcal{F}_t)$, play important roles in dealing with stochastic processes. The *processes* $\{\mathbb{P}(\cdot \mid \mathcal{F}_t)\}$ and $\{E(\cdot \mid \mathcal{F}_t)\}$ record how our assessments of probabilities and expectations evolve over time. Thus, for an $\mathcal{F}_T$-measurable random variable X_T we regard $\{E(X_T \mid \mathcal{F}_t)\}_{0 \leq t \leq T}$ itself as a stochastic process that shows how the expectation progresses with new information. When applying these tools to real financial markets we must keep in mind that the nestedness of the filtration structure, $\mathcal{F}_s \subseteq \mathcal{F}_t$ for $s < t$, implies that nothing is forgotten as time evolves. A joint implication of this feature and the tower property of conditional expectation, (2.50), is that

$$E\left[E(X_T \mid \mathcal{F}_t) \mid \mathcal{F}_s\right] = E(X_T \mid \mathcal{F}_s), \quad 0 \leq s \leq t \leq T.$$

The practical interpretation is that the best guess now (time s) of what our expectation will be at t is our current expectation. For example, today's best forecast of the weather seven days from now is our best forecast of *tomorrow's* forecast of the weather six days hence.

We will often use the following abbreviated notation for conditional expectation and conditional probability in the context of stochastic processes:

$$\begin{aligned} E_t(\cdot) &\equiv E(\cdot|\mathcal{F}_t) \\ \mathbb{P}_t(\cdot) &\equiv \mathbb{P}(\cdot|\mathcal{F}_t). \end{aligned}$$

This has the virtue of saving space, but it does require one to keep in mind that conditioning is on the entire history and not just on the current state. The more elaborate notation will be used when this fact needs special emphasis and when space allows. When no conditioning is indicated, as $E(\cdot)$ and $\mathbb{P}(\cdot)$, it is implicit that conditioning is on the trivial field $\{\emptyset, \Omega\}$. Unless otherwise specified, we shall always regard the initial σ-field $\mathcal{F}_0$ as trivial. This means that the initial value of a stochastic process is considered to be deterministic. Thus, unless otherwise stated,

$$\begin{aligned} E(\cdot) &\equiv E_0(\cdot) \equiv E(\cdot|\mathcal{F}_0) \\ \mathbb{P}(\cdot) &\equiv \mathbb{P}_0(\cdot) \equiv \mathbb{P}(\cdot|\mathcal{F}_0). \end{aligned}$$

Two very important classes of stochastic processes are "martingales" and "Markov processes". We will meet many examples of these, and martingales in particular will be of central importance in the study of financial derivatives.

Martingales

An adapted process $\{X_t\}$ on a filtered probability space $(\Omega, \mathcal{F}, \{\mathcal{F}_t\}, \mathbb{P})$ is a martingale if

1. $E|X_t| < \infty$ for all $t \geq 0$ and
2. $E_s X_t = X_s$ for $0 \leq s \leq t$.

The integrability requirement, property **1**, just ensures that the conditional expectation in **2** always exists, since by the tower property $E|X_t| \equiv E_0|X_t| = E_0(E_s|X_t|)$. The really crucial feature, property **2**, is called the "fair-game property". Here is how it gets its name.

Example 26 *Suppose one starts with a fixed amount X_0 in capital and undertakes a sequence of fair bets at times $t \in \mathcal{T} \equiv \{0,1,2,\ldots\}$, the outcomes of which determine the capital available at t, X_t. Let $\{Y_{t+1}\}_{t\in\mathcal{T}}$ be i.i.d. random variables with zero mean that represent the payoffs per unit of currency of the bets placed at t, and let $\mathcal{F}_t$ be the σ-field generated by $Y_0, Y_1, Y_2, \ldots, Y_t$ (where we define $Y_0 \equiv 0$). The individual bets are fair in the sense that their payoffs have zero expected value. Suppose a nonnegative, uniformly bounded amount b_t, $0 \le b_t \le B < \infty$, is wagered at each t and that $b_t \in \mathcal{F}_t$. The last means that b_t is $\mathcal{F}_t$-measurable and therefore that $\{b_t\}$ is adapted to the filtration $\{\mathcal{F}_t\}_{t\in\mathcal{T}}$. This allows the amount of the bet to depend in any way on the sequence of past gains and losses but rules out any knowledge of the future. Since each bet placed at t produces a payoff b_tY_{t+1} at $t+1$, wealth evolves in the sequence of plays as $X_{t+1} = X_t + b_tY_{t+1}, t \in \mathcal{T}$. Clearly, $\{X_t\}$ itself is adapted to $\{\mathcal{F}_t\}$, since the sizes and outcomes of all past bets determine the capital at each stage. Also, the integrability of $\{Y_t\}$ and the boundedness of $\{b_t\}$ imply*

$$E|X_t| \le X_0 + \sum_{j=1}^{t} BE|Y_j| < \infty;$$

and the fact that the payoffs are independent with mean zero implies

$$\begin{aligned} E_sX_t &= E(X_s + b_sY_{s+1} + \cdots + b_{t-1}Y_t|\mathcal{F}_s) \\ &= X_s + b_sE_sY_{s+1} + \cdots + E_s(b_{t-1}E_{t-1}Y_t) \\ &= X_s \end{aligned}$$

for $0 \le s \le t$. Thus, $\{X_t\}_{t\in\mathcal{T}}$ is a discrete-time martingale. The message is that so long as the outcome of each bet is fair, the game itself is fair, provided the amounts wagered are bounded and do not somehow anticipate future outcomes.

Example 27 *If Z is an $\mathcal{F}_T$-measurable random variable with $E|Z| < \infty$ and $X_t = E_tZ \equiv E(Z \mid \mathcal{F}_t)$ for $t \in [0,T]$, then*

$$E|X_t| = E|E_tZ| \le E(E_t|Z|) = E|Z| < \infty$$

and

$$E_s X_t = E_s(E_t Z) = E_s Z = X_s.$$

Thus, the conditional expectations **process** *with respect to a filtration $\{\mathcal{F}_t\}$ is itself a martingale adapted to that filtration.*

Example 28 *Suppose c is a finite, $\mathcal{F}_0$-measurable constant. It is then obviously adapted to $\{\mathcal{F}_t\}$, and it clearly satisfies conditions 1 and 2. Such a constant is therefore a martingale on any filtered probability space.*

Submartingales and supermartingales are, respectively, favorable games ($E_s X_t \geq X_s$) and unfavorable games ($E_s X_t \leq X_s$). Thus, in reality the wealth processes of gambling houses are submartingales, while those of the patrons are supermartingales. Notice that since the inequalities are weak, martingales are also sub- and supermartingales.

Stopping times (optional times, Markov times) are important concepts in connection with both martingales and Markov processes. These are extended random variables $\tau : \Omega \to \Re^+ \cup \{+\infty\}$ for which events of the form $\tau \leq t$ are $\mathcal{F}_t$-measurable. As such, they model decisions that are made in connection with an adapted process $\{X_t\}$. For a gambling process a plan that decides from current and past conditions whether to terminate play at each point is a stopping time. An example is the random variable $\tau_K = \inf\{t : X_t \geq K\}$, which is the decision to stop the first time wealth hits level K. At any time t one knows whether the event $\tau \leq t$ has occurred by looking at the present and past of wealth, both of which are in the information set $\mathcal{F}_t$. What the definition of a stopping time rules out is reliance on information about the future. For example, $\tau'_K = \sup\{t \leq T : X_t \geq K\}$ represents the plan to stop the last time before T at which wealth is at least K. This is not a stopping time, because one cannot know whether to stop at a given time without knowing the future.

In working with martingales stopping times can provide a means of controlling the processes in some way that is advantageous to the purpose at hand. Localization is one such means of control. For example, letting

$$\tau_n = \inf\{t : |X_t| \geq n\} \tag{2.58}$$

be a stopping time for a martingale $\{X_t\}$, then the stopped process $\{X_{t\wedge\tau_n}\}$ becomes a (locally)-bounded martingale (adapted to $\mathcal{F}_{t\wedge\tau_n}$) that simply terminates at the time it first hits $\pm n$. If a process $\{X_t\}_{t\geq 0}$ satisfies the fair-game property but not the integrability property, condition **1**, it can nevertheless be considered a local martingale if there is a sequence of stopping times $\tau_n \uparrow +\infty$ such that $\{X_{t\wedge\tau_n}\}$ is (for each n) a martingale adapted to $\mathcal{F}_{t\wedge\tau_n}$.[29] The requirement that $\tau_n \uparrow +\infty$ as $n \to \infty$ just ensures that the process can be made to evolve stochastically for as long as desired by taking n sufficiently large. The stopping times that control $\{X_t\}$ in this way are said to "reduce" the process. A local martingale $\{X_t\}$ is also a martingale proper if $E\sup_{s\leq t}|X_s| < \infty$ for each $t \geq 0$.

One last variant on the martingale theme will be important in modeling asset prices in continuous time. A "semimartingale" is a process that would be a local martingale if some other adapted process of finite variation were extracted; in other words, it is a process that can be decomposed into an adapted, finite-variation process and a local martingale.[30] As a discrete-time example, the wealth process $\{X_n\}$ of someone playing a sequence of fair games would not be a martingale if some extraneous income process augmented wealth at each play, but $\{X_n\}$ could still be a semimartingale.

An important result that explains much of martingales' central role in modern probability theory is the martingale convergence theorem:

Theorem 7 *If $\{X_t\}$ is a supermartingale with $\sup_t E|X_t| < \infty$, then the limit of X_t as $t \to \infty$ exists a.s. and is finite.*

In short, supermartingales (and therefore martingales) whose expected values are bounded converge to either a finite constant or a random variable; that is, they neither diverge nor fluctuate indefinitely.[31] We note that uniform integrability of $\{X_t\}$ implies $\sup_t E|X_t| < \infty$. While the convergence theorem is crucial for modern proofs of laws of large numbers and related asymptotic results, it is mostly the fair-game property of martingales that accounts for their importance in finance.

---

[29] In a situation where the initial value X_0 itself is not integrable ($\mathcal{F}_0$ now being no longer the trivial σ-field), we would require $\{X_{t\wedge\tau_n}\mathbf{1}_{(0,\infty)}(\tau_n)\}$ for localization, where the indicator prevents an automatic stop at $t = 0$. An example is the wealth process of a gambler who gets to starts the play with a Cauchy-distributed initial stake, X_0. $E|X_t| < \infty$ clearly fails here, yet it remains true (so long as the bets are bounded and the individual plays are fair) that $E_s X_t = X_s$ for $0 \leq s < t$, so localization as $\{X_{t\wedge\tau_n}\mathbf{1}_{(0,\infty)}(\tau_n)\}$ controls the process in the way we want.

[30] A "finite-variation" process is one whose sample paths have finite variation over finite intervals.

[31] For a proof and applications see Williams (1991).

Markov Processes

A Markov process is an adapted stochastic process in either discrete or continuous time that is memoryless, in the sense that future behavior depends only on the present state and not on the past. Specifically, an adapted process $\{X_t\}$ on a filtered probability space $(\Omega, \mathcal{F}, \{\mathcal{F}_t\}, \mathbb{P})$ is Markov if for any Borel set B

$$\mathbb{P}_s(X_t \in B) \equiv \mathbb{P}(X_t \in B \mid \mathcal{F}_s) = \mathbb{P}\left(X_t \in B \mid \sigma(X_s)\right) \equiv \mathbb{P}(X_t \in B \mid X_s).$$

In words, the probability of any event given the entire history to s—that is, given $\mathcal{F}_s$—is the same as the probability conditional on knowledge of X_s alone.

An example of a process that is *not* Markov is one that describes the position over time of a massive object subject to unbalanced forces. In such a case the future trajectory would depend on the current rate of change as well as on the current position. The wealth process $\{X_t\}_{t=1,2,...}$ of the gambler who bets on a sequence of independent plays of a game of chance is not Markov if amounts wagered at each play depend in any nontrivial way on the past sequence of outcomes. It is Markov if the bets depend on current wealth only. Here are other examples of processes that do have the Markov property.

Example 29 *The first-order autoregressive process,* $\{X_t = \rho X_{t-1} + U_t\}_{t=1,2,...}$, *where continuous random variables* $\{U_t\}$ *are i.i.d. with zero mean, is a continuous-state, discrete-time Markov process.*

Example 30 *A branching process is a discrete-state, discrete-time Markov process that has been used to represent, among other things, the dynamics of populations. Starting with an initial population of size* $P_0 > 0$, *each member* $j \in \{1, 2, \ldots, P_0\}$ *produces a random number* Q_{j0} *of offspring and then dies off. Together, these progeny constitute the next generation of* $P_1 = \sum_{j=1}^{P_0} Q_{j0}$ *individuals, who in turn bear offspring and die off. The process continues either indefinitely or until some stage at which all individuals remain childless. Thus, taking* $Q_{0n} \equiv 0$ *to allow for the possibility of extinction at generation* $n-1$, *there are* $P_n = \sum_{j=0}^{P_{n-1}} Q_{jn}$ *individuals in the nth generation, where* $\{Q_{jn}\}_{j\in\{1,2,...,P_{n-1}\},n\in\aleph_0}$ *are i.i.d. and integer-valued.*

Example 31 *The Poisson process* $\{N_t\}_{t\geq 0}$ *is a discrete-state, continuous-time Markov process that will be encountered often in succeeding chapters.*

It is defined by the conditions (i) $N_0 = 0$, (ii) $N_t - N_s$ *independent of* $N_r - N_q$ *for* $q < r \leq s < t$, *and (iii)* $N_{t+u} - N_t$ *distributed as Poisson with parameter* θu*; that is,*

$$\mathbb{P}(N_{t+u} - N_t = n) = e^{-\theta u}(\theta u)^n / n!, n \in \aleph_0.$$

The process takes a Poisson-distributed number of unit jumps during a finite period of time, the mean number being proportional to the elapsed time. The independence of increments in nonoverlapping intervals confers the Markov property.

Example 32 *The Wiener or Brownian-motion process is a continuous-state, continuous-time Markov process with independent, normally distributed increments. Whereas sample paths of the Poisson process are discontinuous, those of Brownian motion are a.s. continuous. The* standard *Wiener process* $\{W_t\}_{t \geq 0}$ *is defined by the conditions (i)* $W_0 = 0$, (ii) $W_t - W_s$ *independent of* $W_r - W_q$ *for* $q < r \leq s < t$, *and (iii)* $W_{t+u} - W_t \sim N(0, u)$.

Diffusion processes are continuous-time Markov processes with continuous sample paths. The Wiener process is an example. We will see shortly that the more general Itô processes—some of which have the Markov property and some of which do not—can be built up from the Wiener process. These continuous processes can be combined in various ways with Poisson processes to model phenomena that are subject to unpredictable, possibly discontinuous changes. To use and construct such models requires at least a basic understanding of stochastic calculus, which is the subject of chapter 3.

3

Tools for Continuous-Time Models

This chapter adds to the tools of analysis and probability theory presented in chapter 2 the concepts needed to value derivatives on assets whose prices evolve in continuous time. We begin by surveying the key properties of Brownian motions, which are the most basic building blocks for continuous-time models of asset prices. Sections 3.2–3.4 then develop the tools of stochastic calculus that permit us to work with such erratic processes. The final section of the chapter is devoted to processes whose sample paths may be discontinuous.[1]

3.1 Wiener Processes

3.1.1 *Definition and Background*

Wiener processes are continuous-time, continuous-state-space versions of the random walk. A celebrated early application was by Einstein (1905) in describing the phenomenon of molecular vibration, by which he explained the incessant, random-seeming motion of microscopic particles observed about 80 years earlier by botanist Robert Brown. Louis Bachelier's application to speculative prices in a 1900 doctoral thesis (published in English

[1]Sources for the technical detail that underlies the material in this chapter include Arnold (1974), Chung and Williams (1983), Durrett (1984, 1996), Elliott (1982), Elliott and Kopp (1999), Karatzas and Shreve (1991), Shreve (2004), and Musiela and Rutkowski (2005). Baxter and Rennie (1996) offer a highly readable and intuitive treatment of Girsanov's theorem (section 3.4.1) and stochastic calculus generally. Karlin and Taylor (1975) give a more technical but non-measure-theoretic account of Brownian motion in particular. Protter (1990) provides a rigorous treatment of stochastic calculus for discontinuous processes.

in Cootner (1964)) predated Einstein's work. The appellation "Wiener" honors the American probabilist Norbert Wiener, a pioneer in the mathematical analysis of these processes. In current literature the terms *Wiener process* and *Brownian motion* are used interchangeably.

The standard Wiener process was defined in chapter 2 as a stochastic process $\{W_t\}_{t\geq 0}$ having the following properties: (i) $W_0 = 0$; (ii) $W_t - W_s$ independent of $W_r - W_q$ for $q < r \leq s < t$; and (iii) $W_{t+u} - W_t \sim N(0, u)$. The process thus has time-stationary and independent increments. It is important to note that u is the *variance* of $W_{t+u} - W_t$. The choice of zero for initial value is just a matter of convention, whereas the independence and normality of increments are crucial features. An implication of (iii) that we will often put to use is that $W_{t+u} - W_t = Z\sqrt{u}$, where $Z \sim N(0, 1)$.[2] Also, (ii) and (iii) together imply that increments $\{W_{t_j} - W_{t_{j-1}}\}_{j=1}^n$ over nonoverlapping intervals $\{[t_{j-1}, t_j)\}_{j=1}^n$ are distributed as $\{Z_j\sqrt{t_j - t_{j-1}}\}_{j=1}^n$, where the $\{Z_j\}$ are independent standard normals.

The definition of the Wiener process extends to higher dimension k simply by regarding W_t as a vector in (i) and (ii) and writing (iii) as (iii′) $W_{t+u} - W_t \sim N(0, u\mathbf{I}_k)$, where $\mathbf{I}_k$ is the identity matrix. The k-dimensional standard form generalizes to allow mean vector proportional to u and an arbitrary positive-definite covariance matrix in place of $\mathbf{I}_k$. However, the more general processes we need can be built from the standard form alone, and for now we consider just the one-dimensional case.

Proofs that processes meeting conditions (i)-(iii) do exist can be found in many texts on stochastic processes; e.g., Durrett (1984) and Revuz and Yor (1991). While the conditions are thus known to be mutually consistent, they do give rise to some strange properties. These will be the main source of distinction between stochastic calculus and deterministic calculus.

3.1.2 *Essential Properties*

With $(\Omega, \mathcal{F}, \{\mathcal{F}_t\}, \mathbb{P})$ as a filtered probability space, we consider a Brownian motion adapted to $\{\mathcal{F}_t\}_{t\geq 0}$. Here are the main features.

1. Wiener processes are Markov. Although there are delicate technical issues here[3] the independence of increments makes this easy to believe.

---

[2]Expressions such as $W_t = Z\sqrt{t}$ and $W_t \sim Z\sqrt{t}$ should be interpreted as "equal in distribution".

[3]For these see Durrett (1984, section 1.2).

For example, one gets the idea in a simple way by calculating the conditional characteristic function of W_t. For any $\zeta \in \Re$ and $s \leq t$ we have

$$\begin{aligned} E_s e^{i\zeta W_t} &= E[e^{i\zeta W_s} e^{i\zeta(W_t - W_s)} \mid \mathcal{F}_s] \\ &= e^{i\zeta W_s} E[e^{i\zeta(W_t - W_s)} \mid \mathcal{F}_s] \\ &= e^{i\zeta W_s - \zeta^2 (t-s)/2} \\ &= E(e^{i\zeta W_t} | W_s). \end{aligned}$$

The uniqueness theorem for c.f.s then implies that $\mathbb{P}(W_t \in B \mid \mathcal{F}_s) = \mathbb{P}(W_t \in B \mid W_s)$ for any Borel set B.

2. Standard Wiener processes are martingales. First, W_t is integrable, since

$$E|W_t| = E|Z|\sqrt{t} < \infty \, \forall t,$$

where $Z \sim N(0,1)$. Next, W_t has the fair-game property:

$$E_s(W_t) = E_s[W_s + (W_t - W_s)] = W_s, \ \ s \leq t.$$

3. $E(W_s W_t) = E(W_s \cdot E_s W_t) = E W_s^2 = s, s \leq t.$

4. Sample paths $\{W_t(\omega)\}_{t \geq 0}$ are a.s. continuous. That is, the occurrence of a discontinuity or jump anywhere in the realization of a Brownian path is an event of zero probability. In symbols, if

$$\mathcal{D} = \{\omega : \{W_t(\omega)\}_{t \geq 0} \text{ is not continuous}\}$$

then $\mathbb{P}(\mathcal{D}) = 0$. A theorem due to Kolmogorov[4] states that sample paths of a stochastic process $\{X_t\}_{t \geq 0}$ are a.s. continuous if there are positive numbers a, b, c such that

$$E|X_t - X_s|^a \leq c(t-s)^{1+b}$$

for each $s < t$. Taking $a = 4$ gives $E(W_t - W_s)^4 = EZ^4(t-s)^2 = 3(t-s)^2$, thus satisfying the requirement.

5. Although they are continuous, sample paths of Brownian motion are a.s. not differentiable anywhere. The following, although not a proof of the statement, gets across the idea. Let $\Delta_n \equiv n(W_{t+1/n} - W_t) \sim N(0, n)$ be the average rate of change of W over $[t, t + 1/n)$. For $K > 0$, $\mathbb{P}(|\Delta_n| > K) = \mathbb{P}(|Z| > Kn^{-1/2}) \to 1$ as $n \to \infty$, and since K is arbitrary it follows that $|\Delta_n| \to \infty$ in probability. This nondifferentiability property has an important implication for models of asset prices built up from

[4]For the proof see Karatzas and Shreve (1991, pp. 53–55). Strictly, what is proved is that there is a continuous *modification* of $\{X_t\}$; that is, an a.s. continuous process $\{X_t'\}$ such that $\mathbb{P}(X_t = X_t')$ for each $t \geq 0$.

Brownian motions: one cannot predict an asset's future value even at the next instant by looking at the current value and the change in the immediate past.

6. Sample paths of Brownian motion are of unbounded variation. This implies that the process covers an infinite distance as it evolves over any time interval, no matter how short! The following argument shows specifically that W has infinite variation over $[0, 1]$ (and extends to any interval through an affine transformation of the time scale). Partitioning $[0, 1]$ into subintervals of length $1/n$, let $\Delta_{kn} \equiv W_{k/n} - W_{(k-1)/n}$ be the net change over the kth of these, and let $\mathbb{V}_n^{[0,1]} W \equiv \sum_{k=1}^{n} |\Delta_{kn}|$ be the total variation over the discrete partition. Then $\mathbb{V}^{[0,1]} W = \sup_n \mathbb{V}_n^{[0,1]} W$. Since the $\{\Delta_{kn}\}_{k=1}^{n}$ are i.i.d. as $N(0, n^{-1})$ we have

$$E|\Delta_{kn}| = n^{-1/2}\sqrt{2/\pi},$$
$$V|\Delta_{kn}| = n^{-1}(1 - 2/\pi),$$

and therefore

$$E\mathbb{V}_n^{[0,1]} W = \sqrt{2n/\pi},$$
$$V\mathbb{V}_n^{[0,1]} W = (1 - 2/\pi).$$

Finally, by the central limit theorem

$$Z_n \equiv (V\mathbb{V}_n^{[0,1]} W)^{-1/2}(\mathbb{V}_n^{[0,1]} W - E\mathbb{V}_n^{[0,1]} W) \rightsquigarrow N(0, 1),$$

so that for any $K > 0$

$$\mathbb{P}(\mathbb{V}_n^{[0,1]} W > K) = \mathbb{P}\left(Z_n > \frac{K - \sqrt{2n/\pi}}{\sqrt{1 - 2/\pi}}\right) \to \mathbb{P}(Z > -\infty) = 1$$

as $n \to \infty$.

7. The net change in Brownian motion over an interval cannot be represented as an integral with respect to Lebesgue measure; that is, there is no density w such that $W_t = \int_0^t w_s \cdot ds$. For such a representation to exist it is necessary and sufficient that W be an absolutely continuous function of t. That is, for any $\varepsilon > 0$ there must exist $\delta > 0$ such that for each finite k

$$\sum_{j=1}^{k} |t_j - t_{j-1}| \leq \delta \Rightarrow \sum_{j=1}^{k} |W_{t_j} - W_{t_{j-1}}| < \varepsilon.$$

However, since W is of unbounded variation the sum on the right can be made arbitrarily large by increasing k and partititioning any given interval of length δ more and more finely. Therefore, W is not absolutely continuous and there can be no conventional integral representation in terms of a density function.

3.2 Itô Integrals and Processes

These properties of Brownian motion indicate that the conventional concepts of integration do not extend to stochastic integrals with respect to W_t. Since sample paths of Brownian motion are continuous, an expression such as $\int b(t) \cdot dW_t$ could not have meaning as a countable sum, as for a Stieltjes integral with respect to a step function. But since W_t is not absolutely continuous, neither could there be the other interpretation as a Lebesgue integral of b times a density with respect to Lebesgue measure. A theory of stochastic integration that does provide a useful interpretation of $\int b(t) \cdot dW_t$ was developed during the 1940s and early 1950s by the Japanese probabilist Kiyosi Itô. Itô integrals and the processes constructed from them have since become essential tools in financial modeling.

3.2.1 *A Motivating Example*

What accounts for the importance of stochastic integrals in finance? Recall example 26 on page 88, which showed that wealth after a sequence of fair bets is a discrete-time martingale. Beginning with a known stake X_0, b_t is bet on the tth play and each unit of currency that is wagered returns Y_{t+1} at $t+1$, so that wealth evolves as $X_{t+1} = X_t + b_t Y_{t+1}$. Moving from the casino to a financial market, we can transform the sequence of games into a sequence of investments at times $t_0, t_1, \ldots, t_{n-1}$, where $t_j = j\Delta t$ and $n\Delta t \equiv T$. Now think of b_{t_j} as the number of units of an asset held at t_j or, more aptly, as a vector of units of various assets—which is to say, a portfolio; and let $Y_{t_{j+1}} = \Delta S_{t_j} \equiv S_{t_{j+1}} - S_{t_j}$ be the corresponding vector of price changes. After $n = T/\Delta t$ periods the total change in wealth is[5]

$$X_T - X_0 = \sum_{j=0}^{n-1} b_{t_j} \Delta S_{t_j}.$$

[5]In this example $\{X_t\}$ is, in general, no longer a martingale in the measure $\mathbb{P}$ that governs the actual behavior of prices, since expected price changes are not generally equal to zero. However, we will see later that there is an equivalent measure under which the martingale property does hold when X_t is appropriately normalized.

Since the dynamics of self-financing portfolios that replicate derivatives will be of just this form, the relevance of expressions like this becomes obvious. And when we move to continuous time and allow portfolio weights and prices to vary continuously, such expressions turn into integrals with respect to the stochastic process $\{S_t\}$:

$$X_T - X_0 = \int_0^T b_t \cdot dS_t. \tag{3.1}$$

Although Brownian motion *per se* is not a good model for the price of a financial asset, it is necessary to define integration with respect to Wiener processes in order to develop integrators that *are* good models. We first undertake to find a meaningful construction of integrals $\int_0^t b_s \cdot dW_s$ for suitable processes $\{b_t\}$, then develop plausible models for asset prices in terms of such integrals. Before beginning, however, it is worthwhile to observe that finding integrals *of* Brownian motions, or of suitable functions of Brownian motions, requires nothing new. Since Brownian paths $\{W_t(\omega)\}_{t\geq 0}$ are a.s. continuous, expressions (random variables) like $\int_0^t W_s \cdot ds$ or $\int_0^t b(W_s) \cdot ds$ can be interpreted, path by path, as ordinary Riemann (or Lebesgue) integrals, provided that $b(\cdot)$ is itself Riemann (or Lebesgue) integrable. The following example provides a specific result that will be needed later in applications.

Example 33 *Let us derive the distribution of $\int_0^t W_s \cdot ds$ from its Riemann construction, as $\lim_{n\to\infty} tn^{-1} \sum_{k=1}^n W_{kt/n}$. Note first that*

$$W_{kt/n} \equiv \sum_{j=1}^{k} (W_{tj/n} - W_{t(j-1)/n}) \sim \sqrt{t/n} \sum_{j=1}^{k} Z_j,$$

where "$\sim$" signifies "equal in distribution" and where the $\{Z_j\}$ are i.i.d. as $N(0,1)$. Therefore,

$$\begin{aligned}\sum_{k=1}^{n} W_{kt/n} &\sim \sqrt{t/n}[nZ_1 + (n-1)Z_2 + \cdots + Z_n] \\ &\sim N\left(0, \frac{t}{n}\sum_{k=1}^{n} k^2\right) \\ &= N[0, t(n+1)(2n+1)/6].\end{aligned}$$

The characteristic function of $\int_0^t W_s \cdot ds$ *is*

$$\begin{aligned} E\exp\left(i\zeta\int_0^t W_s\cdot ds\right) &= E\exp\left[i\zeta\lim_{n\to\infty}\frac{t}{n}\sum_{k=1}^{n} W_{kt/n}\right] \\ &= \lim_{n\to\infty} E\exp\left[i\zeta\frac{t}{n}\sum_{k=1}^{n} W_{kt/n}\right] \\ &= \lim_{n\to\infty}\exp\left[-\frac{\zeta^2 t^2}{2n^2}\cdot\frac{t(n+1)(2n+1)}{6}\right] \\ &= e^{-\zeta^2 t^3/6}, \end{aligned}$$

where the second step follows by dominated convergence. From the uniqueness theorem for c.f.s (section 2.2.3) *we conclude that*

$$\int_0^t W_s\cdot ds \sim N(0,t^3/3).$$

3.2.2 *Integrals with Respect to Brownian Motions*

Defining integrals of adapted processes with respect to Brownian motion on some finite interval $[0,t]$ proceeds in steps similar to those for the Lebesgue integral, working from restricted to successively more general forms of integrands. At each stage the restrictions will guarantee that the process $\{X_t\}_{t\geq 0}$, with $X_t \equiv X_0 + \int_0^t b_s \cdot dW_s$ viewed as a function of the upper limit, is at least locally a martingale. Henceforth, the integrands $\{b_t\}$ we consider are processes adapted to the same filtration $\{\mathcal{F}_t\}_{t\geq 0}$ as the Wiener process $\{W_t\}$. With such integrands the process $\{X_t\}$ thus defined belongs to the class of Itô processes. In studying these processes we will always assume that $X_0 \in \mathcal{F}_0$ and that $\mathcal{F}_0$ is the trivial field, $\{\emptyset, \Omega\}$.

Defining the Integral

Stage one begins with "simple" adapted processes. The process $\{b_{n_t,t}\}_{0\leq t}$ is simple if each sample path is constant (and bounded) on intervals. Thus, $\{b_{n_t,t}\}$ is simple adapted if there exists for each $t > 0$ numbers $0 = t_0 < t_1 < \ldots < t_{n_t-1} < t_{n_t} = t$, a finite constant K, and random variables $\hat{b}_0, \hat{b}_1, \ldots, \hat{b}_{n_t-1}$ with $|\hat{b}_j| \leq K$ a.s. and $\hat{b}_j \in \mathcal{F}_{t_j}$ for each j such that

$$b_{n_t,s} = \begin{cases} \hat{b}_0, & s = 0 \\ \sum_{j=1}^{n_t} \hat{b}_{j-1}\mathbf{1}_{(t_{j-1},t_j]}(s), & 0 < s \leq t \end{cases}.$$

For these simple processes the Itô integral $\int_0^t b_{n_t,s} \cdot dW_s$ is defined as

$$X_{n_t,t} - X_0 \equiv \int_0^t b_{n_t,s} \cdot dW_s = \sum_{j=1}^{n_t} \hat{b}_{j-1}(W_{t_j} - W_{t_{j-1}}). \tag{3.2}$$

In words, the realization of the integral for outcome ω is the weighted sum of the (constant) values of b on the intervals, where the weights are the respective (positive or negative) increments to the Brownian motion. The X_0 in (3.2) is simply an arbitrary $\mathcal{F}_0$-measurable initial value.

Stage two extends to adapted processes $\{b_t\}$ that are well-enough behaved that $E\int_0^t b_s^2 \cdot ds < \infty$. For this it is first established that any such process can be attained as the limit of a sequence of simple processes $\{b_{n_t,s}\}_{n_t \in \aleph}$, in the sense that $E\int_0^t (b_{n_t,s} - b_s)^2 \cdot ds \to 0$ as $n_t \to \infty$. Then $X_t = X_0 + \int_0^t b_s \cdot dW_s$ is defined as a random variable to which the sequence $\{X_{n_t,t}\}$ converges in mean square; that is, the variable such that $\lim_{n_t \to \infty} E(X_t - X_{n_t,t})^2 = 0$. That this is a good definition follows from the fact that a random variable X_t' constructed from some other sequence of simple adapted processes $\{b'_{n'_t,s}\}$ equals X_t a.s. (For proofs of these facts consult the technical references listed in the introduction to this chapter.)

A final extension is to adapted processes for which $\int_0^t b_s^2 \cdot ds < \infty$ a.s. This is a weaker condition than that in step two because $E\int_0^t b_s^2 \cdot ds < \infty$ implies $\int_0^t b_s^2 \cdot ds < \infty$ but not conversely. The extension is made by defining stopping times τ_K that produce a bounded version of the process, for which the expected value of $\int_0^t b_s^2 \cdot ds$ is necessarily finite, and then taking limits of the resulting stochastic integrals of the bounded processes. Thus, for some $K > 0$ set $\tau_K \equiv \inf\{t : \int_0^t b_s^2 \cdot ds \geq K\}$, with $\tau_K \equiv +\infty$ if $\int_0^\infty b_s^2 \cdot ds < K$. This is a stopping time since the fact that $\{b_t\}$ is $\{\mathcal{F}_t\}$-adapted means that the value of $\int_0^t b_s^2 \cdot ds$ is known at time t. Also, being bounded, the truncated process $\{b_{K,t} \equiv b_t \mathbf{1}_{[0,\tau_K)}(t)\}_{t \geq 0}$ clearly satisfies $E\int_0^t b_{K,s}^2 \cdot ds < \infty$. As a result the Itô integral, $X_{K,t} - X_0 = \int_0^t b_{K,s} \cdot dW_s$, say, is defined for each K and each $t > 0$. Since $\tau_K \to \infty$ a.s. as $K \to \infty$, $X_t - X_0 = \int_0^t b_s \cdot dW_s$ can be defined as the random variable to which $\int_0^t (X_s - X_{K,s})^2 \cdot ds$ converges in probability. Again, this random variable can be shown to be essentially unique.

Here are three examples that show how Itô integrals can be constructed from the definition. The first two examples illustrate the fact that the Itô and Riemann-Stieltjes constructions yield the same result when $\{b_t\}$ is a process of finite variation.

Example 34 *Taking* $b_t = 1$, *which is already a simple process,* (3.2) *gives* $X_t - X_0 = \int_0^t dW_s$ *as* $W_t - W_0 = W_t$*—the same as in the Stieltjes construction.*

Example 35 *Taking* $b_t = t$, *let us evaluate* $X_t - X_0 = \int_0^t s \cdot dW_s$. *First, partition* $(0, t]$ *as* $\cup_{j=1}^n (t_{j-1}, t_j]$ *with* $t_j = jt/n$, *and create the simple process* $b_{n,s} \equiv \{\sum_{j=1}^n t_{j-1} \mathbf{1}_{(t_{j-1}, t_j]}(s)\}_{0<s\le t}$. *The approximating sum is then*

$$\begin{aligned}\sum_{j=1}^n t_{j-1}(W_{t_j} - W_{t_{j-1}}) &= \frac{t}{n}\sum_{j=1}^n (j-1)(W_{t_j} - W_{t_{j-1}}) \\ &= \frac{t}{n}[-W_{t_1} - \cdots - W_{t_{n-1}} + (n-1)W_t] \\ &= -\sum_{j=1}^n W_{t_{j-1}}(t_j - t_{j-1}) + \frac{n-1}{n} t W_t.\end{aligned}$$

Taking limits path by path gives

$$\int_0^t s \cdot dW_s = tW_t - \int_0^t W_s \cdot ds.$$

Again, this stochastic counterpart to the integration-by-parts formula is the same as for an ordinary Stieltjes integral.

Example 36 *Taking* $b_t = W_t$, *let us evaluate* $X_t - X_0 \equiv \int_0^t W_s \cdot dW_s$. *This differs from the two previous examples in that both integrand and integrator have unbounded variation. Were* W_t *an ordinary absolutely continuous function, the Stieltjes construction would give* $W_s^2/2|_0^t = W_t^2/2$, *but this time the result will be different because the noise in the integrand and integrator interact. Again partitioning* $(0, t]$ *as* $\cup_{j=1}^n (t_{j-1}, t_j]$ *with* $t_j = jt/n$ *and forming the simple process*

$$b_{n,s} \equiv \sum_{j=1}^n W_{t_{j-1}} \mathbf{1}_{(t_{j-1}, t_j]}(s),$$

the Itô integral will emerge as the random variable to which

$$X_t - X_0 \equiv \sum_{j=1}^n W_{t_{j-1}}(W_{t_j} - W_{t_{j-1}})$$

converges in probability. To find this limit, write the summation as

$$\sum_{j=1}^{n} W_{t_{j-1}} W_{t_j} - \sum_{j=1}^{n} W_{t_{j-1}}^2$$
$$= \sum_{j=1}^{n} W_{t_{j-1}} W_{t_j} - \frac{1}{2}\left(\sum_{j=1}^{n} W_{t_{j-1}}^2 + \sum_{j=1}^{n} W_{t_j}^2 - W_t^2\right)$$
$$= \frac{1}{2}\left[W_t^2 - \sum_{j=1}^{n} \left(W_{t_j} - W_{t_{j-1}}\right)^2\right] \sim \frac{1}{2}\left(W_t^2 - \frac{t}{n}\sum_{j=1}^{n} Z_j^2\right),$$

where the $\{Z_j\}$ *are i.i.d. as* $N(0,1)$. *Since* $\frac{1}{n}\sum_{j=1}^{n} Z_j^2$ *converges a.s. to* $EZ^2 = 1$ *by the strong law of large numbers, we conclude that*

$$\int_0^t W_s \cdot dW_s = W_t^2/2 - t/2. \tag{3.3}$$

In the second example it was not crucial to evaluate the integrand at t_{j-1} in the approximating sum for $\int_0^t s \cdot dW_s$, because the integral would have the same value for any sequence of simple adapted processes that converged to the integrand process, $\{t\}$. In the last example, however, it *is* crucial that the integrand be evaluated at t_{j-1} rather than at some $t_j^* \in (t_{j-1}, t_j]$. This is so because the effect of interaction between $W_{t_j} - W_{t_{j-1}}$ and the change in the integrand over $(t_{j-1}, t_j^*]$, $W_{t_j} - W_{t_j^*}$, does not become negligible as $t_j - t_{j-1}$ approaches zero. To appreciate this point, the reader can show that taking $b_{n,s} = \sum_{j=1}^{n} W_{t_j} \mathbf{1}_{(t_{j-1}, t_j]}(s)$ results in

$$\sum_{j=1}^{n} W_{t_j}(W_{t_j} - W_{t_{j-1}}) \to^{\mathbb{P}} W_t^2/2 + t/2.$$

This feature shows that the Itô and Riemann-Stieltjes constructions do differ fundamentally, despite the superficial similarity.

Properties of the Integral

It is convenient to characterize the Itô integral in terms of the stochastic process $X_t = X_0 + \int_0^t b_s \cdot dW_s$ that it generates. As usual, we assume that X_0 is deterministic. The step-wise development of the integral from simple integrands makes it easy to determine the key properties of this process.

1. $\{X_t\}_{t\geq 0}$ is a continuous process. This is a simple consequence of the continuity of $\{W_t\}_{t\geq 0}$, since

$$\lim_{n\to\infty} |X_{t+1/n} - X_t| = \lim_{n\to\infty} |b_t(W_{t+1/n} - W_t)| = 0.$$

2. $\{X_t\}_{t\geq 0}$ has the fair-game property. With $X_{n_t,t}$ as in (3.2) and $s \in (t_\ell, t_{\ell+1}]$, we have

$$X_{n_t,t} - X_{n_s,s} = \hat{b}_\ell(W_{t_{\ell+1}} - W_s) + \sum_{j=\ell+2}^{n_t-1} \hat{b}_{j-1}(W_{t_j} - W_{t_{j-1}}) + \hat{b}_{n_t-1}(W_t - W_{t_{n_t-1}}).$$

Taking expectations and using the tower property of conditional expectation and the $\mathcal{F}_{t_{j-1}}$-measurability of $\hat{b}_{j-1}$ give $E_s(X_{n_t,t} - X_{n_s,s}) = 0$, showing that $E_s X_t = X_s$. Likewise, $E_0 X_t = X_0$.

3. $\{X_t\}_{t\geq 0}$ is a continuous martingale when $E\int_0^T b_t^2 \cdot dt < \infty$. The fair-game property having been demonstrated, it remains just to verify that $E|X_t| < \infty$. Since $E|X_t| \leq E|X_t - X_0| + E|X_0| = E|X_t - X_0| + |X_0|$, we need show only that $E|X_t - X_0| < \infty$. Working again from the approximating simple process, write $(X_{n,t} - X_0)^2$ as

$$\sum_{j=1}^{n_t-1} \hat{b}_{j-1}^2(W_{t_j} - W_{t_{j-1}})^2 + 2\sum_{j=2}^{n_t-1}\sum_{k=1}^{j-1}[\hat{b}_{j-1}(W_{t_j} - W_{t_{j-1}})\hat{b}_{k-1}(W_{t_k} - W_{t_{k-1}})]. \tag{3.4}$$

Taking expectations, apply the tower property by conditioning on $\mathcal{F}_{t_{j-1}}$ in each summand of the first term and conditioning on $\mathcal{F}_{t_{k-1}}$ in each summand of the second term. Each of the latter summands then vanishes, leaving

$$E(X_{n_t,t} - X_0)^2 = E\sum_{j=1}^{n_t-1} \hat{b}_{j-1}^2(t_j - t_{j-1}),$$

which, on taking limits, gives

$$E(X_{n_t,t} - X_0)^2 \to E\int_0^t b_s^2 \cdot ds < \infty. \tag{3.5}$$

Finally, $E|X_t - X_0| < \infty$ follows from Jensen's inequality.

4. $\{X_t\}_{t\geq 0}$ is a continuous *local* martingale when $\int_0^t b_s^2 \cdot ds < \infty$ but $E\int_0^t b_s^2 \cdot dt = \infty$. In this case $\{X_t\}$ is reduced by the stopping times $\tau_K = \inf\{t : |X_t| \geq K\}$.

5. If $X_0 = 0$ the first two moments of $X_t = \int_0^t b_s \cdot dW_s$ are $EX_t = 0$ and $VX_t = EX_t^2 = E\int_0^t b_s^2 \cdot ds$ when $E\int_0^T b_t^2 \cdot dt < \infty$. The first of these comes directly from the fair-game property, and the second follows in the same way as (3.5).
6. $X_t = X_0 + \int_0^t b_s \cdot dW_s \sim N(X_0, \int_0^t b_s^2 \cdot ds)$ when $\{b_s\}_{s \le t} \in \mathcal{F}_0$. When the integrand process is $\mathcal{F}_0$-measurable, as in example 35, normality follows from the construction of the integral as a linear function of increments of Brownian motion.

Generalization to Higher Dimensions

Taking $\{\mathbf{W}_t\}$ to be a k-dimensional Brownian motion and $\{\mathbf{b}_t\}$ to be an $m \times k$ adapted matrix process leads to a multidimensional extension of the integral that makes $\mathbf{X}_t = \mathbf{X}_0 + \int_0^t \mathbf{b}_s \cdot d\mathbf{W}_s$ an m-vector-valued process. The construction proceeds in the same way, from (i) simple processes $\{\mathbf{b}_t\}$ to (ii) processes with $E\int_0^t \mathbf{b}_s \mathbf{b}_s' \cdot ds < \infty$ to (iii) processes with $\int_0^t \mathbf{b}_s \mathbf{b}_s' \cdot ds < \infty$ a.s. Members of $\mathbf{X}_t$ are martingales under the condition (ii), and when $\mathbf{b}_t$ is nonstochastic the increments are distributed as multivariate normal: $\mathbf{X}_t - \mathbf{X}_0 \sim N(0, \int_0^t \mathbf{b}_s \mathbf{b}_s' \cdot ds)$.

3.2.3 *Itô Processes*

Taking $b_t \equiv 1$ as in example 34 gives $X_t - X_0 = \int_0^t b_s \cdot dW_s = W_t$, which we know to be distributed as $N(0, t)$. As we have seen from the other examples, more interesting integrands produce processes $\{X_t\}$ that are themselves more general, having variances $E\int_0^t b_s^2 \cdot ds$ (when these are finite) and increments that are no longer necessarily normally distributed. In addition, such processes are no longer necessarily Markov since $\{b_t\}_{t \ge 0}$ may depend on the entire past of $\{W_t\}_{t \ge 0}$ as, for example, $b_t = \sup_{0 \le s \le t} |W_s|$. One major restriction remains, however. By the fair-game property of the integral a process of the form $X_t = X_0 + \int_0^t b_s \cdot dW_s$ drifts aimlessly and thus lacks the tendency to upward movement that prices of financial assets usually display over long periods. "Itô processes" are defined more generally to allow for such systematic trend.

Definition and Examples

Let $\{a_t\}_{t \ge 0}$ be an adapted process with $\int_0^t |a_s| \cdot ds < \infty$ almost surely for each finite t and (as heretofore) let $\{b_t\}_{t \ge 0}$ be adapted with $\int_0^t b_s^2 \cdot ds < \infty$

a.s. for each $t < \infty$. Then the process

$$\left\{X_t = X_0 + \int_0^t a_s \cdot ds + \int_0^t b_s \cdot dW_s\right\}_{t\geq 0}$$

with $X_0 \in \mathcal{F}_0$ is called an "Itô process". Notice that $\int_0^t a_s \cdot ds$ is not a stochastic integral but is interpreted pathwise in the Lebesgue sense. Also, note that the definition of the integral implies that the *process* $\{\int_0^t a_s \cdot ds\}_{t\geq 0}$ is one of finite variation.[6] Clearly, unless $a_t = 0$ for (almost) all t the process $\{X_t\}$ lacks the fair-game property and is not a martingale. However, since $\{X_t\}$ does have a local martingale component if $\int_0^t b_s^2 \cdot ds < \infty$ for each t, and since $\{\int_0^t a_s \cdot ds\}$ is a continuous, adapted process of finite variation, $\{X_t\}_{t\geq 0}$ remains a continuous *semi*martingale.

The Itô class is rich enough to provide reasonable models for prices of many commodities and financial assets. Moreover, as we shall see, semimartingales in this class can serve as integrators in other stochastic integrals, which in turn produce even richer models. The following examples—all continuous Markov (i.e., diffusion) processes—have been widely used in financial modeling.

Example 37 *Taking $a_t = \mu X_t$ and $b_t = \sigma X_t$ in the Itô integral, where μ and $\sigma > 0$ are constants, gives the process known as "geometric Brownian motion":*

$$\left\{X_t = X_0 + \int_0^t \mu X_s \cdot ds + \int_0^t \sigma X_s \cdot dW_s\right\}_{t\geq 0}. \tag{3.6}$$

This is the process on which the Black-Scholes model for European options is based, and it is still the benchmark continuous-time model for prices of many financial assets. The parameter μ is called the (proportional) mean drift, and σ is the volatility parameter. Notice that the change in the process over an interval $[t, t + \Delta t)$, namely $X_{t+\Delta t} - X_t = \int_t^{t+\Delta t} \mu X_s \cdot ds + \int_t^{t+\Delta t} \sigma X_s \cdot dW_s$, damps out as the initial value X_t approaches zero. A process that starts off at $X_0 > 0$ thus remains nonnegative, as is consistent with the limited-liability feature of most financial assets. Also, the variance of the relative change in the process over $[t, t+\Delta t)$, $X_{t+\Delta t}/X_t - 1$, is approximately

[6]Under the integrability condition the integral process $\{\int_0^t a_s \cdot ds\}_{t\geq 0}$ is a.s. of finite variation even if the *integrand* process $\{a_t\}_{t\geq 0}$ is not. Recall that the definition of the Lebesgue integral presents $\int_0^t a_s \cdot ds$ as $\int_0^t a_s^+ \cdot ds - \int_0^t (-a_s^-) \cdot ds$, where $a_s^+ = a_s \vee 0$ and $a_s^- = a_s \wedge 0$. Thus, on almost all sample paths the integral process is the difference between two nondecreasing processes.

invariant to the initial price level. This implies that holding period returns over short periods—ratios of terminal to initial prices—are approximately stationary.

A slight variation on the theme is to allow mean drift and volatility to be time-varying yet still $\mathcal{F}_0$-measurable (predictable as of $t = 0$).

Example 38 *Taking $a_t = (\mu - \nu X_t)$, where $\nu > 0$, gives a mean-reverting or "Ornstein-Uhlenbeck" process:*

$$\left\{X_t = X_0 + \int_0^t (\mu - \nu X_s) \cdot ds + \int_0^t b_s \cdot dW_s\right\}_{t \geq 0}. \tag{3.7}$$

This has negative mean drift when $X_t > \mu/\nu$ and positive mean drift when $X_t < \mu/\nu$, so that the process is attracted to μ/ν with strength proportional to ν. The $\{b_t\}$ process can take various forms. Mean-reverting processes have been used extensively in modeling interest rates, which do typically fluctuate about some long-term average value. We encounter mean-reverting processes first in chapter 8 in connection with processes whose volatility coefficients are themselves mean-reverting diffusions.

Example 39 *Taking $b_t = \sigma X_t^\gamma$ for $0 \leq \gamma < 1$ gives a "constant-elasticity-of-variance" (c.e.v.) process:*

$$\left\{X_t = X_0 + \int_0^t a_s \cdot ds + \int_0^t \sigma X_s^\gamma \cdot dW_s\right\}_{t \geq 0}. \tag{3.8}$$

When $\gamma = 1/2$ this is commonly called a "square-root process". The chief contrast with geometric Brownian motion is that the variance of the relative change in the process over $[t, t + \Delta t)$, $X_{t+\Delta t}/X_t - 1$, is now decreasing in the initial value. This corresponds to the empirical observation that returns of high-priced stocks tend to be less variable than those of low-priced stocks. Mean-reverting square-root processes,

$$X_t = X_0 + \int_0^t (\mu - \nu X_s) \cdot ds + \int_0^t \sigma X_s^{1/2} \cdot dW_s, \tag{3.9}$$

have been important models for interest rates, as in Cox, Ingersoll, and Ross (1985); *for stochastic volatility models, as in Heston* (1993); *and in modeling prices of securities subject to default risk. These are referred to variously as "Feller processes" after Feller* (1951) *and as "CIR" processes after* Cox *et al.* (1985).

The Quadratic-Variation Process

The quadratic variation of a semimartingale $\{X_t\}_{t\geq 0}$ over $[0,T]$ is the probability limit as $n\to\infty$ of $\sum_{j=1}^n (X_{t_j}-X_{t_{j-1}})^2$, where $t_0=0$, $t_n=T$, and $\max_j |t_j-t_{j-1}|\to 0$ as $n\to\infty$. The process that tracks the quadratic variation over $[0,t]$ for $0\leq t\leq T$ is called the "(predictable) quadratic-variation process" and is denoted $\{\langle X\rangle_t\}$. Of course, $\langle X\rangle_0=0$. We have seen that Itô process $\{X_t = X_0+\int_0^t a_s\cdot ds+\int_0^t b_s\cdot dW_s\}_{0\leq t\leq T}$ is a semimartingale when $\int_0^T b_s^2\cdot ds<\infty$. We will need to determine its quadratic variation process in order to develop Itô's formula for the change over time in a smooth function of X_t—expression (3.15) below.

To start with the simplest case, the quadratic variation of Brownian motion over $[0,T]$ is

$$\langle W\rangle_T = \lim_{n\to\infty}\sum_{j=1}^n (W_{t_j}-W_{t_{j-1}})^2 = T\cdot\lim_{n\to\infty} n^{-1}\sum_{j=1}^n Z_j^2 = T,$$

where (i) we set $t_j=jT/n$ for $j\in\{0,1,\ldots,n\}$, (ii) the $\{Z_j\}$ are i.i.d. as $N(0,1)$, and (iii) convergence is with probability one by the strong law of large numbers. Likewise $\langle W\rangle_t=t$, so that the quadratic variation process $\{\langle W\rangle_t\}_{0\leq t\leq T}$ for Brownian motion is perfectly predictable. More generally, for an Itô process $\{X_t\}_{0\leq t\leq T}$ with $X_t=X_0+\int_0^t a_s\cdot ds+\int_0^t b_s\cdot dW_s$ we will see that the quadratic variation over $[0,T]$ is $\langle X\rangle_T=\int_0^T b_t^2\cdot dt$, the salient fact being that the drift term makes no contribution. To make this plausible without belaboring the details, we prove it for the special case that $\{a_t\}$ and $\{b_t\}$ are a.s. bounded processes.[7]

Approximating $X_{t_j}-X_{t_{j-1}}=\int_{t_{j-1}}^{t_j} a_s\cdot ds+\int_{t_{j-1}}^{t_j} b_s\cdot dW_s$ as $a_{t_{j-1}}(t_j-t_{j-1})+b_{t_{j-1}}(W_{t_j}-W_{t_{j-1}})$, the quadratic variation of $\{X_t\}$ on the grid $\{t_j\}$ is approximately

$$\sum_{j=1}^n a_{t_{j-1}}^2(t_j-t_{j-1})^2+2\sum_{j=1}^n a_{t_{j-1}}b_{t_{j-1}}(t_j-t_{j-1})(W_{t_j}-W_{t_{j-1}}) \\ +\sum_{j=1}^n b_{t_{j-1}}^2(W_{t_j}-W_{t_{j-1}})^2. \tag{3.10}$$

Again setting $t_j=jT/n$, the first term equals $Tn^{-1}\sum_{j=1}^n a_{t_{j-1}}^2(t_j-t_{j-1})$, which converges a.s. to zero since $\sum_{j=1}^n a_{t_{j-1}}^2(t_j-t_{j-1})\to\int_0^T a_t^2\cdot dt<\infty$

[7] For the general case see Durrett (1996, chapter 2).

as $n \to \infty$ when $\{a_t\}$ is bounded. Squaring the second term and applying the Schwarz inequality, (2.41), give an upper bound proportional to $n^{-1}\sum_{j=1}^{n} a_{t_{j-1}}^2 b_{t_{j-1}}^2 (t_j - t_{j-1}) \cdot \sum_{j=1}^{n}(W_{t_j} - W_{t_{j-1}})^2$. The first sum converges to $\int_0^T a_t^2 b_t^2 \cdot dt < \infty$ and the second converges to T, so this term also converges a.s. to zero. This confirms that $\langle X \rangle_T$ does not depend on $\{a_t\}$. The last term of (3.10) is where the action is. Writing it as

$$\sum_{j=1}^{n} b_{t_{j-1}}^2 (t_j - t_{j-1}) + \sum_{j=1}^{n} b_{t_{j-1}}^2 (t_j - t_{j-1})(Z_j^2 - 1),$$

where the $\{Z_j\}$ are i.i.d. as $N(0,1)$, we see that the first sum converges to $\int_0^T b_t^2 \cdot dt$. Letting J_n represent the second sum and setting $J_0 = 0$, we see that J_n is a discrete-time martingale since

$$E_{t_{n-1}} J_n = J_{n-1} + b_{t_{n-1}}^2 (t_n - t_{n-1}) E_{t_{n-1}} (Z_n^2 - 1) = J_{n-1}.$$

Moreover, $\sup_n E|J_n| = E|Z^2 - 1| \int_0^T Eb_t^2 \cdot dt < \infty$, so theorem 7 (martingale convergence) implies that J_n converges a.s. to some finite quantity. Using the tower property of conditional expectation in the manner of (3.4) gives

$$EJ_n^2 = \sum_{j=1}^{n} Eb_{t_{j-1}}^4 (t_j - t_{j-1}) \cdot 2T/n \to 0,$$

showing that $J_n \to^{a.s.} 0$ and therefore that $\langle X \rangle_T = \int_0^T b_t^2 \cdot dt$.

Integration with Respect to Itô Processes

Given an Itô process $\{X_t = X_0 + \int_0^t a_s \cdot ds + \int_0^t b_s \cdot dW_s\}$ and a process $\{c_t\}$ adapted to the filtration $\{\mathcal{F}_t\}$, assume that $\int_0^t |a_s c_s| \cdot ds < \infty$ and $\int_0^t b_s^2 c_s^2 \cdot ds < \infty$ a.s. for each finite t. Then

$$\int_0^t c_s \cdot dX_s = \int_0^t c_s a_s \cdot ds + \int_0^t c_s b_s \cdot dW_s.$$

With $Y_0 \in \mathcal{F}_0$ the process $\{Y_t\}$ with

$$Y_t = Y_0 + \int_0^t c_s \cdot dX_s = Y_0 + \int_0^t c_s a_s \cdot ds + \int_0^t c_s b_s \cdot dW_s$$

is therefore just another Itô process, and therefore another semimartingale.

Generalizing to the multidimensional case, let $\mathbf{X}_t = \mathbf{X}_0 + \int_0^t \mathbf{a}_s \cdot ds + \int_0^t \mathbf{b}_s \cdot d\mathbf{W}_s$ be an m-vector process, where $\{\mathbf{W}_t\}$ is a k-dimensional Brownian motion and $\{\mathbf{a}_t\}$ and $\{\mathbf{b}_t\}$ are m-vector and $m \times k$-matrix adapted processes, respectively. If $\{\mathbf{c}_t\}$ is an $\ell \times m$ adapted process with $\int_0^t |\mathbf{c}_s \mathbf{a}_s| \cdot ds < \infty$ and $\int_0^t (\mathbf{c}_s \mathbf{b}_s)(\mathbf{b}_s' \mathbf{c}_s') \cdot ds < \infty$ for each finite t, then

$$\mathbf{Y}_t = \mathbf{Y}_0 + \int_0^t \mathbf{c}_s \mathbf{a}_s \cdot ds + \int_0^t \mathbf{c}_s \mathbf{b}_s \cdot d\mathbf{W}_s$$

defines the ℓ-vector process $\{\mathbf{Y}_t = \mathbf{Y}_0 + \int_0^t \mathbf{c}_s \cdot d\mathbf{X}_s\}$.

Stochastic Differentials

The integral expression $X_t - X_0 = \int_0^t a_s \cdot ds + \int_0^t b_s \cdot dW_s$ that defines the Itô process can be written in a compact differential form as

$$dX_t = a_t \cdot dt + b_t \cdot dW_t. \tag{3.11}$$

Although this is just a shorthand, it has the useful intuitive interpretation as the instantaneous change in the process at t—that is, the change over $[t, t + dt)$. Differential forms of (3.6), (3.7), and (3.8) are, respectively,

$$\begin{aligned} dX_t &= \mu X_t \cdot dt + \sigma X_t \cdot dW_t \\ dX_t &= (\mu - \nu X_t) \cdot dt + b_t \cdot dW_t \\ dX_t &= a_t \cdot dt + \sigma X_t^\gamma \cdot dW_t. \end{aligned} \tag{3.12}$$

The differential form of geometric Brownian motion is often written as

$$dX_t / X_t = \mu \cdot dt + \sigma \cdot dW_t. \tag{3.13}$$

In the special case that a_t and b_t depend just on time and the current value of the process, as $a_t = a(t, X_t)$, $b_t = b(t, X_t)$, (3.11) represents a stochastic differential equation (s.d.e.). The solution, up to an arbitrary $\mathcal{F}_0$-measurable X_0, is the process X_t that satisfies[8]

$$X_t - X_0 = \int_0^t a(s, X_s) \cdot ds + \int_0^t b(s, X_s) \cdot dW_s.$$

The solution to (3.12) can be characterized explicitly by applying Itô's formula, which we now develop.

---

[8]Conditions on $a(\cdot,\cdot)$ and $b(\cdot,\cdot)$ such that the s.d.e. has a unique solution are given by Karatzas and Shreve (1991, pp. 286–291). Examples of s.d.e.s that meet these conditions are (3.6) and (3.7). Duffie (1996, pp. 291–293) gives weaker conditions that accommodate the constant-elasticity-of-variance process when $\gamma \geq 1/2$.

3.3 Itô's Formula

If $f(t)$ is a continuously differentiable function of time, then the change in the function over $[0, t]$ can of course be expressed as an ordinary Riemann integral,

$$f(t) - f(0) = \int_0^t f_t(s) \cdot ds, \tag{3.14}$$

where f_t denotes the derivative. In differential form this would be $df(t) = f_t \cdot dt$. The Itô formula gives corresponding expressions for the change over time in a smooth function of an Itô process. These now take the form of stochastic integrals or stochastic differentials. What we find, most conveniently, is that smooth functions of Itô processes are themselves Itô processes and therefore semimartingales. This simply stated (though not so simply proven) fact is the primary reason that Itô processes—and, for that matter, continuous-time stochastic models built up from Brownian motions—have been so useful in finance and other fields.

3.3.1 *The Result, and Some Intuition*

With $X_t - X_0 = \int_0^t a_s \cdot ds + \int_0^t b_s \cdot dW_s$ and assuming $f : \Re \to \Re$ to be twice continuously differentiable, we will see that

$$f(X_t) - f(X_0) = \int_0^t f_X(X_s) \cdot dX_s + \frac{1}{2} \int_0^t f_{XX}(X_s) \cdot d\langle X_s \rangle, \tag{3.15}$$

where f_X and f_{XX} are the first two derivatives and

$$d\langle X \rangle_t = d \int_0^t b_s^2 \cdot ds = b_t^2 \cdot dt.$$

In abbreviated, differential form this is

$$df(X_t) = f_X \cdot dX_t + \frac{1}{2} f_{XX} \cdot d\langle X \rangle_t. \tag{3.16}$$

Going a little farther to write out dX_t as $a_t \cdot dt + b_t \cdot dW_t$ and collect terms gives

$$df(X_t) = (f_X a_t + f_{XX} b_t^2/2) \cdot dt + f_X b_t \cdot dW_t, \tag{3.17}$$

showing most clearly that smooth functions of Itô processes are, indeed, still Itô processes.

The key feature of (3.15), and what sets it apart from (3.14), is the second-order term in f_{XX}. Before getting into the details it will be helpful to grasp the intuition for this distinction. For a smooth function of time itself, $f(t)$, we can think of $df = f_t \cdot dt$ as approximating the motion of the function over $[t, t+\Delta t)$ to order Δt—that is, as an $O(\Delta t)$ approximation to $\Delta f(t) = f(t+\Delta t) - f(t)$. In the integral form, the fundamental theorem of calculus tells us that the next order term involving f_{tt} does not contribute to the change in f because the effects of these $o(\Delta t)$ components in the sum that converges to the integral become negligible as $\Delta t \to 0$. Why, then, does it not work this way for functions of an Itô process? Very simply, because the motion of X_t itself over $[t, t+\Delta t)$ is stochastically of order larger than Δt as $\Delta t \to 0$—meaning that on average it goes to zero more slowly than Δt. Recalling that $V(W_{t+\Delta t} - W_t) = \Delta t$, it follows that $E|W_{t+\Delta t} - W_t| = O(\sqrt{\Delta t})$. This extreme variability of Brownian motion, which is what causes it to be nondifferentiable, carries over to the more general Itô processes as well. Therefore, to get an $O(\Delta t)$ approximation to $\Delta f(X_t)$ requires taking the Taylor expansion of f out one more term to include $f_{XX}(\Delta X_t)^2$. That is precisely what Itô's formula does.

3.3.2 *Outline of Proof*

We now sketch the derivation of the Itô formula. To keep it simple, we assume that $\{a_t\}$, $\{b_t\}$, and f_{XX} are a.s. bounded.[9] The development is similar to that for the quadratic variation of $\{X_t\}$. Begin with the usual grid $\{t_j = jt/n\}_{j=0}^n$ and the identity $f(X_t) - f(X_0) = \sum_{j=1}^n [f(X_{t_j}) - f(X_{t_{j-1}})]$. Expanding each $f(X_{t_j})$ to the second order about $X_{t_{j-1}}$, we have

$$\begin{aligned} f(X_t) - f(X_0) &= \sum_{j=1}^n f_X(X_{t_{j-1}})(X_{t_j} - X_{t_{j-1}}) \\ &\quad + \frac{1}{2}\sum_{j=1}^n f_{XX}(X_{t_j^*})(X_{t_j} - X_{t_{j-1}})^2. \end{aligned}$$

Notice that the continuity of $\{X_t\}$ assures that there is indeed a $t_j^* \in (t_{j-1}, t_j)$ such that $X_{t_j^*} \in (X_{t_{j-1}}, X_{t_j})$. Sending $n \to \infty$ turns the first term on the right into $\int_0^t f_X(X_s) \cdot dX_s$. Concentrating on the second term, approximating $X_{t_j} - X_{t_{j-1}}$ as $a_{t_{j-1}}(t_j - t_{j-1}) + b_{t_{j-1}}(W_{t_j} - W_{t_{j-1}})$ and squaring leads to an expression involving three terms. We find their limits one by one.

[9] For the general treatment see Durrett (1996) or Karatzas and Shreve (1991).

1.

$$
\begin{aligned}
&\frac{1}{2}\sum_{j=1}^{n} f_{XX}(X_{t_j^*})a_{t_{j-1}}^2(t_j - t_{j-1})^2 \\
&= \frac{t}{2n}\sum_{j=1}^{n} f_{XX}(X_{t_j^*})a_{t_{j-1}}^2(t_j - t_{j-1}) \\
&\to 0.
\end{aligned}
$$

2.

$$
\begin{aligned}
&\sum_{j=1}^{n} f_{XX}(X_{t_j^*})a_{t_{j-1}}b_{t_{j-1}}(t_j - t_{j-1})(W_{t_j} - W_{t_{j-1}}) \\
&\le \sqrt{tn^{-1}\sum_{j=1}^{n} f_{XX}(X_{t_j^*})^2 a_{t_{j-1}}^2 b_{t_{j-1}}^2 (t_j - t_{j-1})\sum_{j=1}^{n}(W_{t_j} - W_{t_{j-1}})^2} \\
&= \sqrt{tn^{-1}\sum_{j=1}^{n} f_{XX}(X_{t_j^*})^2 a_{t_{j-1}}^2 b_{t_{j-1}}^2 (t_j - t_{j-1}) \cdot tn^{-1}\sum_{j=1}^{n} Z_j^2} \\
&\to 0,
\end{aligned}
$$

where the Schwarz inequality is used in the second line.

3.

$$
\begin{aligned}
&\frac{1}{2}\sum_{j=1}^{n} f_{XX}(X_{t_j^*})b_{t_{j-1}}^2(W_{t_j} - W_{t_{j-1}})^2 \\
&= \frac{1}{2}\sum_{j=1}^{n} f_{XX}(X_{t_{j-1}})b_{t_{j-1}}^2(W_{t_j} - W_{t_{j-1}})^2 \\
&\quad + \frac{1}{2}\sum_{j=1}^{n}[f_{XX}(X_{t_j^*}) - f_{XX}(X_{t_{j-1}})]b_{t_{j-1}}^2(W_{t_j} - W_{t_{j-1}})^2.
\end{aligned}
$$

The first term converges to $\frac{1}{2}\int_0^t f_{XX}(X_s)b_s^2 \cdot ds \equiv \frac{1}{2}\int_0^t f_{XX}(X_s)\cdot d\langle X\rangle_s$ by the same argument used to show that $\langle X\rangle_t = \int_0^t b_s^2 \cdot ds$. Invoking the continuity of f_{XX} and the fact that $X_{t_j^*} - X_{t_{j-1}} \to 0$ as $n \to \infty$, a similar argument implies that the second term converges to zero.

Assembling the pieces leaves us with (3.15).

3.3.3 *Functions of Time and an Itô Process*

The Itô formula extends as follows for a function $f(t, X_t)$ that depends independently on time and an Itô process and is once continuously differentiable in t and twice continuously differentiable in X:

$$f(t, X_t) - f(0, X_0) = \int_0^t f_t(s, X_s) \cdot ds + \int_0^t f_X(s, X_s) \cdot dX_s + \frac{1}{2} \int_0^t f_{XX}(s, X_s) \cdot d\langle X \rangle_s. \tag{3.18}$$

Note that the existence of f_{tX} is not required. The differential form is

$$\begin{aligned} df(t, X_t) &= f_t \cdot dt + f_X \cdot dX_t + \frac{1}{2} f_{XX} \cdot d\langle X \rangle_t \qquad (3.19) \\ &= \left(f_t + f_X a_t + \frac{1}{2} f_{XX} b_t^2 \right) \cdot dt + f_X b_t \cdot dW_t. \end{aligned}$$

3.3.4 *Illustrations*

When integrating ordinary functions of a real variable, we typically just use differentiation formulas to find antiderivatives, rather than work out the result from the definition of the integral as a limiting sum. Itô's formula affords a similar way to find stochastic integrals and solve stochastic differential equations. Here are some examples.

Example 40 *Example* 36 *used the definition of the Itô integral to show that*

$$\int_0^t W_s \cdot dW_s = W_t^2/2 - t/2. \tag{3.20}$$

Since $W_0 = 0$, *this can be written as* $W_t^2 - W_0^2 = 2\int_0^t W_s \cdot dW_s + \int_0^t ds$ *or, in differential form, as* $dW_t^2 = 2W_t \cdot dW_t + dt$. *Setting* $X_t = W_t$ *and* $f(W_t) = W_t^2$ *and applying* (3.16) *with* $d\langle W \rangle_t = dt$ *yield the same result:* $dW_t^2 = 2W_t \cdot dW_t + \frac{1}{2}(2) \cdot dt = 2W_t \cdot dW_t + dt$. *Applying the formula in reverse gives* $W_t \cdot dW_t = dW_t^2/2 - dt/2$.

Example 41 *Example* 35 *showed that*

$$\int_0^t s \cdot dW_s = tW_t - \int_0^t W_s \cdot ds,$$

which has the differential form

$$d(tW_t) = t \cdot dW_t + W_t \cdot dt.$$

With $X_t = W_t$ and $f(t, W_t) = tW_t$ (3.19) gives the same result.

Example 42 *From expression (3.13) a differential form of geometric Brownian motion is $dX_t/X_t = \mu \cdot dt + \sigma \cdot dW_t$. We can propose a solution to this s.d.e. and use Itô's formula to check it. We are used to thinking that $dx/x = d\ln x$, so our first guess might be $X_t = f^*(t, W_t) \equiv X_0 e^{\mu t + \sigma W_t}$. This gives $f_t^* = \mu X_t$, $f_W^* = \sigma X_t$, $f_{WW}^* = \sigma^2 X_t$ and*

$$\begin{aligned} dX_t/X_t &= \mu \cdot dt + \sigma \cdot dW_t + \frac{1}{2}\sigma^2 \cdot dt \\ &= (\mu + \sigma^2/2) \cdot dt + \sigma_t \cdot dW_t. \end{aligned}$$

This misses—and incidentally tells us that $d\ln X_t \neq dX_t/X_t$ in stochastic calculus, but it shows that we need just to subtract $\sigma^2/2$ from μ to get the right solution:

$$X_t = X_0 e^{(\mu - \sigma^2/2)t + \sigma W_t}. \tag{3.21}$$

Given the importance of geometric Brownian motion as a model for assets' prices, (3.21) will be an extremely useful result, to which we will refer repeatedly in later chapters. It implies that

$$\ln X_t = \ln X_0 + (\mu - \sigma^2/2)t + \sigma W_t, \tag{3.22}$$

so that,

$$\ln X_t \sim N[\ln X_0 + (\mu - \sigma^2/2)t, \sigma^2 t]. \tag{3.23}$$

Thus, if X_t represents the price of a financial asset, its future value after t units of time is lognormally distributed. Moreover, the log-price process $\{\ln X_t\}$ is a Gaussian process with independent increments and with autocorrelation function given by

$$\begin{aligned} Corr(\ln X_s, \ln X_t) &= Cov(\ln X_s, \ln X_t)/\sqrt{V\ln X_s \cdot V\ln X_t} \\ &= \sqrt{s/t}, s \le t. \end{aligned}$$

Example 43 *Applying Itô with $dX_t = a_t \cdot dt + b_t \cdot dW_t$ and $f(X_t) = \ln X_t$ gives*

$$\begin{aligned} d\ln X_t &= X_t^{-1} \cdot dX_t - \frac{1}{2} X_t^{-2} \cdot d\langle X \rangle_t \\ &= \left(X_t^{-1} a_t - \frac{1}{2} X_t^{-2} b_t^2 \right) \cdot dt + X_t^{-1} b_t \cdot dW_t. \end{aligned}$$

When $a_t = \mu X_t$ and $b_t = \sigma X_t$, this is

$$d\ln X_t = (\mu - \sigma^2/2) \cdot dt + \sigma \cdot dW_t,$$

in agreement with (3.22).

The next example introduces two concepts whose importance will be seen in section 3.4 in connection with Girsanov's theorem.

Example 44 *Consider the trendless Itô process* $dX_t = \theta_t \cdot dW_t$, *where* $\int_0^t \theta_s^2 \cdot ds < \infty$ *for each finite* t. *Define a new process* $dQ_t = Q_t \cdot dX_t = Q_t\theta_t \cdot dW_t$. *By analogy with example 42 we guess (and can verify using Itô's formula) that solutions to this s.d.e. are of the form*

$$Q_t = Q_0 e^{X_t - \langle X \rangle_t/2} = Q_0 \exp\left(\int_0^t \theta_s \cdot dW_s - \frac{1}{2}\int_0^t \theta_s^2 \cdot ds\right)$$

for arbitrary Q_0. *Taking* $Q_0 = 1$ *gives the "Doléans exponential" or "stochastic exponential" of* $\{X_t\}$. *It can be shown that* $\{Q_t\}$ *is a continuous local martingale.*[10] *Moreover, under the Novikov condition,*

$$E \exp\left(\frac{1}{2}\int_0^t \theta_s^2 \cdot ds\right) < \infty, \tag{3.24}$$

Q_t *is integrable and therefore a continuous proper martingale, with* $EQ_t = 1$. *The Novikov condition holds trivially when* $\{\theta_t\}_{t \geq 0} \in \mathcal{F}_0$ *and* θ_t^2 *is integrable—a case that covers many of our applications.*

3.3.5 *Functions of Higher-Dimensional Processes*

It is often necessary to consider the behavior of functions of several Itô processes. Let $\{\mathbf{X}_t = \mathbf{X}_0 + \int_0^t \mathbf{a}_s \cdot ds + \int_0^t \mathbf{b}_s \cdot d\mathbf{W}_s\}$ be an m-dimensional Itô process, where $\{\mathbf{W}_t\}$ is a k-dimensional standard Brownian motion and $\{\mathbf{a}_t\}$ and $\{\mathbf{b}_t\}$ are an m-vector and an $m \times k$ matrix of adapted processes. The quadratic covariation of $\{X_{js}\}$ and $\{X_{\ell s}\}$ on $[0, t]$ is the probability limit as $n \to \infty$ of $\sum_{i=1}^n (X_{jt_i} - X_{jt_{i-1}})(X_{\ell t_i} - X_{\ell t_{i-1}})$ on a grid $\{t_i = it/n\}$. This equals

$$\langle X_j, X_\ell \rangle_t = \int_0^t \mathbf{b}_{js}\mathbf{b}_{\ell s}' \cdot ds,$$

[10]This is easily seen in two special cases. Write $Q_t = Q_s \exp[\int_s^t \theta_u \cdot dW_u - \frac{1}{2}\int_s^t \theta_u^2 \cdot du]$ for $0 \leq s < t$, and suppose, as the first case, that $\{\theta_t\}$ is $\mathcal{F}_0$-measurable and $\int_0^T \theta_t^2 \cdot dt < \infty$. Then $\int_s^t \theta_u \cdot dW_u \sim N(0, \int_s^t \theta_u^2 \cdot du)$, and $E_s Q_t = Q_s$ follows at once from the form of the normal moment generating function. Considering as the second case *simple* processes $\theta_u = \sum \hat{\theta}_{j-1} \mathbf{1}_{[t_{j-1}, t_j)}(u)$ the result follows easily from the independence of increments to Brownian motion. In both cases the Novikov condition, (3.24), clearly holds, so that $\{Q_t\}$ is in fact a martingale. For the more general case that $\theta_t = \langle X \rangle_t$ and $\{X_t\}$ is a continuous local martingale see Karatzas and Shreve (1991, section 3.5).

where $\mathbf{b}_{js}$ and $\mathbf{b}_{\ell s}$ are the jth and ℓth rows of $\mathbf{b}_s$ and $\mathbf{b}_{js}\mathbf{b}'_{\ell s}$ is the inner product. Of course, when $j = \ell$

$$\langle X_j, X_\ell \rangle_t = \langle X_j \rangle_t = \int_0^t \mathbf{b}_{js}\mathbf{b}'_{js} \cdot ds.$$

Note that $\langle X_j, X_\ell \rangle_t = 0$ when the subsets of $\{W_{1t}, W_{2t}, \ldots, W_{kt}\}$ on which X_{jt} and $X_{\ell t}$ depend have no elements in common, in which case $\mathbf{b}_{js}\mathbf{b}'_{\ell s} = 0, s \in [0, t]$. For a differentiable function $f(t, \mathbf{X}_t)$ the differential form of Itô's formula is

$$df(t, \mathbf{X}_t) = f_t \cdot dt + \sum_{j=1}^{m} f_{X_j} \cdot dX_{jt} + \frac{1}{2} \sum_{j=1}^{m} \sum_{\ell=1}^{m} f_{X_j X_\ell} \cdot d\langle X_j, X_\ell \rangle_t, \quad (3.25)$$

or, written out,

$$df(t, \mathbf{X}_t) = \left(f_t + \sum_{j=1}^{m} f_{X_j} a_{jt} + \frac{1}{2} \sum_{j=1}^{m} \sum_{\ell=1}^{m} f_{X_j X_\ell} \mathbf{b}_{jt}\mathbf{b}'_{\ell t} \right) \cdot dt + \sum_{j=1}^{m} f_{X_j} \mathbf{b}_{jt} \cdot d\mathbf{W}_t.$$

The next two examples develop special cases that will be important in the applications.

Example 45 *Let $\{X_{1t}\}$ and $\{X_{2t}\}$ be Itô processes with*

$$dX_{jt} = a_{jt} \cdot dt + \mathbf{b}_{jt} \cdot d\mathbf{W}_t$$

for $j = 1, 2$, where $\{\mathbf{W}_t\}$ is a k-dimensional standard Brownian motion. Taking $f(\mathbf{X}_t) = X_{1t}X_{2t}$ formula (3.25) gives

$$df(X_{1t}X_{2t}) = X_{2t} \cdot dX_{1t} + X_{1t} \cdot dX_{2t} + d\langle X_1, X_2 \rangle_t. \quad (3.26)$$

Since $d\langle X_1, X_2 \rangle_t = \mathbf{b}_{1t}\mathbf{b}'_{2t} \cdot dt$, this is

$$df(X_{1t}X_{2t}) = (X_{2t}a_{1t} + X_{1t}a_{2t} + \mathbf{b}_{1t}\mathbf{b}'_{2t}) \cdot dt + (X_{2t}\mathbf{b}_{1t} + X_{1t}\mathbf{b}_{2t}) \cdot d\mathbf{W}_t.$$

Notice that if either or both of $\{X_{1t}\}, \{X_{2t}\}$ is of finite variation—in which case $\mathbf{b}_{1t} = \mathbf{0}$ and/or $\mathbf{b}_{2t} = \mathbf{0}$—then the quadratic-covariation process is identically zero, and (3.26) reduces to the ordinary product formula for differentiation:

$$df(X_{1t}X_{2t}) = X_{2t} \cdot dX_{1t} + X_{1t} \cdot dX_{2t}. \quad (3.27)$$

Example 46 *Let* $\{\mathbf{p}_t\}$ *and* $\{\mathbf{S}_t\}$ *be two* m*-dimensional Itô processes,* $m \geq 2$*, as*

$$d\mathbf{p}_t = \boldsymbol{\pi}_t \cdot dt + \boldsymbol{\delta}_t \cdot d\mathbf{W}_t$$
$$d\mathbf{S}_t = \boldsymbol{\mu}_t \cdot dt + \boldsymbol{\sigma}_t \cdot d\mathbf{W}_t,$$

where $\boldsymbol{\pi}_t$ *and* $\boldsymbol{\mu}_t$ *are* m *vectors and* $\boldsymbol{\delta}_t$ *and* $\boldsymbol{\sigma}_t$ *are* $m \times k$ *matrices, all* $\{\mathcal{F}_t\}$*-adapted. Let*

$$P_t = f(\mathbf{p}_t, \mathbf{S}_t) \equiv \mathbf{p}_t'\mathbf{S}_t = \sum_{j=1}^{m} p_{jt}S_{jt}.$$

Then (3.25) *gives*

$$\begin{aligned} dP_t &= \mathbf{S}_t' \cdot d\mathbf{p}_t + \mathbf{p}_t' \cdot d\mathbf{S}_t + d\langle \mathbf{p}_t, \mathbf{S}_t\rangle_t \qquad (3.28)\\ &= \sum_{j=1}^{m}[S_{jt} \cdot dp_{jt} + p_{jt} \cdot dS_{jt} + \boldsymbol{\delta}_{jt}\boldsymbol{\sigma}_{jt}' \cdot dt] \\ &= \sum_{j=1}^{m}[(S_{jt}\pi_{jt} + p_{jt}\mu_{jt} + \boldsymbol{\delta}_{jt}\boldsymbol{\sigma}_{jt}') \cdot dt + (S_{jt}\boldsymbol{\delta}_{jt} + p_{jt}\boldsymbol{\sigma}_{jt}) \cdot d\mathbf{W}_t], \end{aligned}$$

where $\boldsymbol{\delta}_{jt}$ *and* $\boldsymbol{\sigma}_{jt}$ *are the* j*th rows of* $\boldsymbol{\delta}_t$ *and* $\boldsymbol{\sigma}_t$*.*

3.3.6 *Self-Financing Portfolios in Continuous Time*

Working now toward applications, think of $\mathbf{p}_t$ and $\mathbf{S}_t$ in the last example as portfolio and asset-price vectors; that is, component p_{jt} is the number of units of asset j held at t and S_{jt} is its price. With $P_0 = \mathbf{p}_0'\mathbf{S}_0$ given, $P_t = \mathbf{p}_t'\mathbf{S}_t$ is then the value of the portfolio at t—the sum of shares times prices. However, the $\{\mathbf{p}_t\}$ process needs to be restricted if this interpretation is to make sense. Since instantaneous changes in the portfolio should be under the investor's control—hence, predictable—we will need to set $\boldsymbol{\delta}_t = \mathbf{0}$. This change makes $\{\mathbf{p}_t\}$ of finite variation, with derivative $\{d\mathbf{p}_t/dt = \boldsymbol{\pi}_t\}$ an $\{\mathcal{F}_t\}$-adapted process. In this case (3.27) applies, and (3.28) reduces to

$$dP_t = \mathbf{S}_t' \cdot d\mathbf{p}_t + \mathbf{p}_t' \cdot d\mathbf{S}_t.$$

The first component represents the instantaneous change in P_t that is attributable to purchases and sales of assets at the current prices, while the second component is the change attributable to changes in prices of assets already in the portfolio.

Now let us make the portfolio self-financing by insisting that instantaneous purchases and sales balance out. This is done by restricting $\boldsymbol{\pi}_t$ such that $\mathbf{S}_t'\boldsymbol{\pi}_t = \mathbf{S}_t' \cdot d\mathbf{p}_t/dt = 0$, thereby eliminating the first component of dP_t. With these controls

$$dP_t = \mathbf{p}_t' \cdot d\mathbf{S}_t, \tag{3.29}$$

or in integral form

$$P_T - P_0 = \int_0^T \mathbf{p}_t' \cdot d\mathbf{S}_t. \tag{3.30}$$

3.4 Tools for Martingale Pricing

Previous sections of the chapter have shown how to construct Itô integrals and have established that integrals of suitable adapted processes with respect to Brownian motions are continuous local martingales. Thus, given a Wiener process $\{W_t\}$ on a filtered probability space $(\Omega, \mathcal{F}, \{\mathcal{F}_t\}, \mathbb{P})$, the process $\{X_t = X_0 + \int_0^t b_s \cdot dW_s\}_{0 \le t \le T}$ is a continuous local martingale on $[0, T]$ when $b_t \in \mathcal{F}_t$ for each t and $\int_0^T b_t^2 \cdot dt < \infty$. Moreover, $\{X_t\}$ is a proper martingale when $E\int_0^T b_t^2 \cdot dt < \infty$. On the other hand, the more general Itô processes that allow for systematic trend, as $\{X_t = X_0 + \int_0^t a_s \cdot ds + \int_0^t b_s \cdot dW_s\}$, are *not* $\mathbb{P}$-martingales but, rather, are *semi*martingales. Girsanov's theorem will show us how to find a new measure in which a process with trend does become a proper martingale.

3.4.1 *Girsanov's Theorem and Changes of Measure*

Turning Itô processes that are *semi*martingales into Itô processes that are local martingales is just a matter of removing the drift term. Suppose it were possible to replace the standard Brownian motion in $dX_t = a_t \cdot dt + b_t \cdot dW_t$ with a new process $\{\hat{W}_t\}_{t \ge 0}$ with $\hat{W}_t = W_t + \int_0^t (a_s/b_s) \cdot ds$. This would give a formal representation for the stochastic differential of X_t as

$$dX_t = a_t \cdot dt + b_t[d\hat{W}_t - (a_t/b_t) \cdot dt] = b_t \cdot d\hat{W}_t.$$

Obviously, this would require the extra restriction $b_t > 0$, but even granting this one quickly realizes that this sleight of hand accomplishes nothing. $\{\hat{W}_t\}$ itself has systematic trend and is no longer a standard Brownian motion under measure $\mathbb{P}$, so the stochastic integral $\int_0^t b_s \cdot d\hat{W}_s$ is still not a local $\mathbb{P}$-martingale. But suppose there were a different measure $\hat{\mathbb{P}}$ under which $\{\hat{W}_t\}$ actually *was* a Brownian motion. Then the process $\{X_t = X_0 + \int_0^t b_s \cdot d\hat{W}_s\}_{0 \le t \le T}$ would indeed be a local martingale under $\hat{\mathbb{P}}$—and a proper martingale, too, if $E\int_0^T b_t^2 \cdot dt < \infty$. So how could one change the measure so as to bring this about?

Clearly, the Radon-Nikodym theorem will be involved here. A hint toward the answer comes by considering how Radon-Nikodym would be used to change a normally distributed random variable with nonzero mean into one with zero mean. For this, suppose $Z \sim N(0,1)$ on a probability space $(\Omega, \mathcal{F}, \mathbb{P})$, and let $\mathbb{P}_Z$ be the measure on the space $(\Re, \mathcal{B})$ induced by Z. This is defined as

$$\mathbb{P}_Z(B) = \int_B \frac{1}{\sqrt{2\pi}} e^{-z^2/2} \cdot dz, \quad B \in \mathcal{B}.$$

Let $\hat{Z} = Z + \theta \sim N(\theta, 1)$ under $\mathbb{P}$, with $\theta \neq 0$. Then with $\phi(z) = (2\pi)^{-1/2} e^{-z^2/2}$ as the normal p.d.f., introduce the nonnegative random variable $Q(Z) = \exp(-\theta Z - \theta^2/2)$ and define the set function $\hat{\mathbb{P}}_Z$ as

$$\begin{aligned} \hat{\mathbb{P}}_Z(B) &\equiv EQ(Z)\mathbf{1}_B(Z) = \int_B Q(z)\phi(z) \cdot dz \\ &= \int_B \frac{1}{\sqrt{2\pi}} e^{-(z+\theta)^2/2} \cdot dz, \quad B \in \mathcal{B}. \end{aligned} \tag{3.31}$$

Thus defined, $\hat{\mathbb{P}}_Z$ is obviously nonnegative and countably additive, and since $\hat{\mathbb{P}}_Z(\Re) = EQ(Z)\mathbf{1}_\Re(Z) = EQ(Z) = 1$ it follows that $\hat{\mathbb{P}}_Z(\cdot)$ is a probability measure on $(\Re, \mathcal{B})$. As the form of the integral in (3.31) shows, this corresponds to some new measure $\hat{\mathbb{P}}$ on $(\Omega, \mathcal{F})$ under which $Z \sim N(-\theta, 1)$. Since the mean of Z is $-\theta$ under $\hat{\mathbb{P}}$ it is not surprising that $\hat{Z} = Z + \theta \sim N(0,1)$ under the new measure. This is shown explicitly as follows. Letting $\hat{E}$ denote expectation under $\hat{\mathbb{P}}$, the c.f. of $\hat{Z}$ under $\hat{\mathbb{P}}$ is

$$\hat{E}e^{i\varsigma\hat{Z}} = \int_\Re e^{i\varsigma(z+\theta)} \frac{1}{\sqrt{2\pi}} e^{-(z+\theta)^2/2} \cdot dz = \int_\Re \frac{1}{\sqrt{2\pi}} e^{i\varsigma\hat{z} - \hat{z}^2/2} \cdot d\hat{z} = e^{-\varsigma^2/2},$$

showing that $\hat{Z}$ is indeed distributed as $N(0,1)$ under this new measure.

Now let us try to extend this idea to Brownian motions. Considering a finite time horizon T, we work with the filtered probability space $(\Omega, \mathcal{F}_T, \{\mathcal{F}_t\}_{0 \le t \le T}, \mathbb{P})$. Recalling example 44, let

$$Q_t = \exp\left(-\int_0^t \theta_s \cdot dW_s - \frac{1}{2}\int_0^t \theta_s^2 \cdot ds\right) \tag{3.32}$$

be the Doléans exponential of $-\int_0^t \theta_s \cdot dW_s$, where $\{\theta_t\}$ is an adapted process on $[0,T]$ with $\int_0^T \theta_t^2 \cdot dt < \infty$ and such that $EQ_T = 1$. Recall that the Novikov condition,

$$E \exp\left(\frac{1}{2}\int_0^T \theta_t^2 \cdot dt\right) < \infty,$$

is sufficient for $\{Q_t\}_{0\le t\le T}$ to be a $\mathbb{P}$ martingale. Now defining the measure $\hat{\mathbb{P}}$ on $\mathcal{F}_T$ as

$$\hat{\mathbb{P}}(A) = EQ_T\mathbf{1}_A = \int_A Q_T(\omega)\cdot d\mathbb{P}(\omega), \quad A \in \mathcal{F}_T, \tag{3.33}$$

where $Q_T = d\hat{\mathbb{P}}/d\mathbb{P}$ is the Radon-Nikodym derivative, we have the conclusion of Girsanov's theorem:

Theorem 8 (*Girsanov*) *For fixed $T \in [0,\infty)$ the process*

$$\left\{\hat{W}_t = W_t + \int_0^t \theta_s \cdot ds\right\}_{0\le t\le T}$$

is a Brownian motion on $(\Omega, F_T, \{\mathcal{F}_t\}_{0\le t\le T}, \hat{\mathbb{P}})$, where $\hat{\mathbb{P}}$ is defined by (3.33) and (3.32).

For a multidimensional version simply interpret $\{\hat{W}_t\}$, $\{W_t\}$, and $\{\theta_t\}$ as k-dimensional processes and θ_t^2 as scalar product $\boldsymbol{\theta}_t'\boldsymbol{\theta}_t$, making $\{\hat{\mathbf{W}}_t = \mathbf{W}_t + \int_0^t \boldsymbol{\theta}_s \cdot ds\}_{0\le t\le T}$ a k-dimensional Brownian motion under $\hat{\mathbb{P}}$, with

$$\frac{d\hat{\mathbb{P}}}{d\mathbb{P}} = \exp\left(-\int_0^T \boldsymbol{\theta}_s \cdot d\mathbf{W}_s - \frac{1}{2}\int_0^t \boldsymbol{\theta}_s'\boldsymbol{\theta}_s \cdot ds\right).$$

Notice that the martingale property of $\{Q_t\}$ implies that when $A \in \mathcal{F}_t$

$$\hat{\mathbb{P}}(A) = E[\mathbf{1}_A E(Q_T \mid \mathcal{F}_t)] = EQ_t\mathbf{1}_A,$$

Thus, $EQ_t\mathbf{1}_A$ is the restriction of $\hat{\mathbb{P}}$ to $\mathcal{F}_t$ in the sense that $\hat{\mathbb{P}}(A) = EQ_t\mathbf{1}_A$ when $A \in \mathcal{F}_t$. Notice, too, the correspondence between how the measure is changed to produce a Brownian motion without trend and how the measure is changed to produce a normal variate with zero mean.

The proof here is much more involved, however, because the behavior of the entire process $\{\hat{W}_t\}_{0\le t\le T}$ must be established. The standard approach relies on a characterization theorem of Paul Lévy, which states that if $\{X_t\}$ is a continuous martingale with $X_0 = 0$ and if $\{X_t^2 - t\}$ is also a martingale, then $\{X_t\}$ is a Brownian motion. We will use Lévy's characterization to prove a simple one-dimensional version of Girsanov in which $\theta_t = \theta$, a constant. In this case $Q_T = \exp(-\theta W_T - \theta^2 T/2)$ and $\hat{W}_t = W_t + \theta t$.

We need a preliminary result; namely, that if $\{Y_t\}_{0\leq t\leq T}$ is $\{\mathcal{F}_t\}_{0\leq t\leq T}$ adapted and $0 \leq s < t \leq T$, then

$$\hat{E}_s Y_t = E_s(Y_t Q_t / Q_s), \tag{3.34}$$

where $\hat{E}_s$ and E_s are expectations conditional on $\mathcal{F}_s$ with respect to $\hat{\mathbb{P}}$ and $\mathbb{P}$, respectively. Given that $\{Q_t\}$ is a continuous $\mathbb{P}$ martingale, this follows at once from Bayes' rule for conditional expectations, (2.51). Expression (3.34) tells us that the right variable by which to scale an $\mathcal{F}_t$-measurable variable when conditioning on $\mathcal{F}_s$ is

$$\frac{Q_t}{Q_s} = \exp\left[-\int_s^t \theta_u \cdot dW_u - \frac{1}{2}\int_s^t \theta_u^2 \cdot du\right].$$

Also, since $Q_0 = 1$, it follows that $\hat{E}Y_t = EY_tQ_t$.

Now let us use Lévy's characterization theorem to prove Girsanov for a constant process $\{\theta_t = \theta\}$. This requires showing that

$$\hat{E}_s \hat{W}_t^2 - t - (\hat{W}_s^2 - s) = 0.$$

First, write

$$\begin{aligned}\hat{E}_s \hat{W}_t^2 - t - (\hat{W}_s^2 - s) &= \hat{E}_s[2\hat{W}_s(\hat{W}_t - \hat{W}_s) + (\hat{W}_t - \hat{W}_s)^2] - (t - s)\\ &= E_s[2\hat{W}_s(\hat{W}_t - \hat{W}_s) + (\hat{W}_t - \hat{W}_s)^2]Q_t/Q_s - (t - s),\end{aligned}$$

then set $\hat{W}_t = W_t + \theta t$, $\hat{W}_s = W_s + \theta s$, and $Q_t/Q_s = \exp[\theta(W_t - W_s) - \theta^2(t-s)/2]$. Finally, expressing $W_t - W_s$ as $Z\sqrt{t-s}$, where $Z \sim N(0,1)$, and making a further change of variables reduces this to

$$2\hat{W}_s\sqrt{t-s}\int_{\Re} z\phi(z)\cdot dz + (t-s)\int_{\Re} z^2\phi(z)\cdot dz - (t-s) = 0,$$

thus establishing the conclusion of Girsanov in this special case.

Returning to the problem posed at the beginning of this section, let us now look specifically at how to apply Girsanov's theorem. If there is a measure $\hat{\mathbb{P}}$ under which $\{\hat{W}_t = W_t + \int_0^t (a_s/b_s)\cdot ds\}$ is a Brownian motion, then under this measure the Itô process $\{X_t\}$ with $dX_t = a_t \cdot dt + b_t \cdot dW_t = b_t \cdot d\hat{W}_t$ has zero drift. Taking $\theta_t = a_t/b_t$, we conclude from Girsanov that such $\hat{\mathbb{P}}$ exists provided $\{Q_t = \exp[-\int_0^t \theta_s \cdot dW_s - \frac{1}{2}\int_0^t \theta_s^2 \cdot ds]\}$ (with $Q_0 = 1$) is a martingale under the original measure $\mathbb{P}$. The Novikov condition is sufficient for this, and clearly $b_t \neq 0$ is necessary.

3.4.2 *Representation of Martingales*

From what we have seen Itô processes are extremely resilient: smooth functions of Itô processes are still Itô processes, and integrals of suitable adapted processes with respect to Itô processes are still Itô processes. Given this remarkable fungibility, it is not surprising that it is possible to represent either of two different Itô processes in terms of the other if they are adapted to the same filtration. Suppose there are two trendless Itô processes,

$$\left\{X_t = X_0 + \int_0^t a_s \cdot dW_s\right\}_{0\le t\le T}$$

$$\left\{Y_t = Y_0 + \int_0^t b_s \cdot dW_s\right\}_{0\le t\le T},$$

where a_t is a.s. nonzero for each t and $\int_0^T b_s^2 \cdot ds < \infty$. Then setting $c_t = b_t/a_t$ shows that there exists a process $\{c_t\}$ such that $Y_t = Y_0 + \int_0^t c_s \cdot dX_s$ and $\int_0^T a_t^2 c_t^2 \cdot dt < \infty$ a.s. The first condition follows just from algebra:

$$Y_0 + \int_0^t c_s \cdot dX_s = Y_0 + \int_0^t (b_s/a_s)a_s \cdot dW_s = Y_t;$$

and the almost-certain finiteness of $\int_0^T a_t^2 c_t^2 \cdot dt = \int_0^T b_t^2 \cdot dt$ comes from the fact that $\{Y_t\}$ was defined to be an Itô process.

Taking this one step farther, suppose there are two local martingales $\{X_t\}_{0\le t\le T}$ and $\{Y_t\}_{0\le t\le T}$ whose quadratic variation processes $\{\langle X\rangle_t\}$ and $\{\langle Y\rangle_t\}$ are absolutely continuous functions of t. Then it can be shown[11] that there exist adapted processes $\{a_t\}$ and $\{b_t\}$ with $\int_0^T a_t^2 \cdot dt < \infty$ and $\int_0^T b_t^2 \cdot dt < \infty$ such that $X_t = X_0 + \int_0^t a_s \cdot dW_s$ and $Y_t = Y_0 + \int_0^t b_s \cdot dW_s$. In other words, subject to the restriction on quadratic variations, it is always possible to represent local martingales as Itô processes. Piecing together these facts then gives the following result:

Theorem 9 (*Martingale Representation*) *For local martingales $\{X_t\}_{0\le t\le T}$ and $\{Y_t\}_{0\le t\le T}$ such that $a_t^2 \equiv d\langle X\rangle_t/dt > 0$ a.s. there exists an adapted process $\{c_t\}_{0\le t\le T}$ such that $Y_t = Y_0 + \int_0^t c_s \cdot dX_s$ for $t \in [0,T]$ and $\int_0^T a_t^2 c_t^2 \cdot dt < \infty$ a.s.*

---

[11]See Karatzas and Shreve (1991, theorem 4.2).

3.4.3 *Numeraires, Changes of Numeraire, and Changes of Measure*

Prices of domestic assets are normally specified in units of domestic currency—dollars, pounds, Euros, etc. If we allow trades only among domestic assets, there is no need to think about how the currency's value relates to values of foreign currencies. However, even when confining attention to domestic markets, finding arbitrage-free prices of derivatives will require us to work with other numeraires than units of currency. The reason is that currency *per se* is dominated by riskless, interest-bearing assets such as money-market funds and default-free bonds. In arbitrage-free markets containing such (admittedly, idealized) riskless alternatives, currency would simply not be held as an asset. Therefore, to explore arbitrage-free economies, we need to express prices relative to the price of some asset or portfolio of assets that is a viable means of storing wealth. Any such price process that remains always positive can serve as numeraire. Specifically,

Definition 3 *A numeraire is an almost-surely positive price process.*

We refer to the ratio of an asset's price to a numeraire as a "normalized" price process. While the idealized money-market fund introduced in section 4.1—essentially a savings account—will usually serve our purpose as numeraire, others will be more convenient in some applications, particularly in chapter 10, where we treat derivatives on interest-sensitive assets. This section shows how normalized price processes are changed by switching from one numeraire to another. If a particular normalized price process is a martingale in some measure, we will see that it is no longer so after a change of numeraire, but we will also see how to change measures so as to restore the martingale property.

For some finite horizon T let the price process $\{S_t\}_{0\le t\le T}$ of a traded asset evolve as $dS_t = s_t S_t \cdot dt + \boldsymbol{\sigma}_t' S_t \cdot d\mathbf{W}_t$, where $\{\boldsymbol{\sigma}_t\}$ is a k-dimensional process and $\{\mathbf{W}_t\}$ is a vector-valued standard Brownian motion on $(\Omega, \mathcal{F}, \{\mathcal{F}_t\}_{0\le t\le T}, \mathbb{P})$; and let $\{M_t\}$ and $\{L_t\}$ be other, strictly positive price processes evolving as

$$\begin{aligned} dM_t &= m_t M_t \cdot dt \\ dL_t &= l_t L_t \cdot dt + \boldsymbol{\lambda}_t' L_t \cdot d\mathbf{W}_t. \end{aligned}$$

These will be our numeraires. Note that $\{M_t\}$ is represented as a process of finite variation, with $M_t = M_0 \exp(\int_0^t m_s \cdot ds)$. We take $\{s_t, m_t, l_t, \boldsymbol{\sigma}_t, \boldsymbol{\lambda}_t\}$

to be adapted processes with $\int_0^T \boldsymbol{\sigma}_t'\boldsymbol{\sigma}_t \cdot dt < \infty$ and $\int_0^T \boldsymbol{\lambda}_t'\boldsymbol{\lambda}_t \cdot dt < \infty$. Applying Itô's formula to $S_t^L \equiv S_t/L_t = f(S_t, L_t)$ and to $S_t^M \equiv S_t/M_t$ gives

$$\begin{aligned}
dS_t^L &= f_S \cdot dS_t + f_L \cdot dL_t + f_{SL} \cdot d\langle S, L\rangle_t + \frac{1}{2} f_{LL} \cdot d\langle L\rangle_t \\
&= \frac{dS_t}{L_t} - S_t \frac{dL_t}{L_t^2} - \frac{S_t L_t \boldsymbol{\sigma}_t' \boldsymbol{\lambda}_t \cdot dt}{L_t^2} + \frac{S_t L_t^2 \boldsymbol{\lambda}_t' \boldsymbol{\lambda}_t \cdot dt}{L_t^3} \\
&= S_t^L \left[\frac{dS_t}{S_t} - \frac{dL_t}{L_t} - (\boldsymbol{\sigma}_t - \boldsymbol{\lambda}_t)' \boldsymbol{\lambda}_t \cdot dt \right] \\
&= \left[s_t - l_t - (\boldsymbol{\sigma}_t - \boldsymbol{\lambda}_t)' \boldsymbol{\lambda}_t \right] S_t^L \cdot dt + (\boldsymbol{\sigma}_t - \boldsymbol{\lambda}_t)' S_t^L \cdot d\mathbf{W}_t \\
dS_t^M &= (s_t - m_t) S_t^M \cdot dt + \boldsymbol{\sigma}_t' S_t^M \cdot d\mathbf{W}_t
\end{aligned}$$

Note that the volatility vector of $\{S_t^L\}$ is the difference between the volatility vectors of $\{S_t\}$ and of $\{L_t\}$ and that the drift of $\{S_t^L\}$ is influenced by both the trend and the volatility of $\{L_t\}$, whereas the change from $\{S_t\}$ to $\{S_t^M\}$ is in the drift term only since the volatility of $\{M_t\}$ is null. However, since the volatilities of $\{L_t\}$ and $\{M_t\}$ differ, the *change* of numeraire from $\{L_t\}$ to $\{M_t\}$ (or *vice versa*) alters both the drift and volatility of the normalized version of $\{S_t\}$.

Now suppose there exists a measure $\mathbb{P}^M$ under which both $\{S_t^M\}$ and $\{L_t^M \equiv L_t/M_t\}$ are martingales. By the martingale representation theorem there are under this measure a k-vector Brownian motion $\{\mathbf{W}_t^M\}$ and adapted processes $\{\boldsymbol{\sigma}_t^M, \boldsymbol{\lambda}_t^M\}$ such that

$$\begin{aligned}
dS_t^M &= (\boldsymbol{\sigma}_t^M)' S_t^M \cdot d\mathbf{W}_t^M \\
dL_t^M &= (\boldsymbol{\lambda}_t^M)' L_t^M \cdot d\mathbf{W}_t^M.
\end{aligned}$$

However, if we change to numeraire $\{L_t\}$, we have for $S_t^L \equiv S_t/L_t = S_t^M/L_t^M$

$$\begin{aligned}
dS_t^L &= -(\boldsymbol{\sigma}_t^M - \boldsymbol{\lambda}_t^M)' \boldsymbol{\lambda}_t^M S_t^L \cdot dt + (\boldsymbol{\sigma}_t^M - \boldsymbol{\lambda}_t^M)' S_t^L \cdot d\mathbf{W}_t^M \qquad (3.35) \\
&= (\boldsymbol{\sigma}_t^M - \boldsymbol{\lambda}_t^M)' S_t^L \cdot d\left(\mathbf{W}_t^M - \int_0^t \boldsymbol{\lambda}_s^M \cdot ds \right).
\end{aligned}$$

There is now a drift term, so $\{S_t^L\}$ is not a martingale under $\mathbb{P}^M$; but we can ask whether there is a new measure $\mathbb{P}^L$ under which the martingale property is restored. Girsanov's theorem tells us that the answer is 'yes' if $\{\boldsymbol{\lambda}_t\}$ is such that $\{Q_t \equiv \exp(-\int_0^t \boldsymbol{\lambda}_s' \cdot d\mathbf{W}_t^M - \frac{1}{2}\int_0^t \boldsymbol{\lambda}_s'\boldsymbol{\lambda}_s \cdot ds)\}_{0 \le t \le T}$ is a martingale under $\mathbb{P}^M$. In that case $\{\mathbf{W}_t^L \equiv \mathbf{W}_t^M - \int_0^t \boldsymbol{\lambda}_s^M \cdot ds\}$ is a k-dimensional standard Brownian motion in the measure defined by $\mathbb{P}^L(A) = \int_A Q_T \cdot d\mathbb{P}^M$

$= \int_A \frac{d\mathbb{P}^A}{d\mathbb{P}^M} \cdot d\mathbb{P}^M$, $A \in \mathcal{F}_T$; and with E^M, E^L denoting expectation under $\mathbb{P}^M, \mathbb{P}^L$, we have $E_t^L S_T^L \equiv E^L(S_T^L \mid \mathcal{F}_t) = E_t^M S_T^L Q_T / Q_t = S_t^L$. Notice that the change of *measure* has altered only the trend of $\{S_t^L\}$, not its volatility process.

To summarize,

- Changing from one *numeraire* to another that has different drift and volatility alters both the drift and the volatility of an asset's normalized price.
- Changing *measures* without changing numeraires changes the drift of the asset's normalized price but not its volatility.

Conveniently, random variable $d\mathbb{P}^L/d\mathbb{P}^M$ can be represented in terms of observables. As above, let A be any $\mathcal{F}_T$-measurable event, and for $t \in [0, T]$ set $S_t \equiv M_t E_t^M \mathbf{1}_A = E^M(M_t \mathbf{1}_A \mid \mathcal{F}_t)$. In particular, $S_T = M_T \mathbf{1}_A$ since $A \in \mathcal{F}_T$. Then $\{S_t^M\}_{0 \le t \le T}$ is a conditional expectations process adapted to $\{\mathcal{F}_t\}$ and therefore a $\mathbb{P}^M$ martingale, with

$$S_0 = M_0 E^M S_T^M = M_0 E^M(M_T \mathbf{1}_A / M_T) = \int_A M_0 \cdot d\mathbb{P}^M.$$

But by hypothesis $\{S_t^L\}$ is a $\mathbb{P}^L$ martingale, so

$$S_0 = L_0 E^L S_T^L = L_0 E^L(M_T \mathbf{1}_A / L_T) = \int_A L_0 \frac{M_T}{L_T} \frac{d\mathbb{P}^L}{d\mathbb{P}^M} \cdot d\mathbb{P}^M.$$

Since the two integral expressions agree for each $A \in \mathcal{F}_T$, we conclude that

$$\frac{d\mathbb{P}^L}{d\mathbb{P}^M} = \frac{L_T/L_0}{M_T/M_0}, \ \mathbb{P}^M \text{ a.s.} \tag{3.36}$$

3.5 Tools for Discontinuous Processes

This section describes certain continuous-time processes with discontinuous sample paths and introduces the applicable tools of stochastic calculus. The applications of these processes will come in chapter 9. We begin with models for a simple class of pure-jump processes having no continuous component, then turn to more general forms.

3.5.1 *J Processes*

Given a filtered probability space $(\Omega, \mathcal{F}, \{\mathcal{F}_t\}_{t \ge 0}, \mathbb{P})$, an increasing sequence of stopping times $\{\tau_k\}_{k=1}^\infty$, and a collection of random variables $\{U_k\}_{k=1}^\infty$

with $U_k \in \mathcal{F}_{\tau_k}$, we introduce a class of pure-jump processes $\{J_t\}_{t\geq 0}$ having the representation

$$J_t = J_0 + \sum_{k=1}^{\infty} U_k \mathbf{1}_{[\tau_k, \tau_{k+1})}(t), \tag{3.37}$$

where $J_0 \in \mathcal{F}_0$. We shall refer to processes of this type as J processes. They are clearly right-continuous step functions, each sample path of which takes the constant value $U_k(\omega)$ on $[\tau_k(\omega), \tau_{k+1}(\omega))$.

Example 47 *The Poisson process, $\{N_t\}_{t\geq 0}$, was introduced in chapter 2 as an example of a continuous-time Markov process. Recall that $\{N_t\}$ is an adapted process on $(\Omega, \mathcal{F}, \{\mathcal{F}_t\}_{t\geq 0}, \mathbb{P})$ with the properties (i) $N_0 = 0$; (ii) $N_t - N_s$ independent of $N_r - N_q$ for $q < r \leq s < t$; and (iii) $N_{t+u} - N_t \sim P(\theta u)$, with $\mathbb{P}(N_{t+u} - N_t = n) = (\theta u)^n e^{-\theta u}/n!$ for $n \in \aleph_0$ and$\theta, u > 0$. Parameter θ is called the "intensity" or the "arrival rate". Taking $\tau_k = \inf\{t : N_t \geq k\}$ and $U_k = k$ gives a representation in the form (3.37) as $N_t = \sum_{k=1}^{\infty} k \mathbf{1}_{[\tau_k, \tau_{k+1})}(t)$; equivalently, $N_t = \sum_{k=1}^{\infty} \mathbf{1}_{[\tau_k, \infty)}(t)$. The stopping times $\{\tau_k\}$ are called the "arrival times".*

Here are some features of the Poisson process that will be needed in applications. Since $EN_t = VN_t = \theta t$, it is clear from the independent-increments property that $\{N_t - \theta t\}$ and $\{(N_t - \theta t)^2 - \theta t\}$ are martingales. Likewise, the form of the probability generating function,

$$\Pi(\zeta) = E\zeta^{N_t} = \exp[\theta t(\zeta - 1)],$$

shows that $\{\xi^{N_t} e^{-\theta t(\xi - 1)}\}$ is a martingale.

Example 48 *Taking $\{\tau_k\}_{k=1}^{\infty}$ as Poisson arrival times, let $\{U_k\}_{k=1}^{\infty}$ be i.i.d. random variables independent of $\{N_t\}$ (and therefore of the $\{\tau_k\}$), and set $U_0 = 0$. With $J_0 \in \mathcal{F}_0$, we then have the following representations of a "compound-Poisson" process $\{J_t\}$:*

$$\begin{aligned}
J_t &= J_0 + \sum_{k=1}^{\infty} \left(\sum_{j=1}^{k} U_j \right) \mathbf{1}_{[\tau_k, \tau_{k+1})}(t) \\
&= J_0 + \sum_{k=1}^{\infty} U_k \mathbf{1}_{[\tau_k, \infty)}(t) \\
&= J_0 + \sum_{k=0}^{N_t} U_k.
\end{aligned}$$

As the sum of a Poisson distributed number of independent random variables, $\{J_t - J_0\}$ *takes a jump of random size* U_k *at the* k *th Poisson arrival. Clearly,* $\{J_t\}_{t\geq 0}$ *is a martingale if the jump sizes have mean zero. For later reference, if* $J_0 = 0$ *and the jumps have characteristic function* $\Psi_U(\zeta) \equiv Ee^{i\zeta U}$*, then the c.f. of* J_t *is*

$$\begin{aligned}\Psi_{J_t}(\zeta) &= E\left\{E\left[\exp\left(i\zeta\sum_{k=0}^{N_t} U_k\right) \mid N_t\right]\right\} \\ &= E\exp[\Psi_U(\zeta)N_t] \\ &= \exp\{\theta t[\Psi_U(\zeta) - 1]\}. \end{aligned} \quad (3.38)$$

Integration with Respect to J Processes

A special theory for integration with respect to Brownian motion—and Itô processes generally—is required because Brownian motion is of unbounded variation. By contrast, sample paths of J processes necessarily have finite variation on finite time intervals, since step functions can be represented as differences between nondecreasing functions. This means that integrals of continuous functions with respect to J processes can be defined pathwise as ordinary Riemann-Stieltjes integrals. That is, if $\{c_t\}_{t\geq 0}$ is a.s. continuous, then we can interpret $\int_0^t c_s \cdot dJ_s$ on a sample path ω as $\lim_{n\to\infty}\sum_{j=1}^n c_{t_j^*}(J_{t_j} - J_{t_{j-1}})$ for any $t_j^* \in [t_{j-1}, t_j]$, whenever $\max_j |t_j - t_{j-1}| \to 0$ as $n \to \infty$. The following example shows that, contrary to the case for the Itô integral, c_t does not have to be evaluated at the left bound of each interval in the approximating sum if c is continuous.

Example 49 *Taking* $J_t = N_t$, $c_t = t$, $t_j = jt/n$ *and*

$$\Delta_n \equiv \sum_{j=1}^n c_{t_j}(J_{t_j} - J_{t_{j-1}}) - \sum_{j=1}^n c_{t_{j-1}}(J_{t_j} - J_{t_{j-1}}),$$

we have

$$\begin{aligned}\lim_{n\to\infty}\Delta_n &= t\lim_{n\to\infty} n^{-1}\sum_{j=1}^n (N_{t_j} - N_{t_{j-1}}) \\ &= t\lim_{n\to\infty} n^{-1}N_t \\ &= 0.\end{aligned}$$

On the other hand, integrals with respect to $\{J_t\}$ of a process $\{c_t\}$ that is merely right-continuous (such as another J process) must be defined more restrictively if the result is to be unambiguous.

Example 50 *Taking* $J_t = N_t = c_t$ *and* $t_j = jt/n$ *gives*

$$\begin{aligned}
\lim_{n\to\infty} \Delta_n &= \lim_{n\to\infty} \sum_{j=1}^{n} N_{t_j}(N_{t_j} - N_{t_{j-1}}) - \lim_{n\to\infty} \sum_{j=1}^{n} N_{t_{j-1}}(N_{t_j} - N_{t_{j-1}}) \\
&= \sum_{j=1}^{N_t} j - \sum_{j=1}^{N_t} (j-1) \\
&= N_t.
\end{aligned}$$

Adopting as the definition the left-end construction (we motivate this choice momentarily) gives

$$\int_{(0,t]} c_{s-} \cdot dJ_s = \lim_{n\to\infty} \sum_{j=1}^{n} c_{t_{j-1}}(J_{t_j} - J_{t_{j-1}}) = \sum_q c_{q-}(J_q - J_{q-}),$$

where c_{s-} is the left hand limit (that is, $\lim_{n\to\infty} c_{s-1/n}$) and the last summation is over the arrival times $\{\tau_k\}$.

Regarding the stochastic integral with respect to a J process as a function of the upper limit leads to a new J process, $\{K_t\}$ say, with

$$K_t = K_0 + \int_{(0,t]} c_{s-} \cdot dJ_s,$$

for some $K_0 \in \mathcal{F}_0$. The differential shorthand is then

$$dK_t = c_{t-} \cdot dJ_t.$$

If there is an absolutely continuous function[12] ϑ_t such that $\{J'_t \equiv J_t - \vartheta_t\}$ is a martingale adapted to $\{\mathcal{F}_t\}$, then this left-end construction makes the integral with respect to $\{J'_t\}$ a martingale also, for suitably restricted functions c. Thus, for $K_t - K_0 = \int_{(0,t]} c_{s-} \cdot dJ'_s$ and a bounded c,

$$\begin{aligned}
E(K_t - K_0) &= E \lim_{n\to\infty} \sum_{j=1}^{n} c_{t_{j-1}}(J'_{t_j} - J'_{t_{j-1}}) \\
&= \lim_{n\to\infty} E \sum_{j=1}^{n} c_{t_{j-1}}(J'_{t_j} - J'_{t_{j-1}})
\end{aligned}$$

[12]More generally, a function having finite variation on finite intervals.

$$\begin{aligned} &= \lim_{n\to\infty} E \sum_{j=1}^{n} c_{t_{j-1}} E_{t_{j-1}} (J'_{t_j} - J'_{t_{j-1}}) \\ &= 0. \end{aligned}$$

For example, with $c_t = e^{-N_t}$, $J_t = N_t$, $\vartheta_t = \theta t$, and $K_0 \in \mathcal{F}_0$ the process $\{K_t\}$ with

$$K_t = K_0 + \int_{(0,t]} e^{-N_{s-}} \cdot d(N_s - \theta s) = K_0 + \int_{(0,t]} e^{-N_{s-}} \cdot dN_s - \theta \int_0^t e^{-N_{s-}} \cdot ds$$

is an $\{\mathcal{F}_t\}$ martingale.

Integrals with respect to compound-Poisson processes can be written explicitly as

$$K_t - K_0 = \sum_q c_{q-} U \cdot (N_q - N_{q-}) = \sum_{k=0}^{N_t} c_{\tau_k -} U_k, \tag{3.39}$$

where the first summation is over jump times $\tau_1, \tau_2, \ldots$. In differential form this is

$$dK_t = c_{t-} \cdot dJ_t = c_{t-} U \cdot dN_t.$$

Of course, $U \equiv 1$ for a standard Poisson process.

Differentials of Functions of J Processes

Itô's formula provides representations of smooth functions of Itô processes as stochastic integrals and of changes in these functions as stochastic differentials. Obtaining these representations for functions of J processes is straightforward. For any measurable $f : [0, \infty) \times \Re \to \Re$ we have as an identity for $f(t, J_t) - f(0, J_0)$ the expression

$$\sum_{j=1}^{n} [f(t_j, J_{t_j}) - f(t_{j-1}, J_{t_j})] + \sum_{j=1}^{n} [f(t_{j-1}, J_{t_j}) - f(t_{j-1}, J_{t_{j-1}})],$$

where $0 = t_0 < t_1 < \ldots < t_n = t$. Assuming that f is continuously differentiable in the time argument, apply the mean value theorem to express the first summation as

$$\sum_{j=1}^{n} f_t(t_j^*, J_{t_j})(t_j - t_{j-1}),$$

where f_t is the time derivative and $t_j^* \in (t_{j-1}, t_j)$. In the limit as $\max_j |t_j - t_{j-1}|$ approaches zero this gives

$$f(t, J_t) - f(0, J_0) = \int_0^t f_t(s, J_s) \cdot ds + \sum_q [f(q, J_q) - f(q, J_{q-})], \quad (3.40)$$

where the summation is over jump times $\tau_1, \tau_2, \ldots$. The differential form is

$$df(t, J_t) = f_t(t, J_t) \cdot dt + [f(t, J_t) - f(t, J_{t-})]. \quad (3.41)$$

When J_t is an integral with respect to a compound-Poisson process, as in (3.39), (3.40) is

$$f(t, J_t) - f(0, J_0) = \int_0^t f_t(s, J_s) \cdot ds + \sum_{k=0}^{N_t} [f(\tau_k, J_{\tau_k -} + c_{\tau_k -} U_k) - f(\tau_k, J_{\tau_k -})].$$

Consider next a function $f(t, Y_t, J_t)$ that depends on both a J process, $\{J_t\}_{t \geq 0}$, and an Itô process, $\{Y_t\}_{t \geq 0}$, with

$$Y_t = Y_0 + \int_0^t a_s \cdot ds + \int_0^t b_s \cdot dW_s. \quad (3.42)$$

If f is continuously differentiable with respect to t and twice continuously differentiable with respect to Y, then we have the following extension of Itô's formula, generalizing (3.18) and (3.40):

$$\begin{aligned} f(t, Y_t, J_t) - f(0, Y_0, J_0) = & \int_0^t f_s \cdot ds + \int_0^t f_Y \cdot dY_s + \frac{1}{2} \int_0^t f_{YY} \cdot d\langle Y \rangle_s \\ & + \sum_q [f(q, Y_q, J_q) - f(q, Y_q, J_{q-})]. \end{aligned}$$

Here the derivatives are evaluated at J_{s-}; for example, $f_Y = f_Y(s, Y_s, J_{s-})$. Of course the continuity of $\{Y_t\}_{t \geq 0}$ makes $Y_{t-} = Y_t$ a.s. This has the differential form

$$df = f_t \cdot dt + f_Y \cdot dY_t + \frac{f_{YY}}{2} \cdot d\langle Y \rangle_t + f(t, J_t, Y_t) - f(t, J_{t-}, Y_t) \quad (3.43)$$

or equivalently

$$df = \left(f_t + a_t f_Y + \frac{b_t^2 f_{YY}}{2} \right) \cdot dt + b_t f_Y \cdot dW_t + f(t, J_t, Y_t) - f(t, J_{t-}, Y_t).$$

3.5.2 *More General Processes*

We now consider generalizations of J processes that are subject to discontinuous movements but are not necessarily pure jump.

Mixed Itô and J Processes

The discontinuous models most widely applied in finance have been simple combinations of Itô processes and J processes built up from the compound Poisson. These allow for continuous, stochastic movement of unbounded variation on finite intervals, punctuated by randomly timed breaks of random size. With Y_t as in (3.42), K_t as in (3.39), and $X_0 = Y_0 + K_0 \in \mathcal{F}_0$, define the new process $\{X_t\}_{t\geq 0}$ with

$$\begin{aligned} X_t &= X_0 + \int_{(0,t]} (dY_s + dK_s) \qquad (3.44) \\ &= X_0 + \int_0^t a_s \cdot ds + \int_0^t b_s \cdot dW_s + \int_0^t c_{s-} U \cdot dN_s, \end{aligned}$$

where $\{W_t\}_{t\geq 0}$ and $\{N_t\}_{t\geq 0}$ are independent of each other and of the $\{U_k\}$. $\{X_t\}_{t\geq 0}$ has continuous paths governed by the Itô part until the independent Poisson process triggers a jump of size $c_{t-}U$. Therefore, $X_t - X_{t-} = c_{t-}U(N_t - N_{t-})$. For a differentiable function $f(t, X_t)$ the Itô formula for the mixed process becomes

$$\begin{aligned} df &= f_t \cdot dt + f_X \cdot dY_t + \frac{f_{XX}}{2} \cdot d\langle Y \rangle_t + [f(t, X_t) - f(t, X_{t-})] \qquad (3.45) \\ &= (f_t + a_t f_X + b_t^2 f_{XX}/2) \cdot dt + b_t f_X \cdot dW_t + [f(t, X_t) - f(t, X_{t-})], \end{aligned}$$

where the derivatives are evaluated at (t, X_{t-}).

We also require the following more general version of Itô's formula for functions of right-continuous semimartingales:[13]

$$df = f_t \cdot dt + f_X \cdot dX_t + \frac{f_{XX}}{2} \cdot d\langle X \rangle_t^c + [f(t, X_t) - f(t, X_{t-}) - f_X \Delta X_t]. \qquad (3.46)$$

Here, $\langle X \rangle_t^c$ is the quadratic variation of the continuous part of X_t; the derivatives are evaluated at (t, X_{t-}); and $\Delta X_t \equiv X_t - X_{t-}$. When $X_t = Y_t + K_t$, where $\{Y_t\}$ is Itô and $\{K_t\}$ is pure jump, then $\Delta X_t = \Delta K_t$, $f_X \cdot dX_t = f_X \cdot dY_t + f_X \Delta K_t$, and $\langle X \rangle_t^c = \langle Y \rangle_t$, which agrees with (3.45).

Example 51 *Constructing a process that will be important in the applications, take* $a_t = (\alpha + \beta^2/2)X_{t-}$, $b_t = \beta X_{t-}$, *and* $c_t = X_t$ *and assume that* $\mathbb{P}(1+U > 0) = 1$. *This results in a geometric Brownian motion punctuated by Poisson-directed, multiplicative, positive shocks* $1 + U$. *With* $R_t \equiv \ln X_t - \ln X_0$ *and* $R_0 = 0$ *formula* (3.45) *gives*

$$dR_t = \alpha \cdot dt + \beta \cdot dW_t + \ln(1 + U) \cdot dN_t \qquad (3.47)$$

[13] For the proof, consult Protter (1990, pp. 71–74).

or

$$R_t = \alpha t + \beta W_t + \sum_{k=0}^{N_t} \ln(1 + U_k) \tag{3.48}$$

along a sample path with N_t jumps. $\{R_t\}$ is therefore the sum of a Brownian motion and a compound-Poisson process. Letting $\Psi_U(\zeta) = E(1+U)^{i\zeta}$ be the characteristic function of $\ln(1+U)$, the natural log of the c.f. of R_t is (for later reference)

$$\ln \Psi_R(\zeta) = t\{i\zeta\alpha - \zeta^2\beta^2/2 + \theta[\Psi_U(\zeta) - 1]\} \tag{3.49}$$

when the Poisson process has intensity θ.

Lévy Processes

A Lévy process is an adapted, right-continuous stochastic process $\{X_t\}_{t\geq 0}$ with the following features:[14]

1. $X_0 = 0$
2. $X_t - X_s$ independent of $X_r - X_q$ for $q < r \leq s < t$
3. $\mathbb{P}(X_{t+u} - X_t \in B)$ independent of t for each $t \geq 0, u > 0$ and $B \in \mathcal{B}$.

Lévy processes are thus right-continuous processes with stationary, independent increments. It is clear that Brownian motions, Poisson processes, and compound-Poisson processes all belong to this class, as does the process $\{R_t\}$ from (3.48); however, Itô processes, $\{Y_t = Y_0 + \int_0^t a_s \cdot ds + \int_0^t b_s \cdot dW_s\}$, are not, in general, Lévy.

Together, the second and third properties impose a surprising degree of structure on Lévy processes. Letting $\Psi_t(\zeta) = Ee^{i\zeta X_t}$ denote the c.f., stationarity and independence of increments implies that $\Psi_t(\zeta) = \Psi_{t/n}(\zeta)^n$ for each positive integer n and each $t \geq 0$. Thus, for each t the distribution of X_t is infinitely divisible. As such, as was shown in section 2.2.7, the c.f. cannot equal zero, so we can take logs and write

$$\ln \Psi_{t+1/n}(\zeta) - \ln \Psi_t(\zeta) = \ln \Psi_{1/n}(\zeta) = n^{-1} \ln \Psi_1(\zeta)$$

for each $n > 0$. Multiplying by n and letting $n \to \infty$ shows that

$$d \ln \Psi_t(\zeta)/dt = \ln \Psi_1(\zeta)$$

[14]Strictly, it is possible to show that there is a right-continuous version of a process with these properties—e.g., Protter (1990, chapter I, theorem 30). We always deal with such a version.

and, therefore, (by property 1) that

$$\Psi_t(\zeta) = e^{t\Theta(\zeta)} \tag{3.50}$$

for some continuous, complex-valued function $\Theta(\zeta) = \ln \Psi_1(\zeta)$. For standard Brownian motions, Poisson processes, and compound-Poisson processes we have, respectively, $\Theta(\zeta) = -\zeta^2/2$, $\Theta(\zeta) = \theta(e^{i\zeta} - 1)$, and $\Theta(\zeta) = \theta[\Psi_U(\zeta) - 1]$, while

$$\Theta(\zeta) = i\zeta\alpha - \zeta^2\beta^2/2 + \theta[\Psi_U(\zeta) - 1] \tag{3.51}$$

for the mixed process $\{R_t\}$ in (3.48).

Observe that any infinitely divisible distribution with c.f. $\Psi_1(\zeta)$ generates a Lévy process as the collection of random variables $\{X_t\}_{t\geq 0}$, where X_t has c.f. $\Psi_t(\zeta) = \exp[t \ln \Psi_1(\zeta)] = \Psi_1(\zeta)^t$. Assuming that $E|X_t| < \infty$, we can appeal to the Lévy-Khintchine theorem (section 2.2.7) for a canonical representation of $\Psi_t(\zeta)$ as (3.50) with

$$\Theta(\zeta) = i\zeta\mu + \int_{\Re} \frac{e^{i\zeta x} - 1 - i\zeta x}{x^2} M(dx), \tag{3.52}$$

where $\mu = EX_1$ and M is a σ-finite measure on $(\Re, \mathcal{B})$. The first two examples below correspond to those in section 2.2.7.

Example 52 *With* $M(\{0\}) = \sigma^2 \geq 0$ *and* $M(\Re\backslash\{0\}) = 0$ *we have* $\Theta(\zeta) = i\zeta\mu - \zeta^2\sigma^2/2$. *If* $\sigma^2 > 0$ *this corresponds to a Brownian motion with mean rate of drift* μ *and variance rate* σ^2*; otherwise, it corresponds to a deterministic process.*

Example 53 *With* $\mu = M(\{1\}) = \theta > 0$ *and* $M(\Re\backslash\{1\}) = 0$ *we have* $\Theta(\zeta) = \theta(e^{i\zeta} - 1)$*, the Poisson process.*

Example 54 *Let* F *be the c.d.f. of a random variable* U *with c.f.* $\Psi_U(\zeta)$ *and mean* $EU = \upsilon$*, and set* $\mu = \theta\upsilon$. *Taking* $M(\{0\}) = 0$ *and setting* $M(dx) = \theta x^2 \cdot dF(x)$ *give* $\Theta(\zeta) = \theta[\Psi_U(\zeta) - 1]$*, which corresponds to compound-Poisson process* (3.38). *In this case* $M(\Re\backslash\{0\}) = \theta \int x^2 \cdot dF(x)$*, which is finite if and only if* VU *exists.*

Example 55 *Let* F *be as in the previous example but take* $M(\{0\}) = \beta^2 > 0$*, so that* $M(\Re\backslash\{0\}) = \theta \int x^2 \cdot dF(x)$. *Then putting* $\mu = \alpha - \theta\upsilon$ *gives* (3.51).

Example 56 *Taking $M(dx) = xe^{-x} \cdot dx$ for $x > 0$, $M((-\infty, 0]) = 0$, and $\mu = 1$ gives*

$$\begin{aligned}\Theta(\zeta) &= i\zeta + \int_0^\infty \frac{e^{i\zeta x} - 1}{x} e^{-x} \cdot dx - i\zeta \int_0^\infty e^{-x} \cdot dx \\ &= \int_0^\infty \frac{e^{i\zeta x} - 1}{x} e^{-x} \cdot dx.\end{aligned}$$

Differentiating with respect to ζ and recognizing the integral in the resulting expression as the c.f. of the exponential distribution give

$$\Theta'(\zeta) = i \int_0^\infty e^{i\zeta x} e^{-x} \cdot dx = i(1 - i\zeta)^{-1},$$

whence $\Theta(\zeta) = \ln(1 - i\zeta)^{-1}$ and $\Psi_t(\zeta) = (1 - i\zeta)^{-t}$. $\{X_t\}_{t \geq 0}$ is now a gamma process, X_t having the density

$$f_t(x) = \Gamma(t)^{-1} x^{t-1} e^{-x}$$

for $t > 0$ and $x > 0$.

Notice that with $M(\Re \backslash \{0\}) = 0$, as in the first example, we get a continuous process, whereas $M(\Re \backslash \{0\}) > 0$ produces either a pure-jump process or a process with both continuous motion and jumps, according as $M(\{0\}) = 0$ or $M(\{0\}) > 0$. We will see that the gamma process in the last example is indeed pure jump, although not a member of the J class.

The applications we study will employ a variant of (3.52) involving a different measure. For a set of real numbers $B \in \mathcal{B}$ that is bounded away from the origin (meaning that $B \cap [-\varepsilon, \varepsilon] = \emptyset$ for some $\varepsilon > 0$) define $\Lambda(B) = \int_B x^{-2} M(dx)$. Λ is then a σ-finite measure on the Borel sets of $\Re \backslash \{0\}$—the "Lévy measure" of the Lévy process $\{X_t\}_{t \geq 0}$. Setting $M(\{0\}) = \sigma^2 \geq 0$ gives the canonical representation

$$\Theta(\zeta) = i\zeta\mu - \zeta^2\sigma^2/2 + \int_{\Re \backslash \{0\}} (e^{i\zeta x} - 1 - i\zeta x)\Lambda(dx). \tag{3.53}$$

Expression (3.53) shows that Lévy processes can be decomposed into three pure types and combinations thereof: (i) a deterministic trend if $\sigma = 0$ and $\Lambda = 0$; (ii) a trendless Brownian motion if $\mu = 0$ and $\Lambda = 0$; and (iii) a pure-jump process if $\mu = \sigma = 0$. If $\Lambda(\Re \backslash \{0\}) = 0$, as in example 52, then $\Theta(\zeta) = i\zeta\mu - \zeta^2\sigma^2/2$, so that $\{X_t\}_{t \geq 0}$ is either a deterministic trend (if $\sigma^2 = 0$) or a Brownian motion. In either case the sample paths are a.s. continuous. The process will be discontinuous whenever $\Lambda(\Re \backslash \{0\}) > 0$—being pure jump if $\sigma^2 = 0$ and $\mu = \int_{\Re \backslash \{0\}} x\Lambda(dx)$ and otherwise the sum of a Brownian motion and an independent pure-jump process. The nature of the jump part depends on whether $\Lambda(\Re\backslash\{0\})$ is finite. If $\Lambda(\Re\backslash\{0\}) = \theta$, where $0 < \theta < \infty$, then $\theta^{-1}\Lambda$ is a probability measure on $\Re\backslash\{0\}$ and $\{X_t\}$ is then a J process—either Poisson or compound-Poisson according as the entire mass θ is concentrated at a single point (as in example 53) or not (as in example 54). The remaining case, that $\Lambda(\Re\backslash\{0\}) = \infty$, is illustrated in example 56, where the integral

$$\int_{\Re\backslash\{0\}} \Lambda(dx) = \lim_{\varepsilon\downarrow 0} \int_{\varepsilon}^{\infty} x^{-1}e^{-x} \cdot dx \tag{3.54}$$

has no finite value. Since, however, Λ is σ-finite, it is possible to partition $\Re\backslash\{0\}$ into countably many sets of finite measure; for example, as $\cup_{j=-\infty}^{\infty} B_j$, where $B_j = [2^{-j-1}, 2^{-j}) \cup (-2^{-j}, -2^{-j-1}]$. Letting $\Lambda(B_j) = \theta_j$ and $dF_j(x) = \theta_j^{-1}\Lambda(dx)\mathbf{1}_{B_j}(x)$ then gives

$$\begin{aligned}\int_{\Re\backslash\{0\}} (e^{i\zeta x} - 1 - i\zeta x)\Lambda(dx) &= \sum_{j=-\infty}^{\infty} \int_{B_j} \theta_j(e^{i\zeta x} - 1 - i\zeta x) \cdot dF_j(x) \\ &= \sum_{j=-\infty}^{\infty} \{\theta_j[\Psi_j(\zeta) - 1] - i\zeta v_j\}.\end{aligned}$$

This shows that a pure-jump Lévy process with $\Lambda(\Re\backslash\{0\}) = \infty$ can be represented as an infinite sum of independent compound-Poisson processes. Since the original measure M is finite whenever X_t has finite variance, we see that $\int_B \Lambda(dx) = \int_B x^{-2}M(dx)$ is always finite whenever B is bounded away from the origin.

To gain an understanding of just what it is that Λ measures, let us begin with example 53, the Poisson process, where $\Lambda(\Re\backslash\{0\}) = \Lambda(\{1\}) = M(\{1\}) = \theta$. Here the measure is concentrated at the value (unity) that is the size of jumps in the Poisson process and $\Lambda(\{1\})$ equals the expected number of jumps per unit time. Generalizing, suppose Λ is concentrated on a countable set $\{x_j\}_{j=1}^{\infty}$, with $\Lambda(\{x_j\}) = \theta_j$. Then

$$\Theta(\zeta) = \sum_{j=1}^{\infty} \theta_j(e^{i\zeta x_j} - 1 - i\zeta x_j),$$

so that $\Psi_t(\zeta) = e^{t\Theta(\zeta)}$ is the c.f. of the sum of independent scaled and centered Poisson processes with jump sizes $\{x_j\}$ and expected frequencies θ_j. If the $\{x_j\}$ are bounded away from the origin, then $\sum_{j=1}^{\infty} \theta_j < \infty$, so that the expected number of jumps per unit time is finite. In general, we interpret $\Lambda(B)$ as measuring the expected frequency of jumps of size in B, whenever B is a set of real numbers whose closure excludes the origin. The expected number of jumps during any finite interval of time is infinite when

$$\Lambda(\Re\backslash\{0\}) = \lim_{\varepsilon\uparrow 0}\Lambda((-\infty,-\varepsilon]) + \lim_{\varepsilon\downarrow 0}\Lambda([\varepsilon,+\infty)) = \infty,$$

as is the case for the gamma process.[15]

Subordinated Processes

Let $\{X_t\}_{t\geq 0}$ be a Lévy process with c.f. $\Psi_{X_t}(\zeta) = e^{t\Theta(\zeta)}$, and let $\{\mathbb{T}_t\}_{t\geq 0}$ be an a.s. nondecreasing process, independent of $\{X_t\}$, with $\mathbb{T}_0 = 0$. Let us think of $\mathbb{T}_t$ as representing the "operational" time that has elapsed during the interval $[0,t]$ of calendar time. The composite process $\{X_{\mathbb{T}_t}\}_{t\geq 0}$, representing the evolution of X in operational time, is called a "subordinated process", and $\{\mathbb{T}_t\}_{t\geq 0}$ is referred to as a "directing process" or "subordinator". Since time itself is infinitely divisible, consistency requires that the distribution of $\mathbb{T}_1$ be infinitely divisible as well, and therefore that $\{\mathbb{T}_t\}$ be a Lévy process. Since the sample paths are nondecreasing, there can be no Brownian component, and the associated Lévy measure $\Lambda_{\mathbb{T}}$ must be concentrated on $\Re^+$. Letting $e^{t\Upsilon(\zeta)}$ be the c.f. of $\mathbb{T}_t$ and μ the mean of $\mathbb{T}_1$, the canonical representation is[16]

$$\Upsilon(\zeta) = i\zeta\mu + \int_{\Re^+}(e^{i\zeta x} - 1 - i\zeta x)\Lambda_{\mathbb{T}}(dx).$$

Examples of subordinators are the Poisson and gamma processes. The c.f. of the subordinated process is then

$$\begin{aligned}\Psi_{X_{\mathbb{T}}}(\zeta) &= E\{E[e^{i\zeta X_{\mathbb{T}_t}} \mid \mathbb{T}_t]\}\\ &= Ee^{\mathbb{T}_t\Theta(\zeta)}\\ &= \exp\{t\Upsilon[-i\Theta(\zeta)]\},\end{aligned}$$

which shows that $\{X_{\mathbb{T}_t}\}_{t\geq 0}$ is again a Lévy process. Thus, the family of Lévy processes is closed under subordination.

Example 57 *Let $\{W_t\}_{t\geq 0}$ and $\{N_t\}_{t\geq 0}$ be, respectively, a standard Brownian motion and a Poisson process with intensity θ, and take $\mathbb{T}_t = t + N_t$.*

---

[15] Protter (1990) gives further details of Lévy measure and Lévy processes generally, and Bertoin (1996) provides an extensive technical treatment.

[16] For the proof see Karatzas and Shreve (1991, pp. 405–408).

Then $Ee^{i\zeta\mathbb{T}_t} = \exp[i\zeta t + \theta t(e^{i\zeta} - 1)]$*, and composition* $W_{\mathbb{T}_t}$ *has c.f. corresponding to the sum of a Brownian motion and an independent compound Poisson:*

$$\Psi_{W_{\mathbb{T}}}(\zeta) = \exp\{t[-\zeta^2/2 + \theta(e^{-\zeta^2/2} - 1)]\}. \qquad (3.55)$$

Thus $\{W_{\mathbb{T}_t}\}$ *has the same finite-dimensional distributions as the process* $\{W_t + \sum_{j=0}^{N_t} Z_j\}$*, where* $\{N_t\}_{t\geq 0}$ *is a Poisson process,* $Z_0 = 0$*, and* $\{Z_j\}_{j=1}^{\infty}$ *are i.i.d. as standard normal.*

Example 58 *Let us try to replicate the mixed process (3.48) via a time change of Brownian motion when* $\{\ln(1 + U_k) \equiv \sigma Z_k\}_{k=1}^{\infty}$ *are i.i.d. as* $N(0, \sigma^2)$ *and* $\beta \neq 0$*. Representing the value of the mixed process at time* t *as*

$$R_t = \alpha t + \beta W_t + \sigma \sum_{k=0}^{N_t} Z_k,$$

the log of the c.f. is

$$\ln \Psi_R(\zeta) = t\{i\zeta\alpha - \zeta^2\beta^2/2 + \theta[e^{-\zeta^2\sigma^2/2} - 1]\}.$$

Setting $\mathbb{T}_t = t + N_t\sigma^2/\beta^2$ *and* $Y_t = \alpha t + \beta W_{\mathbb{T}_t}$*, calculation of the c.f. of* Y_t *shows that* $\{R_t\}$ *and* $\{Y_t\}$ *do have the same finite-dimensional distributions.*

PART II

PRICING THEORY

4

Dynamics-Free Pricing

The work of pricing derivatives begins in this chapter. Here, using the simplest of mathematical tools, we look at what can be said about the valuation of forward contracts, futures contracts, and options without explicit models for the dynamics of the prices of the underlying assets. This is done by relying just on static replication for the arbitrage arguments. Recall from chapter 1 that static replication involves duplicating a derivative's payoffs with a portfolio that does not have to be rebalanced. We also consider here what the absence of arbitrage opportunities implies about the prices of riskless bonds. In both cases the only assumption made about the dynamics is that assets' prices and interest rates are bounded below by zero. However, the arguments do rely on the **P**erfect-**M**arket conditions that were referred to generally in chapter 1 and which we now spell out in detail, as follows.

PM1 It is costless to trade and to hold assets. Thus, there are no commissions to brokers, no transfer fees, no bid-asked spreads to compensate market makers, no margin requirements, no property taxes, and no taxes on capital gains, dividends, or interest.

PM2 Trades can be made instantaneously and at any time. Besides the obvious meaning, this implies that transactions in different markets—for example, those for options and stocks—can be made simultaneously.

PM3 Primary and derivative assets can be sold short without restriction. Thus, there are no special tick rules governing when a short sale can be consummated, and proceeds of sales are not escrowed but are available for other transactions. This also implies that riskless bonds can be issued as well as bought, so that individuals can, in effect, borrow at riskless rates.

PM4 Assets can be held in any quantity whatsoever. One can therefore buy and sell fractional shares.

We also make the following additional assumption regarding what is known about the underlying asset:

CCK Primary assets' explicit **C**osts of **C**arry over the life of a derivative are **K**nown in advance.

While PM1-PM4 are self-explanatory, condition CCK requires special comment. Actually holding an asset for some period of time, as opposed to taking steps to secure ownership at a later date, can confer both positive costs and benefits (negative costs). Contributing on the positive side are the direct expenses to store and insure the asset, plus any losses from depreciation or deterioration. On the negative side are any receipts to which the owner of the asset is entitled. For financial assets these include dividend and interest payments and perhaps intangibles such as voting rights. For commodities like wheat or petroleum that can be used either directly or indirectly in consumption there is a convenience yield, discussed at length below. This is of much less significance for sterile commodities like gold that are in nearly fixed supply and are held primarily for investment. Whatever the source of the costs and benefits, condition CCK requires that their values be known in advance over the lifetime of a derivative asset—either in actual currency units or as proportions of the evolving market value of the underlying asset or commodity. Notice that the *explicit* cost of carry excludes the implicit opportunity cost associated with tying up investment resources.

While there is no real market in which these five conditions strictly hold, the conditions are nevertheless useful idealizations. They are useful from a methodological standpoint because they permit simple arguments to generate unambiguous predictions about prices of derivatives. They are useful in practice because observation confirms that violations of the implied pricing relations in some real markets are neither persistent nor extreme. One suspects that this is true largely because in those markets the conditions are more restrictive than needed for the results (although not for the arguments). For example, the tick rules and escrow requirements that do in fact inhibit short sales of common stocks in some markets do not restrict sales by those who already own these assets. Likewise, while some of us may pay high commissions that would make otherwise profitable trades unprofitable, it suffices that there are those with ample resources who can trade very cheaply. On the other hand, we will see that there are prominent situations in which the predictions based on these conditions—and CCK in particular—are not well borne out in practice and in which naïve arbitrage arguments alone lead to less useful results.

Since the ability to borrow and lend risklessly is fundamental to arbitrage arguments, we begin with a discussion of interest rates and their relation to spot and forward prices of default-free bonds. Section 4.2 then takes up forward and futures contracts. Because values of forward positions depend linearly on prices of traded primary assets, they can be replicated with static, buy-and-hold portfolios of the underlying and bonds. Since no subsequent trading is involved, valuing forward positions does not require a model for the underlying price dynamics. It is possible also to value futures positions and determine arbitrage-free futures prices without such models, although this does require restrictions on the evolution of bond prices. The final section of the chapter deals with options. Because their values at expiration are nonlinear functions of the underlying prices, arguments based on static replication suffice merely to bound their pre-expiration values or restrict the relative prices of such claims. These bounds and restrictions will, however, guide us in later chapters as we adopt specific models for the underlying dynamics and apply dynamic replication to develop sharp estimates of arbitrage-free prices.

4.1 Bond Prices and Interest Rates

Payoffs of derivative assets are typically combinations of sure claims or obligations and contingent claims or obligations. For example, we saw in chapter 1 that the payoff from contracting for forward delivery of a commodity at future date T is $S_T - \mathsf{f}$. In such a deal one simply trades a deferred obligation to pay the sure amount f, the forward price, for a deferred claim on a commodity worth an uncertain amount S_T. To replicate and price such derivatives requires replicating and pricing the sure claims as well as the contingent ones. Riskless discount bonds are just such sure claims, delivering a single guaranteed receipt of the principal amount at expiration. In the United States, U.S. Treasury bills and Treasury strips come close to this idealized concept. For simplicity we apply the term *bonds* to all such obligations, regardless of their maturity. Thus, a 3-month Treasury bill will be referred to as a "bond" that matures in 0.25 years. Only in the next section and in chapter 10 will we have occasion to consider coupon bonds that make periodic payments to the holder before maturity. The terms *bond, riskless bond*, and *unit bond* (see below) without further qualification always refer to the discount variety.

Interest rates are merely secondary constructs that describe the temporal behavior of bond prices. We will see how average and instantaneous spot rates and forward rates can be deduced, in principle, from a collection of bonds having a continuous range of maturities. Our convention throughout is to measure time to maturity in years and to express interest rates *per annum*. The riskless bonds we consider are riskless only in the sense that the nominal value of their contractual payoffs is certain; that is, they are default-free. Purchasing-power risk arising from uncertainty in goods' prices may still be present. We also deal only with nominal interest rates, not real interest rates adjusted for anticipated changes in the purchasing power of the currency.

4.1.1 *Spot Bond Prices and Rates*

Throughout, $B(t,T)$ represents the price at t of an idealized discount bond that matures and pays the holder one unit of currency for sure at $T \geq t$; thus, $B(t,t) = B(T,T) = 1$. We sometimes refer to these as "unit" bonds to emphasize that they pay a single currency unit at maturity. Of course, no one really issues bonds with unitary "principal", or "face", value; but we can simply think of $B(t,T)$ as the price per unit principal. As the price of a unit bond, $B(t,T)$ is the appropriate discount factor for valuing at t any sure cash amount that is to be received at T. For example, a sure claim on a cash amount c at T would be worth $B(t,T)c$ at time t.[1] The time- t value of a default-free coupon bond that promises to pay the bearer c currency units at each of dates $T_1,\ldots,T_n$ and deliver also a final one-unit principal payment at T_n would thus be

$$B(t,c;T_1,\ldots,T_n) = c\sum_{j=1}^{n} B(t,T_j) + B(t,T_n).$$

Table 4.1 illustrates the term structure of prices of discount bonds through a sample of bid and asked quotations for U.S. Treasury strips at one-year maturity intervals, all in the month of August.[2] Treasury strips

[1]This familiar fact is itself based on an arbitrage argument. Since one could secure the same receipt c by buying c of the unit, T-maturing bonds, a disparity between the price of the claim and the cost of the bonds could be arbitraged by buying the cheaper of the two and selling the other.

[2]Source: *Wall Street Journal*, 8 June 2006.

Table 4.1. Price quotations for U.S. Treasury strips.

Maturity	Bid	Asked
2006	99:03	99:04
2007	94:09	94:10
2008	89:26	89:27
2009	85:16	85:17
2010	81:24	81:25
2011	77:19	77:19
2012	73:29	73:29
2013	70:07	70:08
2014	66:09	66:09
2015	62:29	62:30
2016	59:16	59:17

are discounted (non-coupon paying) claims backed by portfolios of coupon-paying Treasury notes and long-term bonds. Prices are quoted in percentage points of maturity value, with digits after the colon representing 32ds of one percentage point.

Average and Instantaneous Rates

We know from first principles that an initial sum S_0 invested at a continuously compounded average rate of return r per annum would be worth $S_\tau = S_0 e^{r\tau}$ after τ years. Equivalently, S_0, S_τ, and τ together determine the average continuously compounded return, as $r = \tau^{-1} \ln(S_\tau / S_0)$. Thus, $r\tau$ equals the logarithm of the "total return", which is the terminal value of a unit initial investment. Applying these relations to default-free bonds, a unit bond purchased at $t < T$ for an amount $B(t,T)$ and held the $T - t$ years to maturity gives total return $B(T,T)/B(t,T) = B(t,T)^{-1}$ and average continuously compounded *rate* of return

$$r(t,T) = (T-t)^{-1}[\ln B(T,T) - \ln B(t,T)] = -(T-t)^{-1} \ln B(t,T).$$

Since buying a bond at its current or "spot" price is the same as making a loan of $B(t,T)$ for $T - t$ years, $r(t,T)$ represents the average spot rate for lending from t to T. It is also known as the bond's (average) "yield to maturity". Current prices of bonds of various maturities being known, the corresponding yields or average spot rates for those maturities can be readily calculated. The "yield curve" represents a plot of these yields against time to maturity, as of a point in time. Of course, a bond's price can be expressed in terms of its yield as

$$B(t,T) = e^{-r(t,T)(T-t)}. \tag{4.1}$$

Now consider a bond that is just on the verge of maturing. Letting $t-\Delta t$ be the current time and t the maturity date, the average continuously compounded spot rate is

$$r(t-\Delta t, t) = (\Delta t)^{-1}[\ln B(t,t) - \ln B(t-\Delta t, t)].$$

Assuming now that $B(t,T)$ is differentiable with respect to both arguments, letting $\Delta t \downarrow 0$ in the above gives the derivative of $\ln B$ with respect to the current date. This limit is the "instantaneous spot rate" at t:

$$r_t \equiv \lim_{\Delta t \downarrow 0} r(t-\Delta t, t) = \left.\frac{\partial \ln B(t,T)}{\partial t}\right|_{T=t}. \tag{4.2}$$

For brevity we will often refer to the instantaneous spot rate as the "short rate". Conversely, taking the derivative with respect to the maturity date gives

$$\lim_{\Delta t \downarrow 0} (\Delta t)^{-1}[\ln B(t, t+\Delta t) - \ln B(t,t)] \equiv \left.\frac{\partial \ln B(t,T)}{\partial T}\right|_{T=t}.$$

Since

$$\left.\frac{\partial \ln B(t,T)}{\partial t}\right|_{T=t} \cdot dt + \left.\frac{\partial \ln B(t,T)}{\partial T}\right|_{T=t} \cdot dT = d\ln B(t,t) = d\ln(1) = 0,$$

the short rate is also given by

$$r_t = -\left.\frac{\partial \ln B(t,T)}{\partial T}\right|_{T=t}. \tag{4.3}$$

Since a unit of currency at a future date T could always be secured simply by acquiring one now and holding it, a unit bond priced above unity would afford an opportunity for arbitrage unless the explicit cost of carry for cash money were positive. Given the advanced monetary and financial institutions of modern developed nations, we are justified in regarding the explicit cost of carrying cash as zero. Under this condition arbitrage then assures that $B(t,T) \le 1$ for each $t \le T$ or, equivalently, that the average spot rate, $r(t,T)$, and the short rate, r_t, are always nonnegative.[3] Since

[3]Short-term U.S. Treasury bills did sell at prices slightly above maturity value during the late 1930s, perhaps because of weak confidence in depression-era banks and the low coverage limits in the early years of federal deposit insurance. Other factors that could contribute to positive cost of carry are the costs of safekeeping currency and property taxes levied on deposits at financial institutions by some governmental units.

obtaining sure cash for free is (by definition) an arbitrage, it follows that prices of bonds of any finite maturity are positive and therefore that average spot rates at all maturities are finite. Henceforth, then, we take for granted that bond prices and instantaneous spot rates satisfy the following condition:

RB Interest **R**ates are nonnegative, and bond prices are **B**ounded:

$$\begin{aligned} 0 < B(t,T) \leq 1, &\quad 0 \leq t \leq T \\ 0 \leq r_t < \infty, &\quad t \geq 0 \end{aligned} \tag{4.4}$$

The Money Fund

Some of the arguments we make in pricing derivative assets rely on the concept of an idealized savings account or money-market fund, whose value per share is often used as a numeraire. This "money fund", as we shall usually call it, always holds bonds that are just approaching maturity, continuously rolling over the proceeds from one cohort of maturing bonds into the next. Since this strategy earns the short rate on the fund's assets, the share price at t, M_t, grows at instantaneous rate r_t, so that

$$M_t = M_0 \exp\left(\int_0^t r_s \cdot ds\right), \quad t \geq 0.$$

The initial value of one share, M_0, can be specified arbitrarily.

4.1.2 *Forward Bond Prices and Rates*

Paying $B(t,T)$ for a T-maturing unit bond is a direct way of securing one currency unit at T. The currency unit can also be secured indirectly through a forward agreement. For this, one contracts forward at t to buy a T-maturing unit bond at a date $t' \in (t,T)$ at some negotiated forward price, $B_t(t',T)$. Here, subscript t indicates the time at which the forward price is quoted, and the arguments indicate the initiation and maturity dates of the loan. Upon making the forward commitment, one also buys in the spot market $B_t(t',T)$ units of t'-maturing bonds, each of which costs $B(t,t')$. This spot transaction at t thus provides the $B_t(t',T)$ units of cash that is to be swapped at t' for the T-maturing bond that will in turn provide the desired unit payoff at T. Arbitrage will equate the costs of securing the currency unit in these direct and indirect ways. Assuming that the explicit cost of carrying discount bonds is zero, it follows that the time-t forward price of a T-maturing bond at t' must be

$$B_t(t',T) = B(t,t')^{-1}B(t,T).$$

Notice that the forward price depends just on current spot prices and that condition RB assures that such a price always exists and is positive. Current spot prices of discount bonds over a span of maturities thus determine implicitly the prices of forward commitments to lend and borrow over that span of time.

Corresponding to the average spot rate, the average time-t forward rate over (t',T) is

$$\begin{aligned} r_t(t',T) &= -(T-t')^{-1}\ln B_t(t',T) \\ &= (T-t')^{-1}[\ln B(t,t') - \ln B(t,T)]. \end{aligned}$$

If $t' = t$ this is the same as the average spot rate, $r(t,T)$. Taking $T = t' + \Delta t$ and letting $\Delta t \downarrow 0$ give the time-t value of the instantaneous forward rate at time t':

$$r_t(t') \equiv \lim_{\Delta t \downarrow 0} r_t(t', t' + \Delta t) = - \left. \frac{\partial \ln B(t,T)}{\partial T} \right|_{T=t'}.$$

If $t' = t$, we have $r_t(t) = -\partial \ln B(t,T)/\partial T\mid_{T=t} = r_t$, which is just the short rate at t. The assumption that cash money has zero cost of carry implies that average and instantaneous forward rates are necessarily nonnegative, just as for spot rates.

The relations below follow trivially from the definitions of forward bond prices, average forward rates, and instantaneous forward rates:

$$B_t(t',T) = \exp\left(-\int_{t'}^{T} r_t(s)\cdot ds\right)$$

$$B_t(t',T) = e^{-r_t(t',T)(T-t')}$$

$$r_t(t',T) = (T-t')^{-1}\int_{t'}^{T} r_t(s)\cdot ds.$$

Example 59 *Based on the asked quotes in table 4.1 the implied forward price as of June 2006 for August 2010 delivery of a two-year unit bond was*

$$B_t(t+4,t+6) = \frac{B(t,6)}{B(t,4)} = \frac{(73:29)/100}{(81:25)/100} \doteq 0.9037,$$

and the corresponding average forward rate was $-\ln(0.9037)/2 \doteq 0.0506$.

4.1.3 *Uncertainty in Future Bond Prices and Rates*

Although the values of default-free bonds are known both currently and at maturity, we obviously do not know what they will sell for between now and then. That is, while both $B(t,T)$ and $B(T,T)$ are known at t, the value of $B(t',T)$ at $t' \in (t,T)$ will depend on market conditions during $(t,t']$. It follows that the average continuously compounded rate of return from holding a T-maturing bond from t to $t' < T$—namely $(t'-t)^{-1} \ln [B(t',T)/B(t,T)]$—is uncertain, and likewise the instantaneous rate of return at t',

$$\lim_{\Delta t \to 0} (\Delta t)^{-1} [\ln B(t',T) - \ln B(t' - \Delta t, T)] .$$

Likewise, current short rate r_t is known, but its value at $t' > t$ is not.

Although future spot bond prices and interest rates are indeed uncertain, this uncertainty turns out to be of secondary importance in pricing short-term derivatives on volatile underlying assets whose prices are not linked directly to fixed-income instruments; for example, common stocks, stock indexes, and currencies. Since strategies for replicating a derivative's payoffs are much simpler when there is a single source of risk, we will usually assume that future short rates are known when treating derivatives on such volatile assets. Since we refer to this assumption often in this and later chapters, we set it out as a separate condition:

RK The course of the short **R**ate (instantaneous spot rate) of interest is **K**nown over the life of the derivative asset.

In other words, for a derivative initiated at $t = 0$ and expiring at $t = T$, the assumption is that the process $\{r_t\}_{0 \le t \le T}$ is $\mathcal{F}_0$-measurable. Of course, if future spot rates are known they must equal the future forward rates to prevent spot-forward arbitrages; that is, it must be true under RK that $r_{t'} = r_t(t')$ for each $t' \ge t$. Likewise, if future spot rates are known, then so are future bond prices, so that $\{B(t,t')\}_{0 \le t \le t' \le T} \in \mathcal{F}_0$. Moreover, future bond prices must equal the current forward prices as a condition for markets' being arbitrage-free. Therefore, condition RK also implies that

$$B(t',T) = B_t(t',T) = B(t,t')^{-1} B(t,T)$$

for $0 \le t \le t' \le T$. Finally, since current bond prices can always be expressed in terms of instantaneous forward rates, as $B(t,T) = \exp(-\int_t^T r_t(s) \cdot ds)$, they can also be expressed in terms of future spot rates under RK; i.e., $B(t,T) = \exp(-\int_t^T r_s \cdot ds)$. We shall use this simplification frequently in what follows, but one must remember that it is just a simplification. Assumption RK is dispensed with in chapter 10, where we consider stochastic models for interest rates and see how to price derivatives on interest-sensitive assets.

4.2 Forwards and Futures

In the previous section an arbitrage argument was used to express the price at t of a forward bond, $B_t(t',T)$, in terms of known ($\mathcal{F}_t$-measurable) prices of bonds that mature at later dates t' and T. The conclusion that $B_t(t',T) = B(t,T)/B(t,t')$ depended on the perfect-market assumptions PM1-PM4, on the known-cost-of-carry assumption CCK, and on regularity condition RB, but it did not require assumption RK. That is, we did not need to know the future prices of bonds in order to determine the arbitrage-free *forward* prices. One is then led to ask whether forward prices for other assets than default-free bonds are also fully determined in markets with no arbitrage. We will see that under the same conditions arbitrage considerations alone do indeed determine the precise values of forward prices (and of forward contracts as well) without specific models for the dynamics of either underlying prices or discount bonds. The expression for the forward bond price thus turns out to be a special case of a more general result. Under the additional assumption RK, we will see that futures prices for interest-insensitive assets are determined also. We will go through the arguments for forwards, then turn to futures, and finally discuss an important class of cases in which the manifest failure of condition CCK (known cost of carry) leaves us just with bounds on values and prices rather than exact relations.

Here are the conventions and notation to be used. The spot price of the underlying asset or commodity at $t \geq 0$ is S_t. This is the price that would be paid for immediate delivery to a specified location. The party to a forward or futures contract who is obligated to take delivery will be said to have a "long" position—for example, to be "long the futures"—while the party who must make delivery is said to be "short". We will deal with forward contracts initiated at $t = 0$ and expiring at T. Recalling the notation introduced in chapter 1, $\mathfrak{F}(S_t, T-t; \mathsf{f})$ represents the value at $t \in [0,T]$ of a forward contract initiated at forward price $\mathsf{f} \equiv \mathsf{f}(0,T)$, which of course does not change over the life of the contract. "Value" is from the perspective of the party who will take delivery—i.e., the one with the long position. The futures price, which *does* change through time, is represented by $\mathsf{F}(t,T)$. When t denotes a specific time, it is the current time.

4.2.1 *Forward Prices and Values of Forward Contracts*

We use static replication to determine the value at any $t \in [0, T]$ of a forward contract initiated at some price $\mathsf{f} = \mathsf{f}(0, T)$. Since the actual forward price in any contract is such that $\mathfrak{F}(S_0, T; \mathsf{f}) = 0$, finding the value function $\mathfrak{F}(S_t, T - t; \mathsf{f})$ for arbitrary f and each $t \in [0, T]$ will enable us to solve for $\mathsf{f}(0, T)$ as well.

In a forward contract cash and the commodity change hands only at expiration (time T). At that time the value (from the viewpoint of the party who will take delivery) is $\mathfrak{F}(S_T, 0; \mathsf{f}) = S_T - \mathsf{f}$. To replicate this payoff at time t one would need to do three things: (i) buy one unit of the asset; (ii) finance the explicit cost of carrying it to T;and (iii) borrow $B(t, T)\mathsf{f}$ of cash by selling f unit bonds each requiring the repayment of one currency unit at T. The net cost to replicate is therefore $S_t + K(t, T) - B(t, T)\mathsf{f}$, where $K(t, T)$ is the explicit cost of carry. As will be shown presently, under condition CCK $K(t, T)$ can itself be replicated with a portfolio of the underlying asset and/or riskless bonds—which to use depending on whether the cost of carry is known in absolute terms or varies as a known proportion of the value of the underlying. Therefore, $S_t + K(t, T) - B(t, T)\mathsf{f}$ is also the value of a portfolio of underlying asset and bonds. Since this portfolio replicates the payoff of the forward contract, the two will have the same value if there are no opportunities for arbitrage, and so

$$\mathfrak{F}(S_t, T - t; \mathsf{f}) = S_t + K(t, T) - B(t, T)\mathsf{f} \tag{4.5}$$

for $t \in [0, T]$. If interests in such forward contracts sold for more than this, one could secure a riskless profit by taking the short side of the contract (contracting to deliver) and buying the replicating portfolio. If interests sold for less, one could take the opposite positions. We can now use (4.5) to determine the forward price itself. Setting $\mathfrak{F}(S_0, T; \mathsf{f}) = 0$ and solving for $\mathsf{f}(0, T)$ give[4]

$$\mathsf{f}(0, T) = B(0, T)^{-1} [S_0 + K(0, T)] . \tag{4.6}$$

What remains is to see how to replicate the explicit cost of carry. If the payments or receipts associated with holding the asset are of known cash amounts and at known dates, the obligations can be replicated with

[4]In the case that the underlying commodity is a sure claim on one currency unit at T', then the current price is $S_0 = B(0, T')$. There being no explicit cost of carry for such a claim, (4.6) gives $B(0, T)^{-1}B(0, T')$ as the forward price for delivery of the claim at $T \leq T'$, which coincides with the expression for the time-0 price of a forward discount bond, $B_0(T, T')$.

a portfolio of riskless bonds. For example, known payments $c_{t_1}, \ldots, c_{t_n}$ for storage and safekeeping at times $t_1, \ldots, t_n \in [t, T]$ would be financed by acquiring a portfolio of unit bonds maturing at these times and now worth

$$K(t,T) = \sum_{j=1}^{n} B(t,t_j)c_{t_j}.$$

Receipts of dividends or interest would correspond to negative values of the $\{c_{t_j}\}$. These receipts could be used to redeem $\{t_j\}_{j=1}^n$-maturing bonds sold at t for $|K(t,T)|$. If the payments/receipts were of unknown cash value but were known proportions of the future value of the asset, then the carrying cost could be financed through sales/purchases of the asset itself at the times $\{t_j\}_{j=1}^n$. For example, the funds needed to pay $c_{t_j}S_{t_j} > 0$ at t_j could be generated then by selling $c_{t_j} > 0$ units of the asset for each one held, and a receipt of $-c_{t_j}S_{t_j} > 0$ could be used to purchase $-c_{t_j} > 0$ units. Since for each unit held at t one would have $\prod_{j=1}^n (1 - c_{t_j})$ units after n such transactions, one would need to start with $\prod_{j=1}^n (1 - c_{t_j})^{-1}$ units in order to wind up with one unit at T, as is required to replicate the forward commitment. Therefore, the total cost at t to assure the availability of one unit at T would be

$$S_t + K(t,T) = S_t \prod_{j=1}^{n} (1 - c_{t_j})^{-1},$$

and the value of the forward contract would be

$$\mathfrak{F}(S_t, T - t; \mathsf{f}) = S_t \prod_{j=1}^{n} (1 - c_{t_j})^{-1} - B(t,T)\mathsf{f}. \tag{4.7}$$

Notice that the timing of the payments does not matter when they are known proportions of the price at whatever times the payments are to be made.

In some cases it is appropriate to think of such proportional amounts as being paid, received, or accruing continuously through time. For example, an investment in a default-free foreign certificate of deposit grows in value continuously at a known rate in proportion to the (unknown) future value of the currency. Extending (4.7) to allow for such continuous flows, let $t_j - t_{j-1} = (T-t)/n$ and make each c_{t_j} proportional to the time since the previous payment, as $c_{t_j} = \kappa(T-t)/n$. Then $\prod_{j=1}^n (1 - c_{t_j})^{-1} \to e^{\kappa(T-t)}$ as $n \to \infty$ and

$$\mathfrak{F}(S_t, T - t; \mathsf{f}) = S_t e^{\kappa(T-t)} - B(t,T)\mathsf{f}. \tag{4.8}$$

More generally, if instantaneous rates of payment vary through time in such a way that $\{\kappa_s\}_{0\leq s\leq T}$ is integrable and $\mathcal{F}_0$-measurable, then for each $t \in [0,T]$

$$\mathfrak{F}(S_t, T-t; \mathsf{f}) = S_t \exp\left(\int_t^T \kappa_s \cdot ds\right) - B(t,T)\mathsf{f}, \tag{4.9}$$

and initial value $\mathfrak{F}(S_0, T; \mathsf{f}) = 0$ when

$$\mathsf{f}(0,T) = B(0,T)^{-1} S_0 \exp\left(\int_0^T \kappa_s \cdot ds\right). \tag{4.10}$$

4.2.2 *Determining Futures Prices*

Turning now to futures, recall that the parties to such contracts have obligations that differ slightly from those of parties to forward contracts. One who goes long the futures at time $t = 0$ when the futures price is $\mathsf{F}(0,T)$ will actually wind up paying the time-T spot price, S_T, to take delivery; but in the meantime the net price difference, $S_T - \mathsf{F}(0,T)$, will have been posted to the buyer's account through a succession of daily marks to market.[5] The net result differs from that of a forward contract initiated at the same price, $\mathsf{f}(0,T) = \mathsf{F}(0,T)$, because the intervening additions to the futures account could be reinvested and the intervening withdrawals would have to be financed. Despite this difference in timing we will show that futures prices would nevertheless be identical to forward prices under the perfect-market and cost-of-carry conditions if, in addition, spot interest rates over the life of the contract were known in advance—condition RK. Once this is demonstrated, we can draw some qualitative conclusions as to how forward and futures prices should differ given that bond prices are, in reality, uncertain. More than that cannot be done without explicit models for $\{S_t\}_{t\geq 0}$ and $\{r_t\}_{t\geq 0}$, of which we shall see some examples in chapter 10.

Assume now that the course of the short rate during $[0,T]$ is known at $t = 0$, which implies that bond prices, $B(t,t')$, are known at $t = 0$ for all t and t' such that $0 \leq t \leq t' \leq T$. This is the first of many applications of the strategic assumption RK that was discussed in section 4.1.3. Under this

[5]Our discussion pertains to forward and futures contracts that are identical except for provisions of marking to market. For example, we ignore the fact that some futures contracts allow a range of delivery dates.

condition it turns out that dynamic adjustments to a futures position initiated at price $\mathsf{F}(0,T)$ can offset the effect of marking to market and deliver a terminal value of $S_T - \mathsf{F}(0,T)$. This implies that arbitrages between forwards and futures will equate the time-0 futures prices and forward prices, $\mathsf{F}(0,T)$ and $\mathsf{f}(0,T)$, on contracts for delivery at T.

Let $t_1 < t_2 < \cdots < t_{n-1}$ be the times at which a position in T-expiring futures is marked to market, and set $t_0 = 0$ and $t_n = T$. Begin with a long position in $B_0(t_1,T) = B(0,T)/B(0,t_1) < 1$ contracts at futures price $\mathsf{F}(0,T)$. At $t = t_1$ when the futures price is $\mathsf{F}(t_1,T)$, marking to market will generate a (positive or negative) receipt $[\mathsf{F}(t_1,T) - \mathsf{F}(0,T)]B_0(t_1,T)$. Investing this in (financing it with) T-expiring unit bonds, each costing $B(t_1,T)$, will add

$$\frac{B_0(t_1,T)}{B(t_1,T)}[\mathsf{F}(t_1,T) - \mathsf{F}(0,T)] = \mathsf{F}(t_1,T) - \mathsf{F}(0,T)$$

to the value of the account at T. The equality holds because future bond prices, being known at $t = 0$, must equal the current forward prices, so that $B(t_1,T) = B_0(t_1,T)$. Besides investing or financing the capital gain or loss at t_1, we also increase the futures position by the factor $B(t_1,t_2)^{-1}$, giving $B(t_1,T)/B(t_1,t_2) = B_{t_1}(t_2,T)$ contracts in all. Under condition PM1 this requires no cash outlay. Marking to market at t_2 posts an additional change

$$B_{t_1}(t_2,T)[\mathsf{F}(t_2,T) - \mathsf{F}(t_1,T)]$$

to the account, and investing or financing this with T-expiring bonds each worth $B(t_2,T)$ adds

$$\frac{B_{t_1}(t_2,T)}{B(t_2,T)}[\mathsf{F}(t_2,T) - \mathsf{F}(t_1,T)] = \mathsf{F}(t_2,T) - \mathsf{F}(t_1,T)$$

to the final value.

Continuing, arranging to hold at each t_{j-1} a total of $B_{t_{j-1}}(t_j,T)$ contracts and reinvesting at t_j the subsequent the one-period gain,

$$B_{t_{j-1}}(t_j,T)[\mathsf{F}(t_j,T) - \mathsf{F}(t_{j-1},T)],$$

in T-maturing bonds each worth $B(t_j,T)$ adds $\mathsf{F}(t_j,T) - \mathsf{F}(t_{j-1},T)$ to the value of the account at T. The total of the receipts generated by this strategy is

$$\sum_{j=1}^{n}[\mathsf{F}(t_j,T) - \mathsf{F}(t_{j-1},T)] = \mathsf{F}(T,T) - \mathsf{F}(0,T) = S_T - \mathsf{F}(0,T).$$

As we have argued, this implies that futures and forward prices for contracts of the same specifications must agree. The key requirement for this result is that the current forward price of a T-expiring bond acquired at the next marking to market be equal to the actual, realized price at that future time—a requirement that is met under condition RK. As futures price $\mathsf{F}(t,T)$ evolves up to T it must therefore track the current forward prices for new T-expiring contracts having the same specifications. Specifically, in the case of continuous carrying costs we have

$$\mathsf{F}(t,T) = B(t,T)^{-1}S_t \exp\left(\int_t^T \kappa_s \cdot ds\right) = S_t \exp\left[\int_t^T (r_s + \kappa_s) \cdot ds\right] \tag{4.11}$$

under condition RK.

Without explicitly modeling bond prices and $\{S_t\}_{t\geq 0}$ it is not possible to say precisely how forward and futures prices should correspond when RK fails to hold. However, when price movements in the underlying and bonds are strongly related, equilibrium arguments do at least support some educated guesses. First, if the correlation is strongly negative, so that the term structure of bond prices usually shifts in a direction opposite to that of the spot price, then one expects futures prices to be higher than forward prices, for the following reason. Positive changes in S_t lead typically to increases in $\mathsf{F}(t,T)$ and gains for those who are long the futures. If accompanied by a decline in bond prices, these gains purchase more of the cheaper T-maturing bonds and thereby generate larger cash receipts at T than in the absence of negative correlation. This makes long positions in futures more attractive than long positions in forwards when $\mathsf{F}(t,T) = \mathsf{f}(t,T)$. Therefore, the futures price needs to be higher to offset this advantage and equilibrate the two markets. Conversely, if the correlation between movements in prices of bonds and the underlying is strongly positive, then gains on long futures positions usually buy fewer bonds and secure lower terminal receipts than in the absence of such correlation. In that case one expects futures prices to be lower than forward prices. However, it is not so clear what to expect when bond prices are uncertain but only weakly linked to the underlying. We come back to this issue in chapter 10, where we show precisely how futures and forward prices are related under a particular model for the instantaneous forward rate and the underlying asset.

4.2.3 *Illustrations and Caveats*

If the short rate is positive and an asset's explicit cost of carry is nonnegative, so that $\int_t^T (r_s + \kappa_s) \cdot ds > 0$ for each $t \in [0, T)$, then it is clear from (4.11) that futures prices should increase with the remaining time until expiration. This is, in fact, what we usually observe for sterile assets like gold, which generate no income but are nevertheless held largely for investment purposes as hedges against inflation. To illustrate, table 4.2 shows that settlement prices for gold futures traded on the COMEX (a division of the New York Mercantile Exchange) as of t = 7 June 2006 do increase monotonically with time to delivery.[6]

On the other hand, price would decrease monotonically with time to delivery if the cost of carry were sufficiently negative. For example, this would be true of currency futures if foreign interest rates were higher than domestic rates. U.S. dollar prices of Mexican peso futures on 8 June 2006 (table 4.3) illustrate such a monotonic decline, as is consistent with the difference in average spot rates on Mexican and U.S. government securities. (For comparison, the last column shows the futures prices predicted from the simple cost-of-carry relation (4.11) under assumption RK, given a spot price of $S_t = .0882$ \$/peso.)

However, the situation is more complicated for commodities held importantly as inventories for direct consumption or for the production of final goods, such as industrial metals, petroleum products, and agricultural products. Although these generate no explicit receipts and are costly to store, futures prices are often poorly explained by naïve cost-of-carry arguments.

Table 4.2. Gold futures prices vs. length of contract.

Expiration, T	$F(t, T)$(\$/oz.)
Jun 06	627.40
Aug 06	632.60
Dec 06	645.20
Feb 07	651.50
Jun 07	663.80
Dec 07	682.30

[6]Sources for tables 4.2–4.4: *Wall Street Journal*; Chicago Mercantile Exchange; Banco de Mexico (CETES rates); U.S. Treasury (government securities yield curve); Chicago Board of Trade: all as of 8 June 2006. Settlement prices are those officially determined by the exchange in order to mark traders' positions to market.

Table 4.3. Dollar/peso futures prices vs. length of contract.

Expiration, T	$\mathsf{F}(t,T)$($/peso)	$r(t,T)_{MX}$	$r(t,T)_{US}$	$S_t e^{(r_{MX}-r_{US})(T-t)}$
Jun 06	0.0878			
Aug 06	0.0872	0.0701	0.0480	0.0878
Oct 06	0.0869	0.0728	0.0486	0.0874
Dec 06	0.0866	0.0750	0.0505	0.0871
Feb 07	0.0862			
Apr 07	0.0857			
Jun 07	0.0851	0.0756	0.0503	0.0859

Table 4.4. Live cattle futures prices vs. length of contract.

Expiration, T	$\mathsf{F}(t,T)$($.01/lb.)
Jun 06	79.875
Aug 06	80.625
Oct 06	84.100
Dec 06	86.125
Feb 07	88.525
Apr 07	86.275
Jun 07	81.700

For example, futures prices for commodities in seasonal supply or demand are often not monotonic in time to delivery. This is illustrated in table 4.4, which reports futures prices for live cattle on the Chicago Board of Trade as of 8 June 2006. Prices increase progressively for delivery dates from June through February then decline for deliveries during spring 2007.

Why is it that the arguments that apply to investment assets seem not to apply here? Two things are involved. First, there is a latent benefit to holding stocks of such commodities; namely, that inventories afford opportunities for immediate use in consumption or production. Often described as "convenience yields", such opportunities are more accurately regarded as "real" (as opposed to financial) options. Thus, a beef processor who sold April 2007 futures (to deliver then out of current inventory) and simultaneously contracted to take delivery in June 2007 at a lower price would forego some options to market beef during that interval. Second, the fact that existing quantities of the commodity fluctuate as stocks are continually being used up or augmented by new production makes the value of the utilization option variable and hard to determine. Thus, the value of maintaining inventories of live cattle for immediate use depends on the uncertain domestic and foreign demand for beef, on changes in export and import policies, and on cattlemen's evolving plans to deplete or build up their herds. Contrary to what condition CCK requires, there is no way to assign a value to the utilization option—and thus no way to assess the cost of carry—without a model for the dynamics of the spot price. The best we can do with static replication for such consumable commodities is to set upper bounds on futures prices by taking account of just the known components of cost of carry. Clearly, a situation in which the futures price was above the explicit and opportunity cost of carrying live cattle to T would offer a readily exploitable opportunity for arbitrage: short the futures and buy the cows.

4.2.4 *A Preview of Martingale Pricing*

Let us now develop a useful interpretation of our findings about the values of forward contracts and about forward and futures prices. Let $M_t = M_0 \exp(\int_0^t r_s \cdot ds)$ be the value at t of an initial investment of M_0 in a money-market fund earning the short rate continuously, and let $S_t^* = M_t^{-1}[S_t + K(t,T)]$ be the relative value of the underlying asset plus the explicit cost of carry during $[t,T]$. Now suppose there exists a probability measure $\hat{\mathbb{P}}$ and a filtration $\{\mathcal{F}_t\}_{0 \le t \le T}$ such that $\{S_t^*\}$ is a martingale adapted to $\{\mathcal{F}_t\}$. We refer to such $\hat{\mathbb{P}}$ as a "martingale measure". Then by the fair-game property of martingales

$$\hat{E}_t S_T^* \equiv \hat{E}(S_T^* \mid \mathcal{F}_t) = S_t^*,$$

where the carat on E denotes expectation under $\hat{\mathbb{P}}$. Assuming that $K(T,T) = 0$, this implies that

$$M_t^{-1}[S_t + K(t,T)] = M_T^{-1} \hat{E}_t S_T$$

or

$$\frac{\hat{E}_t S_T}{S_t + K(t,T)} = \frac{M_T}{M_t} = \frac{B(T,T)}{B(t,T)} = B(t,T)^{-1},$$

where the equality connecting money fund and bond prices holds under assumption RK. Under the martingale measure, therefore, the expected total return from holding the risky asset during $[t,T]$ is the same as the sure total return of a riskless bond or, equivalently, of an investment in the money fund. This same condition would hold in equilibrium if individuals were risk neutral, since no risk-neutral person would willingly hold bonds or money fund if $B(t,T)^{-1} = M_T/M_t < \hat{E}_t S_T/[S_t + K(t,T)]$, and none would willingly hold the risky asset if the inequality went the other way. Therefore, we can refer to $\hat{\mathbb{P}}$ also as a "risk-neutral" measure.

Applying (4.5), the value of a T-expiring forward contract can be expressed in terms of the risk-neutral expectation as

$$\begin{aligned}\mathfrak{F}(S_t, T-t; \mathsf{f}) &= S_t + K(t,T) - B(t,T)\mathsf{f} \\ &= B(t,T)[\hat{E}_t S_T - \mathsf{f}(0,T)] \\ &= B(t,T)\hat{E}_t \mathfrak{F}(S_T, 0; \mathsf{f}).\end{aligned}$$

Letting $\mathfrak{F}^*(S_t, T-t; \mathsf{f}) \equiv M_t^{-1}\mathfrak{F}(S_t, T-t; \mathsf{f})$ and noting that $B(t,T) = M_t/M_T$ under assumption RK, we see that the above implies

$$\mathfrak{F}^*(S_t, T-t; \mathsf{f}) = \hat{E}_t \mathfrak{F}^*(S_T, 0; \mathsf{f}).$$

Thus, the process representing the evolving normalized value of the forward position is also a martingale under $\hat{\mathbb{P}}$. Finally, the forward price that sets $\mathfrak{F}(S_t, T-t; \mathsf{f}) = 0$ at t would be $\mathsf{f}(t,T) = B(t,T)^{-1}[S_t + K(t,T)] = \hat{E}_t S_T$; and, still abstracting from uncertainty about future bond prices, the evolving futures price $\mathsf{F}(t,T)$ would be the same. Since $\mathsf{f}(T,T) = \mathsf{F}(T,T) = S_T$, this means that processes $\{\mathsf{f}(t,T)\}_{0\le t\le T}$ and $\{\mathsf{F}(t,T)\}_{0\le t\le T}$ are themselves martingales under $\hat{\mathbb{P}}$.

These are remarkable and, as we shall come to see, very significant results. One way of expressing them is that in the absence of possibilities for arbitrage the present value of a forward contract is the discounted value of its expected terminal value in a risk-neutral market; and the forward price on a new contract is the risk-neutral expectation of the future spot. Put another way, in the absence of arbitrage there is a probability measure such that processes representing normalized values of the derivative and the underlying are martingales. We shall see that such a martingale measure does exist in the absence of arbitrage possibilities. We shall also see that the measure is unique when it is possible to replicate the derivative's payoffs with a portfolio of traded assets. Under these conditions we have a simple recipe for pricing derivatives.

4.3 Options

Payoffs of options are harder to replicate than payoffs of forward contracts and futures because they are nonlinear functions of the underlying price. Replication, if possible at all, has to be done dynamically, and whether and how this can be done depends on the price dynamics. Without a model for these, arbitrage arguments can merely establish loose bounds on prices and determine qualitatively how prices depend on specific provisions of the option contract, such as the strike price and time to expiration. On the other hand, under the perfect-market and known cost-of-carry conditions (PM1-PM4, CCK) static replication arguments alone do impose a tight parity relation between prices of puts and calls with the same strike and term. Before developing these parity relations and bounds, it will be useful to get a feel for the risks associated with option positions by relating the probability distributions of option payoffs to the conditional distribution of the price of the underlying. Throughout the section the terms *option*, *put*, and *call* refer to the ordinary ("vanilla") European or American varieties.[7]

4.3.1 *Payoff Distributions for European Options*

We know from chapter 1 that the payoffs of European puts and calls struck at X and expiring at T are $P^E(S_T, 0) = (X - S_T)^+$ and $C^E(S_T, 0) = (S_T - X)^+$, respectively, where S_T is the underlying price at T. It is obvious, then, that the put will wind up in the money with the same probability that S_T is less than X and that the call will end in the money with the same probability that S_T is greater than X. More generally, it is easy to represent the distributions of $C^E(S_T, 0)$ and $P^E(S_T, 0)$ in terms of the c.d.f. of S_T. Regarding $\{S_t\}_{t\geq 0}$ as a stochastic process adapted to the filtered probability space $(\Omega, \mathcal{F}, \{\mathcal{F}_t\}, \mathbb{P})$, it is clear that for each $t \leq T$ and $p, c \in \Re$

$$\begin{aligned}
\mathbb{P}_t[P^E(S_T, 0) \leq p] &\equiv \mathbb{P}[P^E(S_T, 0) \leq p \mid \mathcal{F}_t] = \mathbb{P}_t[S_T \geq X - p] \\
\mathbb{P}_t[C^E(S_T, 0) \leq c] &\equiv \mathbb{P}[C^E(S_T, 0) \leq c \mid \mathcal{F}_t] = \mathbb{P}_t[S_T \leq X + c].
\end{aligned}$$

Therefore, letting $F_S(s) \equiv \mathbb{P}_t(S_T \leq s)$, the conditional c.d.f.s of $C^E(S_T, 0)$ and $P^E(S_T, 0)$ are

$$F_C(c) = \begin{cases} 0, & c < 0 \\ F_S(X + c), & c \geq 0 \end{cases}$$

$$F_P(p) = \begin{cases} 0, & p < 0 \\ 1 - F_S((X - p)-), & 0 \leq p < X \\ 1, & p \geq X \end{cases},$$

[7]The European put-call parity relation was first pointed out by Stoll (1969). Merton's seminal (1973) paper presented many of the arbitrage relationships derived here. Cox and Rubinstein (1985) give a highly readable account.

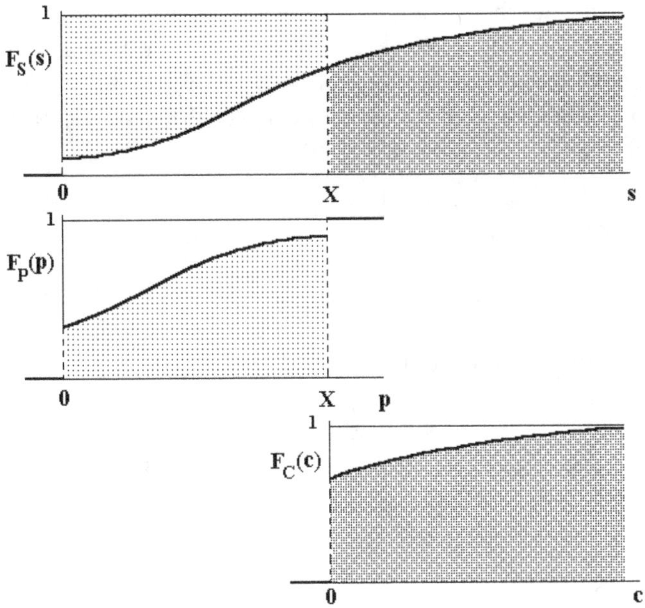

Fig. 4.1. C.d.f.s of terminal values of underlying, put, and call.

where $(X - p)-$ is the left-hand limit. C.d.f.s for European puts and calls at expiration can therefore be read off the c.d.f. of S_T, as illustrated in figure 4.1.[8]

It is also possible to infer from the graphs the mathematical expectations of the options' terminal values. Applying (2.40), $E_t P^E(S_T, 0)$ and $E_t C^E(S_T, 0)$ are represented by the areas between the unit lines and the respective c.d.f.s.

4.3.2 *Put-Call Parity*

Since being long a call gives the right to buy at X and being short a put with the same strike gives the contingent obligation to buy at X, holding the two positions together is clearly akin to having a forward commitment

[8]With one exception these payoff distributions would *not* apply to American options conditional on their being held to term, because that condition would restrict the allowable sample paths of $\{S_t\}_{0 \le t \le T}$ and alter the conditional distribution of S_T. The exceptional case is that of an American call on an underlying asset with nonnegative explicit cost of carry. As we show below, such an option would never be exercised early.

to take delivery at the price f $= X$. Indeed, for European puts and calls with the same term the correspondence is exact, and this imposes a strict relation between the prices of the options before maturity. There is also a relation of this sort for American options, although the early-exercise feature makes it a bit more complicated.

European Options

Being long a T-expiring European call and simultaneously short a put on the same underlying asset and with the same term and same strike, X, produces a payoff at T equal to

$$C^E(S_T, 0) - P^E(S_T, 0) = (S_T - X)^+ - (X - S_T)^+ = S_T - X,$$

identical to that of a forward contract to buy at X. Under the perfect-market and known-cost-of-carry conditions arbitrage will therefore align the prices of the three claims at each $t \in [0, T]$ so that

$$\begin{aligned} C^E(S_t, T-t) - P^E(S_t, T-t) &= \mathfrak{F}(S_t, T-t; X) \\ &= S_t + K(t,T) - B(t,T)X, \end{aligned} \tag{4.12}$$

where $K(t, T)$ is the explicit cost of carrying the underlying asset from t to T. Of course, the explicit cost of carry is typically zero or negative for financial assets. For example, suppose the underlying is a share of stock that generates sure "lump-sum" dividends (that is, cash payments of fixed sums) at known times during $[t, T]$, and let $\Delta(t, T)$ be the present value of the dividends at t. Thus, if the owner of the stock receives a single cash payment of Δ at $t^* \in [t, T]$, then $\Delta(t, T) = B(t, t^*)\Delta$. The cost of carry is then the negative of $\Delta(t, T)$, and

$$C^E(S_t, T-t) - P^E(S_t, T-t) = S_t - \Delta(t, T) + B(t, T)X.$$

If the underlying is a currency or stock index that is regarded as paying a continuous proportional dividend at constant rate $\delta \geq 0$, then setting $\kappa = -\delta$ in (4.8) gives[9]

$$C^E(S_t, T-t) - P^E(S_t, T-t) = S_t e^{-\delta(T-t)} - B(t, T)X. \tag{4.13}$$

[9] If the payout rate varies with time but is known in advance then the right side of (4.13) is simply $S_t \exp(-\int_t^T \delta_s \cdot ds) - B(t, T)X$.

Finally, if the underlying is a futures price or a no-dividend stock, then there is no explicit cost of carry at all, and

$$C^E(S_t, T-t) - P^E(S_t, T-t) = S_t - B(t,T)X. \tag{4.14}$$

Notice that when interest rates are positive (4.14) implies that at-the-money European calls are worth more than the corresponding puts even when the underlying generates no receipts, since

$$C^E(X, T-t) - P^E(X, T-t) = X - B(t,T)X > 0.$$

The higher value of at-the-money calls is sometimes mistakenly attributed to a bullish bias in the market, yet we see that arbitrage would bring this about even if the underlying asset's expected rate of return (under "natural" measure $\mathbb{P}$) were negative.

American Options

Because of the possibility of early exercise, the best one can do in the case of American options is to provide bounds—albeit usually tight ones—on the difference between the values of calls and puts. In the case of known lump-sum dividends with present value $\Delta(t,T)$ at t, the bounds are

$$S_t - \Delta(t,T) - X \leq C^A(S_t, T-t) - P^A(S_t, T-t) \leq S_t - B(t,T)X. \tag{4.15}$$

We shall show that the violation of either bound presents an opportunity for arbitrage, referring to the underlying as a "stock" as we go through the arguments. To establish the lower bound, consider a portfolio that is long the call, short the put, short the stock, and has $X + \Delta(t,T)$ invested in riskless securities. The amount X is placed in shares of the money fund, now worth M_t each, that grow in value at short rate r_t; while $\Delta(t,T)$ is distributed among discount bonds that mature at the stock's dividend dates. If the lower bound is violated, this portfolio can be bought for

$$C^A(S_t, T-t) - P^A(S_t, T-t) - S_t + X + \Delta(t,T) < 0.$$

Since the put may be exercised at a time not under our control, we must be prepared to unwind all the positions when and if that occurs. Therefore, in analyzing the payoffs there are two relevant states: "put not exercised" and "put exercised". Values of the portfolio in each state are laid out and explained in table 4.5.

Table 4.5. Cash flows attainable from violation of lower bound, American put-call parity.

State	Put Not Exercised by T	Put Exercised at $t^* \in [t,T]$
Call	$S_T - X$	$C^A(S_{t^*}, T - t^*)$
Put	0	$-X + S_{t^*}$
Stock	$-S_T$	$-S_{t^*}$
Money fund	XM_T/M_t	XM_{t^*}/M_t
Bonds	0	$\Delta(t^*, T)$
Net	$X(M_T/M_t - 1) \geq 0$	$C^A(S_{t^*}, T - t^*) + X(M_{t^*}/M_t - 1)$ $+\Delta(t^*, T) \geq 0$

If the put is not exercised by T it must then be out of the money, in which case the call can be exercised to cover the shorted stock, paying the strike price out of the money-fund balance. Each stock dividend is repaid to the owner as it comes due using the principal from the bonds of that maturity. What is left is the residual money fund balance, $X(M_T/M_t - 1)$, and this is positive unless the short rate was always zero during $[t,T]$. (Recall that condition RB rules out negative short rates.) If the put is exercised at some t^*, one uses the stock that has to be purchased to cover the short position, again paying the strike price out of the money account. Besides the nonnegative residual balance, $X(M_{t^*}/M_t - 1)$, one still has bonds worth $\Delta(t^*, T)$ that were dedicated to the remaining dividends, plus a call with remaining life $T - t^*$. Since the call can always be thrown away, its value is clearly nonnegative. The net value of the portfolio being at least zero whether the put is exercised or not, this strategy is indeed an arbitrage.

Violation of the upper bound of (4.15) affords such an opportunity as well. Take a portfolio short the call, long the put, long the stock, and short X units of T-maturing bonds worth $B(t,T)X$. Again, this portfolio can be had for a negative sum. Since it is short the call, the relevant states are now "call not exercised" and "call exercised". As dividends are paid, invest them in the money fund. Letting $M(t,t')$ be the resulting balance at time $t' \in [t,T]$, table 4.6 shows that the portfolio's value in each case is positive or zero.

If the underlying is a currency or other asset that generates a continuous proportional receipt at rate δ, then a similar argument shows that the parity inequality is

$$S_t e^{-\delta(T-t)} - X \leq C^A(S_t, T-t) - P^A(S_t, T-t) \leq S_t - B(t,T)X. \quad (4.16)$$

To give some idea of the width of these bounds, the difference between prices of a 3-month at-the-money call and a corresponding put on a stock

Table 4.6. Cash flows attainable from violation of upper bound, American put-call parity.

State	Call Not Exercised by T	Call Exercised at $t^* \in [t, T]$
Call	0	$-S_{t^*} + X$
Put	$X - S_T$	$P^A(S_{t^*}, T - t^*)$
Stock	S_T	S_{t^*}
Money fund	$M(t, T)$	$M(t, t^*)$
Bonds	$-X$	$-B(t^*, T)X$
Net	$M(t, T) \geq 0$	$P^A(S_{t^*}, T - t^*) + X[1 - B(t^*, T)]$ $+ M(t, t^*) \geq 0$

index paying 2% per annum would be between about −0.5% and +1.0% of the index price if the 3-month average spot rate were 4% per annum.

Since the explicit cost of carry for a financial asset paying lump-sum dividends is

$$K(t, T) = -\Delta(t, T) \tag{4.17}$$

and is

$$K(t, T) = S_t[e^{-\delta(T-t)} - 1] \tag{4.18}$$

for an asset paying continuous proportional dividends at constant rate $\delta > 0$, the two expressions (4.15) and (4.16) are both contained in

$$S_t + K(t, T) - X \leq C^A(S_t, T - t) - P^A(S_t, T - t) \leq S_t - B(t, T)X. \tag{4.19}$$

4.3.3 *Bounds on Option Prices*

While arbitrage keeps the difference between prices of comparable calls and puts within fairly narrow limits, the bounds on prices of individual options remain quite wide unless structure is imposed on the prices of the underlying and of default-free bonds. If we assume just that S_t is nonnegative, then under perfect-market conditions PM1-PM4, the known-cost-of-carry assumption (CCK), and the bounded-rate assumption (RB) arbitrage constrains prices of puts and calls as shown in table 4.7. By applying either (4.17) or (4.18), this covers the two cases of known lump-sum dividends and continuous proportional dividends. In the table K stands for $K(t, T)$, the explicit cost of carry, which is assumed to be negative or zero depending on whether or not the underlying generates cash receipts; B stands for $B(t, T)$, the current price of a T-maturing bond; and each option is struck at X.

Table 4.7. Arbitrage bounds on prices of options.

Bound	Lower	Upper
European call, $C^E(S_t, T-t)$	$\max\{0, S_t + K - BX\}$	$S_t + K$
American call, $C^A(S_t, T-t)$	$\max\{0, S_t + K - BX, S_t - X\}$	S_t
European put, $P^E(S_t, T-t)$	$\max\{0, BX - S_t - K\}$	BX
American put, $P^A(S_t, T-t)$	$\max\{0, BX - S_t - K, X - S_t\}$	X

Here are the arguments, starting with the upper bounds. How to arbitrage the condition $C^E(S_t, T-t) > S_t + K$ depends on how the dividends are paid. If they are lump-sum, buy one share of stock and borrow a total of $-K = \Delta(t, T)$, matching the maturities of the loans with the various dividend dates and repaying the loans as the dividends are received. If dividends are proportional to the price of the stock, just buy $S_t + K$ worth of stock and invest the dividends in stock as they are received. In both cases the initial net cost of the purchases would be more than offset by the proceeds from the sale of one call. If the call is exercised, deliver the stock in return for the strike price. If it is not exercised, one is left with stock worth $S_T \geq 0$, thus at least breaking even either way. If an American call sells for more than S_t, buy the stock and sell a covered call, pocketing the difference. If the call is exercised at or before T, deliver the stock at no cost. If it is not exercised, sell the stock. (If the stock does pay a dividend, the upper bound is a little higher for the American call because it could be exercised before any dividends are received.) If the European put sells for more than BX, sell it and put that amount into T-maturing bonds. If the American put sells for more than X, then sell it and put the proceeds in the money fund. If either put is exercised, the respective bond or money fund balances will finance the payment of the strike, and the stock received in exchange can be worth no less than zero.

Moving to the lower bounds, none of the options can have negative value since, as we have argued before, they can always be thrown away. If either of the calls is worth less than $S_t + K - BX$, so that $C(S_t, T-t) + BX < S_t + K$, then buy the call and BX worth of bonds and short the stock. Since we have the option of holding the American call all the way to T, we can consider just that possibility. If the dividends are lump-sum, with $K = -\Delta(t, T)$, we short one share of stock and buy bonds worth $BX + \Delta(t, T)$, distributing the maturities of those covering the dividends to match the payment dates and using principal values to repay the dividends on the shorted stock as they come due. If the dividends are proportional to the stock's price, we short stock worth $S_t + K$ (i.e., $1 + K/S_t$ shares) and finance the dividends as

Table 4.8. Cash flows attainable from violation of 2nd lower bound for price of American call.

State	$S_T \leq X$	$S_T > X$
Call	0	$S_T - X$
Stock	$-S_T$	$-S_T$
Bonds	X	X
Net	$X - S_T \geq 0$	0

they come due by further sales. For example, if proportional dividends were paid continuously at rate δ, we would short $e^{-\delta(T-t)}$ shares and continue to sell short at rate δ until time T. Either way, we wind up at T short one share and with X in cash. If $S_T \leq X$ the call is worthless and we cover the stock and come out ahead by $X - S_T$. If $S_T > X$ we use the bond principal to exercise the call, then repay the borrowed stock and come out even. Table 4.8 summarizes.

If $0 \leq C^A(S_t, T-t) < S_t - X$, violating the third lower bound for the American call, buy the call and exercise it immediately. Of course, this is not possible for the European option. The bounds for puts can be established by similar arguments or, for European options, by applying put-call parity and the bounds on the call.

The bounds on prices of American calls, although very wide when S_t and X are positive, nevertheless have an important consequence for pricing these instruments. Notice that if the stock pays no dividend (has zero explicit cost of carry), the second lower bound dominates the third so long as $t < T$, since in that case $B(t,T) \leq 1$ and

$$S_t - B(t,T)X \geq S_t - X.$$

Therefore, $C^A(S_t, T-t) \geq S_t - X$ when $t < T$, and the inequality is strict if interest rates are strictly positive. This means that an American call on a no-dividend stock is always worth at least as much as its intrinsic value—its value if exercised immediately, and therefore it will not (rationally) be exercised early.[10] It follows that when no dividends are to be paid during the option's life an American call must have the same value as an otherwise identical European call. This is important because, as we shall see in later chapters, it is much easier to value European options than American options. On the other hand, when the underlying is to pay a lump-sum dividend during the term of the option, it may well be optimal to exercise just before the stock begins to trade *ex* dividend. Moreover, when

[10]This is also true, *a fortiori*, if the explicit cost of carry is positive.

dividends are paid continuously, early exercise is possible at any time. All this is explained in chapters 5 and 7 as we consider ways to value American options on dividend-paying stocks in discrete and continuous time.

The situation is entirely different for the early exercise of American puts. Prior to expiration the third lower bound, $X - S_t$, dominates the second, $BX - S_t - K$, unless there is a dividend. This means that early exercise of an American put could well be optimal. One can see at once that, if the stock's price fell to zero (the lower bound) at t, nothing could be gained by waiting, and interest on X would be lost. Since the early-exercise feature has value, an American put is usually worth more than an otherwise identical European put.

The bounds on American and European options on underlying financial assets become sharp at certain extreme values of the underlying and strike, as summarized in table 4.9. (We take for granted that $K(t, T) \leq 0$ for financial assets.)

Notice that when $X = 0$ a call—even a European call on a no-dividend stock—effectively gives the stock for the taking. In fact, it is sometimes helpful to think of a stock or other underlying asset as a call with a strike price of zero.

The bounds for some of the options also become sharp as the time to expiration approaches infinity. Interestingly, the limiting behavior as $T \to \infty$ depends critically on whether dividends are lump-sum or proportional to the stock's price. Table 4.10 summarizes the results under the

Table 4.9. Values of options at extremes of underlying price and strike.

Case	$S_t = 0$	$X = 0$
European call	$C^E(S_t, T-t) = 0$	$C^E(S_t, T-t) = S_t + K$
American call	$C^A(S_t, T-t) = 0$	$C^A(S_t, T-t) = S_t$
European put	$P^E(S_t, T-t) = BX$	$P^E(S_t, T-t) = 0$
American put	$P^A(S_t, T-t) = X$	$P^A(S_t, T-t) = 0$

Table 4.10. Values and bounds of perpetual options.

	Type of Dividend	
Option	Lump-Sum	Proportional
European call	$C^E(S_t, \infty) = S_t$	$C^E(S_t, \infty) = 0$
American call	$C^A(S_t, \infty) = S_t$	$S_t - X \leq C^A(S_t, \infty) \leq S_t$
European put	$P^E(S_t, \infty) = 0$	$P^E(S_t, \infty) = 0$
American put	$X - S_t \leq P^A(S_t, \infty) \leq X$	$X - S_t \leq P^A(S_t, \infty) \leq X$

assumptions that both the average spot rate, $r(t,T)$, and the dividend proportions are positive and nonvanishing (that is, bounded away from zero) and that the present value of the lump-sum dividend stream approaches zero. For example, for the European call when dividends are lump-sum, both $K(t,T)$ and $B(t,T)$ approach zero under these assumptions, so that the lower and upper bounds in table 4.7 coincide asymptotically. If, on the other hand, dividends are proportional to price, positive, and nonvanishing, then $S_t + K(t,T) \to 0$. For example, if dividends are paid continuously at constant rate $\delta > 0$, then $S_t + K(t,T) = S_t e^{-\delta(T-t)}$. In this case as $T \to \infty$ we can replicate the call's maximum possible time-T payoff with a vanishingly small fractional share.

4.3.4 *How Prices Vary with T, X, and S_t*

Let us now see what static-replication arguments can establish about the relations of option prices to time to expiration, strike price, and the current price of the underlying.

Variation with Term

A simple argument shows that the value of any vanilla American-style put or call must be a nondecreasing function of the time to expiration. For example, a call expiring at $T_2 > T_1$ is equivalent to a pair of options—a T_1-expiring call plus a T_2-expiring call that kicks in at T_1 if the first call was not exercised. Since the second option cannot have negative value, it follows that $C^A(S_t, T_2 - t) \geq C^A(S_t, T_1 - t)$and, likewise, $P^A(S_t, T_2 - t) \geq P^A(S_t, T_1 - t)$. Obviously, this argument does not apply to European options, since the one expiring at T_2 cannot be exercised before T_2. Of course, if the underlying is a financial asset that pays no dividend (and hence has zero cost of carry) then we know that American and European calls have the same prices. In this one case we can conclude that $C^E(S_t, T_2 - t) \geq C^E(S_t, T_1 - t)$ when $T_2 > T_1$; otherwise, without imposing specific structure on $\{S_t\}$ the change in the price of a European option with term is of ambiguous sign.

Variation with Strike Price

Arbitrage does enforce very specific relations of option prices to the strike, for European as well as American varieties. We will see that calls are

Table 4.11. Cash flows attainable if call values increased with strike.

Position	State		
	$S_T \leq X_1$	$X_1 < S_T \leq X_2$	$S_T > X_2$
Long call	0	$S_T - X_1$	$S_T - X_1$
Short call	0	0	$X_2 - S_T$
Net	0	$S_T - X_1 > 0$	$X_2 - X_1 > 0$

nonincreasing functions of X, that puts are nondecreasing functions, that the absolute values of rates of change of puts and calls alike are no more than unity, and that both $C(\cdot,\cdot;X)$ and $P(\cdot,\cdot;X)$ are convex functions. Since all comparisons are for the same values of S_t and T, we simplify notation by writing $C(X)$ and $P(X)$, omitting superscripts for relations that apply to both European and American options.

We first establish that $C(X_2) - C(X_1) \leq 0$ and $P(X_2) - P(X_1) \geq 0$ when $X_2 > X_1$. Starting with the call, if $C(X_2) > C(X_1)$ we could sell the call struck at X_2, buy the one struck at X_1, and have money left over. In the European case, or if the options are American and the X_2 call is not exercised before T, then the payoffs of the arbitrage portfolio in the three relevant terminal price states will be as shown in table 4.11. If the options are American and the shorted put is exercised at some $t^* \in [0,T)$, then the long put would also be in the money and exercising it at t^* would produce the same positive payoff as that in the last column. The same kind of argument shows that the condition $P(X_2) - P(X_1) < 0$ would also present an opportunity for arbitrage.

Having signed the slopes of the put and call functions, we can now verify that their absolute values are bounded above by unity; that is, $\Delta C/\Delta X \geq -1$ and $\Delta P/\Delta X \leq 1$. Starting with American calls we show that $C^A(X_2) - C^A(X_1) < -(X_2 - X_1)$ presents us with an arbitrage. Buying the X_2 call, shorting the X_1 call, and putting $\Delta X \equiv X_2 - X_1$ in the money fund brings in a positive amount now and leads to the time-T payoffs shown in table 4.12 if the shorted call is not exercised early. If the X_1 call is exercised at some $t^* < T$, then the payoffs are as in the last two columns of the table but with t^* replacing T throughout. Either way there is an arbitrage.

For European calls the bound on $|\Delta C^E/\Delta X|$ can actually be sharpened a bit, to $\Delta C^E/\Delta X \geq -B(t,T)$. If $C^E(X_2) - C^E(X_1) < -B(t,T)(X_2 - X_1)$ we can buy the X_2 call and sell the X_1 call as before, but since early exercise is no longer a worry we can lock in a return on the net take from the calls as

Table 4.12. Cash flows attainable from violation of lower bound on slope of price of American call.

Position	State		
	$S_T \leq X_1$	$X_1 < S_T \leq X_2$	$S_T > X_2$
Long call	0	0	$S_T - X_2$
Short call	0	$X_1 - S_T$	$X_1 - S_T$
Money fund	$\frac{M_T}{M_t}\Delta X$	$\frac{M_T}{M_t}\Delta X$	$\frac{M_T}{M_t}\Delta X$
Net	$\frac{M_T}{M_t}\Delta X \geq 0$	$X_2 - S_T +$ $(\frac{M_T}{M_t} - 1)\Delta X \geq 0$	$(\frac{M_T}{M_t} - 1)\Delta X \geq 0$

Table 4.13. Cash flows attainable from violation of lower bound on slope of price of European call.

Position	State		
	$S_T \leq X_1$	$X_1 < S_T \leq X_2$	$S_T > X_2$
Long call	0	0	$S_T - X_2$
Short call	0	$X_1 - S_T$	$X_1 - S_T$
Bonds	$\frac{\Delta C^E}{B} + \Delta X$	$\frac{\Delta C^E}{B} + \Delta X$	$\frac{\Delta C^E}{B} + \Delta X$
Net	$\frac{\Delta C^E}{B} + \Delta X > 0$	$\frac{\Delta C^E}{B} + X_2 - S_T > 0$	$\frac{\Delta C^E}{B} > 0$

well as from the difference in the strikes. This would be done by buying T -maturing bonds worth $\Delta C^E + B(t,T)\Delta X$. While this leaves us with zero cash at time t, table 4.13 shows that it buys strictly positive payoffs in all states at T. Similar arguments for American and European puts show that $\Delta P^A/\Delta X \leq 1$ and $\Delta P^E/\Delta X \leq B(t,T)$.

Not only does arbitrage bound the slopes of calls and puts with respect to X, it requires $C(X)$ and $P(X)$ to be (weakly) convex functions. We now show that for each $X_1 > 0$, each $X_3 > X_1$, and each $p \in (0,1)$ and $q = 1-p$

$$C(pX_1 + qX_3) \leq pC(X_1) + qC(X_3) \tag{4.20}$$

$$P(pX_1 + qX_3) \leq pP(X_1) + qP(X_3) \tag{4.21}$$

for both American and European options. For this let $X_2 = pX_1 + qX_3$, so that $p = (X_3 - X_2)/(X_3 - X_1)$ and $q = (X_2 - X_1)/(X_3 - X_1)$. Starting with the call, suppose, contrary to (4.20), that $C(X_2) > pC(X_1) + qC(X_3)$. Selling one X_2 call and buying p of the X_1 calls and q of the X_3 calls brings in a positive amount of cash now. If the shorted call is not exercised before T then table 4.14 shows that the portfolio has nonnegative value for any outcome $S_T \geq 0$.

Table 4.14. Cash flows attainable if call's price were concave function of strike.

	State			
Position	$S_T \leq X_1$	$X_1 < S_T \leq X_2$	$X_2 < S_T \leq X_3$	$S_T > X_3$
X_1 calls	0	$p(S_T - X_1)$	$p(S_T - X_1)$	$p(S_T - X_1)$
X_2 call	0	0	$-(S_T - X_2)$	$-(S_T - X_2)$
X_3 calls	0	0	0	$q(S_T - X_3)$
Net	0	$p(S_T - X_1) > 0$	$q(X_3 - S_T) \geq 0$	0

Table 4.15. Cash flows attainable if put's price were concave function of strike.

	State			
Position	$S_T \leq X_1$	$X_1 < S_T \leq X_2$	$X_2 < S_T \leq X_3$	$S_T > X_3$
X_1 puts	$p(X_1 - S_T)$	0	0	0
X_2 put	$-(X_2 - S_T)$	$-(X_2 - S_T)$	0	0
X_3 puts	$q(X_3 - S_T)$	$q(X_3 - S_T)$	$q(X_3 - S_T)$	0
Net	0	$p(S_T - X_1) > 0$	$q(X_3 - S_T) \geq 0$	0

This takes care of the European case. If the options are American and the shorted X_2 call is exercised at some $t^* < T$, then the payoffs in the only relevant states, $X_2 < S_{t^*} \leq X_3$ and $S_{t^*} > X_3$, will be as in the last two columns with t^* in place of T. Arbitraging the condition $P(X_2) > pP(X_1) + qP(X_3)$ proceeds in the same way by buying the portfolio of X_1 and X_3 puts and shorting the X_2 option. If the shorted put is not exercised before T the payoffs are as shown in table 4.15, and if it is exercised at $t^* < T$ the payoffs are as in the first two columns with T replaced by t^*.

Homogeneity of Option Prices in S_t and X

On any price path $\{S_t\}_{t \geq 0}$ that produces a positive payoff of a vanilla European or American option the payoff will depend linearly on the underlying price and strike, as $S_{t^*} - X$ or $X - S_{t^*}$ for calls and puts, respectively. The obvious implication of this is that call and put prices at any $t \in [0, T]$ will be denominated in the same units as S_t and X. If we choose to measure all the quantities in units of the strike price, then the options' payoffs become $S_{t^*}/X - 1$ and $1 - S_{t^*}/X$, and the prices at t become $C(S_t, T - t; X)/X$ and $P(S_t, T - t; X)/X$. It follows that

$$C(S_t, T - t; X) = XC(S_t/X, T - t; 1) \tag{4.22}$$

$$P(S_t, T - t; X) = XP(S_t/X, T - t; 1). \tag{4.23}$$

These relations will be of value in later chapters when we develop explicit expressions for the prices of puts and calls and when we approximate them numerically. For example, values of $C(S_t, T - t; X)$ on a two-dimensional grid of price and strike values can be found by evaluating (4.22) on a one-dimensional grid of price/strike ratios. The relations are also useful in helping to spot erroneous pricing formulas. However, one must be careful not to misinterpret them. For example, (4.22) should not be taken to mean that an at-the-money equity call when the stock's price is $S_t = 10$ currency units has value ten times that of an at-the-money call when $S_t = 1$ currency unit. The reason is that the dynamics of $\{S_u\}_{u \geq t}$—that is, the nature of the possible price paths emanating from (t, S_t)—may well depend on the initial price level.

With this *caveat* in mind, one should be able to spot the flaw in the following

FALSE Theorem: Under conditions PM1-PM4, and given just that $S_t \geq 0$ for all $t \geq 0$, the price of a European or American put is a strictly decreasing function of the current price of the underlying.

The "proof" of this statement might run as follows. Taking $u > 1$, (4.23) gives

$$P(uS_t, T - t; X) = uP(S_t, T - t; X/u).$$

Applying (4.21) with $X_3 = X$, $X_2 = X/u$, and $X_1 = 0$ then gives

$$\begin{aligned} P(uS_t, T - t; X) &\leq \frac{u-1}{u} P(S_t, T - t; 0) + \frac{1}{u} P(S_t, T - t; X) \\ &= \frac{1}{u} P(S_t, T - t; X) \\ &< P(S_t, T - t; X). \end{aligned}$$

Bergman *et al.* (1996) give specific examples of price processes for which this sort of reasoning fails. We present one of their examples in section 8.2.1 to show that call prices do not necessarily increase monotonically with S_t. In sum, without further restriction on the dynamics of the underlying price nothing can be said about how option prices depend on S_t.

5

Pricing under Bernoulli Dynamics

In chapter 4, static replication was used to value forward commitments to buy or sell an investment asset. This involved finding a position in traded assets—the underlying itself plus riskless bonds—that produces the same terminal payoff as the forward contract, $S_T - \mathsf{f}(0,T)$. Because this payoff is linear in the price of the underlying asset, and because values of portfolios are also linear in the prices, replicating the forward contract in this way requires just a static portfolio—one that does not have to be adjusted as time passes. Also, since it is static, the replicating portfolio is necessarily self-financing, in that no funds need be added or withdrawn along the way. Having found a self-financing replicating portfolio, we could apply the law of one price to value the forward contract and, ultimately, to deduce the initial arbitrage-free forward price, $\mathsf{f}(0,T)$. The linearity of the contract's terminal payoff makes the process very easy.

Imagine, however, trying to follow this program to determine the value at time $t < T$ of a T-expiring European call option, $C^E(S_t, T-t)$. If the strike is X the option's only potential payoff is $C^E(S_T, 0) = (S_T - X)^+$ at T. But since $(S_T - X)^+$ is not linear in S_T, it will not ordinarily be possible to construct a static replicating portfolio. If it turns out that $S_T > X$ we would need to be long a full unit of the asset and short X currency units worth of bonds at T, whereas if $S_T \leq X$ we would want to have no net position in either asset. As we approached time T, the replicating portfolio would have to be adjusted continually toward one or the other of these extreme positions, depending on the path taken by S_t. Pricing a call by arbitrage therefore requires a recipe for constructing a dynamic replicating portfolio. And since the call position that is being replicated requires no additions or withdrawals of cash after the initial purchase, the replicating portfolio must also be self-financing once it is put in place, with subsequent purchases of one asset being financed by sales of the other.

This chapter shows how to construct dynamic, self-financing replicating portfolios for options and other nonlinear derivatives under a very simple Bernoulli model for the underlying price. The method yields what are called "binomial" estimates of arbitrage-free prices. The central features of Bernoulli dynamics are that time is measured discretely and that assets' prices after any span of time are limited to a finite number of possible states. Although the Bernoulli model will at first seem highly unrealistic, it does have two very desirable features. First, it helps in understanding the continuous-time, continuous-state-space models taken up in later chapters, for which the underlying mathematics is far more difficult. Second, despite its apparent lack of realism, the binomial method turns out to be capable of producing very useful quantitative estimates of prices.

We begin by explaining the motivation for and structure of the Bernoulli model and the conditions for portfolios to be self-financing in this discrete-time setup. Section 5.2 then gives stylized applications of the binomial method to European-style derivatives, using an underlying asset and riskless bonds to form the dynamic replicating portfolio. Although the method is presented first as a recursive procedure, we will see that simple pricing formulas can be found for European-style derivatives. Section 5.3 shows that these formulas—and binomial estimates generally—can be given two different interpretations. One interpretation is that they are solutions to difference equations that govern how the price of the derivative evolves over time and with the underlying price. The other interpretation is that binomial estimates are discounted expected values of the derivative's payoffs, relative to a special probability distribution. The latter viewpoint is that of equivalent-martingale or risk-neutral pricing. Both methods have counterparts in the continuous-time framework of later chapters, but they are much easier to understand in the Bernoulli context.

After these basic concepts are presented, we turn in section 5.4 to specific applications: European options on stocks that pay no dividends, futures prices and derivatives on futures, American-style derivatives, and derivatives on assets that yield cash receipts, such as dividend-paying stocks and positions in foreign currencies. Section 5.5 provides further details on implementation, demonstrates that binomial valuations of European options can approximate the Black-Scholes formulas from continuous time, and suggests efficient computational methods. Of course, like all other models, the standard binomial setup has its limitations. The final section shows how it can be adapted to better fit the observed market prices of traded options.

5.1 The Structure of Bernoulli Dynamics

Let us consider the effort to replicate at some initial time $t = 0$ a generic European-style derivative whose payoff at time T, $D(S_T, 0)$, depends just on the price of a single underlying traded asset. We have seen that it is not generally possible to construct a static replicating portfolio from just the underlying asset and riskless bonds unless D is linear in S_T . However, it is clear that any proper function is linear on a domain that comprises just two points! For example, $f(s) = \exp(s)$ is linear if restricted to the domain $\{s', s''\}$. Thus, the payoff of a European call would also be linear in the terminal value, S_T, if the dynamics were such that $\mathbb{P}(S_T \in \{s', s''\}) = 1$. This simple observation is the basis of the binomial model for derivatives pricing, proposed by Cox , Ross, and Rubinstein (1979).

Of course, it is hardly realistic to propose that the price of a stock, a foreign currency, or a precious metal, etc., could have one of just two values after an interval of a month, a week, or even a day. However, suppose the time interval $[0, T)$ on which the derivative lives is subdivided arbitrarily finely, into say n equal subperiods of length T/n. Letting $s_j \equiv S_{jT/n}$ be the price at time jT/n for $j \in \{0, 1, \ldots, n-1\}$, model s_{j+1} as $s_{j+1} = s_j R_{j+1}$, where price relatives $R_1, R_2, \ldots, R_n$ are independent random variables with the same (translated and rescaled) Bernoulli distribution:

$$\mathbb{P}(R_j = u) = \pi = 1 - \mathbb{P}(R_j = d)$$

for positive numbers $u > d$. In fact, that the distributions be the same for each time step is not essential, and we will later consider schemes that allow for deterministic variation. Although the notation at this point does not indicate it, the uptick and downtick values of the price relatives, u and d, will certainly have to depend on n, since prices are less variable over short periods than over long. Starting with initial value $s_0 = S_0$, there are two possible price states at $t = T/n$ $(s_0 u, s_0 d)$, three states at $t = 2T/n$ $(s_0 u^2, s_0 ud, s_0 d^2)$, and $n + 1$ states at $t = T$ $(s_0 u^n, s_0 u^{n-1} d, \ldots, s_0 ud^{n-1}, s_0 d^n)$. The multiplicative setup with positive u and d has the virtue of keeping s_j positive, which is appropriate for most financial assets. Clearly, we must have $u > 1 > d$ in order to allow for both positive and negative changes in price.

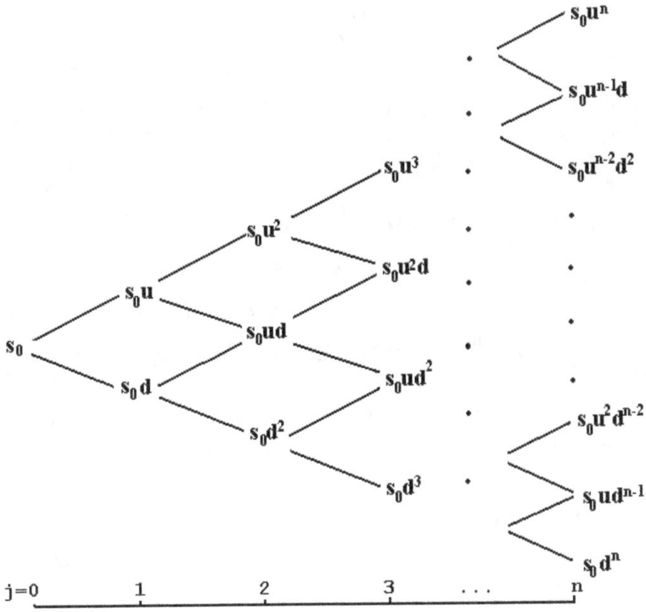

Fig. 5.1. A recombinant binomial tree.

The tree diagram in figure 5.1 depicts the state space after n periods. Positions from which the two branches emanate—the attainable price states at a given stage—will be called the "nodes". Notice that with this arrangement the downtick from a given node at any step j leads to the same price level at $j+1$ as does an uptick from the node below it; for example, at step 2 we have $s_0 u \cdot d = s_0 d \cdot u$. A tree with this property is said to "recombine". Later, we will encounter situations leading to nonrecombinant trees. The total number of paths after n steps is 2^n whether the tree recombines or not. If it does fully recombine there are $n+1$ terminal nodes; if not, there can be up to 2^n.

Figure 5.1 depicts a stochastic process $\{s_j\}$ adapted to a filtered probability space, $(\Omega, \mathcal{F}, \{\mathcal{F}_j\}_{j=0}^n, \mathbb{P})$. An element ω of Ω is one of the 2^n sample paths of $\{s_j\}$, and $\mathcal{F} = \mathcal{F}_n$ is the field of events generated by $\{s_j\}_{j=0}^n$. Likewise, $\mathcal{F}_j$ conveys the whole history of the process through stage j, and we take $\mathcal{F}_0$ to be the trivial field, $\{\emptyset, \Omega\}$. The process $\{s_j\}$ is clearly Markov in view of the independence of the one-step price relatives. Finally, we note that the probability associated with any attainable price level at stage j has a binomial form. That is, since the event $s_j = s_0 u^{j-k} d^k$ occurs when and only when there are k downticks in j independent trials, we have for $k \in \{0, 1, \ldots, j\}$

$$\mathbb{P}(s_j = s_0 u^{j-k} d^k) \equiv \mathbb{P}(\{\omega : s_j(\omega) = s_0 u^{j-k} d^k\}) = \binom{j}{k} \pi^{j-k}(1-\pi)^k$$

and, for $j \in \{0, 1, \ldots, n\}$ and $k \in \{0, 1, \ldots, n-j\}$,

$$\mathbb{P}(s_n = s_j u^{n-j-k} d^k | \mathcal{F}_j) = \binom{n-j}{k} \pi^{n-j-k}(1-\pi)^k.$$

Before beginning to construct self-financing, replicating portfolios that will enable us to value derivatives, we should make very clear what it means for a portfolio to be self-financing in the discrete-time setup. Consider a portfolio of just two assets, worth $P_j = p_j s_j + q_j s'_j$ at time j. At $j+1$ two things will ordinarily have happened: the prices will have changed, and the portfolio weights will have been adjusted. The change in value of the portfolio between j and $j+1$, $P_{j+1} - P_j$, can be decomposed into these two effects, as

$$\begin{aligned}
&(p_{j+1}s_{j+1} + q_{j+1}s'_{j+1}) - (p_j s_j + q_j s'_j) \\
&\quad = [(p_{j+1} - p_j)s_{j+1} + (q_{j+1} - q_j)s'_{j+1}] \\
&\qquad + [p_j(s_{j+1} - s_j) + q_j(s'_{j+1} - s'_j)] \\
&\quad = (\Delta P \text{ from portfolio adjustment}) + (\Delta P \text{ from price change}).
\end{aligned}$$

The portfolio-adjustment term would be zero for a self-financing portfolio. In this case purchases of one asset are financed by sales of the other or, to put it the other way, proceeds of sales of one asset are reinvested in the other. Since no cash is either added or withdrawn, the change in the portfolio's value between j and $j+1$ would be $\Delta P_j = p_j \Delta s_j + q_j \Delta s'_j$. Maintaining this discipline over n trading intervals, the net change in the portfolio's value from $j = 0$ to $j = n$ will be

$$P_n - P_0 = \sum_{j=0}^{n-1} (p_j \Delta s_j + q_j \Delta s'_j). \tag{5.1}$$

More generally, the net change in value of a self-financing portfolio $\mathbf{p}_j = (p_{1j}, p_{2j}, \ldots, p_{kj})'$ chosen from k assets with step-j prices $\mathbf{s}_j = (s_{1j}, s_{2j}, \ldots, s_{kj})'$ would be

$$P_n - P_0 = \sum_{j=0}^{n-1} \mathbf{p}'_j \Delta \mathbf{s}_j. \tag{5.2}$$

This is the discrete counterpart to expression (3.30), which applies in continuous time.

5.2 Replication and Binomial Pricing

To see how the Bernoulli setup makes it possible to construct dynamic, self-financing replicating portfolios, let us replicate a generic European-style derivative that pays the holder $D(S_T, 0)$ upon expiration at stage n. Continuing to measure time in steps, we write $D(s_j, n-j) \equiv D(S_{jT/n}, T - jT/n)$ for the value of the derivative at step j when the underlying price is s_j and there are $n-j$ steps to go. With this convention $D(s_n, 0) = D(S_T, 0)$ at step n. Similarly, for $0 \leq j \leq k \leq n$ $B(j,k) \equiv B(jT/n, kT/n)$ represents the price at step j of a unit discount bond that matures at step k. Invoking assumption RK, current and future prices of bonds maturing at step n or earlier are supposed to be known at step 0. We refer to the underlying asset as a "stock" and assume for now that it pays no dividend and has no explicit cost of carry. The objective is to determine the arbitrage-free value of the derivative at the initial step, $D(s_0, n)$, given knowledge of the initial price of one share of the stock, s_0.

As a second asset needed for replication we could use a unit discount bond maturing at step n or later, or else we could manage by rolling over bonds of shorter lives; however, it is a little easier notationally to use the "money" fund. Recall that this is like a savings account that rolls over investments in riskless bonds that are just on the verge of maturing. In the discrete-time setup these are bonds that mature after one time step, each being worth $B(j, j+1)$ at step j and $B(j+1, j+1) = 1$ at $j+1$. Thus, the value of a unit of the money fund grows from $m_j \equiv M_{jT/n}$ at step j to $m_{j+1} = m_j B(j, j+1)^{-1} \equiv m_j(1+r_j)$ at step $j+1$, where r_j is the discrete-time counterpart of the short rate at step j. To keep the notation as simple as possible for now, we fix $r_j = r$ for each j; and, as for u and d, we do not make explicit its dependence on n. Starting off the money fund at some arbitrary initial value m_0 per share, the value at stage j is $m_j = m_0(1+r)^j$. Under assumption RK the entire sequence of values $\{m_j\}_{j=0}^{n}$ is known as of step 0, and by assumption RB it is nondecreasing in j. Also, in accord with assumptions PM1-PM4, trading and holding stock and money fund are costless, and any positive or negative number of shares of either can be held.

With this setup the value of the derivative asset can be replicated at each time step by working backwards through the tree. We work *backwards* because the terminal value of the derivative at step n is a known, contractually specified function of the underlying price, whereas the values at earlier stages are what we are attempting to determine. At the final stage there are $n+1$ possible values of the stock and up to $n+1$ corresponding known values of the derivative, $D(s_n, 0)$. We begin the program at stage $n-1$ by finding $D(s_{n-1}, 1)$ at each of the n possible realizations of s_{n-1}. Given any such s_{n-1} the stock will be worth either $s_{n-1}u$ or $s_{n-1}d$ next period, the derivative will be worth either $D(s_{n-1}u, 0)$ or $D(s_{n-1}d, 0)$, and the value of the money-market share will be $m_n = m_{n-1}(1+r)$. The goal is to find a portfolio of the stock and money fund that can be bought at $n-1$ and yet, without our adding or withdrawing funds, will be sure to have the same value as the derivative at stage n. $D(s_{n-1}, 1)$, the arbitrage-free value of the derivative at $n-1$ given s_{n-1}, will then equal the value of this portfolio.

If the replicating portfolio contains some p_{n-1} shares of the stock and q_{n-1} units of the money fund, its value at stage $n-1$ is

$$P_{n-1} = p_{n-1}s_{n-1} + q_{n-1}m_{n-1}.$$

If it is self-financing, (5.1) implies that its value at stage n will be

$$\begin{aligned} P_n &= P_{n-1} + p_{n-1}(s_n - s_{n-1}) + q_{n-1}(m_n - m_{n-1}) \\ &= p_{n-1}s_n + q_{n-1}m_n. \end{aligned}$$

Choosing portfolio shares to replicate the derivative requires

$$P_n^u \equiv p_{n-1}s_{n-1}u + q_{n-1}m_n = D(s_{n-1}u, 0) \tag{5.3}$$

$$P_n^d \equiv p_{n-1}s_{n-1}d + q_{n-1}m_n = D(s_{n-1}d, 0), \tag{5.4}$$

or

$$p_{n-1} = \frac{D(s_{n-1}u, 0) - D(s_{n-1}d, 0)}{s_{n-1}u - s_{n-1}d} \tag{5.5}$$

$$q_{n-1} = \frac{D(s_n, 0)}{m_n} - p_{n-1}\frac{s_n}{m_n}. \tag{5.6}$$

In (5.6) we express the step-n values of the derivative and the stock in terms of the $\mathcal{F}_n$-measurable random variable s_n since q_{n-1} is the same for $s_n = s_{n-1}u$ and $s_n = s_{n-1}d$.

Given the realization of s_{n-1} at stage $n-1$, the portfolio $\{p_{n-1}, q_{n-1}\}$ does replicate the derivative when it has one period to go, and it does so without having to add or withdraw funds. The value of this portfolio, and therefore the arbitrage-free value of the derivative itself, is then $D(s_{n-1}, 1) = p_{n-1}s_{n-1} + q_{n-1}m_{n-1}$, a function of the $\mathcal{F}_{n-1}$-measurable quantities s_{n-1}, u, d, and r. Once the value of $D(s_{n-1}, 1)$ is determined in this way at each of the n possible realizations of s_{n-1} we can back up to stage $n-2$ and value $D(s_{n-2}, 2)$ at each of the $n-1$ realizations of s_{n-2}.

Continuing in this way leads ultimately to the value of $D(s_0, n)$ at the known initial value s_0. This is the essence of the binomial method.

It will be helpful to characterize this recursive process more generally before giving an example. Suppose we are at any stage $j \in \{0, 1, \ldots, n-1\}$ with s_j known and the next-stage value function, $D(\cdot, n-j-1)$, having already been found. Simplifying notation still further, let $D_j = D(s_j, n-j)$ be the value of the derivative at step j (which is to be found) and $D_{j+1}^u = D(s_j u, n-j-1)$ and $D_{j+1}^d = D(s_j d, n-j-1)$ be the known future values in the two states that can be reached from s_j. Forming a portfolio worth

$$P_j = p_j s_j + q_j m_j \tag{5.7}$$

now and

$$P_{j+1} = p_j s_{j+1} + q_j m_{j+1} \tag{5.8}$$

next period requires that p_j, q_j satisfy

$$\begin{aligned} P_{j+1}^u &\equiv p_j s_j u + q_j m_{j+1} = D_{j+1}^u \\ P_{j+1}^d &\equiv p_j s_j d + q_j m_{j+1} = D_{j+1}^d. \end{aligned}$$

Substituting solutions

$$\begin{aligned} p_j &= \frac{D_{j+1}^u - D_{j+1}^d}{s_j u - s_j d} \\ q_j &= \frac{D_{j+1}}{m_{j+1}} - p_j \frac{s_{j+1}}{m_{j+1}} = \frac{1}{m_{j+1}} \left(\frac{u D_{j+1}^d - d D_{j+1}^u}{u-d} \right) \end{aligned}$$

into (5.7) and simplifying give as the derivative's arbitrage-free price

$$D_j = \frac{m_j}{m_{j+1}} \left[\left(\frac{m_{j+1}/m_j - d}{u-d} \right) D_{j+1}^u + \left(\frac{u - m_{j+1}/m_j}{u-d} \right) D_{j+1}^d \right]. \tag{5.9}$$

Finally, by setting $m_{j+1}/m_j = 1 + r$ and

$$\hat{\pi} \equiv \frac{1 + r - d}{u - d}$$

we obtain the simple expression

$$D_j = (1+r)^{-1} \left[\hat{\pi} D_{j+1}^u + (1-\hat{\pi}) D_{j+1}^d \right], \quad j \in \{0, 1, \ldots, n-1\}. \tag{5.10}$$

Both the forms (5.9) and (5.10) will prove useful later on, and the weights $\hat{\pi}$ and $1 - \hat{\pi}$ will be seen to have special significance. The notation deliberately suggests a parallel to the state probabilities π and $1-\pi$, but it is evident that neither $\hat{\pi}$ nor the binomial estimate D_j depends directly on those actual probabilities. Of course, behind the scenes is a market that has presumably taken those probabilities into account in determining the interest rate, the stock's current price, and the distribution of price relatives, all of which are taken now as given.

Example 60 *Pricing a European call struck at* $X = 10.0$ *on an underlying stock worth* $s_0 = 10.0$, *let us divide the option's life into* $n = 3$ *subperiods and choose* $u = 1.05$, $d = 0.97$, *and* $r = 0.01$, *so that* $\hat{\pi} = \frac{1}{2}$. *(Later we will see how to determine the move sizes in realistic applications.) Figure 5.2 summarizes the steps to find the call's replicating portfolio at each stage and its arbitrage-free value at* $t = 0$. *The first step is to determine the value of the option at the terminal nodes. At stage 3 the possible realizations of the stock's price are* $10.0(1.05^3) \doteq 11.576$, $10.0(1.05^2)(.97) \doteq 10.694$, $10.0(1.05)(.97^2) \doteq 9.879$, *and* $10.0(.97^3) \doteq 9.127$. *Taking* $m_0 = 1$, *the money fund is worth* $m_3 = 1.01^3 \doteq 1.030$ *in all four states. The call is out of the money at the two bottom nodes, and is worth about* $11.576 - 10.0 = 1.576$ *and* $10.694 - 10.0 = 0.694$ *at the two upper nodes. Now proceed to value the call recursively at each preceding stage. At the top node of stage 2 with the*

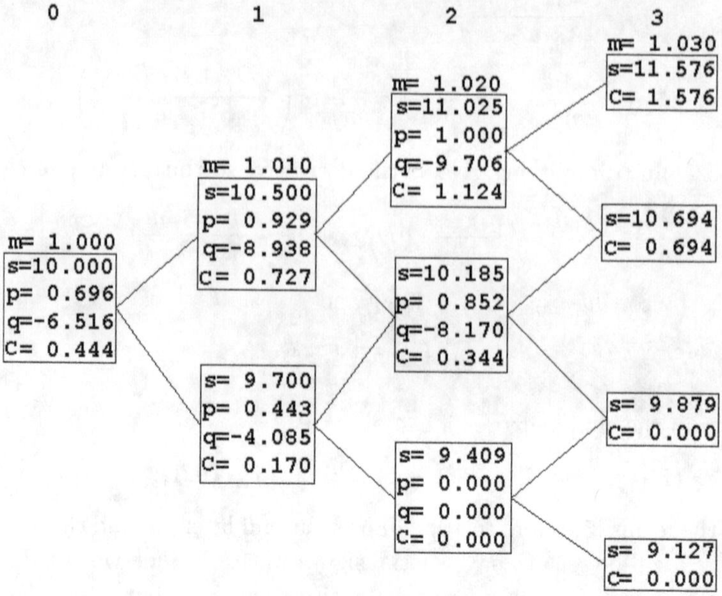

Fig. 5.2. Binomial pricing of a European call.

stock worth $10.0(1.05^2) = 11.025$, *the replicating portfolio contains*

$$p_2 = \frac{1.576 - 0.694}{11.576 - 10.694} = 1.0$$

shares of stock and

$$q_2 = \frac{1.576 - 11.576}{1.01^3} = -\frac{10.0}{1.01^3}$$

units of the money-market fund. (Notice that it makes sense to hold a full unit of the stock, since from this position the call must wind up in the money at step 3.) The values of the replicating portfolio and call are then

$$\begin{aligned} P_2 &= p_2 s_2 + q_2 m_2 \\ &= (1.0)10.0(1.05^2) - \frac{10.0}{1.01^3}(1.01^2) \\ &\doteq 1.124. \end{aligned}$$

Alternatively, the value of the portfolio and call can be found directly from (5.10) *as*

$$\begin{aligned} C^E(11.025, 1) &= \frac{\hat{\pi} C^E(11.576, 0) + (1 - \hat{\pi}) C^E(10.694, 0)}{1 + r} \\ &\doteq \frac{(.5)\,1.576 + (.5)\,0.694}{1.01} \\ &\doteq 1.124. \end{aligned}$$

At the middle node of stage 2, *where the stock is worth* $10.0(1.05)(0.97) = 10.185$, *the call is no longer certain to wind up in the money. The replicating portfolio here contains only*

$$p_2 = \frac{0.694 - 0.0}{10.694 - 9.878} \doteq 0.852$$

shares of stock and

$$q_2 = \frac{0.694 - (0.852) \cdot 10.694}{1.01^3} \doteq -8.170$$

units of the fund. The value of the portfolio and the call is then

$$C^E(10.185, 1) = \frac{(.5)0.694 + (.5)0.0}{1.01} \doteq 0.344.$$

From the bottom node of stage 2 *the call has no chance to finish in the money, so the replicating portfolio is null* ($p_2 = q_2 = 0$) *and the call is worthless. Similar calculations at stage* 1 *give call values of about* 0.727 *and* 0.170 *in the two states. Finally, at stage* 0 *the replicating portfolio has*

$$p_0 = \frac{0.727 - 0.170}{10.5 - 9.7} \doteq 0.696$$

shares of stock and

$$q_0 = \frac{0.727 - (0.696)10.5}{1.01} \doteq -6.516$$

units of the money fund, with total value $P_0 \doteq 0.444$. *Starting out with this portfolio and making the adjustments shown at each node in the figure delivers a portfolio that has the same value as the expiring call in every state at stage* 3. *Accordingly, the call's initial arbitrage-free price is* $C^E(10.0, 3) \doteq 0.444$.

Although the solution was found recursively in the example, the initial value of a European-style derivative can be expressed directly in terms of s_0 by means of the following formula:

$$D(s_0, n) = (1+r)^{-n} \sum_{k=0}^{n} \binom{n}{k} \hat{\pi}^{n-k}(1-\hat{\pi})^k D(s_0 u^{n-k} d^k, 0). \qquad (5.11)$$

To prove this by induction, setting $n = 1$ in (5.10) gives

$$D(s_0, 1) = (1+r)^{-1}[\hat{\pi} D(s_0 u, 0) + (1-\hat{\pi}) D(s_0 d, 0)],$$

which verifies (5.11) for that case. Now supposing (5.11) holds for any $n - 1 \in \aleph$, set $j = 0$ in (5.10) to get

$$\begin{aligned} D(s_0, n) &= (1+r)^{-1}[\hat{\pi} D(s_0 u, n-1) + (1-\hat{\pi}) D(s_0 d, n-1)] \\ &= (1+r)^{-n} \left[\sum_{k=0}^{n-1} \binom{n-1}{k} \hat{\pi}^{n-k}(1-\hat{\pi})^k D(s_0 u^{n-k} d^k, 0) \right. \\ &\quad \left. + \sum_{j=0}^{n-1} \binom{n-1}{j} \hat{\pi}^{n-1-j}(1-\hat{\pi})^{j+1} D(s_0 u^{n-1-j} d^{j+1}, 0) \right]. \end{aligned}$$

Noting from (2.6) on page 24 that $\binom{n-1}{n} = \binom{n-1}{-1} = 0$, one can extend the first sum to n and change variables in the second sum as $k = j + 1$ to get

$$D(s_0, n) = (1+r)^{-n} \sum_{k=0}^{n} \left[\binom{n-1}{k} + \binom{n-1}{k-1} \right] \hat{\pi}^{n-k}(1-\hat{\pi})^k D(s_0 u^{n-k} d^k, 0),$$

from which (5.11) follows by identity (2.7). In the next section we shall discover a much simpler way to show this.

How do things change if short rates are allowed to vary over time but are still known in advance, so that $m_{j+1}/m_j = B(j, j+1)^{-1} = 1 + r_j$, and $m_n = m_0(1 + r_0), \ldots, (1 + r_{n-1})$? The same replicating argument leads to

$$D_j = (1 + r_j)^{-1}[\hat{\pi}_j D^u_{j+1} + (1 - \hat{\pi}_j)D^d_{j+1}], \tag{5.12}$$

where now

$$\hat{\pi}_j \equiv \frac{1 + r_j - d}{u - d}. \tag{5.13}$$

The recursive procedure works almost as easily as before despite the time dependence of $\hat{\pi}$, but there is no longer such a simple formula for the value of a European derivative. Let

$$\widehat{\Pi}(n, k) \equiv \sum_{j_0+\cdots+j_{n-1}=k} \hat{\pi}_0^{1-j_0}(1 - \hat{\pi}_0)^{j_0} \cdot \cdots \cdot \hat{\pi}_{n-1}^{1-j_{n-1}}(1 - \hat{\pi}_{n-1})^{j_{n-1}}, \tag{5.14}$$

where the summation is over values of $j_0, \ldots, j_{n-1} \in \{0, 1\}$ such that $j_0 + \cdots + j_{n-1} = k$. Then (5.11) becomes

$$D(s_0, n) = \frac{m_0}{m_n} \sum_{k=0}^{n} \widehat{\Pi}(n, k) D(s_0 u^{n-k} d^k, 0).$$

This is equivalent to

$$D(s_0, n) = B(0, n) \sum_{j=0}^{n} \widehat{\Pi}(n, k) D(s_0 u^{n-k} d^k, 0), \tag{5.15}$$

since when interest rates are known in advance m_0/m_n must equal the price of a discount bond maturing at step n.

If we think of the $\{\hat{\pi}_j\}$ as pseudo-probabilities, then $\widehat{\Pi}(n, k)$ equals the probability of k down moves in n steps. When the $\hat{\pi}$'s are time-invariant, this has the binomial form $\binom{n}{k}\hat{\pi}^{n-k}(1 - \hat{\pi})^k$. We are about to see that this probabilistic interpretation is both justifiable and useful.

5.3 Interpreting the Binomial Solution

Whether obtained recursively or directly as in (5.11) or (5.15), binomial solutions can be interpreted in two useful ways. Viewed as solutions to a certain difference equation, they are the discrete-time counterparts of solutions to partial differential equations that arise in the continuous-time framework adopted in later chapters. Alternatively, thinking of the $\hat{\pi}$'s as pseudo-probabilities, binomial solutions appear to be mathematical expectations of the derivatives' normalized payoffs. This interpretation, which also has its counterpart in continuous time, corresponds to risk-neutral or martingale pricing. The recursive formula (5.12) is a good place from which to start in order to attain both perspectives.

5.3.1 *The P.D.E. Interpretation*

Formula (5.12) really amounts to a partial difference equation that governs the variation of D with respect to the discrete-time counter and the price of the underlying asset. To see this, apply some algebra and reduce (5.12) to

$$0 = -r_j D_j + r_j s_j \left(\frac{D_{j+1}^u - D_{j+1}^d}{s_j u - s_j d} \right) + \left[\frac{(u-1)(D_{j+1}^d - D_j) + (1-d)(D_{j+1}^u - D_j)}{(u-1)+(1-d)} \right].$$

The factor in parentheses in the second term is $\Delta D/\Delta s$, the instantaneous rate of change of D at step $j+1$ with respect to s_{j+1}, holding time constant. The third term, a weighted average of the state-dependent changes in D with time, amounts to the partial difference with respect to t, or $\Delta D/\Delta t$. With these interpretations (5.12) can be written as the partial difference equation

$$-rD + rs \cdot \Delta D/\Delta s + \Delta D/\Delta t = 0.$$

The binomial estimate of $D(s_0, n)$ solves this equation subject to the terminal condition imposed by the derivative's contractual relation to the stock's price at expiration, $D(s_n, 0)$. Equations (5.11) and (5.15) are explicit solutions for a European-style derivative.

5.3.2 *The Risk-Neutral or Martingale Interpretation*

Gaining an understanding of the equivalent-martingale interpretation is a bit more involved, but this is by far the easiest setting in which to do it. It is also well worth the cost. The first step is to see that the ability to replicate a derivative's payoffs within the Bernoulli framework implies the existence of a pseudo-probability measure, with respect to which normalized prices of the derivative, the underlying asset, and the money fund are all martingales.

The Martingale Measure

Begin by recognizing that weighting factors $\hat{\pi}_j = (1 + r_j - d)/(u - d)$ and $1 - \hat{\pi}_j = (u - 1 - r_j)/(u - d)$ in (5.12) must be strictly positive in an arbitrage-free market. Weight $\hat{\pi}_j$ could be zero or negative only if $1 + r_j \leq d$. In this case there would be an arbitrage against riskless bonds, since shorting bonds (borrowing) and buying the stock could produce a gain but could not produce a loss. Likewise, $1 - \hat{\pi}_j \leq 0$ would require $u \leq 1 + r_j$ and set up an arbitrage against the stock. The implication is that $\hat{\pi}_j \in (0, 1)$ at each step j. Since it thus possesses all the features of a Bernoulli probability, let us construct a new pseudo-probability space $(\Omega, \mathcal{F}, \{\mathcal{F}_j\}, \hat{\mathbb{P}})$. This has the same information structure as the "real" one in this model but contains a new measure $\hat{\mathbb{P}}$ such that

$$\hat{\mathbb{P}}(s_n = s_j u^{n-j-k} d^k \mid \mathcal{F}_j) = \widehat{\Pi}(n-j, k), k \in \{0, 1, \ldots, n-j\} \tag{5.16}$$

for each s_j and each $j \in \{0, 1, \ldots, n-1\}$. Again, $\{s_j\}$ is an adapted process in this new filtered space.

It is clear that $\hat{\mathbb{P}}$ and $\mathbb{P}$ are equivalent measures since $\mathbb{P}$ assigns a positive value to a given sample path if and only if $\hat{\mathbb{P}}$ does. By the Radon-Nikodym theorem (section 2.2.4) there is then at each step $j \in \{0, 1, \ldots, n-1\}$ a unique random variable Q_j such that $\hat{\mathbb{P}}(A \mid \mathcal{F}_j) = \int_A Q_j(\omega) \cdot d\mathbb{P}(\omega \mid \mathcal{F}_j)$ for each $A \in \mathcal{F}$. The construction of Q_j is very simple in this discrete Bernoulli framework. For each step j a sample path ω that passes through s_j and terminates at step n is defined by a sequence of upticks and downticks. Under $\mathbb{P}$ this path has probability

$$\mathbb{P}(\omega \mid \mathcal{F}_j) = \pi_j^{1-i_j}(1-\pi_j)^{i_j} \cdot \ldots \cdot \pi_{n-1}^{1-i_{n-1}}(1-\pi_{n-1})^{i_{n-1}},$$

where each index $i_j, i_{j+1}, \ldots, i_{n-1}$ takes a value in $\{0, 1\}$ according as the path moves down or up at that stage. The expression for $\hat{\mathbb{P}}(\omega \mid \mathcal{F}_j)$, the path probability under $\hat{\mathbb{P}}$, is the same except that $\hat{\pi}$'s replace π's. Then Q_j is defined by

$$Q_j(\omega) = \frac{\hat{\mathbb{P}}(\omega \mid \mathcal{F}_j)}{\mathbb{P}(\omega \mid \mathcal{F}_j)}.$$

For example, taking A_{jk} to be the set of sample paths emanating from s_j and having k downticks between stages j and n, we have

$$\begin{aligned}
\hat{\mathbb{P}}(A_{jk} \mid \mathcal{F}_j) &\equiv \hat{\mathbb{P}}(s_n = s_j u^{n-j-k} d^k \mid \mathcal{F}_j) \\
&= \widehat{\Pi}(n-j, k) \\
&= \sum_{\omega \in A_{jk}} Q_j(\omega) \mathbb{P}(\omega \mid \mathcal{F}_j).
\end{aligned}$$

Here summing over $\omega \in A_{jk}$ is equivalent to summing over all $\binom{n-j}{k}$ indices $i_j, i_{j+1}, \ldots, i_{n-1}$ such that $i_j + i_{j+1} + \cdots + i_{n-1} = k$.

Let us now see that the new measure $\hat{\mathbb{P}}$ has the special property of turning suitably normalized values of s_j and m_j—and of any self-financing portfolio of these—into martingales. Normalize the prices of both assets by the current per-unit value of the money fund, which is thus treated as numeraire. With "$*$" denoting normalized values, we have $s_j^* = s_j/m_j$ and $m_j^* = 1$, both quantities being dimensionless. Of course, $\{m_j^*\}$ is trivially a martingale under both $\mathbb{P}$ and $\hat{\mathbb{P}}$. As for $\{s_j^*\}$, taking conditional expectations in the new measure gives

$$\begin{aligned}
\hat{E}_j s_{j+1}^* &= \hat{\pi}_j \frac{s_j u}{m_{j+1}} + (1-\hat{\pi}_j)\frac{s_j d}{m_{j+1}} \\
&= \left(\frac{m_{j+1}/m_j - d}{u-d}\right)\frac{s_j u}{m_{j+1}} + \left(\frac{u - m_{j+1}/m_j}{u-d}\right)\frac{s_j d}{m_{j+1}} \\
&= \frac{s_j}{m_j} \\
&= s_j^*.
\end{aligned}$$

This with the tower property of conditional expectation then implies that

$$\hat{E}|s_n^*| = \hat{E}s_n^* = s_0^* + \hat{E}\left[\sum_{j=0}^{n-1} \hat{E}_j(s_{j+1}^* - s_j^*)\right] = s_0^*$$

for all $n \in \aleph$. Therefore, $\{s_j^*\}$ is also a martingale adapted to $\{\mathcal{F}_j\}$.[1] The appropriateness of the term "equivalent martingale measure" for $\hat{\mathbb{P}}$ is now evident.

Martingale/Risk-Neutral Pricing

Now comes the crucial implication for pricing a derivative on s. Given that the derivative can be replicated at each step with a self-financing portfolio of the underlying and the money fund, its own normalized value at $j+1$, $D_{j+1}^* \equiv D_{j+1}/m_{j+1}$, is a linear function of s_{j+1}^* with weights that are $\mathcal{F}_j$-measurable. Therefore, the martingale property of s^* and m^* under $\hat{\mathbb{P}}$ carries over to the normalized derivative also. That is, if at step j there are known portfolio weights p_j and q_j such that $p_j s_j + q_j m_j = D_j$ and $p_j s_{j+1} + q_j m_{j+1} = D_{j+1}$ for sure, it follows that

$$\hat{E}_j D_{j+1}^* = \hat{E}_j(p_j s_{j+1}^* + q_j) = p_j s_j^* + q_j = D_j^*.$$

[1]Since $\{s_j^*\}$ is a Markov process in the Bernoulli setup, one could as well write $\hat{E}_j(s_{j+1}^*) \equiv \hat{E}(s_{j+1}^* \mid \mathcal{F}_j)$ as $\hat{E}(s_{j+1}^* \mid s_j^*)$.

Indeed, in the Bernoulli setup one sees directly that the derivative's normalized price is a martingale under $\hat{\mathbb{P}}$, since by (5.12)

$$\begin{aligned} D_j^* &= \hat{\pi}_j \frac{D_{j+1}^u}{m_{j+1}} + (1 - \hat{\pi}_j) \frac{D_{j+1}^d}{m_{j+1}} \\ &= \hat{E}_j D_{j+1}^* \end{aligned}$$

for each $j \in \{0, 1, \ldots, n-1\}$.

The martingale property of $\{D_j^*\}$ can be exploited to find the derivative's arbitrage-free price at any step j by writing

$$D_j^* = \hat{E}_j D_n^*. \tag{5.17}$$

Since D_n^* is a known function of s_n and m_n, this expectation can be evaluated—either recursively or directly. Next, multiply both sides of (5.17) by the current value of the money-market fund, m_j, to restore to currency units:

$$D_j = m_j D_j^* = m_j \hat{E}_j \left(\frac{D_n}{m_n} \right). \tag{5.18}$$

For the initial price at $j = 0$ this gives

$$D_0 = m_0 \hat{E}(D_n / m_n),$$

or, since $\{m_j\}_{j=0}^n$ are all $\mathcal{F}_0$-measurable under assumption RK,

$$D_0 = \frac{m_0}{m_n} \hat{E} D_n = B(0, n) \hat{E} D_n \tag{5.19}$$

under that condition. Finally, using (5.16) to evaluate $\hat{E} D_n$ gives directly the explicit formula that was proved earlier by brute-force induction:

$$D_0 = B(0, n) \sum_{k=0}^{n} \hat{\Pi}(n, k) D(s_0 u^{n-k} d^k, 0). \tag{5.20}$$

These last expressions show that martingale pricing can be thought of equivalently as "risk-neutral" pricing when the price of the money fund is numeraire. Since $s_0 = B(0, n) \hat{E} s_n$ and $D_0 = B(0, n) \hat{E} D_n$ under assumption RK, current prices of both the underlying and the derivative are just what they would be in an equilibrium with risk-neutral agents who happened to regard $\hat{\mathbb{P}}$ as the actual probability measure. For this reason $\hat{\mathbb{P}}$ is often called the "risk-neutral measure" as well as an "equivalent-martingale measure". As we shall see later, this distinguishes $\hat{\mathbb{P}}$ from other martingale measures that apply for other numeraires. By switching to this new measure, assets can be priced by pretending that the economy is made up of individuals who are indifferent to risk. However, one must be careful not to misinterpret this procedure. Individuals in the Bernoulli world would really regard measure $\mathbb{P}$ as driving the dynamics of s, and there is no reason whatever to think that $\mathbb{P}$ and $\hat{\mathbb{P}}$ are the same. The crucial point is that the validity of risk-neutral pricing does not depend on the assumption of risk neutrality. Rather, given that replication is possible, the method is justified by the law of one price and the observation that markets do not afford systematic opportunities for arbitrage.

It is helpful to consider why pricing by expectation requires a change of measure. Notice that, if individuals were *really* risk-neutral, prices would satisfy $s_0 = B(0,n)Es_n$ and $D_0 = B(0,n)ED_n$, since expected returns from both positions under $\mathbb{P}$ would then equal the sure return from holding a riskless bond to maturity or equivalently (under assumption RK) investing in the money fund. However, since we think that most people display risk aversion in their investment decisions, this is not the right discount factor to use when expectations are taken with respect to $\mathbb{P}$. To find the factor B^* such that $D_0 = B^* ED_n$ would require measuring the derivative's risk and determining how much compensation the representative investor demands to bear it—facts about which reasonable people (and plausible models) are apt to disagree.[2] By contrast, if a self-financing, replicating portfolio exists, then arbitrage will drive the prices of the derivative and the portfolio together and tell us what the derivative is worth—regardless of peoples' tastes for risk.

We are now in a position to state in more detail the recipe for martingale pricing that was given in chapter 1. To find the arbitrage-free value of a derivative that makes a single payoff at some future date contingent on the price of an underlying asset,

- Verify that there exists a self-financing portfolio that replicates the derivative's payoff;
- Find a measure that (i) is equivalent to the real measure governing the price of the underlying and (ii) makes the underlying price a martingale after normalization by some numeraire;
- Relying on the fair-game property of martingales, calculate the current normalized price of the derivative as the conditional expectation of its normalized payoff under the new measure; and then
- Multiply by the current value of the numeraire to obtain the arbitrage-free price in currency units.

---

[2]Of course, once D_0 is found by martingale methods B^* can be found indirectly as $B^* = B\hat{E}D_n/ED_n$, assuming that we know enough to determine ED_n in measure $\mathbb{P}$.

When the price of the money fund is used as numeraire and assumption RK applies, the last two steps together are equivalent to finding the expectation of the derivative's unnormalized payoff under $\hat{\mathbb{P}}$ and discounting back to the present at the riskless rate. Discounting is equivalent to multiplying by the current price of a riskless unit bond that matures on the date of payoff. If there are contingent payoffs at multiple dates, price each of them individually in this way and add them up.[3]

Martingale Measures and Arbitrage-Free Markets

Within the Bernoulli framework it is easy to see that the existence of a martingale measure implies and is implied by the absence of opportunities for arbitrage, a statement that is now regarded as the *Fundamental Theorem of Asset Pricing.* Consider the simple one-period setup depicted in figure 5.3. An asset costs v_0 initially, and its price can take values v_1^u and v_1^d at $t = 1$ with probabilities $\mathbb{P}(v_1 = v_1^u) = \pi > 0$ and $\mathbb{P}(v_1 = v_1^d) = 1 - \pi > 0$. Any equivalent measure $\hat{\mathbb{P}}$ will assign positive probabilities $\hat{\pi}$ and $1 - \hat{\pi}$ to these same two outcomes. In this framework there are two canonical conditions

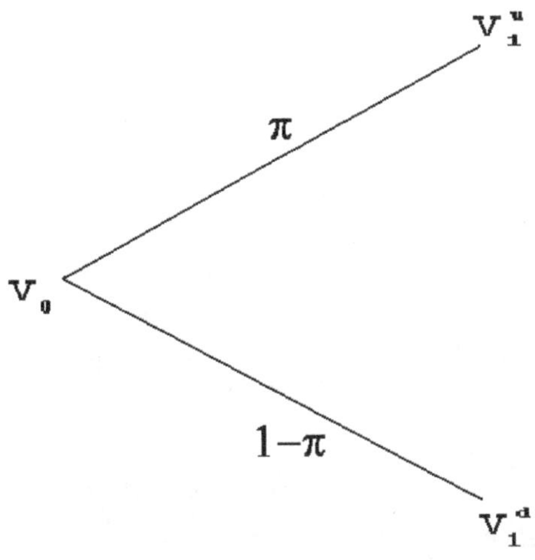

Fig. 5.3. A simple Bernoulli tree.

[3]In the continuous-time setting of chapter 10 we will distinguish between the "spot" martingale measure, $\hat{\mathbb{P}}$, for numeraire $\{M_t\}$ and the "time-T forward measure", $\mathbb{P}^T$, for numeraire $\{B(t,T)\}$. When the expectation is taken under $\mathbb{P}^T$, relation $D(S_t, T-t) = B(t,T)E^T[D(S_T, 0) \mid \mathcal{F}_t]$ holds even when future short rates are uncertain. When RK does hold, $\hat{\mathbb{P}}$ and $\mathbb{P}^T$ coincide.

for an arbitrage:

$$v_0 < 0, \quad v_1^u \geq 0, \quad v_1^d \geq 0; \tag{5.21}$$

$$v_0 = 0, \quad v_1^u > 0, \quad v_1^d \geq 0. \tag{5.22}$$

Under the first condition the asset delivers a positive initial payout (has negative cost), yet can be sold for a nonnegative amount in the future. An example would be the opportunity at $t = 0$ to exchange one portfolio for a cheaper one that was sure to be worth at least as much at $t = 1$. Under the second condition the asset costs nothing initially but has a chance to produce a positive payoff. For example, portfolios could be exchanged for even money with the result that future value might be greater and could not be less. In either case a rational person who prefers more wealth to less would work the trades on the highest possible scale. Now if (5.21) holds we will have

$$Ev_1 \geq 0 > v_0$$

under $\mathbb{P}$ and under any measure equivalent to $\mathbb{P}$, while if (5.22) holds we will have

$$Ev_1 > 0 = v_0.$$

In either case the martingale condition $Ev_1 = v_0$ fails to hold, showing that the existence of an arbitrage rules out the possibility of a martingale measure that is equivalent to $\mathbb{P}$.

Conversely, the existence of a martingale measure is inconsistent with both forms of arbitrage; for if $Ev_1 - v_0 = 0$ then either $v_1^u - v_0 > 0 > v_1^d - v_0$ in violation of (5.21) or else $v_1^u - v_0 = v_1^d - v_0 = 0$, violating both conditions. While we have been thinking here of prices denominated in some constant unit of exchange, it is obvious that these conclusions hold also for any numeraire w (numeraires being positive by definition) and corresponding normalized process, $\{v_j^* = v_j/w_j\}$. Moreover, the condition

$$\frac{v_1^d}{w_1^d} \leq \frac{v_0}{w_0} \leq \frac{v_1^u}{w_1^u} \tag{5.23}$$

implies the existence of at least one measure equivalent to $\mathbb{P}$ that makes the normalized price process a martingale, and when it is violated there is an opportunity for arbitrage.[4]

---

[4]That is, when (5.23) holds, there is necessarily a $\hat{\pi}$ such that $v_0/w_0 = \hat{\pi} v_1^u/w_1^u + (1-\hat{\pi})v_1^d/w_1^d$. When (5.23) does not hold, then either $v_1^d/w_1^d \leq v_1^u/w_1^u < v_0/w_0$ or $v_0/w_0 < v_1^d/w_1^d \leq v_1^u/w_1^u$. The first case implies $v_1^d/v_0 < w_1^d/w_0$ and $v_1^u/v_0 < w_1^u/w_0$, so that shorting v and investing the proceeds in w gives a costless, certain gain. The second case is exploited by the opposite trade.

It is not crucial that the price of the money fund be the numeraire. In fact, under assumption RK that future interest rates (and therefore bond and money-fund prices) are known, we can price an n-expiring derivative in exactly the same way using $\{B(j,n)\}_{j=0}^{n}$ as numeraire.[5] That is,

$$\left\{\frac{s_j}{B(j,n)} = \frac{m_n s_j}{m_j} = m_n s_j^*\right\}, \quad \left\{\frac{m_j}{B(j,n)} = m_n = m_n m_j^*\right\}$$

are still martingales under $\hat{\mathbb{P}}$, and so

$$D_0 = B(0,n)\hat{E}\frac{D_n}{B(n,n)} = B(0,n)\hat{E}D_n$$

as in (5.19). We can also normalize instead by $\{s_j\}$, creating $\{s_j^{**} = 1\}$, $\{m_j^{**} = m_j/s_j\}$, and $\{D_j^{**} = D_j/s_j\}$. Of course $\{s_j^{**}\}$ is automatically a martingale in any measure equivalent to $\mathbb{P}$ and $\hat{\mathbb{P}}$, but for the others to be martingales a new measure is required. Defining $\widetilde{\mathbb{P}}$ such that

$$\widetilde{\mathbb{P}}(s_{j+1} = s_j u \mid \mathcal{F}_j) = \tilde{\pi}_j = \frac{m_j/m_{j+1} - d^{-1}}{u^{-1} - d^{-1}},$$

we have

$$\begin{aligned}\tilde{E}(m_{j+1}^{**}|\mathcal{F}_j) &= \left(\frac{m_j/m_{j+1} - d^{-1}}{u^{-1} - d^{-1}}\right)\frac{m_{j+1}}{s_j u} + \left(\frac{u^{-1} - m_j/m_{j+1}}{u^{-1} - d^{-1}}\right)\frac{m_{j+1}}{s_j d} \\ &= m_j/s_j \\ &= m_j^{**}\end{aligned}$$

Likewise, using (5.9) it is easy to show that $\tilde{E}(D_{j+1}^{**}|\mathcal{F}_j) = D_j^{**}$. However, we would no longer refer to measure $\widetilde{\mathbb{P}}$ as a "risk-neutral measure", since $s_0 \neq B(0,n)\tilde{E}s_n$. Thus, while different martingale measures are, in general, associated with different numeraires, the price of any asset in the replicating portfolio—indeed, of any traded asset—can serve as the normalizing factor, so long as it is strictly positive.

Complete Markets

While the existence of a martingale measure is tied to the absence of arbitrage, the uniqueness of the measure for any given choice of numeraire is tied to the completeness of markets. To see what *completeness* means within the Bernoulli framework, let the market's information structure evolve over

[5]See footnote 3 in this connection.

a finite span of n periods according to the filtration $\{\mathcal{F}_j\}_{j=0}^n$. Traded in this market is a collection of $k+1$ assets with prices $\mathbf{s}_j = (s_{0j}, s_{1j}, \ldots, s_{kj})'$ at step j. The market is complete relative to the information structure $\{\mathcal{F}_j\}_{j=0}^n$ if each adapted payoff process $\{y_j\}_{j=1}^n$ can be replicated by a self-financing portfolio process $\{\mathbf{p}_j = (p_{0j}, p_{1j}, \ldots, p_{kj})'\}_{j=0}^{n-1}$. This means that, given an appropriate initial endowment, there exists a portfolio that generates any arbitrary stream of payoffs, $y_1, y_2, \ldots, y_n$.

Consider such a complete market. For any $j \in \{1, 2, \ldots, n\}$ let $A_j \subset \Omega$ be an arbitrary $\mathcal{F}_j$-measurable set of outcomes. Completeness implies that there exists at least one portfolio of assets that pays one currency unit at j when A_j occurs, and nothing otherwise. (Take the adapted process $\{y_i\}_{i=1}^n$ to be $y_i(\omega) = 1$ for $i = j$ and $\omega \in A_j$, and $y_i(\omega) = 0$ otherwise.) If there are no opportunities for arbitrage, then corresponding to whatever particular price is chosen as numeraire there exists at least one measure such that the normalized value of this claim is a martingale. Suppose there are two such (equivalent) measures, $\hat{\mathbb{P}}^j$ and $\widetilde{\mathbb{P}}^j$, when the price of a j-maturing unit bond is numeraire.[6] The current value of the claim is then either

$$\hat{v}_0(A_j) \equiv B(0,j)\hat{E}\mathbf{1}_{A_j} = B(0,j)\int_{A_j} d\hat{\mathbb{P}}^j(\omega) = B(0,j)\hat{\mathbb{P}}^j(A_j)$$

or

$$\tilde{v}_0(A_j) \equiv B(0,j)\tilde{E}\mathbf{1}_{A_j} = B(0,j)\int_{A_j} d\widetilde{\mathbb{P}}^j(\omega) = B(0,j)\widetilde{\mathbb{P}}^j(A_j).$$

Since there is no arbitrage, $\hat{v}_0(A_j)$ and $\tilde{v}_0(A_j)$ must be equal, and since such an equality holds for each j and each measurable A_j it follows that the measures $\hat{\mathbb{P}}$ and $\widetilde{\mathbb{P}}$ are identical. Thus, in a complete, arbitrage-free market there corresponds to each numeraire a unique martingale measure.

Completeness in the Bernoulli Framework

Let us now consider what it takes to make the martingale measure unique when the underlying price for our derivatives follows Bernoulli dynamics. Since uniqueness holds in complete markets, we can ask under what condition a market with a "stock" whose price, s, follows a Bernoulli process is complete relative to the information structure $\{\mathcal{F}_j\}_{j=0}^n$, where $\mathcal{F}_j \equiv \sigma(s_0, s_1, \ldots, s_j)$ is the field of events generated by $s_0, s_1, \ldots, s_j$.

---

[6] Under assumption RK, we can remove the "j" and consider $\hat{\mathbb{P}}$ and $\widetilde{\mathbb{P}}$ as alternative spot measures.

In order to create a portfolio that reproduces any adapted process $\{y_j\}_{j=1}^n$, it suffices to find a portfolio that pays y_j for any single $j \in \{1, 2, \ldots, n\}$ and zero at other times, since a collection of n such portfolios generates the entire process. Now since y_j is $\mathcal{F}_j$-measurable, there exists f such that $y_j = f(s_0, s_1, \ldots, s_j)$. Notice that y_j may depend on the entire history of the stock's price through stage j and thus on the path by which s_j is reached.

To replicate y_j let us try to work backwards in the usual fashion, beginning at stage $j - 1$ with $s_0, s_1, \ldots, s_{j-1}$ known. Given the Bernoulli dynamics, there are two possible realizations of y_j; namely, $y_j^u \equiv f(s_0, \ldots, s_{j-1}, s_{j-1}u)$ and $y_j^d \equiv f(s_0, \ldots, s_{j-1}, s_{j-1}d)$. We require a portfolio now worth some amount P_{j-1} and worth $P_j^u = y_j^u$ in the up state and $P_j^d = y_j^d$ in the down state next period. Since y_j^u and y_j^d are arbitrary, we clearly need a portfolio such that (P_j^d, P_j^u) can take on any value in $\Re_2$. To form such a portfolio requires, besides the stock itself, an asset whose price, v_j, is adapted to our market information structure, $\{\mathcal{F}_j = \sigma(s_0, s_1, \ldots, s_j)\}_{j=0}^n$, and which, together with the stock, spans $\Re_2$. In other words, we require the payoff matrix of the two assets,

$$\begin{pmatrix} s_{j-1}d & v_j^d \\ s_{j-1}u & v_j^u \end{pmatrix}$$

to have rank two or, equivalently, that the vectors (d, u) and (v_j^d, v_j^u) be linearly independent. This in turn requires that $u/d \neq v_j^u/v_j^d$. Figure 5.4

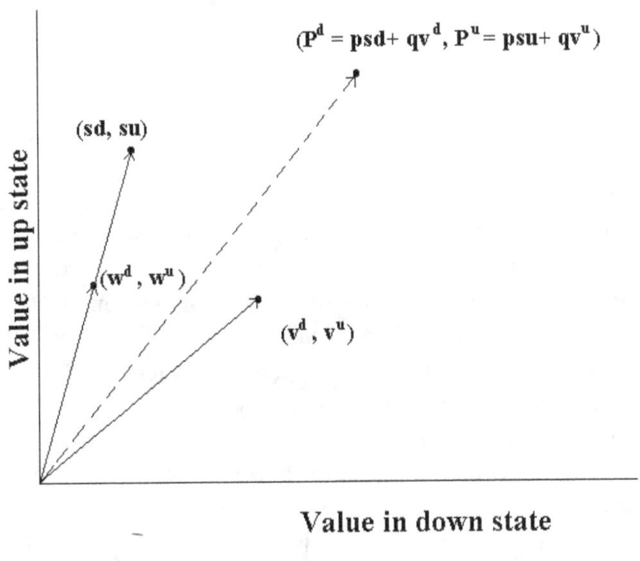

Fig. 5.4. Spanning the Bernoulli state space.

depicts state prices (v^d, v^u) that meet this condition and prices (w^d, w^u) that do not. The figure also shows how an appropriate portfolio (p, q) of the two assets with linearly independent prices can be used to replicate an arbitrary payoff $(y^d, y^u) = (P^d, P^u)$.

Suppose, besides a stock that follows Bernoulli dynamics, the market contains also the riskless money fund, a unit of which will be worth $m_j^u = m_j^d = m_{j-1}(1 + r_{j-1})$ in both the up and down states next period. This degenerate random variable m_j is linearly independent of the stock's price, since $m_j^u / m_j^d = 1 < u/d$. Therefore, under Bernoulli dynamics an arbitrage-free market consisting of the stock and the money fund is indeed complete with respect to the filtration generated by $\{s_j\}$, and so a unique martingale measure does correspond to whatever numeraire is chosen.

5.4 Specific Applications

This section treats binomial pricing of specific derivatives. To develop the intuition and some parallels with the continuous-time models in later chapters, we start with the simple case of European options on assets with zero explicit cost of carry, then move on to futures prices, derivatives on futures, American-style derivatives, and derivatives on assets that pay dividends. To focus on the main ideas, we continue to take as given the move sizes and one-period interest rates that define the Bernoulli dynamics. Ways to choose these in real applications are described in the next section.

5.4.1 *European Stock Options*

Pricing a derivative recursively by working backward through a binomial tree gives estimates of its value at all the nodes. Taken all together, the nodal values at any single time step represent the derivative's value function on a discrete domain of points; that is, the functional relation between the derivative's price and that of the underlying. Thus, after j steps backward in time the $n - j + 1$ nodal values of an n-step tree constitute the value function for a derivative with j steps to go before expiration. It is instructive to see how the value function evolves as one steps backward in the tree, and figure 5.5 illustrates the process for a European call. The bold kinked line in the top panel is the option's piecewise-linear terminal payoff function, drawn to connect the dots corresponding to the $n + 1$ nodal stock prices at step n. The dots in the next panel of the figure represent the value function at stage $n - 1$ on the lattice of n values of s_{n-1}. Taking the short rate as constant and setting $B = (1 + r)^{-1}$ as the one-period discount factor,

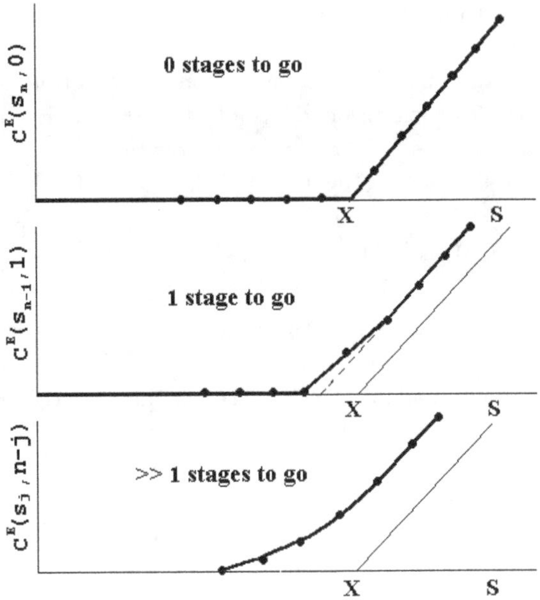

Fig. 5.5. Recursive evolution of binomial call function.

the call's value at each point is $C^E(s_{n-1}, 1) = B[\hat{\pi}C^E(s_{n-1}u, 0) + (1 - \hat{\pi})C^E(s_{n-1}d, 0)]$. At high values of s_{n-1} from which the call is sure to end up in the money this is

$$\begin{aligned} C^E(s_{n-1}, 1) &= B[\hat{\pi}(s_{n-1}u - X) + (1 - \hat{\pi})(s_{n-1}d - X)] \\ &= s_{n-1} - BX, \end{aligned}$$

where X is the strike price. These points lie along a line parallel to but above the terminal payoff, $s - X$. Skipping to the low values of s_{n-1}, dots along the S axis correspond to points from which the call must finish out of the money. The interest centers on the one value of s_{n-1} for which $s_{n-1}u - X > 0$ and $s_{n-1}d - X \leq 0$, where the call is worth $C^E(s_{n-1}, 1) = B\hat{\pi}(s_{n-1}u - X)$. This is positive and so lies above the axis, and it is also above the extension of the (dashed) line $s_{n-1} - BX$. This point thus puts an additional kink in the trace of call values at step $n - 1$. Another kink appears at step $n - 2$, and so on. Although the value function is observed at progressively fewer points as the tree reduces, one still perceives that the function becomes smoother as the process continues. We will soon be in a position to do the math and show that with the right specification of move sizes the value function approaches a smooth, convex curve as the number of remaining stages approaches infinity.

To get additional insight into the binomial valuation of a European option, apply the explicit pricing formula (5.20) to a European call. To make things easier, regard interest rates as constant over time, so that the simpler (5.11) applies to give for the call's initial value

$$C^E(s_0, n) = B(0, n) \sum_{k=0}^{n} \binom{n}{k} \hat{\pi}^{n-k} (1 - \hat{\pi})^k (s_0 u^{n-k} d^k - X)^+. \tag{5.24}$$

Reducing this further, let $k_{X,n}$ be the largest number of downticks such that the call finishes in the money. Then $(s_0 u^{n-k} d^k - X)^+$ equals $s_0 u^{n-k} d^k - X$ when $k \leq k_{X,n}$ and equals zero otherwise, so that the right side of (5.24) becomes

$$B(0, n) s_0 \sum_{k=0}^{k_{X,n}} \binom{n}{k} (\hat{\pi} u)^{n-k} [(1 - \hat{\pi}) d]^k - B(0, n) X \sum_{k=0}^{k_{X,n}} \binom{n}{k} \hat{\pi}^{n-k} (1 - \hat{\pi})^k.$$

The sum in the second term is $\hat{\mathbb{P}}(s_n > X)$. Expressing the bond's price as

$$B(0, n) = (1 + r)^{-n} = (1 + r)^{-n+k} (1 + r)^{-k},$$

the first term equals

$$s_0 \sum_{k=0}^{k_{X,n}} \binom{n}{k} \left[\frac{\hat{\pi} u}{1 + r} \right]^{n-k} \left[\frac{(1 - \hat{\pi}) d}{1 + r} \right]^k.$$

But since $\hat{\pi} u + (1 - \hat{\pi}) d = 1 + r$, the two bracketed factors sum to unity. Therefore, letting $\widetilde{\mathbb{P}}$ be a new measure equivalent to $\hat{\mathbb{P}}$ and $\mathbb{P}$ but with uptick probability $\tilde{\pi} = \hat{\pi} u / (1 + r)$, the call's value can be expressed finally as

$$C^E(s_0, n) = s_0 \widetilde{\mathbb{P}}(s_n > X) - B(0, n) X \hat{\mathbb{P}}(s_n > X). \tag{5.25}$$

The European put can be developed in the same way as

$$P^E(s_0, n) = B(0, n) X \hat{\mathbb{P}}(s_n < X) - s_0 \widetilde{\mathbb{P}}(s_n < X). \tag{5.26}$$

The formulas relate the prices of European calls and puts to the probabilities (in two different pseudo-measures) that the options expire in the money. It is clear immediately that the formulas are consistent with European put-call parity for options on an asset with zero explicit cost of carry: $C^E(s_0, n) - P^E(s_0, n) = s_0 - B(0, n) X$. Expressions (5.25) and (5.26) are very similar to the Black-Scholes formulas from the continuous-time theory, which are given in section 5.5 as (5.45) and (5.46). Indeed, with appropriate choices of u and d we will see that they actually converge to those expressions as $n \to \infty$. Thus, while (5.25) and (5.26) dominate the slower recursive procedure in pricing European options, their value is strictly pedagogical. On the other hand, we will see that the recursive method comes into its own in pricing more complicated derivatives, such as American options, for which exact computational formulas are not available.

5.4.2 *Futures and Futures Options*

The static replication arguments in chapter 4 established that the arbitrage-free forward price for time-T delivery of a commodity with zero explicit cost of carry is

$$\mathsf{f}(0, T) = B(0, T)^{-1} S_0. \tag{5.27}$$

We also saw that under condition RK the initial futures price $\mathsf{F}(0, T)$ must be the same as the forward price. Reassuringly, one can confirm these answers in the Bernoulli framework by applying martingale pricing formula (5.19); but to do so one must be clear about what the derivative asset actually *is* in connection with futures or forward contracts. The futures price itself is not the price of a derivative asset. It merely sets the terms for a later transaction, and no asset actually trades at the futures price until the contract expires, at which time the futures price is just the spot price. The real derivative asset whose discounted value is a martingale under the risk-neutral measure is the futures *position*, which is to say the value of the futures account. Using the time-step notation from the Bernoulli setup and assuming that the futures position is marked to market at each time step, the value at step $j+1$ of a long position initiated at step j is $\mathsf{F}_{j+1} - \mathsf{F}_j$ per commodity unit. Applying (5.19) gives

$$0 = \frac{\mathsf{F}_j - \mathsf{F}_j}{m_j} = \hat{E}_j \left(\frac{\mathsf{F}_{j+1} - \mathsf{F}_j}{m_{j+1}} \right). \tag{5.28}$$

which (since $m_{j+1} \in \mathcal{F}_0 \subset \mathcal{F}_j$ by assumption RK) implies that $\mathsf{F}_j = \hat{E}_j \mathsf{F}_{j+1}$, for $j \in 0, 1, \ldots, n-1$. The fact that the normalized value of the futures *position* is a martingale thus does imply that the futures price itself is a martingale under $\hat{\mathbb{P}}$, provided we abstract from uncertainty about future interest rates.

With this result we can now see that the forward and futures prices do coincide and that they satisfy cost-of-carry relation (5.27). Since $\mathsf{F}_n = s_n \equiv S_T$ and since $\{s_j/m_j\}_{j=0}^n$ is a martingale under $\hat{\mathbb{P}}$, repeated application of (5.28) gives $\mathsf{F}_0 = \hat{E}s_n = s_0(m_n/m_0) = B(0,n)^{-1}s_0 = \mathsf{f}_0$ or, in the continuous-time notation, $\mathsf{F}(0,T) = B(0,T)^{-1}S_0 = \mathsf{f}(0,T)$.

Having developed some background, let us now consider derivatives whose payoffs are linked to a futures price. An example is futures options, which on exercise yield cash payments plus positions in the futures. Specifically, one who exercises a European futures call option struck at X receives $\mathsf{F}(T,T') - X$ in cash when the option expires at T, plus a long position in a futures contract that expires at some $T' > T$. Since the futures position has no value when it is first assumed, the option would be exercised only if $\mathsf{F}(T,T') > X$, and the call's terminal value is therefore $[\mathsf{F}(T,T') - X]^+$. Similarly, the value of an expiring put on the futures is $[X - \mathsf{F}(T,T')]^+$. If exercised at T, it would convey a cash payment of $X - \mathsf{F}(T,T')$ plus a short position in the futures. In a frictionless market these long and short positions in the underlying futures can be closed out immediately at no cost by taking offsetting short or long positions.

Modeling the futures price itself as a Bernoulli process, with $\mathsf{F}^u_{j+1} = \mathsf{F}_j\mathsf{u}_j$ and $\mathsf{F}^d_{j+1} = \mathsf{F}_j\mathsf{d}_j$, we will try to form a self-financing, dynamic replicating portfolio for a European-style futures derivative using futures contracts and investments in the money fund. (We will see presently why it might be appropriate to allow time-dependent tick sizes, as the notation suggests.) Things will work a little differently than for derivatives on stocks and other primary assets. Because a futures contract has zero value at the time it is acquired, the step-j value of a replicating portfolio comprising p_j contracts and q_j units of the money fund is just $D_j = q_j m_j$. Since the contracts acquire positive or negative value at the next time step as the futures price moves, the replicating shares for futures and money fund at step j must solve

$$\begin{aligned} p_j(\mathsf{F}^u_{j+1} - \mathsf{F}_j) + q_j m_{j+1} &= D^u_{j+1} \\ p_j(\mathsf{F}^d_{j+1} - \mathsf{F}_j) + q_j m_{j+1} &= D^d_{j+1}, \end{aligned}$$

where D^u_{j+1} and D^d_{j+1} are uptick and downtick values of the derivative. The solutions are

$$\begin{aligned} p_j &= \frac{D^u_{j+1} - D^d_{j+1}}{\mathsf{F}^u_{j+1} - \mathsf{F}^d_{j+1}} \\ q_j &= \frac{D_{j+1} - p_j(\mathsf{F}_{j+1} - \mathsf{F}_j)}{m_{j+1}}. \end{aligned}$$

A little algebra then shows that the value of the replicating portfolio at step j equals

$$\begin{aligned} q_j m_j &= \frac{m_j}{m_{j+1}} \frac{(\mathsf{F}_j - \mathsf{F}_{j+1}^d) D_{j+1}^u + (\mathsf{F}_{j+1}^u - \mathsf{F}_j) D_{j+1}^d}{\mathsf{F}_{j+1}^u - \mathsf{F}_{j+1}^d} \\ &= \frac{m_j}{m_{j+1}} \left[\tilde{\pi}_j D_{j+1}^u + (1 - \tilde{\pi}_j) D_{j+1}^d \right], \end{aligned}$$

with

$$\tilde{\pi}_j = \frac{1 - \mathsf{d}_j}{\mathsf{u}_j - \mathsf{d}_j}. \tag{5.29}$$

The pseudo-probabilities $\{\tilde{\pi}_j, 1 - \tilde{\pi}_j\}$ are the basis for a new measure $\widetilde{\mathbb{P}}$ that depends on the move sizes for the futures price. At step $j+1$ the futures position from step j is closed out, the profits (losses) are reinvested in (made up from) the money fund, and new futures commitments are made to replicate at the next step. In this way the replicating portfolio is forced to be self-financing. As usual, the normalized price of the replicating portfolio, and therefore of the derivative itself, is a martingale in the corresponding measure $\widetilde{\mathbb{P}}$; that is,

$$\begin{aligned} \frac{D_j}{m_j} &= \tilde{\pi}_j \frac{D_{j+1}^u}{m_{j+1}} + (1 - \tilde{\pi}_j) \frac{D_{j+1}^d}{m_{j+1}} \\ &= \tilde{E}_j \left(\frac{D_{j+1}}{m_{j+1}} \right) \end{aligned}$$

and, as in (5.19),

$$D_0 = \frac{m_0}{m_n} \tilde{E} D_n = B(0, n) \tilde{E} D_n. \tag{5.30}$$

This parallels the development for derivatives on primary assets, such as the one that underlies the futures, yet it seems that we have wound up with a different risk-neutral probability—$\tilde{\pi}_j$ instead of the $\hat{\pi}_j$ that drives the underlying price itself—and a different martingale measure. In fact, $\tilde{\pi}_j$ and $\hat{\pi}_j$ turn out to be the same if move sizes for the futures are made consistent with those for the underlying primary asset. This would have to be true since the dynamics of the primary asset are driving everything here, but we can see it formally as follows. Impose the usual Bernoulli dynamics with move sizes u and d for the primary asset, and suppose there are N time steps until the futures contract expires and $n < N$ steps in the life of the futures derivative. Since the futures price itself is a martingale under

$\hat{\mathbb{P}}$, we have $\mathsf{F}_j = \hat{E}_j s_N = (m_N/m_j)s_j$, so that

$$\mathsf{F}^u_{j+1} = \frac{m_N}{m_{j+1}} s_j u = \left(\frac{m_N}{m_j} s_j\right) \frac{m_j}{m_{j+1}} u = \mathsf{F}_j \frac{m_j}{m_{j+1}} u,$$

and likewise

$$\mathsf{F}^d_{j+1} = \mathsf{F}_j \frac{m_j}{m_{j+1}} d.$$

Thus, if the dynamics for the futures are to be consistent with those for the underlying, the move sizes for the two prices must be related. For a traded asset with zero cost of carry the relation is that $\mathsf{u}_j/u = \mathsf{d}_j/d = m_j/m_{j+1} = (1+r_j)^{-1}$. Recognizing this connection, we see that $\tilde{\pi}_j$ and $\hat{\pi}_j$ are indeed the same. Of course, if we work in terms of futures prices rather than prices of the underlying commodity, the measures $\widetilde{\mathbb{P}}$ and $\hat{\mathbb{P}}$ are indeed different, because the state spaces differ. In applications it is common to ignore the connection between (u, d) and (u, d) and to estimate the move sizes for futures directly from observations on futures prices rather than prices of the underlying.

Example 61 *Consider replicating and pricing a 3-period European futures call with strike $X = 10.0$ when the initial futures price is $\mathsf{F}_0 = 10.0$. Take r to be constant at* 0.01 *and $\mathsf{u} = 1.05/1.01$ and $\mathsf{d} = 0.97/1.01$ in each period, which would be consistent with move sizes of $u = 1.05$ and $d = 0.97$ for the price of the underlying commodity. Then* (5.29) *gives $\tilde{\pi} = \frac{1}{2}$. Figure* 5.6 *illustrates recursive valuation and replication of the call. Applying* (5.30)*, the initial value can also be obtained directly as*

$$C^E(10.0, 3) = \frac{1}{1.01^3} \sum_{k=0}^{3} \frac{\binom{3}{k}}{2^3} \left[10.0 \left(\frac{1.05}{1.01}\right)^{3-k} \left(\frac{0.97}{1.01}\right)^{k} - 10.0\right]^+ .$$

5.4.3 *American-Style Derivatives*

Ordinary or "vanilla" American-style derivatives can be exercised at any time before expiration and, upon exercise, deliver payoffs that depend just on the current value of the underlying. We can consider just as easily slightly more general forms whose payoffs can depend on time as well. For this purpose $D^X_j \equiv D^X_j(s_j)$ represents the derivative's value at the (s_j, j) node if it is not held beyond step j. Since the derivative could fail to be held (by someone) only if it were exercised or discarded, and since disposal is costless, it follows that $D^X_j \geq 0$. We refer to this (somewhat imprecisely)

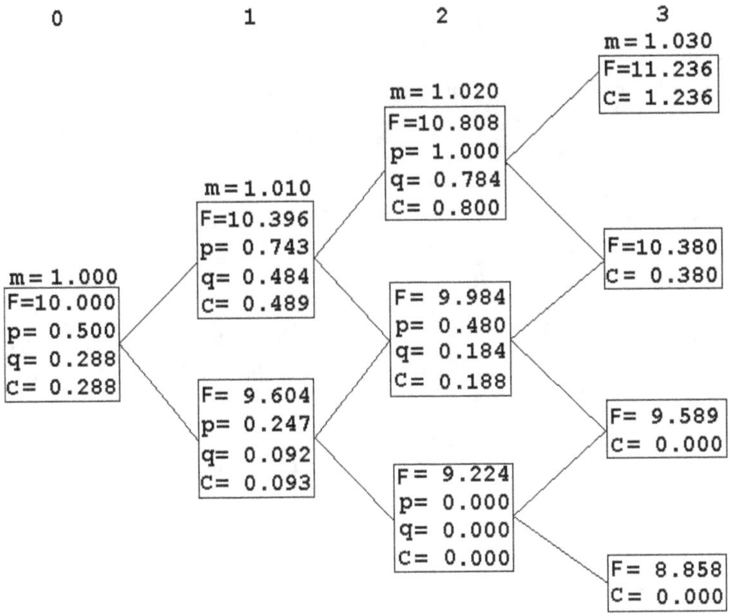

Fig. 5.6. Pricing a European call on a futures.

as the derivative's "exercise value". As usual, D_j without the "X" represents the derivative's arbitrage-free price at j. Obviously $D_j \geq D_j^X$, since the holder has the right to exercise (or discard) at any time, and exercise occurs if and only if $D_j = D_j^X > 0$. The tasks are to discover at each step and each underlying price whether exercise is optimal and, thereby, ultimately to determine the derivative's current price, D_0. We first treat derivatives on primary assets ("stocks") and then turn to American-style futures derivatives. Throughout this section we continue to assume that the underlying has no explicit cost of carry, meaning that one who holds the underlying receives no dividends or other emoluments and incurs no explicit costs. These complications are considered in section 5.4.4.

Derivatives on Primary Assets

Begin by valuing an American derivative at an arbitrary node (s_j, j) of an n-step binomial tree, where n corresponds to the expiration time. As usual, there is nothing to do at the terminal nodes, since values there are known functions of s_n. When $j < n$ the holder of the derivative has two relevant options: (i) hold until at least $j+1$ or (ii) exercise (or discard) immediately. As for European-style derivatives, the self-financing portfolio that replicates under the "hold" option contains

$$p_j = \frac{D_{j+1}^u - D_{j+1}^d}{s_j u - s_j d}$$

shares of the stock and

$$q_j = \frac{D_{j+1}}{m_{j+1}} - p_j \frac{s_{j+1}}{m_{j+1}}$$

units of the money fund. The current value of the replicating portfolio, which equals the derivative's arbitrage-free value under the hold option, is then

$$\begin{aligned} p_j s_j + q_j m_j &= \frac{m_j}{m_{j+1}} \left[\hat{\pi}_j D_{j+1}^u + (1-\hat{\pi}) D_{j+1}^d \right] \\ &= \frac{m_j}{m_{j+1}} \hat{E}_j D_{j+1}. \end{aligned} \tag{5.31}$$

Since the derivative is worth D_j^X if not held, and since holding is discretionary, the solution for D_j at the (s_j, j) node is

$$D_j = \frac{m_j}{m_{j+1}} \hat{E}_j D_{j+1} \vee D_j^X, \tag{5.32}$$

where "$\vee$" signifies "greater of" and D_{j+1} represents the value at $j+1$ if the derivative is not exercised at step j.

The procedure for pricing an American-style derivative is then as follows. One begins at step n, where the derivative's terminal value is known, and works backward through the tree. At each node of each step $j < n$, (5.31) is used to calculate the value under the hold option from the two state-dependent values at the next step. This value is compared with the value under the exercise option, D_j^X, and then D_j is set equal to the greater of these. When prices at all nodes of step j have been found in this way, one moves back to step $j-1$, repeats the process, and then continues in this manner step by step back to stage 0.

The next section shows how to modify this procedure when the underlying pays a dividend or has some other positive or negative explicit carrying cost. When there are no such costs, we have seen that there are explicit solutions for prices of European-style derivatives and that it is not necessary to work through the entire tree to price them. But, with one exception, the recursive method is essential for American derivatives, since at each step the values at the *next* future time step must be known in order to decide whether or not to hold. The exceptional case is that of an American call on a no-dividend stock, for which early exercise would never be optimal. On the other hand, an American put may well be exercised before expiration, regardless of whether the underlying pays dividends. As a result, American puts are typically worth more than their European counterparts, and they must always be priced recursively.

Example 62 *Figure* 5.7 *illustrates binomial pricing of a* 4*-period vanilla American put struck at* $X = 10.0$ *on an underlying stock with initial price* $s_0 = 10.0$. *Taking* $u = 1.05$, $d = 0.95$, *and* $r = 0.01$ *gives* $\hat{\pi} = (1.01 - 0.95) \div (1.05 - 0.95) = 0.6$. *The put's value if not held beyond* j *is* $P_j^X(s_j) = (X - s_j)^+$, *independent of time. The values* p *and* q *at each node in the figure are the number of shares of stock and money fund that replicate the put if it is held to the next period. A shaded block at a node indicates that the put would be exercised there. As a sample of the calculations, look at the bottom node at step* 2, *which is marked with a check* ($\checkmark$) *in the figure. The put's*

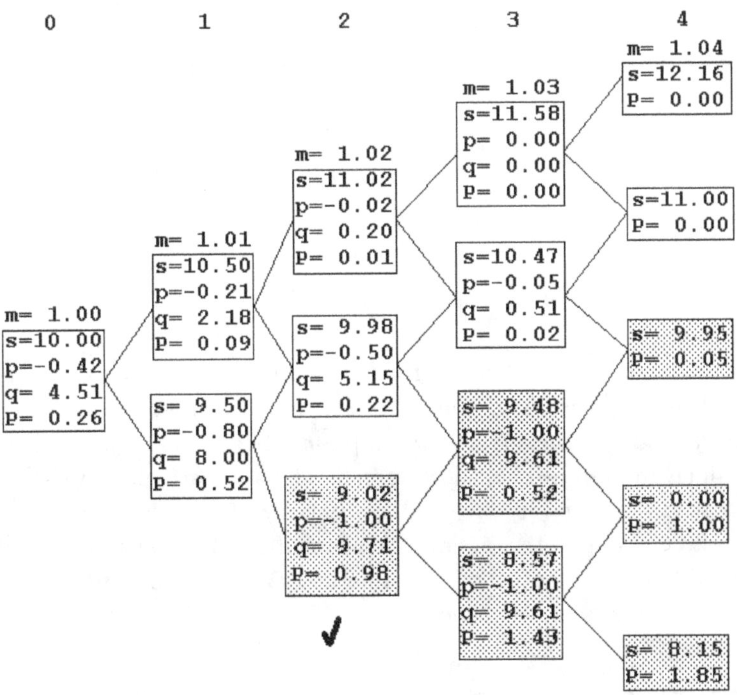

Fig. 5.7. Binomial pricing of American put.

value if held is calculated via (5.31) *as* $[.6(0.524) + .4(1.426)] \div 1.01 \doteq 0.876$, *and the exercise value is* $(10.000 - 9.025) = 0.975$; *hence, the value of the American put at this node is* $P_2^A = 0.876 \vee 0.975 \doteq 0.98$.

Futures Derivatives

Consider now American options on futures. Once the move sizes are determined appropriately to model the futures price, the binomial method works in the same way as for American options on primary assets. However, in one instance there is an important qualitative difference in the results: even though a dividend is never associated with a futures price, an American call on the futures might still be exercised early. Accordingly, American calls on futures will ordinarily be worth more than otherwise identical European calls and must be priced recursively.

The possibility of rational early exercise for the American call on futures can be demonstrated easily in a one-step binomial tree. When the call expires at step 1, it will be worth $(\mathsf{F}_1 - X)^+$, where $\mathsf{F}_1 = \mathsf{F}_0\mathsf{u}$ in the up state and $\mathsf{F}_1 = \mathsf{F}_0\mathsf{d}$ in the down state. Suppose $\mathsf{F}_1 - X > 0$ in *both* states. Since the call is certain to wind up in the money from the initial F_0, it must be in the money at step 0 also. Then the current value of the American call is

$$\begin{aligned}
C^A(\mathsf{F}_0, 1) &= (1+r)^{-1}\hat{E}(\mathsf{F}_1 - X)^+ \vee C_0^X \\
&= (1+r)^{-1}\hat{E}(\mathsf{F}_1 - X) \vee (\mathsf{F}_0 - X) \\
&= (1+r)^{-1}(\mathsf{F}_0 - X) \vee (\mathsf{F}_0 - X) \\
&= (\mathsf{F}_0 - X).
\end{aligned}$$

Here, the third equality follows from the martingale property of the undiscounted futures price in the measure $\hat{\mathbb{P}}$, as in (5.28) with $n = 1$, and the last equality holds since $r \geq 0$ (assumption RB). This shows that it would be optimal to exercise the call at step 0. The same argument would apply at any step j of an n-step tree and any initial F_j from which the call would be sure to wind up in the money.[7] Notice that it is the fact that the futures price is a $\hat{\mathbb{P}}$ martingale that makes American calls on futures fundamentally different from calls on no-dividend stocks. Following through the steps of the

[7]The certainty of finishing in the money is merely sufficient for early exercise, however. It is not a necessary condition.

example with a one-period equity call instead of a call on the futures gives

$$\begin{aligned} C^A(s_0, 1) &= (1+r)^{-1}\hat{E}(S_1 - X)^+ \vee (S_0 - X) \\ &= (1+r)^{-1}\hat{E}(S_1 - X) \vee (S_0 - X) \\ &= (1+r)^{-1}[S_0(1+r) - X] \vee (S_0 - X) \\ &= S_0 - (1+r)^{-1}X. \end{aligned}$$

Consistent with the arguments in chapter 4, American calls on no-dividend stocks would be kept to expiration, even if they were certain to end up in the money. In the next section we shall see that this is no longer always true when the stock pays dividends.

Another View of American Derivatives

While the mechanics of valuation in the Bernoulli framework are clear, the method is not particularly insightful. There is another perspective from which to view the arbitrage-free price of an American-style derivative. The frontier of nodes at which early exercise is optimal represents the "exercise boundary". There are always such critical exercise boundaries for American options, whether they be vanilla puts or calls or more exotic varieties. For the put in example 62 this is the boundary between the unshaded and shaded nodes in figure 5.7. In general, for vanilla puts the exercise boundary is a function of the time step, $\mathfrak{B} : \{0, 1, \ldots, n\} \to \Re^+$, such that exercise occurs at step j whenever $s_j < \mathfrak{B}_j$. A given path followed by the stock's price from an initial point above the boundary produces an outcome of positive value for the put when and only when it crosses this boundary for the first time. The put then generates a cash receipt, and (since exercise has occurred) the remainder of the path becomes irrelevant. Paths that never cross the exercise boundary contribute nothing to the option's value. Those that do cross contribute to the value of the derivative an amount equal to the product of the receipt itself times the risk-neutral probability of the path times the discount factor that converts the future receipt to present value. The value of an American derivative can be regarded as the sum of these contributions. For example, in figure 5.7 there are six such paths to a first crossing. Each path is shown in table 5.1 as a string of d's and u's, along with its contribution to value. These sum to about 0.26, as in the figure.

This path-by-path view suggests formal ways to characterize values of American-style derivatives. Letting "ω" designate a sample path of the underlying price and assuming that it is not initially optimal to exercise,

Table 5.1. Sample paths of underlying contributing to value of American put.

Path	Probability		Receipt		Time Discount		Value
dd	0.4^2	×	0.975	÷	1.01^2	=	0.153
dud	$0.4^2(0.6)$	×	0.524	÷	1.01^3	=	0.049
udd	$0.4^2(0.6)$	×	0.524	÷	1.01^3	=	0.049
uudd	$0.4^2(0.6^2)$	×	0.050	÷	1.01^4	=	0.003
udud	$0.4^2(0.6^2)$	×	0.050	÷	1.01^4	=	0.003
duud	$0.4^2(0.6^2)$	×	0.050	÷	1.01^4	=	0.003
					Total value:		0.260

define

$$J(\omega) = \min\{1 \leq j \leq n : D_j^X \geq D_j\}.$$

This represents the number of the first step on path ω at which the value of the derivative is not greater than its exercise value. If ω crosses the exercise boundary before n, $J(\omega)$ will be the time step of the first crossing. If there is no crossing before n, then $J(\omega) = n$, since in that case the value and exercise value will coincide for the first time at expiration. (If the derivative expires out of the money, both D_n and D_n^X are zero.) With this convention the initial value of the American derivative is

$$\hat{E}\left[\frac{m_0}{m_J} D_J^X(s_J)\right] = \hat{E}[B(0, J) D_J^X(s_J)].$$

This amounts to doing just what was done in the table, except that it averages in the paths that contribute nothing to value along with those that do.

The value of a particular derivative can be also stated directly in terms of its positive payoffs at the exercise boundary and the times of first passage. The first-passage time is defined as

$$J^*(\omega) = \min\{1 \leq j \leq n : s_j < \mathfrak{B}_j\}$$

if this set is not empty (that is, if the crossing does occur by step n) and $J^*(\omega) = n + 1$ otherwise. Note that events $J(\omega) = j$ and $J^*(\omega) = j$ are measurable with respect to $\mathcal{F}_j$, since one always knows at stage j whether the condition for exercise is met. Thus, J and J^* are stopping times with respect to the filtered space $(\Omega, \mathcal{F}, \{\mathcal{F}_j\}, \mathbb{P})$. For the American put, which has exercise value $P_j^X(s_j) = X - \mathfrak{B}_j$ at the boundary, the initial value is

$$P^A(s_0, n) = \sum_{j=1}^{n} B(0, j)(X - \mathfrak{B}_j)\hat{\mathbb{P}}(J^* = j),$$

provided the stock's initial price s_0 is above $\mathfrak{B}_0$. If it is not, then $P^A(s_0, n) = X - s_0$.

5.4.4 *Derivatives on Assets That Pay Dividends*

The prospect of a dividend payment may either trigger or delay the exercise of an American option and typically affects the values of derivatives of all types. In this context the term *dividends* applies to any receipts of positive monetary value that (i) accrue to the holder of an asset, (ii) alter the dynamics of its price, and (iii) are not automatically adjusted for in the contractual terms of a derivative asset. Dividends are the negative components of an underlying asset's explicit cost of carry, and for financial assets they are typically the only such components. Examples of dividends in the broad sense (as the term is used here) include the interest accrued on a long position in a money-market fund denominated in a foreign currency, coupon interest received from a portfolio of bonds, and cash dividends on common or preferred stock. The definition does exclude stock dividends, which are pro-rata disbursements of shares of the underlying stock itself, when considering derivatives (such as exchange-traded options) whose terms are adjusted for the price effects of such dilutions of value.[8]

As described in chapter 4, dividends can sometimes be modeled accurately as proportional to the value of the underlying asset at the time payments are made. This would be true, for example, of interest payments at a known rate on a foreign money fund or discount bond, whose value in domestic currency units would depend on the exchange rate. In other cases it is reasonable to model dividends as lump-sum cash amounts that are completely unrelated to the value of the underlying. Interest payments on default-free coupon bonds are examples, and cash dividends paid by common stocks are often assumed to be lump-sum as well. In line with the assumption that cost of carry is known (condition CCK), if the dividends

[8]For exchange-traded options both stock dividends and stock splits trigger adjustments to the strike price and either the contract size or the number of contracts. Letting ρ be the split factor (the number of shares outstanding per initial share after the split or dividend), strike prices are adjusted as X/ρ. If ρ is an integer (as in a two-for-one split or 100% stock dividend), then the number of option contracts is multiplied by ρ. If ρ is not an integer, then the number shares per option contract (normally 100) is increased by that factor.

However, such contractual adjustments do not insulate derivatives' prices from the volatility changes that usually accompany splits and stock dividends. In pricing the derivatives such effects have to be captured by the appropriate modeling of volatility.

paid by the underlying are proportional to price, we assume that the rates (proportions) over the life of the derivative are known; if they are lump-sum, we assume that cash amounts and dates of receipt are known.

The frequency with which common stocks pay dividends varies from country to country. For example, in the U.S. they are typically paid quarterly, while annual payments are the standard in Germany. Because it takes time to disburse the payments, a stock begins to trade "*ex* dividend" a short time before the payments are made, and those who buy the stock after the *ex* date do not receive the dividend. On average, the price of a share drops by about the amount of the dividend on the *ex* date. The short delay before payment is received and the tax disadvantage of dividends *vs.* capital gains in the U.S. both slightly diminish the price effect for U.S. stocks. For purposes of pricing derivatives on stocks it is sensible just to define the dividend as the price effect that results from the disbursement.

The proper treatment of dividends in the binomial scheme depends on the frequency of payments and whether they are proportional to value of the underlying or of fixed amount. We will look at three cases, in increasing order of difficulty: (i) proportional dividends paid continuously, (ii) proportional dividends paid at discrete intervals, and (iii) fixed (lump-sum) cash amounts paid at discrete intervals. In each case we will work out the replicating portfolio and martingale measure for pricing an American-style derivative on an underlying asset (which we continue to call a "stock"), and will indicate briefly how the procedure is to be modified for European-style derivatives.

Continuous, Proportional Dividends

In the Bernoulli framework *continuous* means occurring at each time step. Working with the usual n-step binomial tree, the holder of the underlying stock receives a dividend δs_j, proportional to price, at each node $(s_j, j)_{j=1}^n$. (Allowing for time dependence of δ would just require adding a subscript.) A key observation here is that it is not possible without portfolio adjustments to maintain a desired position in the underlying stock alone, because it generates cash payments at each step. One thing that can be done is to reinvest the dividend in the stock each time it is received, using the payment δs_j to purchase δ additional shares. Alternatively, some or all of the cash could be invested the money market fund. We show below that the result would be the same, but for now we consider just reinvestment in the stock. Let $s_j^c = s_j(1+\delta)^j$ be the value at step j of one unit of this *cum*-dividend position that begins at step 0 with one actual share of stock and collects additional units as dividends are received and reinvested. This *position* is the asset that must be used for replication—and the asset whose discounted value should be a martingale under the risk-neutral measure.

The usual Bernoulli dynamics for the price of the stock itself, as $s_{j+1} = s_j R_{j+1}$ for $R_{j+1} \in \{d, u\}$, give $s^c_{j+1} = s^c_j(1+\delta)R_{j+1}$ for the value of the *cum*-dividend position. Let $D_{j+1}(s_{j+1}) \equiv D(s_{j+1}, n-j-1)$ be the value of the derivative asset at $j+1$, when it has $n-j-1$ periods to go. Notice that $D_{j+1}(s_{j+1})$ depends on the actual market price of the stock rather than on s^c_{j+1} since it is the former that determines exercise value. The replicating portfolio (p_j, q_j) that sets

$$p_j s^c_j(1+\delta)u + q_j m_{j+1} = D_{j+1}(s_j u) \tag{5.33a}$$

$$p_j s^c_j(1+\delta)d + q_j m_{j+1} = D_{j+1}(s_j d) \tag{5.33b}$$

is

$$p_j = \frac{D_{j+1}(s_j u) - D_{j+1}(s_j d)}{s^c_j(1+\delta)(u-d)},$$

$$q_j = \frac{D_{j+1}(s_j d)\cdot u - D_{j+1}(s_j u)\cdot d}{m_{j+1}(u-d)}.$$

Setting $m_{j+1} = m_j(1+r)$, some algebra gives the derivative's value if held as

$$p_j s^c_j + q_j m_j = \frac{1}{1+r}[\hat{\pi} D_{j+1}(s_j u) + (1-\hat{\pi}) D_{j+1}(s_j d)], \tag{5.34}$$

where

$$\hat{\pi} = \frac{(1+r)/(1+\delta) - d}{u-d}. \tag{5.35}$$

Therefore, if the current exercise value is $D^X_j(s_j)$ the derivative's current value at step j is

$$D_j(s_j) = \frac{1}{1+r}\hat{E}_j D_{j+1}(s_{j+1}) \vee D^X_j(s_j).$$

Notice that, for actual calculations, accounting for continuous, proportional dividends is a simple matter of dividing $1+r$ by $1+\delta$ in the expression for $\hat{\pi}$ and proceeding as if there were no dividends. That is why the continuous-dividend case is the simplest to handle.

It is worth pausing a moment to see a more direct way to get the same result and to appreciate its intuitive content. The pseudo-probability $\hat{\pi}$ in

(5.35) could have been found right away by determining the probability that makes the process $\{s_j^c/m_j\}$ a martingale; that is, by solving

$$\begin{aligned}\frac{s_j^c}{m_j} &= \hat{E}_j \frac{s_{j+1}^c}{m_{j+1}} \\ &= \frac{s_j^c}{m_j}\left(\frac{1+\delta}{1+r}\right)[\hat{\pi}u + (1-\hat{\pi})d]\,. \end{aligned} \tag{5.36}$$

Once this was done we could have seen that it was possible to construct a self-financing, replicating portfolio by just writing down (5.33a) and (5.33b) without having to solve for p_j and q_j. With the existence of a martingale measure and a replicating portfolio thus verified, the right side of (5.34) could have been obtained immediately to value the derivative. As to the intuitive content, notice that the expected one-period return of the stock from capital gain alone (in the martingale measure) is

$$\hat{E}_j(s_{j+1}/s_j) = \frac{1+r}{1+\delta}.$$

This corresponds to the fact that in a risk-neutral market the expected value of the *cum*-dividend, one-period return, $s_{j+1}(1+\delta)/s_j$, would be the same as the sure return on the money fund, $1+r$.

Had all dividends been invested in the money fund rather than in the stock, it is the normalized value of the stock-fund *position* that would be a martingale. Letting v_j be the position's value at step j, it is easy to see that $v_{j+1} = s_j R_{j+1}(1+\delta) + (v_j - s_j)m_{j+1}/m_j$ and hence that

$$\frac{v_{j+1}}{m_{j+1}} = \frac{v_j}{m_j} + \left[\frac{s_{j+1}}{m_{j+1}}(1+\delta) - \frac{s_j}{m_j}\right].$$

Accordingly, the same value of $\hat{\pi}$ equates $\hat{E}_j v_{j+1}/m_{j+1}$ to v_j/m, so that the martingale measure is independent of the reinvestment choice for dividends.

To summarize, in pricing a European-style derivative on an asset with continuous proportional dividends at rate δ, one merely divides $(1+r)$ or $(1+r_j)$ by $(1+\delta)$ in calculating the $\hat{\pi}$ or $\hat{\pi}_j$ that enters expression (5.11) or (5.15). Moreover, if r and δ are small, $(1+r)/(1+\delta)$ can be approximated as $1+r-\delta$, an expression for which we will see parallels in the continuous-time treatment of chapter 6.

A Symmetry Relation for Puts and Calls

We shall see later that uptick and downtick values are inversely related in a standard implementation of Bernoulli dynamics; i.e., $d = u^{-1}$. When this holds, there is a useful symmetry between the values of puts and calls—European and American alike. Consider first European puts and calls on an underlying asset that pays proportional dividends continuously at rate δ. Representing the values of the options in terms of underlying price, strike price, dividend rate, and the short rate, the call's value at step $n-1$ is

$$\begin{aligned} C^E_{n-1}(s_{n-1}; X, \delta, r) &= (1+r)^{-1}\hat{E}_{n-1}(s_n - X)^+ \\ &= \left(\frac{\frac{1}{1+\delta} - \frac{d}{1+r}}{u-d}\right)(s_{n-1}u - X)^+ \\ &\quad + \left(\frac{\frac{u}{1+r} - \frac{1}{1+\delta}}{u-d}\right)(s_{n-1}d - X)^+, \end{aligned}$$

and the put's value is

$$\begin{aligned} P^E_{n-1}(s_{n-1}; X, \delta, r) &= (1+r)^{-1}\hat{E}_{n-1}(X - s_n)^+ \\ &= \left(\frac{\frac{1}{1+\delta} - \frac{d}{1+r}}{u-d}\right)(X - s_{n-1}u)^+ \\ &\quad + \left(\frac{\frac{u}{1+r} - \frac{1}{1+\delta}}{u-d}\right)(X - s_{n-1}d)^+. \end{aligned}$$

In the last expression for the put interchange s_{n-1} and X and also δ and r to obtain

$$\left(\frac{\frac{1}{1+r} - \frac{d}{1+\delta}}{u-d}\right)(s_{n-1} - Xu)^+ + \left(\frac{\frac{u}{1+\delta} - \frac{1}{1+r}}{u-d}\right)(s_{n-1} - Xd)^+.$$

Then if $u = d^{-1}$ this becomes

$$\begin{aligned} &\left(\frac{\frac{1}{1+r} - \frac{d}{1+\delta}}{u-d}\right)\frac{(s_{n-1}d - X)^+}{d} + \left(\frac{\frac{u}{1+\delta} - \frac{1}{1+r}}{u-d}\right)\frac{(s_{n-1}u - X)^+}{u} \\ &= \left(\frac{\frac{1/d}{1+r} - \frac{1}{1+\delta}}{u-d}\right)(s_{n-1}d - X)^+ + \left(\frac{\frac{1}{1+\delta} - \frac{1/u}{1+r}}{u-d}\right)(s_{n-1}u - X)^+ \\ &= \left(\frac{\frac{u}{1+r} - \frac{1}{1+\delta}}{u-d}\right)(s_{n-1}d - X)^+ + \left(\frac{\frac{1}{1+\delta} - \frac{d}{1+r}}{u-d}\right)(s_{n-1}u - X)^+ \\ &= C^E_{n-1}(s_{n-1}; X, \delta, r). \end{aligned}$$

Thus, permuting the arguments for the European put with one step to go produces the corresponding expression for the call, and *vice versa*. Moreover, the symmetry persists as we back up through the tree. At step j the expression for the European put with permuted arguments is

$$\begin{aligned}P_j^E(X; s_j, r, \delta) &= \left(\frac{\frac{u}{1+r} - \frac{1}{1+\delta}}{u-d}\right) P_{j+1}^E(Xu; s_j, r, \delta) \\ &\quad + \left(\frac{\frac{1}{1+\delta} - \frac{d}{1+r}}{u-d}\right) P_{j+1}^E(Xd; s_j, r, \delta) \\ &= C_j^E(s_j; X, \delta, r).\end{aligned}$$

Since the corresponding American options at each stage j are valued as

$$\begin{aligned}C_j^A(s_j; X, \delta, r) &= \frac{1}{1+r}\hat{E}_j C_{j+1}^A(s_{j+1}; X, \delta, r) \vee (s_j - X)^+ \\ P_j^A(s_j; X, \delta, r) &= \frac{1}{1+r}\hat{E}_j P_{j+1}^A(s_{j+1}; X, \delta, r) \vee (X - s_j)^+,\end{aligned}$$

permuting the arguments has the same effect for American puts and calls as well. Thus, under the usual specification of Bernoulli dynamics reversing the roles of S and X and of δ and r turns puts into calls and calls into puts.

Proportional Dividends at Discrete Intervals

Continuing to explore the treatment of dividends, consider now a derivative on a stock that pays at a single time step, $k > 0$, a dividend equal to a proportion δ of its price. The stock's observed price up to k is a *cum*-dividend price, s_j^c, since (ignoring the lag between *ex*-dividend and payment dates) whoever owns the stock prior to k will get the dividend. The dividend payment itself is $s_k^c\delta$. From k onwards what we observe is an *ex*-dividend price, s_j; but the relation $s_k = s_k^c(1-\delta)$ makes it possible to construct a continuation of the *cum*-dividend sequence for $j \in \{k, k+1, \ldots, n\}$, as $s_j^c = s_j/(1-\delta)$. Again, it is the (discounted) *cum*-dividend asset that should be a martingale under the risk-neutral measure. The martingale probability that solves

$$\begin{aligned}s_j^c &= \frac{m_j}{m_{j+1}}\hat{E}_j s_{j+1}^c \\ &= \frac{1}{1+r}\left[\hat{\pi} s_j^c u + (1-\hat{\pi}) s_j^c d\right],\end{aligned} \tag{5.37}$$

is the standard one:

$$\hat{\pi} = \frac{1+r-d}{u-d}. \tag{5.38}$$

Up through step $k-2$ both the stock's current and next-period prices are *cum*-dividend, so the derivative is priced off the observed *cum*-dividend price as

$$D_j(s_j^c) = \frac{1}{1+r}\hat{E}_j D_{j+1}(s_{j+1}^c) \vee D_j^X(s_j^c). \tag{5.39}$$

From k onward the derivative is priced off the observed *ex*-dividend price as

$$\begin{aligned} D_j(s_j) &= \frac{1}{1+r}\hat{E}_j D_{j+1}(s_{j+1}) \vee D_j^X(s_j) \\ &= \frac{1}{1+r}\hat{E}_j D_{j+1}\left[s_{j+1}^c(1-\delta)\right] \vee D_j^X(s_j). \end{aligned} \tag{5.40}$$

At $k-1$, the critical stage just before the *ex* date, one compares the derivative's current exercise value, which is based on the *cum*-dividend price, with its value if held until the stock's price reacts to the dividend:

$$D_{k-1}(s_{k-1}) = \frac{1}{1+r}\hat{E}_{k-1} D_k[s_k^c(1-\delta)] \vee D_{k-1}^X(s_{k-1}^c).$$

An example will show what this means in applications.

Example 63 *Figure 5.8 illustrates pricing a 3-period American call struck at* $X = 10.0$ *when the initial* cum*-dividend price is* $s_0^c = 10.0$ *and a 5% dividend is paid at step 2. The dynamics are* $u = 1.05$, $d = 0.95$, *and* $r = 1.01$, *giving* $\hat{\pi} = 0.6$. *At step 2 the* cum*-dividend prices in the three states are* $s_0^c uu = 11.025$, $s_0^c ud = 9.975$, *and* $s_0^c dd = 9.025$. *Each declines by 5% to* $s_0^c uu(1-\delta) \doteq 10.474$, $s_0^c ud(1-\delta) \doteq 9.476$, *and* $s_0^c dd(1-\delta) \doteq 8.574$, *and these then produce the four values shown in the final period. Following the calculations along the top-most path of the tree, the call ends up in the money at the top terminal node with value* $s_0^c uu(1-\delta)u - X = 0.997$. *At step 2, with the stock trading* ex *dividend for the first time, the call's value is*

$$C^A(10.474, 1) = \frac{.6(0.997) + .4(0.0)}{1.01} \vee (10.474 - 10.0) = 0.592.$$

At step 1, just before the ex *date, the value is*

$$C^A(10.500, 2) = \frac{.6(0.592) + .4(0.0)}{1.01} \vee (10.500 - 10.0) = 0.500,$$

which is attained by early exercise. This illustrates the general result that an American call on a primary asset will be exercised, if ever, just before it begins to trade ex *dividend. Finally, the initial value is*

$$C^A(10.0, 3) = \frac{.6(0.500) + .4(0.0)}{1.01} \vee (10.0 - 10.0) = 0.297.$$

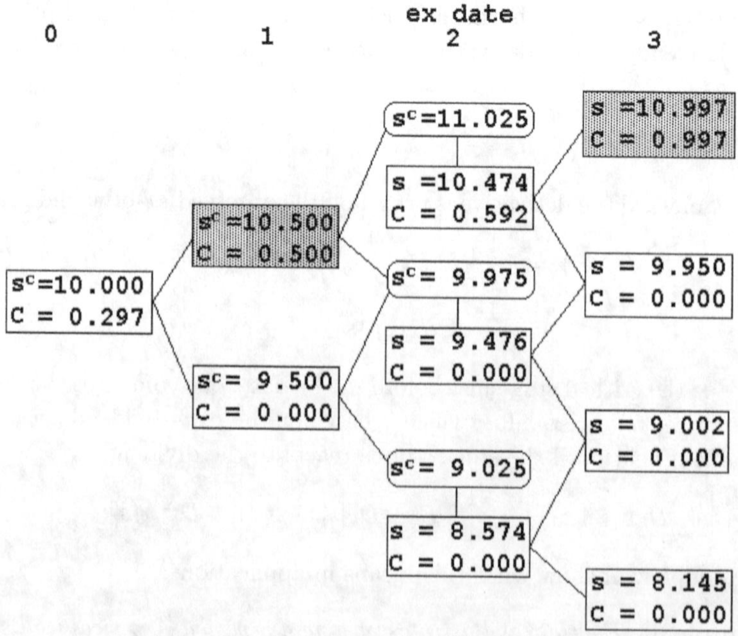

Fig. 5.8. Binomial pricing of American call on stock paying proportional dividend.

Pricing a European derivative on an asset that makes a single dividend payment proportional to price just involves adjusting the initial, *cum*-dividend price by the dividend factor, since

$$D_0[s_0^c(1-\delta)] = B(0,n)\hat{E}D_n[s_n^c(1-\delta)] = B(0,n)\hat{E}D_n(s_n).$$

Multiple proportionate payments at rates $\{\delta_k\}_{k=1}^K$ during the life of the derivative would simply require correcting s_0^c by the factor $\prod_{k=1}^K(1-\delta_k)$.

Fixed Dividends at Discrete Intervals

This case is the one that best fits short-term derivatives on equities, since cash amounts paid as dividends during periods of a few months are highly predictable and (therefore) not very sensitive to changes in the underlying price. It is also the hardest case to implement in the binomial setup. Let us again suppose that a single dividend is paid at step $k > 0$, but now fix the amount at Δ units of currency rather than at δ units of stock. As before, it is the *cum*-dividend price that is observed prior to k. Two methods of pricing a derivative on this asset can be considered.

First suppose that the *cum*-dividend price of the stock follows the usual Bernoulli dynamics during the entire life of the derivative, so that $s_{j+1}^c = s_j^c R_{j+1}$ with $R_{j+1} \in \{d, u\}$. For this to make sense the dividend must have been reinvested in the stock at stage k to purchase an additional Δ/s_k shares. Thus, $s_j^c = s_j(1 + \Delta/s_k)$ for $j \in \{k, k+1, \ldots, n-1\}$. This means that the *ex*-dividend price, $s_j = s_j^c/(1 + \Delta/s_k)$, follows the usual Bernoulli dynamics also from stage k onward. With this construction of s_j^c (5.37) is solved as before to give the same $\hat{\pi}$ as in (5.38). Up to step $k-2$ the derivative is priced off the *cum*-dividend price, s_j^c, as in (5.39), and from k onward it is based on the *ex*-dividend price, s_j, as in (5.40). At step $k-1$, just before the dividend, it depends on both prices, as

$$D_{k-1}(s_{k-1}) = \frac{1}{1+r}\hat{E}_{k-1}D_k(s_k^c - \Delta) \vee D_{k-1}^X(s_{k-1}^c).$$

What is hard about this? The difficulty here is not conceptual but computational. The problem can be seen at once by looking at the tree diagram in figure 5.9, which illustrates the pricing of an American call with the same

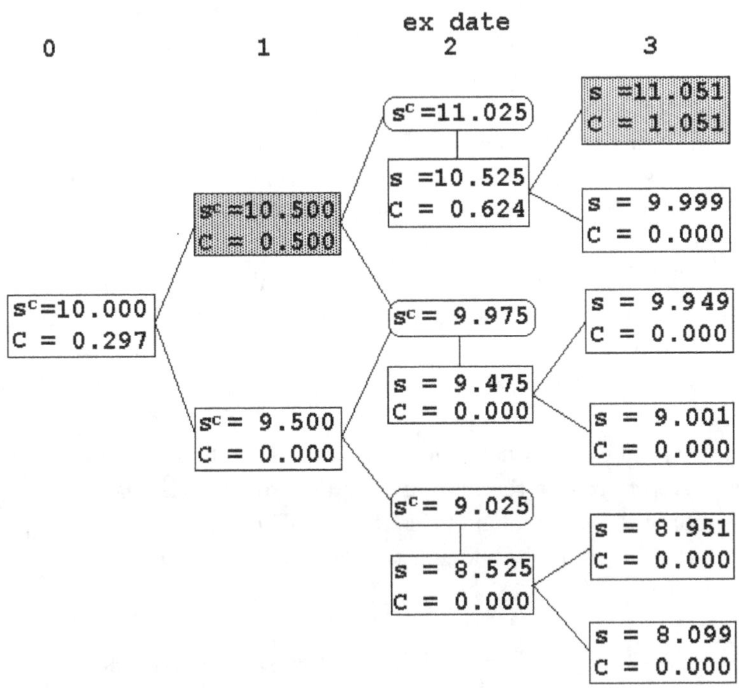

Fig. 5.9. Binomial pricing of American call on stock paying lump-sum dividend.

data as figure 5.8 but with a fixed dividend of $\Delta = 0.50$ currency units at step 2. Following the fixed dividend the price paths no longer recombine as they did with proportional dividends. From the top node at step 2 the *ex*-dividend price of $s_2 = s_0^c uu - \Delta$ becomes in the down state at step 3 $(s_0^c uu - \Delta)d$, whereas from the *ex*-dividend price of $s_2 = s_0^c ud - \Delta$ at the middle node at step 2 there is a move to $(s_0^c ud - \Delta)u$ in the up state at step 3. The difference is $\Delta(u-d)$. The effect is that an entirely new tree has to be constructed at each node on the *ex*-dividend date. There will thus be $k+1$ new $(n-k)$-step trees and $(k+1)(n-k+1)$ terminal nodes. This is not difficult to manage if the binomial scheme is programmed in modular form so that it can be executed repeatedly with different input data. In that case one merely starts off the routine and values the derivative in an $(n-k)$-step tree at each of the $k+1$ *ex*-dividend prices, then uses those prices as terminal values in a second-stage k-step tree that starts from the initial *cum*-dividend price. However, the procedure is much slower, particularly so if the dividend is paid near the midpoint in the derivative's life. The number of nodes at which the derivative must be valued reaches a maximum of $(n/2+1)^2$ at $k = n/2$, versus the usual $n+1$ in a recombining tree.

There is a way to treat fixed dividends that avoids the nonrecombinant tree, but it has the disadvantage of introducing some inconsistency in the modeling of the stock dynamics. We refer to this as the "set-aside" method. The assumption is that in anticipation of paying the dividend the firm sets aside a money-fund deposit equal to the present value of the payment obligation. Equivalently (under assumption RK), the firm purchases discount bonds of face value equal to the promised dividend and maturing on the *ex* date. This portion of the firm's assets contributes to the share price but does not follow the usual Bernoulli dynamics generated by the firm's other ("real") assets. Paying the dividend removes this component of value and leaves just the residual real component, which continues to follow the same dynamics as before.

Here, specifically, is how the idea works. At all time steps before the *ex* date, step k, regard the *cum*-dividend share price that is then observed as the sum of the *ex*-dividend price (the "real" component) and the present value of the anticipated dividend; that is, as

$$s_j^c = s_j + \Delta(1+r)^{j-k} \tag{5.41}$$

for $j \in \{0, 1, \ldots, k-1\}$. The *ex*-dividend price s_j follows the usual dynamics, $s_{j+1} = s_j R_{j+1}$, $R_{j+1} \in \{d, u\}$. From step k onward this is the price we observe, but a comparable *cum*-dividend asset can be constructed simply by reinvesting the dividend in the money-market account. This done, (5.41) applies for each $j \in \{0, 1, \ldots, n\}$. Again, the martingale probability $\hat{\pi}$ solves $s_j^c = \frac{m_j}{m_{j+1}} \hat{E}_j s_{j+1}^c$, or

$$s_j + \Delta(1+r)^{j-k} = \frac{\hat{\pi}}{1+r}[s_j u + \Delta(1+r)^{j-k+1}] + \frac{1-\hat{\pi}}{1+r}[s_j d + \Delta(1+r)^{j-k+1}],$$

and this has the usual form (5.38). As before, the derivative is priced off the *cum*-dividend price prior to step $k-1$, as

$$D_j(s_j^c) = \frac{1}{1+r} \hat{E}_j D_{j+1}(s_{j+1}^c) \vee D_j^X(s_j^c),$$

and off the *ex*-dividend price from k onward, as

$$D_j(s_j) = \frac{1}{1+r} \hat{E}_j D_{j+1}(s_{j+1}) \vee D_j^X(s_j);$$

while at $k-1$

$$D_{k-1}(s_{k-1}) = \frac{1}{1+r} \hat{E}_j D_k(s_k^c - \Delta) \vee D_{k-1}^X(s_{k-1}^c).$$

For a European-style derivative one can simply evaluate (5.11) or (5.15) at the current *ex*-dividend price, which is the observed price minus the present value of the dividend, $s_0 = s_0^c - B(0,k)\Delta$.

Figure 5.10 shows how the set-aside assumption avoids nonrecombinant trees, illustrating once again the pricing of an American call with $n = 3$ time steps, dividend payment at step $k = 2$, $s_0^c = X = 10.0$, $u = 1.05$, $d = 0.95$, $r = 1.01$ and $\Delta = 0.50$. The observed price of $s_0^c = 10.0$ at step 0 equals the money-market deposit of $\Delta(1+r)^{-2} \doteq 0.490$ (which finances the dividend) plus the residual real component, $s_0 \doteq 9.510$. At step 1 in the up state the real component grows to $s_0 u \doteq 9.985$ and the money fund deposit grows to $\Delta(1+r)^{-1} \doteq 0.495$, giving an observed price of $s_1^c \doteq 10.480$. At step 2 in the upper-most state the *ex*-dividend price is $s_2 = s_0 uu \doteq 10.485$, which drops to $s_0 uud \doteq 9.961$ in the down state at the second node from the top in step 3. The same point is reached after an uptick from the middle node at step 2, where the *ex*-dividend price is $s_2 = s_0 ud \doteq 9.486$

Unfortunately, while this arrangement does make the tree recombine and cuts computational time, the dynamic structure does not quite make sense. The problem is that we start out conveniently at step 0 with the set-aside for the step-k dividend already in place, but no similar provision is made at step k for the next dividend to be paid after the derivative expires. Doing so would produce the same nonrecombinant tree that was to be avoided. Thus, this method should be thought of just as a shortcut rather

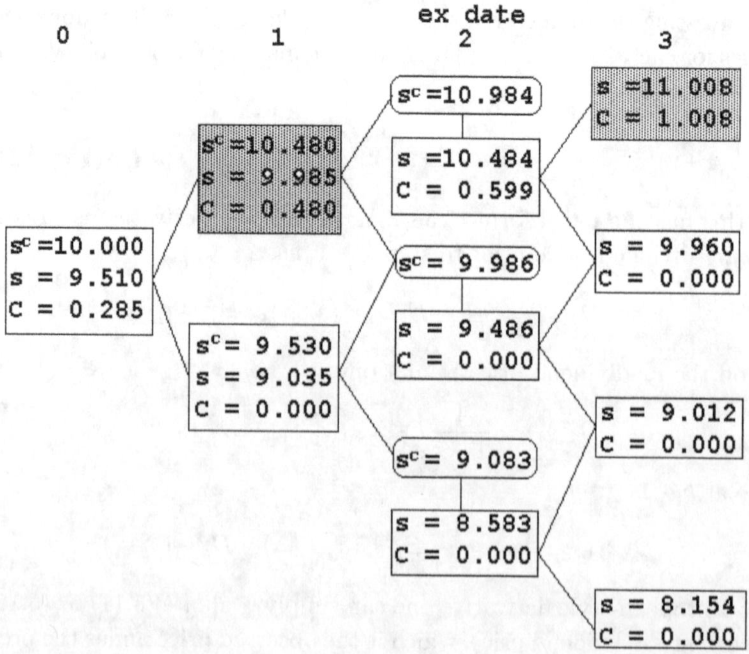

Fig. 5.10. Binomial pricing of American call on stock with lump-sum, set-aside dividend.

than as a completely satisfactory solution. Notice that the final answers for the call price in figures 5.9 and 5.10 do differ slightly. The set-aside makes the call worth a little less because there is less volatility and less potential for growth in share prices.

5.5 Implementing the Binomial Method

This section covers the details of how to put the binomial method into practice. The main issues are (i) how to choose move sizes u and d to afford a satisfactory approximation to the dynamics of the underlying asset, and (ii) how to do the recursive calculations so as to use computational resources to best advantage.

5.5.1 *Modeling the Dynamics*

The Bernoulli model for the price of an asset clearly becomes more plausible the shorter is the interval of calendar time that corresponds to each step—that is, the larger is n. Moreover, we would have little confidence in the binomial method unless the sequence of estimates converged to some definite, finite value as n increased. Since the move sizes must themselves depend on n for this to happen, we write them now as u_n, d_n. The most basic approach for pricing derivatives on stocks, stock indexes, and financial futures is to choose u_n and d_n so that the n-step distribution of terminal price converges to the lognormal. We will now see how to do this. Since the limiting lognormal distribution for terminal price is the same as that implied by the continuous-time model for $\{S_t\}$ that underlies the Black-Scholes theory (that is, geometric Brownian motion), it will not be surprising to see that the n-step binomial prices for European options converge to those given by the Black-Scholes formulas. Our first look at those formulas will come in this section.

Specifying Move Sizes to Achieve Convergence

We work first with dynamics for prices of primary assets ("stocks"), treating futures prices below. For now we assume that there are no dividends or other explicit carrying costs. Begin as usual by dividing the T-year life of the derivative into n equal intervals $\{[t_j, t_{j+1})\}_{j=0}^{n-1}$ with $t_j = jT/n$. On $[t_j, t_{j+1})$ the price of a share in the money fund grows by the proportion $m_{j+1}/m_j = 1 + r_j$. Under assumption RK this equals the sure return $B(t_j, t_{j+1})^{-1}$ on a one-period discount bond, and the common value of these for each j is known as of $t = 0$. Since the goal is to approximate continuous-time dynamics, we will express $B(t_j, t_{j+1})^{-1}$ in terms of the average continuously compounded spot rate, as $B(t_j, t_{j+1})^{-1} \equiv e^{\bar{r}_j T/n}$, where we set $\bar{r}_j \equiv r(t_j, t_{j+1})$ for brevity. Of course, under RK $\bar{r}_j$ also equals the average forward rate as of $t = 0$, $r_0(t_j, t_{j+1})$. Since $e^{\bar{r}_j T/n}$ corresponds to $(1 + r_j)$ in the discrete-time notation, the martingale probability on the jth interval can now be written as

$$\hat{\pi}_{jn} = \frac{e^{\bar{r}_j T/n} - d_n}{u_n - d_n}. \tag{5.42}$$

We now derive the limiting distribution as $n \to \infty$ of the stock's terminal price after n steps, given the following choices of u_n and d_n:

$$\begin{aligned} u_n &= e^{\sigma\sqrt{T/n}}, \\ d_n &= u_n^{-1}. \end{aligned} \tag{5.43}$$

Here σ is a positive parameter not depending on n, whose interpretation will be given later.

The limiting argument requires making large-n approximations to various quantities. The first thing we need is an approximation for $\hat{\pi}_{jn}$. This is

$$\begin{aligned}\hat{\pi}_{jn} &= \frac{e^{\bar{r}_j T/n} - e^{-\sigma\sqrt{T/n}}}{e^{\sigma\sqrt{T/n}} - e^{-\sigma\sqrt{T/n}}}\\ &= \frac{e^{\bar{r}_j T/n+\sigma\sqrt{T/n}} - 1}{e^{2\sigma\sqrt{T/n}} - 1}\\ &= \frac{\sigma\sqrt{T/n} + (\bar{r}_j + \sigma^2/2)T/n + o(n^{-1})}{2\sigma\sqrt{T/n} + 2\sigma^2 T/n + o(n^{-1})}\\ &= \frac{\frac{1}{2} + \frac{\bar{r}_j+\sigma^2/2}{2\sigma}\sqrt{T/n} + o(n^{-1/2})}{1 + \sigma\sqrt{T/n} + o(n^{-1/2})}\\ &= \frac{1}{2} + \frac{\bar{r}_j - \sigma^2/2}{2\sigma}\sqrt{T/n} + o(n^{-1/2}).\end{aligned}$$

Now let U_n be the number of upticks in the n time steps, and set

$$R_n \equiv \ln s_n - \ln s_0 = \ln S_T - \ln S_0 \equiv R(0,T),$$

which is the stock's continuously compounded return over $[0,T]$. (Recall that we assume for now that there are no dividends.) Since $s_n = s_0 u_n^{U_n} d_n^{n-U_n}$, we have

$$\begin{aligned}R_n &= n \ln d_n + U_n \ln(u_n/d_n)\\ &= (2U_n - n)\ln u_n\\ &= (2U_n - n)\sigma\sqrt{T/n}.\end{aligned}$$

Being the number of upticks, U_n is the sum of independent (but not identically distributed) Bernoulli variates with parameters $\hat{\pi}_{0n}, \hat{\pi}_{1n}, \ldots, \hat{\pi}_{n-1,n}$. The jth of these has characteristic function

$$\Psi_j(\zeta) = 1 + \hat{\pi}_{jn}(e^{i\zeta} - 1)$$

for $\zeta \in \Re$ and $i = \sqrt{-1}$, so that the sum has c.f.

$$\Psi_{U_n}(\zeta) = \prod_{j=0}^{n-1}[1 + \hat{\pi}_{jn}(e^{i\zeta} - 1)].$$

Accordingly, the c.f. of R_n is

$$\Psi_{R_n}(\zeta) = e^{-i\zeta\sigma\sqrt{Tn}} \prod_{j=0}^{n-1}[1 + \hat{\pi}_{jn}(e^{2i\zeta\sigma\sqrt{T/n}} - 1)].$$

Taking the logarithm of Ψ_{R_n} and then expanding $e^{2i\zeta\sigma\sqrt{T/n}}$ in Taylor series in powers of $n^{-1/2}$ give for $\ln \Psi_{R_n}(\zeta)$

$$
\begin{aligned}
&-i\zeta\sigma\sqrt{Tn} + \sum_{j=0}^{n-1} \ln[1 + \hat{\pi}_{jn}(e^{2i\zeta\sigma\sqrt{T/n}} - 1)] \\
&\quad = -i\zeta\sigma\sqrt{Tn} + \sum_{j=0}^{n-1} \ln\{1 + \hat{\pi}_{jn}[2i\zeta\sigma\sqrt{T/n} - 2\zeta^2\sigma^2 T/n + o(n^{-1})]\}.
\end{aligned}
$$

Now use the $O(n^{-1/2})$ approximation for $\hat{\pi}_{jn}$, simplify, and expand the logarithm in Taylor series to write $\ln \Psi_{R_n}(\zeta)$ as

$$
\begin{aligned}
&-i\zeta\sigma\sqrt{Tn} + \sum_{j=0}^{n-1} \ln\left[1 + i\zeta\sigma\sqrt{\frac{T}{n}} + i\zeta(\bar{r}_j - \sigma^2/2)\frac{T}{n} - \zeta^2\sigma^2\frac{T}{n} + o(n^{-1})\right] \\
&\quad = -i\zeta\sigma\sqrt{Tn} + \sum_{j=0}^{n-1} \Big[i\zeta\sigma\sqrt{T/n} + i\zeta(\bar{r}_j - \sigma^2/2)T/n \\
&\qquad - \frac{\zeta^2\sigma^2}{2}T/n + o\left(n^{-1}\right)\Big] \\
&\quad = i\zeta\left(n^{-1}\sum_{j=0}^{n-1}\bar{r}_j - \sigma^2/2\right)T - \zeta^2\sigma^2 T/2 + o(1) \\
&\quad = i\zeta[r(0,T) - \sigma^2/2]T - \zeta^2\sigma^2 T/2 + o(1).
\end{aligned}
$$

The last step follows from the fact that the log of the total return on a T-period bond at $t = 0$ is

$$
\sum_{j=0}^{n-1} \bar{r}_j T/n = \ln \prod_{j=0}^{n-1} B(t_j, t_{j+1})^{-1} = \ln B(0,T)^{-1} = r(0,T)T
$$

under assumption RK. Letting $n \to \infty$ gives

$$
\begin{aligned}
\Psi_{R_n}(\zeta) &\to \Psi_{R(0,T)}(\zeta) \\
&= \exp\{i\zeta[r(0,T) - \sigma^2/2]T - \zeta^2\sigma^2 T/2\}.
\end{aligned}
$$

Now apply the continuity theorem for characteristic functions (theorem 4 on page 74) to conclude that the limiting distribution of R_n is normal with mean $r(0,T)T - \sigma^2 T/2$ and variance $\sigma^2 T$. This implies that terminal price $S_T = S_0 e^{R(0,T)}$ is distributed under the martingale measure as

$$
S_0 \exp\left[r(0,T)T - \sigma^2 T/2 + Z\sigma\sqrt{T}\right], \tag{5.44}
$$

where $Z \sim N(0,1)$. The terminal price is therefore lognormally distributed. This shows also that σ^2 governs the rate at which the variance of the log of the asset's total return grows with time, and is therefore the variance of log return over a unit interval. For this reason σ is called the "volatility parameter".

Other choices of uptick and downtick parameters than those in (5.43) can lead to the same lognormal limiting distribution; for example,

$$\begin{aligned} u_{jn} &= \exp\left[(\bar{r}_j - \sigma^2/2)T/n + \sigma\sqrt{T/n}\right] \\ d_{jn} &= \exp\left[(\bar{r}_j - \sigma^2/2)T/n - \sigma\sqrt{T/n}\right]. \end{aligned}$$

Here, with time-varying spot rates the move sizes themselves are time-varying, but the risk-neutral probabilities,

$$\hat{\pi}_n = \frac{e^{\bar{r}_j T/n} - d_{jn}}{u_{jn} - d_{jn}} = \frac{e^{\sigma^2 T/(2n)} - e^{-\sigma\sqrt{T/n}}}{e^{\sigma\sqrt{T/n}} - e^{-\sigma\sqrt{T/n}}},$$

are constant. Although they are a bit more complicated, these expressions do have the advantage of revealing a more direct link to the comparable continuous-time model, geometric Brownian motion.

Since the distribution of $R(0,T)$ was derived from risk-neutral Bernoulli dynamics, one expects it to be consistent with risk neutrality over the period $[0,T]$. It is so. From the expression for the m.g.f. of a normal variate we have $E\exp(Z\sigma\sqrt{T}) = \exp(\sigma^2 T/2)$, so that $\hat{E}_0 S_T = S_0 e^{r(0,T)} = S_0 B(0,T)^{-1}$ and $S_0 = B(0,T)\hat{E}_0 S_T$.

In pricing European options we have seen that the recursive procedure can be sidestepped and the arbitrage-free price written as a simple formula; for example, as (5.24) or (5.25) for call options. These formulas result from evaluating $\hat{E}C^E(s_n, 0)$ in the martingale measure that applies to the Bernoulli dynamics, using the fact that $\ln s_n$ is a linear function of a binomial variate. We have just seen that the limiting distribution of s_n as $n \to \infty$ is lognormal, so it is not surprising that evaluating $\hat{E}C^E(S_T, 0) = \hat{E}(S_T - X)^+$ with S_T as in (5.44) leads to a related formula,

$$C^E(S_0,T) = S_0\Phi\left[\frac{\ln(\frac{S_0}{BX}) + \sigma^2 T/2}{\sigma\sqrt{T}}\right] - BX\Phi\left[\frac{\ln(\frac{S_0}{BX}) - \sigma^2 T/2}{\sigma\sqrt{T}}\right], \tag{5.45}$$

where $\Phi(\cdot)$ is the standard normal c.d.f. and $B \equiv B(0,T)$. Likewise, evaluating $\hat{E}P^E(S_T, 0) = \hat{E}(X - S_T)^+$ gives for the European put

$$P^E(S_0,T) = BX\Phi\left[\frac{\ln(\frac{BX}{S_0}) + \sigma^2 T/2}{\sigma\sqrt{T}}\right] - S_0\Phi\left[\frac{\ln(\frac{BX}{S_0}) - \sigma^2 T/2}{\sigma\sqrt{T}}\right]. \tag{5.46}$$

These are in fact the famous Black-Scholes (1973) formulas, of which we give two derivations in chapter 6.

Modifications for Futures Prices

In section 5.4.2 move sizes of the futures price, u_{jn} and d_{jn}, were shown to be related to the move sizes of the price, s_j, of the primary asset that underlies the futures. Multiplying u_{jn} and d_{jn} by one-period discount factor $m_j/m_{j+1} = (1+r_j)^{-1} = e^{-\bar{r}_j T/n} = B(t_{j-1}, t_j)$ eliminates the average positive trend in $\{s_j\}$ under the risk-neutral measure and makes the futures price itself a $\hat{\mathbb{P}}$ martingale. In moving from stage j to stage $j+1$ of the n-stage binomial tree on $[0, T]$ we thus take $\mathsf{F}^u_{j+1} = \mathsf{F}_j B(t_j, t_{j+1}) u_n$ and $\mathsf{F}^d_{j+1} = \mathsf{F}_j B(t_j, t_{j+1}) d_n$ in the up and down states, respectively, where $\mathsf{F}_j \equiv \mathsf{F}(t_j, T')$ is the futures price at t_j on a contract expiring at $T' \geq T$. With this convention the equivalent-martingale probabilities remain as in (5.42).

How does this alteration of move sizes affect the limiting distribution of F_n as $n \to \infty$? Recall that the terminal value of the underlying can be represented as $s_n = s_0 u_n^{U_n} d_n^{n-U_n}$, where U_n is the number of upticks in the n stages. Since up and down moves in the futures at stage j are both proportional to $B(t_j, t_{j+1})$, the terminal futures price can be expressed as

$$\mathsf{F}_n = \prod_{j=0}^{n} B(t_j, t_{j+1}) \mathsf{F}_0 u_n^{U_n} d_n^{n-U_n} = B(0, T) \mathsf{F}_0 u_n^{U_n} d_n^{n-U_n}.$$

Since the continuously compounded n-period rate of return on the underlying is

$$R(0, T) = R_n \equiv \ln(s_n/s_0) = \ln(u_n^{U_n} d_n^{n-U_n}),$$

the corresponding n-period log return on the futures price is

$$\begin{aligned} \mathsf{R}_n &\equiv \ln(\mathsf{F}_n/\mathsf{F}_0) = \ln B(0, T) + R_n \\ &= -r(0, T)T + R_n, \end{aligned}$$

where $r(0, T)$ is the average spot rate for T-year loans initiated at $t = 0$. Since the limiting distribution of R_n is normal with mean $[r(0, T) - \sigma^2/2]T$ and variance $\sigma^2 T$, it follows that R_n is asymptotically normal with the same variance and mean $-\sigma^2 T/2$. The limiting distribution of $\mathsf{F}_n/\mathsf{F}_0$ is therefore lognormal with these two parameters. Consistent with the fact that the futures price is a martingale under the risk-neutral Bernoulli dynamics, the limiting result implies that the $\hat{\mathbb{P}}$-expected gain from the futures position over $[0, T]$ is

$$\hat{E}_0[\mathsf{F}(T, T') - \mathsf{F}(0, T)] = \mathsf{F}(0, T)e^{-\sigma^2 T/2 + \sigma^2 T/2} - \mathsf{F}(0, T) = 0.$$

An alternative to modeling the dynamics of the underlying commodity or asset is to model the futures price directly. This would clearly be appropriate for futures on a commodity used in consumption and not held primarily as an investment asset, which could not in practice be used to construct a replicating portfolio. Taking the move sizes for the futures price as

$$\mathsf{u}_n = e^{\sigma\sqrt{T/n}}, \quad \mathsf{d}_n = e^{-\sigma\sqrt{T/n}}$$

(the same in each subperiod) and setting

$$\hat{\pi}_n = \frac{1 - \mathsf{d}_n}{\mathsf{u}_n - \mathsf{d}_n}$$

again makes $\mathsf{R}_n \equiv \ln(\mathsf{F}_n/\mathsf{F}_0)$ converge in distribution to normal with mean $-\sigma^2 T/2$ and variance $\sigma^2 T$. The interpretation of σ^2 as the variance of the log return over a unit interval remains the same.

Modifications for Assets Paying Continuous Dividends

Moving from derivatives on futures back to derivatives on stocks, currencies, or other primary assets, let us see how the payment of continuous, proportional dividends can be specified so as to provide the correct limiting distribution of $R(0, T)$. Suppose the underlying asset pays a continuous proportional dividend at known constant rate δ over $[0, T]$. With reinvestment of the dividends one share bought for S_0 at $t = 0$ would be worth $S_0 e^{\delta T}$ at T, apart from capital gains. For the asset to be priced like a riskless bond, as it would be in the risk-neutral measure, the expected total return from capital gains must be reduced by the factor $e^{-\delta T}$. This requires setting the mean parameter in the normal distribution of $\ln(S_T/S_0)$ to $[r(0, T) - \delta - \sigma^2/2]T$. Simply subtracting δ from the average forward rate in each subinterval makes the binomial setup produce this result in the limit. We therefore take

$$\hat{\pi}_{jn} = \frac{e^{(\bar{r}_j - \delta)T/n} - d_n}{u_n - d_n}.$$

To allow for time-varying (but still deterministic) dividend rates, just replace δ by δ_j. For derivatives on a foreign currency the foreign interest rate is the continuous dividend, since it governs the rate at which an initial investment in foreign-denominated bonds or money deposits grows over time. To apply binomial pricing in this case, just deduct from $\bar{r}_j$ the average period-j forward rate in the foreign country.

By contrast, dividends or storage costs on an asset or commodity underlying a futures contract do not affect the expected gain on a futures position under $\hat{\mathbb{P}}$, which always equals zero. These receipts or costs require no adjustment of $\hat{\pi}_n$ or the move sizes in pricing derivatives on futures.

Choosing Binomial Parameters in Applications

To apply the binomial method one must specify the number of time steps, n, and determine the move sizes, $u_n = \exp(\sigma\sqrt{T/n})$ and $d_n = u_n^{-1}$, and the martingale probability $\hat{\pi}_{jn}$. The probability for step j depends on the move sizes, the period-j discount factor, $B(t_j, t_{j+1})$, and (if the underlying pays continuous dividends) the dividend rate, δ. The volatility, σ, is the key parameter because it is the only ingredient that is not observable or cannot readily be approximated in terms of observables. We consider this first.

In principle, estimating σ should be a simple matter. The Bernoulli dynamics and the limiting lognormal model imply that log returns of the underlying asset in nonoverlapping periods of equal length are independent and identically distributed. Since σ is just the standard deviation of the log return over a unit time interval, the obvious thing to do is to calculate the sample standard deviation of returns in an historical sample. For stocks we might calculate the log returns from dividend-corrected daily closing prices, as $\{R_\tau = \ln(S_{\tau+1} + \Delta_{\tau+1}) - \ln S_\tau\}_{\tau=1}^{M}$. Here τ counts trading days, and $\Delta_{\tau+1}$ is the dividend if the stock goes *ex* dividend on day $\tau + 1$, and zero otherwise. For futures the calculations are usually made with daily settlement prices—the prices officially determined by the exchange in order to mark traders' positions to market. The unbiased sample variance from a sample of size M is

$$S_M^2 = (M-1)^{-1} \sum_{\tau=1}^{M} (R_\tau - \bar{R})^2,$$

where $\bar{R}$ is the sample mean. Once S_M^2 has been calculated, one converts it to the same time units used for interest rates and time to expiration. For example, if the basic time unit is years, S_M^2 from daily returns is multiplied by the number of days per year.[9] Under the limiting lognormal form of

[9]Alternatively, the variance might be scaled by the number of trading days per year or by a weighted average of calendar and trading days. Averaging can be justified by the fact that volatility over non-trading intervals is greater than zero but less than that over comparable trading spells.

the Bernoulli model S_M^2 is a sufficient statistic for σ^2 and therefore has minimum sampling variance in the class of unbiased estimators. Since it is also strongly consistent, it seems that we could get as good an estimate of σ as desired by just taking M to be large enough. What could possibly be wrong with this simple plan?

The problem is that for most securities the i.i.d.-normal model for log returns is a poor description of the actual data. The empirical marginal distributions typically exhibit much thicker tails than the normal, especially for short holding periods, such as one day. This indicates that large absolute deviations from the mean occur much more often than they would for a normal variate with the same variance. Moreover, returns are not independent: the conditional distribution of tomorrow's return manifestly depends on past experience. Although the past record does not help much, on average, in forecasting tomorrow's actual return, it does provide good forecasts of the square of the return or the squared difference from the mean. Shocks that raise volatility in one period seem to persist and keep it high for a time.[10] To apply binomial pricing we need to know, at the very least, the average volatility over the life of the derivative, and the sample standard deviation gives a poor estimate of this because it treats all historical observations as conveying the same information about the future.

Of course, if σ fluctuates through time, there is clearly a flaw in binomial-pricing models based on the Bernoulli dynamics thus far considered, and the same flaw is present in continuous-time models like Black-Scholes that depend on the limiting lognormal case. In later chapters we introduce more elaborate and more realistic continuous-time models, and in the final section of the present chapter we discuss a way of incorporating time- and price-dependent volatility into the binomial framework. However, despite their inconsistency with empirical evidence about the underlying dynamics, the standard binomial and Black-Scholes models remain the benchmarks against which others are judged because of their simplicity and tractability. The trick in using them is to obtain a better estimate of the average volatility over the derivative's lifetime than is provided by the historical sample standard deviation. There are various ways to do this,

[10] Not only is volatility persistent, for stock returns it has regular weekly and annual cycles. For the evidence see Campbell, Lo, and MacKinlay (1997). Temporal variation and dependence in volatility can explain some of the observed thickness of the tails in the empirical marginal distributions of daily returns, but the usual models of time-varying volatility do not fully account for it. Even conditionally on the past, the normal model gives a poor description of daily log returns.

some sophisticated and some *ad hoc*. A common *ad hoc* approach that often works very well is to use a value of σ that puts estimates for comparable traded derivatives close to their observed prices. This "implicit" volatility that can be deduced from prices of traded derivatives will be discussed in section 5.6 below and again in connection with the Black-Scholes model in chapter 6. Also in common use are more sophisticated statistical techniques for estimating σ from historical data. Nonlinear time-series models such as ARCH (for autoregressive, conditional heteroskedasticity) and its many generalizations (GARCH, EGARCH, etc.) characterize the persistence in volatility parametrically. Having estimated the parameters by maximum likelihood, one can use the models to generate conditional forecasts of average volatility over future periods.[11]

Once σ has been estimated and u_n and d_n have been calculated, we still need the one-period interest rate or price of a one-period bond at each step j in order to find the martingale probabilities:

$$\hat{\pi}_{jn} = \frac{\exp[r(t_j, t_{j+1})T/n] - d_n}{u_n - d_n} = \frac{B(t_j, t_{j+1})^{-1} - d_n}{u_n - d_n}.$$

Since in actual practice these future one-period rates and prices are not known in advance, they must be replaced by the corresponding forward rates and prices. This, too, is not entirely straightforward. If spot prices of discount bonds could be observed at a continuum of maturities, the forward prices could just be calculated as $B_0(t_j, t_{j+1}) = B(0, t_{j+1})/B(0, t_j)$, but we must make do with the limited number of maturities that are available. A direct way to estimate $B(0, t)$ for arbitrary t is to interpolate between prices of nearby bonds. A better way is to interpolate along the yield curve between the annualized spot rates for the available maturities, $r(0, t) = -\ln B(0, t)^{1/t}$, then back out the corresponding t_j-maturing spot bond price. In the U.S. there is an active secondary market for Treasury bills and Treasury strips of a wide range of maturities, from which the spot bond prices and spot rates can be calculated in this way.

Prices of most equity and index derivatives are not highly sensitive to variation in interest rates. If the yield curve is relatively flat over the derivative's life, it is reasonable to treat one-period interest rates as constant and equal to the average spot rate on a riskless bond with about the same maturity, $r(0, T)$. In this case the risk-neutral probabilities are the same for

[11]Campbell *et al.* (1997, chapter 12) give a full account of the procedure.

each time step:

$$\hat{\pi}_n = \frac{\exp[r(0,T)T/n] - d_n}{u_n - d_n} = \frac{B(0,T)^{-1/n} - d_n}{u_n - d_n}.$$

If the underlying pays a continuous dividend at rate δ, this is subtracted from $r_0(t_j, t_{j+1})$ or $r(0,T)$ in the above formulas. (Dividends paid at discrete intervals are handled as explained in section 5.4.4 and require no adjustment of $\hat{\pi}$.)

Example 64 *As an illustration of the whole process, let us construct the move sizes and risk-neutral uptick probability for pricing a 3-month (.25 -year) option on a no-dividend common stock using a binomial tree with* $n = 100$ *time steps. Given the estimate* $\hat{\sigma} = .30$ *for annualized volatility and an annualized yield of* $r_0(0, .25) = .05$ *on 3-month Treasury bills, we have*

$$u_n = e^{.30\sqrt{.25/100}} \doteq 1.015$$

$$d_n = u_n^{-1} \doteq 0.985$$

$$\hat{\pi}_n = \frac{e^{.05(.25/100)} - 0.985}{1.015 - 0.985} \doteq 0.504.$$

5.5.2 *Efficient Calculation*

With specifications for u_n, d_n, and $\hat{\pi}_{jn}$ that cause s_n to converge in distribution to lognormal, one expects binomial price estimates to improve in accuracy as n increases. How large n should be in a particular application depends on the degree of accuracy required, the nature of the derivative being priced, the costliness of computational time, the limitations of computer memory, and the computational design. This section presents some design suggestions for making efficient use of computer resources. We begin with some techniques of general applicability that economize on memory and computational time, then describe some smoothing techniques that greatly accelerate convergence for certain derivative assets.

Economizing on Memory

Memory requirements for binomial computations quickly get out of hand even for moderate values of n unless storage is used efficiently. The recombinant trees we have used to illustrate the technique have been two-dimensional arrays over time and price, involving $n+1$ terminal price states after n time steps. One's first thought in designing a program is to construct one or perhaps two $n \times (n+1)$ arrays—one for the underlying price and one for the derivative. However, all that is really needed is just a single vector array of dimension $n+1$. Indeed, it is possible to get by with even less than that.

First, here is how to do it with an $(n+1)$-dimensional array. For concreteness, consider an American-style derivative on a no-dividend stock that is worth $D_j^X \equiv D_j^X(s_j)$ upon exercise at some node (s_j, j). After n time steps the binomial tree will have $n+1$ terminal nodes, corresponding to underlying prices ranging from $s_0 d^n$ to $s_0 u^n$. In the program, as in the numerical examples, one starts at step n when the option expires and works backward successively to steps $n-1, n-2, \ldots$ and ultimately to step 0, where the desired initial value is finally obtained. We work with an $(n+1)$-dimensional vector $\mathbb{D}$, which can be envisioned as a columnar array. At step n fill $\mathbb{D}$ with the derivative's terminal values, $\{D(s_n, 0) = D(s_0 u^{n-k} d^k, 0)\}_{k=0}^n$, storing $D(s_0 u^n, 0)$ at the top as $\mathbb{D}_n$ and $D(s_0 d^n, 0)$ at the bottom as $\mathbb{D}_0$. For a put struck at X these would be calculated as $D(s_0 u^{n-k} d^k, 0) \equiv (X - s_0 u^{n-k} d^k)^+$. Working down from the top with underlying price $s_0 u^n$, each successive value of s_n can be obtained recursively from its predecessor by multiplying by d/u. Now moving back to the $n-1$ step, calculate new derivative values, $\{D(s_0 u^{n-1-k} d^k, 1)\}_{k=0}^{n-1}$ and store them in the same $\mathbb{D}$ array, as follows. Starting at the top, with $B_{n-1} \equiv B(t_{n-1}, t_n \equiv T)$ as the one-step discount factor for step $n-1$, calculate $D(s_0 u^{n-1}, 1)$ as

$$B_{n-1}[\hat{\pi}_{n-1} D(s_0 u^n, 0) + (1 - \hat{\pi}_{n-1}) D(s_0 u^{n-1} d, 0)] \vee D_{n-1}^X(s_0 u^{n-1})$$

and store as $\mathbb{D}_n$, writing over the old value $D(s_0 u^n, 0)$. Continue down the array, evaluating the derivative at position k, $D(s_0 u^{n-1-k} d^k, 1)$, as

$$\begin{aligned} &B_{n-1}[\hat{\pi}_{n-1} D(s_0 u^{n-k} d^k, 0) + (1 - \hat{\pi}_{n-1}) D(s_0 u^{n-k-1} d^{k+1}, 0)] \\ &\vee D_{n-1}^X(s_0 u^{n-1-k} d^k) \end{aligned} \tag{5.47}$$

and storing as $\mathbb{D}_{n-k}$, for $k = 1, 2, \ldots, n-1$. These n operations replace just the top n elements of $\mathbb{D}$, leaving $\mathbb{D}_0$ unchanged as $D(s_0 d^n, 0)$. This is not needed after the n step. At the $n-2$ step the top $n-1$ elements of $\mathbb{D}$ will be modified while leaving $\mathbb{D}_1$ and $\mathbb{D}_0$ unchanged, and so on. In general, at each step j just the top $j+1$ elements, $\mathbb{D}_n, \mathbb{D}_{n-1}, \ldots, \mathbb{D}_{n-j}$, will be calculated. In particular, at step 0 only $\mathbb{D}_n$ is changed, and this represents initial value $D(s_0, n) \equiv D(S_0, T)$.

Now here is how to manage with even fewer than $n+1$ storage positions. This is done by storing only nonzero values of the derivative in the $\mathbb{D}$ array, these being the only ones that will affect its initial value. To make the notation more compact, we illustrate specifically for a European call option. Let n_X be the maximum number of downticks for which the call is in the money at the n step; that is, n_X is such that $s_0u^{n-k}d^k - X > 0$ for $k = 0, 1, \ldots, n_X$ and $(s_0u^{n-k}d^k - X)^+ = 0$ for $k > n_X$. We will need a $\mathbb{D}$ array with only $n_X + 1$ elements. At step n, fill $\mathbb{D}$ with the nonzero option values, beginning with $\mathbb{D}_{n_X} = C^E(s_0u^n, 0) = s_0u^n - X$ at the top and continuing through $\mathbb{D}_0 = C^E(s_0u^{n-n_X}d^{n_X}, 0) = s_0u^{n-n_X}d^{n_X} - X$ at the bottom. Moving to step $n-1$, recalculate the top n_X elements, $\mathbb{D}_{n_X}, \mathbb{D}_{n_X-1}, \ldots, \mathbb{D}_1$, as

$$\begin{aligned}\mathbb{D}_{n_X-k} &= C^E(s_0u^{n-1-k}d^k, 1)\\ &= B_{n-1}[\hat{\pi}_{n-1}C^E(s_0u^{n-k}d^k, 0) + (1-\hat{\pi}_{n-1})C^E(s_0u^{n-k-1}d^{k+1}, 0)]\end{aligned}$$

for $k = 0, 1, \ldots, n_X - 1$. The bottom element of the array, $\mathbb{D}_0$, will contain

$$C^E(s_0u^{n-1-n_X}d^{n_X}, 1) = B_{n-1}\hat{\pi}_{n-1}C^E(s_0u^{n-n_X}d^{n_X}, 0).$$

This depends just on the uptick value at step n, since the downtick value is zero. The process is repeated at steps $j = n-2, n-3, \ldots, n_X$. From the bottom $j+1-n_X$ nodes of the complete tree at each such step it would be impossible to attain a terminal node corresponding to a positive value of the call. Thus, at each such j the element $\mathbb{D}_0$ will depend on just the bottom element of the previous step. Once step $n_X - 1$ is reached, however, the complete tree will have only n_X nodes, and from even the lowest of these, where $s_{n_X-1} = s_0d^{n_X-1}$, it is possible to attain positive terminal values on downtick as well as uptick. For each of steps $j = n_X - 1, n_X - 2, \ldots, 0$, therefore, each of the $j+1$ elements $\{\mathbb{D}_{n_X-k} = C^E(s_0u^{j-k}d^k, n-j)\}_{k=0}^{j}$ is computed as

$$B_j[\hat{\pi}_jC^E(s_0u^{j-k+1}d^k, n-j-1) + (1-\hat{\pi}_j)C^E(s_0u^{j-k}d^{k+1}, n-j-1)],$$

just as if the full $(n+1)$-dimensional array had been used. We wind up with the desired initial value at the top, as before.

The procedure for $n = 4$ and $n_X = 3$ is illustrated in table 5.2, where $\mathbb{D}_k^j$ stands for the element in the kth position at the jth step and β and $\bar{\beta}$ represent $B\hat{\pi}$ and $B(1-\hat{\pi})$, respectively. We start at step 4 of the tree with the array $\mathbb{D}^4$ of the derivative's nonzero terminal values and work backward to step 0. The entries show how elements of $\mathbb{D}$ are calculated at

Table 5.2. Value arrays for memory-efficient binomial calculations.

$\mathbb{D}^0$	$\mathbb{D}^1$	$\mathbb{D}^2$	$\mathbb{D}^3$	$\mathbb{D}^4$
$\beta\mathbb{D}_3^1 + \bar{\beta}\mathbb{D}_2^1$	$\beta\mathbb{D}_3^2 + \bar{\beta}\mathbb{D}_2^2$	$\beta\mathbb{D}_3^3 + \bar{\beta}\mathbb{D}_2^3$	$\beta\mathbb{D}_3^4 + \bar{\beta}\mathbb{D}_2^4$	$\mathbb{D}_3^4$
$[\beta\mathbb{D}_2^2 + \bar{\beta}\mathbb{D}_1^2]$	$\beta\mathbb{D}_2^2 + \bar{\beta}\mathbb{D}_1^2$	$\beta\mathbb{D}_2^3 + \bar{\beta}\mathbb{D}_1^3$	$\beta\mathbb{D}_2^4 + \bar{\beta}\mathbb{D}_1^4$	$\mathbb{D}_2^4$
$[\beta\mathbb{D}_1^3 + \bar{\beta}\mathbb{D}_0^3]$	$[\beta\mathbb{D}_1^3 + \bar{\beta}\mathbb{D}_0^3]$	$\beta\mathbb{D}_1^3 + \bar{\beta}\mathbb{D}_0^3$	$\beta\mathbb{D}_1^4 + \bar{\beta}\mathbb{D}_0^4$	$\mathbb{D}_1^4$
$[\beta\mathbb{D}_0^4]$	$[\beta\mathbb{D}_0^4]$	$[\beta\mathbb{D}_0^4]$	$\beta\mathbb{D}_0^4$	$\mathbb{D}_0^4$

each successive step. The bracketed elements at the bottom are those that get carried over from the previous step and are not used again.

In pricing path-independent European derivatives it is unnecessary to know the values of the asset except at the terminal nodes, but they are needed for American derivatives in order to judge whether to exercise early. Rather than store the asset values, it is efficient to compute them recursively as one moves down the $\mathbb{D}$ array at each step, just as we described doing at the n step. That is, at the j step start with s_0u^j (the highest price) and multiply each element by d/u to obtain the next lower price in the array.

Program `BINOMIAL` on the accompanying CD carries out binomial pricing of European, American, and extendable puts and calls on various underlying assets, allowing for various types of dividends. (Extendable options are discussed below.)

Economizing on Time

Two time-saving numerical methods that work in a number of contexts, including binomial pricing, are (i) the use of control variates and (ii) Richardson extrapolation.

A control variate is an artificial variable with known properties that is constructed in order to reduce the estimation error in a quantity whose properties are not well understood. Suppose, for example, that we have estimated the value of an American put option using an n-step binomial tree that is set up to converge to geometric Brownian motion as $n \to \infty$. The estimate, P_n^A, is the sum of the true value, P^A, plus an unknown estimation error, a_n. The goal is to obtain a new estimate, $\hat{P}_n^A$, with smaller error. For this purpose we would construct as a control a binomial estimate, P_n^E, of an otherwise identical European put from the same tree. The estimation error in the control, $e_n = P_n^E - P^E$, can be determined with high accuracy by comparing the binomial estimate with that obtained from the Black-Scholes formula. Presuming that the errors a_n and e_n are closely related, we would expect to improve the estimate for the American put by subtracting e_n from it. The new estimate would then be $\hat{P}_n^A = P_n^A - e_n = P^A + (a_n - e_n)$. If a_n and e_n were random variables having about the same mean and variance, then we could say that the mean squared error of $\hat{P}^A$ would be smaller than that of P_n^A provided the correlation between a_n and e_n exceeds 0.5.[12]

Example 65 *Consider 3-month (.25-year) puts struck at* $X = 10.0$ *on an underlying no-dividend stock priced at* $S_0 = 10.0$ *and with volatility* $\sigma = 0.4$. *With 3-month unit discount bonds selling for* $B(0, .25) = 0.9850$, *program* `BINOMIAL` *yields for an* $n = 25$*-stage tree* $P_{25}^A \doteq 0.7384$ *and* $P_{25}^E \doteq 0.7257$, *while the Black-Scholes estimate from* (5.46) *is* $P^E \doteq 0.7178$. *The error,* $e_{25} \doteq 0.7257 - 0.7178 = 0.0079$, *yields control-variate estimate* $\hat{P}_{25}^A \doteq 0.7384 - 0.0079 = 0.7305$. *For comparison, a 1000-stage binomial tree delivers* $P_{1000}^A \doteq 0.7300$.

Turn now to the second error-reduction technique, Richardson extrapolation. Broadly, the idea is to estimate the price of a derivative for each of a short sequence of relatively small but increasing values of n—for $n_1, n_2, \ldots, n_m$, say—and, in effect, to extrapolate from these to $n = \infty$. For example, we could estimate for $n = 10, 20, 30$ and extrapolate from there. We will see just how to do this directly, but first consider the logic of it. The number of calculations in the standard binomial solution of n steps is the same as the number of nodes in the tree, or $(n+1)(n+2)/2$ (somewhat less if the compact solution is used). Getting estimates for $n = 10, 20, 30$ requires a total of 793 calculations. If extrapolating from these gives accuracy comparable to one estimate with $n = 50$, we will have saved 533 calculations; if it is comparable to one with $n = 100$, the saving is 4,358 calculations!

Being thus motivated, let us see how it works. Instead of extrapolating a sequence of functions of n out to $n = \infty$, we actually extrapolate down to zero a sequence of functions of $1/n$. These functions are the binomial estimates indexed by the time interval between steps, T/n. For a derivative expiring after T units of time, let $h_k = T/n_k$ be the interval corresponding to an n_k-step tree and let $D(h_k)$ be the corresponding binomial estimate.

---

[12] A refinement of the control-variate estimator is $\hat{P}_n^A(\alpha) = P_n^A - \alpha e_n$, where α is chosen to minimize the variance of $\hat{P}_n^A(\alpha)$. The optimal value, $\alpha^* = cov(P_n^A, e_n)/Ve_n$, can be determined by evaluating the covariance and variance in the terminal (binomial) distribution implied by the tree. Details and applications of this idea in the Monte Carlo context are given in chapter 11.

The hope is that $D(h)$ is smooth enough that it can be approximated by a few terms of a Taylor expansion about $h = 0$, as $D(h) = a_0 + a_1 h + a_2 h^2 + \cdots + a_{m-1}h^{m-1}$. If this is so, then obtaining estimates $D(h_1), \ldots, D(h_m)$ for m values of h gives m equations in m unknowns, and one of these is the value $a_0 = D(0)$ that corresponds to $n = \infty$. For example, taking just a linear approximation with $m = 2$ gives $D(h_1) = a_0 + a_1 h_1$, $D(h_2) = a_0 + a_1 h_2$, and the solution

$$D(0) = \frac{h_1 D(h_2) - h_2 D(h_1)}{h_1 - h_2}. \tag{5.48}$$

With $m = 3$ the estimate is

$$D(0) = \frac{h_1 h_2 (h_1 - h_2) D(h_3) - h_1 h_3 (h_1 - h_3) D(h_2) + h_2 h_3 (h_2 - h_3) D(h_1)}{h_1^2 h_2 + h_1 h_3^2 + h_2^2 h_3 - h_2 h_3^2 - h_1 h_2^2 - h_1^2 h_3}.$$

Even the linear approximation can be very effective.

Example 66 *Consider pricing a one-year European put struck at* $X = 10.0$ *on a stock with* $S_0 = 10.0$ *and volatility* $\sigma = 0.30$, *given an annual discount factor of* $B(0, 1) = 0.95$. *With* $n_1 = 10$ $(h_1 = 0.10)$ *and* $n_2 = 20$ $(h_2 = 0.05)$ *program* `BINOMIAL` *yields* $D(0.10) \doteq 0.900$ *and* $D(0.05) \doteq 0.915$. *Plugging into* (5.48) *gives* $D(0) \doteq 0.930$. *This is closer to the Black-Scholes solution of* 0.929 *than the* 0.926 *obtained with a* 100*-step tree, which took* 4,853 *more calculations (counting the extrapolation itself as one).*

Special Smoothing Techniques

While Richardson extrapolation can indeed produce impressive gains in efficiency, there is cause for concern that sequences of binomial estimates may lack sufficient smoothness for the method to work reliably in some applications. We now look at the cause of this problem and then consider some ways of smoothing the estimates. We have seen that Bernoulli dynamics leads to an explicit solution for arbitrage-free prices of European-style derivatives. For example, when the short rate is considered to be constant out to T the initial price of a European call is

$$C^E(S_0, T) = B(0, T) \sum_{j=0}^{n} \binom{n}{j} \hat{\pi}_n^j (1 - \hat{\pi}_n)^{n-j} (S_0 u_n^j d_n^{n-j} - X)^+.$$

The top panel of figure 5.11 depicts what is involved in calculating the summation; namely, adding the $n + 1$ products of binomial weights (represented by the vertical bars) times the corresponding terminal payoffs.

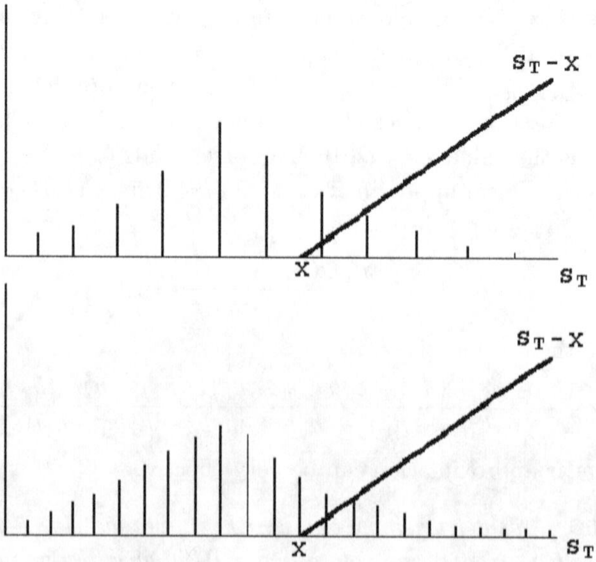

Fig. 5.11. Binomial pricing of European call as a discrete expectation.

The bottom panel shows what happens as the number of time steps, n, is increased. This increases the number of terminal nodes at which there are atoms of probability mass, both extending the range of the support of S_T and, since u_n and d_n shrink toward unity as n grows, also decreasing the spacing between the atoms. In the process there is a net transfer of probability mass from one side to the other of the kink in the terminal payoff function. Since the payoff is zero to the left of the kink and positive to the right, this shift of mass can appreciably affect the value of the summation. If one plots a sequence of binomial estimates based on $n_1, n_2, \ldots$ time steps, the result is a slightly ragged curve, the estimates oscillating back and forth as they tend toward the limiting Black-Scholes value. The kink typically has more effect for small n since the individual probability weights around the strike price are usually larger when n is small. Although the mechanism is more complicated for American options, the piecewise linearity of the payoff function still has a similar effect.

The problem arising from lack of smoothness in the terminal payoff function of an American option can be reduced by the following simple trick. Backing up to stage $n-1$ of the tree, the retention value of the derivative is just that of a European option with remaining life T/n. This can be priced by Black-Scholes. As we shall see in chapter 6, the Black-Scholes price (prior to expiration) is a smooth, convex function of the price of the underlying, so starting off an $(n-1)$-step tree with Black-Scholes prices eliminates the kink in the value function. The smaller n is and the larger is T/n, the greater is the gain in smoothness and the greater is the improvement in the binomial estimates. Reviewing a number of approaches for pricing American options, Broadie and Detemple (1996) conclude that this enhanced version of the binomial method is quite competitive in terms of accuracy and speed, particularly when coupled with Richardson extrapolation.

Lack of smoothness in the payoff function is an even greater problem when pricing what are called "extendable" options. If not in the money at each of a sequence of specific dates prior to expiration, the life of an extendable option is extended to the next date in the sequence. If it is in the money at any of these dates, it must be exercised. For example, a put option might expire after six months if then in the money, but otherwise be extended for one additional six-month period. Using program `BINOMIAL` to price these options, one can see that convergence as n increases is extremely slow. This is illustrated in figure 5.12. The three bold, jagged curves are plots of binomial estimates of prices of three European-style extendable

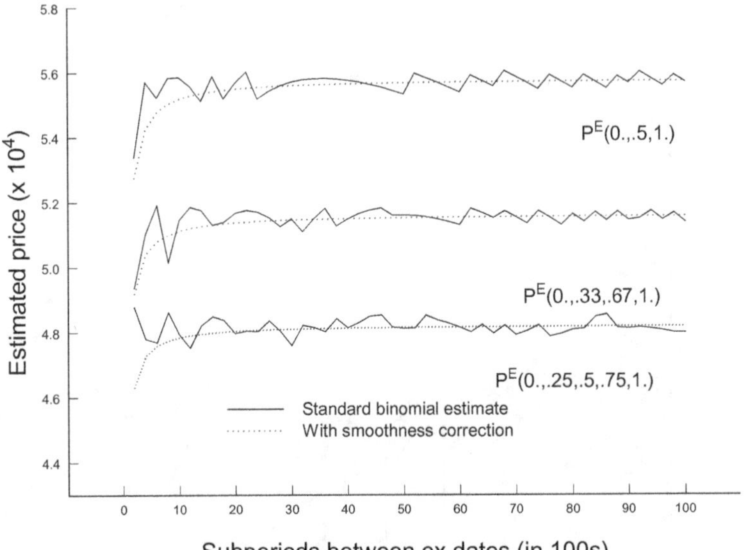

Fig. 5.12. Binomial estimates of extendable puts vs. number of time steps between exercise dates.

puts *vs.* the number of time steps between dates. The top-most curve is for a once-extendable 6-month (half-year) put, which is denoted $P^E(0, .5, 1)$. The other curves are for 4-month and 3-month puts with up to two and three extensions, respectively. For all three the oscillations continue to be well above the graphic threshold, even out to 10,000 steps between exercise dates.

To see why convergence is so slow, consider valuing at $t = 0$ a European put that can be extended once at $t = t^*$ and then expires at $T > t^*$. At t^* the put is worth $X - S_{t^*}$ if in the money; otherwise its value is that of an ordinary, nonextendable European put with remaining life $T - t^*$. The initial value can be represented as the discounted expectation of this time-t^* payoff function in the martingale measure; that is, as

$$P^E(S_0, t^*, T) = B(0, t^*)\hat{E}[(X - S_{t^*})^+ + P^E(S_{t^*}, T - t^*)\mathbf{1}_{[X,\infty)}(S_{t^*})].$$

The sawtooth curve in figure 5.13 depicts the value at t^* as a function of S_{t^*}, with a discrete probability mass function overlaid. The value function is discontinuous at X because the payoff approaches zero as S_{t^*} approaches the strike price from below, whereas its value at X is that of an at-the-money European put with $T - t^*$ to go. What happens as n increases is similar to what was described for an ordinary nonextendable option, but the discontinuity in the t^* value function makes the effect much more dramatic. Probability mass shifts back and forth across the discontinuity, causing the binomial estimates to oscillate wildly and persistently.

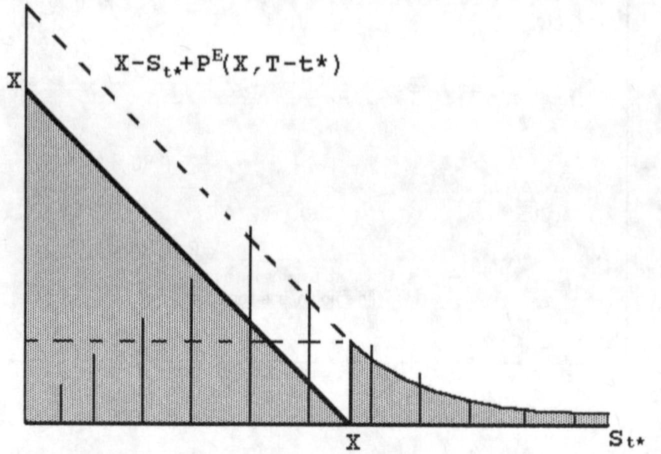

Fig. 5.13. Binomial price of extendable put as discrete expectation.

Dealing with this problem is a little more complicated. Suppose the payoff function were shifted upward in the interval $[0, X)$ so as to eliminate the discontinuity, as represented by the dashed diagonal line in figure 5.13. The required shift would equal the value of an at-the-money vanilla European put with $T - t^*$ to go, or $P^E(X, T - t^*)$, and this can be computed from the Black-Scholes formula. The value of a derivative with this quite smooth, continuized payoff function can be estimated relatively accurately by the binomial method. Its terminal (t^*) value would be defined for the n step of a binomial recursion as $X - s_n + P^E(X, T - t^*)$ for $s_n \in [0, X)$ and as $P^E(s_n, T - t^*)$ for $s_n \geq X$. Of course, the value of this displaced time-t^* payoff would overstate that of the extendable put by the discounted expected value of $P^E(X, T - t^*)$ times an indicator function of the interval $[0, X)$; namely,

$$B(0, t^*)\hat{E}[P^E(X, T - t^*)\mathbf{1}_{[0,X)}(S_{t^*})].$$

This is the value of a "digital" option that pays $P^E(X, T - t^*)$ whenever $S_{t^*} < X$. Since $P^E(X, T - t^*)$ is a known constant, however, this overstatement is easily calculated as

$$B(0, t^*)P^E(X, T - t^*)\hat{\mathbb{P}}(S_{t^*} < X),$$

and it can simply be subtracted from the binomial estimate of the displaced payoff function. The probability can be calculated from the limiting lognormal distribution for S_{t^*} given S_0.[13] The smooth, dotted curves in figure 5.12 are plots of these continuized estimates.

5.6 Inferring Trees from Prices of Traded Options

In the standard binomial setup considered thus far the move sizes, u_n and d_n, are the same for all nodes. While the risk-neutral probabilities do vary with the time step if the short rate of interest is time varying, the values of $\hat{\pi}_j$ at step j are nevertheless the same at all $j + 1$ nodes. A more general model would allow move sizes and probabilities alike to depend on both time

[13]This corresponds to the factor $\Phi[\cdot]$ in the first term of the Black-Scholes put formula, (5.46). With $T = t^*$ as the time of expiration this is

$$\widehat{\mathbb{P}}(S_{t^*} < X) = \Phi\left[\frac{\ln(BX/S_0) + \sigma^2 t^*/2}{\sigma\sqrt{t^*}}\right].$$

Details of the smoothing method and extensions to multiple exercise dates are given in Epps *et al.* (1996).

and price. In effect, since move probabilities are calculated from interest rates and move sizes, and since move sizes are calculated from the volatility parameter (as $u_n = e^{\sigma\sqrt{T/n}}$ in the usual setup), such an arrangement would amount to having price- and time-dependent volatilities. Whereas the log of price approaches normality as $n \to \infty$ in the usual constant-volatility Bernoulli setup, an appropriate price-dependent specification would lead to marginal distributions with thicker tails than the normal and could account for some degree of intertemporal dependence of volatilities. As discussed at the end of section 5.5.1, prices of financial assets usually exhibit both of these features. We will see in chapter 8 that continuous-time models with price-dependent volatilities can also generate option prices whose implicit volatilities vary systematically with the strike price. While this is contrary to both the Black-Scholes and standard binomial models, it, too, is consistent with empirical observation.

Just as the constant σ in the standard model can be inferred from a single option price, it is possible to infer $\sigma(t, S_t)$ as a function of time and price from prices of a suitable collection of derivatives. Methods for doing this within the Bernoulli framework have been proposed by Derman and Kani (1994) and Rubinstein (1994).[14] Here we describe the simpler Rubinstein procedure, as modified by Jackwerth and Rubinstein (1996), which uses a collection of options all of the same maturity to determine node-dependent move sizes and probabilities. The method requires knowing current prices of a collection of European-style options at different strikes but with the same expiration date, T. Because of European put-call parity either a put or call at each strike suffices, and nothing is gained from having both. The method is therefore limited to underlying assets on which European options with a large range of strikes do trade, such as the S&P 500 index.[15]

5.6.1 *Assessing the Implicit Risk-Neutral Distribution of S_T*

Letting $(\Omega, \mathcal{F}, \{\mathcal{F}_j\}, \mathbb{P})$ be the actual filtered probability space on which the Bernoulli price process is defined, our task will be to find an equivalent martingale measure $\widetilde{\mathbb{P}}$ that correctly prices a collection of T-expiring,

[14]Dupire (1994) presented a related idea in the context of diffusion models and trinomial dynamics. Working also in the continuous-time framework, Dumas *et al.* (1998) use option prices to fit a flexible quadratic model for σ. We describe their findings in chapter 8.

[15]In principle, it would suffice to have a large number of T-expiring American calls on an underlying stock that paid no dividends before T.

traded, European-style derivatives. The first step is to build a standard n-step tree with time steps T/n, using a guess as to what an appropriate constant σ would be. This might be an average of implicit volatilities deduced from options close to the money or an estimate from historical returns. This trial value of σ is then used to calculate values of move sizes, u and d, in the standard way. From the move sizes one then determines prices of the underlying at the terminal nodes, as $\{s_{n,k}\}_{k=0}^{n} = \{s_0 u^{n-k} d^k\}_{k=0}^{n}$, and a corresponding vector of risk-neutral probabilities, $\{\hat{\pi}_k\}_{k=0}^{n-1}$. Together, the move sizes and probabilities determine a benchmark measure, $\hat{\mathbb{P}}$, of precisely the form we have worked with heretofore. Assuming a constant interest rate, the move probabilities will be the same at each step—that is, $\hat{\pi}$ and $1 - \hat{\pi}$—and the nodal probabilities under $\hat{\mathbb{P}}$ will just be binomial, as

$$\widehat{\Pi}_{n,k} \equiv \hat{\mathbb{P}}(S_T = s_0 u^{n-k} d^k) = \binom{n}{k} \hat{\pi}^{n-k}(1-\hat{\pi})^k, k = 0, 1, \ldots, n.$$

If the asset pays a continuous proportional dividend at rate δ, then S_T is interpreted as a *cum*-dividend price and $\hat{\pi}$ is calculated as in (5.35).

The goal is to replace these benchmark probabilities with new ones, $\widetilde{\Pi}_{n,k} \equiv \widetilde{\mathbb{P}}(S_T = s_0 u^{n-k} d^k)$, that do a better job of explaining the observed prices of derivatives. Ultimately, all of the interior nodes of the tree will be moved around as well in the process of developing distinct move sizes for each node, but the next task is just to find the new probabilities for the terminal nodes. Finding the new vector of probabilities, $\widetilde{\mathbf{\Pi}}_n \equiv \{\widetilde{\Pi}_{n,k}\}_{k=0}^{n}$, can be posed as a constrained optimization problem. The idea is to minimize a certain criterion function, $\mathcal{C}(\widetilde{\mathbf{\Pi}}_n)$, subject to the following constraints on the elements of $\widetilde{\mathbf{\Pi}}_n$:[16]

- That they be nonnegative and sum to unity;
- That they preserve the martingale property for the (cum-dividend) normalized process $\{s_j/m_j\} = \{(1+r)^{-j} s_j/m_0\}$; and
- That they correctly price all the options in the working sample.

Here, "correctly price" means that the implied prices fall within the current bid and asked quotes at the time the option sample is taken. Thus, if $\{D(s_0; X_m)^q\}, m \in \{1, 2, \ldots, M\}, q \in \{b, a\}$ are quotes on the collection of

[16]Rubinstein (1994) does not impose the second constraint, which requires him to redefine the interest rate (an observable!) in order to restore the martingale property.

M puts and calls expiring at T, the constraints would be

$$\sum_{k=0}^{n} \widetilde{\Pi}_{n,k} = 1, \widetilde{\Pi}_{n,k} \geq 0, k \in \{0, 1, \ldots, n\} \tag{5.49}$$

$$\left(\frac{1+\delta}{1+r}\right)^n \sum_{k=0}^{n} \widetilde{\Pi}_{n,k} s_{n,k} = s_0 \tag{5.50}$$

and

$$(1+r)^{-n} \sum_{k=0}^{n} \widetilde{\Pi}_{n,k} D(s_{nk}; X_m) \in \left[D(s_0; X_m)^b, D(s_0, X_m)^a\right], \quad m \in \{1, 2, \ldots, M\}. \tag{5.51}$$

In these expressions the initial price of the underlying, s_0, is defined as the average of the current bid and asked, $(s_0^b + s_0^a)/2$. Rubinstein (1994) takes the vector of standard binomial probabilities, $\widehat{\mathbf{\Pi}}_n$, as the benchmark and uses the squared Euclidean distance of $\widetilde{\mathbf{\Pi}}_n$ from $\widehat{\mathbf{\Pi}}_n$ as criterion function:

$$\mathcal{C}\left(\widetilde{\mathbf{\Pi}}_n; \widehat{\mathbf{\Pi}}_n\right) = \sum_{k=0}^{n} \left(\widetilde{\Pi}_{n,k} - \widehat{\Pi}_{n,k}\right)^2.$$

Jackwerth and Rubinstein (1996) consider several other criteria, including the Kullback-Leibler information,[17]

$$\mathcal{C}(\widetilde{\mathbf{\Pi}}_n; \widehat{\mathbf{\Pi}}_n) = \sum_{k=0}^{n} \widetilde{\Pi}_{n,k} \ln\left(\widetilde{\Pi}_{n,k}/\widehat{\Pi}_{n,k}\right), \tag{5.52}$$

but they wind up favoring a simple smoothness criterion that does not involve the original binomial probabilities at all; namely,

$$\mathcal{C}\left(\widetilde{\mathbf{\Pi}}_n\right) = \sum_{k=0}^{n} \left[\left(\widetilde{\Pi}_{n,k+1} - \widetilde{\Pi}_{n,k}\right) - \left(\widetilde{\Pi}_{n,k} - \widetilde{\Pi}_{n,k-1}\right)\right]^2.$$

Notice that the tree must have more than $m+2$ time steps if the optimization is to be feasible, since (5.49)-(5.51) impose a total of $m+2$ constraints on the $n+1$ elements of $\widetilde{\mathbf{\Pi}}_n$.

[17]The K-L information or entropy function is a measure of distance between two probability measures that takes the form (5.52) in the discrete case. It is nonnegative and equals zero if and only if the two measures are the same. For an overview of properties see the article by Kullback (1983). Stutzer (1996) uses the entropy function in a clever nonparametric scheme to estimate the risk-neutral distribution over a holding period of length T.

5.6.2 *Building the Tree*

As usual, the tree branches out from initial price s_0 through a sequence of up and down moves. What is different now is that the move sizes u and d and the uptick and downtick probabilities will vary from node to node. To handle this notationally, we distinguish the probabilities in the same way as the prices; thus, $\tilde{\pi}_0, u_0, d_0$ are the initial uptick probability and move sizes from s_0, and $\tilde{\pi}_{j,k}, u_{j,k}, d_{j,k}$ are the probability and moves from price $s_{j,k}$, where j counts time steps and $k \in \{0, 1, \ldots, j\}$ indexes the nodes. Since k corresponds to the number of downticks from the initial price, the prices $\{s_{j,k}\}_{k=0}^{j}$ are ranked as $s_{j,0} > s_{j,1} > \cdots > s_{j,j}$.

The following assumptions provide the additional structure needed to guarantee a nice solution to the problem of finding equivalent measure $\widetilde{\mathbb{P}}$.

1. The tree is recombining. This means that the product of a sequence of $j - k$ up moves and k down moves leading to a step-j price is invariant under a permutation of the moves. For example, there are $2^2 = 4$ paths leading to the prices at step two, but the uptick-downtick and downtick-uptick paths both lead to the same place: $s_0 u_0 d_{1,0} = s_0 d_0 u_{1,1}$. An n-step tree will therefore have just $n + 1$ terminal prices, as usual.
2. All paths leading to any given node are equally likely. Of the 2^n paths in an n-step tree, $\binom{n}{k}$ lead to the kth terminal node, where the nodal probability is $\widetilde{\Pi}_{n,k}$. The assumption implies that each of these paths has the same probability, $\tilde{P}_{n,k} \equiv \widetilde{\Pi}_{n,k}/\binom{n}{k}$. Combined with assumption 1, this tells us that all paths with given numbers of up and down moves are equally likely. For example, the path probabilities at the three nodes at step two are $\tilde{P}_{2,0} = \widetilde{\Pi}_{2,0}$, $\tilde{P}_{2,1} = \widetilde{\Pi}_{2,1}/2$, $\tilde{P}_{2,2} = \widetilde{\Pi}_{2,2}$.

Within this framework let us see how to solve for the move probabilities and move sizes at any node. The $n+1$ terminal prices at step n have already been determined, along with corresponding nodal probabilities $\{\widetilde{\Pi}_{n,k}\}_{k=0}^{n}$. Let us take a position at any node $k \in \{0, 1, \ldots, n - 1\}$ on the $n - 1$ step, at which the price (to be determined) will be $s_{n-1,k}$ and from which the prices $s_{n,k}$ and $s_{n,k+1}$ can be reached at step n. The probabilities of the paths leading to $s_{n,k}$ and $s_{n,k+1}$ through $s_{n-1,k}$ are $\tilde{P}_{n,k} = \widetilde{\Pi}_{n,k}/\binom{n}{k}$ and $\tilde{P}_{n,k+1} = \widetilde{\Pi}_{n,k+1}/\binom{n}{k+1}$, respectively. Since all paths through $s_{n-1,k}$ must go to one of these nodes, the probability for any single path through $s_{n-1,k}$ is $\tilde{P}_{n-1,k} = \tilde{P}_{n,k} + \tilde{P}_{n,k+1}$, and by assumption 2 all such paths are equally likely. Given the path probabilities, we can now find the uptick probability at this node, $\tilde{\pi}_{n-1,k}$. Since this is the conditional probability of getting to $s_{n,k}$ via $s_{n-1,k}$ given that we have arrived at $s_{n-1,k}$, it is clear that $\tilde{\pi}_{n-1,k} = \tilde{P}_{n,k}/\tilde{P}_{n-1,k}$. Now it remains just to find $s_{n-1,k}$ itself. This is done by imposing the martingale property of the $\widetilde{\mathbb{P}}$ measure, as

$$s_{n-1,k} = \frac{1+\delta}{1+r}[\tilde{\pi}_{n-1,k}s_{n,k} + (1-\tilde{\pi}_{n-1,k})s_{n,k-1}],$$

where δ is the proportionate dividend received each period and all the prices are interpreted as *cum*-dividend. Of course, $s_{n-1,k}$, $s_{n,k}$, and $s_{n,k-1}$ implicitly determine the move sizes, as $u_{n-1,k} = s_{n,k}/s_{n-1,k}$ and $d_{n-1,k} = s_{n,k-1}/s_{n-1,k}$, but these are now of no direct relevance.

Once $s_{n-1,k}$ is found for each $k \in \{0, 1, \ldots, n-1\}$, the process is repeated at step $n-2$. In general, at any step $j \in \{0, 1, \ldots, n-1\}$ and node $k \in \{0, 1, \ldots, j\}$ we

1. Find the probability of each path passing through the $s_{j,k}$ node, as $\tilde{P}_{j,k} = \tilde{P}_{j+1,k} + \tilde{P}_{j+1,k+1}$; then
2. Calculate the uptick probability as $\tilde{\pi}_{j,k} = \tilde{P}_{j+1,k}/\tilde{P}_{j,k}$; and, finally,
3. Determine the price level as

$$s_{j,k} = \frac{1+\delta}{1+r}\left[\tilde{\pi}_{j,k}s_{j+1,k} + (1-\tilde{\pi}_{j,k})\, s_{j+1,k-1}\right].$$

Example 67 *Figure 5.14 illustrates the procedure for a tree with* $n = 3$ *steps whose terminal prices match those in figure 5.2. The data used for the original construction* ($s_0 = 10.0$, $r = 0.01$, $\delta = 0.0$, $u = 1.05$, $d = 0.97$, $\hat{\pi} = 1/2$) *give as terminal prices and benchmark nodal probabilities*

$$\{s_{3,k}\}_{k=0}^{3} = \{11.576, 10.694, 9.875, 9.127\}$$
$$\{\widehat{\Pi}_{3,k}\}_{k=0}^{3} = \{0.125, 0.375, 0.375, 0.125\}.$$

Suppose that an optimization subject to constraints (5.49)–(5.51) *has produced the new risk-neutral probabilities*

$$\{\widetilde{\Pi}_{3,k}\}_{k=0}^{3} = \{0.136, 0.393, 0.303, 0.168\}.$$

At each node of the tree are shown the path probabilities, uptick probabilities, and prices of the underlying asset and of a European call struck at $X = 10.0$. *At the top of step* 2, *for example, the path probability is calculated as*

$$\tilde{P}_{2,0} = \tilde{P}_{3,0} + \tilde{P}_{3,1} = 0.136 + 0.131 = 0.267,$$

and the move probability is

$$\tilde{\pi}_{2,0} = \tilde{P}_{3,0}/\tilde{P}_{2,0} = 0.136/0.267 \doteq 0.509.$$

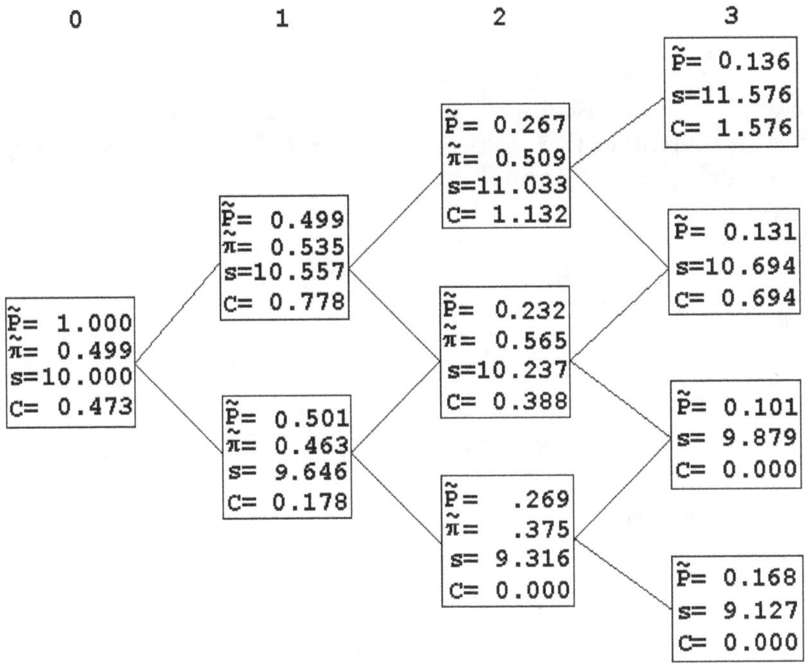

Fig. 5.14. Pricing a call from an inferred tree.

The implied prices for the underlying and the call are then

$$\begin{aligned} s_{2,0} &= (1+r)^{-1}[(\tilde{\pi}_{2,0})s_{3,0} + (1-\tilde{\pi}_{2,0})s_{3,1}] \\ &= 1.01^{-1}[(.509)11.576 + (.491)10.694] \\ &\doteq 11.033 \end{aligned}$$

and

$$\begin{aligned} C^E_{2,0} &= (1+r)^{-1}[(\tilde{\pi}_{2,0})C^E_{3,0} + (1-\tilde{\pi}_{2,0})C^E_{3,1}] \\ &= 1.01^{-1}[(.509)1.576 + (.491)0.694] \\ &\doteq 1.132. \end{aligned}$$

The path probability at stage 0 *works out to be unity, and the underlying price works out to be* $s_0 = 10.0$, *consistent with the original construction. Notice that the price of the European call is a little higher than in figure* 5.2, *as is consistent with the fact that* $\widetilde{\mathbb{P}}$ *puts more probability mass in the tails than does* $\hat{\mathbb{P}}$.

5.6.3 *Appraisal*

Although we chose to price a European call in order to keep the example simple, it is unnecessary to construct the entire tree if the purpose is to price just European-style derivatives, since terminal probabilities $\{\widetilde{\Pi}_{n,k}\}_{k=0}^{n}$ alone suffice for this. Thus, in the example the call's price can be found directly as

$$\begin{aligned} C^E(s_0, 3) &= (1+r)^{-3} \sum_{k=0}^{3} \widetilde{\Pi}_{3,k}(s_{3,k} - X)^+ \\ &= 1.01^{-3}[(.136)1.576 + (.393).694] \\ &= 0.473. \end{aligned}$$

The real reason for building the tree is to price American-style options or exotic derivatives whose payoffs depend on the entire price path. The premise is that these prices will be more accurate if derived in a framework that is consistent with the revealed prices of traded European derivatives. While this seems reasonable, one does question whether accurate inferences about the intertemporal behavior of price under the martingale measure can really be drawn from a single cohort of European-style options.[18] A relevant consideration in this regard is how closely implicit volatility structures $\{\sigma(t, S_t)\}$ correspond when constructed from cohorts of options having different terms. Some measure of such correspondence might turn out to a better criterion for optimizing the $\{\widetilde{\Pi}_{n,k}\}$ than just smoothness or conformity with log-binomial priors.

[18]The Derman-Kani (1994) method does simultaneously fit options having different terms as well as different strikes, but it requires interpolating prices on what is ordinarily a coarse grid of times and strikes.

6

Black-Scholes Dynamics

We now take up the valuation of derivative assets when prices of underlying primary assets are considered to evolve in continuous time. This chapter and the next treat the classic geometric Brownian motion dynamics that were the basis for the pioneering option-pricing theory of Black and Scholes (1973) and Merton (1973). The first section describes the slight generalization of geometric Brownian motion that we will use and lays out the assumptions about the behavior of the money-market fund in continuous time. The price of the money fund still serves as a convenient numeraire, as it did in the discrete-time Bernoulli framework of chapter 5. Section 6.2 introduces the two pricing strategies that are available in this setting—solving the partial differential equation that gives the law of motion of a financial derivative, and evaluating expectations of normalized payoffs in the risk-neutral measure. Section 6.3 first shows how the two approaches are applied in valuing futures and European puts and calls on an underlying primary asset that can pay continuous dividends, then extends to options on stocks paying lump-sum dividends and options on stock indexes and futures. The last section works out the properties of arbitrage-free prices of European options within the Black-Scholes framework. There we (i) see how Black-Scholes prices respond in a comparative-static sense to changes in the parameters that define the options and the underlying assets, (ii) determine the implied instantaneous risk-reward trade-offs from positions in European options, and (iii) express the expected return (in natural measure $\mathbb{P}$) from holding European options over arbitrary periods. Pricing American options and other derivatives with more complicated payoff structures under Black-Scholes dynamics is the subject of chapter 7.

6.1 The Structure of Black-Scholes Dynamics

Consider a primary asset (a "stock") whose price starts off at a known value S_0 and evolves over time in accord with the stochastic differential equation

$$dS_t = \mu_t S_t \cdot dt + \sigma_t S_t \cdot dW_t. \tag{6.1}$$

Here $\{W_t\}_{t\geq 0}$ is a standard Brownian motion on $(\Omega, \mathcal{F}, \{\mathcal{F}_t\}, \mathbb{P})$, and $\{\mu_t\}_{t\geq 0}$ and $\{\sigma_t\}_{t\geq 0}$ are deterministic ($\mathcal{F}_0$-measurable) processes with $\int_0^t |\mu_s| \cdot ds < \infty$, $\int_0^t \sigma_s^2 \cdot ds < \infty$, and $\sigma_t > 0$ for each t. The drift process, $\{\mu_t\}_{t\geq 0}$, will be essentially irrelevant until section 6.4.5, but the volatility process, $\{\sigma_t\}_{t\geq 0}$, is of key importance throughout. When these are constant, as assumed in the classic Black-Scholes-Merton theory of option pricing, the model is geometric Brownian motion; but extending to allow for deterministic variation in volatility with time, as we do here, is almost costless. With this addition we refer to the model as "Black-Scholes dynamics".

Applying Itô's formula to $\ln S_t$ gives

$$\ln S_t - \ln S_s = \int_s^t (\mu_u - \sigma_u^2/2) \cdot du + \int_s^t \sigma_u \cdot dW_u.$$

From the definition of the stochastic integral the change in the logarithm of the stock's price over $[s,t]$ is therefore normal with mean $\int_s^t (\mu_u - \sigma_u^2/2) \cdot du$ and variance $\int_s^t \sigma_u^2 \cdot du$. The stock-price process itself is Markov in view of the independence of the increments of Brownian motion, and $\mathcal{F}_t$ can be regarded equivalently as the σ-field generated by the history of either W or S through time t. Accordingly, for $0 \leq s \leq t$ we can write

$$S_t \sim S_s \exp\left[\int_s^t (\mu_u - \sigma_u^2/2) \cdot du + Z\sqrt{\int_s^t \sigma_u^2 \cdot du}\right] \tag{6.2}$$

where $Z \sim N(0,1)$, concluding that S_t is lognormally distributed conditional on $\sigma(S_s)$ (the σ-field generated by S_s). Since $E(e^{a+bZ}) = e^{a+b^2/2}$, the first moment is $E_s S_t \equiv E(S_t \mid \mathcal{F}_s) = E(S_t \mid S_s) = S_s \exp(\int_s^t \mu_u \cdot du)$. These facts are all we need to value derivatives on this underlying primary asset.

We shall also consider futures and derivatives on futures under the assumption that the futures price follows the same Black-Scholes dynamics as (6.1); that is,

$$d\mathsf{F}_t = \mu_t \mathsf{F}_t \cdot dt + \sigma_t \mathsf{F}_t \cdot dW_t.$$

Next, we need to recall the definitions and assumptions about the behavior of interest rates and bond prices, which in the present setting evolve in continuous time. As usual, $B(t,T)$ represents the price at t of a riskless zero-coupon bond paying one currency unit at T, and

$$r_t = \left.\frac{\partial \ln B(t,T)}{\partial t}\right|_{T=t} = -\left.\frac{\partial \ln B(t,T)}{\partial T}\right|_{T=t}$$

is the short or instantaneous spot rate of interest. As in chapter 5 we continue to invoke assumption RK, which is that short rates are known over the life of each derivative that we price. The numeraire is again the price of the money fund that rolls over instantaneous riskless spot loans. Its per-unit value therefore grows continuously through time at deterministic rates $\{r_t\}$, as

$$dM_t/dt = r_t M_t. \tag{6.3}$$

Taking the initial value to be M_0, this implies that the process

$$\left\{M_t = M_0 \exp\left(\int_0^t r_s \cdot ds\right)\right\}_{t\geq 0}$$

is itself $\mathcal{F}_0$-measurable.

If the stock that underlies our derivative asset pays proportional dividends continuously at $\mathcal{F}_0$-measurable instantaneous rates $\{\delta_t\}$, it will be necessary also to model the market value of the position that comprises the stock and its dividend stream. There are two possibilities, according as the dividends are used to purchase more stock or are invested in the money fund. In the former case, we have a *cum-* dividend process,

$$S_t^c = S_t \exp\left(\int_0^t \delta_u \cdot du\right), \tag{6.4}$$

which represents the value of a position that begins with one share at $t = 0$ and reinvests the dividends continuously. Applying Itô's formula to (6.4) shows that the dynamics are given by

$$dS_t^c = (\mu_t + \delta_t) S_t^c \cdot dt + \sigma_t S_t^c \cdot dW_t. \tag{6.5}$$

Alternatively, if the dividends are invested in the money fund, the value at t of a position that starts with one share of stock at $t = 0$ is

$$V_t = S_t + M_t \int_0^t \delta_s S_s / M_s \cdot ds, \tag{6.6}$$

with

$$dV_t = r_t V_t \cdot dt + [\mu_t - (r_t - \delta_t)] S_t \cdot dt + \sigma_t S_t \cdot dW_t. \tag{6.7}$$

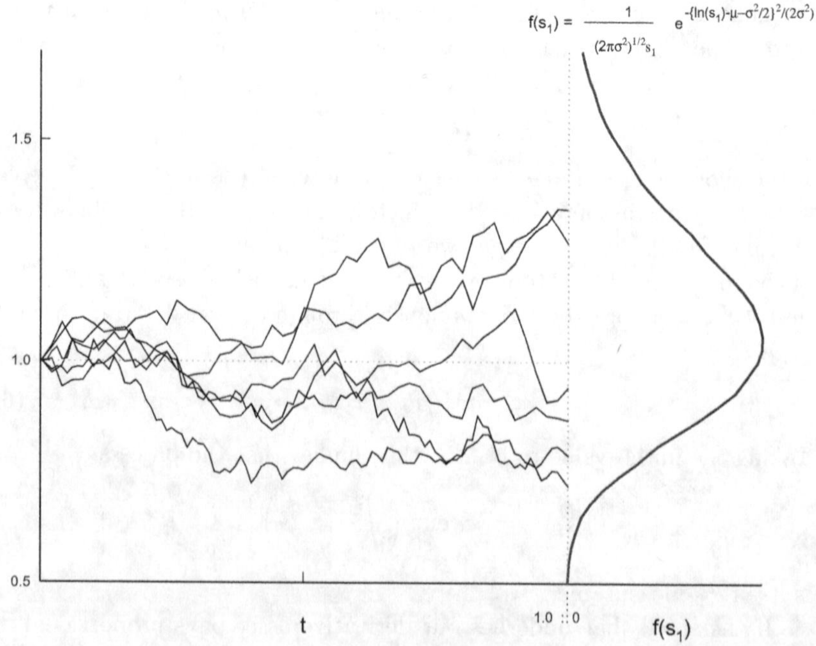

Fig. 6.1. Simulated sample paths of geometric Brownian motion and p.d.f. of S_1 ($S_0 = 1$, $\mu = .07$, $\sigma = .21$).

By allowing $\delta_t < 0$, these expressions apply also to positions in investment commodities for which the explicit cost of carry is positive and deterministic.

Figure 6.1 shows five simulated realizations of sample paths of geometric Brownian motion for $t \in [0, 1]$, all with initial value $S_0 = 1$. The lognormal p.d.f. of S_1 is juxtaposed.

6.2 Approaches to Arbitrage-Free Pricing

The market value of a derivative security in an arbitrage-free market can be determined uniquely if there exists a self-financing portfolio of traded assets that replicates the derivative's payoffs. One way to value a derivative is just to set out to construct such a portfolio. Considering path-independent derivatives whose values at each moment depend just on the contemporaneous price of the underlying asset, it is usually possible to find such a replicating portfolio if the derivative's value is a smooth function of time and the underlying price. In that case the conditions for replication will be seen to imply a law of motion for the value of the derivative, which takes the form of a partial differential equation. Solving this p.d.e. subject to initial conditions (and, typically, boundary conditions) gives the derivative's value in terms of the observables, t and S_t. This is the differential-equation approach to valuation.

The standard alternative to this is the equivalent-martingale or risk-neutral approach. It works like this. One first finds a measure equivalent to $\mathbb{P}$ under which the value of a position in the underlying, normalized by some numeraire, is a martingale. One then just verifies that a self-financing, replicating portfolio for the derivative exists—without having to construct it. The martingale representation theorem (section 3.4.2) will make this possible. This done, one can conclude that the portfolio's normalized value is itself a martingale under the new measure. Exploiting the fair-game property of martingales then leads to an expression for the derivative's current value in terms of the conditional expected value of its future payoffs.

This section explains both approaches. We consider derivatives on both primary assets, such as stocks and foreign currencies, and on futures prices. We limit attention to European-style derivatives, each of which makes a single state/time-contingent payoff $\bar{D}$ at some future date T. Of course, derivatives with payoffs at more than one date can be valued as the sum of the parts.

6.2.1 *The Differential-Equation Approach*

Consider first a T-expiring European-style derivative on an underlying primary asset (a "stock") that can pay dividends continuously at deterministic rates $\{\delta_t\}$ per unit time. Representing the value function at $t \in [0, T]$ as $D(S_t, T-t)$, let us try to construct a self-financing replicating portfolio of the stock and money fund. If the plan is to hold p_t shares of the *cum*-dividend stock position and q_t units of the money fund, we thus want

$$D(S_t, T-t) = p_t S_t^c + q_t M_t$$

at each (t, S_t) (in order to replicate) and

$$\begin{aligned} dD(S_t, T-t) &= p_t \cdot dS_t^c + q_t \cdot dM_t \\ &= p_t \cdot dS_t^c + [D(S_t, T-t) - p_t S_t^c] \cdot dM_t / M_t \qquad (6.8) \end{aligned}$$

(to make the position self-financing). Notice that the value of the derivative itself depends on the *ex*-dividend value of the underlying, whereas one works with the *cum*-dividend position in order to replicate. Using (6.5) and (6.3), the right side of (6.8) is

$$[D(S_t, T-t)r_t + p_t(\mu_t - r_t + \delta_t)S_t^c] \cdot dt + p_t\sigma_t S_t^c \cdot dW_t.$$

Assuming that the value function is smooth enough to apply Itô's formula—that is, continuously differentiable once with t and twice with S_t, the left side of (6.8) is

$$(-D_{T-t} + D_S\mu_t S_t + D_{SS}\sigma_t^2 S_t^2/2) \cdot dt + D_S\sigma_t S_t \cdot dW_t,$$

where subscripts on D denote partial derivatives. (Note that D_{T-t} and $-D_{T-t}$ are the partial derivatives of D with respect to $T-t$ and t, respectively.) Equating the two sides shows that the portfolio shares must satisfy

$$\begin{aligned} 0 = \int_t^T & [-D_{T-u} + D_S\mu_u S_u + D_{SS}\sigma_u^2 S_u^2/2 \\ & - Dr_u - p_u(\mu_u - r_u + \delta_u)S_u^c] \cdot du \end{aligned} \tag{6.9}$$

and

$$0 = \int_t^T (D_S S_u - p_u S_u^c)\sigma_u \cdot dW_u$$

for each $t \in [0, T]$. The second integral is zero for sure if and only if

$$0 = E\left[\int_t^T (D_S S_u - p_u S_u^c)\sigma_u \cdot dW_u\right]^2 = \int_t^T E[(D_S S_u - p_u S_u^c)^2\sigma_u^2] \cdot du,$$

which requires that $p_u = D_S S_u / S_u^c$ for almost all u.[1] Making this substitution in (6.9) shows that $D(\cdot, \cdot)$ must satisfy the partial differential equation

$$0 = -D_{T-t} + D_S(r_t - \delta_t)S_t + D_{SS}\sigma_t^2 S_t^2/2 - Dr_t. \tag{6.10}$$

Equation (6.10) is the "Black-Scholes" or "fundamental" p.d.e. for valuing contingent claims. Solving this subject to the terminal condition $D(S_T, 0) = \bar{D}(S_T)$ and applicable boundary conditions gives the value of the derivative at each (t, S_t). For example, boundary conditions for a European call struck at X on a no-dividend stock are $S_t - B(t,T)X \leq C^E(S_t, T-t) \leq S_t$. (See table 4.7 on page 166.) The "terminal" condition at $t = T$ for calendar time corresponds to an *initial* condition at $T - t = 0$ for time to expiration, and is sometimes referred to as such. The solution of (6.10) also implicitly

---

[1]Recall that $\{S_t\}_{0 \leq t \leq T}$ and $\{S_t^c\}_{0 \leq t \leq T}$ are a.s. positive under geometric Brownian motion, assuming that $\{\delta_t\}_{0 \leq t \leq T}$ is nonnegative.

determines the composition of the replicating portfolio, as

$$p_t = D_S(S_t, T-t)S_t/S_t^c$$
$$q_t = [D(S_t, T-t) - p_t S_t^c]/M_t.$$

If dividends are invested in the money fund rather than used to buy more stock, one can replicate using stock-savings combination V_t in (6.6) instead of S_t^c. In this case it is easy to see that the replicating portfolio comprises $p_t = D_S$ units of the stock-fund position and $q_t = (D - V_t)/M_t$ additional units of the money fund. Of course, (6.10) still applies.

Now consider a T-expiring derivative on a futures price, $\{\mathsf{F}_t\}$. Since futures contracts have no initial value, replicating a derivative worth $D(\mathsf{F}_t, T-t)$ with q_t units of the money fund and p_t futures contracts requires that $q_t M_t = D(\mathsf{F}_t, T-t)$. To be self-financing, the portfolio must satisfy

$$\begin{aligned} dD(\mathsf{F}_t, T-t) &= p_t \cdot d\mathsf{F}_t + q_t \cdot dM_t \\ &= p_t \cdot d\mathsf{F}_t + D(\mathsf{F}_t, T-t) \cdot dM_t/M_t \\ &= p_t(\mu_t \mathsf{F}_t \cdot dt + \sigma_t \mathsf{F}_t \cdot dW_t) + D(\mathsf{F}_t, T-t) r_t \cdot dt. \end{aligned}$$

Expressing $dD(\mathsf{F}_t, T-t)$ with Itô's formula as before, the last equality is seen to require that $p_t = D_{\mathsf{F}}$ and

$$0 = -D_{T-t} + D_{\mathsf{FF}}\sigma_t^2 \mathsf{F}_t^2/2 - Dr_t. \tag{6.11}$$

This is solved subject to terminal condition $D(\mathsf{F}_T, 0) = \bar{D}(\mathsf{F}_T)$ (and applicable boundary conditions) to obtain the value function at arbitrary $(\mathsf{F}_t, T-t)$. Note that solutions to (6.10) with the same side conditions also work for (6.11) upon setting $\delta_t = r_t$.

Equations (6.10) and (6.11) are linear second-order p.d.e.s of the parabolic type. With appropriate changes of variables each can be reduced to the simplest case of the classic heat equation from physics, $u_t = \kappa u_{xx}$, for which solutions are well known.[2] In the physical application the solution $u(x,t)$ relates the temperature in an idealized, insulated, one-dimensional rod to the time (t) and distance (x) from an origin, given an initial distribution of temperature, $u(\cdot, 0) = h(\cdot)$. The constant $\kappa > 0$ characterizes the rod's physical conductivity properties. In section 6.3 we use Fourier methods to work out a solution to (6.10) subject to $D(S_T, 0) = (X - S_T)^+$, thereby deriving the Black-Scholes formula for the value of a European put.

[2]Cannon (1984, section 1.2) gives transformations that reduce various linear, second-order p.d.e.s to the heat equation. Wilmott *et al.* (1995, section 5.4) provide such a reduction for (6.10) specifically.

6.2.2 *The Equivalent-Martingale Approach*

In an arbitrage-free market there exists a probability measure with respect to which normalized values of traded assets and of self-financing portfolios of assets are martingales. Within the Bernoulli framework it was possible to determine this measure explicitly by finding the pseudo-probabilities of the up and down states at each time step, $\hat{\pi}_j$ and $1 - \hat{\pi}_j$. Things obviously will not be so simple in the continuous-time Black-Scholes environment, where prices of underlying assets are Itô processes built up from Brownian motions. In order to be a martingale an Itô process must have zero mean drift, yet prices of risky assets typically exhibit systematic trends through time—even after they are normalized by values of bonds or money-market fund. Girsanov's theorem is the tool that will enable us to find an alternative measure in which normalized price processes are, in fact, trendless. We consider first how things work for underlying primary assets that can pay dividends continuously at deterministic rates. We shall see how to find the measure in which normalized asset prices are martingales, how to verify the existence of self-financing replicating portfolios for suitable derivatives, and how to use the fair-game property of martingales to find the derivatives' arbitrage-free prices. Once all this is done, it will be easy to see how to modify the procedure for derivatives on futures.

Finding Martingale Measures in the Black-Scholes Setting

Consider an underlying primary asset (a "stock") with deterministic, proportional dividend-rate process $\{\delta_t\}_{0 \le t \le T}$, where T is some fixed future date. Let us suppose first that the dividends are reinvested in the stock as they are received, so that the *cum*-dividend process evolves as $dS_t^c = (\mu_t + \delta_t)S_t^c \cdot dt + \sigma_t S_t^c \cdot dW_t$ with $\sigma_t > 0$ for $t \in [0, T]$. With $\{M_t\}_{0 \le t \le T}$ as numeraire, Itô's formula shows that the normalized process $\{S_t^{c*} \equiv S_t^c / M_t\}_{0 \le t \le T}$ evolves as

$$dS_t^{c*} = (\mu_t - r_t + \delta_t)S_t^{c*} \cdot dt + \sigma_t S_t^{c*} \cdot dW_t.$$

Notice that since $\{M_t\}$ is of finite variation, normalization alters the drift but not the volatility, as was explained in section 3.4.3. This normalized process can be a martingale under $\mathbb{P}$ only if it happens that $\mu_t - r_t + \delta_t = 0$ for all t. However, suppose there exists a measure $\hat{\mathbb{P}}$, equivalent to $\mathbb{P}$, with respect to which the process

$$\hat{W}_t = W_t + \int_0^t \sigma_u^{-1}(\mu_u - r_u + \delta_u) \cdot du \tag{6.12}$$

is a Brownian motion on $[0, T]$.[3] In this new measure the drift term for S_t^{c*} vanishes, since

$$\begin{aligned} dS_t^{c*} &= (\mu_t - r_t + \delta_t)S_t^{c*} \cdot dt + \sigma_t S_t^{c*}[d\hat{W}_t - \sigma_t^{-1}(\mu_t - r_t + \delta_t) \cdot dt] \\ &= \sigma_t S_t^{c*} \cdot d\hat{W}_t. \end{aligned} \tag{6.13}$$

Removing the trend in this way makes $\{S_t^{c*}\}_{0 \le t \le T}$ a local martingale under $\hat{\mathbb{P}}$. That it is also a martingale proper (requiring that $\hat{E}|S_t^{c*}| < \infty$ for all $t \ge 0$) is confirmed by Novikov's condition, since the $\mathcal{F}_0$ -measurability and integrability of $\{\sigma_t^2\}$ clearly guarantee that $\hat{E}\exp(\frac{1}{2}\int_0^T \sigma_t^2 \cdot dt) < \infty$.[4] Girsanov's theorem (section 3.4.1) verifies that such a $\hat{\mathbb{P}}$ does exist and shows that it can be constructed as

$$\hat{\mathbb{P}}(A) = E\mathbf{1}_A Q_T = \int_A Q_T(\omega) \cdot d\mathbb{P}(\omega), \quad A \in \mathcal{F},$$

where

$$Q_T = \exp\left[-\int_0^T \sigma_t^{-1}(\mu_t - r_t + \delta_t) \cdot dW_t - \frac{1}{2}\int_0^T \sigma_t^{-2}(\mu_t - r_t + \delta_t)^2 \cdot dt\right].$$

All this is strictly mechanical, but there is a very intuitive way to interpret the change of measure. Referring again to figure 6.1, think of this as depicting sample paths of a normalized process $\{S_t^{c^*}\}_{0 \le t \le 1}$ with initial value $S_0^{c^*} = 1$ and a drift process $\{\mu_t - (r_t - \delta_t)\}$ that is positive for all t. Then, as in the simulation that actually generated the data, these paths have an upward trend, on average under natural measure $\mathbb{P}$, so that $\int S_1^{c^*}(\omega) \cdot d\mathbb{P}(\omega) > 1$. Now when we change the measure, the potential paths themselves do not change. We merely change the weights attached to them, giving less weight to those for which $S_1^{c^*}(\omega) > 1$ and more to those with $S_1^{c^*}(\omega) < 1$ so as to make

$$\int S_1^{c^*}(\omega)Q_1(\omega) \cdot d\mathbb{P}(\omega) = \int S_1^{c^*}(\omega) \cdot d\hat{\mathbb{P}}(\omega) = 1.$$

Now let us consider what happens if the dividends are invested in the money fund instead of being used to buy more stock. In this case the value

[3] The reason for assuming that $\sigma_t > 0$ becomes is now evident.

[4] $\mathcal{F}_0$-measurability and integrability under $\mathbb{P}$ imply measurability and integrability under $\hat{\mathbb{P}}$ by equivalence. The martingale property of $\{S_t^{c*}\}$ under $\hat{\mathbb{P}}$ can also be verified directly from the solution to s.d.e. (6.13),

$$S_t^{c*} = S_0^{c*} \exp\left[\int_0^t \sigma_s \cdot d\hat{W}_s - \int_0^t \sigma_s^2/2 \cdot ds\right].$$

at t of the position in stock and fund is given by (6.6), and the normalized process $\{V_t^* = V_t/M_t\}_{0\le t\le T}$ evolves as

$$\begin{aligned} dV_t^* &= S_t^*[\mu_t - (r_t - \delta_t)] \cdot dt + \sigma_t S_t^* \cdot dW_t \\ &= \sigma_t S_t^* \cdot d\hat{W}_t. \end{aligned}$$

Thus, $\{V_t^*\}$ is also a martingale under the same measure $\hat{\mathbb{P}}$.[5]

Notice that the unnormalized, *ex*-dividend price process $\{S_t\}$ evolves under $\hat{\mathbb{P}}$ as

$$\begin{aligned} dS_t &= \mu_t S_t \cdot dt + \sigma_t S_t [d\hat{W}_t - \sigma_t^{-1}(\mu_t - r_t + \delta_t) \cdot dt] \\ &= (r_t - \delta_t) S_t \cdot dt + \sigma_t S_t \cdot d\hat{W}_t. \end{aligned} \tag{6.14}$$

This implies that

$$\hat{E}_t S_T = S_t \exp\left[\int_t^T (r_u - \delta_u) \cdot du\right] \tag{6.15}$$

and therefore $\hat{E}_t S_T^c = S_t^c \exp(\int_t^T r_u \cdot du)$, so that the value of the *cum*-dividend stock position has the same expected growth rate as does the value of the money fund. Likewise, if dividends are invested in the money fund, the unnormalized value of the stock-fund position evolves under $\hat{\mathbb{P}}$ as

$$dV_t = r_t V_t \cdot dt + \sigma_t S_t \cdot d\hat{W}_t,$$

and $\hat{E}_t V_T = V_t \exp(\int_t^T r_u \cdot du)$. Since this is precisely how the values of these positions would behave in a risk-neutral market, $\hat{\mathbb{P}}$ is also the risk-neutral measure.

So long as we abstract from uncertainty about future interest rates (as we do until chapter 10) nothing is changed if we use $\{B(t,T)\}$ as numeraire instead of $\{M_t\}$, since under assumption RK

$$\{S_t^c / B(t,T) \equiv S_t^c B(0,T)/M_t\}_{0\le t\le T}$$

is also a martingale on $(\Omega, \mathcal{F}, \{\mathcal{F}_t\}, \hat{\mathbb{P}})$.[6] In general, however, different martingale measures do correspond to different numeraires. For example, using $\{S_t^c\}$ as numeraire and letting $M_t^{**} = M_t/S_t^c$, we have

$$dM_t^{**} = (r_t - \delta_t + \sigma_t^2 - \mu_t) M_t^{**} \cdot dt - \sigma_t M_t^{**} \cdot dW_t.$$

---

[5]Write $dV_t^*/V_t^* = \sigma_t S_t/V_t \cdot d\hat{W}_t$. Assuming $\int_0^t \delta_s \cdot ds \ge 0$ for each $t \in [0,T]$, we have $0 \le S_t/V_t \le 1$, $\hat{\mathbb{P}}$ a.s. Hence, $\hat{E}\int_0^T \exp(\frac{1}{2}\sigma_t^2 S_t^2/V_t^2) \cdot dt < \infty$ and $\{V_t^*\}$ is indeed a proper martingale under $\hat{\mathbb{P}}$.

[6]When we introduce stochastic interest-rate processes in chapter 10 we will see that "spot" and T-forward measures $\hat{\mathbb{P}}$ and $\mathbb{P}^T$ associated with numeraires $\{M_t\}$ and $\{B(t,T)\}$ do in fact differ.

With

$$Q'_T = \exp\left[\int_0^T \left(\frac{r_t - \delta_t + \sigma_t^2 - \mu_t}{\sigma_t}\right) \cdot dW_t \right.$$
$$\left. -\frac{1}{2}\int_0^T \left(\frac{r_t - \delta_t + \sigma_t^2 - \mu_t}{\sigma_t^2}\right)^2 \cdot dt\right]$$

and

$$\widetilde{\mathbb{P}}(A) = E\mathbf{1}_A Q'_T = \int_A Q'_T(\omega) \cdot d\mathbb{P}(\omega), \ A \in \mathcal{F}, \tag{6.16}$$

we have a measure under which $\widetilde{W}_t = W_t - \int_0^t \sigma_u^{-1}(r_u - \delta_u + \sigma_u^2 - \mu_u) \cdot du$ is a Brownian motion and M_t^{**} becomes trendless, with $dM_t^{**} = -\sigma_t M_t^{**} \cdot d\widetilde{W}_t$.

Verifying a Self-financing Replicating Strategy for D

The martingale representation theorem (theorem 9 on page 122) makes it possible to confirm the existence of a self-financing replicating strategy within the Black-Scholes framework without actually constructing it. Applying the theorem requires restricting the class of derivative assets to those whose payoffs have finite expected value under $\hat{\mathbb{P}}$, but this limitation is of no practical importance under Black-Scholes dynamics.

Here is how the process works. When $\hat{E}|\bar{D}(S_T)| < \infty$, the derivative's normalized terminal value gives rise to a conditional-expectation process, $\{\hat{E}_t[\bar{D}(S_T)/M_T] \equiv D^*(S_t, T-t)\}_{0 \le t \le T}$ say, that is a $\hat{\mathbb{P}}$ martingale adapted to the same filtration as price process $\{S_t\}_{0 \le t \le T}$.[7] The representation theorem tells us that any such $\hat{\mathbb{P}}$ martingale can be represented as an Itô process and therefore as a stochastic integral with respect to a Brownian motion under $\hat{\mathbb{P}}$. This means that there exists a process $\{\mathfrak{D}_t\}_{0 \le t \le T}$ with $\int_0^T \mathfrak{D}_s^2 \cdot ds < \infty$ such that

$$D^*(S_t, T-t) = D^*(S_0, T) + \int_0^t \mathfrak{D}_s \cdot d\hat{W}_s.$$

Moreover, the theorem implies that $D^*(S_t, T-t)$ can also be represented as a stochastic integral with respect to the normalized price of the underlying. That is, since $dS_s^{c*} = \sigma_s S_s^{c*} \cdot d\hat{W}_s$ and both σ_s and S_s^{c*} are positive, taking

[7]See example 27 on page 88.

$p_s \equiv (\sigma_s S_s^{c*})^{-1}\mathfrak{D}_s$ for $s \geq 0$ gives

$$D^*(S_t, T-t) = D^*(S_0, T) + \int_0^t p_s \cdot dS_s^{c*}$$

or $dD^*(S_t, T-t) = p_t \cdot dS_t^{c*}$. Now define $D(S_t, T-t) \equiv M_t D^*(S_t, T-t)$ and set $q_t = D^*(S_t, T-t) - p_t S_t^{c*}$, so that $D(S_t, T-t) = p_t S_t^c + q_t M_t$. This will be the current value of the replicating portfolio. It does replicate the payoff, since $p_T S_T^c + q_T M_T = M_T D^*(S_T, 0) = M_T \hat{E}_T[\bar{D}(S_T)/M_T] = \bar{D}(S_T)$. Moreover, it is self-financing, since

$$\begin{aligned} dD(S_t, T-t) &= M_t \cdot dD^*(S_t, T-t) + D^*(S_t, T-t) \cdot dM_t \\ &= M_t p_t \cdot dS_t^{c*} + (p_t S_t^{c*} + q_t) \cdot dM_t \\ &= p_t \cdot d(M_t S_t^{c*}) + q_t \cdot dM_t \\ &= p_t \cdot dS_t^c + q_t \cdot dM_t. \end{aligned}$$

With replication thus verified, we now know that the measure $\hat{\mathbb{P}}$ that prices derivatives on the underlying asset is unique.

Valuing the Derivative

While a replicating portfolio is thus shown to exist, we do not know explicitly the numbers of shares, p_t and q_t, from which to calculate the derivative's value at any $t < T$. This value can be found directly, however, by exploiting the martingale property of $D^*(S_t, T-t)$ under $\hat{\mathbb{P}}$. Since $D^*(S_t, T-t) = \hat{E}_t D^*(S_T, 0) = \hat{E}_t[\bar{D}(S_T)/M_T]$, we have under assumption RK

$$\begin{aligned} D(S_t, T-t) &= M_t \hat{E}_t[\bar{D}(S_T)/M_T] \\ &= \frac{M_t}{M_T} \hat{E}_t \bar{D}(S_T) \\ &= \exp\left(-\int_t^T r_u \cdot du\right) \hat{E}_t \bar{D}(S_T) \end{aligned}$$

or

$$D(S_t, T-t) = B(t, T) \hat{E}_t \bar{D}(S_T). \tag{6.17}$$

The derivative's initial value is then

$$\begin{aligned} D(S_0, T) &= \exp\left(-\int_0^T r_u \cdot du\right) \hat{E} \bar{D}(S_T) \\ &= B(0, T) \hat{E} \bar{D}(S_T). \end{aligned}$$

Under $\hat{\mathbb{P}}$ the derivative itself is therefore valued as it would be in an equilibrium within a risk-neutral market, just as was true of the underlying asset itself. Under the new measure the price of the underlying behaves precisely as it did under $\mathbb{P}$ except that the drift term, μ_t, in (6.1) is replaced by the short rate minus the dividend rate, $r_t - \delta_t$, as in (6.14). This means that under $\hat{\mathbb{P}}$ the conditional distribution of S_T given S_t still has the lognormal form

$$S_T \sim S_t \exp\left[\int_t^T (r_u - \delta_u - \sigma_u^2/2) \cdot du + Z\sqrt{\int_t^T \sigma_u^2 \cdot du}\right], \tag{6.18}$$

where $Z \sim N(0,1)$. With this result pricing a European-style derivative under Black-Scholes dynamics reduces to the purely mechanical process of evaluating the expectation of a function of a standard normal variate.

Futures Prices under the Risk-Neutral Measure

Modeling the evolution of a futures price over $[0,T]$ under the natural measure $\mathbb{P}$ as

$$d\mathsf{F}_t = \mu_t \mathsf{F}_t \cdot dt + \sigma_t \mathsf{F}_t \cdot dW_t, \tag{6.19}$$

let us consider the implications for the stochastic behavior of $\{\mathfrak{F}_t\}$, the value of a futures position that was initiated at $t = 0$. Supposing this to have been marked to market at times $0 < t_1 < \cdots < t_n \le t$, with gains (losses) being invested in (financed by) shares of the money fund, we have

$$\mathfrak{F}_t = (\mathsf{F}_{t_1} - \mathsf{F}_0)\frac{M_t}{M_{t_1}} + (\mathsf{F}_{t_2} - \mathsf{F}_{t_1})\frac{M_t}{M_{t_2}} + \cdots + (\mathsf{F}_{t_n} - \mathsf{F}_{t_{n-1}})\frac{M_t}{M_{t_n}} + (\mathsf{F}_t - \mathsf{F}_{t_n}).$$

Now as an approximation imagine that the marking to market has taken place continuously. Taking limits in the above as $\max\{t_j - t_{j-1}\}_{j=1}^n \to 0$ gives $\mathfrak{F}_t \doteq M_t \int_0^t M_s^{-1} \cdot d\mathsf{F}_s$ and, upon dividing by M_t to normalize,

$$d\mathfrak{F}_t^* \equiv d(\mathfrak{F}_t/M_t) = M_t^{-1} \cdot d\mathsf{F}_t = \mu_t \mathsf{F}_t^* \cdot dt + \sigma_t \mathsf{F}_t^* \cdot dW_t.$$

Since it is the futures position rather than the futures price that has the status of an asset, it is $\mathfrak{F}_t^*$ rather than F_t^* that should be a martingale under the appropriate change of measure. Applying Girsanov's theorem with

$$Q_T = \exp\left(-\int_0^T \sigma_t^{-1}\mu_t \cdot dW_t - \frac{1}{2}\int_0^T \sigma_t^{-2}\mu_t^2 \cdot dt\right)$$

produces the measure $\hat{\mathbb{P}}$ that makes $\{\hat{W}_t = W_t + \int_0^t \sigma_s^{-1}\mu_s \cdot ds\}$ a Brownian motion and thereby removes the trend in $\mathfrak{F}_t^*$, as $d\mathfrak{F}_t^* = \sigma_t \mathsf{F}_t^* \cdot d\hat{W}_t$. Referring to (6.19), we see that under this measure $d\mathsf{F}_t = \sigma_t \mathsf{F}_t \cdot d\hat{W}_t$, so that the *unnormalized* futures price is itself a trendless Itô process motion, and hence a martingale. Accordingly, the time-T futures price has the following distribution conditional on what is known at $t \leq T$:

$$\mathsf{F}_T \sim \mathsf{F}_t \exp\left(-\frac{1}{2}\int_t^T \sigma_u^2 \cdot du + Z\sqrt{\int_t^T \sigma_u^2 \cdot du}\right), \quad Z \sim N(0,1).$$

With this result valuing European options or other European-style derivatives on the futures is again reduced to finding the discounted expectation of a function of a normally distributed random variable.

Overview of Martingale Pricing

To summarize, here is how the equivalent-martingale/risk-neutral method is used within the Black-Scholes framework to value a European-style derivative whose payoff depends just on time and the price of an underlying asset at some future date T.

- First, define the underlying asset in such a way that it can be used to replicate the derivative. For a primary asset with zero explicit cost of carry, such as a common stock that pays no dividends during the life of the derivative, ordinary shares can be held and used along with the money fund in replication. For derivatives on dividend-paying assets, however, one must take positions in either *cum*-dividend shares or portfolios of shares and money fund; and for derivatives on futures one has to work with futures positions rather than futures prices. In each case, so long as we are within the Black-Scholes framework the instantaneous change in the value of this underlying asset will be given by an s.d.e. that corresponds to an Itô process with either constant or time-dependent (but deterministic) volatility. Either way, the distribution of the asset's future value will be lognormal, conditional on the current value.
- Next, set the drift term in the s.d.e. for the underlying to what it would be if the asset were priced under conditions of risk neutrality. For a primary asset paying dividends continuously at rates $\{\delta_t\}_{0 \leq t \leq T}$ this amounts to setting the mean proportional drift at t equal to $r_t - \delta_t$. For a futures price just set the mean proportional drift to zero, or else treat the underlying as a primary asset with dividend rate equal to r_t.

- Finally, to find the derivative's arbitrage-free value at $t \in [0, T)$, find the expected value of the price-dependent payoff in the risk-neutral measure, as $\hat{E}_t\bar{D}(S_T)$, $\hat{E}_t\bar{D}(\mathsf{F}_T)$, etc., and weight it by the riskless discount factor, $B(t, T) = M_t M_T^{-1} = \exp(-\int_t^T r_u \cdot du)$. If price-dependent payoffs can occur at more than one time, weight each of their expected values by the appropriate discount factor and sum them.

6.3 Applications

Let us now begin to price some specific derivatives. Simple forward contracts are a good place to start since their arbitrage-free values are already known from the static replication arguments of chapter 4 and since risk-neutral pricing is trivially easy. For options we will work out the prices both by solving the fundamental p.d.e. and by following the steps for martingale pricing. Then we will stand back and reflect a bit on why these two very different procedures lead to the same result.

6.3.1 *Forward Contracts*

Our derivative is a forward contract expiring at T with pre-established forward price f. At expiration it is worth $\mathfrak{F}(S_T, 0; \mathsf{f}) = S_T - \mathsf{f}$ to the long side, and the goal is to find the contract's value at any $t \in [0, T)$. The underlying asset either pays a continuous dividend at rate $\delta_t \geq 0$ or else incurs an explicit cost of carry at rate $-\delta_t = \kappa_t \geq 0$. We already know that the *cum*-dividend/*ex*-cost position can be used for replication by continuously reinvesting the receipts or selling down to finance the outlays. Take the underlying price to evolve as $dS_t = \mu_t S_t \cdot dt + \sigma_t S_t \cdot dW_t$ under measure $\mathbb{P}$, where $\{\mu_t, \sigma_t\}_{0 \leq t \leq T} \in \mathcal{F}_0$. Changing to $\hat{\mathbb{P}}$, we set $\mu_t = r_t - \delta_t$, write $dS_t = (r_t - \delta_t)S_t \cdot dt + \sigma_t S_t \cdot d\hat{W}_t$, and solve the s.d.e. to get

$$S_T = S_t \exp \int_t^T (r_u - \delta_u - \sigma_u^2/2) \cdot du + \int_t^T \sigma_u \cdot d\hat{W}_u. \qquad (6.20)$$

The risk-neutral value of the contract is then found as

$$\begin{aligned}
\mathfrak{F}(S_t, T - t; \mathsf{f}) &= B(t, T)\hat{E}_t(S_T - \mathsf{f}) \\
&= B(t, T)S_t \exp\left[\int_t^T (r_u - \delta_u) \cdot du\right] - B(t, T)\mathsf{f} \\
&= S_t e^{-\int_t^T \delta_u \cdot du} - B(t, T)\mathsf{f}.
\end{aligned}$$

This clearly satisfies terminal condition $\mathfrak{F}(S_T, 0; \mathsf{f}) = S_T - \mathsf{f}$, and with $\delta_u = -\kappa_u$ it corresponds to our dynamics-free solution (4.9) on page 153. Checking that it also solves the fundamental p.d.e., we have

$$\mathfrak{F}_t = \partial\mathfrak{F}/\partial t = \delta_t S_t e^{-\int_t^T \delta_u \cdot du} - r_t B(t,T)\mathsf{f}$$
$$\mathfrak{F}_S = \partial\mathfrak{F}/\partial S_t = e^{-\int_t^T \delta_u \cdot du}$$
$$\mathfrak{F}_{SS} = \partial^2\mathfrak{F}/\partial S_t^2 = 0$$

and

$$\mathfrak{F}_t + \mathfrak{F}_S(r_t - \delta_t)S_t + \mathfrak{F}_{SS}\sigma_t^2 S_t^2/2 - \mathfrak{F}r_t = 0.$$

6.3.2 *European Options on Primary Assets*

Consider now European options on primary assets such as stocks and currency positions. To illustrate both techniques, we price a European put by solving the Black-Scholes p.d.e., then use martingale methods to value a European call.

Solving the Black-Scholes P.D.E.

Letting $P^E(S_t, T-t)$ represent the price at t of a T-expiring European put and setting $D(S_t, T-t) = P^E(S_t, T-t)$ in (6.10), the task is to solve the p.d.e. subject to terminal condition $P^E(S_T, 0) = \bar{P}(S_T) \equiv (X - S_T)^+$. This is most easily done using Fourier transforms, which convert the p.d.e. to an ordinary differential equation in the time variable alone. To simplify notation let us regard the volatility parameter (σ), the continuous dividend rate (δ), and the short rate (r) as time invariant. Later we modify the final result to let them vary deterministically.

Recall from chapter 2 that if $g : \Re \rightarrow \Re$ is integrable, then its Fourier integral transform is the complex-valued function

$$\mathcal{F}g(\cdot) = \int_{-\infty}^{\infty} e^{i\zeta x} g(x) \cdot dx \equiv f(\zeta),$$

where $\zeta \in \Re$ and $i = \sqrt{-1}$. For present purposes we need to generalize slightly to functions $g : \Re_2 \rightarrow \Re$, allowing g to depend on a parameter t that is independent of x. Then

$$\mathcal{F}g(\cdot, t) = \int_{-\infty}^{\infty} e^{i\zeta x} g(x,t) \cdot dx \equiv f(\zeta, t),$$

and the inversion and convolution formulas, (2.31) and (2.32), become

$$g(x,t) = \mathcal{F}^{-1}f(\zeta,t) \equiv \frac{1}{2\pi}\int_{-\infty}^{\infty} e^{-i\zeta x}f(\zeta,t)\cdot d\zeta$$
$$\mathcal{F}[g(\cdot,\tau) * h(\cdot,\tau)] = \mathcal{F}g(\cdot,\tau)\cdot\mathcal{F}h(\cdot,\tau).$$

The first three relations below correspond to properties **1–3** on page 46, while the last expresses the transform of the derivative of g with respect to t:

(i) $\mathcal{F}[bg(\cdot,t) + ch(\cdot,t)] = b\mathcal{F}g(\cdot,t) + c\mathcal{F}h(\cdot,t)$, where b, c are constants.
(ii) $\mathcal{F}g_x(\cdot,t) = -i\zeta\cdot f(\zeta,t)$ if $\lim_{|x|\to\infty} g(x,t) = 0$
(iii) $\mathcal{F}g_{xx}(\cdot,t) = -\zeta^2\cdot f(\zeta,t)$ if $\lim_{|x|\to\infty} g_x(x,t) = 0$
(iv) $\mathcal{F}g_t(\cdot,t) = f_t(\zeta,t)$.

The first step in solving (6.10) is to simplify with a normalization and a change of variables. Recalling the homogeneity of the put's value with respect to S and X, we can normalize by the strike price, putting $s \equiv \ln(S_t/X)$, $\tau \equiv T-t$, and $p(s,\tau) \equiv P^E(Xe^s,\tau)/X$. With these substitutions (6.10) becomes

$$0 = -p_\tau + p_s(r - \delta - \sigma^2/2) + p_{ss}\sigma^2/2 - pr, \tag{6.21}$$

and the terminal condition becomes an initial condition, $p(s,0) = (1-e^s)^+$. From the static replication arguments of chapter 4 (table 4.7 on page 166) we also recognize the following boundary conditions:

$$(e^{-r\tau} - e^{s-\delta\tau})^+ \le p(s,\tau) \le e^{-r\tau}.$$

However, it will turn out that these are automatically satisfied by the solution from the initial condition only.

Now we would like to Fourier transform $p(s,\tau)$ with respect to s, thus eliminating that variable, and use relations (ii)–(iv) above to express the derivatives in (6.21). $p(s,\tau)$ itself is bounded and therefore integrable, so its transform does exist, but the condition for (ii) fails because $\lim_{s\to-\infty} p(s,\tau) = e^{-r\tau} > 0$. We circumvent this problem by working not with $p(s,\tau)$ but with the still-integrable function $q(s,\tau) \equiv e^s p(s,\tau)$, for which the required conditions in (ii) and (iii) do hold. Expressing p and its derivatives in terms of q and simplifying give the new p.d.e.

$$0 = -q_\tau + q_s(\nu - \sigma^2) + q_{ss}\sigma^2/2 - q(r + \nu - \sigma^2/2),$$

where $\nu \equiv r - \delta - \sigma^2/2$. Now we can work the transform, putting $f(\zeta,\tau) = \mathcal{F}q(\cdot,\tau) = \int_{-\infty}^{\infty} e^{i\zeta s}q(s,\tau)\cdot ds$ and using (ii)–(iv) to get what is now an ordinary differential equation:

$$0 = -f_\tau - i\zeta(\nu - \sigma^2)f - \zeta^2\sigma^2 f/2 - (r + \nu - \sigma^2/2)f.$$

Writing as

$$\frac{f_\tau}{f} = -i\zeta(\nu - \sigma^2) - \zeta^2\sigma^2/2 - (r + \nu - \sigma^2/2)$$

and solving give

$$f(\zeta, \tau) = e^{-(r+\nu-\sigma^2/2)\tau} f(\zeta, 0) \exp[-i\zeta(\nu - \sigma^2)\tau - \zeta^2\sigma^2\tau/2].$$

The task now is to invert the transform. Assuming that $f(\zeta, 0) = \mathcal{F}q(\cdot, 0)$, and noting that the second exponential factor in the solution for $f(\zeta, \tau)$ is the c.f. of a normal random variable with mean $-(\nu - \sigma^2)\tau$ and variance $\sigma^2\tau$, we can write $f(\zeta, \tau) \equiv \mathcal{F}q(\cdot, \tau)$ as

$$f(\zeta, \tau) = e^{-(r+\nu-\sigma^2/2)\tau} \mathcal{F}q(\cdot, 0) \mathcal{F}\left\{ \frac{1}{\sqrt{2\pi\sigma^2\tau}} e^{-[(\cdot)+(\nu-\sigma^2)\tau]^2/(2\sigma^2\tau)} \right\}.$$

In words, the Fourier transform of the function we seek is proportional to the product of transforms of these two other functions. It follows from the convolution formula that $q(s, t)$ is proportional to the convolution of those functions and so equal to

$$e^{-(r+\nu-\sigma^2/2)\tau} \int_{-\infty}^{\infty} q(w, 0) \frac{1}{\sqrt{2\pi\sigma^2\tau}} \exp\left\{ -\frac{[w - s - (v - \sigma^2)\tau]^2}{2\sigma^2\tau} \right\} \cdot dw. \tag{6.22}$$

Writing $q(w, 0) = e^w(1 - e^w)^+$ in recognition of the initial condition and noting that this is zero for $w > 0$, we are led to evaluate

$$e^{-(r+\nu-\sigma^2/2)\tau} \int_{-\infty}^{0} (e^w - e^{2w}) \frac{1}{\sqrt{2\pi\sigma^2\tau}} \exp\left\{ -\frac{[w - s - (v - \sigma^2)\tau]^2}{2\sigma^2\tau} \right\} \cdot dw.$$

Integrals

$$\int_{-\infty}^{0} e^{kw} \frac{1}{\sqrt{2\pi\sigma^2\tau}} \exp\left\{ -\frac{[w - s - (v - \sigma^2)\tau]^2}{2\sigma^2\tau} \right\} \cdot dw$$

can be reduced to

$$e^{sk+\nu k\tau+\sigma^2(k^2/2-k)\tau} \Phi\left(-\frac{s + [\nu + (k-1)\sigma^2]\,\tau}{\sigma\sqrt{\tau}} \right)$$

($\Phi(\cdot)$ being the standard normal c.d.f.) by completing the square in the exponent and changing variables as

$$z = \frac{w - s - [\nu + (k-1)\sigma^2]\tau}{\sigma\sqrt{\tau}}.$$

Evaluating at $k \in \{1, 2\}$ and using $\nu \equiv r - \delta - \sigma^2/2$, one obtains for $q(s, \tau)$

$$e^{s-r\tau}\Phi\left(-\frac{s+(r-\delta-\sigma^2/2)\tau}{\sigma\sqrt{\tau}}\right) - e^{2s-\delta\tau}\Phi\left(-\frac{s+(r-\delta+\sigma^2/2)\tau}{\sigma\sqrt{\tau}}\right).$$

Finally, on recovering $p(s, \tau)$ as $e^{-s}q(s, \tau)$, restating in terms of the original variables, and putting $B \equiv B(t, T) = e^{-r(T-t)}$, we can express $P^E(S_t, T - t)$ as

$$BX\Phi\left[\frac{\ln\left(\frac{BX}{S_te^{-\delta(T-t)}}\right)+\frac{\sigma^2(T-t)}{2}}{\sigma\sqrt{T-t}}\right]$$
$$-S_te^{-\delta(T-t)}\Phi\left[\frac{\ln\left(\frac{BX}{S_te^{-\delta(T-t)}}\right)-\frac{\sigma^2(T-t)}{2}}{\sigma\sqrt{T-t}}\right]$$

or

$$\begin{aligned} P^E(S_t, T-t) &= BX\Phi\left[q^+\left(\frac{BX}{S_te^{-\delta(T-t)}}\right)\right] \\ &\quad - S_te^{-\delta(T-t)}\Phi\left[q^-\left(\frac{BX}{S_te^{-\delta(T-t)}}\right)\right]. \end{aligned} \tag{6.23}$$

With $\delta = 0$ this is the usual Black-Scholes formula for a European put on an underlying asset that pays no dividends and has no other explicit cost of carry. We show in section 6.4.2 that (6.23) is consistent with terminal condition $P^E(S_T, 0) = (X - S_T)^+$ and that it satisfies the differentiability conditions that were assumed in deriving the fundamental p.d.e.

The shorthand notation

$$q^{\pm}(x) \equiv \frac{\ln x \pm \sigma^2(T-t)/2}{\sigma\sqrt{T-t}} \tag{6.24}$$

will often appear in pricing formulas derived within the Black-Scholes framework. A second argument will sometimes be included, as $q^{\pm}(x; T - t)$ or $q^{\pm}(x; \sigma)$, if more than one such feature is involved in a particular problem. One can convert between the + and − versions as

$$q^-(x) = q^+(x) - \sigma\sqrt{T-t} = -q^+(x^{-1}). \tag{6.25}$$

While this shows that a single symbol "q" would really do, it will be convenient to use both.

Since the T-forward price of an underlying asset that pays continuous proportional dividends at rate δ is

$$\mathsf{f} \equiv \mathsf{f}(t, T) = B(t, T)^{-1}S_te^{-\delta(T-t)}, \tag{6.26}$$

formula (6.23) can be written also as

$$P^E(S_t, T-t) = BX\Phi[q^+(X/\mathsf{f})] - B\mathsf{f}\Phi[q^-(X/\mathsf{f})]. \tag{6.27}$$

A similar exercise for the European call, solving (6.10) subject to terminal condition $C^E(S_T, 0) = (S_T - X)^+$, leads to

$$C^E(S_t, T-t) = B\mathsf{f}\Phi[q^+(\mathsf{f}/X)] - BX\Phi[q^-(\mathsf{f}/X)]. \tag{6.28}$$

Below, we derive this expression via the martingale approach.

Deterministic variation in σ, δ, and r is accommodated in these formulas just by setting

$$r = (T-t)^{-1}\int_t^T r_u \cdot du$$

$$\sigma^2 = (T-t)^{-1}\int_t^T \sigma_u^2 \cdot du$$

$$\delta = (T-t)^{-1}\int_t^T \delta_u \cdot du.$$

This amounts to regarding r, σ^2, and δ as averages of the time-varying values over the interval $[t, T]$. The fact that the average values are all that matter for valuing European options in this setting will be of importance in chapter 8, where we look at dynamics in which volatility of the underlying varies stochastically.

Risk-Neutral Valuation of European Options

Let us now derive the European call formula by martingale methods. Taking the solution for S_T given S_t to be as in (6.20) and recognizing $C^E(S_T, 0) = \bar{C}(S_T) \equiv (S_T - X)^+$ as the call's terminal value, we will apply (6.17) to find the current value as

$$C^E(S_t, T-t) = B(t,T)\hat{E}_t(S_T - X)^+. \tag{6.29}$$

For this purpose it is helpful to have a simple representation for the conditional c.d.f. of S_T in the risk-neutral measure—namely, $\hat{F}_t(s) \equiv \hat{\mathbb{P}}(S_T \le s \mid \mathcal{F}_t)$. Obtaining this will be our first step. Continuing to take r, σ^2, and δ as the time averages of these deterministic processes, (6.20) implies that S_T is distributed under $\hat{\mathbb{P}}$ as $S_t \exp[(r - \delta - \sigma^2/2)(T-t) + Z\sigma\sqrt{T-t}]$, where $Z \sim N(0,1)$. It follows that $\hat{F}_t(s) = 0$ for

$s \leq 0$, while for positive s

$$\begin{aligned}\hat{F}_t(s) &= \hat{\mathbb{P}}[\ln(S_T/S_t) \leq \ln(s/S_t) \mid \mathcal{F}_t] \\ &= \hat{\mathbb{P}}[(r - \delta - \sigma^2/2)(T-t) + Z\sigma\sqrt{T-t} \leq \ln(s/S_t)] \\ &= \Phi\left[\frac{\ln(s/S_t) - (r - \delta - \sigma^2/2)(T-t)}{\sigma\sqrt{T-t}}\right].\end{aligned}$$

Finally, using again the shorthand

$$q^{\pm}(x) \equiv \frac{\ln x \pm \sigma^2(T-t)/2}{\sigma\sqrt{T-t}}$$

and taking the forward price $\mathsf{f} = \mathsf{f}(t,T)$ from (6.26), $\hat{F}_t(\cdot)$ can be expressed compactly as

$$\hat{F}_t(s) = \Phi[q^+(s/\mathsf{f})]. \tag{6.30}$$

Having found $\hat{F}_t$, we are not far from a solution for the arbitrage-free price of the call. Applying (6.29) with $B \equiv B(t,T)$, we have

$$\begin{aligned}C^E(S_t, T-t) &= B\int_0^\infty (s - X)^+ \cdot d\hat{F}_t(s) \\ &= B\int_X^\infty s \cdot d\hat{F}_t(s) - BX\int_X^\infty d\hat{F}_t(s) \\ &= B\int_X^\infty s \cdot d\Phi[q^+(s/\mathsf{f})] - BX\int_X^\infty d\Phi[q^+(s/\mathsf{f})].\end{aligned}$$

Changing variables in the first integral as $z = q^+(s/\mathsf{f}) - \sigma\sqrt{T-t} = q^-(s/\mathsf{f})$ gives

$$B\mathsf{f}\int_{q^-(X/\mathsf{f})}^\infty d\Phi(z) = B\mathsf{f}\Phi[q^+(\mathsf{f}/X)],$$

and setting $z = q^+(s/\mathsf{f})$ in the second integral gives

$$-BX\int_{q^+(X/\mathsf{f})}^\infty d\Phi(z) = -BX\Phi[q^-(\mathsf{f}/X)].$$

Assembling the results gives (6.28).

We could have got the same solution by working not with risk-neutral measure $\hat{\mathbb{P}}$ but with the martingale measure $\widetilde{\mathbb{P}}$ that corresponds to numeraire $\{S_t\}$, as in (6.16). In the present setting, with σ, r, and δ still as constants or averages over $[t,T]$, we find that S_T is distributed under $\widetilde{\mathbb{P}}$ (conditional on $\mathcal{F}_t$) as $S_T = S_t \exp[(r - \delta + \sigma^2/2)(T-t) + Z\sigma\sqrt{T-t}]$, and from this we deduce that $\widetilde{F}_t(s) = \Phi[q^-(s/\mathsf{f})]$. Under this measure it is now

$\{C^E(S_t, T-t)/S_t\}_{0 \le t \le T}$ rather than $\{C^E(S_t, T-t)/M_t\}_{0 \le t \le T}$ that is a martingale. Thus,

$$C^E(S_t, T-t)/S_t = \widetilde{E}_t[(S_T - X)^+/S_T]$$

and

$$C^E(S_t, T-t) = S_t \int_0^\infty s^{-1}(s-X)^+ \cdot d\widetilde{F}_t(s) = S_t \int_X^\infty (1 - X/s) \cdot d\widetilde{F}_t(s).$$

Working out the integral leads again to (6.28).

Notice that we can write (6.28) using relation (6.25) as

$$\begin{aligned} C^E(S_t, T-t) &= B\mathsf{f}\Phi[-q^-(\mathsf{f}/X)] - BX\Phi[-q^+(\mathsf{f}/X)] \\ &= B\mathsf{f}[1 - \widetilde{F}_t(X)] - BX[1 - \widehat{F}_t(X)] \\ &= B\mathsf{f}\widetilde{\mathbb{P}}_t(S_T > X) - BX\hat{\mathbb{P}}_t(S_T > X). \end{aligned}$$

Likewise, the solution for $P^E(S_t, T-t)$ can be expressed as

$$P^E(S_t, T-t) = BX\widetilde{\mathbb{P}}_t(S_T < X) - B\mathsf{f}\hat{\mathbb{P}}_t(S_T < X).$$

One will recall the corresponding expressions (5.25) and (5.26) from Bernoulli dynamics (page 198). Thus, the values of the call and put depend on the probabilities of finishing in the money under the distinct martingale measures corresponding to numeraires $\{M_t\}$ and $\{S_t\}$.

The Feynman-Kac Solution

We have presented two very different derivations of the Black-Scholes formulas. The differential-equation method required solving the p.d.e. that represents the derivative's law of motion, while the martingale method involved calculating the mathematical expectation of the option's terminal payoff in the risk-neutral distribution or other martingale measure. It seems astonishing at first that calculating such an expectation turns out to deliver a solution to the financial p.d.e. One wonders how the two methods are connected and whether there is something special about derivative securities that makes it possible to solve their equations of motion by means of probabilistic arguments. In fact, probabilistic methods are applicable to p.d.e.s that describe a wide variety of phenomena. Just as was demonstrated for the call option, the probabilistic approach delivers solutions in the form of expectations of functions of Brownian motion. Attribution is often made to Feynman (1948) and Kac (1949), although their work focused just on a specific version of the heat equation. Durrett (1996) describes many other applications.

To get the flavor of the probabilistic approach, consider its application to the classic heat equation, which was described on page 253:

$$\begin{aligned} u_t &= \kappa u_{xx} \\ u(x,0) &= h(x). \end{aligned} \tag{6.31}$$

Here t and x represent time and position, κ is a positive constant, and h is a given function that describes the initial distribution of temperature. The Feynman-Kac method hinges on the following observation: If $\{W_s\}_{0\le s\le t}$ is a Brownian motion on $(\Omega, \mathcal{F}, \{\mathcal{F}_s\}, \mathbb{P})$ and if $\{U_s = u(W_s\sqrt{2\kappa}, t-s)\}_{0\le s\le t}$ is uniformly integrable, then $\{U_s\}$ is a martingale adapted to $\{\mathcal{F}_s\}$ when (6.31) is satisfied. This follows because Itô's formula together with (6.31) imply that $\{U_s\}_{0\le s\le t}$ is trendless, since

$$\begin{aligned} du(W_s\sqrt{2\kappa}, t-s) &= (-u_t + \kappa u_{xx}) \cdot ds + u_x\sqrt{2\kappa} \cdot dW_s \\ &= u_x\sqrt{2\kappa} \cdot dW_s, \end{aligned}$$

and because uniform integrability assures that $\sup_{0\le s\le t} E|U_s| < \infty$. Martingale convergence (theorem 7) thus applies, from which we conclude that $\lim_{s\to t} U_s = h(W_t\sqrt{2\kappa})$, and so $U_s = E_s h(W_t\sqrt{2\kappa})$. Therefore, taking $W_0 = x/\sqrt{2\kappa}$ to meet the initial condition gives as the solution to the p.d.e. $u(x,t) = Eh(W_t\sqrt{2\kappa})$.

Applied to the problem of solving

$$0 = -D_{T-t} + D_S(r-\delta)S_t + D_{SS}\sigma^2 S_t^2/2 - Dr \tag{6.32}$$

subject to $D(S_T, 0) = \bar{D}(S_T)$, the method requires exactly the same steps as in the equivalent-martingale approach. Under risk-neutral measure $\hat{\mathbb{P}}$ we have

$$dS_t = (r-\delta)S_t \cdot dt + \sigma S_t \cdot d\hat{W}_t$$

with $\{\hat{W}_t\}$ a Brownian motion, so applying Itô's formula to $D^*(S_t, T-t) \equiv M_t^{-1}D(S_t, T-t)$ and invoking (6.32) give

$$\begin{aligned} dD^* &= \frac{-D_{T-t} + D_S(r-\delta)S_t + D_{SS}\sigma^2 S_t^2/2 - Dr}{M_t} \cdot dt + \frac{D_S S_t \sigma}{M_t} \cdot d\hat{W}_t \\ &= M_t^{-1} D_S S_t \sigma \cdot d\hat{W}_t. \end{aligned}$$

Thus, assuming uniform integrability (which is guaranteed for vanilla options by the dynamics-free bounds in chapter 4), we conclude that

$\{D^*(S_t, T-t)\}$ is a martingale under $\hat{\mathbb{P}}$, with $D^*(S_t, T-t) = \hat{E}_t D^*(S_T, 0)$ and

$$D(S_t, T-t) = M_t M_T^{-1} \hat{E}_t \bar{D}(S_T) = e^{-r(T-t)} \hat{E}_t \bar{D}(S_T).$$

6.3.3 *Extensions of the Black-Scholes Theory*

By making some approximations the Black-Scholes formulas can be extended to European options on stocks that pay deterministic lump-sum dividends and to options on stock indexes. There is also a straightforward extension to futures, due to Black (1976b).

Handling Lump-Sum Dividends

The formulas derived thus far allow for the underlying asset to pay dividends continuously at a deterministic rate. This is appropriate for modeling options on currencies, which could be held in the form of foreign discount bonds, and is a common approximation for options on diversified stock indexes like the S&P 500. For ordinary stock options there is also a way to adapt the Black-Scholes formulas to allow for predictable lump-sum dividends. This corresponds to the "set-aside" method that was introduced in chapter 5 to avoid a nonrecombining binomial tree. Like that method, it is merely an approximation rather than a thoroughly satisfactory solution.

To see how it works, suppose first that a single lump-sum, per-share payment $\Delta_{t'}$ is to be made at $t' \in (t, T]$. The firm can finance this payment by holding $\Delta_{t'}$ unit, t'-maturing bonds per share of stock. If this is done, the value of the stock up to t' has two components: (i) the bonds that provide for the dividend payment, which are worth $\Delta(t, t') \equiv B(t, t')\Delta_{t'}$, and (ii) the ownership claim on the firm's real assets, which is worth $S_t - \Delta(t, t')$. Assuming that this latter quantity follows Black-Scholes dynamics, the continuous-dividend formula still holds with $S_t - \Delta(t, t')$ in place of $e^{-\delta(T-t)}S_t$. To extend to multiple dividend payments beginning at t', just take $\Delta(t, t')$ to be their aggregate present value at t. Notice that the forward price for delivery at T of a stock that pays lump-sum dividends is $\mathsf{f}(t, T) = B(t, T)^{-1}[S_t - \Delta(t, t')]$, corresponding to expression (4.6) with $K(t, T) = -\Delta(t, t')$ as the explicit cost of carry. Therefore, formulas (6.27) and (6.28) already accommodate the lump-sum case.

As explained in chapter 5, the conceptual problem with this approach is that it makes no allowance for the financing of dividends to be paid after the option expires.

Options on Stock Indexes

The value at t of a value-weighted index like the S&P 500 is just a weighted sum of the prices of n individual stocks:

$$I_t = \mathbf{v}'\mathbf{S}_t = \sum_{j=1}^{n} v_j S_{jt}.$$

Weights for the S&P in particular are computed as 10 times the ratio of the number of shares outstanding to the average aggregate value of the firms during a reference period. This scheme automatically adjusts for splits and stock dividends. Let us consider how the index itself behaves if each individual price follows geometric Brownian motion. For this it is easiest to regard the weights as constants and figure the prices themselves to be split adjusted—that is, multiplied by the cumulative split factor. Suppose each $\{S_{jt}\}$ is modeled as

$$\frac{dS_{jt}}{S_{jt}} = \mu_{jt} \cdot dt + \boldsymbol{\sigma}_j' \cdot d\mathbf{W}_t, \quad j = 1, 2, \ldots, n,$$

where (i) $\mathbf{W}_t \equiv (W_{1t}, \ldots, W_{Nt})'$ is N-dimensional standard Brownian motion, with $N \geq n$; (ii) $\boldsymbol{\sigma}_j \equiv (\sigma_{j1}, \ldots, \sigma_{jN})'$; and $\boldsymbol{\sigma}_j'\mathbf{W}_t = \sum_{k=1}^{N} \sigma_{jk} W_{kt}$ is the inner product. Introducing the multidimensional process allows the returns of different stocks to be imperfectly correlated, but returns can also be expressed in one-dimensional form as

$$dS_{jt} = \mu_{jt} S_{jt} \cdot dt + \theta_j S_{jt} \cdot dW_{jt}^{\sigma},$$

where $\theta_j \equiv \sqrt{\boldsymbol{\sigma}_j'\boldsymbol{\sigma}_j}$ and $W_{jt}^{\sigma} \equiv \theta_j^{-1}\boldsymbol{\sigma}_j'\mathbf{W}_t$. The index itself then satisfies the stochastic differential equations

$$\begin{aligned} dI_t &= \sum_{j=1}^{n} v_j S_{jt} \mu_{jt} \cdot dt + \sum_{j=1}^{n} v_j S_{jt} \boldsymbol{\sigma}_j' \cdot d\mathbf{W}_t \\ &= \sum_{j=1}^{n} v_j S_{jt} \mu_{jt} \cdot dt + \sum_{j=1}^{n} v_j S_{jt} \theta_j \cdot dW_{jt}^{\sigma}. \end{aligned}$$

Momentarily ignoring dividends from the individual stocks, each μ_{jt} is replaced by short rate r_t when we move to risk-neutral measure $\hat{\mathbb{P}}$, so that

$$\frac{dI_t}{I_t} = r_t \cdot dt + \boldsymbol{\Sigma}_t' \cdot d\hat{\mathbf{W}}_t,$$

where

$$\boldsymbol{\Sigma}_t \equiv \frac{\sum_{j=1}^n v_j S_{jt} \boldsymbol{\sigma}_j}{\sum_{j=1}^n v_j S_{jt}} = I_t^{-1} \sum_{j=1}^n v_j S_{jt} \boldsymbol{\sigma}_j.$$

Here subscript t signifies the time/state dependence that enters through prices $\{S_{jt}\}$. Now admitting dividends on the individual stocks but approximating their joint effect as a continuous proportional dividend on the index, we can write

$$\begin{aligned} dI_t/I_t &= (r_t - \delta_t) \cdot dt + \boldsymbol{\Sigma}_t' \cdot d\hat{\mathbf{W}}_t \\ &= (r_t - \delta_t) \cdot dt + \theta_{I_t} \cdot d\hat{W}_t^I, \end{aligned}$$

where $\theta_{I_t} \equiv \sqrt{\boldsymbol{\Sigma}_t' \boldsymbol{\Sigma}_t}$ and $\hat{W}_t^I \equiv \theta_{I_t}^{-1} \boldsymbol{\Sigma}_t' \hat{\mathbf{W}}_t$. It is now clear that the index itself does not follow Black-Scholes dynamics under $\hat{\mathbb{P}}$, since $\boldsymbol{\Sigma}_t$ and θ_{I_t} are not generally deterministic. They would be deterministic—and constant—if and only if volatility vectors $\{\boldsymbol{\sigma}_j\}_{j=1}^n$ were all the same, but this would mean that returns on all stocks were perfectly correlated. Still, Black-Scholes dynamics may serve as a reasonable empirical approximation for the index process. In discrete time this is equivalent to approximating the distribution of a weighted sum of correlated lognormals as lognormal, and this can provide a very satisfactory fit if the weights are not too disparate. For this reason, formulas (6.28) and (6.27) have often been used in practice to price European options on diversified stock indexes.[8]

Options on Futures

We have seen that (subject to an integrability condition) futures prices are themselves martingales in the risk-neutral measure, without the need for normalization by the value of the money fund. If a futures price $\{\mathsf{F}_t\}$ follows Black-Scholes dynamics under $\hat{\mathbb{P}}$, it thus solves the s.d.e. $d\mathsf{F}_t = \sigma_t \mathsf{F}_t \cdot d\hat{W}_t$

[8]Chapter 11 shows how to avoid the lognormal approximation by using simulation to price "basket" options on portfolios of assets.

under $\hat{\mathbb{P}}$, so that

$$\begin{aligned} \mathsf{F}_T &= \mathsf{F}_t \exp\left[-\frac{1}{2}\int_t^T \sigma_s^2 \cdot ds + \int_t^T \sigma_s \cdot d\hat{W}_s\right] \\ &\sim \mathsf{F}_t \exp[-\sigma^2(T-t)/2 + Z\sigma\sqrt{T-t}], \end{aligned}$$

where $\sigma^2 \equiv (T-t)^{-1}\int_t^T \sigma_s^2 \cdot ds$ and $Z \sim N(0,1)$. Given the implied conditional lognormality of the futures price, it is easy to adapt the Black-Scholes formulas to price European options on futures, as done originally by Black (1976b). For this, consider T-expiring options on futures prices for delivery at $T' > T$. Since the futures price evolves under $\hat{\mathbb{P}}$ just like a primary asset that has no net trend, one simply sets $r_t - \delta_t = 0$ and $S_t = \mathsf{F}_t$ in the derivations of the previous section. This has the effect of replacing forward price $\mathsf{f}(t,T)$ in (6.27) and (6.28) with futures price F_t. Black's formulas for arbitrage-free prices of European options on the futures are thus

$$P^E(\mathsf{F}_t, T-t) = BX\Phi[q^+(X/\mathsf{F}_t)] - B\mathsf{F}_t\Phi[q^-(X/\mathsf{F}_t)] \tag{6.33a}$$

$$C^E(\mathsf{F}_t, T-t) = B\mathsf{F}_t\Phi[q^+(\mathsf{F}_t/X)] - BX\Phi[q^-(\mathsf{F}_t/X)], \tag{6.33b}$$

where $q^{\pm}(\cdot)$ is given by (6.24) and $B \equiv B(t,T)$.

For reference in chapter 7 when discussing American-style options, notice that when $t < T$ and $r > 0$ the ratio $C^E(\mathsf{F}_t, T-t)/(\mathsf{F}_t - X) \to B < 1$ as $\mathsf{F}_t \to \infty$. Since the value of the live option is thus ultimately below intrinsic value, early exercise of a call on the futures will at some point be desirable. Therefore, as in the Bernoulli model so also in the Black-Scholes framework: American calls on futures may well be exercised early. This is easy to remember if one recalls that the futures price behaves just like the price of a primary asset with dividend rate equal to the short rate, since we already know that calls on dividend-paying assets are subject to early exercise.

Routine `B-S` on the CD calculates Black-Scholes prices for European puts and calls on stocks paying lump-sum dividends, on indexes paying continuous dividends, on currency deposits appreciating at the foreign interest rate, and on futures. Program `doB_S` handles the input and output conveniently for single calculations. This is also implemented in the `ContinuousTime.xls` spreadsheet.

6.4 Properties of Black-Scholes Formulas

This section documents a number of important features of the Black-Scholes formulas. We show that Black-Scholes prices satisfy put-call parity, that they have the right terminal values, and that they fall within the no-arbitrage bounds worked out in chapter 4. We also see precisely how they vary with the current underlying price, the strike price, and other parameters. The last four parts of the section treat "implicit" volatility, hedging and the construction of "synthetic" derivatives, and instantaneous and holding-period returns to positions in European options.

6.4.1 *Symmetry and Put-Call Parity*

Not surprisingly, there are symmetries in the Black-Scholes formulas for calls and puts like those found in the binomial approximations. If the price of the call in (6.28) is written symbolically as a function of $B\mathsf{f}$ and BX, as

$$C^E(S_t, T-t) = f(B\mathsf{f}, BX),$$

then the corresponding expression for the put in (6.27) is obtained just by permuting the arguments; that is,

$$P^E(S_t, T-t) = f(BX, B\mathsf{f}). \tag{6.34}$$

Of course, something like this must be true since the terminal payoffs of puts and calls are symmetric in the same way. Less obvious is the connection between δ and r. If the underlying asset pays continuous proportional dividends at rate δ, then $\mathsf{f}(t,T) = S_t e^{(r-\delta)(T-t)}$, and the option formulas can be written as

$$\begin{aligned} C^E(S_t, T-t) &= f(S_t e^{-\delta(T-t)}, Xe^{-r(T-t)}) \\ P^E(S_t, T-t) &= f(Xe^{-r(T-t)}, S_t e^{-\delta(T-t)}). \end{aligned} \tag{6.35}$$

This shows that the price of the European put can be expressed in terms of the European call, and *vice versa*, by interchanging S_t and X, δ and r. We saw in chapter 5 that this was true for both European and American options under Bernoulli dynamics, and we will see in chapter 7 that it applies also for American options in the Black-Scholes framework.[9]

---

[9]While put-call symmetry holds more generally than under Black-Scholes dynamics, it is not model-independent. Williams (2001) shows that if the distribution of $Y \equiv \ln[S_T/\mathsf{f}(t,T)]$ is independent of S_t, r, and δ, then it is necessary and sufficient for symmetry that $dF(y) = e^{-y}dF(-y)$. The condition does hold in the Black-Scholes model, where the conditional density of Y is proportional to $\exp\{-[(y + \sigma^2(T-t)]^2/2\}$.

To verify that the Black-Scholes formulas are consistent with European put-call parity, notice that

$$\begin{aligned}\Phi[q^+(\mathsf{f}/X)] &\equiv \Phi\left[\frac{\ln(\mathsf{f}/X)+\sigma^2(T-t)/2}{\sigma\sqrt{T-t}}\right]\\ &= 1-\Phi\left[\frac{\ln(X/\mathsf{f})-\sigma^2(T-t)/2}{\sigma\sqrt{T-t}}\right]\\ &\equiv 1-\Phi[q^-(X/\mathsf{f})],\end{aligned}$$

which from (6.27) and (6.28) implies that

$$C^E(S_t,T-t)-P^E(S_t,T-t)=B(t,T)[\mathsf{f}(t,T)-X]. \tag{6.36}$$

This is the put-call parity relation stated in terms of the forward price for an underlying asset with arbitrary (but deterministic) explicit cost of carry. If there are neither dividends nor positive carrying costs, then $\mathsf{f}(t,T) = S_t/B(t,T)$ and

$$C^E(S_t,T-t)-P^E(S_t,T-t)=S_t-B(t,T)X.$$

6.4.2 *Extreme Values and Comparative Statics*

Let us now verify that the Black-Scholes formulas satisfy the terminal value conditions and the boundary conditions worked out in chapter 4, then see how the prices depend on stock and option parameters.

Calls

We present call formula (6.28) again here for convenience as

$$C^E(S_t,T-t)=B\mathsf{f}\Phi[q^+(\mathsf{f}/X)]-BX\Phi[q^-(\mathsf{f}/X)], \tag{6.37}$$

where once again

$$q^{\pm}(x)\equiv\frac{\ln x\pm\sigma^2(T-t)/2}{\sigma\sqrt{T-t}},$$

$B\equiv B(t,T)$ is the price of a discount T-maturing bond, and $\mathsf{f}\equiv\mathsf{f}(t,T)$ is the T-forward price as of time t.

To check that this satisfies terminal value condition $C^E(S_T,0)=(S_T-X)^+$, begin with the obvious facts that $B\to B(T,T)=1$ and $\mathsf{f}\to\mathsf{f}(T,T)=S_T$ as $t\to T$. If $S_T>X$ then $q^{\pm}(\mathsf{f}/X)\to+\infty$ as $t\to T$ and $C^E(S_t,T-t)\to S_T-X$, whereas if $S_T\le X$ then $q^{\pm}(\mathsf{f}/X)\to-\infty$ and $C^E(S_t,T-t)\to 0$.

Thus, the call formula does deliver at expiration the correct piecewise-linear payoff function, $(S_T - X)^+$.

Now fixing $t < T$, consider how the value function behaves as S_t approaches its extreme values. As S_t (and therefore f) approaches zero, it is apparent that $q^{\pm}(\mathsf{f}/X) \to -\infty$ and $C^E(S_t, T-t) \to 0$. Also,

$$C^E(S_t, T-t) - B(\mathsf{f} - X) = B\mathsf{f}\{\Phi[q^+(\mathsf{f}/X)] - 1\} - BX\{\Phi[q^-(\mathsf{f}/X)] - 1\}$$

approaches zero as $S_t \to \infty$. Thus, the value function starts at the origin and approaches asymptotically the line $B(\mathsf{f} - X) = S_t e^{-\delta(T-t)} - BX$ as $S_t \to \infty$. When there is no dividend, this is the line $S_t - BX$, as in figure 6.2. Notice that the fact that $C^E(S_t, T-t) > S_t - BX \geq S_t - X$ when $t < T$ is consistent with the general conclusion in chapter 4 that there would be no incentive to exercise a call on a no-dividend stock before expiration.

It is obvious from (6.37) that $C^E(S_t, T-t) \leq S_t e^{-\delta(T-t)}$, which is the upper bound in table 4.7 that we derived from static replicating arguments. To see that it satisfies lower bound $C^E(S_t, T-t) \geq (S_t e^{-\delta(T-t)} - BX)^+$, set $X = 1$ as a normalization and put

$$c(\mathsf{f}) \equiv C^E(S_t, T-t)/B = \mathsf{f}\Phi[q^+(\mathsf{f})] - \Phi[q^-(\mathsf{f})].$$

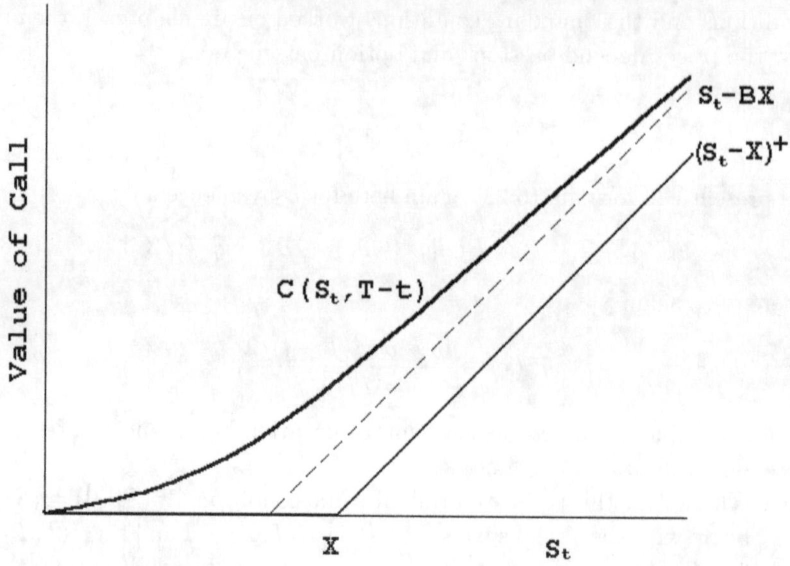

Fig. 6.2. Shape of Black-Scholes call function.

With these conventions the bound becomes $c(\mathsf{f}) \geq (\mathsf{f}-1)^+$. Differentiating $c(\mathsf{f})$ gives

$$c'(\mathsf{f}) = \Phi[q^+(\mathsf{f})] + \mathsf{f}\phi[q^+(\mathsf{f})] \cdot dq^+/d\mathsf{f} - \phi[q^-(\mathsf{f})] \cdot dq^-/d\mathsf{f},$$

where ϕ is the standard normal p.d.f. This reduces to $c'(\mathsf{f}) = \Phi[q^+(\mathsf{f})]$ on expressing the last two terms. Thus, $c'(\mathsf{f}) > 0$ when $\mathsf{f} > 0$, and since $c(0) = 0$ it follows that $c(\mathsf{f}) \geq 0$ for all $\mathsf{f} \geq 0$. On the other hand, we have the stronger bound $c(\mathsf{f}) \geq (\mathsf{f}-1)$ when $\mathsf{f} \geq 1$, since

$$\begin{aligned}\frac{c(\mathsf{f}) - (\mathsf{f}-1)}{\mathsf{f}} &= \{\Phi[q^+(\mathsf{f})] - 1\} - \mathsf{f}^{-1}\{\Phi[q^-(\mathsf{f})] - 1\} \\ &= \mathsf{f}^{-1}\Phi[q^+(\mathsf{f}^{-1})] - \Phi[q^-(\mathsf{f}^{-1})] \\ &= c(\mathsf{f}^{-1}),\end{aligned}$$

which is nonnegative by the previous result.

Figure 6.2 depicts the call function as monotonic and convex in S_t. To verify this, we can sign the first two derivatives. In the standard notation the derivative with respect to the current price of the underlying, which is called the "delta" of the call, is

$$C_S^E = e^{-\delta(T-t)}\Phi[q^+(\mathsf{f})] = e^{-\delta(T-t)}\Phi\left[q^+\left(\frac{S_t e^{-\delta(T-t)}}{BX}\right)\right] > 0.$$

Interestingly, this is the same expression one would get on differentiating (6.37) with respect to S_t but overlooking its presence in the arguments of the c.d.f.s. The second derivative with respect to S_t, the "gamma", is also positive. Specifically,

$$C_{SS}^E \propto \frac{\phi[q^+(\mathsf{f})]}{\mathsf{f}\sigma\sqrt{T-t}} > 0,$$

confirming that the Black-Scholes call function is indeed increasing and convex in S_t.

Differentiating with respect to the strike price gives

$$C_X^E = -B\Phi[q^-(\mathsf{f})] < 0.$$

This and the expression for C_S^E show that a call on a no-dividend stock can be represented as

$$C^E(S_t, T-t) = C_S^E S_t + C_X^E X, \tag{6.38}$$

a decomposition that will be found useful as we proceed.

Here are other comparative statics for European calls in the no-dividend case ($\delta = 0$), where subscripts denote partial derivatives and the "Greeks" in parentheses are the nicknames applied by practitioners:

1. $C_t^E = -C_{T-t}^E < 0$ ("theta"). Value declines as remaining life diminishes, since there is less potential to get deeply into the money, and since the present value of the liability to pay X upon exercise increases as $t \to T$.
2. $C_\sigma^E > 0$ ("vega"). Greater volatility adds potential to be in the money at expiration.
3. $C_r^E > 0$ ("rho"). Higher interest rates reduce the present value of the contingent liability to pay X in order to exercise the call.
4. $C_{XX}^E > 0$. Convexity of C^E with respect to X follows from convexity of P^E with respect to S_t (see below) and identity (6.34).

Puts

The corresponding results for European puts can be deduced most easily using European put-call parity, expression (6.36). For the terminal condition, $P^E(S_t, T-t) \to (S_T - X)^+ - (S_T - X) = (X - S_T)^+$ as $t \to T$. At the extremes of S_t, we have $P^E(S_t, T-t) \to B(t,T)X$ as $S_t \to 0$ and $P^E(S_t, T-t) \to 0$ as $S_t \to \infty$. The bounds are

$$[B(t,T)X - S_t e^{-\delta(T-t)}]^+ = B(t,T)(X - \mathsf{f})^+ \le P^E(S_t, T-t) \le B(t,T)X,$$

in line with the findings in chapter 4. For the comparative statics, the delta for a European put is

$$P_S^E = -e^{-\delta(T-t)}\Phi\left[q^-\left(\frac{BX}{S_t e^{-\delta(T-t)}}\right)\right] < 0 \tag{6.39}$$

and the gamma, P_{SS}^E, is positive, so that Black-Scholes values of European puts are also convex functions of the underlying price, decreasing from $B(t,T)X$ to 0 as S_t ranges from 0 to ∞. These features are depicted in figure 6.3. Notice from the figure that $P^E(S_t, T-t) < X - S_t$ when S_t is sufficiently small, indicating that the holder of a put would sometimes prefer to exercise before expiration.

The derivative of P^E with respect to X is

$$P_X^E = B\Phi\left[q^+\left(\frac{BX}{S_t e^{-\delta(T-t)}}\right)\right] > 0.$$

This and the expression for P_S^E show that the put's value can be decomposed into terms involving P_X^E and its delta. When $\delta = 0$ the result is

$$P^E(S_t, T-t) = P_S^E S_t + P_X^E X. \tag{6.40}$$

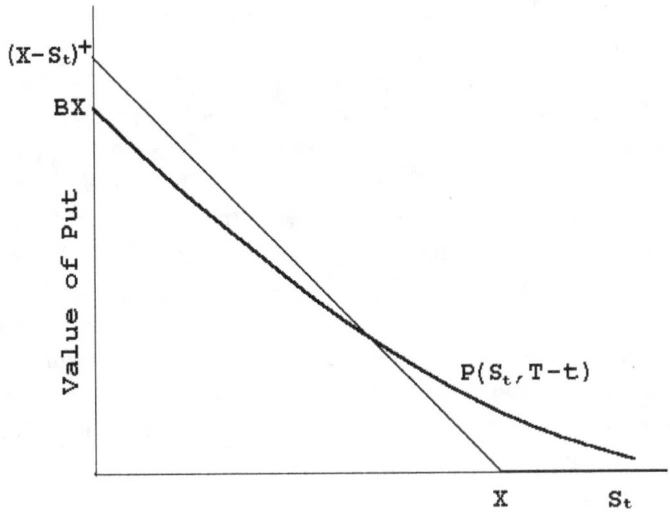

Fig. 6.3. Shape of Black-Scholes put function.

Other comparative statics when $\delta = 0$ are

1. $P_t^E = -P_{T-t}^E$ (theta) has indeterminate sign. Value rises as time to maturity falls if the S_t is low enough that early exercise would be desirable; otherwise, longer life adds value by increasing the chance of getting into the money.
2. $P_\sigma^E > 0$ ("vega"). Greater volatility adds potential value.
3. $P_r^E < 0$ (rho). Higher interest rates reduce the present value of whatever is received at exercise.
4. $P_{XX}^E > 0$. Convexity of the call in S_t implies convexity of the put in X.

6.4.3 *Implicit Volatility*

Since the "vegas" of vanilla European puts and calls are strictly positive, the options' values are strictly increasing in σ. Also, it is easy to see from the formulas that Black-Scholes option prices range from their lower to their upper arbitrage-free bounds as σ ranges from zero to infinity; specifically, the prices of calls and puts increase from $B(\mathsf{f}-X)^+$ to $B\mathsf{f}$ and from $B(X-\mathsf{f})^+$ to BX, respectively. It follows that one and only one nonnegative value of σ corresponds to each feasible call value and likewise to each feasible put value. Given the market price of a European put or call (along with the prices of the underlying asset and the riskless bond), it is therefore possible to find a unique value of volatility that corresponds. This is the volatility that is said to be "implied by", or "implicit in", the Black-Scholes formulas.

Under the strict assumptions of Black and Scholes, including that volatility is constant through time, the volatilities implicit in any collection of European options on the same underlying asset should be precisely the same, regardless of expiration date, strike price, and even the current calendar time. Under the less restrictive assumption that volatility may vary predictably through time, puts and calls with different strike prices but the same expiration date should still have the same implicit volatilities. Of course, one expects some variation empirically because of transaction costs, the discreteness of observed prices, and the fact that price quotes on the option and the underlying may not be simultaneous. Even so, before the pronounced market reversal of October 1987 differences in implicit volatilities across strike prices were relatively minor, albeit somewhat higher for options far away from the money. After 1987 things changed radically in many markets. Implicit volatilities at low strikes have been substantially higher than those at strikes close to and above the current underlying price. This has been particularly pronounced for options on stock indexes. Figure 6.4 illustrates a typical relation between implicit volatility and the

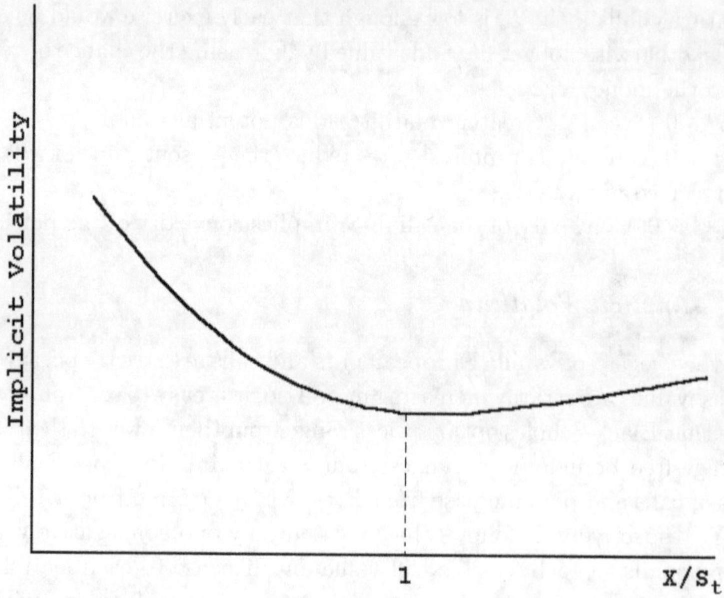

Fig. 6.4. An implicit-volatility "smile" curve.

"moneyness" of options, X/S_t. The shape of the curve suggests the common terms for the phenomenon: volatility "smile" and volatility "smirk".

To give a specific example of what the curve implies, take as benchmark the implicit volatility of at-the-money options, $\hat{\sigma}(1)$ say, and use this to calculate Black-Scholes prices for far-out-of-money puts, for which $X/S_t \ll 1$. Because of the monotonic relation between price and volatility, the high implicit volatility of these out-of-money puts indicates that their actual market prices are higher than those calculated from the Black-Scholes formula with $\sigma = \hat{\sigma}(1)$. In other words, the market places a higher relative value on out-of-money puts than on those at or in the money. In-the-money calls are also valued higher, because the put-call parity relation fixes the difference between prices of puts and calls with the same strike and expiration. Explaining such systematic anomalies has been—and is—a primary research objective. Chapters 8 and 9 treat alternative models based on richer dynamics than geometric Brownian motion that yield predictions more in accord with current experience.

6.4.4 *Delta Hedging and Synthetic Options*

The derivation of the Black-Scholes formulas relies on the ability to replicate payoffs of European options with dynamic, self-financing portfolios of the underlying asset and money fund or riskless bonds. The formulas themselves actually provide the exact recipes for such replicating portfolios. For example, (6.40) shows that replicating a put on a no-dividend stock requires holding $P_X^E X = X\Phi[q^+(BX/S_t)]$ worth of the money fund and $P_S^E = -\Phi[q^-(BX/S_t)]$ units of the stock at time t. That is, one holds a number of shares of stock equal to the option's delta, together with $P_X^E X/M_t$ units of the fund. Since $P_S^E < 0$ and $P_X^E > 0$, this requires a short position in the stock and a long position in the fund.

As t, S_t, and $P_S^E(S_t, T-t)$ change, the portfolio shares have to be adjusted, but it must be possible to make the adjustments without cash infusions or withdrawals if the replicating portfolio is to be self-financing. To see that this is so, we must show that

$$dP^E(S_t, T-t) = P_S^E \cdot dS_t + \frac{P_X^E X}{M_t} \cdot dM_t.$$

Applying Itô's formula to $P^E(S_t, T-t)$ gives

$$dP^E(S_t, T-t) = P_S^E \cdot dS_t + (P_{SS}^E \sigma^2 S_t^2/2 - P_{T-t}^E) \cdot dt;$$

and fundamental p.d.e. (6.10) implies (with $\delta = 0$)

$$\begin{aligned} P_{SS}^E \sigma^2 S_t^2/2 - P_{T-t}^E &= r_t[P^E(S_t, T-t) - P_S^E S_t] \\ &= r_t P_X^E X \\ &= \frac{P_X^E X}{M_t} \cdot dM_t, \end{aligned}$$

which gives the desired result.

Actually replicating puts or other derivatives by following this recipe is an example of "delta hedging", which derives its name from the fact that the derivative's delta is the required position in the underlying asset. An institution that sold a put to a client would typically hedge the exposure by replicating a long position in the option.[10] In hedging the put, consider what happens as the price of the underlying declines. Because P^E is convex in S_t, P_S^E decreases (increases in absolute value) as S_t falls. This means that more stock must be sold short and that proceeds of the sale have to be added to the money-fund balance. As the share price rises, the money fund balance is drawn down to purchase stock and reduce the short position. By the time of expiration both positions will have been closed out entirely if the put is out of the money. If it is in the money, the account ends up short one share of the underlying asset and with X units of cash. Figure 6.5 gives a visual illustration of the replication process. The slope of the tangent line at S_t (the put's delta) indicates the required (short) position in the underlying, and the vertical intercept of the tangent line (XP_X) indicates the value of the replicating position in the money fund. The put's value shows up on the vertical axis as the sum of XP_X and the negative quantity $S_t P_S$. By moving the tangent line appropriately, one can track how the positions vary with S_t. Since $P_S^E \downarrow -1$ and $P_X^E \uparrow B(t,T)$ as $S_t \downarrow 0$, the portfolio would be long BX/M_t units of the money fund and short one share of the stock should the stock become worthless, and both positions would ultimately be closed out as $S_t \to \infty$.

Other than to hedge a short position arising from selling a put, why would anyone want to replicate one? Owning a put on an asset provides a floor on the loss that could be sustained if the price falls. For example, the maximum potential loss as of time T from buying a share of (no-dividend) stock and a T-expiring put at $t = 0$ is $S_0 + P(S_0, T) - X < S_0$, as compared

[10]Of course, institutions actually worry just about hedging their net exposures rather than the individual positions.

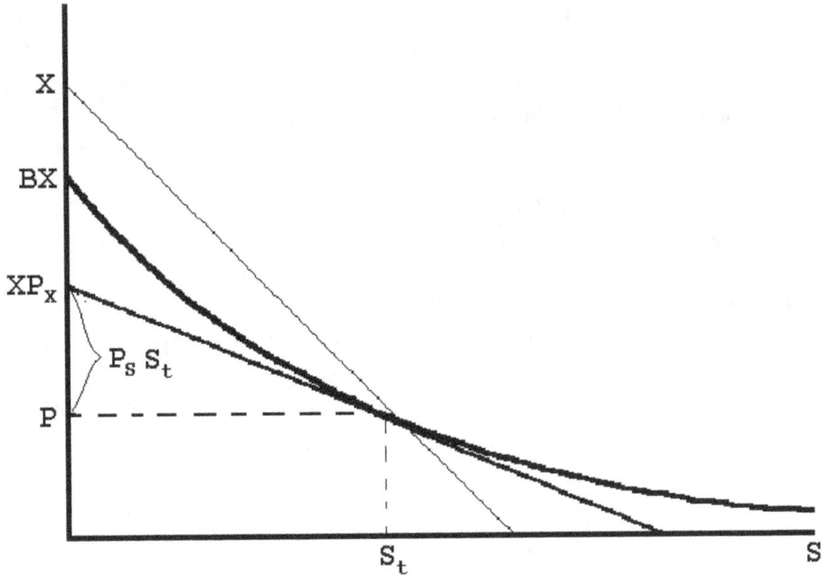

Fig. 6.5. Replicating portfolio for European put.

with maximum loss S_0 from an unhedged position in the stock alone. The put thus serves as a form of insurance, with initial cost $P(S_0, T)$ as the insurance premium. Indeed, one presumes that the high implicit volatilities for out-of-money puts since 1987 are attributable to the high valuations the market now places on such insurance. However, one cannot always find an exchange-traded put with the precise X, T, underlying asset, and contract size that one desires. Moreover, if the goal is to insure the value of a particular portfolio, it would be wasteful to insure each separate position, even if that were possible. This is where delta hedging comes in. Delta hedging can be used to create "synthetic" puts of arbitrary expiration date and strike price and on individual assets or portfolios of assets for which options do not actively trade. Creating synthetic puts on diverse portfolios is in fact a common means of creating "portfolio insurance".[11] Thus, to start at time t

[11]... but a far less common means than before the 1987 crash, to which its widespread use during the 1980s was a prime contributor. The concept works well on a scale small enough to prevent substantial feedback to the underlying price, but on a large scale the program of dumping shares into a declining market and scrambling to get them back on upticks is clearly destabilizing. It is interesting and ironical that the Black-Scholes-Merton model is less useful today because of its triumphant early success.

to insure out to T a portfolio worth S_t (and assumed to follow Black-Scholes dynamics), one would sell off an amount worth $|P_S(S_t, T-t)|S_t$ and buy $P_X(S_t, T-t)X$ worth of riskless bonds—the net outlay (an increasing function of the strike) being the cost of insuring.

6.4.5 *Instantaneous Risks and Expected Returns of European Options*

Under risk-neutral measure $\hat{\mathbb{P}}$ all primary assets and derivatives have instantaneous expected rate of return (proportional mean drift) equal to the short rate, r_t, just as do riskless savings accounts. Of course, in the real world (that is, in measure $\mathbb{P}$) one expects returns of risky assets to compensate for risk on average over time. Looking in turn at European calls and European puts, we will see what the Black-Scholes formulas imply about their actual instantaneous mean returns and risks. To simplify the expressions, we continue to work with options on no-dividend assets and to treat underlying volatilities as constants.

Calls

Modeling the evolution of the underlying price as $dS_t/S_t = \mu_{S_t} \cdot dt + \sigma_S \cdot dW_t$, where $\{\mu_{S_t}\}$ is a deterministic process with $\int_0^T |\mu_{S_t}| \cdot dt < \infty$, Itô's lemma applied to $C^E(S_t, T-t)$ gives

$$\begin{aligned}\frac{dC^E}{C^E} &= \frac{-C^E_{T-t} + \mu_{S_t} C^E_S S_t + C^E_{SS} S_t^2 \sigma_S^2/2}{C^E} \cdot dt + \frac{\sigma_S C^E_S S_t}{C^E} \cdot dW_t \\ &\equiv \mu_{C_t} \cdot dt + \sigma_{C_t} \cdot dW_t. \end{aligned} \tag{6.41}$$

Notice that while the option's price is an Itô process, the volatility, σ_{C_t}, is not deterministic since it depends on S_t—directly and through C^E and C^E_S. Letting $\eta_{C_t} \equiv C^E_S S_t / C^E = \sigma_{C_t}/\sigma_S$ be the call's price elasticity with respect to S_t and using (6.38) to express $C^E(S_t, T-t)$, one sees that

$$\eta_{C_t} = \frac{C^E_S S_t}{C^E_S S_t + C^E_X X}.$$

Since $C^E_X < 0$, we have $\eta_{C_t} > 1$ and thus $\sigma_{C_t} > \sigma_S$. Thus, the call is always more volatile than the stock. Moreover, $\eta_{C_t} \to 1$ as $S_t \to \infty$ and $\eta_{C_t} \to +\infty$

as $S_t \to 0$. [12] From (6.41) the call's mean proportional drift is

$$\mu_{C_t} = \frac{-C^E_{T-t} + r_t C^E_S S_t + C^E_{SS} S_t^2 \sigma_S^2/2}{C^E} + \eta_{C_t}(\mu_{S_t} - r_t).$$

and applying the fundamental p.d.e. reduces this to

$$\mu_{C_t} - r_t = \eta_{C_t}(\mu_{S_t} - r_t).$$

Thus, since $\eta_{C_t} > 1$, if the stock's instantaneous excess rate of return is positive, then the call's excess return is even larger, in compensation for its extra volatility.

Now let us suppose that the stock's instantaneous rate of return is related to the excess return of the "market" portfolio in the manner predicted by Capital Asset Pricing Model (CAPM).[13] To set up the CAPM in continuous time, model the price of asset $j \in \{1, 2, \ldots, n\}$ as

$$dS_{jt}/S_{jt} = \mu_{jt} \cdot dt + \boldsymbol{\sigma}_j' \cdot d\mathbf{W}_t,$$

where $\mathbf{W}_t$ is N-dimensional standard Brownian motion with $N > n$. We also need the equivalent one-dimensional form,

$$dS_{jt}/S_{jt} = \mu_{jt} \cdot dt + \theta_j \cdot dW^\sigma_{jt},$$

where $\theta_j \equiv \sqrt{\boldsymbol{\sigma}_j' \boldsymbol{\sigma}_j}$ and $W^\sigma_{jt} \equiv \theta_j^{-1} \boldsymbol{\sigma}_j' \mathbf{W}_t$. Then if $\mathbf{S}_t$ is the vector of asset prices and $\mathbf{v} \equiv (\nu_1, \ldots, \nu_n)'$ is the vector of numbers of shares (regarded as

---

[12]This is true because $\frac{C_X X}{C_S S_t} \to -1$ as $S_t \to 0$. To show this, apply l'Hospital's rule to get

$$\lim_{S_t \to 0} \frac{C_X X}{C_S S_t} = \lim_{S_t \to 0} \frac{X C_{XS}}{C_S + S_t C_{SS}}.$$

Expressing C_{XS} and C_{SS} in terms of normal p.d.f. ϕ and C_S in terms of normal c.d.f. Φ, setting $\sigma_S \sqrt{T-t} = 1$ without loss of generality, and letting $y \equiv S/(BX)$ shows that

$$\lim_{S_t \to 0} \frac{C_X X}{C_S S_t} = -\lim_{y \to 0} \frac{\frac{1}{y} \cdot \phi(\ln y - \frac{1}{2})}{\Phi(\ln y + \frac{1}{2}) + \phi(\ln y + \frac{1}{2})}.$$

The numerator simplifies to $\phi(\ln y + \frac{1}{2})$. Dividing numerator and denominator by this and letting $z \equiv \ln y + \frac{1}{2}$ give

$$\lim_{S_t \to 0} \frac{C_X X}{C_S S_t} = -[1 + \lim_{z \to -\infty} \Phi(z)/\phi(z)]^{-1}.$$

Finally, applying l'Hospital again shows that $\lim_{z \to -\infty} \Phi(z)/\phi(z) = \lim_{z \to -\infty} -z^{-1} = 0$.

[13]Under the Sharpe (1964)-Lintner (1965)-Mossin (1966) version of the CAPM an asset's equilibrium expected excess rate of return is proportional to that of the market portfolio, as $ER_t - r_t = \beta(ER_{\mathcal{M}} - r_t)$, where $\beta = cov(R, R_{\mathcal{M}})/VR_{\mathcal{M}}$.

constants), the value of the market portfolio is $\mathcal{M}_t = \mathbf{v}'\mathbf{S}_t$ and

$$d\mathcal{M}_t/\mathcal{M}_t = \mu_{\mathcal{M}_t} \cdot dt + \boldsymbol{\Sigma}'_{\mathcal{M}_t} \cdot d\mathbf{W}_t \equiv \mu_{\mathcal{M}_t} \cdot dt + \theta_{\mathcal{M}_t} \cdot dW_{\mathcal{M}_t},$$

where

$$\boldsymbol{\Sigma}_{\mathcal{M}_t} \equiv \frac{\sum_{j=1}^n v_j S_{jt} \boldsymbol{\sigma}_j}{\sum_{j=1}^n v_j S_{jt}},$$

$$\theta^2_{\mathcal{M}_t} \equiv \boldsymbol{\Sigma}'_{\mathcal{M}_t} \boldsymbol{\Sigma}_{\mathcal{M}_t},$$

and $W_{\mathcal{M}_t} \equiv \boldsymbol{\Sigma}'_{\mathcal{M}_t} \mathbf{W}_t$. The instantaneous correlation between the returns of stock j and the market is

$$\rho_{j\mathcal{M}_t} \equiv \frac{d\langle \mathcal{M}, S_j \rangle_t}{\sqrt{d\langle \mathcal{M}\rangle_t d\langle S_j \rangle_t}} = \frac{\boldsymbol{\sigma}'_j \boldsymbol{\Sigma}_{\mathcal{M}_t}}{\theta_j \theta_{\mathcal{M}_t}},$$

and the instantaneous "beta" coefficient is

$$\beta_{S_j t} = \rho_{j\mathcal{M}_t} \theta_j / \theta_{\mathcal{M}_t}.$$

In this model beta represents the asset's systematic risk—its exposure to the forces that cause aggregate wealth to vary unpredictably. Because individuals seek to avoid this risk the stock's excess return varies in proportion to its beta, as $\mu_{jt} - r_t = \beta_{S_j t}(\mu_{\mathcal{M}_t} - r_t)$. (We take for granted that $\mu_{\mathcal{M}_t} > r_t$.)

Focusing on some particular asset $j \in \{1, 2, \ldots, n\}$ now but dropping the subscript, the volatility of the call on the asset can be expressed as $\sigma_{C_t} = \eta_{C_t} \theta$, so that the call's beta is $\beta_{C_t} = \eta_{C_t} \beta_{S_t}$. If the instantaneous CAPM does hold for the stock, we find that it holds for the call as well, since

$$\begin{aligned} \mu_{C_t} - r_t &= \eta_{C_t}(\mu_{S_t} - r_t) \\ &= \eta_{C_t} \beta_{S_t}(\mu_{\mathcal{M}_t} - r_t) \\ &\equiv \beta_{C_t}(\mu_{\mathcal{M}_t} - r_t). \end{aligned}$$

Moreover, since $\eta_{C_t} > 1$ the call has higher systematic risk than the stock whenever the stock's own systematic risk is positive.

Puts

The situation for puts is a bit more complicated. Corresponding expressions do hold; that is,

$$\begin{aligned} \mu_{P_t} - r_t &= \eta_{P_t}(\mu_{S_t} - r_t) \\ &= \eta_{P_t} \beta_{S_t}(\mu_{\mathcal{M}_t} - r_t) \\ &= \beta_{P_t}(\mu_{\mathcal{M}_t} - r_t). \end{aligned}$$

However, since $\eta_{P_t} \equiv P_S^E S_t / P^E < 0$, the put offers negative excess return whenever the underlying has positive systematic risk. That is, within the CAPM framework $\mu_{P_t} - r_t < 0$ when $\beta_{S_t} > 0$. The intuition is that being long a put reduces the overall systematic risk of one's portfolio (recall its insurance value for portfolios of primary assets), and this makes it worthwhile to hold the put even though it returns less on average than do riskless bonds.

Unlike calls, puts do not necessarily have higher volatility than the underlying asset in the Black-Scholes framework. Writing

$$|\eta_{P_t}| = \left| \frac{P_S^E S_t}{P_S^E S_t + P_X^E X} \right|,$$

shows that the absolute price elasticity is greater than, equal to, or less than unity as $-2P_S^E S_t \gtreqless P_X^E X$, or as

$$2\Phi \left[\frac{\ln(BX/S_t) - \sigma_S^2(T-t)/2}{\sigma_S \sqrt{T-t}} \right] S_t \gtreqless \Phi \left[\frac{\ln(BX/S_t) + \sigma_S^2(T-t)/2}{\sigma_S \sqrt{T-t}} \right] BX.$$

Letting $\mathsf{f} \equiv S_t / BX$, this is equivalent to

$$2\Phi \left[\frac{\ln \mathsf{f}^{-1} - \sigma_S^2(T-t)/2}{\sigma_S \sqrt{T-t}} \right] \gtreqless \mathsf{f}^{-1} \Phi \left[\frac{\ln \mathsf{f}^{-1} + \sigma_S^2(T-t)/2}{\sigma_S \sqrt{T-t}} \right].$$

The right side will certainly be greater than the left when the option is far enough in the money that $\mathsf{f} < 1/2$. Thus, $|\eta_{P_t}| < 1$ for puts that are deep in the money, whereas $|\eta_{P_t}| > 1$ for puts that are far enough out. The location of the threshold of unit elasticity depends on the volatility and time to expiration. When $\mathsf{f} = 1$ some algebra shows that $|\eta_{P_t}| \gtreqless 1$ according as $\Phi(\sigma_S \sqrt{T-t}/2) \lesseqgtr \frac{2}{3}$. As expiration nears, $\Phi(\sigma_S \sqrt{T-t}/2) \to \frac{1}{2}$, so that absolute elasticity is certainly greater than unity for at-the-money puts that are about to expire; however, $|\eta_{P_t}| < 1$ when volatility and time to expiration are sufficiently large.

Summing up, the Black-Scholes formulas imply that European calls are always more volatile than the underlying stocks, but that puts are more volatile only when far out of the money or far from expiration.

6.4.6 *Holding-Period Returns for European Options*

Having just seen what the Black-Scholes formulas imply for options' instantaneous expected returns, let us now look at expected returns over arbitrary holding periods. It is clear that in risk-neutral measure $\hat{\mathbb{P}}$ expected holding-period returns of all primary assets equal the sure return from holding a default-free bond to maturity. Under assumption RK this is the same as the return of the money fund and the same as the return from holding a bond to a time *before* maturity. Thus, for both an underlying asset with initial price S_0 and a T-expiring derivative worth $D(S_0, T)$ the $\hat{\mathbb{P}}$-expected total return during $[0, t]$ is

$$\frac{\hat{E}S_t}{S_0} = \frac{\hat{E}D(S_t, T-t)}{D(S_0, T)} = \frac{1}{B(0,t)} = \frac{M_t}{M_0} = \frac{B(t,T)}{B(0,T)} = e^{\int_0^t r_s \cdot ds} \tag{6.42}$$

for $t \le T$. However, under actual measure $\mathbb{P}$, wherein

$$dS_t/S_t = \mu_t \cdot dt + \sigma \cdot dW_t$$

with deterministic $\{\mu_t\}$, we have $E(S_t/S_0) = \exp(\int_0^t \mu_s \cdot ds)$. Can we also calculate the expected holding-period returns of European puts and calls under $\mathbb{P}$ for arbitrary paths of μ_s? In fact, we can thanks to the following relation (proved in section 2.2.7):

$$\int_{-\infty}^{\infty} \Phi(\alpha \pm \beta x) \cdot d\Phi(x) = \Phi(\alpha/\sqrt{1+\beta^2}). \tag{6.43}$$

Following Rubinstein (1984) we shall use (6.43) to work out the expected holding-period return from 0 to $t \le T$ of a T-expiring European call on a no-dividend stock. The c.d.f. of S_t under $\mathbb{P}$, given what is known at $t = 0$, is

$$\begin{aligned} F(s) &\equiv \mathbb{P}(S_t \le s) \\ &= \mathbb{P}(S_0 e^{\int_0^t \mu_s \cdot ds - \sigma^2 t/2 + \sigma W_t} \le s) \\ &= \mathbb{P}(ES_t \cdot e^{-\sigma^2 t/2 + \sigma W_t} \le s) \\ &= \Phi[q^+(s/ES_t; t)], \end{aligned}$$

where

$$q^{\pm}(x;t) \equiv \frac{\ln x \pm \sigma^2 t/2}{\sigma\sqrt{t}}.$$

The time-zero expectation of the European call's value at t is then

$$\begin{aligned} EC^E(S_t, T-t) = &\int_0^{\infty} s\Phi\left[q^+\left(\frac{s}{BX}; T-t\right)\right] \cdot d\Phi\left[q^+\left(\frac{s}{ES_t}; t\right)\right] \\ &-BX \int_0^{\infty} \Phi\left[q^-\left(\frac{s}{BX}; T-t\right)\right] \cdot d\Phi\left[q^+\left(\frac{s}{ES_t}; t\right)\right], \end{aligned}$$

where $B \equiv B(t, T)$. Changing variables as

$$z \leftarrow \frac{\ln(s/ES_t) - \sigma^2 t/2}{\sigma\sqrt{t}} = q^+\left(\frac{s}{ES_t}; t\right) - \sigma\sqrt{t}/2 = q^-\left(\frac{s}{ES_t}; t\right)$$

in the first integral and as

$$z \leftarrow \frac{\ln(s/ES_t) + \sigma^2 t/2}{\sigma\sqrt{t}} = q^+\left(\frac{s}{ES_t}; t\right)$$

in the second gives, after simplifying,

$$\begin{aligned} EC^E(S_t, T-t) = ES_t \int_{-\infty}^{\infty} \Phi\left[\frac{\ln\left(\frac{ES_t}{BX}\right) + \sigma^2 T/2}{\sigma\sqrt{T-t}} + z\sqrt{\frac{t}{T-t}}\right] \cdot d\Phi(z) \\ -BX \int_{-\infty}^{\infty} \Phi\left[\frac{\ln\left(\frac{ES_t}{BX}\right) - \sigma^2 T/2}{\sigma\sqrt{T-t}} + z\sqrt{\frac{t}{T-t}}\right] \cdot d\Phi(z). \end{aligned}$$

Since $B \equiv B(t,T) = B(0,T)/B(0,t)$ under assumption RK, we can write

$$\frac{ES_t}{B(t,T)X} = \frac{B(0,t)ES_t}{B(0,T)X}$$

in the above. Formula (6.43) then gives for $EC^E(S_t, T-t)$ the expression

$$ES_t\Phi\left[q^+\left(\frac{B(0,t)ES_t}{B(0,T)X}; T\right)\right] - B(t,T)X\Phi\left[q^-\left(\frac{B(0,t)ES_t}{B(0,T)X}; T\right)\right].$$

Finally, multiplying by $B(0,t)$ shows that

$$B(0,t)EC^E(S_t, T-t) = C^E[B(0,t)ES_t, T]$$

and gives the following remarkable formula for the call's expected holding-period return under $\mathbb{P}$:

$$\frac{EC^E(S_t, T-t)}{C^E(S_0, T)} = \frac{C^E[B(0,t)ES_t, T]}{B(0,t)C^E(S_0, T)}. \tag{6.44}$$

Notice that when $\mathbb{P} = \hat{\mathbb{P}}$ the right side reverts to

$$\frac{C^E[B(0,t)\hat{E}S_t, T]}{B(0,t)C^E(S_0, T)} = \frac{C^E(S_0, T)}{B(0,t)C^E(S_0, T)} = B(0,t)^{-1},$$

in line with (6.42).

While the numerator on the right side of (6.44) increases with the $\mathbb{P}$-expected future price of the stock, the current value of the call in the denominator does not change, since it based on measure $\hat{\mathbb{P}}$. An increase in ES_t therefore raises the call's mean holding-period return, as one would expect. Things work in the opposite direction for puts. Figures 6.6 and 6.7 show, for several values of volatility, how mean holding-period returns of European options vary with the stock's average proportional drift over the holding period, $\bar{\mu}(0,t) \equiv t^{-1}\int_0^t \mu_s \cdot ds$. The figures plot the expected annualized, continuously compounded rates of return of calls and puts *vs.* $\bar{\mu}(0,t)$

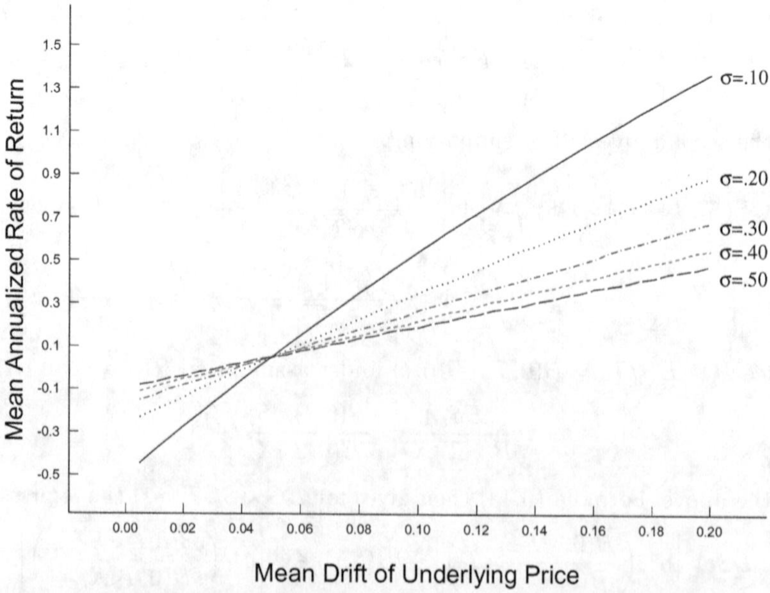

Fig. 6.6. Expected holding-period returns of call vs. mean drift of stock.

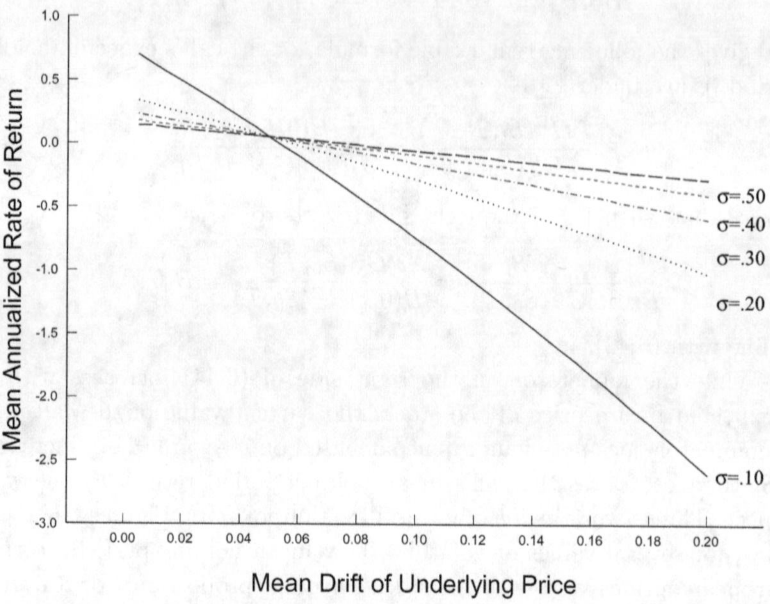

Fig. 6.7. Expected holding-period return of put vs. mean drift of stock.

for each of five values of σ. (Annualized, continuously compounded rates of return are natural logs of holding-period returns divided by the length of the holding period in years.) In each case $t = 0.5, T = 1.0$, $S_0 = X = 10.0$, and the average short rate is $r = 0.05$. The curves for all values of σ intersect at $\bar{\mu} = r = 0.05$, since here the actual and risk-neutral values coincide. Notice that the options' mean returns are especially sensitive to variations in the stock's mean drift when volatility is low, because in that case they have little chance of being in the money at expiration unless the trend is favorable.

7

American Options and "Exotics"

This chapter takes up the pricing of derivatives that have more complicated payoff structures than European-style options: American options with both finite and indefinite lives; "compound" options, for which the underlying assets themselves are (or depend on) options; European-style options that can be extended finitely or indefinitely many times; options with terminal payoff functions besides just $(S_T - X)^+$ or $(X - S_T)^+$; and options that let the holder choose among alternative underlying assets or that let the buyer decide after the purchase whether the option is to be a put or a call. We also consider various path-dependent derivatives such as barrier options, Asian options, and lookbacks, all of whose payoffs depend on the price of the underlying at more than one point in time. While finite-lived American options are traded on the organized exchanges—and, indeed, are the most common types of exchange-traded options—the various exotics are traded primarily over the counter, being created by financial firms to meet specific hedging demands of their clients.

Throughout this chapter we adhere to the standard Black-Scholes framework, taking the dynamics for the underlying as geometric Brownian motion with volatility σ and regarding the short rate of interest as a constant, r.

7.1 American Options

American calls on no-dividend stocks are valued just as European calls, because their always-positive time value prior to expiration rules out early exercise. This was established in chapter 4 by simple static replicating arguments. On the other hand, unless they are dividend-protected by a contractual adjustment of the strike price, calls on stocks that do pay dividends may well be exercised early. Likewise, early exercise may well occur for American calls on foreign currencies and, as explained on page 273, for American calls on futures. Moreover, American puts on any of these underlying assets are subject to early exercise. Obviously, one would want to exercise a put if the price of the underlying were so low that foregone interest on the strike exceeded the possible gain from further decline.

Valuing finite-lived American puts and finite-lived puts and calls on futures or on assets paying continuous dividends is complicated by the path dependence of the date of exercise. While this early-exercise option is easily handled in the backwards-recursive binomial setup, no exact computational formula like that for European options has ever been found within the Black-Scholes framework. This is one reason that the binomial method remains a useful tool in derivatives pricing. On the other hand, in the continuous-time environment it is relatively easy to price finitely lived calls on stocks that pay known dividends at discrete times and, oddly enough, American options that never expire. We begin with the first of these cases, then treat the harder case of puts and calls on assets that may pay continuous dividends, then conclude by developing simple valuation formulas for options with no definite termination dates.

7.1.1 *Calls on Stocks Paying Lump-Sum Dividends*

Consider an American call expiring at T on a stock that pays a single known, cash dividend to holders of record up to some date $t^* \in (t, T)$, the *ex*-dividend date. On this date the price will drop below its *cum*-dividend value by some amount Δ, which we *define* to be the value of the cash dividend.[1] The same argument that rules out early exercise of a call on a no-dividend stock shows that exercise in this case would occur, if at all, just prior to t^*. Let t^{*-} denote this last instant before t^*. In order to value the option at t, we must determine its value at t^{*-}. The holder of the option would exercise it at that time if intrinsic value $S_{t^{*-}} - X$ exceeded its live value on the *ex*-dividend asset. Since by assumption only one dividend is paid before T, the option's *ex*-dividend value is the value of a European call with remaining life $T - t^*$ on a stock worth $S_{t^{*-}} - \Delta$. Thus, the value of the American call at t^{*-} is the greater of $S_{t^{*-}} - X$ and $C^E(S_{t^{*-}} - \Delta, T - t^*) = C^E(S_{t^*}, T - t^*)$.

Figure 7.1 illustrates. We know that the value of a call—either European or American—on a no-dividend stock would approach asymptotically the

[1]See the discussion of tax and timing issues for dividends in section 5.4.4.

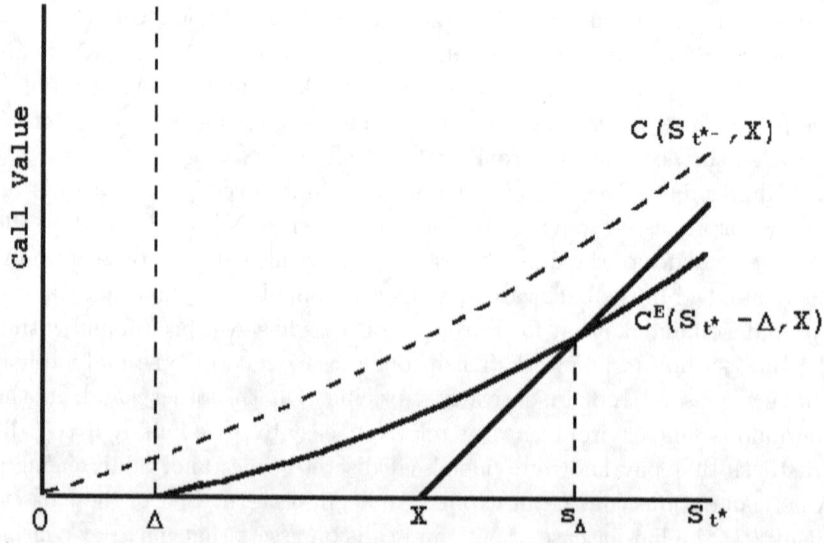

Fig. 7.1. Early exercise condition for American call on stock paying dividend Δ at t^*.

line $S_{t^*} - B(t^*, T)X$ as $S_{t^*} \to \infty$. This lies above the $S_{t^*} - X$ line by an amount $X[1 - B(t^*, T)]$. This difference increases with the average short rate of interest over the option's remaining life and, so long as short rates are sometimes positive, with $T - t^*$. This no-dividend case is depicted by the dashed curve emanating from the origin. That the value function lies always above the intrinsic value line prior to expiration indicates that early exercise would never be advantageous in the absence of a dividend. Now consider how the payment of a dividend alters the picture. Since the stock's price will drop at once to $S_{t^*} = S_{t^*-} - \Delta$ on the *ex* date, a *cum*-dividend price of Δ at t^{*-} would correspond to an *ex*-dividend price of zero at t^*. Once the dividend is out of the way, the American call is again valued like a European call with $T - t^*$ to go. Thus, its value at t^* as a function of *ex*-dividend price S_{t^*} is simply the value function of a European call at S_{t^*-} shifted to the right by Δ. This translation is depicted by the bold curve in the figure. If the curve intersects the line $(S_{t^*} - X)^+$, as it does in the figure at price s_Δ, then it would be advantageous to exercise the American call at t^{*-} whenever $S_{t^*-} > s_\Delta$. The critical issue, then, is whether such an intersection occurs, and this depends on the size of the dividend and on $B(t^*, T)$—that is, on the average short rate and on how long the option has yet to live.

To see whether such a critical s_Δ exists and to determine its value just requires a numerical search for a solution to

$$S_{t^*} - X - C^E(S_{t^*} - \Delta, T - t^*) = 0,$$

using the Black-Scholes formula to value the call. If there is no solution, the value of the American call at t will be that of a European call on a dividend-paying stock. This can be approximated by the set-aside approach discussed in section 6.3.3. If there is a solution, then one has to do some work, as follows. Applying the risk-neutral paradigm, express the call's current value, denoted $C^A(S_t, T-t; t^*, \Delta)$, as

$$B(t,t^*) \int_0^\infty \max\left[s_{t^*} - X, C^E(s_{t^*} - \Delta, T - t^*)\right] \cdot d\hat{F}_t(s_{t^*})$$

or

$$B(t,t^*) \int_\Delta^{s_\Delta} C^E(s_{t^*} - \Delta, T-t^*) \cdot d\hat{F}_t(s_{t^*}) + B(t,t^*) \int_{s_\Delta}^\infty (s_{t^*} - X) \cdot d\hat{F}_t(s_{t^*}),$$

where $\hat{F}_t(s_{t^*}) \equiv \hat{\mathbb{P}}_t(S_{t^*} \leq s_{t^*})$ is the conditional c.d.f. at t of the *cum-dividend* price at t^* in the risk-neutral measure. Look first at the second term. If the strike price of the option were s_Δ rather than X, the value of this term would be given by the Black-Scholes formula with s_Δ in place of X. As it is, X just takes its usual place as the coefficient of the second term in the formula, with s_Δ appearing in place of X everywhere else; that is, setting $\tau_1 \equiv t^* - t$ and $B_1 \equiv B(t, t^*)$ for brevity,

$$B_1 \int_{s_\Delta}^\infty (s_{t^*} - X) \cdot d\hat{F}_t(s_{t^*}) = S_t \Phi\left[q^+\left(\frac{S_t}{B_1 s_\Delta}; \tau_1\right)\right] - B_1 X \Phi\left[q^-\left(\frac{S_t}{B_1 s_\Delta}; \tau_1\right)\right].$$

To work on the first term in the expression for $C^A(S_t, T-t; t^*, \Delta)$, let $\tau_2 \equiv T - t^*$ and restate $\hat{F}_t$ in terms of Φ to get

$$B_1 \int_\Delta^{s_\Delta} C^E(s_{t^*} - \Delta, \tau_2) \cdot d\Phi\left[q^+\left(\frac{B_1 s_{t^*}}{S_t}; \tau_1\right)\right].$$

Now change variables as

$$z \leftarrow q^+\left(\frac{B_1 s_{t^*}}{S_t}; \tau_1\right)$$

$$s_{t^*} \rightarrow \frac{S_t}{B_1} \exp\left(z\sigma\sqrt{\tau_1} - \sigma^2 \tau_1 / 2\right),$$

and calculate this as

$$B_1 \int_{z_\Delta}^{z_{s_\Delta}} C^E\left[\frac{S_t}{B_1} \exp\left(z\sigma\sqrt{\tau_1} - \sigma^2\tau_1/2\right) - \Delta, \tau_2\right] \cdot d\Phi(z),$$

where the limits of integration are

$$z_\Delta \equiv q^+\left(\frac{B_1\Delta}{S_t};\tau_1\right)$$
$$z_{s_\Delta} \equiv q^+\left(\frac{B_1 s_\Delta}{S_t};\tau_1\right).$$

Alternatively, expressing $C^E(\cdot,\tau_2)$ gives

$$\int_{z_\Delta}^{z_{s_\Delta}}\left\{\left(S_t e^{z\sigma\sqrt{\tau_1}-\sigma^2\tau_1/2}-B_1\Delta\right)\Phi\left[q^+\left(\frac{S_t e^{z\sigma\sqrt{\tau_1}-\sigma^2\tau_1/2}-B_1\Delta}{BX};\tau_2\right)\right]\right.$$
$$\left.-BX\Phi\left[q^-\left(\frac{S_t e^{z\sigma\sqrt{\tau_1}-\sigma^2\tau_1/2}-B_1\Delta}{BX};\tau_2\right)\right]\right\}\cdot d\Phi(z),$$

where $B \equiv B_1 B(t^*,T) \equiv B(t,T)$. This integral can be computed numerically using, for example, routine GAUSSINT on the accompanying CD. The method can be adapted for a dividend that is a fixed proportion of S_{t^*} or, for that matter, a dividend that is any well-behaved function of the underlying price.

7.1.2 *Options on Assets Paying Continuous Dividends*

This section treats American options on an underlying asset that follows geometric Brownian motion and pays continuous proportional dividends at constant, deterministic rate δ. The dynamics under the risk-neutral measure are thus

$$dS_t = (r-\delta)S_t \cdot dt + \sigma S_t \cdot d\hat{W}_t,$$

where $\{\hat{W}\}_{t\ge 0}$ is a Brownian motion under $\hat{\mathbb{P}}$. We refer to the underlying asset as a "stock", but the analysis applies also to options on currencies, on diversified stock portfolios (indexes), and (taking $\delta = r$) even on futures, so long as it is appropriate to model their price processes as geometric Brownian motion. The plan is to deal first with the American put and then extend to calls by exploiting the symmetry between these two types. We consider a T-expiring put with strike X and terminal value (if not previously exercised) equal to $P^A(S_T,0) = (X-S_T)^+$, with $P^A(S_t,T-t)$ denoting the value at $t\in[0,T]$. Although there is no simple analytical formula for $P^A(S_t,T-t)$, there are many implicit characterizations, some of which lead to useful numerical approximations. The binomial model considered in chapter 5 gives one such approximation, but standard binomial solutions are generally slower than those we now develop.[2] We shall describe procedures based on three characterizations of $P^A(S_t, T-t)$. Since all of these somehow involve the critical exercise boundary, $\{\mathfrak{B}_t\}_{0\le t\le T}$, we begin by describing this function.

The Put's Critical Exercise Boundary

The exercise boundary for the put is a function $\mathfrak{B} : [0, T] \to \Re^+$ such that exercise is optimal at any $t \in [0, T]$ whenever $S_t < \mathfrak{B}_t$. Given the parameters that define the put and the dynamics of the underlying (that is, X, T, r, δ, and σ), this boundary has to be found numerically by some process that is capable of valuing the put itself. We saw examples of this in chapter 5, where we discovered the exercise boundary in the process of valuing the option in the binomial lattice. Nevertheless, there are several features of $\{\mathfrak{B}_t\}$ that can be deduced analytically. First, it is clear that $\mathfrak{B}_t \le X$ for $t \in [0, T]$, since it would never be optimal to exercise an out-of-money option. Second, it must be the case that $\mathfrak{B}_T = X$, since an in-the-money put would be let to expire unexercised if $\mathfrak{B}_T < X$ and $\mathfrak{B}_T \le S_T < X$. Third, if $\sigma = 0$ and $r \ge 0$ then $\mathfrak{B}_t = X$ for all $t \in [0, T]$, since with nonnegative mean drift and zero volatility there would be no chance that the option could ever get further in the money. Fourth, $\mathfrak{B}_t$ is nondecreasing with time; that is, as we move closer to expiration the threshold below which exercise is desirable cannot decrease. The intuition for this is that by exercising at t one trades the put's time value—the value associated with the contingency of further declines in S—for the sure intrinsic value $X - \mathfrak{B}_t$. Since the time value cannot increase as $t \to T$, neither can $X - \mathfrak{B}_t$. Indeed, with $r > 0$ and $\sigma > 0$ (as we assume in the remainder of this discussion) the option's time value strictly decreases as $t \to T$, and $\mathfrak{B}_t$ must strictly increase. Fifth, since there is nothing that could introduce discontinuities in the put's time value or its derivatives, $\mathfrak{B}_t$ itself should be a smooth, differentiable function. And, finally, the values of P^A and $\partial P^A/\partial S \equiv P^A_S$ at the exercise boundary are known;[3] namely,

$$P^A(\mathfrak{B}_t, T-t) = X - \mathfrak{B}_t \tag{7.1}$$

$$P^A_S(\mathfrak{B}_t, T-t) = -1. \tag{7.2}$$

---

[2]On the other hand, we will see that the smoothing technique of Broadie and Detemple (1996) coupled with Richardson extrapolation is competitive with the best of these approximations.

[3]See van Moerbeke (1976) for proofs of these properties.

Although the "high-contact" condition, (7.2), looks a bit strange, it is not hard to develop some intuition for it. Clearly, the put's delta would be negative unity at the point of exercise, since a replicating portfolio would then be short precisely one share of the underlying. More formally, letting $P^A(S_t, T-t; \mathfrak{b}_t)$ represent the put's value at t when the critical exercise price is $\mathfrak{b}_t$, the fact that $\mathfrak{B}_t$ is the optimal threshold suggests that $\partial P^A(S_t, T-t; \mathfrak{b}_t)/\partial \mathfrak{b}_t = 0$ at $\mathfrak{b}_t = \mathfrak{B}_t$. Now differentiating (7.1) gives

$$\begin{aligned}-1 &= dP^A(\mathfrak{B}_t, T-t; \mathfrak{B}_t)/d\mathfrak{B}_t \\ &= \partial P^A(S_t, T-t; \mathfrak{B}_t)/\partial S_t \mid_{S_t=\mathfrak{B}_t} + \partial P^A(\mathfrak{B}_t, T-t; \mathfrak{b}_t)/\partial \mathfrak{b}_t \mid_{\mathfrak{b}_t=\mathfrak{B}_t} \\ &= P_S^A(\mathfrak{B}_t, T-t; \mathfrak{B}_t) + 0.\end{aligned}$$

Characterizations of $P^A(S_t, T-t)$

We saw in chapter 5 that the value of an American put can be represented as the discounted expected value of the receipt generated when underlying price first passes the exercise boundary. Specifically, in the discrete-time Bernoulli setup we expressed the put's initial value as

$$P_0^A = \sum_{j=1}^{n} B(0,j)(X - \mathfrak{B}_j)\hat{\mathbb{P}}(J^* = j),$$

where stopping time J^* is the first step at which $s_j < \mathfrak{B}_j$ and $B(0,j)$ is the discrete-time notation for the price of a bond expiring at step j. The applicable boundary function $\{\mathfrak{B}_j\}$ maximizes the value of this expression. Corresponding expressions in continuous time are

$$P^A(S_0, T) = \max_{\mathfrak{B}} \int_0^T e^{-ru}(X - \mathfrak{B}_u) \cdot d\hat{\mathbb{P}}(U \le u)$$

and

$$P^A(S_0, T) = \max_{\mathfrak{B}} \hat{E} e^{-rU}(X - S_U), \tag{7.3}$$

where stopping time $U = \inf\{t : S_t < \mathfrak{B}_t\}$. The characterization (7.3) will be used below in developing a symmetry relation for American puts and calls and again in chapter 11 to value options by simulation. However, neither of these expressions has been found helpful in devising analytical approximations, because distributions of first-passage times of Brownian motions through arbitrary boundaries are not easily described, and because the boundary itself has to be found along with the value of the option.

More fruitful for analytical results is the representation of $P^A(S_t, T - t)$ as solution to the Black-Scholes p.d.e. Merton (1973) showed that the American put satisfies

$$-P^A_{T-t} + P^A_S(r - \delta)S_t + P^A_{SS}\sigma^2 S_t^2/2 - P^A r = 0 \tag{7.4}$$

subject to the terminal condition $P^A(S_T, 0) = (X - S_T)^+$ and boundary conditions

$$P^A(S_t, T - t) \in [(X - S_t)^+, X] \tag{7.5}$$

$$\lim_{S_t \to \infty} P^A(S_t, T - t) = 0. \tag{7.6}$$

There are also conditions (7.1) and (7.2). Brennan and Schwartz (1977) presented an algorithm that yields numerical solutions for this problem, using discretized state spaces for both time and the underlying price. These finite-difference methods for solving p.d.e.s are described in chapter 12, which gives specific examples of their application to American options.[4] Later in this section we work out an analytical solution to the p.d.e. in the case $T = \infty$, which turns out to be a much easier problem.

A third characterization of the American put involves discretizing time alone. Letting $P_n(S_0, T)$ denote the value of a put that can be exercised just at times $t_j \equiv jT/n$ for $j \in \{0, 1, \ldots, n\}$, the value of the standard American put is approached in the limit as $n \to \infty$. We develop below the following recursion, which, in principle, permits valuation for arbitrary n:

$$\begin{aligned} P_n(S_0, T) = {} & B(0, t_1)\hat{E}(X - S_{t_1})\mathbf{1}_{[0, \mathfrak{B}_{t_1}]}(S_{t_1}) \\ & + B(0, t_1)\hat{E}P_{n-1}(S_{t_1}, T - t_1)\mathbf{1}_{(\mathfrak{B}_{t_1}, \infty)}(S_{t_1}). \end{aligned} \tag{7.7}$$

Here, $\mathbf{1}_A$ is the usual indicator notation and $\mathfrak{B}_{t_1}$ is the critical exercise value at t_1. Implementing this procedure gives an approximation proposed by Geske and Johnson (1984).

Finally, the value of the American put can be expressed as the sum of the value of a corresponding European option and an early-exercise premium:

$$P^A(S_t, T - t) = P^E(S_t, T - t) + \Pi(S_t, T - t). \tag{7.8}$$

There are a number of ways to represent the premium and to exploit this relationship in order to approximate the value of the put. We describe methods proposed by MacMillan (1986) and Ju (1998).

[4] Wilmott *et al.* (1993, 1995) give extensive treatment of this subject.

The Geske-Johnson Approximation

Abstracting from the fact that markets do not operate around the clock, an American put can in principle be exercised at any time before expiration. However, one can imagine discrete versions that could be exercised only at specific times up to and including T. Indeed, there is an active over-the-counter market in "Bermudan" options that have precisely this feature. Now consider a sequence of Bermudan puts indexed by the number of equally spaced exercise dates, with prices $\{P_n(S_0, T)\}_{n=1}^{\infty}$. Given the continuity of underlying process $\{S_t\}_{t\geq 0}$, one expects such a sequence of prices to converge to the arbitrage-free price of a vanilla American put. The idea behind the Geske-Johnson (1984) approach is to value each of a short sequence of these puts, then apply Richardson extrapolation to extend to an infinite number of dates. The process starts with $P_1(S_0, T)$, which is nothing but the Black-Scholes value of a European put. Next comes $P_2(S_0, T)$, the value of an option that can be exercised at $t_1 = T/2$ and at $t_2 = T$. Geske and Johnson represent this in terms of the discounted risk-neutral expectations of the contingent receipts at the two times:

$$\begin{aligned} P_2(S_0, T) = B(0, t_1)\hat{E}(X - S_{t_1})\mathbf{1}_{[0, \mathfrak{B}_{t_1})}(S_{t_1}) \\ + B(0, T)\hat{E}(X - S_T)^+\mathbf{1}_{[\mathfrak{B}_{t_1}, \infty)}(S_{t_1}). \end{aligned}$$

With some manipulation the expectation in the second term, which involves the correlated lognormals S_{t_1} and S_T, can be expressed as the expectation of a function of bivariate normals and evaluated numerically. Alternatively, conditioning on what is known at t_1 and writing $B(0, T)\hat{E}(X - S_T)^+$ in the second term as

$$B(0, t_1)\hat{E}[B(t_1, T)\hat{E}_{t_1}(X - S_T)^+] = B(0, t_1)\hat{E}P_1(S_{t_1}, T - t_1)$$

gives a recursive expression that makes the computations easier to program and leads naturally to the next stage in the sequence:

$$\begin{aligned} P_2(S_0, T) &= B(0, t_1)\hat{E}[(X - S_{t_1})\mathbf{1}_{[0, \mathfrak{B}_{t_1})}(S_{t_1}) \\ &\quad + P_1(S_{t_1}, T - t_1)\mathbf{1}_{[\mathfrak{B}_{t_1}, \infty)}(S_{t_1})] \\ &= B(0, t_1)\hat{E}\max\{X - S_{t_1}, P_1(S_{t_1}, T - t_1)\}. \end{aligned} \tag{7.9}$$

This shows that one who owns the put at t_1 chooses either to receive $X - S_{t_1}$ in cash or to hold on to what becomes a European put that dies at T. Critical exercise price $\mathfrak{B}_{t_1}$ equates the values of these two claims at t_1 and can therefore be found by solving $X - \mathfrak{B}_{t_1} = P_1(\mathfrak{B}_{t_1}, T - t_1)$ numerically. (Of

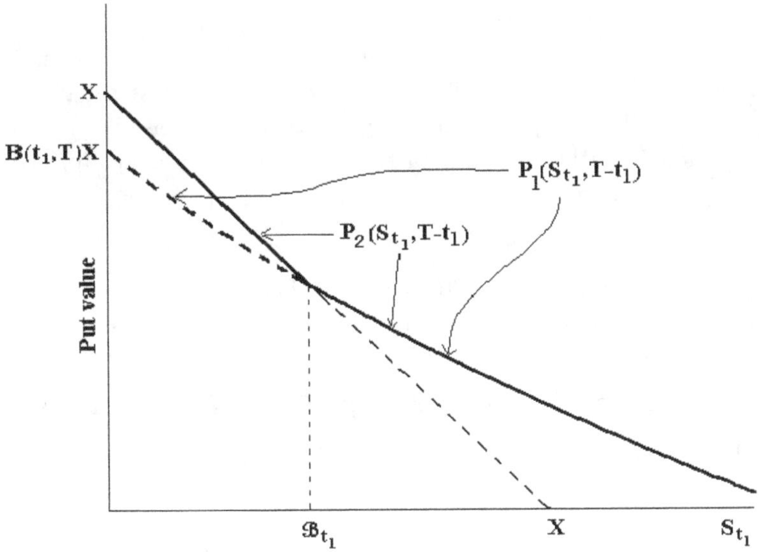

Fig. 7.2. Value function at t_1 of Bermudan put exercisable at t_1 and T.

course, this is the critical price for the Bermudan approximation rather than for a genuine American put.) The solid, kinked curve in figure 7.2 depicts the value of the two-step Bermudan option at t_1 as a function of S_{t_1}.

Proceeding to the third stage, where exercise can occur at $t_1 = T/3$, $t_2 = 2T/3$, or $t_3 = T$, the recursion progresses as

$$\begin{aligned} P_3(S_0, T) &= B(0, t_1)\hat{E}\left[(X - S_{t_1})\mathbf{1}_{[0,\mathfrak{B}_{t_1}]}(S_{t_1}) \right. \\ &\qquad \left. + P_2(S_{t_1}, T - t_1)\mathbf{1}_{[\mathfrak{B}_{t_1},\infty)}(S_{t_1})\right] \\ &= B(0, t_1)\hat{E}\max\{X - S_{t_1}, P_2(S_{t_1}, T - t_1)\}, \qquad (7.10) \end{aligned}$$

where $\mathfrak{B}_{t_1}$ now solves $X - \mathfrak{B}_{t_1} = P_2(\mathfrak{B}_{t_1}, T - t_1)$—and so on to (7.7) at stage n.

Evaluating the discounted expectation of the first terms in (7.9) and (7.10) is straightforward since

$$\hat{E}(X - S_{t_1})\mathbf{1}_{[0,\mathfrak{B}_{t_1}]}(S_{t_1}) = X\Phi[q^+(\mathfrak{B}_{t_1}/\mathsf{f})] - \mathsf{f}\Phi[q^-(\mathfrak{B}_{t_1}/\mathsf{f})],$$

where Φ is the standard normal c.d.f., $\mathsf{f} = S_0 e^{-\delta t_1}/B(0, t_1)$, and $q^{\pm}(x) = (\ln x)/(\sigma\sqrt{t_1}) \pm \sigma\sqrt{t_1}/2$. However, the computations get more difficult as one proceeds, since at each new stage beyond the first calculating

$\hat{E}P_{n-1}(S_{t_1}, T-t_1)\mathbf{1}_{[\mathfrak{B}_{t_1},\infty)}(S_{t_1})$ requires an additional iteration of numerical integration. Geske and Johnson suggest using Richardson extrapolation to extend to P_∞ from a short sequence of estimates. Letting $h_j = T/j$ for $j \in \{1, 2, \ldots, n\}$ and representing $P_j(S_0, T)$ temporarily as $\hat{P}(h_j)$, we recall from section 5.5.2 that Richardson extrapolation proceeds by approximating $\hat{P}(h)$ as a polynomial,

$$\hat{P}(h) = a_0 + a_1 h + a_2 h^2 + \cdots + a_{n-1} h^{n-1},$$

in which a_0 corresponds to $\hat{P}(0)$ or P_∞. This is found by solving the n equations corresponding to $\hat{P}(h_1)$ through $\hat{P}(h_n)$. The linear approximation $(n = 2)$ works out to be

$$P_\infty(S_0, T) \doteq 2\hat{P}(h_2) - \hat{P}(h_1) \equiv 2P_2(S_0, T) - P_1(S_0, T),$$

and the quadratic approximation is

$$P_\infty(S_0, T) \doteq 9P_3(S_0, T)/2 - 4P_2(S_0, T) + P_1(S_0, T)/2.$$

Computations beyond $n = 3$ are very slow because of the need to compute $P_4, P_5, \ldots$, but (depending on σ and T) the results can be quite accurate even for $n = 2$. While the Geske-Johnson method is usually faster than a standard binomial tree of the same accuracy, Broadie and Detemple (1996) and Ju (1998) find that a still better speed/accuracy frontier is attained by applying Richardson extrapolation to binomial estimates that start with Black-Scholes values at the terminal nodes, as described in section 5.5.2.

The MacMillan Approximation

In a 1986 paper L.W. MacMillan proposed a very fast procedure that exploits the decomposition (7.8) to produce a p.d.e. whose solution approximates the value of the early-exercise premium. Barone-Adesi and Whaley (1987) extend the method to allow for continuous dividends and (in their 1988 paper) for dividends at discrete times. We consider just the case of continuous dividends.

The idea is based on the observation that, since prices of both the European and American puts satisfy (7.4), so must early-exercise premium $\Pi(S_t, T-t)$ satisfy

$$-\Pi_{T-t} + \Pi_S(r-\delta)S_t + \Pi_{SS}\sigma^2 S_t^2/2 - \Pi r = 0 \qquad (7.11)$$

subject to certain boundary conditions. However, since P^A and P^E share the same terminal condition, $\overline{P}(S_T) = (X - S_T)^+$, the corresponding condition for the premium is simply $\Pi(S_T, 0) = 0$. The plan is to simplify still more by a change of variable and an approximation that reduces the problem to solving an ordinary differential equation in S_t alone.

Begin with the change of variable. Introducing the function $K(T-t) \equiv 1 - e^{-r(T-t)}$, write

$$\Pi(S_t, T-t) = f(S_t, K) \cdot K,$$

thus defining a new function f. With this substitution, and using

$$K' \equiv dK(T-t)/d(T-t) = r(1-K),$$

(7.11) becomes

$$-r(1-K)Kf_K - fr + Kf_S(r-\delta)S_t + Kf_{SS}\sigma^2 S_t^2/2 = 0. \qquad (7.12)$$

The time-dependence of f is expressed here explicitly only through the first term, and this term is small under two circumstances: (i) when $T-t \doteq 0$, in which case $1-K \doteq 0$, or (ii) when $T-t$ is large, in which case f is relatively insensitive to time dimension and $f_K \doteq 0$. MacMillan's approximation is simply to drop the first term. This leaves just the ordinary differential equation

$$-fr + Kf_S(r-\delta)S_t + Kf_{SS}\sigma^2 S_t^2/2 = 0, \qquad (7.13)$$

which is to be solved for $f(\cdot, K)$. The relevant boundary conditions are (i) $f \geq 0$, since an American put can be worth no less than a European; (ii) $\lim_{S_t \to \infty} f = 0$, since both American and European options are worthless in the limit; and (iii) that f is bounded on $[\mathfrak{B}_t, \infty)$. Terminal condition $\Pi(S_T, 0) = 0$ is automatically satisfied by the specification of K and imposes no condition on f. When $K \neq 0$ (7.13) is a second-order ordinary differential equation with variable coefficients, of which any function of the form $f = cS_t^{-u}$ is a solution.[5] Substituting, one finds that there are two roots,

$$u^{\pm}(K) = \frac{1}{2}\left\{-1 + 2(r-\delta)/\sigma^2 \pm \sqrt{[1 - 2(r-\delta)/\sigma^2]^2 + 8r/(K\sigma^2)}\right\},$$

so that the general solution is

$$f(S, K) = c^+(K)S_t^{-u^+(K)} + c^-(K)S_t^{-u^-(K)}.$$

[5] $K = 0$ requires either $T-t = 0$ or $r = 0$. In the first case the option's value is given by the terminal condition. The other case, which is for practical purposes irrelevant, is treated by MacMillan (1986) in an appendix.

Boundary condition (ii) rules out negative root u^-, whereas condition (iii) still allows positive root u^+. Dropping the "+" and letting $u(K), c(K)$ correspond to the positive root, one sees that all three conditions are met by $f(S_t, K) = c(K)S_t^{-u(K)}$, so that

$$\hat{P}^A(S_t, T-t) = \begin{cases} X - S_t, & S_t < \mathfrak{B}_t \\ P^E(S_t, T-t) + Kc(K)S_t^{-u(K)}, & S_t \geq \mathfrak{B}_t \end{cases}. \tag{7.14}$$

Coefficient $c(K)$ and critical exercise price $\mathfrak{B}_t$ are found from conditions imposed by (7.1) and (7.2):

$$P^E(\mathfrak{B}_t, T-t) + Kc(K)\mathfrak{B}_t^{-u(K)} = X - \mathfrak{B}_t \tag{7.15}$$

$$P_S^E(\mathfrak{B}_t, T-t) - Kc(K)u(K)\mathfrak{B}_t^{-u(K)-1} = -1. \tag{7.16}$$

Eliminating $c(K)$ gives a nonlinear equation for $\mathfrak{B}_t$,

$$\mathfrak{B}_t = \frac{u(K)\left[X - P^E(\mathfrak{B}_t, T-t)\right]}{1 + u(K) + P_S^E(\mathfrak{B}_t, T-t)},$$

which must be solved numerically. For this, $P^E(\mathfrak{B}_t, T-t)$ is calculated from the Black-Scholes formula,

$$P^E(\mathfrak{B}_t, T-t) = BX\Phi\left[q^+\left(\frac{BX}{\mathfrak{B}_t e^{-\delta(T-t)}}\right)\right] - \mathfrak{B}_t e^{-\delta(T-t)}\Phi\left[q^-\left(\frac{BX}{\mathfrak{B}_t e^{-\delta(T-t)}}\right)\right],$$

where q^+ and q^- are given in (6.24), and

$$P_S^E(\mathfrak{B}_t, T-t) = -\Phi\left[q^-\left(\frac{BX}{\mathfrak{B}_t e^{-\delta(T-t)}}\right)\right].$$

With $\mathfrak{B}_t$ determined equations (7.15) and (7.16) give an explicit solution for $c(K)$:

$$c(K) = \mathfrak{B}_t^{1+u(K)}\frac{1 + P_S^E(\mathfrak{B}_t, T-t)}{Ku(K)}.$$

These values can be calculated very quickly, and findings reported by MacMillan (1986) and Barone-Adesi and Whaley (1987) indicate that their accuracy compares favorably over an extensive range of parameter values with 3-stage Geske-Johnson (1984) estimates and numerical solutions after Brennan and Schwartz (1977).

It is interesting that the approximation (7.14) becomes exact as $T \to \infty$ and $K \to 1$, since (7.12) and (7.13) then coincide. Since $\lim_{T\to\infty} P^E(\mathfrak{B}_t, T-t) = \lim_{T\to\infty} P_S^E(\mathfrak{B}_t, T-t) = 0$, (7.15) and (7.16) yield explicit solutions:

$$\mathfrak{B}_t = \frac{u(1)X}{1+u(1)}$$

$$c(K) = \frac{X}{[1+u(1)]}\mathfrak{B}_t^{u(1)}$$

and

$$P^A(S_t, T-t) = \frac{X}{1+u(1)}\left(\frac{S_t}{\mathfrak{B}_t}\right)^{-u(1)}.$$

Since $u(1) > 0$, the value function is hyperbolic above the critical exercise price. We will get the same result later by solving directly the infinite-horizon version of (7.4).

The Multipiece Exponential Approximation

It appears that the current state of the art in pricing American puts under Black-Scholes dynamics is an approach proposed by Nengjiu Ju (1998). This is based on the following characterization of the early-exercise premium, developed by Kim (1990), Carr *et al.* (1992), and others:

$$\Pi(S_t, T-t) = \hat{E}_t\left[\int_t^T e^{-r(u-t)}\left(rX - \delta S_u\right)\mathbf{1}_{[0,\mathfrak{B}_u)}(S_u)\cdot du\right]. \qquad (7.17)$$

This amounts to the difference between the value of the earnings from investing the strike price after early exercise and the value of the dividends forgone by putting the stock. Carr *et al.* (1992) give an intuitive explanation, involving a strategy that effectively converts the American put into a European by stripping away the potential gain from early exercise.

The idea is easiest to grasp if one begins with a degenerate case in which both interest rate and dividend rate are zero. Start at t by buying a "live" American option at price $P^A(S_t, T-t)$—that is, one for which S_t is above the exercise boundary. So long as S stays above the boundary, continue to hold the option. If it crosses the boundary, exercise the option, borrowing the share of stock that is put to the writer and receiving the strike price X in cash. This involves no net outlay. If the stock's price recrosses the boundary, use the cash/stock account to repurchase the option, again for no net outlay. In this way one holds the option when S is above the exercise boundary, and one holds an account worth $X - S$ when it is below the boundary. At $t = T$ the position has the same value as a European option, $(X - S_T)^+$. Since the strategy replicates the European put, is self-financing, and costs $P^A(S_t, T - t)$ initially, we conclude that American and European puts have the same value if $r = \delta = 0$.

Now try to follow the same strategy in the relevant case that $r > 0$ and δ is arbitrary. Again, start by buying a live American option for $P^A(S_t, T-t)$, and again exercise if S crosses the boundary, shorting a share of stock and investing the strike in riskless bonds. The difference now is that the value of this account is no longer $X - S$ because interest accrues on the bonds and dividends have to be constantly repaid to the lender of the borrowed stock. During an interval $[u, u+du)$ in which S is below the boundary these effects change the value of the account by $(rX - \delta S_u) \cdot du$. The present value of this prospective increment at t is $e^{-r(u-t)}(rX - \delta S_u) \cdot du$. The strategy still yields the European put's terminal payoff, $(X - S_T)^+$, on any path, but a given path $\{S_u(\omega)\}_{t \le u \le T}$ generates in addition a cash stream worth

$$\int_t^T e^{-r(u-t)}[rX - \delta S_u(\omega)]\mathbf{1}_{[0,\mathfrak{B}_u)}[S_u(\omega)] \cdot du.$$

Averaging over paths (in the risk-neutral measure) gives (7.17) as the additional value of the American put over the European. Working out the integral yields a relatively simple expression for the exercise premium:

$$\Pi(S_t, T-t) = \int_t^T e^{-r(u-t)} rX\Phi[q^+(e^{-(r-\delta)(u-t)}\mathfrak{B}_u/S_t)] \cdot du$$
$$- \int_t^T e^{-\delta(u-t)}\delta S_t\Phi[q^-(e^{-(r-\delta)(u-t)}\mathfrak{B}_u/S_t)] \cdot du,$$

where

$$q^{\pm}(x) \equiv \frac{\ln x \pm \sigma^2(u-t)/2}{\sigma\sqrt{u-t}}.$$

This is easily calculated once the boundary function $\{\mathfrak{B}_u\}$ is known. While $\{\mathfrak{B}_u\}$ can be solved for in principle from the equation

$$P^A(\mathfrak{B}_t, T-t) = \mathfrak{B}_t - X = P^E(\mathfrak{B}_t, T-t) + \Pi(\mathfrak{B}_t, T-t),$$

this is very difficult to do in practice.

Ju (1998) proposes to break the time to expiration into subintervals $\{[t_j, t_{j+1})\}_{j=0}^{n-1}$ with $t_0 = t$ and $t_n \equiv T$ and to approximate $\{\mathfrak{B}_u\}$ within $[t_j, t_{j+1})$ by an exponential, $\mathfrak{B}_u = \alpha_j e^{\beta_j u}$. When $\mathfrak{B}_u$ is of this form, the integrals involved in $\Pi(S_t, T-t)$ can be expressed explicitly in terms of univariate-normal c.d.f.s, which can be calculated very quickly. The constants α_j and β_j are found by applying $P^A(\mathfrak{B}_t, T-t) = \mathfrak{B}_t - X$ and high-contact condition (7.2) at t_j. One may consult the original article for the lengthy formulas involved in the approach. Ju gives extensive comparisons with other methods, including Geske-Johnson, the standard binomial, and Broadie and Detemple's (1996) smoothed binomial that starts with Black-Scholes prices at the terminal nodes. Using a three-piece exponential approximation to $\mathfrak{B}_t$, Ju's method seems to dominate in terms of speed and accuracy.[6]

Extension to American Calls

A neat symmetry relation immediately extends to American calls all the methods that apply to American puts on assets that may pay continuous, proportional dividends. In chapter 6 we demonstrated that under Black-Scholes dynamics the price of either the European put or the call could be obtained from the other by permuting arguments S_t and X, r and δ. Specifically, if we write $C^E(S_t, T-t; X, r, \delta)$ and $P^E(S_t, T-t; X, r, \delta)$ for T-expiring European options struck at X on an asset paying dividends at rate δ, then

$$C^E(S_t, T-t; X, r, \delta) = P^E(X, T-t; S_t, \delta, r).$$

We now show formally that the same relation holds for American options C^A and P^A. Also, representing the critical exercise boundaries for calls and puts as $\mathfrak{B}_t^C = \beta_t^C X$ and $\mathfrak{B}_t^P = \beta_t^P X$, it will turn out that $\beta_t^C = 1/\beta_t^P$.

Start with the characterization of American puts and calls as

$$\begin{aligned} P^A(S_t, T-t; X, r, \delta) &= \max_{\beta^P} \hat{E}[e^{-rU^P}(X - S_{U^P})^+ \mid \mathcal{F}_t], \\ U^P &\equiv \inf\{u \in [t, T] : S_u < \beta_u^P X\}, \end{aligned}$$

[6]The ongoing search for better approximations to values of American options under Black-Scholes dynamics is, without question, a worthy intellectual challenge; and fast and accurate methods are clearly of value to institutions that must price and hedge many of these instruments in the course of a business day. However, one wonders if intellectual capital might not now be better spent elsewhere, given the limited descriptive accuracy of Black-Scholes dynamics for financial assets. Within this framework, the simple smoothed-binomial scheme of Broadie and Detemple (1996) appears perfectly adequate for most purposes, as Ju himself suggests in the following statement (p. 641):

> "When the requirement of accuracy is not too stringent, [the Broadie-Detemple scheme] could be the choice of method in many applications because it is simple, fast, and easy to program."

and

$$C^A(S_t, T-t; X, r, \delta) = \max_{\beta^C} \hat{E}[e^{-rU^C}(S_{U^C} - X)^+ \mid \mathcal{F}_t],$$
$$U^C \equiv \inf\{u \in [t,T] : S_u > \beta_u^C X\},$$

where $U^P = T$, $U^C = T$ if the respective boundaries are not crossed by T. These represent the prices of American options as solutions to the problem of finding the optimal exercise boundary. Let $\{Y_u \equiv S_u e^{(\delta-r)u}\}_{t \le u \le T}$, noting that this is a martingale under $\hat{\mathbb{P}}$ since $\hat{E}|Y_u| < \infty$ and $\hat{E}_t Y_u = S_t e^{(\delta-r)t} = Y_t$ for $u \ge t$. Then

$$P^A(Y_t, T-t; X, r, \delta) = \max_{\beta^P} \hat{E}_t(e^{-rU^P}X - e^{-\delta U^P}Y_{U^P})^+,$$
$$U^P = \inf\{u \in [t,T] : Y_u < \beta_u^P X e^{(\delta-r)u}\}.$$

Permuting the arguments,

$$P^A(X, T-t; Y_t, \delta, r) = \max_{\beta^P} \hat{E}_t(e^{-\delta U^P}Y_{U^P} - e^{-rU^P}X)^+,$$
$$U^P = \inf\{u \in [t,T] : X < \beta_u^P Y_u e^{(r-\delta)u}\}$$
$$= \inf\{u \in [t,T] : Y_u e^{(r-\delta)u} \ge X/\beta_u^P\},$$

or

$$P^A(X, T-t; Y_t, \delta, r) = \max_{\beta^C} \hat{E}e^{-rU^C}(S_{U^C} - X)^+,$$
$$U^C = \inf\{u \in [t,T] : Y_u e^{(r-\delta)u} \equiv S_u \ge \beta_u^C X\},$$

with $\beta^C = 1/\beta^P$. Thus, as promised,

$$C^A(S_t, T-t; X, r, \delta) = P^A(X, T-t; S_t, \delta, r). \qquad (7.18)$$

7.1.3 *Indefinitely Lived American Options*

Paradoxically, the problem of valuing American-style derivatives is often easier when there is no fixed termination date. When the underlying follows geometric Brownian motion with time-invariant volatility, and when the instantaneous spot rate and dividend rates are also constant through time, the value of the derivative itself becomes time-invariant. The fundamental p.d.e. that describes its dynamics then becomes an ordinary differential equation. If the boundary conditions for the particular derivative are not too difficult, there may be an elementary solution. We demonstrate the techniques through several examples, which are of interest in their own right. In each case the short rate and dividend rate are constants, r and δ, and the dynamics of the underlying under the risk-neutral measure are

$$dS_t = (r - \delta)S_t \cdot dt + \sigma S_t \cdot d\hat{W}_t.$$

1. Let $P^A(S_t, \infty)$ be the value of an indefinitely lived American put struck at X. Since this does not depend on time independently of S_t, the Black-Scholes p.d.e. becomes

$$P^A_S(r - \delta)S_t + P^A_{SS}\sigma^2 S_t^2/2 - P^A r = 0, \tag{7.19}$$

which is a second-order ordinary differential equation with variable coefficients. The American put will be exercised when the stock's price falls to a critical level $\mathfrak{B}$ that is now time-invariant. This imposes a lower boundary condition, $P^A(\mathfrak{B}, \infty) = X - \mathfrak{B}$, and there is the usual upper bound, $\lim_{S_t \to \infty} P^A(S_t, \infty) = 0$. A solution to (7.19) will be a function of the form $P^A(S_t, \infty) = cS_t^{-u}$ for appropriate constants c and u. Substituting and simplifying shows that there are two roots,

$$u, u' = \frac{r - \delta - \sigma^2/2 \pm \sqrt{(r - \delta - \sigma^2/2)^2 + 2r\sigma^2}}{\sigma^2},$$

with $u' < 0$ and u positive or zero according as r is positive or zero. The general solution is therefore

$$P^A(S_t, \infty) = cS_t^{-u} + c'S_t^{-u'},$$

but since the upper boundary condition is inconsistent with the negative root, we conclude that $c' = 0$ and $P^A(S_t, \infty) = cS_t^{-u}$. Applying the lower boundary condition gives $c = (X - \mathfrak{B})\mathfrak{B}^u$, whence

$$P^A(S_t, \infty) = (X - \mathfrak{B})(S_t/\mathfrak{B})^{-u}.$$

The maximum value of P^A is attained at

$$\mathfrak{B} = \frac{Xu}{1+u}.$$

Notice that this is consistent with high-contact condition $P^A_S(\mathfrak{B}, \infty) = -1$. Assuming that the put is live to begin with (i.e., that $S_t \geq \mathfrak{B}$) the solution for the put's current value is

$$P^A(S_t, \infty) = \frac{X}{1+u}(S_t/\mathfrak{B})^{-u}. \tag{7.20}$$

Recall that this is the same result obtained in the limit as $T \to \infty$ from MacMillan 's (1986) approximation for the finite-lived put. Notice, too, that $X - S_t \leq P^A(S_t, \infty) \leq X$, consistent with the bounds derived in section 4.3.3.

2. Let $C^A(S_t, \infty)$ be the value of a perpetual American call with strike X. Lower and upper boundary conditions are $C^A(0, \infty) = 0$ and

$$C^A(\mathfrak{B}, \infty) = \mathfrak{B} - X,$$

where $\mathfrak{B}$ is the critical exercise price. This satisfies an ordinary differential equation of the same form as (7.19),

$$C_S^A(r-\delta)S_t + C_{SS}^A \sigma^2 S_t^2/2 - C^A r = 0, \tag{7.21}$$

of which the general solution is

$$C^A(S_t, \infty) = cS_t^u + c' S_t^{u'},$$

with

$$u, u' = \frac{-r + \delta + \sigma^2/2 \pm \sqrt{(r - \delta - \sigma^2/2)^2 + 2r\sigma^2}}{\sigma^2}.$$

The lower and upper boundary conditions require $c' = 0$ and $c = (\mathfrak{B} - X)\mathfrak{B}^{-u}$, respectively, so that, assuming $S_t \le \mathfrak{B}$,

$$C^A(S_t, \infty) = (\mathfrak{B} - X)(S_t/\mathfrak{B})^u. \tag{7.22}$$

Algebra shows that $u > 1$ when $\delta > 0$ and that $u = 1$ when $\delta = 0$. When $\delta > 0$ and $X > 0$, $C^A(S_t, \infty)$ is maximized by taking

$$\mathfrak{B} = \frac{u}{u-1}X,$$

which is the same value inferred from high-contact condition $C_S^A(\mathfrak{B}, \infty) = 1$. When $\delta = 0$ there is no unique maximum; instead, $C^A(S_t, \infty) \uparrow S_t$ as $\mathfrak{B} \to \infty$. This tells us that a perpetual American call on a no-dividend stock will never be exercised and that its value equals that of the stock regardless of the strike price.

A little algebra shows that switching X for S_t and δ for r in (7.22) gives (7.20), so that

$$C^A(X, \infty; S_t, \delta, r) = P^A(S_t, \infty; X, r, \delta),$$

corresponding to the symmetry relation for finite-lived American options in (7.18).

3. Let $C(S_t, \infty; K)$ be an indefinitely lived "capped" call option with strike X and cap $K > X$. The capped call pays $K - X$ and terminates as soon as S hits the cap, but it cannot be exercised voluntarily. The problem is the same as for the infinite American call except that the upper boundary condition is $C(K, \infty; K) = K - X$. Thus, assuming $S_t < K$ initially,

$$C(S_t, \infty; K) = (K - X)(S_t/K)^u,$$

where u is the same as in the previous example.

4. Let $C^A(S_t, \infty; L)$ be an indefinitely lived American "down-and-out" call with strike X and knockout price $L \in (0, X)$. This option can be exercised at any time, but it terminates and becomes worthless when S_t first hits L. The boundary conditions are now $C^A(L, \infty; L) = 0$ and $C^A(\mathfrak{B}, \infty; L) = \mathfrak{B} - X$ for some critical exercise price $\mathfrak{B}$. Again, the general solution to (7.21) is

$$C^A(S_t, \infty; L) = cS_t^u + c'S_t^{u'},$$

where u and u' are as in the second example. Here the boundary conditions do not rule out either term, but they determine c and c' as solutions to

$$\begin{aligned} cL^u + c'L^{u'} &= 0 \\ c\mathfrak{B}^u + c\mathfrak{B}^{u'} &= \mathfrak{B} - X. \end{aligned}$$

When $S_t > L > 0$ the particular solution is

$$C^A(S_t, \infty; L) = \left(\frac{\mathfrak{B} - X}{\mathfrak{B}^u L^{u'} - \mathfrak{B}^{u'} L^u} \right) (L^{u'} S_t^u - L^u S_t^{u'}).$$

To complete the problem, the first factor is maximized numerically with respect to the optimal exercise price, $\mathfrak{B}$.

7.2 Compound and Extendable Options

This section treats European-style derivatives with special features that make their valuation more difficult: (i) options written on underlying financial assets that are either options themselves or portfolios of options and real assets; and (ii) options whose lives are extended one or more times if not in the money at an expiry date. In the simple framework of Black-Scholes dynamics it is often possible to develop tractable computational formulas even in such nonstandard cases.

7.2.1 *Options on Options*

Black and Scholes (1973) pointed out that a common stock itself has option-like characteristics. Take for example a stylized, one-period firm that has debt (i.e., liabilities) in the form of zero-coupon bonds paying face value L at T. Let A_T represent the value of the firm's assets at T. The aggregate value of the common is then $S_T = (A_T - L)^+$. That is, having limited liability, shareholders simply turn assets over to the bondholders if the firm is insolvent; otherwise, they pay off the debt and claim the residual. This makes the firm's stock a call option on the assets, and it makes an option on the stock an option on an option, or a "compound option". We develop an analytical formula for the value of a compound call like that first presented by Geske (1979) and then give a corresponding expression for a put on a call.

A related pricing problem has attracted interest because of the technology-driven boom in U.S. stocks during the 1990s. During that period Microsoft, Intel, and several other successful firms began to take large positions in options on their own stocks. The primary strategy involved selling puts, the goal being to enhance earnings by collecting premiums on options that would expire worthless if the stocks' prices continued to rise. Of course, the firms were obligated to buy their own stock at above-market prices if and when the puts were exercised. Just as happens when a firm incurs debt, issuing these "inside" puts creates opportunities and liabilities that change the dynamics of the price of its stock. Similar issues arise when firms issue warrants (long-term calls on their own stock), when they compensate executives with inside call options, and when they use derivatives either to hedge exposure to various business risks or to take affirmative positions. In the second part of this section we will see how the writing of inside European puts affects the price of the firm's stock and the prices of other derivatives.

Compound Options

Following Geske (1979) we consider first a European-style call with strike X and expiration date t^* on the common stock of a firm having debt L that comes due at some $T \geq t^*$. We view this as a compound call, whose value, $C^C(\cdot, \cdot)$, is to be determined. Assume now that the process representing the value of the firm's assets, $\{A_t\}$, follows geometric Brownian motion and that the firm pays no dividend on the stock. Both L and A are expressed per share of common. To simplify notation, take the current time to be $t = 0$. Then under the risk-neutral measure the distribution of A_{t^*} is lognormal with c.d.f.

$$\hat{F}(a_{t^*}) = \Phi[q^+(a_{t^*}e^{-rt^*}/A_0; t^*)], \tag{7.23}$$

where for arbitrary positive x and τ

$$q^{\pm}(x; \tau) \equiv \frac{\ln x \pm \sigma^2 \tau/2}{\sigma\sqrt{\tau}}.$$

Regarding each share of stock as a European call with strike L on the firm's assets, its value at t^* is given by Black-Scholes as

$$\begin{aligned} S_{t^*} &= C^E(A_{t^*}, T - t^*) \\ &= A_{t^*}\Phi[q^+(A_{t^*}e^{r(T-t^*)}/L; T - t^*)] \\ &\quad - e^{-r(T-t^*)}L\Phi[q^-(A_{t^*}e^{r(T-t^*)}/L; T - t^*)]. \end{aligned}$$

Let a_X be the value of assets that puts the *compound* call just at the money at t^*. That is, a_X satisfies $C^E(a_X, T - t^*) = X$. Assuming that $X > 0$, a solution necessarily exists and is unique because C^E increases monotonically from zero with the price of the underlying. Martingale pricing then sets the value of the compound call at

$$C^C(A_0, t^*) = e^{-rt^*}\int_{a_X}^{\infty}[C^E(a_{t^*}, T - t^*) - X] \cdot d\hat{F}(a_{t^*}). \tag{7.24}$$

Notice that if $t^* = T$, then $C^E(a_X, T - t^*) = (a_X - L)^+$ so that $a_X = L + X$. In this case (7.24) becomes

$$e^{-rT}\int_{L+X}^{\infty}[a_T - (L + X)] \cdot d\hat{F}(a_T) = e^{-rT}\hat{E}[A_T - (L + X)]^+,$$

which is given by the Black-Scholes formula with $L + X$ as the strike price. We concentrate on the case $t^* < T$ to avoid this triviality.

Breaking (7.24) into two terms and expressing $e^{-rt^*}X[1 - \hat{F}(a_X)]$ using (7.23) give

$$e^{-rt^*}\int_{a_X}^{\infty} C^E(a_{t^*}, T - t^*) \cdot d\hat{F}(a_{t^*}) - e^{-rt^*}X\Phi[q^-(A_0e^{rt^*}/a_X; t^*)]. \tag{7.25}$$

For the first term, express $C^E(a_{t^*}, T - t^*)$ via the Black-Scholes formula to get

$$\begin{aligned} &e^{-rt^*}\int_{a_X}^{\infty} a_{t^*}\Phi[q^+(a_{t^*}e^{-r(T-t^*)}/L; T - t^*)] \cdot d\Phi[q^+(a_{t^*}e^{-rt^*}/A_0; t^*)] \\ &- e^{-rT}L\int_{a_X}^{\infty}\Phi[q^-(a_{t^*}e^{-r(T-t^*)}/L; T - t^*)] \cdot d\Phi[q^+(a_{t^*}e^{-rt^*}/A_0; t^*)]. \end{aligned}$$

Changing variables in the first integral as

$$a_{t^*} \to A_0 \exp\left(rt^* + \sigma^2 t^*/2 - z\sigma\sqrt{t^*}\right)$$

and in the second as

$$a_{t^*} \to A_0 \exp\left(rt^* - \sigma^2 t^*/2 - z\sigma\sqrt{t^*}\right)$$

reduces the above to

$$A_0 \int_{-\infty}^{q^+\left(\frac{A_0 e^{rt^*}}{a_X};t^*\right)} \Phi\left[\frac{\ln\left(\frac{A_0 e^{rT}}{L}\right) + \frac{\sigma^2 T}{2}}{\sigma\sqrt{T-t^*}} - z\sqrt{\frac{t^*}{T-t^*}}\right] \cdot d\Phi(z)$$

$$- \frac{L}{e^{rT}} \int_{-\infty}^{q^-\left(\frac{A_0 e^{rt^*}}{a_X};t^*\right)} \Phi\left[\frac{\ln\left(\frac{A_0 e^{rT}}{L}\right) - \frac{\sigma^2 T}{2}}{\sigma\sqrt{T-t^*}} - z\sqrt{\frac{t^*}{T-t^*}}\right] \cdot d\Phi(z) \qquad (7.26)$$

This form is computationally easy using a routine like GAUSSINT that finds expectations of functions of standard normals, but there is a better representation in terms of a bivariate normal c.d.f. For this we need the result from section 2.2.7 that for arbitrary constants α, β, γ

$$\int_{-\infty}^{\gamma} \Phi(\alpha - \beta y) \cdot d\Phi(y) = \Phi\left(\gamma, \alpha/\sqrt{1+\beta^2}; \beta/\sqrt{1+\beta^2}\right), \qquad (7.27)$$

where $\Phi(x, y; \rho)$ represents the bivariate normal c.d.f. Using (7.27) to express the integrals in the expression for the first term of (7.25) gives for the value of the compound call

$$\begin{aligned} C^C(A_0, t^*) = A_0 \Phi &\left[q^+(A_0 e^{rt^*}/a_X; t^*), q^+(A_0 e^{rT}/L; T); \sqrt{t^*/T}\right] \\ &- e^{-rT} L\Phi\left[q^-(A_0 e^{rt^*}/a_X; t^*), q^-(A_0 e^{rT}/L; T); \sqrt{t^*/T}\right] \\ &- e^{-rt^*} X\Phi[q^-(A_0 e^{rt^*}/a_X; t^*)]. \end{aligned}$$

The arbitrage-free price of a put option on an underlying call comes from put-call parity. One who is long the compound call at t^* and short the corresponding compound put receives for sure the difference between the vanilla European call on the assets and the strike; therefore,

$$e^{-rt^*} \hat{E}[C^C(A_{t^*}, 0) - P^C(A_{t^*}, 0)] = e^{-rt^*} \hat{E}[C^E(A_{t^*}, T - t^*) - X].$$

Since normalized values of all these instruments are martingales under $\hat{\mathbb{P}}$, it follows that

$$C^C(A_0, t^*) - P^C(A_0, t^*) = C^E(A_0, T) - e^{-rt^*} X,$$

so that an exact parity relation holds for these compound options, just as for European options. Written out, the solution for $P^C(A_0, t^*)$ is

$$\begin{aligned} e^{-rT}L\Bigg\{\Phi\left[q^-\left(\frac{A_0e^{rT}}{L};T\right)\right] \\ -\Phi\left[q^-\left(\frac{A_0e^{rt^*}}{a_X};t^*\right), q^-\left(\frac{A_0e^{rT}}{L};T\right);\sqrt{\frac{t^*}{T}}\right]\Bigg\} \\ -A_0\Bigg\{\Phi\left[q^+\left(\frac{A_0e^{rT}}{L};T\right)\right] \\ -\Phi\left[q^+\left(\frac{A_0e^{rt^*}}{a_X};t^*\right), q^+\left(\frac{A_0e^{rT}}{L};T\right);\sqrt{\frac{t^*}{T}}\right]\Bigg\} \\ +e^{-rt^*}X\Phi\left[q^+\left(\frac{a_Xe^{-rt^*}}{A_0};t^*\right)\right]. \end{aligned}$$

Good approximations are available that make it easy to evaluate the bivariate normal c.d.f. that appears in these formulas. Mee and Owen (1983) show that the following delivers mean absolute error over $\Re_2$ of less than 1.0% when $|\rho|$ is less than about 0.9:

$$\Phi(x, y; \rho) \doteq \Phi(x)\Phi\left(\frac{y-\alpha}{\beta}\right), \tag{7.28}$$

where

$$\begin{aligned} \alpha &= -\rho\phi(x)/\Phi(x), \\ \beta &= 1 + \rho x\alpha - \alpha^2. \end{aligned}$$

While this should be adequate for most financial applications, the error may be unacceptable when $|\rho|$ is close to unity, as it would be in the present case when $t^* \doteq T$. For such cases Moskowitz and Tsai (1989) have developed a procedure based on the decomposition of the bivariate normal as the product of conditional and marginal distributions, using second- or third-degree polynomials to approximate the conditional c.d.f. Both quadratic and cubic versions deliver lower mean absolute error than (7.28) for $|\rho| > 0.6$ and have particularly high accuracy for $|\rho|$ near unity.

"Inside" Options

It is clear that writing puts on a firm's own stock is a strategy that enhances volatility. When the firm does well and the price of its stock rises, extra revenue will have been gained from the sale of out-of-money puts that ultimately expire worthless. On the other hand, a firm with declining revenues and stock prices might have to buy back its own shares at a premium over market. The practice of writing inside puts (also called put "warrants") also gives managers some perverse incentives, such as delaying the release of negative information or altering dividend policy. Abstracting from such potential manipulative effects, let us see how the writing of European puts affects the dynamics of the firm's equity and the prices of derivatives.

To simplify, consider a firm having no debt and no initial position in options, whose equity consists entirely of common stock, on which it pays no dividends. Under these conditions the price of a share of stock equals the market value of assets per share; that is, $S_t = A_t$. At time t^* the firm sells a number of European puts amounting to $p < 1$ options per share of common. (Selling more would be infeasible, since exercise would require the firm to repurchase all its outstanding shares.) The new inside puts are struck at X, expire at T, and have initial value $P^I(A_{t^*}, T - t^*; X)$. The transaction has no immediate effect on shareholders' equity, since the acquired short position in puts is exactly compensated by the premiums received. From then on, at any $t \in [t^*, T]$, the value of a share of stock equals the value of assets per share plus the proceeds from selling the puts minus their current value:

$$\begin{aligned} S_t &= A_t + p[P^I(S_{t^*}, T - t^*; X) - P^I(S_t, T - t; X)] \\ &= A_t^+ - pP^I(S_t, T - t; X), \end{aligned} \tag{7.29}$$

where $A_t^+ \equiv A_t + pP^I(S_{t^*}, T - t^*; X)$ is the augmented value of assets per share.

Table 7.1 shows the values of the put and the stock at T for various values of A_T^+. To explain the entries, if $A_T^+ < pX$ the firm will expend all its assets toward meeting its obligation to pay the strike price, leaving nothing for shareholders. When $A_T^+ \in [pX, X]$ the assets will fully satisfy

Table 7.1. Values of inside put and stock *vs.* values of augmented assets.

Range of A_T^+	$P^I(S_T, 0; X)$	S_T
$[0, pX)$	$\frac{A_T^+}{p}$	0
$[pX, X]$	$\frac{X - A_T^+}{1-p}$	$\frac{A_T^+ - pX}{1-p}$
(X, ∞)	0	A_T^+

the obligation, in which case

$$\begin{aligned} P^I(S_T, 0; X) &= X - [A_T^+ - pP^I(S_T, 0; X)] \\ &= \frac{X - A_T^+}{1-p} \end{aligned}$$

and

$$\begin{aligned} S_T &= A_T^+ - pP^I(S_T, 0; X) \\ &= \frac{A_T^+ - pX}{1-p}. \end{aligned}$$

Notice that exercise of the inside puts has opposing effects on the price of the stock: the expenditure of assets reduces the aggregate value of stock by pX times the original number of shares outstanding, but the retiring of shares moderates the decline in price per share. Finally, when $A_T^+ > X$ the puts expire worthless, giving shareholders sole claim to the augmented assets. Since the sign of $A_T^+ - X$ is the same as that of $S_T - X$, the condition for exercise can be stated equivalently in terms of A_T^+ or S_T.

Assume now that the proceeds from selling the puts are invested in the core business and that this additional investment does not alter the stochastic process for the value of assets, which we take to be geometric Brownian motion; that is,

$$dA_t^+/A_t^+ = dA_t/A_t = \mu_t \cdot dt + \sigma \cdot dW_t.$$

The fact that the firm's augmented assets can be still replicated by a portfolio of the common stock and riskless bonds justifies changing to the equivalent martingale measure, replacing μ_t with r_t and W_t with $\hat{W}_t$. The conditional c.d.f. of A_T^+ under $\hat{\mathbb{P}}$ is therefore

$$\hat{F}_t(a) = \Phi[q^+(Ba/A_t^+)] \equiv \Phi\left[\frac{\ln(Ba/A_t^+) + \sigma^2(T-t)/2}{\sigma\sqrt{T-t}}\right],$$

where $B \equiv B(t,T)$.

The put's value at $t \in [t^*, T]$ is then

$$\begin{aligned} P^I(S_t, T-t; X) &= B \cdot \hat{E}_t P^I(S_T, 0; X) \\ &= \frac{B}{p}\int_0^{pX} a \cdot d\hat{F}_t(a) + \frac{B}{1-p}\int_{pX}^{X} (X-a) \cdot d\hat{F}_t(a). \end{aligned}$$

A bit of algebra shows that this can be written in terms of ordinary outside European puts on the augmented assets:

$$P^I(S_t, T-t; X) = \frac{1}{1-p}\left[P^E(A_t^+, T-t; X) - \frac{1}{p}P^E(A_t^+, T-t; pX)\right]. \tag{7.30}$$

Note that the convexity of the Black-Scholes put price with respect to X (recall that $P_{XX}^E > 0$) guarantees that $P^I(S_t, T-t; X) \geq 0$, with equality if and only if $X = 0$.[7]

The intuitive content of (7.30) is as follows. The put on the stock itself differs from a put on the assets in two respects. First, if the puts are exercised the value of the stock declines, since shareholders have residual claim after put holders. This feature adds value to the inside put on the stock relative to a put on the assets themselves, accounting for the factor $(1-p)^{-1}$ in (7.30). Second, the stockholders have a trick up their sleeves that reduces the value of the inside put. The trick is that their liability to the put holders is limited to the value of the firms' assets should that value fall below pX. In this sense, shareholders themselves have an option to discharge their obligation by putting the assets to holders of the inside put at strike price pX. Holders of the inside put are, in effect, short this secondary option of shareholders, since these two parties are the only claimants to assets. The value of this secondary option (per unit of inside puts) is precisely the second term inside the brackets.

Applying risk-neutral pricing to the stock itself,

$$\begin{aligned}
S_t &= B\hat{E}_t[A_T^+ - pP^E(S_T, 0; X)] \\
&= B\int_{pX}^{X} \frac{a - pX}{1-p} d\hat{F}_t(a) + B\int_X^\infty a \cdot d\hat{F}_t(a) \\
&= B\int_0^\infty a \cdot d\hat{F}_t(a) \\
&\quad - \frac{B}{1-p}\left[p\int_0^X (X-a)\cdot d\hat{F}_t(a) - \int_0^{pX}(pX-a)\cdot d\hat{F}_t(a)\right] \\
&= A_t^+ - pP^E(S_t, T-t; X),
\end{aligned}$$

which agrees with (7.29).

[7]By Jensen's inequality (suppressing arguments of $P^E(\cdot, \cdot; \cdot)$ other than the strike), $P^E(pX) = P^E[(1-p)\cdot 0 + p \cdot X] \leq (1-p)P^E(0) + pP^E(X) = pP^E(X)$, so that $p^{-1}P^E(pX) \leq P^E(X)$.

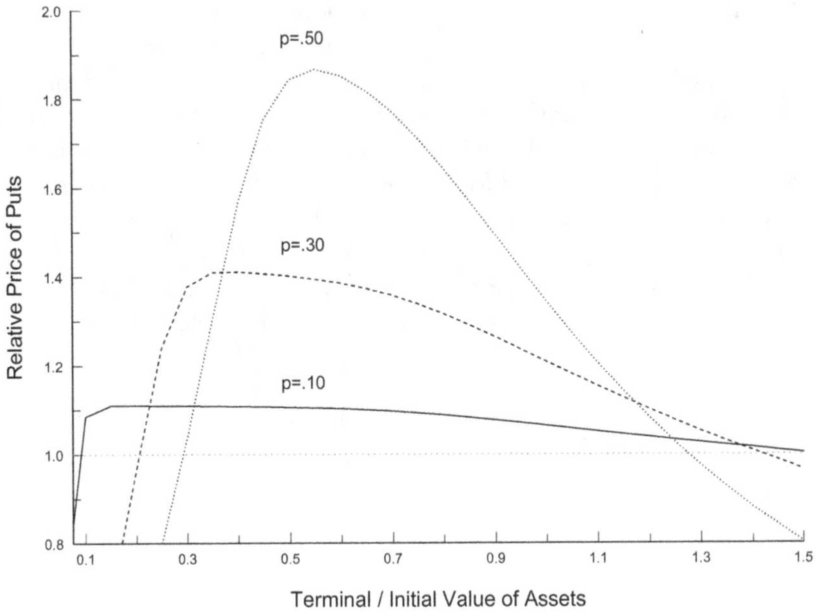

Fig. 7.3. Relative values, inside/outside puts, *vs.* terminal/initial value of assets.

Figure 7.3 illustrates the relation between values of inside and outside puts having the same strike price and time to expiration. It was constructed in the following way. At time t^*, having assets worth $A_{t^*} = 1.1$ currency units per share, the firm was assumed to have sold p puts per share of stock, each struck at $X = 1.0$ and having $T - t^* = 0.5$ years to maturity. The proceeds of these sales for each of $p = 0.1, 0.3$, and 0.5 were calculated from (7.30) (with $S_{t^*} = 1.1$, $T - t^* = 0.5$, $X = 1.0$, $r = 0.05$, $\sigma = 0.30$) to be 0.0043 , 0.0167, and 0.0390 per share, respectively. Adding each of these in turn to the initial value of assets to create $A_{t^*}^+$, time was advanced to $t = t^* + .25$. For values of A_t^+ ranging from $0.05A_{t^*}^+$ to $1.5A_{t^*}^+$ the values of the inside puts, $P^I(S_t, 0.25; 1.0)$, were calculated from (7.30). Likewise for values of A_t ranging from $0.25A_{t^*}$ to $1.5A_{t^*}$ values of outside European puts on the per-share value of assets, $P^E(A_t, 0.25; 1.0)$, were calculated from the Black-Scholes formula (again with $r = 0.05$ and $\sigma = 0.30$). The plot for each value of p shows how the ratio $P^I(S_t, 0.25; 1.0) \div P^E(A_t, 0.25; 1.0)$ varies with the relative change in asset values over the 0.25-year interval from t^* to t.

Figure 7.3 shows that an inside put is ordinarily worth more than one written by an outsider in the absence of such activity by the firm. This can be attributed to the increase in volatility of the stock when the firm takes such positions, as documented below. There are, however, two exceptional circumstances under which the inside puts are worth less. The first is when the firm's assets have fallen near enough to pX that put holders may not receive full intrinsic value at expiration. At the other extreme, when assets grow very rapidly the additional funds raised by selling the puts multiply to the point that the options have much less chance of expiring in the money.

To see how the sale of puts affects the stock's volatility, apply Itô's formula to S_t to obtain

$$dS_t/S_t = r \cdot dt + \sigma_S(A_t^+, t) \cdot dW_t,$$

where

$$\sigma_S(A_t^+, t)/\sigma = \frac{A_t^+ - \frac{p}{1-p} A_t^+ P_A^E(A_t^+, T-t; X) + \frac{1}{1-p} A_t^+ P_A^E(A_t^+, T-t; pX)}{A_t^+ - \frac{p}{1-p} P^E(A_t^+, T-t; X) + \frac{1}{1-p} P^E(A_t^+, T-t; pX)}$$

and subscripts on the P's indicate partial derivatives. Next, express the values of P^E in the denominator using the identity $P^E(A, \tau; X) = P_A^E A + P_X^E X$. Finally, noting that

$$\begin{aligned} \partial P^E(A_t^+, T-t; X)/\partial X &= \Phi[q^+(BX/A_t^+)] \\ &> \Phi[q^+(BpX/A_t^+)] \\ &= \partial P^E(A_t^+, T-t; pX)/\partial(pX), \end{aligned}$$

one sees that $\sigma_S(A_t^+, t) > \sigma$ for each $A_t^+ > 0$ and each $t \in (t^*, T)$. Therefore, the practice of writing inside puts does indeed enhance volatility. Bergman *et al.* (1996) derive inequalities for prices of European options on assets whose prices follow diffusion processes with volatilities depending on the underlying price and time. They show that an increase in the volatility function necessarily raises prices of all European options. Therefore, under the conditions maintained here we conclude that a firm's strategy of selling puts on its own stock increases the prices of outside derivatives.

The price, $D(S_t, T_O - t)$, of some such generic outside European-style derivative expiring at $T_O \leq T$ can be evaluated as

$$\begin{aligned} D(S_t, T_O - t) &= B(t, T_O)\hat{E}_t D(S_{T_O}, 0) \\ &= B(t, T_O)\hat{E}_t D[A_{T_O}^+ - pP^I(S_{T_O}, T-t; X), 0]. \end{aligned}$$

For example a European call struck at X_O would be valued as

$$B(t,T_O)\hat{E}_t\left[A^+_{T_O} - \frac{pP^E(A^+_{T_O},T-t;X)}{1-p} + \frac{P^E(A^+_{T_O},T-t;pX)}{1-p} - X_O\right]^+ .$$

Explicit expressions can be found for values of outside derivatives that expire at the same time as the inside puts. Following Qiu (2004), the value of an outside European-style put written on the stock of a firm with inside puts having the same strike is

$$P^E(S_t,T-t;X) = B(t,T)\left[\int_0^{pX} X\cdot d\hat{F}_t(a) + \int_{pX}^{X}\frac{X-a}{1-p}\cdot d\hat{F}_t(a)\right],$$

which reduces to

$$P^E(S_t,T-t;X) = (1-p)^{-1}[P^E(A^+_t,T-t;X) - P^E(A^+_t,T-t;pX)]. \quad (7.31)$$

Comparing with (7.30) shows that the value of the outside put exceeds that of the inside option by

$$P^E(S_t,T-t;X) - P^I(S_t,T-t;X) = p^{-1}P^E(A^+_t,T-t;pX).$$

Generalizing (7.31) to allow different strike prices, X_I and X_O,

$$P^E(S_t,T-t;X_O) = P^E(A^+_t,T-t;X_O) + \frac{p}{1-p}P^E(A^+_t,T-t;X_I\wedge X_O)$$
$$-\frac{1}{1-p}P^E(A^+_t,T-t;pX_I) + \frac{p}{1-p}(X_I-X_O)^+\Delta(A^+_t,T-t;X_O),$$

where $\Delta(A^+_t,T-t;X_O)$ is the value of a digital option that pays one currency unit at T if $A^+_T < X_O$.

7.2.2 *Options with Extra Lives*

The discussion of compound options involved a stylized firm with debt L in the form of discount bonds all maturing at time T. Stockholders in such a firm will exercise their collective European call option on the assets if $A_T - L > 0$, paying off the bondholders and retaining the residual; otherwise, they will default on the loan and transfer the assets to creditors in partial payment of what is owed. Let us consider how to develop a security that effectively insures the bondholders against such default. Such a derivative would pay $L - A_T$ per share if the firm were insolvent at T, otherwise nothing. Of course, this payment of $(L - A_T)^+$ is the same as that of a put option struck at L. Therefore, an insurer of the firm's creditors would, in effect, be writing a European put.

This is essentially the position of the Federal Deposit Insurance Corporation (FDIC) when it insures deposits at U.S. commercial banks. As the bank's creditors, depositors are endowed by the FDIC with a put option struck at the nominal value of their deposits. For this continuing protection of its depositors the bank pays a recurring fee. Typically the FDIC (or other authority) assesses a bank's condition twice per year, closing any bank found to be insolvent and paying off depositors, thus forcing the exercise of the put. Options for banks that are solvent are simply extended until the next examination. Valuing deposit insurance for some fixed term over which there are one or more intermediate examinations is like valuing a finitely extendable, European-style put option. We have already seen how to value finitely extendable puts within the Bernoulli framework. Our first task in this section, following Longstaff (1990), will be to develop a computational formula for the value of a once-extendable put under Black-Scholes dynamics. In principle, the methods could be applied to options with any finite number of extensions, although in practice the computations quickly get very hard. Moreover, thinking again of the application to deposit insurance, there is a related problem that these methods cannot handle at all. Taking the longer view of the FDIC's situation, it actually has an indefinite commitment to insure a member bank's deposits, in that it must continue to provide the insurance so long as the bank is found solvent at the periodic examinations. Having this commitment is like being short an infinitely extendable put option on the bank's assets—one that will be exercised if and only if it is in the money at any future periodic examination. The second part of the section describes a computational procedure for valuing such an indefinitely extendable option.

Pricing Finitely Extendable Puts

Consider a European-style put that pays $X - S_{t^*}$ if in the money at t^* and is otherwise extended once to expire unconditionally at $T > t^*$. Taking $t = 0$ as the current time, let $P^X(S_0, t^*)$ denote the current value of this once-extendable option. We will price this by finding the discounted expectation under $\hat{\mathbb{P}}$ of the option's value at t^*. The value at that time is easily determined. It is $X - S_{t^*}$ when S_{t^*} is less than X; otherwise, it is the value of an ordinary European put with remaining life $T - t^*$. That is,

$$P^X(S_{t^*}, 0) = (X - S_{t^*})^+ + P^E(S_{t^*}, T - t^*)\mathbf{1}_{[X,\infty)}(S_{t^*}).$$

Relying on the martingale property under $\hat{\mathbb{P}}$, we find that

$$P^X(S_0, t^*) = e^{-rt^*}\hat{E}(X - S_{t^*})^+ + e^{-rt^*}\hat{E}P^E(S_{t^*}, T - t^*)\mathbf{1}_{[X,\infty)}(S_{t^*}). \tag{7.32}$$

The first term is just the value of a t^*-expiring European put, $P^E(S_0, t^*)$. The second term, which is the extra value of the extendable feature, is

$$e^{-rt^*}\int_X^\infty P^E(s_{t^*}, T - t^*) \cdot d\hat{F}(s_{t^*}), \tag{7.33}$$

where

$$\hat{F}(s_{t^*}) = \Phi[q^+(e^{-rt^*}s_{t^*}/S_0; t^*)]$$

and

$$q^\pm(x;\tau) = \frac{\ln x \pm \sigma^2\tau/2}{\sigma\sqrt{\tau}}.$$

Evaluating (7.33) is the part that takes some work, although it proceeds much as for compound options. First, write out $P^E(s_{t^*}, T - t^*)$ from the Black-Scholes formula and express $\hat{F}$ in terms of $\Phi(q^+)$:

$$e^{-rT}X\int_X^\infty \Phi[q^+(Xe^{-r(T-t^*)}/s_{t^*}; T - t^*)] \cdot d\Phi[q^+(s_{t^*}e^{-rt^*}/S_0; t^*)]$$
$$-e^{-rt^*}\int_X^\infty s_{t^*}\Phi[q^-(Xe^{-r(T-t^*)}/s_{t^*}; T - t^*)] \cdot d\Phi[q^+(s_{t^*}e^{-rt^*}/S_0; t^*)].$$

Next, change variables in the first integral as

$$s_{t^*} \to S_0\exp(rt^* - \sigma^2t^*/2 - z\sigma\sqrt{t^*})$$

and in the second integral as

$$s_{t^*} \to S_0\exp(rt^* + \sigma^2t^*/2 - z\sigma\sqrt{t^*}).$$

This gives

$$e^{-rT}X\int_{-\infty}^{q^-(S_0e^{rt^*}/X;t^*)} \Phi\left(\frac{\ln(Xe^{-rT}/S_0) + \sigma^2T/2}{\sigma\sqrt{T - t^*}} + z\sqrt{\frac{t^*}{T - t^*}}\right) \cdot d\Phi(z)$$
$$-S_0\int_{-\infty}^{q^+(S_0e^{rt^*}/X;t^*)} \Phi\left(\frac{\ln(Xe^{-rT}/S_0) - \sigma^2T/2}{\sigma\sqrt{T - t^*}} + z\sqrt{\frac{t^*}{T - t^*}}\right) \cdot d\Phi(z).$$

As with compound options, these are reduced to expressions involving the bivariate c.d.f. using

$$\int_{-\infty}^{\gamma} \Phi(\alpha - \beta y) \cdot d\Phi(y) = \Phi\left(\gamma, \alpha/\sqrt{1 + \beta^2}; \beta/\sqrt{1 + \beta^2}\right).$$

Substituting for γ, α, β using symmetry relation $\Phi(-x, y; -\rho) = \Phi(x, y; \rho)$ and assembling all the pieces from (7.32) give as the arbitrage-free value of the once-extendable European put

$$\begin{aligned} P^X(S_0, t^*) &= e^{-rt^*} X\Phi[q^+(Xe^{-rt^*}/S_0; t^*)] - S_0\Phi[q^-(Xe^{-rt^*}/S_0; t^*)] \\ &\quad + e^{-rT} X\Phi\left[q^+(Xe^{-rt^*}/S_0; t^*), q^+(Xe^{-rT}/S_0; T); \sqrt{t^*/T}\right] \\ &\quad - S_0\Phi\left[q^-(Xe^{-rt^*}/S_0; t^*), q^-(Xe^{-rT}/S_0; T); \sqrt{t^*/T}\right]. \end{aligned}$$

Pricing the Indefinitely Extendable Put

Figure 7.4 illustrates the mechanics of an indefinitely extendable put option and gives the first glimmer of an idea of how to price it. Depicted there are four possible price paths for the underlying asset, starting off at $t = 0$ from an out-of-money initial value S_0. The height of each vertical bar is the strike price, X. The bars are spaced at intervals equal to the period between extensions, which we take as the unit of time. The bars thus appear as hurdles that the price path must clear at unit intervals if the option is not to be terminated. S_t can dip arbitrarily far below X between exercise dates so long as it does not touch the absorbing barrier at the origin, but the option remains alive at any t such that $S_j \geq X$ for each $j \in \{0, 1, \ldots, [t]\}$. (The brackets denote "greatest integer".) When $S_j < X$ at some j, however, then the option pays $X - S_j$ and dies.

An obvious way to price the put is by simulating price paths corresponding to the risk-neutral version of S_t, following each one either until it hits a hurdle and generates a receipt or to a time sufficiently remote that a hit could produce a payoff of only negligible discounted value. Averaging the discounted values of receipts generated by a large sample of such paths gives an estimate of the option's initial value as the discounted $\hat{\mathbb{P}}$-expected payoff. We discuss simulation techniques at length in chapter 11 and encounter other applications in chapters 8–10, where we take up more complicated dynamics. Simulation is a feasible technique for valuation in many applications, but, being slow, it is almost always a last resort. In this case contemplating those evenly spaced bars in figure 7.4 suggests a better way.

Represent the value of the put at an arbitrary $t \geq 0$ as $P^\infty(S_t, [t+1]-t)$, the second argument showing the time that remains until the next exercise date. For now we concentrate on pricing the put at some integral value of t—that is, at one of the exercise dates. The put's current value is then $P^\infty(S_t, 1)$, and at the next exercise date it will be worth $P^\infty(S_{t+1}, 1)$.

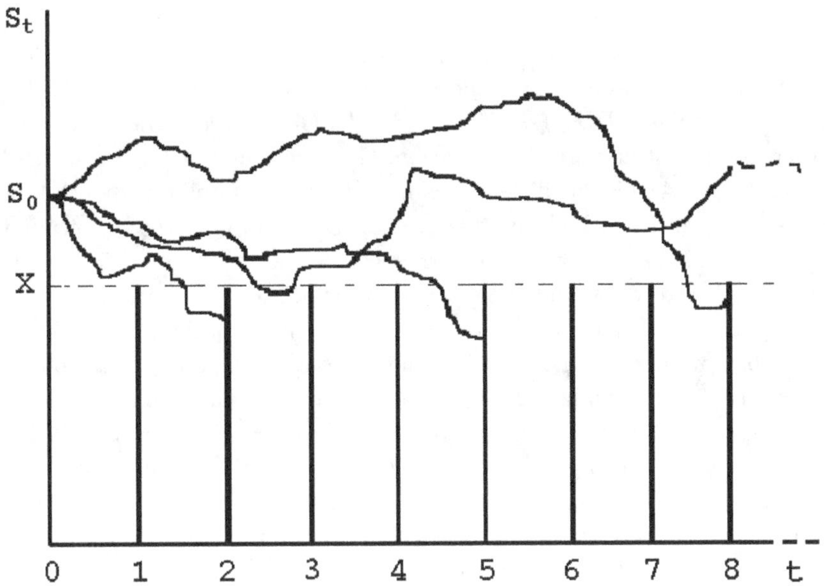

Fig. 7.4. Price paths *vs.* strike hurdles for indefinitely extendable puts.

Now here is the key point. Assuming constant volatility, interest rates, dividend rate, and strike price, the Markov property of geometric Brownian motion guarantees that the value at t will be the same *function* of the stock's price as at $t+1$. This observation and the martingale paradigm then lead to the following recursive expression for the put's value at any exercise date—that is, at any $t \in \aleph_0$:

$$P^{\infty}(S_t, 1) = (X - S_t)^+ + B\hat{E}P^{\infty}(S_{t+1}, 1) \cdot \mathbf{1}_{[X,\infty)}(S_t),$$

where $B \equiv B(t, t+1) = e^{-r}$ is the discount factor. In words, the put is worth its present intrinsic value if in the money; else, its value is the discounted expectation under $\hat{\mathbb{P}}$ of its value one period hence.

To save notation, we will value the put as of $t = 0$. If the underlying pays dividends continuously at rate δ, the value then can be expressed as

$$P^{\infty}(S_0, 1) = (X - S_0)^+ + BEP^{\infty}(S_0 e^{r-\delta-\sigma^2/2+\sigma Z}, 1) \cdot \mathbf{1}_{[X,\infty)}(S_0),$$

where Z is distributed as standard normal. Since the put's value necessarily lies between 0 and X, this is a contraction mapping from $[0, X]$ into itself. This guarantees that value function $P^{\infty}(\cdot, 1)$ at any exercise date can be discovered by recursive programming.

The method is as follows. First, construct a trial function $\hat{P}_0(\cdot)$ on a discrete domain of points $s \in \mathcal{S}$, arrayed as a vector. The trial value should be $X - s$ for any $s < X$, but values at other points can be chosen freely. For example, one can use Black-Scholes values of one-period vanilla European puts. Then, starting at the first out-of-money value in $\mathcal{S}$ (the first $s \geq X$), calculate $BE\hat{P}_0\ (se^{r-\delta-\sigma^2/2+\sigma Z})$ by Gaussian quadrature or some other method of numerical integration. Gaussian quadrature (Press *et al.*, 1992, chapter 4.5) is an efficient technique that works well for smooth functions.[8] Any such method involves evaluating Z on a discrete set of points $\mathcal{Z} \equiv \{z_1, \ldots, z_n\}$, say. Since the corresponding values of $se^{r-\delta-\sigma^2/2+\sigma z}$ for $z \in \mathcal{Z}$ will rarely be among the set $\mathcal{S}$ on which $\hat{P}_0$ was initially defined, values of $\hat{P}_0$ at those points must be found by interpolation. Having found the expectation at s, one then has the value at that point of an updated function, $\hat{P}_1$. Continuing, $\hat{P}_1$ is evaluated at each of the remaining out-of-money points of $\mathcal{S}$. Then, as for $\hat{P}_0$, one sets $\hat{P}_1(s) = X - s$ for $s < X$. This process is repeated to produce $\hat{P}_2$, $\hat{P}_3$, and so on until at some stage m the maximum difference between $\hat{P}_m$ and $\hat{P}_{m-1}$ reaches some desired tolerance. The boundedness of the put function assures eventual convergence.

Figure 7.5 illustrates the results of such calculations using $r = \delta = 0.0$, $X = 1.0$, extensions at intervals of 0.5 years, and five values of σ (expressed

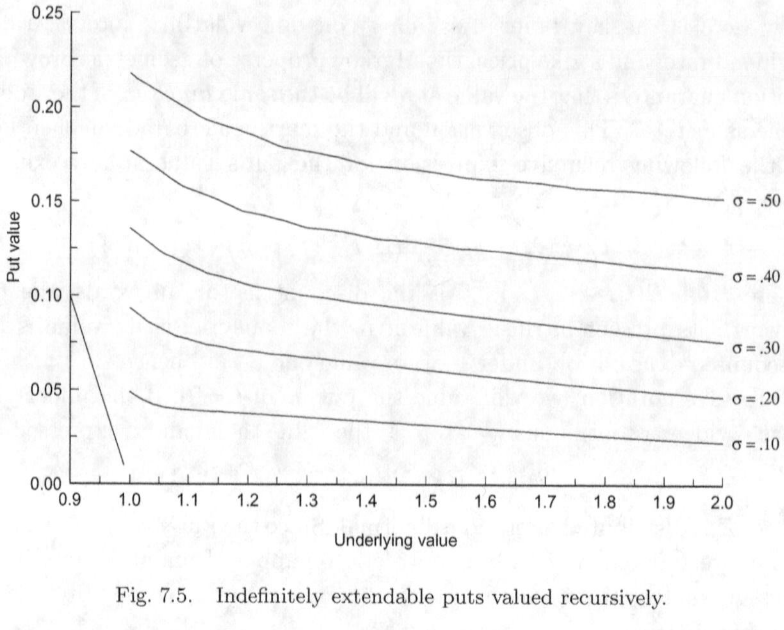

Fig. 7.5. Indefinitely extendable puts valued recursively.

[8]We will see shortly that some smoothing technique must be applied for this to work well here.

in annual time units). Each curve depicts the put's value as a function of the initial value of the underlying asset, S_0. Notice that these functions are *not* smooth. Indeed, they are discontinuous at $S_0 = X$, because $\lim_{s\uparrow X} P^{\infty}(s,1) = 0$ whereas $\lim_{s\downarrow X} P^{\infty}(s,1)$ is the positive value of a still-viable instrument that could wind up in the money at a future exercise date. Given that any method of numerical integration involves averaging the function over a discrete domain, it is important to use a continuization method like that applied in section 5.5.2 to value finitely extendable options in the binomial framework. The figure was produced using that method in conjunction with Gaussian quadrature.

Having found the value function at an extension date, we can value it at a nonintegral value of t as

$$P^{\infty}(S_t, [t+1]-t) = e^{-r([t+1]-t)}\hat{E}_t P^{\infty}(S_{[t+1]}, 1),$$

where the integration would again be done numerically after continuizing $P^{\infty}(S_{[t+1]}, 1)$.

7.3 Other Path-Independent Claims

This section looks at certain other "exotic" derivative products whose payoffs depend on prices of one or more underlying assets at a finite number of points in time. These are distinguished from path-dependent derivatives (treated in the next section), whose payoffs cannot be determined without knowing the entire price path of the underlying. We continue to assume that the underlying follows geometric Brownian motion with constant volatility σ and pays dividends (if at all) continuously at constant rate δ. The instantaneous spot rate of interest is also a constant, r. However, the results in this section are easily generalized to allow σ, δ, and r to be time-varying but deterministic. Prices for options on futures are obtained by setting $\delta = r$, since the futures price has zero mean drift under the risk-neutral measure. Since the goal is to exploit the relative simplicity of the Black-Scholes setting to derive explicit pricing formulas, we consider only European-style versions of these exotic options. It is straightforward to get approximate prices of American versions via the binomial method.

7.3.1 *Digital Options*

Standard digital calls and puts pay a fixed number of currency units conditional on the event that the price of the underlying is in some stated range on the expiration date. For example, payoffs of standard digital calls and puts struck at X have the form

$$C^D(S_T, 0; K) = K \cdot \mathbf{1}_{[X,\infty)}(S_T)$$
$$P^D(S_T, 0; K) = K \cdot \mathbf{1}_{[0,X]}(S_T),$$

where K is the number of currency units contingently paid. A variety of complex terminal payoffs can be constructed by combining digital options that have different strike prices and contingent payoffs.

Martingale pricing of European digitals is very simple in the Black-Scholes framework. We have

$$\begin{aligned} C^D(S_t, T-t; X, K) &= e^{-r(T-t)} K \cdot \hat{\mathbb{P}}_t(S_T \geq X) \\ &= B(t,T) K \Phi \left[\frac{\ln(\mathsf{f}/X) - \sigma^2(T-t)/2}{\sigma\sqrt{T-t}} \right] \\ &= B(t,T) K \Phi[q^-(\mathsf{f}/X)] \\ P^D(S_t, T-t; X, K) &= B(t,T) K \Phi[q^+(X/\mathsf{f})], \end{aligned}$$

where $\mathsf{f} = \mathsf{f}(t,T)$ is the time-t price for forward delivery of the underlying at T. The simple put-call parity relation for European digitals is

$$C^D(S_t, T-t; X, K) - P^D(S_t, T-t; X, K) = B(t,T)K.$$

7.3.2 *Threshold Options*

Threshold puts and calls struck at X deliver the usual payoffs $(S_T - X)^+$ and $(X - S_T)^+$, but only if they are sufficiently far in the money. For example, a standard arrangement is

$$C^T(S_T, 0; X, K) = (S_T - X) \cdot \mathbf{1}_{[K,\infty)}(S_T), \quad K > X$$
$$P^T(S_T, 0; X, K) = (X - S_T) \cdot \mathbf{1}_{[0,K)}(S_T), \quad K < X,$$

which requires the options to be in the money by at least $|K - X|$. The appeal of these is that they allow hedging of forward positions in the underlying at lower premiums than vanilla options struck at X, but give greater protection (at higher cost) than vanilla options struck at K. Pricing these threshold options is easy because they are replicable with a static portfolio of European vanilla options and digitals. Writing

$$\begin{aligned} C^T(S_T, 0; X, K) &= [(S_T - K) + (K - X)] \cdot \mathbf{1}_{[K,\infty)}(S_T) \\ &= (S_T - K)^+ + (K - X) \cdot \mathbf{1}_{[K,\infty)}(S_T), \quad K > X \\ P^T(S_T, 0; X, K) &= (K - S_T)^+ + (X - K) \cdot \mathbf{1}_{[0,K)}(S_T), \quad K < X \end{aligned}$$

shows that

$$\begin{aligned} C^T(S_t, T-t; X, K) &= C^E(S_t, T-t; K) + C^D(S_t, T-t; K, K-X) \\ P^T(S_t, T-t; X, K) &= P^E(S_t, T-t; K) + P^D(S_t, T-t; K, X-K), \end{aligned}$$

where P^E and C^E are European puts and calls and P^D and C^D are the digitals.

7.3.3 "As-You-Like-It" or "Chooser" Options

The holder of a "chooser" option at some time t gets to decide at a future date t^* whether the option will be a T-expiring put or a T-expiring call. When t^* arrives and the decision has been made, it becomes a vanilla option of the specified type; therefore, the interest is in pricing it at $t < t^*$ while the option to choose the type is still alive. Such an instrument might appeal to one who anticipates the occurrence of an upcoming event that could affect the price of the underlying either positively or negatively—for example, an earnings announcement that could be either above or below analysts' expectations. An alternative play on such information would be a T-expiring straddle, which is a simultaneous long position in a put and a call that have the same strike. We will see that the chooser is cheaper than the straddle by a margin that diminishes with $T - t^*$.

Pricing the European chooser is made simple by an application of put-call parity. Taking $t = 0$ as the present time, the current price, denoted[9] $H(S_0, t^*, T; X)$, is

$$\begin{aligned} &e^{-rt*}\hat{E}\max[C^E(S_{t^*}, T-t^*; X), P^E(S_{t^*}, T-t^*; X)] \\ &\quad = e^{-rt*}\hat{E}\{P^E(S_{t^*}, T-t^*; X) + [S_{t*}e^{-\delta(T-t^*)} - e^{-r(T-t^*)}X]^+\} \\ &\quad = P^E(S_0, T; X) + e^{-rt^*}\hat{E}[S_{t*}e^{-\delta(T-t^*)} - e^{-r(T-t^*)}X]^+. \end{aligned}$$

Here, the second equality follows upon using put-call parity to express C^E, and the last equality holds because the normalized value of $\{P^E\}$ is itself a martingale under $\hat{\mathbb{P}}$. Thus,

$$H(S_0, t^*, T; X) = P^E(S_0, T; X) + C^E(S_0 e^{-\delta(T-t^*)}, t^*; e^{-r(T-t^*)}X).$$

Equivalently, using put-call parity to express P^E instead of C^E gives

$$H(S_0, t^*, T; X) = C^E(S_0, T; X) + P^E(S_0 e^{-\delta(T-t^*)}, t^*; e^{-r(T-t^*)}X).$$

---

[9] "*Hermaphroditic*" seems an apt descriptor for this derivative!

The price difference between a T-expiring straddle and a comparable chooser—$\Delta_{SH}(t^*, T)$, say—is

$$\begin{aligned}\Delta_{SH}(t^*,T) &\equiv C^E(S_0,T;X) + P^E(S_0,T;X) - H(S_0,t^*,T;X)\\ &= C^E(S_0,T;X) - C^E(S_0e^{-\delta(T-t^*)},t^*;e^{-r(T-t^*)}X)\\ &= P^E(S_0,T;X) - P^E(S_0e^{-\delta(T-t^*)},t^*;e^{-r(T-t^*)}X).\end{aligned}$$

Expressing either

$$C^E(S_0e^{-\delta(T-t^*)},t^*;e^{-r(T-t^*)}X)$$

or

$$P^E(S_0e^{-\delta(T-t^*)},t^*;e^{-r(T-t^*)}X)$$

via the Black-Scholes formula shows that it is equal to the price of a T -expiring option of the same type with volatility $\sigma\sqrt{t^*/T}$. Since Black-Scholes prices increase monotonically in volatility, it follows that $\Delta_{SH}(t^*,T) > 0$ for $t^* < T$ and that $\Delta_{SH}(t^*,T) \downarrow 0$ as $t^* \to T$. This makes sense because owning the straddle lets one decide right at the last moment whether the instrument to be exercised shall be a put or a call, whereas with the chooser one must commit in advance.

7.3.4 *Forward-Start Options*

A European forward-start put or call begins its life at some future time t^* as an at-the-money vanilla put or call that expires at a still later date T. That is, at t^* it becomes a European option struck at S_{t^*} and having $T-t^*$ to go. Since from t^* forward it is an ordinary European option, the interest is in pricing it at $t < t^*$. Once more, this is easy to do. Again taking $t = 0$ as the current date, the price of a forward-start call is

$$\begin{aligned}C^F(S_0,t^*,T) &= e^{-rt^*}\hat{E}C^E(S_{t^*},T-t^*;S_{t^*})\\ &= e^{-rt^*}\hat{E}[S_{t^*}C^E(1,T-t^*;1)]\\ &= e^{-\delta t^*}S_0C^E(1,T-t^*;1)\\ &= C^E(e^{-\delta t^*}S_0,T-t^*;e^{-\delta t^*}S_0).\end{aligned}$$

The second and last equalities follow from the homogeneity of the Black-Scholes call formula in the price of the underlying and the strike, and the third equality follows from the martingale property of the normalized, *cum*-dividend value of S under $\hat{\mathbb{P}}$. Similarly, the arbitrage-free price of a forward-start put is

$$P^F(S_0,t^*,T) = P^E(e^{-\delta t^*}S_0,T-t^*;e^{-\delta t^*}S_0).$$

Notice that $e^{-\delta t^*}S_0$ is the current value of an *ex* -dividend stock position held to t^*. Thus, options that commence at the money at t^* and expire at T are equivalent to $(T - t^*)$-expiring vanilla options that are at the money at current *ex* -dividend values of the underlying.

7.3.5 *Options on the Max or Min*

Given two assets with price processes $\{S_{1t}\}_{t\geq 0}$ and $\{S_{2t}\}_{t\geq 0}$ there are several ways one can construct derivatives whose payoffs depend on the greater or lesser of these prices at a future date. Options on the max or min are one approach. For a given strike X and expiration date T we could have (i) a call on the maximum, with terminal payout $(S_{1T} \vee S_{2T} - X)^+$; (ii) a call on the minimum, with payout $(S_{1T} \wedge S_{2T} - X)^+$; (iii) a put on the maximum, paying $(X - S_{1T} \vee S_{2T})^+$; or (iv) a put on the minimum, worth $(X - S_{1T} \wedge S_{2T})^+$ at T. Stultz (1982) gave the first systematic account of options of this sort under Black-Scholes dynamics. Another scheme produces what are called "best-of" options. Consider two primary assets whose initial prices are S_{10} and S_{20}. An investment of one currency unit in the first will be worth S_{1T}/S_{10} at T, while a like investment in the second will be worth S_{2T}/S_{20}. For each unit of principal a "best-of-assets {1, 2}" instrument gives the holder the better of these returns or unity, whichever is greater. Thus, the payoff at T per unit of principal is

$$\begin{aligned}\max(S_{1T}/S_{10}, S_{2T}/S_{20}, 1) &= 1 + (S_{1T}/S_{10} - 1)^+ \vee (S_{2T}/S_{20} - 1)^+ \\ &= 1 + (S_{1T}/S_{10} \vee S_{2T}/S_{20} - 1)^+.\end{aligned}$$

Clearly, if we can value a call on the maximum at unit strike price and for unit initial values of the assets, then adding unity to it will give the value of a unit position in the best-of {1, 2}. We will work out the pricing formula for a call on the maximum, present corresponding results for calls on the minimum, and then show how what amounts to a put-call parity relation can be used to generate the put prices from these.

Pricing a Call on the Max

Martingale pricing of any of these instruments requires developing an expression for the conditional distribution of an extremum of the two prices. Working still within the Black-Scholes framework, we assume that prices evolve under $\hat{\mathbb{P}}$ as

$$dS_{1t} = (r - \delta_1)S_{1t} \cdot dt + \sigma_1 S_{1t} \cdot d\hat{W}_{1t}$$
$$dS_{2t} = (r - \delta_2)S_{2t} \cdot dt + \sigma_2 S_{2t} \cdot d\hat{W}_{2t},$$

where $\{\hat{W}_{1t}\}$ and $\{\hat{W}_{2t}\}$ are Brownian motions with $\hat{E}\hat{W}_{1t}\hat{W}_{2t} = \rho t$. The terminal value of a call on the maximum thus depends on the maximum of two correlated lognormal variates. The call's price at t can be expressed as

$$C^{\max}(S_{1t}, S_{2t}; \tau) = e^{-r\tau}\hat{E}_t[(S_{1T} \vee S_{2T} - X)\mathbf{1}_{(X,\infty)}(S_{1T} \vee S_{2T})], \quad (7.34)$$

where $\tau = T - t$. A nice computational expression for this can be found by working out the conditional density of $\ln(S_{1T} \vee S_{2T})$.

Letting $\hat{\mathbb{P}}_t(\cdot)$ denote $\hat{\mathbb{P}}(\cdot \mid \mathcal{F}_t)$ as usual, begin by observing that for $h > 0$ and $\eta = \ln h$

$$\begin{aligned}\hat{\mathbb{P}}_t(S_{1T} \vee S_{2T} \le h) &= \hat{\mathbb{P}}_t(\ln S_{1T} \le \eta, \ln S_{2T} \le \eta) \\ &= \hat{\mathbb{P}}[\mu_1 + \sigma_1 \hat{W}_{1\tau} \le \eta, \mu_2 + \sigma_2 \hat{W}_{2\tau} \le \eta] \\ &= \Phi\left(\frac{\eta - \mu_1}{\sigma_1\sqrt{\tau}}, \frac{\eta - \mu_2}{\sigma_2\sqrt{\tau}}; \rho\right), \end{aligned} \quad (7.35)$$

where

$$\mu_j \equiv \ln S_{jt} + (r - \delta_j - \sigma_j^2/2)\tau \quad (7.36)$$

for $j \in \{1, 2\}$ and $\Phi(\cdot, \cdot; \rho)$ is the c.d.f. of bivariate standard normals with correlation ρ. Likewise, we use $\phi(\cdot, \cdot; \rho)$ to represent the p.d.f. The right side of expression (7.34) can now be written as

$$e^{-r\tau} \int_{\ln X}^{\infty} e^{\eta} \hat{f}_t(\eta) \cdot d\eta - e^{-r\tau} X\{1 - \hat{\mathbb{P}}_t[\ln(S_{1T} \vee S_{2T}) \le \ln X]\}, \quad (7.37)$$

where $\hat{f}_t$ is the conditional p.d.f. of $\ln S_{1T} \vee S_{2T}$ under $\hat{\mathbb{P}}$. Since the probability in the second term comes directly from (7.35) evaluated at $\eta = \ln X$, only the first term requires attention. Letting $Z_j(\eta) \equiv (\eta - \mu_j)/(\sigma_j\sqrt{\tau})$ for $j = 1, 2$, we find the density as

$$\begin{aligned}\hat{f}_t(\eta) &= \frac{d}{d\eta}\Phi\left[Z_1(\eta), Z_2(\eta); \rho\right] \\ &= \frac{\partial \Phi}{\partial Z_1} Z_1'(\eta) + \frac{\partial \Phi}{\partial Z_2} Z_2'(\eta) \\ &= \frac{1}{\sigma_1\sqrt{\tau}} \int_{-\infty}^{Z_2(\eta)} \phi[Z_1(\eta), z_2; \rho] dz_2 + \frac{1}{\sigma_2\sqrt{\tau}} \int_{-\infty}^{Z_1(\eta)} \phi[Z_2(\eta), z_1; \rho] dz_1 \\ &\equiv \hat{f}_t^1(\eta) + \hat{f}_t^2(\eta).\end{aligned}$$

Notice that the two terms are identical, modulo an interchange of subscripts.

Decomposing the bivariate p.d.f. into the product of conditional and marginal, as

$$\phi(z_1, z_2; \rho) = \frac{1}{\sqrt{1-\rho^2}}\phi\left(\frac{z_2 - \rho z_1}{\sqrt{1-\rho^2}}\right)\phi(z_1),$$

gives

$$\hat{f}_t^1(\eta) = \frac{1}{\sigma_1\sqrt{\tau}}\Phi\left[\frac{Z_2(\eta) - \rho Z_1(\eta)}{\sqrt{1-\rho^2}}\right]\phi[Z_1(\eta)],$$

where $\phi(\cdot)$ and $\Phi(\cdot)$ are the univariate normal p.d.f. and c.d.f. The same kind of expression holds for $\hat{f}_t^2(\eta)$, just interchanging subscripts. The first term on the right side of (7.37) is then

$$\begin{aligned} e^{-r\tau}\int_{\ln X}^{\infty} e^{\eta}\hat{f}_t(\eta)\cdot d\eta &= e^{-r\tau}\int_{\ln X}^{\infty} e^{\eta}\hat{f}_t^1(\eta)\cdot d\eta + e^{-r\tau}\int_{\ln X}^{\infty} e^{\eta}\hat{f}_t^2(\eta)\cdot d\eta \\ &\equiv e^{-r\tau}(I_1 + I_2). \end{aligned} \tag{7.38}$$

Only the first integral needs to be developed, since the second will have the same form. Writing out $e^{\eta}\phi[Z_1(\eta)]$ and completing the square in the exponent give

$$I_1 = \frac{e^{\mu_1+\sigma_1^2\tau/2}}{\sqrt{2\pi\sigma_1^2\tau}}\int_{\ln X}^{\infty}\Phi\left[\frac{Z_2(\eta) - \rho Z_1(\eta)}{\sqrt{1-\rho^2}}\right]\exp\left\{-\frac{\left[\eta - (\mu_1 + \sigma_1^2\tau)\right]^2}{2\sigma_1^2\tau}\right\}\cdot d\eta.$$

Now change variables, as

$$y \leftarrow -\frac{\eta - (\mu_1 + \sigma_1^2\tau)}{\sigma_1\sqrt{\tau}}$$

and write out $Z_1(\eta)$, $Z_2(\eta)$, and the $\{\mu_j\}$ to get

$$I_1 = \mathsf{f}_{1t}\int_{-\infty}^{q^+(\mathsf{f}_{1t}/X;\sigma_1)}\Phi(a_{12} - b_{12}y)\cdot d\Phi(y), \tag{7.39}$$

where

$$\begin{aligned}
\mathsf{f}_{jt} &= S_{jt}e^{(r-\delta_j)\tau} \text{ (the forward price)} \\
q^{\pm}(x;\theta) &= \frac{\ln x \pm \theta^2\tau/2}{\theta\sqrt{\tau}} \\
a_{12} &= \frac{\ln(\mathsf{f}_{1t}/\mathsf{f}_{2t}) + \sigma^2\tau/2}{\sqrt{1-\rho^2}\sigma_2\sqrt{\tau}} && (7.40) \\
b_{12} &= \frac{\sigma_1 - \rho\sigma_2}{\sigma_2\sqrt{1-\rho^2}} && (7.41) \\
\sigma^2 &= \sigma_1^2 - 2\rho\sigma_1\sigma_2 + \sigma_2^2.
\end{aligned}$$

This can be tidied up by expressing (7.39) in terms of a bivariate normal c.d.f. For this, apply (2.55) to write

$$\int_{-\infty}^{\gamma} \Phi(a_{12} - b_{12}y) \cdot d\Phi(y) = \Phi\left(\gamma, a_{12}/\sqrt{1+b_{12}^2}; b_{12}/\sqrt{1+b_{12}^2}\right).$$

Since $a_{12}/\sqrt{1+b_{12}^2} = q^+(\mathsf{f}_{1t}/\mathsf{f}_{2t};\sigma)$ and $b_{12}/\sqrt{1+b_{12}^2} = (\sigma_1 - \rho\sigma_2)/\sigma$, this gives

$$I_1 = \mathsf{f}_{1t}\Phi[q^+(\mathsf{f}_{1t}/X;\sigma_1), q^+(\mathsf{f}_{1t}/\mathsf{f}_{2t};\sigma); (\sigma_1 - \rho\sigma_2)/\sigma].$$

An interchange of subscripts then gives I_2, the second integral in (7.38). Finally, assembling the pieces of (7.37) yields this final expression for the call on the max, $C^{\max}(S_{1t}, S_2; \tau)$:

$$\begin{aligned}
&e^{-r\tau}\sum_{j=1}^{2} \mathsf{f}_{jt}\Phi[q^+(\mathsf{f}_{jt}/X;\sigma_j), q^+(\mathsf{f}_{jt}/\mathsf{f}_{3-j,t};\sigma); \sigma^{-1}(\sigma_j - \rho\sigma_{3-j})] \quad (7.42) \\
&\quad - Xe^{-r\tau}\{1 - \Phi[q^+(X/\mathsf{f}_{1t};\sigma_1), q^+(X/\mathsf{f}_{2t};\sigma_2); \rho]\},
\end{aligned}$$

where

$$\begin{aligned}
q_j^{\pm}(x;\theta) &= \frac{\ln x \pm \theta^2\tau/2}{\theta\sqrt{\tau}} \\
\sigma^2 &= \sigma_1^2 - 2\rho\sigma_1\sigma_2 + \sigma_2^2 \\
\mathsf{f}_{jt} &= S_{jt}e^{(r-\delta_j)\tau} \\
\tau &= T - t.
\end{aligned}$$

(The $3-j$ subscript in the first line just handles the interchange: $3-1=2$ and $3-2=1$.)

Other Options on the Max or Min

Options on the minimum of two prices can be valued in the same way using

$$\begin{aligned}\hat{\mathbb{P}}(S_{1T} \wedge S_{2T} \leq \ell) &= 1 - \hat{\mathbb{P}}(\ln S_{1T} > \ln \ell, \ln S_{2T} > \ln \ell) \\ &= 1 - \Phi\left(-\frac{\ln \ell - \mu_1}{\sigma_1\sqrt{\tau}}, -\frac{\ln \ell - \mu_2}{\sigma_2\sqrt{\tau}}; \rho\right).\end{aligned}$$

The following expression holds for the call on the min, $C^{\min}(S_{1t}, S_{2t}; \tau)$ [*cf.* Stultz (1982)]:

$$\begin{aligned}&e^{-r\tau}\sum_{j=1}^{2} \mathsf{f}_{jt}\Phi[q^+(\mathsf{f}_{jt}/X;\sigma_j), q^-(\mathsf{f}_{3-j,t}/\mathsf{f}_{jt};\sigma);\sigma^{-1}(\rho\sigma_{3-j}-\sigma_j)] \\ &\quad - Xe^{-r\tau}\Phi[q^-(\mathsf{f}_{1t}/X;\sigma_1), q^-(\mathsf{f}_{2t}/X;\sigma_2);\rho].\end{aligned}$$

To get prices of the puts, we infer from the identities

$$\begin{aligned}C^{\max}(S_{1T}, S_{2T}; 0) - P^{\max}(S_{1T}, S_{2T}; 0) &= S_{1T} \vee S_{2T} - X \\ C^{\min}(S_{1T}, S_{2T}; 0) - P^{\min}(S_{1T}, S_{2T}; 0) &= S_{1T} \wedge S_{2T} - X\end{aligned}$$

that

$$\begin{aligned}P^{\max}(S_{1t}, S_{2t}; \tau) &= C^{\max}(S_{1t}, S_{2t}; \tau) - e^{-r\tau}[\hat{E}_t(S_{1T} \vee S_{2T}) - X] \\ P^{\min}(S_{1t}, S_{2t}; \tau) &= C^{\min}(S_{1t}, S_{2t}; \tau) - e^{-r\tau}[\hat{E}_t(S_{1T} \wedge S_{2T}) - X].\end{aligned}$$

An expression for $\hat{E}_t(S_{1T} \wedge S_{2T})$ is developed by analogy with (7.38) and (7.39) as

$$\begin{aligned}e^{-r\tau}\int_{-\infty}^{\infty} e^{\eta}\hat{f}_t(\eta)\cdot d\eta &= e^{-r\tau}\int_{-\infty}^{\infty} e^{\eta}\hat{f}_t^1(\eta)\cdot d\eta + e^{-r\tau}\int_{-\infty}^{\infty} e^{\eta}\hat{f}_t^2(\eta)\cdot d\eta \\ &= \sum_{j=1}^{2}\mathsf{f}_{jt}\int_{-\infty}^{\infty}\Phi(a_{j,3-j} - b_{j,3-j}y)d\Phi(y).\end{aligned}$$

The integrals are evaluated using

$$\int_{-\infty}^{\infty}\Phi(a - by)\phi(y)\cdot dy = \Phi\left(a/\sqrt{1+b^2}\right)$$

to give

$$\hat{E}_t(S_{1T} \wedge S_{2T}) = \sum_{j=1}^{2}\mathsf{f}_{jt}\Phi[q^+(\mathsf{f}_{jt}/\mathsf{f}_{3-j,t})].$$

7.3.6 *Quantos*

One who buys an asset in a foreign market ordinarily bears two kinds of risk. One is the risk of fluctuations in price as measured in the foreign currency, and the other is the risk of fluctuations in the rate at which foreign currency can be exchanged for domestic currency. Quanto contracts are designed to remove the currency risk from such transactions by providing for foreign asset positions to pay off at a fixed exchange rate. For example, a U.S. investor might buy a quanto forward contract on a German stock with the provision that the price in Euros would convert one-for-one into U.S. dollars. In pricing such derivatives one is tempted to ignore the exchange-rate mechanism entirely and treat the problem just as it would be treated by an investor in the foreign market, whose position in the foreign (non-quanto) version of the same instrument would also be free of currency risk. However, this overlooks an important consideration. The foreign investor could use the derivative and the underlying to replicate a position in foreign-denominated riskless bonds, whereas currency risk makes it impossible for the domestic investor to replicate a position in domestic bonds with the underlying and a quanto, since the underlying is denominated in the foreign currency. Accordingly, arbitrage-free prices of quanto derivatives do wind up depending on exchange-rate risk.

The notation we use to develop the ideas will seem a bit cumbersome, but it has the advantage of making the meanings of the symbols transparent.

$R_t^{d/f}$	domestic-currency value of a unit of foreign currency
$M_t^f = M_0^f e^{r_f t}$	foreign-currency value of the foreign money fund
$M_t^d = M_0^d e^{r_d t}$	domestic-currency value of the domestic money fund
$M_t^{d/f} = M_t^f R_t^{d/f}$	domestic-currency value of foreign money fund
S_t^f	foreign-currency value of foreign underlying asset
$S_t^{d/f} = S_t^f R_t^{d/f}$	domestic-currency value of foreign underlying asset
$\hat{\mathbb{P}}^f$	martingale measure for foreign market
$\hat{\mathbb{P}}^d$	martingale measure for domestic market

Note especially that $M_t^{d/f}$ and $S_t^{d/f}$ are the domestic values of the foreign assets at t. To model the dynamics of these quantities, take the foreign and local short rates, r_f and r_d, to be constants, and assume that the motions of S_t^f and $R_t^{d/f}$ under the actual measure $\mathbb{P}$ are described by

$$\begin{aligned} dS_t^f/S_t^f &= \mu_S \cdot dt + \sigma_S \cdot dW_t^f \\ dR_t^{d/f}/R_t^{d/f} &= \mu_R \cdot dt + \sigma_R \left[\rho \cdot dW_t^f + \sqrt{1-\rho^2} \cdot dW_t^d\right], \end{aligned} \tag{7.43}$$

where W^f and W^d are independent Brownian motions and $\rho \in (-1, 1)$ is a constant. This setup allows for systematic comovements in the exchange rate and the price of the foreign asset while fixing the volatility of $R_t^{d/f}$ at σ_R. (We assume for now that the foreign asset pays no dividend.) One last bit of notation is needed. Let $M_t^{d/f*} \equiv M_t^{d/f}/M_t^d$ and $S_t^{d/f*} \equiv S_t^{d/f}/M_t^d$ denote the normalized domestic values of the foreign assets. Itô's formula gives the dynamics of these as

$$\begin{aligned} dM_t^{d/f*}/M_t^{d/f*} &= (\mu_R + r_f - r_d) \cdot dt + \rho\sigma_R \cdot dW_t^f \\ &\quad + \sigma_R\sqrt{1-\rho^2} \cdot dW_t^d \qquad (7.44) \\ dS_t^{d/f*}/S_t^{d/f*} &= (\mu_R + \mu_S - r_d + \rho\sigma_S\sigma_R) \cdot dt + (\sigma_S + \rho\sigma_R) \cdot dW_t^f \\ &\quad + \sigma_R\sqrt{1-\rho^2} \cdot dW_t^d. \qquad (7.45) \end{aligned}$$

The economics of the problem of pricing a quanto derivative becomes clearer once it is understood how the perspectives of the foreign and domestic investor differ. There exists within the foreign arbitrage-free market a measure $\hat{\mathbb{P}}^f$ under which values of all locally traded assets normalized by $\{M_t^f\}$ are martingales. Likewise, in the domestic market there exists $\hat{\mathbb{P}}^d$ under which the prices of domestically traded assets normalized by $\{M_t^d\}$ are martingales. Since foreign-denominated assets are not traded in the domestic market, there is no reason to expect their normalized prices to be martingales under $\hat{\mathbb{P}}^d$, and yet they must become so upon converting their prices to domestic currency if there is to be no arbitrage. In other words, while neither $\{S_t^f/M_t^d\}$ nor $\{M_t^f/M_t^d\}$ is a martingale under $\hat{\mathbb{P}}^d$ if exchange rates are uncertain, the converted values $\{S_t^{d/f*} = S_t^{d/f}/M_t^d\}$ and $\{M_t^{d/f*} = M_t^{d/f}/M_t^d\}$ *will* be martingales. The key to pricing quanto derivatives on S^f is to discover how S^f itself behaves under the measure that makes $\{S_t^{d/f*}\}$ and $\{M_t^{d/f*}\}$ martingales.

To do this, it is necessary to remove the drift terms from (7.44) and (7.45). Girsanov's theorem confirms the existence of $\hat{\mathbb{P}}^d$ under which both $\{\hat{W}_t^f = W_t^f + \alpha_f t\}$ and $\{\hat{W}_t^d = W_t^d + \alpha_d t\}$ are Brownian motions for any finite constants α_f and α_d. Substituting $\hat{W}_t^f - \alpha_f t$ and $\hat{W}_t^d - \alpha_d t$ for W_t^f and W_t^d in (7.44) and (7.45) shows that the drift terms are eliminated if α_f and α_d solve

$$\begin{aligned} \rho\sigma_R\alpha_f + \sigma_R\sqrt{1-\rho^2}\alpha_d &= \mu_R + r_f - r_d \\ (\rho\sigma_R + \sigma_S)\alpha_f + \sigma_R\sqrt{1-\rho^2}\alpha_d &= \mu_R + \mu_S - r_d + \rho\sigma_S\sigma_R. \end{aligned}$$

However, only α_f is needed for pricing quanto derivatives on S^f, and this is

$$\alpha_f = \frac{\mu_S - r_f + \rho\sigma_R\sigma_S}{\sigma_S}.$$

Since σ_R, σ_S, and ρ are constants, Novikov condition (3.24) is clearly satisfied, so that $\{S_t^{d/f*}\}$ and $\{M_t^{d/f*}\}$ are indeed martingales under $\hat{\mathbb{P}}^d$. Replacing W_t^f by $\hat{W}_t^f - \alpha_f t$ in (7.43), the dynamics of S_t^f under $\hat{\mathbb{P}}^d$ are then

$$\begin{aligned} dS_t^f/S_t^f &= \mu_S \cdot dt + \sigma_S \cdot d(\hat{W}_t^f - \alpha_f t) \\ &= (r_f - \rho\sigma_R\sigma_S) \cdot dt + \sigma_S \cdot d\hat{W}_t^f . \end{aligned}$$

Using this result to price quanto derivatives is made easier and more intuitive by setting $\delta_{d/f} \equiv r_d - r_f + \rho\sigma_R\sigma_S$ and writing

$$dS_t^f/S_t^f = (r_d - \delta_{d/f}) \cdot dt + \sigma_S \cdot d\hat{W}_t^f . \tag{7.46}$$

This makes $\delta_{d/f}$ look like a continuous dividend yield on a domestically traded asset. To interpret it in this context, note that under $\hat{\mathbb{P}}^d$ the mean proportionate drift of the domestic value of a position in the foreign asset—that is, of $\{S_t^{d/f} \equiv S_t^f R_t^{d/f}\}$—is simply r_d. Therefore, $\delta_{d/f}$ can be thought of as the component of the average yield of $S^{d/f}$ that is attributable to changes in the exchange rate. Of this, $r_d - r_f$ is the mean drift of $\{R^{d/f}\}$ itself under $\hat{\mathbb{P}}^d$, and $\rho\sigma_R\sigma_S$ reflects the effect of systematic comovements in $\{R_t^{d/f}\}$ and $\{S_t^f\}$. Expressing dS_t^f/S_t^f as (7.46) makes it possible to calculate arbitrage-free prices of quanto derivatives from the same formulas that apply to derivatives on domestically traded assets that pay continuous dividends.

Here are three particular cases. Since $\hat{E}_t^d S_T^f = S_t^f e^{(r_d - \delta_{d/f})(T-t)}$, the value at t of a T-expiring quanto forward contract at forward price $\mathsf{f}_0^Q \equiv \mathsf{f}^Q(0,T)$ is

$$e^{-r_d(T-t)} \hat{E}_t^d (S_T^f - \mathsf{f}_0^Q) = S_t^f e^{-\delta_{d/f}(T-t)} - e^{-r_d(T-t)} \mathsf{f}_0^Q .$$

The initial forward price that make this value zero is

$$\mathsf{f}^Q(0,T) = S_0^f e^{(r_d - \delta_{d/f})T} = S_0^f e^{(r_f - \rho\sigma_R\sigma_S)T} .$$

Similarly, current values of European-style quanto calls and puts are

$$\begin{aligned} C^Q(S_t^f, T-t) &= e^{-r_d(T-t)} \{\mathsf{f}_t^Q \Phi[q^+(\mathsf{f}_t^Q/X)] - X\Phi[q^-(\mathsf{f}_t^Q/X)]\} \\ P^Q(S_t^f, T-t) &= e^{-r_d(T-t)} \{X\Phi[q^+(X/\mathsf{f}_t^Q)] - \mathsf{f}_t^Q \Phi[q^-(X/\mathsf{f}_t^Q)]\}, \end{aligned}$$

where

$$\mathsf{f}_t^Q \equiv \mathsf{f}^Q(t,T) = S_t^f e^{(r_f - \rho\sigma_R\sigma_S)(T-t)}$$

is the forward price at t for T delivery. In all three cases pricing the quanto product just involves replacing the domestic interest rate by the foreign rate minus the instantaneous covariance between the foreign asset's return and the return on a position in the foreign currency. If the foreign asset itself yields a continuous dividend at rate δ_f (paid in the foreign currency), then δ_f is simply subtracted from r_f.

7.4 Path-Dependent Options

A European-style derivative is said to be "path dependent" if to determine its terminal value requires knowing the entire sample path of the underlying during all or part of the derivative's life. Two examples are lookback options, whose payoffs depend on the extreme values of the underlying—either maxima or minima, and Asian options, whose payoffs depend on the average price. The mechanics of each of these is explained below. While the contractual terms of these instruments usually call for averages and extrema to be determined from a finite set of prices at fixed trading intervals, the pricing problem turns out to be simpler in the Black-Scholes framework if based on the entire continuous record. The simpler formulas derived in this context then serve as approximations for the prices of real-world instruments. Basic to the whole process are some beautiful theoretical results about the distributions of maxima and minima of Brownian paths.

7.4.1 *Extrema of Brownian Paths*

As usual, let us take $S_t = S_0 \exp[(r-\delta-\sigma^2/2)t+\sigma\hat{W}_t]$ as the model for the underlying price under the risk-neutral measure, where δ is the continuous dividend rate, $r \geq 0$ is the short rate, and $\{\hat{W}_t\}_{t\geq 0}$ is a standard Brownian motion. Letting $Y_t \equiv (r-\delta-\sigma^2/2)t+\sigma\hat{W}_t$ and representing the price process as $\{S_0 e^{Y_t}\}_{t\geq 0}$ shows that extrema of price paths are functions of extrema of *non*standard Brownian motions—that is, Brownian motions with arbitrary mean drift and scale. The first step in characterizing the price extrema is to work out distributions of extrema of standard Wiener processes. Although the arguments here will depend crucially on the symmetry of these trendless processes, Girsanov's theorem will make it possible to extend them to the case at hand.

Standard Wiener Processes

Considering a standard Brownian motion $\{W_t\}_{t\geq 0}$ on $(\Omega, \mathcal{F}, \{\mathcal{F}_t\}, \mathbb{P})$ with $W_0 = 0$, let $\overline{W} \equiv \overline{W}_{0,T}$ and $\underline{W} \equiv \underline{W}_{0,T}$ denote the maximum and minimum on the finite interval $[0, T]$. We shall first apply the reflection principle to work out the joint distributions of $\overline{W}, W_T$ and $\underline{W}, W_T$ and then deduce from these the marginal distributions of $\overline{W}$ and $\underline{W}$ themselves.[10]

The first step will be to develop an identity for the joint probability $\mathbb{P}(\overline{W} \geq \overline{w}, W_T \leq w)$ for arbitrary real $\overline{w}, w$. This is where the reflection principle comes in, and figure 7.6 tells the story. Consider a path for which $\overline{W} \geq \overline{w}$ does occur, as depicted by the solid curve that ends up at $W_T = w$. In view of the almost-sure continuity of Brownian paths, there is necessarily a time t^* at which $W_{t^*} = \overline{w}$ along this path. Imagine being at this point in time, observing that $W_{t^*} = \overline{w}$ and thus knowing that the event $\overline{W} \geq \overline{w}$ has occurred, but not knowing the subsequent behavior of W_t beyond t^*. Regarding $\mathcal{F}_{t^*}$ as the information then available, let us try to assess the chance that the path will end up at or below an arbitrary value w—that is, to determine $\mathbb{P}_{t^*}(W_T \leq w) \equiv \mathbb{P}(W_T \leq w \mid \mathcal{F}_{t^*})$. Since the conditional distribution of $W_T - \overline{w}$ is symmetric about zero, it is clear that the path has as much chance of rising by $|\overline{w} - w|$ or more as it does of falling by $|\overline{w} - w|$ or more. That is, corresponding to any path that winds up at or below w

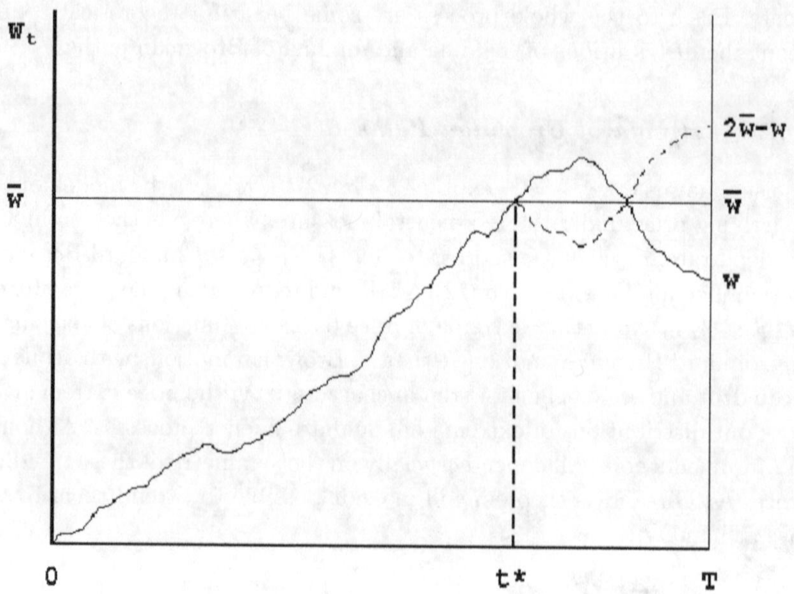

Fig. 7.6. Applying the reflection principle.

[10] Karlin and Taylor (1975) provide a fuller and wonderfully accessible treatment of this and other applications of the reflection principle.

is its reflection that winds up at or above $\overline{w} + (\overline{w} - w)$. The reflected path is depicted by the dashed track in figure 7.6. Therefore,

$$\mathbb{P}_{t^*}(W_T - \overline{w} \geq \overline{w} - w) = \mathbb{P}_{t^*}(W_T - \overline{w} \leq w - \overline{w}),$$

or equivalently

$$\mathbb{P}_{t^*}(W_T \geq 2\overline{w} - w) = \mathbb{P}_{t^*}(W_T \leq w).$$

Since the same relation holds for some such t^* on any path for which $\overline{W} \geq \overline{w}$, we conclude that

$$\mathbb{P}(\overline{W} \geq \overline{w}, W_T \leq w) = \mathbb{P}(\overline{W} \geq \overline{w}, W_T \geq 2\overline{w} - w) \tag{7.47}$$

for all $\overline{w}, w$. The same kind of argument for the minimum, $\underline{W}$, gives

$$\mathbb{P}(\underline{W} \geq \underline{w}, W_T \leq w) = \mathbb{P}(\underline{W} \leq \underline{w}, W_T \geq 2\underline{w} - w). \tag{7.48}$$

Putting (7.47) to work produces two useful results. First, by picking a value $w \leq \overline{w}$ we have $2\overline{w} - w \geq \overline{w}$, so that $W_T \geq 2\overline{w} - w$ implies $W_T \geq \overline{w}$, which in turn implies (is a subset of the event) that $\overline{W} \geq \overline{w}$. Thus,

$$\mathbb{P}(\overline{W} \geq \overline{w}, W_T \leq w) = \mathbb{P}(W_T \geq 2\overline{w} - w) \tag{7.49}$$

whenever $w \leq \overline{w}$. Next, using that[11]

$$\begin{aligned}\mathbb{P}(\overline{W} \geq \overline{w}) &= \mathbb{P}(\overline{W} \geq \overline{w}, W_T \leq w) + \mathbb{P}(\overline{W} \geq \overline{w}, W_T \geq w) \qquad (7.50)\\ &= \mathbb{P}(\overline{W} \geq \overline{w}, W_T \geq 2\overline{w} - w) + \mathbb{P}(\overline{W} \geq \overline{w}, W_T \geq w)\end{aligned}$$

and taking $w = \overline{w}$ give

$$\mathbb{P}(\overline{W} \geq \overline{w}) = 2\mathbb{P}(\overline{W} \geq \overline{w}, W_T \geq \overline{w}).$$

Finally, again using the fact that $W_T \geq \overline{w}$ implies $\overline{W} \geq \overline{w}$ gives the remarkable result that the distribution of the maximum of Brownian motion can be expressed just in terms of the distribution of its terminal value:

$$\mathbb{P}(\overline{W} \geq \overline{w}) = 2\mathbb{P}(W_T \geq \overline{w}). \tag{7.51}$$

Note that this is consistent with $\mathbb{P}(\overline{W} \geq 0) = 1$, which is necessarily true given that $W_0 = 0$.

Likewise, (7.48) (or simply the symmetry of Brownian motion) implies that for $w \geq \underline{w}$

$$\mathbb{P}(\underline{W} \leq \underline{w}, W_T \geq w) = \mathbb{P}(W_T \leq 2\underline{w} - w) \tag{7.52}$$

$$\mathbb{P}(\underline{W} \leq \underline{w}) = 2\mathbb{P}(W_T \leq \underline{w}). \tag{7.53}$$

[11]Observe that event $W_T = w$ is counted twice in (7.50) but has zero probability.

Wiener Processes with Drift

Consider now the trending process $\{Y_t = \mu t + W_t\}$, where $\{W_t\}_{t\geq 0}$ is again a standard Brownian motion and μ is a constant not equal to zero. Although there is nonzero drift, the process still starts at the origin. Adhering to the same notation, with $\overline{Y} \equiv \overline{Y}_{0,T}$ as the maximum value on $[0,T]$, we shall develop results for $\overline{Y}$ that correspond to (7.49) and (7.51).

The hard part is the counterpart of (7.49). Girsanov's theorem implies that taking

$$d\tilde{\mathbb{P}}/d\mathbb{P} = e^{-\mu Y_T + \mu^2 T/2}$$

yields an equivalent measure $\tilde{\mathbb{P}}$ under which $\{Y_t\}_{t\geq 0}$ is a standard Brownian motion on $[0,T]$. Of course, the goal is to find not $\tilde{\mathbb{P}}(\overline{Y} \geq \overline{y}, Y_T \leq y)$ but $\mathbb{P}(\overline{Y} \geq \overline{y}, Y_T \leq y)$. However, since

$$\begin{aligned} \mathbb{P}(\overline{Y} \geq \overline{y}, Y_T \leq y) &= E\mathbf{1}_{[\overline{y},\infty)\times(-\infty,y]}(\overline{Y}, Y_T) \\ &= \tilde{E}[\mathbf{1}_{[\overline{y},\infty)\times(-\infty,y]}(\overline{Y}, Y_T) \cdot d\mathbb{P}/d\tilde{\mathbb{P}}] \\ &= \tilde{E}[\mathbf{1}_{[\overline{y},\infty)\times(-\infty,y]}(\overline{Y}, Y_T) \cdot e^{\mu Y_T - \mu^2 T/2}], \end{aligned} \tag{7.54}$$

the probability we want can be found by evaluating the expectation. $\{Y_t\}_{t\geq 0}$ being a standard Brownian motion under $\tilde{\mathbb{P}}$, (7.49) implies that

$$\tilde{\mathbb{P}}(\overline{Y} \geq \overline{y}, Y_T \leq y) = \tilde{\mathbb{P}}(Y_T \geq 2\overline{y} - y)$$

for $y \leq \overline{y}$. The joint density of $\overline{Y}, Y_T$ under $\tilde{\mathbb{P}}$ is therefore

$$\begin{aligned} \tilde{f}(\overline{y}, y) &= \frac{\partial^2}{\partial \overline{y}\partial y}\tilde{\mathbb{P}}(\overline{Y} \leq \overline{y}, Y_T \leq y) \\ &= \frac{\partial^2}{\partial \overline{y}\partial y}[\tilde{\mathbb{P}}(Y_T \leq y) - \tilde{\mathbb{P}}(\overline{Y} \geq \overline{y}, Y_T \leq y)] \\ &= -\frac{\partial^2}{\partial \overline{y}\partial y}\tilde{\mathbb{P}}(\overline{Y} \geq \overline{y}, Y_T \leq y) \\ &= -\frac{\partial}{\partial y}\left[\frac{\partial}{\partial \overline{y}}\int_{2\overline{y}-y}^{\infty} \frac{1}{\sqrt{2\pi T}} e^{-z^2/(2T)} \cdot dz\right] \\ &= \sqrt{\frac{2}{\pi}}(2\overline{y} - y)T^{-3/2} e^{-(y-2\overline{y})^2/(2T)}, \quad y \leq \overline{y}, \overline{y} > 0. \end{aligned}$$

This makes it possible to evaluate the expectation in (7.54) for $y \leq \overline{y}$ as

$$\begin{aligned} &\int_{-\infty}^{y}\int_{\overline{y}}^{\infty} e^{\mu u - \mu^2 T/2}\, \tilde{f}(\overline{u}, u) \cdot d\overline{u}\, du \\ &= \int_{-\infty}^{y} T^{-1} e^{\mu u - \mu^2 T/2}\left[\int_{\overline{y}}^{\infty} \frac{2(2\overline{u} - u)}{\sqrt{2\pi T}} e^{-(u-2\overline{u})^2/(2T)} \cdot d\overline{u}\right] \cdot du. \end{aligned}$$

Changes of variables and algebra reduce this to

$$e^{2\mu\overline{y}} \int_{-\infty}^{y-2\overline{y}} \frac{1}{\sqrt{2\pi T}} e^{-(z-\mu T)^2/(2T)} \cdot dz,$$

yielding as the counterpart to (7.49)

$$\mathbb{P}(\overline{Y} \geq \overline{y}, Y_T \leq y) = e^{2\mu\overline{y}} \mathbb{P}(Y_T \leq y - 2\overline{y}) \tag{7.55}$$

for $y \leq \overline{y}$ and $\overline{y} > 0$. Since $Y_T = \mu T + W_T$ and $\mathbb{P}(W_T \leq w) = \mathbb{P}(W_T \geq -w)$, this can also be written as

$$\mathbb{P}(\overline{Y} \geq \overline{y}, Y_T \leq y) = e^{2\mu\overline{y}} \mathbb{P}(Y_T \geq 2\overline{y} - y + 2\mu T).$$

A similar process yields for the minimum of $\{Y_t\}_{0 \leq t \leq T}$

$$\mathbb{P}(\underline{Y} \leq \underline{y}, Y_T \geq y) = e^{2\mu\underline{y}} \mathbb{P}(Y_T \geq y - 2\underline{y}) \tag{7.56}$$

for $y \geq \underline{y}$ and $\underline{y} < 0$.

The counterparts of (7.51) for $\overline{Y}$ and of (7.53) for $\underline{Y}$ now come at once. Applying (7.55) and the identity corresponding to (7.50) gives

$$\mathbb{P}(\overline{Y} \geq \overline{y}) = e^{2\mu\overline{y}} \mathbb{P}(Y_T \leq -\overline{y}) + \mathbb{P}(Y_T \geq \overline{y})$$

or

$$\mathbb{P}(\overline{Y} \geq \overline{y}) = e^{2\mu\overline{y}} \mathbb{P}(Y_T \geq \overline{y} + 2\mu T) + \mathbb{P}(Y_T \geq \overline{y}) \tag{7.57}$$

for $\overline{y} > 0$. Likewise, (7.56) and (7.50) imply

$$\mathbb{P}(\underline{Y} \leq \underline{y}) = e^{2\mu\underline{y}} \mathbb{P}(Y_T \geq -\underline{y}) + \mathbb{P}(Y_T \leq \underline{y})$$

or equivalently

$$\mathbb{P}(\underline{Y} \leq \underline{y}) = e^{2\mu\underline{y}} \mathbb{P}(Y_T \leq \underline{y} + 2\mu T) + \mathbb{P}(Y_T \leq \underline{y}) \tag{7.58}$$

for $\underline{y} < 0$. Note that since $\overline{Y}$, $\underline{Y}$, and Y_T are absolutely continuous, all of (7.55), (7.56), (7.57), and (7.58) continue to hold when the weak inequalities are replaced by strict inequalities.

Supplied with these results, we can now price some path-dependent derivatives under Black-Scholes dynamics.

7.4.2 *Lookback Options*

"Fixed-strike" and "variable-strike" options are the two common types of lookbacks. Payoffs of fixed-strike T-expiring European lookbacks issued at $t = 0$ depend on differences between an extremum of the underlying on $[0, T]$ and strike price X. Specifically, fixed-strike lookback calls pay $(\overline{S}_{0,T} - X)^+$, and fixed-strike lookback puts pay $(X - \underline{S}_{0,T})^+$, where overbars and underbars denote maxima and minima, respectively. With variable-strike lookbacks, payoffs depend on the difference between the terminal price and the maximum or minimum; thus, variable-strike puts pay $(\overline{S}_{0,T} - S_T)$ and variable-strike calls pay $(S_T - \underline{S}_{0,T})$. Since the terminal value can be no larger than the maximum and no smaller than the minimum, variable-strike options will be exercised with probability one under Black-Scholes dynamics. Goldman *et al.* (1979) did the pioneering work on European variable-strike lookbacks, demonstrating, among other things, that they are replicable with self-financing portfolios. Conze and Viswanathan (1991) apply Girsanov's theorem to price both variable-strike and fixed-strike lookbacks, and they provide bounds for prices of American lookbacks. We follow their treatment in working out arbitrage-free prices for European variable-strike puts and fixed-strike calls at an arbitrary $t \in [0, T)$. In both cases the main trick is to figure out the conditional distribution of $\overline{S}_{0,T}$ given what is known at t, and the hardest part of that has already been done.

Variable-Strike Options

Letting $P^{VL}(S_t, T - t)$ be the price of the variable-strike lookback put at t, martingale pricing gives

$$\begin{aligned} P^{VL}(S_t, T - t) &= e^{-r(T-t)} \hat{E}_t(\overline{S}_{0,T} - S_T) \\ &= e^{-r(T-t)} \hat{E}_t(\overline{S}_{0,T}) - S_t e^{-\delta(T-t)}. \end{aligned} \tag{7.59}$$

As usual, δ is the continuous dividend rate, $\hat{E}$ denotes expectation under the risk-neutral measure, and $\hat{E}_t(\cdot) \equiv \hat{E}(\cdot \mid \mathcal{F}_t)$. Obviously, one has only to evaluate the expectation in order to price the option. Standing at t, we already know the maximum price up to that point, $\overline{S}_{0,t}$. Therefore, conditional on $\mathcal{F}_t$, the expectation of the maximum price during $[0, T]$ is

$$\begin{aligned} \hat{E}_t(\overline{S}_{0,T}) &= \hat{E}_t(\overline{S}_{0,t} \vee \overline{S}_{t,T}) \\ &= \int_0^\infty \hat{\mathbb{P}}_t(\overline{S}_{0,t} \vee \overline{S}_{t,T} > s) \cdot ds \\ &= \overline{S}_{0,t} + \int_{\overline{S}_{0,t}}^\infty \hat{\mathbb{P}}_t(\overline{S}_{t,T} > s) \cdot ds. \end{aligned} \tag{7.60}$$

Here, $\hat{\mathbb{P}}_t(\cdot) \equiv \hat{\mathbb{P}}(\cdot|\mathcal{F}_t)$; the second equality exploits the identity (2.40) for the expectation; and the third equality follows from the obvious relation

$$\hat{\mathbb{P}}_t(\overline{S}_{0,t} \vee \overline{S}_{t,T} > s) = \begin{cases} 1, & s < \overline{S}_{0,t} \\ \hat{\mathbb{P}}_t(\overline{S}_{t,T} > s), & s \geq \overline{S}_{0,t} \end{cases}.$$

Letting $\gamma \equiv (r - \delta - \sigma^2/2)/\sigma$ and $Y_t \equiv \gamma t + \hat{W}_t$, where $\{\hat{W}_t\}_{t\geq 0}$ is a Brownian motion under $\hat{\mathbb{P}}$, we have $S_t = S_0 e^{\sigma Y_t}$ and $\overline{S}_{t,T} = S_t \exp(\sigma \overline{Y}_{t,T})$. Thus,

$$\hat{\mathbb{P}}_t(\overline{S}_{t,T} > s) = \hat{\mathbb{P}}_t[\overline{Y}_{t,T} > \sigma^{-1} \ln(s/S_t)]$$

and

$$\hat{E}_t(\overline{S}_{0,T}) = \overline{S}_{0,t} + \sigma S_t \int_{\sigma^{-1}\ln(\overline{S}_{0,t}/S_t)}^{\infty} e^{\sigma y} \hat{\mathbb{P}}_t(\overline{Y}_{t,T} > y) \cdot dy. \tag{7.61}$$

Setting $\tau \equiv T - t$ and applying (7.57) with $\hat{\mathbb{P}}_t$ in place of $\mathbb{P}$ give for the integral in (7.61)

$$\int_{\sigma^{-1}\ln(\overline{S}_{0,t}/S_t)}^{\infty} e^{\theta y} \hat{\mathbb{P}}_t(Y_T - Y_t > y + 2\gamma\tau) \cdot dy \tag{7.62}$$
$$+ \int_{\sigma^{-1}\ln(\overline{S}_{0,t}/S_t)}^{\infty} e^{\sigma y} \hat{\mathbb{P}}_t(Y_T - Y_t > y) \cdot dy,$$

where $\theta \equiv 2(r - \delta)/\sigma$. For the time being we assume that $\theta \neq 0$, saving that case, which is relevant for options on futures, until later.

Evaluating the integrals in (7.62) requires the following relation, which integration by parts easily establishes. For arbitrary $\alpha \neq 0$ assume that X is an absolutely continuous random variable such that $e^{\alpha x}\mathbb{P}(X \geq x)$ is integrable. Then

$$\int_{x_0}^{\infty} e^{\alpha x} \mathbb{P}(X > x) \cdot dx = \alpha^{-1} E(e^{\alpha X} - e^{\alpha x_0}) \mathbf{1}_{[x_0,\infty)}(X). \tag{7.63}$$

Attacking the first integral in (7.62), set $Y_t = \gamma t + \hat{W}_t$ and change variables as $u = y + \gamma t$ to get

$$e^{-\theta\gamma\tau} \int_{\sigma^{-1}\ln(\overline{S}_{0,t}/S_t)+\gamma\tau}^{\infty} e^{\theta u} \hat{\mathbb{P}}_t(\hat{W}_T - \hat{W}_t > u) \cdot du.$$

Now applying (7.63) with $\alpha = \theta$ and $x_0 = \sigma^{-1}\ln(\overline{S}_{0,t}/S_t) + \gamma\tau$ gives

$$e^{-\theta\gamma\tau}\theta^{-1}\hat{E}_t[e^{\theta(\hat{W}_T - \hat{W}_t)} - e^{\theta x_0}]\mathbf{1}_{[x_0,\infty)}(\hat{W}_T - \hat{W}_t),$$

or

$$\frac{e^{r'\tau}}{\theta} \int_{x_0}^{\infty} e^{\theta w} \frac{1}{\sqrt{2\pi\tau}} e^{-(w-\theta\tau)^2/(2\tau)} \cdot dw - \frac{1}{\theta}(\overline{S}_{0,t}/S_t)^{\theta/\sigma} \hat{\mathbb{P}}_t(\hat{W}_T - \hat{W}_t > x_0),$$

where $r' \equiv r - \delta$. One more change of variables in the integral leads finally to

$$I_1 \equiv \frac{\sigma}{2r'}\left\{e^{r'\tau}\Phi\left[q^+\left(\frac{S_t e^{r'\tau}}{\overline{S}_{0,t}}\right)\right] - \left(\frac{\overline{S}_{0,t}}{S_t}\right)^{2r'/\sigma^2}\Phi\left[q^+\left(\frac{S_t e^{-r'\tau}}{\overline{S}_{0,t}}\right)\right]\right\} \tag{7.64}$$

as the expression for the first integral in (7.62). As usual,

$$q^{\pm}(x) = \frac{\ln x \pm \sigma^2\tau/2}{\sigma\sqrt{\tau}}.$$

Similar steps involving another application of (7.63) give for the second integral in (7.62)

$$I_2 \equiv \sigma^{-1}\left\{e^{r'\tau}\Phi\left[q^+\left(\frac{S_t e^{r'\tau}}{\overline{S}_{0,t}}\right)\right] - \frac{\overline{S}_{0,t}}{S_t}\Phi\left[q^-\left(\frac{S_t e^{r'\tau}}{\overline{S}_{0,t}}\right)\right]\right\}. \tag{7.65}$$

Finally, assembling the pieces, we have from (7.61) that

$$\hat{E}_t(\overline{S}_{0,T}) = \overline{S}_{0,t} + \sigma S_t(I_1 + I_2),$$

and finally, from (7.59), the arbitrage-free value of the variable-strike lookback put:

$$\begin{aligned}P^{VL}(S_t,\tau) = {} & -S_t e^{-\delta\tau}\Phi\left[q^-\left(\frac{\overline{S}_{0,t}}{S_t e^{r'\tau}}\right)\right] + \overline{S}_{0,t}e^{-r\tau}\Phi\left[q^+\left(\frac{\overline{S}_{0,t}}{S_t e^{r'\tau}}\right)\right] \\ & + \frac{\sigma^2}{2r'}S_t e^{-\delta\tau}\Phi\left[q^+\left(\frac{S_t e^{r'\tau}}{\overline{S}_{0,t}}\right)\right] \\ & - \frac{\sigma^2}{2r'}S_t e^{-r\tau}\left(\frac{\overline{S}_{0,t}}{S_t}\right)^{2r'/\sigma^2}\Phi\left[q^+\left(\frac{S_t e^{-r'\tau}}{\overline{S}_{0,t}}\right)\right],\end{aligned}$$

where $r' \equiv r - \delta$ and $\tau \equiv T - t$.

The variable-strike lookback call option, which is worth $(S_T - \underline{S}_{0,T})$ at expiration, can be valued in much the same way using (7.58) to deduce the distribution of the minimum:

$$\begin{aligned}C^{VL}(S_t,\tau) = {} & S_t e^{-\delta\tau}\Phi\left[q^+\left(\frac{S_t e^{r'\tau}}{\underline{S}_{0,t}}\right)\right] - \underline{S}_{0,t}e^{-r\tau}\Phi\left[q^-\left(\frac{S_t e^{r'\tau}}{\underline{S}_{0,t}}\right)\right] \\ & - \frac{\sigma^2}{2r'}S_t e^{-\delta\tau}\Phi\left[q^-\left(\frac{\underline{S}_{0,t}}{S_t e^{r'\tau}}\right)\right] \\ & + \frac{\sigma^2}{2r'}S_t e^{-r\tau}\left(\frac{\underline{S}_{0,t}}{S_t}\right)^{2r'/\sigma^2}\Phi\left[q^-\left(\frac{\underline{S}_{0,t}}{S_t e^{-r'\tau}}\right)\right].\end{aligned} \tag{7.66}$$

Notice that a variable-strike lookback straddle—a position long in both the call and the put—produces a payoff of $\overline{S}_{0,T} - \underline{S}_{0,T}$, which effectively guarantees a purchase at the low and a sale at the high over the period.

The results differ considerably for lookback options on futures. Since futures price $\{\mathsf{F}_t\}$ is a martingale under $\hat{\mathbb{P}}$, modeling the dynamics requires setting $\delta = r$. In that case $\theta = 0$ and $\gamma = -\sigma/2$ in (7.62), and (7.63) no longer applies to evaluate the first integral in (7.62). However, a direct application of integration by parts ultimately reduces it to

$$\int_{\sigma^{-1}\ln(\overline{\mathsf{F}}_{0,t}/\mathsf{F}_t)}^{\infty} \hat{\mathbb{P}}_t(Y_T - Y_t \geq y - \sigma\tau) \cdot dy = \sqrt{\tau}\left[q^+\Phi(q^+) + \phi(q^+)\right],$$

where $\phi(\cdot)$ is the standard normal p.d.f., $\tau = T - t$, and

$$q^{\pm} \equiv \frac{\ln(\mathsf{F}_t/\overline{\mathsf{F}}_{0,t}) \pm \sigma^2\tau/2}{\sigma\sqrt{\tau}}.$$

The final result for the variable-strike lookback put on the futures is

$$\begin{aligned} P^{VL}(\mathsf{F}_t, \tau) &= \mathsf{F}_t e^{-r\tau}\left\{\Phi(q^+)\left[2 + \sigma\sqrt{\tau}(q^+ + \phi(q^+))\right] - 1\right\} \\ &\quad + \overline{\mathsf{F}}_{0,t} e^{-r\tau}\Phi\left(-q^-\right). \end{aligned}$$

Fixed-Strike Options

Most of the work needed to price fixed-strike lookbacks has already been done. It just remains to show that this is true. To that end we will work out the arbitrage-free price of a fixed-strike European call on an underlying for which $r - \delta \neq 0$. Letting $C^{FL}(S_t, \tau \equiv T - t)$ denote the price at t and recalling that the terminal value is $C^{FL}(S_T, 0) = (\overline{S}_{0,T} - X)^+$, we have

$$\begin{aligned} C^{FL}(S_t, \tau) &= e^{-r\tau}\hat{E}_t(\overline{S}_{0,T} - X)^+ \\ &= e^{-r\tau}\hat{E}_t(\overline{S}_{0,t} \vee \overline{S}_{t,T} - X)^+ \\ &= e^{-r\tau}\int_0^{\infty} \hat{\mathbb{P}}_t[(\overline{S}_{0,t} \vee \overline{S}_{t,T} - X)^+ > u] \cdot du. \end{aligned}$$

Now the probability in the last line is

$$\hat{\mathbb{P}}_t\left[(\overline{S}_{0,t} \vee \overline{S}_{t,T} - X)^+ > u\right] = \begin{cases} 1, & u < (\overline{S}_{0,t} - X)^+ \\ \hat{\mathbb{P}}_t(\overline{S}_{t,T} > X + u), & u \geq (\overline{S}_{0,t} - X)^+ \end{cases}.$$

Therefore,

$$
\begin{aligned}
C^{FL}(S_t,\tau) &= e^{-r\tau}\left[\int_0^{(\overline{S}_{0,t}-X)^+} 1\cdot du + \int_{(\overline{S}_{0,t}-X)^+}^{\infty} \hat{\mathbb{P}}_t\left(\overline{S}_{t,T} > X+u\right)\cdot du\right] \\
&= e^{-r\tau}\left[(\overline{S}_{0,t}-X)^+ + \int_{\overline{S}_{0,t}\vee X}^{\infty} \hat{\mathbb{P}}_t\left(\overline{S}_{t,T} > s\right)\cdot ds\right],
\end{aligned}
$$

where in the second line we have set $s = X + u$ and used the identity $(\overline{S}_{0,t} - X)^+ + X = \overline{S}_{0,t} \vee X$. Now the integral in the last expression is just the same as that in (7.60) except that the lower limit is $\overline{S}_{0,t} \vee X$ instead of $\overline{S}_{0,t}$. Evaluating $C^{FL}(S_t, \tau)$, then, just involves making this substitution in the integrals (7.64) and (7.65) and collecting the results:

$$
\begin{aligned}
C^{FL}(S_t,\tau) &= e^{-r\tau}(\overline{S}_{0,t}-X)^+ + S_t e^{-\delta\tau}\Phi\left[q^+\left(\frac{S_t e^{r'\tau}}{\overline{S}_{0,t}\vee X}\right)\right] \\
&\quad - (\overline{S}_{0,t}\vee X)e^{-r\tau}\Phi\left[q^-\left(\frac{S_t e^{r'\tau}}{\overline{S}_{0,t}\vee X}\right)\right] \\
&\quad + \frac{\sigma^2}{2r'}S_t e^{-\delta\tau}\Phi\left[q^+\left(\frac{S_t e^{r'\tau}}{\overline{S}_{0,t}\vee X}\right)\right] \\
&\quad - \frac{\sigma^2}{2r'}S_t e^{-r\tau}\left(\frac{\overline{S}_{0,t}\vee X}{S_t}\right)^{2r'/\sigma^2}\Phi\left[q^+\left(\frac{S_t e^{-r'\tau}}{\overline{S}_{0,t}\vee X}\right)\right],
\end{aligned}
$$

where $\tau = T - t$, $r' = r - \delta \neq 0$, and

$$
q^{\pm}(x) = \frac{\ln x + \sigma^2\tau/2}{\sigma\sqrt{\tau}}.
$$

Notice that if we hold $S_t < \overline{S}_{0,t} \vee X$ while letting $t \to T$, the value of the call converges to $(\overline{S}_{0,T} - X)^+$, as it should.

7.4.3 *Barrier Options*

Terminal payoffs of variable-strike lookback options depend on both an extremum of price and the terminal value, S_T, but the dependency is strictly linear. For example, the payoff is $\overline{S}_{0,T} - S_T$ for the variable-strike put. Payoffs of European barrier options, however, depend in a more intricate way on terminal price and extremum. These options deliver the same payoffs at T as vanilla call and puts—that is, $(S_T - X)^+$ or $(X - S_T)^+$—but only under the condition that an extremum of price over $[0, T]$ either has or

Table 7.2. Terminal values of various barrier options with strike X and barrier K.

Option	Call	Put
Up & in, $K > X$	$(S_T - X)^+ \mathbf{1}_{[K,\infty)}(\overline{S}_{0,T})$	$(X - S_T)^+ \mathbf{1}_{[K,\infty)}(\overline{S}_{0,T})$
Up & out, $K > X$	$(S_T - X)^+ \mathbf{1}_{[0,K]}(\overline{S}_{0,T})$	$(X - S_T)^+ \mathbf{1}_{[0,K]}(\overline{S}_{0,T})$
Down & in, $K < X$	$(S_T - X)^+ \mathbf{1}_{[0,K]}(\underline{S}_{0,T})$	$(X - S_T)^+ \mathbf{1}_{[0,K]}(\underline{S}_{0,T})$
Down & out, $K < X$	$(S_T - X)^+ \mathbf{1}_{[K,\infty)}(\underline{S}_{0,T})$	$(X - S_T)^+ \mathbf{1}_{[K,\infty)}(\underline{S}_{0,T})$

has not surpassed the pertinent "barrier". Many variations are possible, depending on how the contingency is specified. The payoff formulas for the four most common types of barrier options are listed in table 7.2. In each case X is the strike price and the payoff is contingent on the event that price attains some value K during $[0, T]$. Thus, the up-and-in call pays $S_T - X$ if this is positive and if price has reached $K > X$ during the option's lifetime; otherwise, it pays nothing. Similarly, the down-and-out put pays $X - S_T$ if this is positive and price has not fallen below $K < X$. Clearly, risk-neutral pricing of derivatives with these payoffs requires knowing the joint distributions of terminal price and extrema. Fortunately, these come easily from (7.55) and (7.56).

There is one sense in which the barrier options in table 7.2 are easier to deal with than lookbacks. The price at $t \in (0, T)$ does not depend on the precise value attained by the relevant extremum up to that point—$\overline{S}_{0,t}$ or $\underline{S}_{0,t}$. All that matters is whether the pertinent barrier has been surpassed. If it has, then the option has either been "knocked in", becoming a vanilla European whose price comes directly from Black-Scholes, or else it has been "knocked out" and is worthless. To price a barrier option that has not already been knocked out or in, we can save notation by taking $t = 0$ as the current time. To illustrate the technique, we shall price the up-and-in call and the down-and-out put. Except for the timing convention, the assumptions and notation are the same as for lookback options.

The Up-and-In Call

Letting $C^{UI}(S_0, T)$ denote the current value of the up-and-in call, we have

$$\begin{aligned} C^{UI}(S_0, T) &= e^{-rT} \hat{E} C^{UI}(S_T, 0) \\ &= e^{-rT} \hat{E}(S_T - X)^+ \mathbf{1}_{[K,\infty)}(\overline{S}_{0,T}), \end{aligned}$$

where $K > X > 0$. We also assume that $K > S_0$ to exclude the trivial case that the option is a vanilla European at the outset. We shall derive the p.d.f. of the terminal value by deducing the complementary c.d.f. under $\hat{\mathbb{P}}$. This is given by

$$\hat{\mathbb{P}}[C^{UI}(S_0,T) > c] = \begin{cases} 1, & c < 0 \\ \hat{\mathbb{P}}(S_T > c+X, \overline{S}_{0,T} > K), & 0 \le c \le K - X \\ \hat{\mathbb{P}}(S_T > c+X), & c > K - X \end{cases},$$

where the last expression applies because $S_T > K$ implies $\overline{S}_{0,T} > K$. Consider first the case $c \in [0, K - X]$. Letting $Y_t \equiv \gamma t + \hat{W}_t$ and $\sigma\gamma \equiv r - \delta - \sigma^2/2 \equiv r' - \sigma^2/2$, where $\{\hat{W}_t\}_{t\ge 0}$ is a Brownian motion under $\hat{\mathbb{P}}$, we have $S_T = S_0 e^{\sigma Y_T}$ and $\overline{S}_{0,T} = S_0 \exp(\sigma \overline{Y}_{0,T})$. Therefore,

$$\hat{\mathbb{P}}(S_T > c{+}X, \overline{S}_{0,T} > K) = \hat{\mathbb{P}}\left[Y_T > \sigma^{-1}\ln\left(\frac{c+X}{S_0}\right), \overline{Y}_{0,T} > \sigma^{-1}\ln\left(\frac{K}{S_0}\right)\right].$$

Now (7.55) and (7.57) imply that, for $y \le \overline{y}$,

$$\begin{aligned} \hat{\mathbb{P}}(Y_T > y, \overline{Y}_{0,T} > \overline{y}) &= \hat{\mathbb{P}}(\overline{Y}_{0,T} > \overline{y}) - \hat{\mathbb{P}}(Y_T \le y, \overline{Y}_{0,T} > \overline{y}) \\ &= \hat{\mathbb{P}}(Y_T > \overline{y}) + e^{2\gamma\overline{y}}\hat{\mathbb{P}}(Y_T > \overline{y} + 2\gamma T) - e^{2\gamma\overline{y}}\hat{\mathbb{P}}(Y_T \le y - 2\overline{y}) \\ &= \Phi\left(\frac{-\overline{y} + \gamma T}{\sqrt{T}}\right) + e^{2\gamma\overline{y}}\left[\Phi\left(-\frac{\overline{y} + \gamma T}{\sqrt{T}}\right) - \Phi\left(\frac{y - 2\overline{y} - \gamma T}{\sqrt{T}}\right)\right]. \end{aligned}$$

Substituting $\sigma^{-1}\ln(\frac{c+X}{S_0})$ for y and $\sigma^{-1}\ln(K/S_0)$ for $\overline{y}$ and simplifying give for $\hat{\mathbb{P}}[C^{UI}(S_0,T) > c]$ the expression

$$\begin{aligned} &\left(\frac{K}{S_0}\right)^{2r'/\sigma^2 - 1}\left\{\Phi\left[q^+\left(\frac{S_0}{e^{r'T}K}\right)\right]\right. \\ &\qquad \left. - \Phi\left[q^+\left(\frac{(c+X)S_0}{K^2 e^{r'T}}\right)\right]\right\} + \Phi\left[q^-\left(\frac{S_0 e^{r'T}}{K}\right)\right] \end{aligned}$$

when $0 \le c \le K - X$. For $c > K - X$ it is easy to see that

$$\hat{\mathbb{P}}[C^{UI}(S_0,T) > c] = \hat{\mathbb{P}}(S_T > c + X) = \Phi\left[q^-\left(\frac{S_0 e^{r'T}}{c+X}\right)\right].$$

Differentiating each expression to get the p.d.f. on the corresponding interval and evaluating $e^{-rT}\int_0^\infty c\hat{f}(c)\cdot dc$ piecewise as $e^{-rT}\int_0^{K-X} c\hat{f}(c)\cdot dc +$

$e^{-rT}\int_{K-X}^{\infty} c\hat{f}(c)\cdot dc$, we have

$$
\begin{aligned}
C^{UI}(S_0,T) = S_0e^{-\delta T}\left(\frac{K}{S_0}\right)^{2r'/\sigma^2+1}&\left\{\Phi\left[q^+\left(\frac{K^2e^{r'T}}{S_0X}\right)\right]\right.\\
&\left.-\,\Phi\left[q^+\left(\frac{Ke^{r'T}}{S_0X}\right)\right]\right\}-Xe^{-rT}\left(\frac{K}{S_0}\right)^{2r'/\sigma^2-1}\\
&\times\left\{\Phi\left[q^-\left(\frac{K^2e^{r'T}}{S_0X}\right)\right]-\Phi\left[q^-\left(\frac{Ke^{r'T}}{S_0X}\right)\right]\right\}\\
&+S_0e^{-\delta T}\Phi\left[q^+\left(S_0e^{r'T}/K\right)\right]-Xe^{-rT}\left[q^-\left(S_0e^{r'T}/K\right)\right],
\end{aligned}
$$

where $r' = r - \delta$ and

$$q^{\pm}(x) = \frac{\ln x \pm \sigma^2 T/2}{\sigma\sqrt{T}}.$$

Upon setting $r' = 0$ and $S_0 = \mathsf{F}_0$, the formula applies to a call on a futures price. Notice that as either $K \downarrow S_0$ or $K \downarrow X$ the expression for $C^{UI}(S_0,T)$ reduces to the Black-Scholes formula for a vanilla European call on an asset paying continuous dividends.

The Down-and-Out Put

Letting $P^{DO}(S_0,T)$ denote the initial value of the down-and-out, we have

$$
\begin{aligned}
P^{DO}(S_0,T) &= e^{-rT}\hat{E}P^{DO}(S_T,0)\\
&= e^{-rT}\hat{E}(X-S_T)^+\mathbf{1}_{[K,\infty)}(\underline{S}_{0,T}),
\end{aligned}
$$

where $0 \le K < X$ and (again to exclude the trivial case) $S_0 > K$. We shall first deduce the c.d.f. of $P^{DO}(S_0,T)$ under the martingale measure, then derive the p.d.f. from that, then evaluate the expectation as $\int_0^\infty p\hat{f}(p)\cdot dp$. The c.d.f. of $P^{DO}(S_0,T)$ under $\hat{\mathbb{P}}$ is

$$
\hat{\mathbb{P}}(P \le p) = \begin{cases} 0, & p < 0\\ \hat{\mathbb{P}}(P=0) + \hat{\mathbb{P}}(\underline{S}_{0,T} > K, S_T > X - p), & 0 \le p < X - K\\ 1, & p \ge X - K, \end{cases}
$$

where

$$
\begin{aligned}
\hat{\mathbb{P}}(P=0) &= \hat{\mathbb{P}}[(S_T \ge X)\cup(\underline{S}_{0,T} \le K)]\\
&= \hat{\mathbb{P}}(S_T \ge X) + \hat{\mathbb{P}}(\underline{S}_{0,T} \le K) - \hat{\mathbb{P}}[(S_T \ge X)\cap(\underline{S}_{0,T} \le K)].
\end{aligned}
$$

Again letting $Y_t \equiv \gamma t + \hat{W}_t$ and $\sigma\gamma \equiv r - \delta - \sigma^2/2 \equiv r' - \sigma^2/2$, we have, for $0 \le p < X$,

$$\begin{aligned}
&\hat{\mathbb{P}}(\underline{S}_{0,T} > K, S_T > X - p)\\
&\quad = \hat{\mathbb{P}}\left[\underline{Y}_{0,T} \ge \sigma^{-1}\ln\left(\frac{K}{S_0}\right), Y_T \ge \sigma^{-1}\ln\left(\frac{X-p}{S_0}\right)\right]\\
&\quad = \hat{\mathbb{P}}\left[Y_T \ge \sigma^{-1}\ln\left(\frac{X-p}{S_0}\right)\right]\\
&\qquad - \hat{\mathbb{P}}\left[\underline{Y}_{0,T} < \sigma^{-1}\ln\left(\frac{K}{S_0}\right), Y_T \ge \sigma^{-1}\ln\left(\frac{X-p}{S_0}\right)\right].
\end{aligned}$$

An application of (7.56) reduces this to an expression involving the marginal distribution of Y_T,

$$\begin{aligned}
&\hat{\mathbb{P}}\left[Y_T \ge \sigma^{-1}\ln\left(\frac{X-p}{S_0}\right)\right]\\
&\qquad - e^{2\gamma\sigma^{-1}\ln(K/S_0)}\hat{\mathbb{P}}\left[Y_T \ge \sigma^{-1}\ln\left(\frac{X-p}{S_0}\right) - 2\sigma^{-1}\ln\left(\frac{K}{S_0}\right)\right];
\end{aligned}$$

and this simplifies to give the value of $\hat{\mathbb{P}}(P \le p)$ for $0 < p < X - K$ as

$$\hat{\mathbb{P}}(P = 0) + \Phi\left[q^-\left(\frac{S_0 e^{r'T}}{X-p}\right)\right] - \left(\frac{K}{S_0}\right)^{2r'/\sigma^2 - 1}\Phi\left[q^-\left(\frac{K^2 e^{r'T}/S_0}{X-p}\right)\right].$$

On $(0, X - K)$ the p.d.f. under the martingale measure is

$$\hat{f}(p) = \begin{cases} d\hat{\mathbb{P}}(P \le p)/dp, & 0 < p < X - K\\ 0, & \text{elsewhere} \end{cases}.$$

Differentiating and calculating $e^{-rT}\int_0^{X-K} p\hat{f}(p)\cdot dp$ ultimately give as the value of the down-and-out put, $P^{DO}(S_0, T)$,

$$\begin{aligned}
&Xe^{-rT}\left\{\Phi\left[q^-\left(\frac{S_0e^{r'T}}{K}\right)\right] - \Phi\left[q^-\left(\frac{S_0e^{r'T}}{X}\right)\right]\right\}\\
&\quad - S_0e^{-\delta T}\left\{\Phi\left[q^+\left(\frac{S_0e^{r'T}}{K}\right)\right] - \Phi\left[q^+\left(\frac{S_0e^{r'T}}{X}\right)\right]\right\}\\
&\quad - \left(\frac{K}{S_0}\right)^{2r'/\sigma^2-1} Xe^{-rT}\left\{\Phi\left[q^-\left(\frac{Ke^{r'T}}{S_0}\right)\right] - \Phi\left[q^-\left(\frac{K^2e^{r'T}}{S_0X}\right)\right]\right\}\\
&\quad + \left(\frac{K}{S_0}\right)^{2r'/\sigma^2+1} S_0e^{-\delta T}\left\{\Phi\left[q^+\left(\frac{Ke^{r'T}}{S_0}\right)\right] - \Phi\left[q^+\left(\frac{K^2e^{r'T}}{S_0X}\right)\right]\right\},
\end{aligned}$$

where $r' = r - \delta$ and

$$q^{\pm}(x) = \frac{\ln x \pm \sigma^2 T/2}{\sigma\sqrt{T}}.$$

Setting $r' = 0$ and $S_0 = \mathsf{F}_0$ produces the corresponding formula for a put on a futures price.

7.4.4 *Ladder Options*

Payoffs of barrier options depend on whether the price path of the underlying has at some point crossed the pertinent barrier, but given terminal price S_T there is only one possible terminal payoff. European ladder options are a bit more sophisticated. There are several barriers, called "rungs", and final payoffs depend on how many of these have been crossed, as well as on the terminal price. Consider first a ladder call having strike X and rungs $K_n > K_{n-1} > \cdots > K_1 > X$. For some $j \geq 1$ suppose the jth rung, K_j, is the highest one exceeded by the maximum price during $[0, T)$. Then the payoff at T is the greatest of 0, $S_T - X$, and $K_j - X$. If, however, the price never gets above the first rung, K_1, then the payoff is just the greater of 0 and $S_T - X$. As a notational device, introducing $K_0 \equiv 0$ allows payoffs in both cases to be expressed as

$$C^L(S_T, 0) = (S_T - X)^+ \vee (K_j - X)^+ = (S_T - X)^+ \vee (K_j - X).$$

Figure 7.7 depicts the payoffs of a 2-rung ladder call on two alternative price paths, both of which pass K_1 but fail to reach K_2. On the lower track the terminal price winds up below K_1, so the payoff is $K_1 - X$. On the upper track, where the terminal price is above K_1, the payoff is $S_T - X$. With a last piece of notation, $K_{n+1} \equiv +\infty$, the whole payoff structure can be captured succinctly as

$$C^L(S_T, 0) = \sum_{j=0}^{n} \left[(S_T - X)^+ \vee (K_j - X)\right] \cdot \mathbf{1}_{[K_j, K_{j+1})}(\overline{S}_{0,T}).$$

The ladder put's payoff works symmetrically. Taking $0 \equiv K'_{n+1} < K'_n < K'_{n-1} < \cdots < K'_1 < X$ and $K'_0 = +\infty$, it pays $(X - S_T)^+ \vee (X - K_j)^+$ if price ever drops below rung j, so that[12]

$$P^L(S_T, 0) = \sum_{j=0}^{n} \left[(X - S_T)^+ \vee (X - K'_j)\right] \cdot \mathbf{1}_{(K'_{j+1}, K'_j]}(\underline{S}_{0,T}).$$

[12] For the term $j = n$ note that $\widehat{\mathbb{P}}(\underline{S}_{0,T} = 0) = 0$.

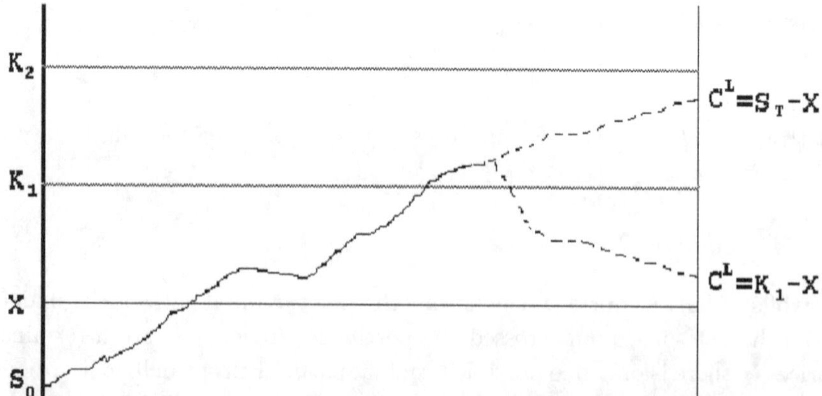

Fig. 7.7. Alternative payoffs of a ladder call.

Pricing a Ladder Call

We shall work out the arbitrage-free price of a call with n rungs to illustrate the application of martingale pricing to ladder options. As usual, we begin by developing the c.d.f. under $\hat{\mathbb{P}}$ from the option's payoff structure. An extra complication here that was avoided with barrier options is that the conditional distribution of the terminal payoff does change substantively as we move through time, since knowing that j rungs have been surpassed by t tells us that the payoff cannot be less than $(K_j - X)^+$. Fortunately, this causes no great difficulty.

Standing at $t \in [0,T)$, we observe current price S_t and the highest surpassed rung. That is, we observe S_t and a $\bar{j}_t \in \{0,1,\ldots,n\}$ such that $\overline{S}_{0,t} \in [K_{\bar{j}_t}, K_{\bar{j}_t+1})$, where we continue to take $K_0 = 0$ and $K_{n+1} = +\infty$. Then a little reflection shows that the conditional c.d.f. of the terminal payoff is $\hat{\mathbb{P}}_t[C^L(S_T,0) \le c] = 0$ for $c < (K_{\bar{j}_t} - X)^+$ and

$$\hat{\mathbb{P}}_t[C^L(S_T,0) \le c] = \hat{\mathbb{P}}_t(S_T \le c + X, \overline{S}_{t,T} < K_{j+1}) \tag{7.67}$$

for $(K_j - X)^+ \le c \le K_{j+1} - X$ and $j \in \{\bar{j}_t, \bar{j}_t + 1, \ldots, n\}$. For example, taking $j = 2$ and picking a c between $K_2 - X$ and $K_3 - X$, the payoff would be *larger* than c if either $S_T > c + X$ or the maximum price from t forward were at least K_3. Notice that the c.d.f. has discontinuities at values of c corresponding to all the rungs. Specifically, for $j \in \{\bar{j}_t, \bar{j}_t + 1, \ldots, n\}$ and $c = (K_j - X)^+$

$$\hat{\mathbb{P}}_t[C^L(S_T, 0) = c] = \hat{\mathbb{P}}_t(S_T \le c + X, K_j \le \overline{S}_{t,T} < K_{j+1}),$$

which is the probability that the jth rung is the highest one passed and that the terminal price ends up below it. (Note that $\hat{\mathbb{P}}_t(S_T = K_j) = 0$.) figure 7.8 illustrates for a 2-rung call.

The arbitrage-free price of the ladder call as of $t \in [0, T)$ is

$$\begin{aligned} C^L(S_t, \tau) &= e^{-r\tau} \hat{E}_t C^L(S_T, 0) \\ &= e^{-r\tau} \int_{[0,\infty)} c \cdot d\hat{\mathbb{P}}_t[C^L(S_T, 0) \le c], \end{aligned}$$

where $\tau = T - t$ and the integral is Lebesgue-Stieltjes. Separating out the atomic points that have positive probability mass, this is written as

$$\begin{aligned} C^L(S_t, \tau) = e^{-r\tau} \sum_{j=\bar{j}_t}^{n} \Big\{ &\int_{(K_j - X)^+}^{K_{j+1} - X} c\hat{\mathbb{P}}_t' \left[C^L(S_T, 0) \le c\right] \cdot dc \\ &+ (K_j - X)^+ \hat{\mathbb{P}}_t \left[C^L(S_T, 0) = (K_j - X)^+\right] \Big\}, \end{aligned} \quad (7.68)$$

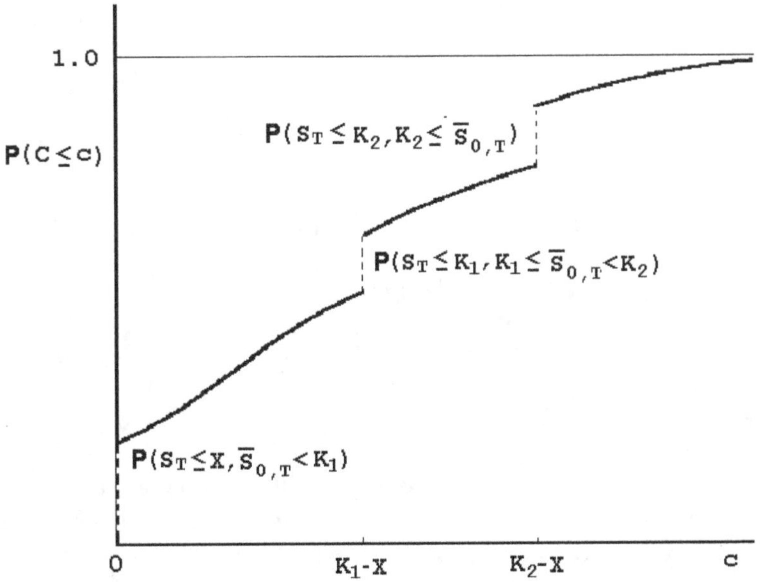

Fig. 7.8. Payoff c.d.f. for ladder call option.

where $\hat{\mathbb{P}}'_t(\cdot) \equiv d\hat{\mathbb{P}}_t(\cdot)/dc$. Now evaluate $\hat{\mathbb{P}}_t[C^L(S_T, 0) \le c]$ at the continuity points using (7.67) and (7.55), and then differentiate to get

$$\hat{\mathbb{P}}'_t[C^L(S_T,0) \le c] = \phi(z) \cdot \frac{dz}{dc} - \left(\frac{K_{j+1}}{S_t}\right)^{2r'/\sigma^2 - 1} \phi(z') \cdot \frac{dz'}{dc}.$$

Here

$$z \equiv q^+\left(\frac{c+X}{\mathsf{f}_t}\right),$$

$$z' \equiv q^+\left(\frac{c+X}{\mathsf{f}_t}\frac{S_t^2}{K_{j+1}^2}\right),$$

$$q^\pm(x) \equiv \frac{\ln x \pm \sigma^2\tau/2}{\sigma\sqrt{\tau}},$$

and $\mathsf{f}_t \equiv \mathsf{f}(t,T) = S_t e^{r'\tau}$ is the underlying's T-delivery forward price at t. Evaluating the integrals and probabilities in (7.68) leads finally to the following expression for $C^L(S_t, \tau)$:

$$\begin{aligned}
e^{-r\tau} \sum_{j=\bar{j}_t}^{n} \Bigg\langle & X\left(\frac{K_{j+1}}{S_t}\right)^{\frac{2r'}{\sigma^2}+1} \left\{\Phi\left[q^+\left(\frac{h^t_{j+1}S_t^2}{K_{j+1}^2}\right)\right] - \Phi\left[q^+\left(\frac{h^t_j S_t^2}{K_{j+1}^2}\right)\right]\right\} \\
& - \mathsf{f}_t\left(\frac{K_{j+1}}{S_t}\right)^{\frac{2r'}{\sigma^2}+1} \left\{\Phi\left[q^-\left(\frac{h^t_{j+1}S_t^2}{K_{j+1}^2}\right)\right] - \Phi\left[q^-\left(\frac{h^t_j S_t^2}{K_{j+1}^2}\right)\right]\right\} \\
& + \mathsf{f}_t\{\Phi[q^-(h^t_{j+1})] - \Phi[q^-(h^t_j)]\} - X\{\Phi[q^+(h^t_{j+1})] - \Phi[q^+(h^t_j)]\} \\
& + (K_j - X)^+ \left\{\left(\frac{K_j}{S_t}\right)^{\frac{2r'}{\sigma^2}-1} \Phi\left[q^+\left(\frac{h^t_j S_t^2}{K_j^2}\right)\right]\right. \\
& \left. - \left(\frac{K_{j+1}}{S_t}\right)^{\frac{2r'}{\sigma^2}-1} \Phi\left[q^+\left(\frac{h^t_j S_t^2}{K_{j+1}^2}\right)\right]\right\}\Bigg\rangle,
\end{aligned}$$

where $r' \equiv r - \delta$ and $h^t_j \equiv (K_j \vee X)/\mathsf{f}_t$.

7.4.5 *Asian Options*

Payoffs of "Asian" or "average-price" options depend on the average price of the underlying over all or part of the option's lifetime. There are two general classes. Payoffs of fixed-strike options depend on the difference between the average price of the underlying and a fixed strike price X; while variable-strike or floating-strike options depend on the difference between terminal

Table 7.3. Terminal values of fixed- and variable-strike Asian options.

Class	Put	Call
Fixed-strike	$(X - A_{[t^*,T]})^+$	$(A_{[t^*,T]} - X)^+$
Variable-strike	$(A_{[t^*,T]} - S_T)^+$	$(S_T - A_{[t^*,T]})^+$

price S_T and the average price. These options are almost always European. Table 7.3 states explicitly the terminal payoffs of the two classes of Asian puts and calls. Here $A_{[t^*,T]}$ represents the average over a specified period extending from some $t^* \geq 0$ up to T. Averaging is typically done arithmetically. (We discuss geometric-averaging in the last part of the section.) Depending on the application the average can be updated as often as daily, or it can be based on as few as 10–15 points over the course of several months.

Asians are attractive vehicles for hedging risks associated with payments that are distributed more or less evenly over time. For example, an exporter making continual sales abroad might find it convenient to hedge with an Asian currency option rather than a vanilla option that depends on the exchange rate at one date only. Asians also take some of the risk out of hedging in options on thinly traded commodities, whose price fluctuations on any one day may be very large. In this sense, too, the lower volatility associated with the average price tends to make Asians cheaper than otherwise identical vanilla options. Since both advantages are more pronounced for the fixed-strike class, they are by far the more common of the two types. We focus entirely on fixed-strike options here. Levy (1992) discusses the pricing of variable-strike Asians.

As we have seen, the Black-Scholes framework is well suited for pricing many path-dependent options, because it lets us exploit known facts about hitting times and extrema of Brownian motion. Unfortunately, modeling underlying prices as geometric Brownian motion does not facilitate the martingale pricing of arithmetic Asian options. The reason is that there is no simple representation for the distribution of a discrete arithmetic average of jointly lognormal variates. (As we shall see, things are much easier in the case of geometric averages.) Inevitably, simple formulas for arbitrage-free prices within the Black-Scholes framework must be approximations of one form or another, and many ways of producing these have been proposed in what is now a large literature on the subject. Our survey begins by introducing notation and deriving some general conditions that arbitrage-free prices of fixed-strike, arithmetic Asians must satisfy. We then summarize some of the solutions that have been put forward before turning to a detailed treatment of one method that seems to offer a particularly good balance among conceptual simplicity, quickness of computation, and accuracy. The final part develops the analytical solution for prices of Asians whose payoffs depend on continuous geometric averages of price.

Notation and Useful Identities

Consider a fixed-strike Asian option issued at $t = 0$, expiring at T, and having terminal payoff depending on $A_{[t^*,T]} = (N+1)^{-1}\sum_{j=0}^{N} S_{t_j}$, which is the arithmetic average of prices at the $N+1$ discrete times $0 \le t^* \equiv t_0 < t_1 < \cdots < t_N \equiv T$. For $t \in [0,T)$ let $N_t = \max\{j : t_j < t\} - 1$. That is, N_t is one fewer than the number of "fixing points" already passed as of time t. Define the average price up to any $t \in [0,T]$ as

$$A_{[t^*,t]} = \begin{cases} 0, & t < t^* \\ (N_t+1)^{-1}\displaystyle\sum_{j=0}^{N_t} S_{t_j}, & t^* \le t \le T. \end{cases}$$

We can then decompose $A_{[t^*,T]}$ into a part that has already been observed and a part that has not:

$$\begin{aligned} A_{[t^*,T]} &= (N+1)^{-1}\left[\sum_{j=0}^{N_t} S_{t_j} + \sum_{j=N_t+1}^{N} S_{t_j}\right] \\ &= \frac{N_t+1}{N+1}A_{[t^*,t]} + \frac{N-N_t}{N+1}A_{(t,T]}. \end{aligned}$$

The price of an Asian fixed-strike call at $t < T$ can, therefore, be written as

$$\begin{aligned} C^{AF}(S_t, A_{[t^*,t]}; T-t) &= e^{-r(T-t)}\hat{E}_t(A_{[t^*,T]} - X)^+ \\ &= e^{-r(T-t)}\hat{E}_t\left(\frac{N_t+1}{N+1}A_{[t^*,t]} + \frac{N-N_t}{N+1}A_{(t,T]} - X\right)^+ \\ &= \frac{N-N_t}{N+1}e^{-r(T-t)}\hat{E}_t(A_{(t,T]} - X_t)^+, \end{aligned} \tag{7.69}$$

where

$$X_t \equiv \frac{N+1}{N-N_t}X - \frac{N_t+1}{N-N_t}A_{[t^*,t]}. \tag{7.70}$$

Thus, at any t the problem boils down to pricing a call on the average price during the option's remaining life.

As Levy (1992) points out, the no-arbitrage condition requires that puts be priced off the call (or *vice versa*) using a version of put-call parity that comes from the following argument. A long position in a put coupled with a short position in a call at the same fixed strike guarantees a cash receipt at expiration equal to

$$X - A_{[t^*,T]} = \left[X - (N+1)^{-1}\sum_{j=0}^{N_t} S_{t_j}\right] + (N+1)^{-1}\sum_{j=N_t+1}^{N} S_{t_j}.$$

This future receipt can be exactly replicated at t by borrowing

$$e^{-r(T-t)}\left[X - (N+1)^{-1}\sum_{j=0}^{N_t} S_{t_j}\right]$$

currency units and entering forward contracts to deliver $(N+1)^{-1}e^{-r(T-t_j)}$ of the underlying "stock" at each $t_j \in \{N_t+1,\ldots,N\}$. To see that this is the right replicating strategy, note that at each such t_j one can meet the obligation to deliver stock by borrowing $(N+1)^{-1}e^{-r(T-t_j)}S_{t_j}$ currency units to finance the purchase, thereby committing to repay $(N+1)^{-1}S_{t_j}$ at time T. The present value of the receipt $X - A_{[t^*,T]}$ is then

$$V(t,T) = e^{-r(T-t)}\left[X - (N+1)^{-1}\sum_{j=0}^{N_t} S_{t_j}\right] + \frac{e^{-r(T-t)}}{N+1}\sum_{j=N_t+1}^{N} \mathsf{f}(t,t_j), \tag{7.71}$$

where $\mathsf{f}(t,t_j) = S_t e^{(r-\delta)(t_j-t)}$ is the forward price at t for t_j delivery, assuming a continuous yield at rate δ. Therefore, $V(t,T)$ must equal the difference in the arbitrage-free prices of the put and call at t.

A Survey of Proposed Pricing Solutions

Having seen that the key to valuing the arithmetic-average Asian call is carrying out the expectation in (7.69), we can now look at some ways that have been proposed to do this. To simplify notation a bit, suppose we stand at $t \in [0,T)$ and that t_j is the time at which the next price enters the average. Set $\tau_j = t_j - t$, $\tau_{j+1} = t_{j+1} - t_j$, and so on, with $\tau_N = T - t_{N-1}$. Thus, τ_j is the time until the next price enters the average, while the remaining τ's are times between successive price fixings. Letting

$Y_k \equiv (r-\delta-\sigma^2/2)\tau_k+\sigma\hat{W}_{\tau_k}$ for $k \in \{j, j+1, \ldots, N\}$, where $\{\hat{W}_t\}$ is a Brownian motion under $\hat{\mathbb{P}}$, we have $S_{t_j} = S_t \exp(Y_j)$, $S_{t_{j+1}} = S_t \exp(Y_j+Y_{j+1}) = S_{t_j} \exp(Y_{j+1})$, and so on, with $S_T = S_t \exp(\sum_{k=j}^{N} Y_k) = S_{t_{N-1}} \exp(Y_N)$. Each price can thus be built up from the previous one by multiplying by an exponentiated normal (that is, a lognormal) variate that is independent of all that have gone before.

This recursive setup lends itself ideally to Monte Carlo simulation. The way it works is very simple. Beginning with initial price S_t, we take a draw Z from a standard normal distribution, scale by $\sigma\sqrt{\tau_j}$, add mean $(r-\delta-\sigma^2/2)\tau_j$, and obtain a realization of S_{t_j}, the first new price to enter the average, as

$$S_{t_j} = S_t \exp[(r-\delta-\sigma^2/2)\tau_j + \sigma\sqrt{\tau_j}Z].$$

Taking another draw and constructing Y_{j+1} we generate $S_{t_{j+1}}$, and so on to the end. Calculating the average of these prices yields one realization of $(A_{(t,T]}-X)^+$. Repeating the procedure $M-1$ more times delivers a sample of M realizations, which are simply averaged to estimate $\hat{E}_t(A_{(t,T]}-X)^+$. With large enough M (and enough computer time) any desired degree of accuracy can be attained. We discuss in chapter 11 some ways to enhance the efficiency of the process, including the use of special control variates for valuing continuously averaged Asian options. Although slower than analytical approximations, simulation provides the standard against which the accuracy of other methods has been judged.

Another computationally intensive procedure, proposed by Carverhill and Clewlow (1990), provides what is essentially an exact numerical solution to the problem. To understand the idea it is best to start with a specific case as an example. Take $j=1$, so that t_1 is the next time at which a price enters the average, and suppose there are in all $N=3$ more prices to come. We then have

$$\begin{aligned} A_{(t,T]} &= (S_{t_1}+S_{t_2}+S_{t_3})/3 \\ &= S_t(e^{Y_1}+e^{Y_1+Y_2}+e^{Y_1+Y_2+Y_3})/3 \\ &= S_t\{e^{Y_1}[1+e^{Y_2}(1+e^{Y_3})]\}/3. \end{aligned}$$

Starting at the end of the chain, the first step is to find the p.d.f. of $Q_2 \equiv \ln(1+e^{Z_3})$, where (to appreciate the recursive pattern) we take $Z_3 \equiv Y_3$. Since Y_3 is distributed under $\hat{\mathbb{P}}$ as $N[(r-\delta-\sigma^2/2)\tau_3, \sigma^2\tau_3]$, finding $f_{Q_2}(\cdot)$

requires just an application of the change-of-variable formula. Thus,

$$f_{Q_2}(q) = \phi[\ln(e^q - 1); (r - \delta - \sigma^2/2)\tau_3, \sigma^2\tau_3] \left| \frac{d}{dq} \ln(e^q - 1) \right|$$
$$= \phi\left[\ln(e^q - 1); (r - \delta - \sigma^2/2)\tau_3, \sigma^2\tau_3\right] \left(\frac{e^q}{e^q - 1}\right),$$

where $\phi(\cdot; \alpha, \beta^2)$ is the normal density with mean α and variance β^2. Evaluating $f_{Q_2}(q)$ on a discrete set of points $q \in \mathcal{Q}$, the characteristic function, $\Psi_{Q_2}(\zeta) = \int_{-\infty}^{\infty} e^{i\zeta q} f_{Q_2}(q) \cdot dq$, is then found on a discrete set $\zeta \in \Xi$ using the fast Fourier transform (f.F.t.), an algorithm that provides an efficient way of doing the calculation.[13] The c.f. of Y_2 can of course be expressed analytically. Taking the product of this and $\Psi_{Q_2}(\zeta)$ yields $\Psi_{Z_2}(\zeta)$, which is the c.f. of

$$Z_2 \equiv Y_2 + Q_2 = \ln[e^{Y_2}(1 + e^{Y_3})].$$

The density of Z_2, which is the convolution of the densities of Y_2 and Q_2, is then evaluated on the grid $\mathcal{Q}$ by inverting $\Psi_{Z_2}(\zeta)$ (again using the f.F.t.), as

$$f_{Z_2}(z) = \frac{1}{2\pi} \int_{-\infty}^{\infty} e^{-i\zeta z} \Psi_{Z_2}(\zeta) \cdot d\zeta.$$

Another application of the change-of-variable formula gives the density of $Q_1 \equiv \ln(1 + e^{Z_2})$, from which its c.f., $\Psi_{Q_1}(\zeta)$, is then found. Taking the product of $\Psi_{Q_1}(\zeta)$ and $\Psi_{Y_1}(\zeta)$ gives $\Psi_{Z_1}(\zeta)$, the c.f. of

$$Z_1 \equiv Y_1 + Q_1 = \ln\{e^{Y_1}[1 + e^{Y_2}(1 + e^{Y_3})]\},$$

which is again inverted to find the density. Finally, one more change of variable yields the density of

$$A_{(t,T]} = S_t e^{Z_1}/3.$$

Having thought through the specific case, it is easy to see how the general procedure goes. Let the next time at which a price enters the average be t_j, and suppose there are in all $N - j + 1$ more prices to enter. Starting off by defining $Z_N \equiv Y_N$, proceed step by step for $k = N - 1, N - 2, \ldots, j$ to find the density of $Q_k = \ln(1 + e^{Z_{k+1}})$ by the change-of-variable formula, and then the density of $Z_k = Y_k + Q_k$ by using the c.f.s to convolve f_{Y_k} and

[13]The f.F.t. is discussed in connection with stochastic volatilty models in chapters 8 and 9. Routine `INVFFT` on the accompanying CD implements the f.F.t. procedure in Press *et al.* (1992) to invert the c.f. of a centered, scaled random variable.

f_{Q_k}. Having worked all the way back to Z_j, find the p.d.f. of $A_{(t,T]}$ from the transformation $(S_t e^{Z_j})/(N-j+1)$ and use it to calculate $\hat{E}_t(A_{(t,T]} - X)^+$.

Since both this approach and Monte Carlo require extensive computation, there have been many attempts to develop fast approximations. In certain cases simplifications can be achieved by treating the averaging as continuous, in which case

$$A_{(t,T]} = (T-t)^{-1} \int_t^T S_u \cdot du.$$

In the special situation that the average to date, $A_{[t^*,t]}$, is such as to assure that the option will finish in the money, Geman and Yor (1992, 1993) and Geman and Eydeland (1995) present an appealing closed-form solution. However, dealing with the usual case that the option's terminal moneyness is not yet known still requires a lot of calculation. Milevsky and Posner (1998) develop a relatively simple pricing formula using the fact that as $T \to \infty$ the reciprocal of $\int_t^T S_u \cdot du$ converges in distribution to the gamma under the condition $r - \delta - \sigma^2/2 < 0$. This condition would obviously hold for futures, but there are many applications in which it would not.

Method-of-Moments Solutions

Several relatively simple techniques for pricing Asian options are based on lognormal approximations to the distribution of $A_{(t,T]}$. Such approximations are often quite good, especially so when the variability of the variates composing the average is not too large. Recall that if X is distributed as lognormal, its distribution is characterized by just two parameters, the mean and variance of $\ln X$. Therefore, the task of fitting a lognormal distribution to $A_{(t,T]}$ comes down to choosing these two parameters. Levy and Turnbull (1992) describe several such methods, of various degrees of complexity. One of the simplest, which nevertheless works quite well when the annualized volatility of the underlying price is no more than 0.30, is a procedure proposed by Levy (1992). We now consider this method in detail.

Levy's approach works much like the classical method of moments in statistical estimation. Begin with the first two moments of $A_{(t,T]}$ about the origin—$\mu_1' = \hat{E}_t A_{(t,T]}$ and $\mu_2' = \hat{E}_t A_{(t,T]}^2$, say—and then find parameters $\alpha = \alpha(t,T)$ and $\beta = \beta(t,T)$ that equate the corresponding lognormal moments to these by solving the equations $\mu_1' = Ee^{\alpha+\beta Z} = e^{\alpha+\beta^2/2}$ and $\mu_2' = E(e^{2\alpha+2\beta Z}) = e^{2\alpha+2\beta^2}$. The solutions are

$$\alpha = 2\ln\mu_1' - \ln\sqrt{\mu_2'}, \quad \beta^2 = \ln\mu_2' - 2\ln\mu_1'. \tag{7.72}$$

To find expressions for μ_1' and μ_2' it is helpful to note first that the price of the underlying at t_j is $S_{t_j} = S_t e^{R(N_t+1,j)}$ for $j \in \{N_t+1, \ldots, N\}$, where

$$R(i,j) \equiv \sum_{\ell=i}^{j} Y_{\tau_\ell} \sim N[(r-\delta-\sigma^2/2)(t_j - t_i), \sigma^2(t_j - t_i)].$$

The first moment is then

$$\begin{aligned}
\mu_1' &= (N-N_t)^{-1} S_t \sum_{j=N_t+1}^{N} \hat{E}_t e^{R(N_t+1,j)} \\
&= (N-N_t)^{-1} S_t \sum_{j=N_t+1}^{N} e^{(r-\delta-\sigma^2/2)(t_j-t)+\sigma^2(t_j-t)/2} \\
&= (N-N_t)^{-1} S_t \sum_{j=N_t+1}^{N} e^{(r-\delta)(t_j-t)}, \qquad (7.73)
\end{aligned}$$

and the second moment is

$$\begin{aligned}
\mu_2' &= \left(\frac{S_t}{N-N_t}\right)^2 \sum_{j=N_t+1}^{N} \sum_{k=N_t+1}^{N} \hat{E}_t e^{R(N_t+1,j)+R(N_t+1,k)} \\
&= \left(\frac{S_t}{N-N_t}\right)^2 \left[\sum_{j=N_t+1}^{N} \hat{E}_t e^{2R(N_t+1,j)} \right. \\
&\qquad \left. + 2 \sum_{j=N_t+1}^{N-1} \sum_{k=j+1}^{N} \hat{E}_t e^{2R(N_t+1,j)+R(j+1,k)} \right].
\end{aligned}$$

Using the independence of $R(N_t+1, j)$ and $R(j+1, k)$, evaluating the expectations, and simplifying give

$$\begin{aligned}
\mu_2' = \left(\frac{S_t}{N-N_t}\right)^2 & \left[\sum_{j=N_t+1}^{N} e^{2(r-\delta+\sigma^2/2)(t_j-t)} \right. \\
& \left. + 2 \sum_{j=N_t+1}^{N-1} \sum_{k=j+1}^{N} e^{(r-\delta+\sigma^2/2)(t_j+t_k-2t)} \right].
\end{aligned}$$

Putting this all together using (7.69) gives as the two-moment approximation to the value $C^{AF}(S_t, A_{[t^*,t]}; T-t)$ of the fixed-strike Asian call the expression

$$\frac{N-N_t}{N+1} e^{-r(T-t)} \left[\mu_1 \Phi\left(\frac{\alpha - \ln X_t}{\beta} + \beta\right) - X_t \Phi\left(\frac{\alpha - \ln X_t}{\beta}\right)\right],$$

where X_t is given in (7.70) and α and β are as in (7.72). The approximate arbitrage-free price of the put would then be determined from the parity relation as

$$P^{AF}(S_t, A_{[t^*,t]}; T-t) = C^{AF}(S_t, A_{[t^*,t]}; T-t) + V(t,T),$$

where $V(t,T)$ is given in (7.71).

Asians Based on Geometric Averages

The fact that products of jointly lognormal variates are themselves lognormal makes the pricing of geometric-average, fixed-strike Asian options much more straightforward. With averaging starting at $t^* = 0$ and N fixings at times $0 = t_0 < t_1 < \cdots < t_{N-1} < t_N = T$ the (time-weighted) geometric mean is

$$G_N[0,T] = \prod_{j=1}^{N} S_{t_j}^{(t_j - t_{j-1})/T} = \exp\left[T^{-1}\sum_{j=1}^{N} \ln S_{t_j}(t_j - t_{j-1})\right].$$

Of course, if the spacing is equal with $t_j = jT/N$, then each factor $(t_j - t_{j-1})/T$ is just N^{-1}. Taking $S_t = S_0 \exp(\nu t + \sigma \hat{W}_t)$, where $\nu \equiv r - \delta - \sigma^2/2$ and $\{\hat{W}_t\}_{t\geq 0}$ is a Brownian motion under $\hat{\mathbb{P}}$, this is

$$G_N[0,T] = S_0 \exp\left[T^{-1}\sum_{j=1}^{N}(\nu t_j + \sigma \hat{W}_{t_j})(t_j - t_{j-1})\right].$$

The distribution of $G_N[0,T]$ is clearly lognormal, and it is a simple matter to work out the mean and variance of $\ln G_N[0,T]$. We consider instead the pricing of puts and calls whose payoffs depend on the continuous mean. This is defined as

$$\begin{aligned} G_{[0,T]} &\equiv \exp\left(T^{-1}\int_0^T \ln S_t \cdot dt\right) \\ &= S_0 \exp\left(\nu T/2 + \sigma T^{-1}\int_0^T \hat{W}_t \cdot dt\right) \end{aligned}$$

for $T > 0$, and $G_{[0,0]} \equiv S_0$.

We first need the distribution of $G_{[0,T]}$ for $T > 0$. Example 33 on page 98 showed that $\int_0^T \hat{W}_t \cdot dt$ is distributed as $N(0, T^3/3)$, so

$$\ln(G_{[0,T]}/S_0) = \nu T/2 + \sigma T^{-1}\int_0^T \hat{W}_t \cdot dt \sim N\left(\nu T/2, \sigma^2\sqrt{T/3}\right).$$

For $t \in [0,T]$ we can therefore write

$$G_{[0,t]} \sim S_0 \exp\left(\nu t/2 + \sigma \tilde{W}_t/\sqrt{3}\right), \tag{7.74}$$

where $\{\tilde{W}_t\}_{t\geq 0}$ is a $\hat{\mathbb{P}}$ Brownian motion. However, to price an option at any $t \in (0,T)$ when S_t and $G_{[0,t]}$ are known requires knowing the conditional distribution of $G_{[0,T]}$ given $\mathcal{F}_t$. Letting $\tau \equiv T - t$, we can express $G_{[0,T]}$ in terms of $G_{[0,t]}$ and S_t as

$$\begin{aligned}
G_{[0,T]} &= G_{[0,t]}^{t/T} G_{[t,T]}^{\tau/T} \\
&= G_{[0,t]}^{t/T} \left\{ S_0 \exp\left[\tau^{-1} \int_t^T (\nu s + \sigma \hat{W}_s) \cdot ds\right]\right\}^{\tau/T} \\
&= G_{[0,t]}^{t/T} \left\{ S_0 e^{\nu t + \sigma \hat{W}_t} \exp\left[\frac{\nu}{\tau}\int_t^T (s-t)\cdot ds + \frac{\sigma}{\tau}\int_t^T (\hat{W}_s - \hat{W}_t)\cdot ds\right]\right\}^{\tau/T} \\
&= G_{[0,t]}^{t/T} \left[S_t \exp\left(\frac{\nu}{\tau}\int_0^\tau u \cdot du + \frac{\sigma}{\tau}\int_0^\tau \overline{W}_u \cdot du\right)\right]^{\tau/T} \\
&= G_{[0,t]}^{t/T} S_t^{\tau/T} \exp\left[\frac{\tau}{T}\left(\nu\tau/2 + \frac{\sigma}{\tau}\int_t^T \overline{W}_u\right)\cdot du\right],
\end{aligned}$$

where $\{\overline{W}_u\}_{u\geq 0} = \{\hat{W}_{t+u} - \hat{W}_t\}_{u \geq 0}$ is a standard $\hat{\mathbb{P}}$ Brownian motion. Using (7.74) gives

$$G_{[0,T]} \sim G_{[0,t]}^{t/T} S_t^{\tau/T} \exp\left[\frac{\tau}{T}\left(\nu\tau/2 + \sigma\tilde{W}_\tau/\sqrt{3}\right)\right]. \tag{7.75}$$

To put this to work in the simplest way, set $Y_t \equiv G_{[0,t]}^{t/T} S_t^{\tau/T}$ so that (7.75) corresponds to

$$Y_T = Y_t \exp[(r - \delta' - \beta^2/2)\tau + \beta\tilde{W}_\tau], \tag{7.76}$$

where $\beta = \sigma\tau/(T\sqrt{3})$ and $\delta' = r - \nu\tau/(2T) - \sigma^2\tau^2/(6T^2)$. The correspondence between (7.75) and (7.76) lets us express prices of continuous geometric-average, fixed-strike Asian options in terms of the Black-Scholes formula on an underlying with price Y_t, volatility β, and dividend rate δ'. For example, the arbitrage-free value, $P^{AF}(S_t, G_{[0,t]}; \tau)$, of a T-expiring put struck at X is

$$BX\Phi\left[q^+\left(\frac{BXe^{\delta'\tau}}{G_{[0,t]}^{t/T} S_t^{\tau/T}}\right)\right] - G_{[0,t]}^{t/T} S_t^{\tau/T} e^{-\delta'\tau}\Phi\left[q^-\left(\frac{BXe^{\delta'\tau}}{G_{[0,t]}^{t/T} S_t^{\tau/T}}\right)\right],$$

where $B \equiv B(t,T)$ is the price at t of a T-maturing riskless bond and $q^{\pm}(x) \equiv (\ln x \pm \beta^2\tau/2)/(\beta\sqrt{\tau})$. When $t = 0$ (in which case $\tau = T$ and $G_{[0,0]} = S_0$), this simplifies to

$$P^{AF}(S_0;T) = BX\Phi\left[q^{+}\left(\frac{BX_0 e^{\delta' T}}{S_0}\right)\right] - S_0 e^{-\delta' T} q^{-}\left(\frac{BX_0 e^{\delta' T}}{S_0}\right), \tag{7.77}$$

with $\delta' = r - \nu/2 - \sigma^2/6 = (r + \delta + \sigma^2/6)/2$.

8

Models with Uncertain Volatility

This chapter and the next treat the pricing of derivatives when the dynamics of underlying prices are modeled in ways that fit the data better than does Black-Scholes. We continue to consider just derivatives on stocks, indexes, and currencies, whose prices are relatively insensitive to stochastic variation in interest rates. Thus, we still treat future **R**ates and bond prices as **K**nown in accord with assumption RK from chapter 4.

8.1 Empirical Motivation

Two sets of observations motivate the search for richer models than Black-Scholes dynamics:

- Empirical findings that underlying prices themselves violate the salient predictions of that model; and
- Systematic discrepancies between actual market prices of derivatives and those predicted within the Black-Scholes framework.

We review the evidence on both counts.

8.1.1 *Brownian Motion Does Not Fit Underlying Prices*

The key predictions of the geometric Brownian motion model are (i) that changes in the logarithm of price over fixed intervals of time are normally distributed, (ii) that increments of log price over nonoverlapping time periods of equal length are independent and identically distributed, and (iii) that price paths are almost surely continuous. There is compelling evidence that prices of financial assets and commodities violate all three.

The nonnormality of marginal distributions of logs of price relatives (that is, of continuously compounded returns) of stocks, indexes, currencies, commodities, and even real estate over fixed intervals is documented in many studies.[1] Empirical distributions of changes in log price are usually found to have thicker tails than the normal, indicating a higher frequency of outliers than would be consistent with a Gaussian law with the same mean and variance. Although the divergence from normality is slight when returns are measured over periods as long as one month, it is very prominent in daily returns. For example, in daily data for common stocks estimates of kurtosis, $\alpha_4 = \mu_4/\sigma^4$, are often in the 5.0–10.0 range *vs.* 3.0 for the normal distribution.

Consistent with the thick tails in the data, but inconsistent with the independent-increments feature of geometric Brownian motion, are the observations that absolute (or squared) returns (e.g., $|\ln(S_t/S_{t-1})|$) are highly predictable from their past and that shocks to the returns process tend to be highly persistent. One interpretation of these findings is that the volatility parameter that pertains at any point in time depends in some way on past price realizations. Evidence of variable and persistent volatility is given in Chou (1988) and Bollerslev *et al.* (1992).[2] There are also indications that the variance of $\ln(S_t/S_{t-1})$ for stocks tends to be inversely related to the initial price level. Often called the "leverage effect" since its early description by Black (1976a) and the studies by Schmalensee and Trippi (1978) and Christie (1982), the standard explanation has been that firms financed partly by debt become more highly leveraged when their stocks' prices fall, making the stocks riskier and their prices more volatile. While leverage *per se* now seems unlikely to be the entire explanation,[3] there is no doubt that volatility does move inversely to the price level for many financial assets.

Temporal variation in dispersion implies that empirical marginal distributions of returns are mixtures—that is, draws from distributions that

---

[1]For example, Mandelbrot (1963), Fama (1965), Officer (1972), Praetz (1972), Clark (1973), Blattberg and Gonedes (1974), Boothe and Glassman (1987), Young and Graff (1995), Mandelbrot and Hudson (2004). McCulloch (1996) and McDonald (1996) provide detailed surveys of this literature.

[2]See Ghysels *et al.* (1996) and Palm (1996) for extensive references that further document the conditional heteroskedasticity in returns and describe efforts to model it.

[3]For example, it appears that volatility of stock prices rises and stays higher after stock splits that involve no substantive change in financial structure. For the evidence see Ohlson and Penman (1985) and Dubofsky (1991).

differ in scale. The finding of excess kurtosis in the marginal distributions of returns is consistent with this; however, the extreme thickness of tails of empirical distributions of daily returns seems too large to be explained by standard models of time-varying volatility with Gaussian errors. Accounting for the extreme price movements that occur occasionally across entire markets (as in U.S. stock markets on 19 October 1987) and much more often for individual assets seems to call for models that allow for discontinuities in price paths.

8.1.2 *Black-Scholes No Longer Fits Option Prices*

Observations of derivatives' prices also indicate the need for richer models of price dynamics. Section 6.4.3 described the anomalous nature of post-1987 Black-Scholes implicit volatilities derived from market prices of stock. The characteristic smile and smirk effects, which are also prominent in prices of options on stock indexes and currencies, are depicted stylistically in figure 6.4. The smile effect is that implicit volatilities of options with strikes far from the current underlying price are higher than for options near the money. This means that Black-Scholes prices for options at the extremes of moneyness are too low relative to market prices when calculated with volatilities backed out from prices of near-the-money options. The effect is qualitatively consistent with the evidence that risk-neutral distributions of logs of terminal prices are thick-tailed, since out-of-money options are more apt to end up in the money when the probability of outliers is high. The smirk effect, manifested in the steeply descending left branch of the plot of implicit volatility vs. X/S_t, indicates that Black-Scholes especially underprices out-of-money puts. The high valuations the market places on these instruments are consistent with an excess of probability mass in the left tail of the risk-neutral distribution of terminal price; that is, with a left-skewed conditional distribution of log price or log return, $\ln(S_T/S_t)$. This effect is consistent with a special fear of market reversals like that of October 1987. It could reflect either the perception that such price drops are likely to reoccur, or else that people put high values on contingent claims that pay off in such states of the world.

Shapes of implicit-volatility curves typically vary over time, and the changes sometimes seem to reflect specific political and economic events. For example, in S&P 500 futures options Kochard (1999) finds evidence of enhanced left skewness in the risk-neutral distribution of $\ln(S_T/S_t)$ around the time of the 1991 Gulf War. He also documents a systematic increase in left skewness following large negative price moves in the futures. There is no obvious way to determine whether these short-term variations come mainly from changes in subjective probabilities or from changes in tastes—subjective valuations.

There is also a term structure associated with implicit volatilities. Smile and smirk effects are most pronounced in prices of options with short maturities, the volatility curve flattening out as time to expiration increases. Also, both implicit volatilities at particular levels of moneyness and averages of volatilities *across* levels may vary systematically with time to expiration. Even these relations can be unstable. Taylor and Xu (1994) find in currency options that the slopes of volatility-to-maturity curves sometimes change sign over time.[4]

This chapter and the next treat models that can capture at least some of the discrepant features observed in prices of underlying assets and derivatives. The present chapter focuses on the modeling of volatility within the framework of continuous Itô processes. We first consider models in which volatility coefficients depend on the current price level or on current and past prices. Since the Wiener process that drives the underlying price is the only source of risk, these models permit dynamic replication of derivatives' payoffs and are therefore consistent with the existence of a unique martingale measure. In this setting, as in the Black-Scholes environment, derivatives prices can be determined from arbitrage considerations alone. However, the equivalent martingale measure is no longer unique when there are other risks that cannot be hedged with traded assets. Section 8.3 considers models in which the stochastic volatility process cannot be so replicated. We shall see that in this situation there may be many preference-dependent prices, all consistent with the absence of arbitrage. The same situation is encountered in chapter 9, where we first consider models of discontinuous price processes.

8.2 Price-Dependent Volatility

This section deals primarily with the following class of models for underlying prices under risk-neutral measure $\hat{\mathbb{P}}$:

$$dS_t/S_t = r_t \cdot dt + \sigma(S_t, t) \cdot d\hat{W}_t, \tag{8.1}$$

[4]Bates (1996b) gives an extensive survey of empirical findings on the nature and evolution of volatility curves.

where short-rate process $\{r_t\}_{t\geq0}$ is deterministic and $\{\hat{W}_t\}_{t\geq0}$ is a Brownian motion. The key feature is that $\sigma(S_t, t)$ depends on the current value of the underlying, while still satisfying the regularity condition $\int_0^T \sigma(S_t, t)^2 \cdot dt < \infty$ a.s. for each finite T. Models of this sort were first proposed by Cox and Ross (1976) to capture the leverage effect, whereby volatility varies inversely with the price level. We begin by examining certain qualitative features of option prices under this general setup, discovering that they have much in common with those of Black-Scholes. Specific computational formulas for prices of European options can be worked out under certain specifications of $\sigma(t, S_t)$. We derive two such formulas, then describe how numerical methods can be used to handle more general cases. Unfortunately, there is evidence that some fairly flexible specifications of $\sigma(t, S_t)$ still fail to predict option prices out of sample better than does an *ad hoc* version of Black-Scholes with strike-specific volatilities. This suggests that a still richer class of models is needed, such as those in the following section that allow for additional stochastic influences on volatility. However, there is an alternative model that is more in the spirit of (8.1). We shall see that time-varying smiles and smirks might be well accounted for without additional risk sources by allowing σ to depend on past as well as present values of the underlying price.

8.2.1 *Qualitative Features of Derivatives Prices*

Most of the qualitative features of derivatives prices in the Black-Scholes setting turn out to extend to general diffusion processes. Recall that diffusion processes are Itô processes with the Markov property, and that property is preserved if volatility depends on time and the current price of the underlying, as in (8.1). Comparing the models' features, option prices remain weakly monotone and convex in the prices of the underlying, and the comparative statics with respect to time to expiration and volatility are also qualitatively the same as for Black-Scholes. Monotonicity and convexity hold as well when volatility depends on some other diffusion process (or processes) whose drift and volatility parameters do not depend on the price of the underlying asset. However, some strange behavior is possible when the latter condition is violated.

These conclusions are due to Bergman *et al.* (1996). Many of them follow from a simple and intuitive result that they call the "no-crossing lemma". Let $\{S_t\}_{0\leq t}$ be a diffusion process on a filtered probability space $(\Omega, \mathcal{F}, \{\mathcal{F}_t\}, \mathbb{P})$, whose motion under the risk-neutral measure is described by (8.1). Of course, to each realization $\{\hat{W}_t(\omega)\}_{t\geq 0}$ of a Brownian path there corresponds a path $\{S_t(\omega)\}_{t\geq 0}$ for price.

Lemma 10 *Suppose $\{S_t'(\omega)\}_{t\geq 0}$ and $\{S_t''(\omega)\}_{t\geq 0}$ are processes with the same drift and volatility parameters that are generated by the same Brownian path $\{\hat{W}_t(\omega)\}_{t\geq 0}$ but from different initial values, $S_0' > S_0''$. Then $S_t'(\omega) \geq S_t''(\omega)$ for all $t \in [0,T]$ and almost all ω.*

The restriction "almost all ω" just means that the exceptional outcomes for which the result fails form a set of $\hat{\mathbb{P}}$ measure zero. The proof exploits the continuity and Markov behavior of diffusion processes. By continuity, $S_t'' > S_t'$ for some t only if the paths crossed at some $\tau \in (0,t)$. But from the Markov property, $S_\tau' = S_\tau''$ implies that $S_t' = S_t''$ for all $t > \tau$, which is a contradiction. The processes could, in fact, merge in this way by hitting an absorbing barrier, such as the origin.

A simple application of this result is to show that the delta of a European-style derivative on an underlying asset whose price follows (8.1) is bounded by the extremes of the slope of the terminal payoff function with respect to S_T. For example, the delta of a European call on a no-dividend stock is bounded by zero and unity. When the terminal payoff function is everywhere differentiable, the proof is a simple consequence of the mean-value theorem, but Bergman *et al.* show that the result extends to payoff functions with jump discontinuities (e.g., those of digital options) and functions whose right and left derivatives exist but may be unequal (e.g., those of ordinary puts and calls).

To see this, let $\Delta S_T(\omega) \equiv S_T'(\omega) - S_T''(\omega)$ be the difference in terminal values of two price paths corresponding to the same Brownian path but with initial values $S_0' > S_0''$, and let $D_S^- \equiv \inf\{D_S(S_T,0) : S_T \geq 0\}$ and $D_S^+ \equiv \sup\{D_S(S_T,0) : S_T \geq 0\}$ be the infimum and supremum of slopes of the terminal payoff function. With $\hat{\mathbb{P}}$ as the risk-neutral measure and a deterministic short-rate process, we have $e^{-\int_t^T r_s \cdot ds}\hat{E}_t D(S_T,0) = D(S_t, T-t)$ and $e^{-\int_t^T r_s \cdot ds}\hat{E}_t \Delta S_T = \Delta S_t \geq 0$, where the inequality follows from the no-crossing lemma. The mean value theorem implies that

$$D(S_T',0) - D(S_T'',0) = D_S(S_T^*,0)\cdot \Delta S_T,$$

for some $S_T^* \in (S_T'', S_T')$, so that

$$D_S^- \cdot \Delta S_T \leq D(S_T',0) - D(S_T'',0) \leq D_S^+ \cdot \Delta S_T.$$

Taking expectations conditional on $\mathcal{F}_t$ and discounting give

$$D_S^- \cdot \Delta S_t \leq D(S_t', T-t) - D(S_t'', T-t) \leq D_S^+ \cdot \Delta S_t,$$

which implies that

$$D_S^- \le D_S(S_t, T-t) \le D_S^+$$

for $t \in [0, T]$.

Similar arguments show that convexity or concavity of the terminal payoff function implies convexity or concavity of $D(S_t, T-t)$. When the underlying follows (8.1), Bergman *et al.* also show (i) that prices of European calls increase with an upward shift in the entire yield curve for default-free bonds, but not necessarily with a change in yields that merely reduces the discounted present value of the strike price; and (ii) that an upward shift in volatility function $\sigma(S_t, t)$ increases current prices of European-style derivatives. We have already applied the latter result in section 7.2.1 in reference to firms that write options on their own stocks. The monotonicity result extends also to cases in which volatility depends on another (possibly vector-valued) diffusion process $\{Y_t\}$, if its drift and diffusion parameter do not depend on $\{S_t\}$. Yet another condition is needed to extend the convexity result.

To show what can go wrong, Bergman *et al.* give the following insightful example of a nonMarkov price process for which the value of a call does not increase monotonically with the current price of the underlying. A firm's managers will be rewarded if the price of its stock is above a threshold $\bar{S}$ at some time t'', and they have an opportunity to revise investment policy at $t' < t''$. The firm is financed entirely by equity capital. If the stock's price exceeds $\bar{S}\exp(-\int_{t'}^{t''} r_s \cdot ds)$ at t', managers will lock in the reward by choosing riskless projects that cause S to grow at the riskless rate. In this case a call option expiring at $T \le t''$ with strike price greater than $S_{t'}\exp(\int_{t'}^{T} r_s \cdot ds)$ will be worthless. However, if $S_{t'} < \bar{S}\exp(-\int_{t'}^{t''} r_s \cdot ds)$ managers will choose risky projects that produce high volatility in S so as to gain some chance of reward. Since this also makes it possible for the call to be in the money at T, it is clear that the call is worth more when $S_{t'}$ is less than $\bar{S}\exp(-\int_{t'}^{t''} r_s \cdot ds)$ than when it is greater. The problem here is precisely that the controlled process $\{S_t\}_{0\le t\le T}$ is not a Markov process. It is not because the volatility at each $t > t'$ depends on the value of the stock at t'.

8.2.2 *Two Specific Models*

Turning now to some specific formulations of price-dependent volatility, we begin with a simplified version of Rubinstein's (1983) "displaced diffusion" model, in which the underlying asset has two components, one risky and the other riskless. In this model the volatility of the underlying rises and falls stochastically as the value of the risky component increases and declines. Although this behavior runs counter to the leverage effect observed in the returns of common stocks, it does characterize portfolios balanced between stocks and short-term government debt. Entities such as trusts and endowments that hold such portfolios sometimes purchase over-the-counter derivative products in order to reduce their risk.[5] In any case the model is interesting pedagogically because it allows prices of European options to be expressed in terms of the Black-Scholes formulas. We also look at the constant-elasticity-of-variance (c.e.v.) model, which does capture the inverse relation between volatility and price level that is observed in stocks' returns.

The Displaced-Diffusion Model

Consider derivatives on an underlying asset that is itself a portfolio. The portfolio comprises (i) a risky component worth A_t at time t; and (ii) b units of T'-maturing riskless bonds each worth $B(t, T')$ at t. Assuming that the risky asset is traded and that its price follows geometric Brownian motion, process $\{A_t\}_{t \geq 0}$ evolves as

$$dA_t/A_t = r_t \cdot dt + \sigma_A \cdot d\hat{W}_t$$

under the risk-neutral measure. Letting $S_t = A_t + bB(t, T')$ be the value of the portfolio at t, we have

$$dS_t = dA_t + b \cdot dB(t, T') = r_t[A_t + bB(t, T')] \cdot dt + \sigma_A A_t \cdot d\hat{W}_t,$$

and

$$dS_t/S_t = r_t \cdot dt + \sigma(S_t, t) \cdot d\hat{W}_t,$$

where

$$\sigma(S_t, t) = \sigma_A[1 - bB(t, T')/S_t].$$

Even though the future volatility of $\{S_t\}$ is unpredictable, it is nevertheless easy to express prices of European-style derivatives. For example,

---

[5]Notice that the polar case of a firm with negative holdings of the riskless asset would correspond to debt financing. However, because of shareholders' limited liability, the compound-option model is more appropriate in this case. Indeed, that model actually belongs to the class (8.1).

consider a European call option that expires at $T \leq T'$. Its arbitrage-free price at t is

$$\begin{aligned} C^E(S_t, T-t) &= B(t,T)\hat{E}_t(S_T - X)^+ \\ &= B(t,T)\hat{E}_t\{A_T - [X - bB(T,T')]\}^+ \end{aligned} \tag{8.2}$$

Since $A_T > 0$ a.s., the option will necessarily expire in the money if $X \leq bB(t,T')$. In this case the "+" symbols in (8.2) are superfluous, and the call's value is that of the portfolio minus the discounted value of the exercise price,

$$\begin{aligned} C^E(S_t, T-t) &= B(t,T)\hat{E}_t(S_T - X) \\ &= S_t - B(t,T)X \\ &= A_t - B(t,T)X + B(t,T')b. \end{aligned}$$

In the less trivial case that $X > bB(t,T')$ one just appeals to the fact that $\{A_t\}_{t\geq 0}$ is geometric Brownian motion and expresses (8.2) by means of the Black-Scholes formula with A_t in place of S_t and $X - bB(T,T')$ in place of X.

Option Pricing under the C.E.V. Model

The constant-elasticity-of-variance process is represented by the stochastic differential equation

$$dS_t/S_t = \mu_t \cdot dt + \sigma_0 S_t^{\gamma-1} \cdot dW_t, \tag{8.3}$$

with $\gamma \in (0,1)$. The name applies because the "elasticity" of the volatility function with respect to the underlying price, $\partial \ln \sigma(S_t,t)/\partial \ln S_t$, equals the constant $\gamma - 1$. Since it appears that s.d.e. (8.3) does not have a unique solution when $\gamma < .5$,[6] applications are normally confined to the case $\gamma \in [.5, 1)$. In this range volatility decreases with the price level, as is consistent with what we observe in returns of common stocks, but $\{S_t\}$ remains almost-surely nonnegative. Among many possible schemes for linking volatility to price, this one has the advantage of supporting computational formulas for prices of European-style options.

W. Feller (1951) has derived the distribution of S_T conditional on S_t in the special case $\gamma = 1/2$, and Cox (1975) has extended to other values.

[6]See Duffie (1996, p. 293).

Following Cox and Ross (1976), we develop a formula for prices of European options on a no-dividend stock whose price follows the "square-root" process, with

$$dS_t/S_t = \mu \cdot dt + \sigma_0 S_t^{-1/2} \cdot dW_t. \tag{8.4}$$

Letting $F_t(s) \equiv \mathbb{P}_t(S_T \leq s)$, Feller's results give

$$\begin{aligned} dF_t(s) &= \sqrt{\theta_t \xi_t} e^{-\xi_t - \theta_t s} s^{-1/2} I_1(2\sqrt{\theta_t \xi_t} s^{1/2}) \cdot ds, s > 0 \\ F_t(0) &= G(\mu; 1) \\ F_t(s) &= 0, s < 0. \end{aligned} \tag{8.5}$$

Here θ_t and ξ_t are given by

$$\begin{aligned} \theta_t &\equiv \frac{2\mu}{\sigma_0^2[e^{\mu(T-t)} - 1]} \\ \xi_t &\equiv S_t \theta_t e^{\mu(T-t)}, \end{aligned}$$

while $I_1(\cdot)$ is the modified Bessel function of the first kind of order one, and G is the complementary gamma c.d.f.:

$$G(x; \alpha) \equiv \int_x^\infty \frac{1}{\Gamma(\alpha)} y^{\alpha-1} e^{-y} \cdot dy.$$

$I_1(\cdot)$ can be approximated using the series expansion (2.24) with $\nu = 1$.

To apply martingale pricing, one changes to the measure $\hat{\mathbb{P}}$ in which $\hat{W}_t = W_t + \sigma_0^{-1} \int_0^t S_s^{1/2}(\mu - r_s + \delta_s) \cdot ds$ is a martingale, where r_t and δ_t are the $\mathcal{F}_0$-measurable instantaneous short rate and dividend rate at t, and replaces μ with $r - \delta \equiv (T-t)^{-1} \int_t^T (r_u - \delta_u) \cdot du$ in the expressions for θ_t and ξ_t. The price of the European call is then $C^E(S_t, T-t) = B(t,T)\hat{E}_t(S_T - X)^+$, where the expectation is given by

$$\int_X^\infty \left[(s - X)\sqrt{\theta_t \xi_t} e^{-\xi_t - \theta_t s} s^{-1/2} \cdot \sqrt{\theta_t \xi_t} s^{1/2} \sum_{k=0}^\infty \frac{1}{k!(k+1)!} (\theta_t \xi_t s)^k \right] \cdot ds.$$

To develop further, rearrange as

$$\begin{aligned} &\frac{\xi_t}{\theta_t} \sum_{k=0}^\infty \frac{e^{-\xi_t} \xi_t^k}{k!} \int_X^\infty \frac{\theta_t^{k+1} s^{k+1} e^{-\theta_t s}}{(k+1)!} \cdot d(\theta_t s) \\ &\quad - X \sum_{k=0}^\infty \frac{e^{-\xi_t} \xi_t^{k+1}}{(k+1)!} \int_X^\infty \frac{\theta_t^k s^k e^{-\theta_t s}}{k!} \cdot d(\theta_t s), \end{aligned}$$

then change variables as $(\theta_t s) \to y$ in the integrals. Finally, writing $\xi_t/\theta_t = S_t e^{(T-t)(r-\delta)} = S_t e^{-\delta(T-t)} B(t,T)^{-1}$ and $g(x; \alpha)$ for the gamma p.d.f., one obtains for the value of the call under the square-root version of the c.e.v. process

$$C^E(S_t, T-t; 1/2) = S_t e^{-\delta(T-t)} \sum_{k=1}^{\infty} g(\xi_t; k) G(\theta_t X; k+1) \qquad (8.6)$$
$$- B(t,T) X \sum_{k=1}^{\infty} g(\xi_t; k+1) G(\theta_t X; k).$$

Cox's (1975) more general expression that applies for $\gamma \in [.5, 1)$ is

$$C^E(S_t, T-t; \gamma)$$
$$= S_t e^{-\delta(T-t)} \sum_{k=1}^{\infty} g(\xi_t'; k) G\left(\theta_t' X^{2-2\gamma}; k + \frac{1}{2-2\gamma}\right) \qquad (8.7)$$
$$- B(t,T) X \sum_{k=1}^{\infty} g\left(\xi_t'; k + \frac{1}{2-2\gamma}\right) G(\theta_t' X^{2-2\gamma}; k),$$

where

$$\theta_t' \equiv \frac{r-\delta}{\sigma_0^2 (1-\gamma)[e^{2(1-\gamma)(r-\delta)(T-t)} - 1]}$$
$$\xi_t' \equiv S_t^{2-2\gamma} \theta_t' e^{2(1-\gamma)(r-\delta)(T-t)}.$$

Schroder (1989) shows that the formula can be expressed in terms of c.d.f.s of noncentral chi-squared distributions and provides some computational shortcuts.

Considering the square-root version of the c.e.v. model, Beckers (1980) finds that (8.6) prices in-the-money calls and out-of-money puts higher than Black-Scholes (evaluated at implicit volatilities of at–the-money options), while the c.e.v. model prices out-of-money calls and in-the money puts lower than Black-Scholes. This is at least qualitatively consistent with the smirk that is often observed in implicit volatilities. Figure 8.1 shows for several values of γ how the difference in prices, c.e.v. minus Black-Scholes, varies with S_t/X for European calls and puts on no-dividend stocks.[7] Note that the differences are positive for $S_t/X > 1$, where puts are out of money and calls in, and they increase as γ declines from unity. Figure 8.2 shows the same information through the Black-Scholes implicit volatilities. Measuring moneyness as X/S_t to correspond to figure 6.4, we can see that the

[7]By put-call parity differences in call prices equal differences in put prices. The calculations use $X = 10.0$, $r = 0.05$, $\delta = 0.0, T - t = 0.5$ and $\sigma_0 = 0.30$, with volatility divided by $S_t^{\gamma-1}$ for c.e.v. to make initial volatilities comparable.

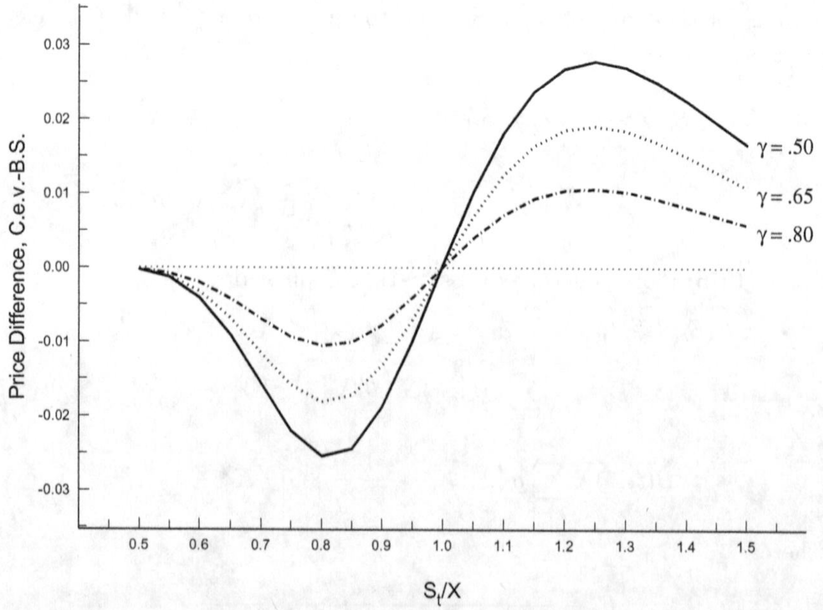

Fig. 8.1. C.e.v. minus Black-Scholes prices vs. moneyness.

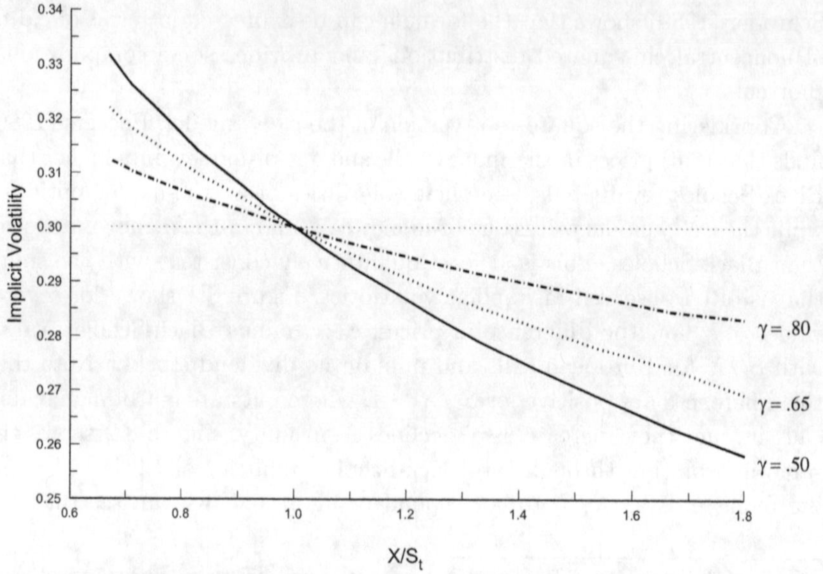

Fig. 8.2. Implicit volatilities for c.e.v. model.

c.e.v. model does introduce the smirk pattern that is commonly observed in prices of equity, index, and currency options. On the other hand, c.e.v. effects by themselves seem incapable of accounting for the other side of the smile curve, corresponding to Black-Scholes' underpricing of in-the-money puts and out-of-money calls. In addition, the c.e.v. model implies an essentially flat term structure of implicit volatilities, in that implicit volatility is insensitive to the length of time to expiration. This is inconsistent with the observation that smile curves tend to flatten out with time.

8.2.3 *Numerical Methods*

Martingale pricing of a derivative requires being able either to express analytically the risk-neutral conditional distribution of the underlying price or to simulate replicates from that distribution. The main attraction of the c.e.v. model is that the distribution of price is known—albeit quite complicated. Under other *a priori* reasonable specifications, such as $\sigma(S_t, t) = \sigma_0 + \sigma_1 S_t^{\gamma-1}$, the distribution is not explicitly known. Simulation, discussed in chapter 11, is often a feasible approach to such a problem, but it is usually faster to solve the partial differential equation that is implied by the dynamic replicating argument (section 6.2.1). While analytical solutions to these p.d.e.s are rarely available in interesting cases, numerical ones obtained by finite-difference methods usually are. These methods also apply when derivatives are subject to early exercise. Chapter 12 presents an introductory description of these techniques. In section 8.4 we shall see that there are simple alternatives for pricing European-style options when one has a computational formula for the characteristic function of the underlying price.

8.2.4 *Limitations of Price-Dependent Volatility*

Although models in the class (8.1) are consistent with preference-free pricing and can account for some moneyness-related variation in implicit volatilities, there are good reasons to look for still more general models. It is unlikely that price-level effects alone could fully account for the observed shapes of implicit volatility curves, nor do they explain the flattening out of the smile as time to expiration increases. Flattening might be captured by allowing explicitly for time dependence, as $\sigma(S_t, T-t) = \sigma_0 + \sigma_1 (T-t)^{-1} S_t^{1-\gamma}$, but this kind of arrangement could still not accommodate the day-to-day variations in the slopes of volatility term structures, nor could it pick up changes over time in the shapes of smile curves.

There are also indications that price/time-dependent volatility models deliver poor price predictions out of sample. This is the conclusion reached by Dumas *et al.* (1998) after extensive efforts to find price/time-volatility forms that do give good predictions. Their approach was, in effect, to use market prices of S&P 500 index options at different strikes to fit pricing models based on the following flexible specification of volatility:

$$\sigma(S_t, T-t) = a_0 + a_1 S_t + a_2 S_t^2 + a_3(T-t) + a_4(T-t)^2 + a_5 S_t(T-t). \quad (8.8)$$

Fitting involved estimating the coefficients $\{a_j\}$ by minimizing the sum of squared deviations of actual option prices from those implied by the model, as calculated by finite-difference methods. Although weekly estimates over several years gave good fits within sample, the coefficients in (8.8) were found to be highly unstable over time. Moreover, weekly out-of-sample prediction errors based on coefficients from the previous week were large and did not improve on errors from an *ad hoc* application of Black-Scholes with strike- and time-specific volatilities. Although one may question the appropriateness and generality of the quadratic specification for σ, these findings do suggest the need for richer models than one-factor diffusions with volatilities that depend just on current prices and time.

8.2.5 *Incorporating Dependence on Past Prices*

A promising direction in which to extend these models is to allow both current and *past* prices of the underlying to drive volatility. For example, volatility might be specified as a deterministic function of the difference between the current price and an average of past prices, all detrended by discounting at the short rate. The idea, similar to that underlying discrete-time ARCH models, is that large deviations of price from a "normal" (under risk-neutrality) trend tend to forecast higher future variation. Such a specification could account for temporal changes in smile curves and volatility term structures as current price and the norm evolve through time; yet limiting the determinants of volatility to time and the underlying price still permits payoff replication and preference-free pricing.

A class of such models was proposed by Hobson and Rogers (1998). One simple version starts with the primitive that the logarithm of the discounted price of a no-dividend asset, $s_t \equiv \ln(e^{-rt} S_t)$, is an Itô process of the form

$$ds_t = \mu(\Delta_t) \cdot dt + \sigma(\Delta_t) \cdot dW_t.$$

Here Δ_t is the deviation from norm, defined as

$$\Delta_t = s_t - \lambda \int_{-\infty}^{t} e^{-\lambda(t-u)} s_u \cdot du,$$

with extrinsic parameter λ governing the rate at which the influence of past prices decays. Note that the time origin is also, in effect, a parameter here. Taking $\Delta_0 = 0$, this setup makes $\{\Delta_t\}_{t\geq 0}$ itself an Itô process adapted to $\{\mathcal{F}_t\}$, with

$$\begin{aligned} d\Delta_t &= ds_t - \lambda\Delta_t \cdot dt \\ &= [\mu(\Delta_t) - \lambda\Delta_t] \cdot dt + \sigma(\Delta_t) \cdot dW_t. \end{aligned}$$

The quadratic specification for σ,

$$\sigma^2(\Delta_t) = \alpha_0 + \alpha_1\Delta_t + \alpha_2\Delta_t^2,$$

is feasible if supplemented (for technical reasons) by an upper bound. This could accommodate different responses of volatility to positive and negative price deviations, depending on the sign of α_1. Defining

$$\theta(\Delta_t) \equiv \mu(\Delta_t)/\sigma(\Delta_t) + \sigma(\Delta_t)/2,$$

a martingale measure $\hat{\mathbb{P}}$ for the discounted price of the underlying can be determined by Girsanov's theorem as

$$\frac{d\hat{\mathbb{P}}}{d\mathbb{P}} = \exp\left[-\int_0^t \theta(\Delta_u) \cdot dW_u - \frac{1}{2}\int_0^t \theta(\Delta_u)^2 \cdot du\right],$$

assuming $\mu(\cdot)$ and $\sigma(\cdot)$ are such that the right-hand side is a martingale under $\mathbb{P}$. Then $\{\hat{W}_t = W_t + \int_0^t \theta(\Delta_u) \cdot du\}$ becomes a $\hat{\mathbb{P}}$-Brownian motion and $\{S_t\}_{t\geq 0}$ and $\{\Delta_t\}_{t\geq 0}$ evolve as,

$$dS_t/S_t = r \cdot dt + \sigma(\Delta_t) \cdot d\hat{W}_t \tag{8.9}$$

$$d\Delta_t = -[\lambda\Delta_t + \sigma(\Delta_t)^2/2] \cdot dt + \sigma(\Delta_t) \cdot d\hat{W}_t. \tag{8.10}$$

Applying the martingale representation theorem (section 3.4.2), a European-style derivative with integrable terminal payoff function, $\bar{D}(S_T)$, that depends on S_T and time alone can be replicated by a self-financing portfolio and valued as

$$D(S_t, \Delta_t, T - t) = e^{-r(T-t)}\hat{E}_t\bar{D}(S_T).$$

Notice that the deviation from norm now enters as another state variable on which the derivative's arbitrage-free price depends.

While Monte Carlo would be an obvious approach to evaluating $\hat{E}_t\bar{D}(S_T)$, Hobson and Rogers (1998) suggest solving for $D(S_t, \Delta_t, T - t)$

from the associated p.d.e. Since $\{M_t^{-1}D(S_t, \Delta_t, T-t)\} = \{M_0^{-1}e^{-rt}D(S_t, \Delta_t, T-t)\}$ is a martingale under $\hat{\mathbb{P}}$ (where M_t is the value of the money-market fund at t) it has zero mean drift. Applying Itô's formula, one obtains

$$e^{rt} \cdot d[e^{-rt}D(S_t, \Delta_t, T-t)] = -(rD + D_{T-t}) \cdot dt + D_S \cdot dS_t + \frac{D_{SS}}{2} \cdot d\langle S \rangle_t + D_{S\Delta} \cdot d\langle S, \Delta \rangle_t + D_\Delta \cdot d\Delta_t + \frac{D_{\Delta\Delta}}{2} \cdot d\langle \Delta \rangle_t.$$

Using (8.9) and (8.10) to express the stochastic differentials and quadratic variations, collecting the drift terms, and equating to zero give the p.d.e.

$$\begin{aligned} 0 &= -D_{T-t} + D_S r S_t - \lambda D_\Delta \Delta_t \\ &\quad + (D_{SS}S_t^2 + 2D_{S\Delta}S_t + D_{\Delta\Delta} - D_\Delta)\sigma(\Delta_t)^2/2 - Dr. \end{aligned}$$

Notice that this reduces to the standard Black-Scholes p.d.e. in the degenerate cases that $\sigma(\Delta_t) = \sigma_0$, a constant, or $\sigma(\Delta_t) = \sigma_t$, a deterministic function of time alone. Allowing for time-varying, deterministic interest rates and continuously paid dividends (at rates that could in principle depend on t, S_t, and Δ_t) would be accomplished just by replacing r with $r_t - \delta_t$. The p.d.e. would be solved by finite-difference methods. Of course, the presence of an additional state variable increases to three the required dimension of the grid. Hobson and Rogers (1998) show that the model can indeed generate a wide variety of smile patterns and volatility term structures.

8.3 Stochastic-Volatility Models

This section treats models in which volatility is itself an Itô process that may be subject to some stochastic influence other than the price of the underlying asset. Specifically, with the underlying price evolving as

$$dS_t = \mu_t S_t \cdot dt + \sigma_t S_t \cdot dW_{1t}, \tag{8.11}$$

$\{\sigma_t\}$ is modeled as

$$d\sigma_t = \xi_t \cdot dt + \gamma_t(\rho_t \cdot dW_{1t} + \bar{\rho}_t \cdot dW_{2t}), \tag{8.12}$$

where $\bar{\rho}_t = \sqrt{1-\rho_t^2}$. Here $\{W_{1t}, W_{2t}\}$ are independent Brownian motions under $\mathbb{P}$, and ξ_t, $\gamma_t \geq 0$, and $\rho_t \in [-1, 1]$ are adapted to the filtration $\{\mathcal{F}_t\}$ generated by $\{W_{1t}, W_{2t}\}$. If $|\rho_t| = 1$ the model reduces to those in section 8.2 where $\sigma_t = \sigma(S_t, t)$. The other degenerate case is that there are other traded assets whose prices are adapted to $\{\mathcal{F}_t\}$ and such that σ_t can be replicated by a portfolio of these. In both cases preference-free pricing is still possible. Otherwise, the market is not complete, and arbitrage arguments alone do not yield a unique solution for the price of the derivative.

The first task in this section is to demonstrate the truth of the last two statements. We show that replication is indeed possible when the uncertainty is "spanned" by prices of traded assets and that there can otherwise be no unique arbitrage-free price for a derivative whose payoffs depend nonlinearly on the price of the underlying. We then see how to price options under particular specifications for $\{\sigma_t\}$, beginning with the case $\rho_t = 0$ in which shocks to volatility and price are independent.

8.3.1 *Nonuniqueness of Arbitrage-Free Prices*

While the absence of arbitrage guarantees the existence of a measure under which normalized prices of traded assets are martingales, it is the completeness of markets—the ability to replicate payoffs with traded assets—that makes the martingale measure unique. Without replication there can be no unique arbitrage-free solution for the price of a derivative. If the risk associated with holding a derivative cannot be hedged away, the derivative's market price, like prices of primary assets, will depend on how individuals respond to risk and how much compensation they require to bear it. We will identify explicit conditions under which preference-free pricing of derivative assets is and is not possible under (8.11) and (8.12). Of course, we know from chapter 4 that derivatives like forwards and futures, whose payoffs depend linearly on the underlying price, can be replicated with static portfolios of the underlying and riskless bonds, regardless of the underlying dynamics. Our discussion therefore pertains to options and other derivatives with nonlinear payoff structures.

Since (8.11) and (8.12) involve two independent sources of risk, it is clear that dynamic replication of a derivative will require at least two risky assets. Consider first the case that there is a second traded asset whose price depends directly on both risk sources, as

$$dS_t' = \mu_t' S_t' \cdot dt + \sigma_t' S_t' (\rho_t' \cdot dW_{1t} + \bar{\rho}_t' \cdot dW_{2t}).$$

Here $\{\sigma_t'\}$ may be stochastic also, provided that it is adapted to the $\{\mathcal{F}_t\}$ with respect to which $\{S_t\}$ and $\{\sigma_t\}$ are measurable. Let $D(S_t, \sigma_t, T - t)$ be the price of a T-expiring European-style derivative whose terminal value depends just on S_T. We will try to replicate the payoff $\bar{D}(S_T)$ with a self-financing portfolio comprising p_t units of the underlying, q_t units of the secondary asset, and m_t units of the money fund. The procedure is like that followed in section 6.2.1 except that we simplify by excluding dividends. Also, σ_t must be recognized now as an additional state variable since it is not a function just of S_t and t. Setting $m_t = [D(S_t, \sigma_t, T-t) - p_t S_t - q_t S'_t,]/M_t$ and using $dM_t/M_t = r_t \cdot dt$, the self-financing property of the portfolio requires

$$dD(S_t, \sigma_t, T-t) = p_t \cdot dS_t + q_t \cdot dS'_t + (D - p_t S_t - q_t S'_t) r_t \cdot dt. \quad (8.13)$$

Using Itô's formula to express the left side (assuming that D has at least two continuous derivatives in S and σ) gives for $dD(S_t, \sigma_t, T-t)$

$$\begin{aligned}(-D_{T-t} &+ D_S \mu_t S_t + D_{SS}\sigma_t^2 S_t^2/2 + D_\sigma \xi_t + D_{\sigma\sigma}\gamma_t^2/2 + D_{S\sigma}\rho_t S_t \sigma_t \gamma_t) \cdot dt \\ &+ (D_S \sigma_t S_t + D_\sigma \rho_t \gamma_t) \cdot dW_{1t} + D_\sigma \bar{\rho}_t \gamma_t \cdot dW_{2t}, \end{aligned} \quad (8.14)$$

while the right side of (8.13) is

$$\begin{aligned}&[Dr_t + p_t(\mu_t - r_t)S_t + q_t(\mu'_t - r_t)S'_t] \cdot dt \\ &\quad + (p_t \sigma_t S_t + q_t \sigma'_t S'_t \rho'_t) \cdot dW_{1t} + q_t \sigma'_t S'_t \bar{\rho}'_t \cdot dW_{2t}. \end{aligned} \quad (8.15)$$

Equating the terms in dW_{1t} and dW_{2t} on the two sides requires

$$q_t = D_\sigma \frac{\bar{\rho}_t \gamma_t}{\sigma'_t S'_t \bar{\rho}'_t}$$

$$p_t = D_S + D_\sigma \frac{\gamma_t}{\sigma_t S_t}(\rho_t - \rho'_t \bar{\rho}_t / \bar{\rho}'_t).$$

Substituting these in (8.15), equating the dt term with that in (8.14), and rearranging give

$$\begin{aligned} 0 = &-D_{T-t} + D_S r_t S_t + D_{SS}\sigma_t^2 S_t^2/2 + D_\sigma \xi_t + D_{\sigma\sigma}\gamma_t^2/2 + D_{S\sigma}\rho_t S_t \sigma_t \gamma_t \\ &- Dr_t - \frac{D_\sigma \gamma_t}{\bar{\rho}'_t}\left[\left(\frac{\mu_t - r_t}{\sigma_t}\right)(\rho_t \bar{\rho}'_t - \rho'_t \bar{\rho}_t) + \left(\frac{\mu'_t - r_t}{\sigma'_t}\right)\bar{\rho}_t\right] \end{aligned} \quad (8.16)$$

as the differential equation that $D(S_t, \sigma_t, T-t)$ must satisfy.

Notice that this involves two preference-dependent parameters, μ_t and μ'_t, that reflect the market's required compensation for the risk of holding the two primary assets. However, assuming there is no arbitrage, there must still be a measure $\hat{\mathbb{P}}$ under which $\{S_t^* \equiv S_t/M_t\}$ and $\{S_t'^* \equiv S'_t/M_t\}$ are martingales. To develop such a measure, introduce adapted processes $\{\theta_{1t}\}$ and $\{\theta_{2t}\}$ satisfying

$$r_t = \mu_t - \sigma_t \theta_{1t} = \mu'_t - \sigma'_t \rho'_t \theta_{1t} - \sigma'_t \bar{\rho}'_t \theta_{2t}$$

and apply Girsanov's theorem to construct $\hat{\mathbb{P}}$ Brownian motions $\hat{W}_{1t} \equiv W_{1t} + \int_0^t \theta_{1s} \cdot ds$ and $\hat{W}_{2t} \equiv W_{2t} + \int_0^t \theta_{2s} \cdot ds$. Girsanov's theorem guarantees that this construction is possible if $\{\theta_{1t}\}$ and $\{\theta_{2t}\}$ are such that

$$\frac{d\hat{\mathbb{P}}}{d\mathbb{P}} = \exp\left[\sum_{j=1}^{2} \int_0^t \left(-\theta_{js} \cdot dW_{js} - \frac{1}{2}\theta_{js}^2 \cdot ds\right)\right] \tag{8.17}$$

is a $\mathbb{P}$-martingale. Under this new measure we have

$$\begin{aligned}
dS_t &= \mu_t S_t \cdot dt + \sigma_t S_t (d\hat{W}_{1t} - \theta_{1t} \cdot dt) \\
&= r_t S_t \cdot dt + \sigma_t S_t \cdot d\hat{W}_{1t} \\
dS_t' &= \mu_t' S_t' \cdot dt + \sigma_t' S_t' [\rho_t' (dW_{1t} - \theta_{1t} \cdot dt) + \bar{\rho}_t' (dW_{2t} - \theta_{2t} \cdot dt)] \\
&= r_t S_t' \cdot dt + \sigma_t' S_t' (\rho_t' \cdot d\hat{W}_{1t} + \bar{\rho}_t' \cdot d\hat{W}_{2t}),
\end{aligned}$$

as the martingale property for S_t^* and $S_t'*$ requires. The time-t values of the adapted processes involved in this transition are

$$\begin{aligned}
\theta_{1t} &= \left(\frac{\mu_t - r_t}{\sigma_t}\right) \\
\theta_{2t} &= \frac{1}{\bar{\rho}_t'}\left(\frac{\mu_t' - r_t}{\sigma_t'}\right) - \frac{\rho_t'}{\bar{\rho}_t'}\left(\frac{\mu_t - r_t}{\sigma_t}\right).
\end{aligned}$$

Under this measure the final term of (8.16) vanishes, and a solution to the p.d.e. subject to the boundary conditions and the terminal condition $D(S_T, \sigma_T, 0) = \bar{D}(S_T)$ gives a preference-free price for the derivative. Alternatively, since the drift of $\{D(S_t, \sigma_t, T-t)/M_t\}$ itself is zero under this measure, we could just exploit the martingale property and price the derivative as

$$D(S_t, \sigma_t, T-t) = \hat{E}_t[\bar{D}(S_T) M_T / M_t] = B(t,T)\hat{E}_t \bar{D}(S_T).$$

What happens when there is no traded asset that completes the market in this way? Since replication still requires two risky assets, a logical idea would be to replicate the derivative of interest using the underlying and another derivative that expires at the same date or later. Since the price of the other derivative would have to respond directly to both risk sources, it could not be a linear instrument like a forward contract, whose arbitrage-free price depends just on prices of the underlying and riskless bonds. When $\rho_t = 0$, so that volatility and price shocks are independent, Bajeux and Rochet (1992) have shown that a European option does complete the market, and the result was extended by Romano and Touzi (1997) (under restrictive conditions on ξ_t and γ_t) to the case $\rho_t \neq 0$. Supposing, then, for the sake of example, that we have chosen a T-expiring European put as the second asset, let us try to follow the same program as before.

Forming a replicating portfolio with p_t units of the underlying, q_t puts each worth $P^E(S_t, \sigma_t, T-t)$, and m_t units of the money fund, we have

$$dD(S_t, \sigma_t, T-t) = p_t \cdot dS_t + q_t \cdot dP^E(S_t, \sigma_t, T-t) + (D - p_t S_t - q_t P^E) r_t \cdot dt.$$

Expressing dD and dP^E by Itô's formula and dS_t as (8.11) then collecting the terms in dW_{1t} and dW_{2t} give $q_t = D_\sigma / P^E_\sigma$ and $p_t = D_S - P^E_S D_\sigma / P^E_\sigma$. With these substitutions, equating the dt terms produces the p.d.e.

$$\begin{aligned} 0 = &\left[-D_{T-t} + D_S r_t S_t + D_{SS} \frac{\sigma_t^2}{2} S_t^2 + D_{\sigma\sigma} \frac{\gamma_t^2}{2} + D_{S\sigma} \rho_t S_t \sigma_t \gamma_t - D r_t \right] \\ &- \frac{D_\sigma}{P^E_\sigma} \left[-P^E_{T-t} + P^E_S r_t S_t + P^E_{SS} \frac{\sigma_t^2}{2} S_t^2 + P^E_{\sigma\sigma} \frac{\gamma_t^2}{2} \right. \\ &\left. + P^E_{S\sigma} \rho_t S_t \sigma_t \gamma_t - P^E r_t \right]. \end{aligned} \tag{8.18}$$

Although there are no preference-dependent parameters here, a moment's reflection shows that there is an essential indeterminacy in the solutions for P^E and D. Letting $\mathcal{L}D$ and $\mathcal{L}P^E$ denote the two bracketed expressions, consider functions $D(S_t, \sigma_t, T-t)$ and $P^E(S_t, \sigma_t, T-t)$ that solve, respectively

$$0 = \mathcal{L}D + D_\sigma \lambda_t \tag{8.19}$$

$$0 = \mathcal{L}P^E + P^E_\sigma \lambda_t. \tag{8.20}$$

Notice that in these expressions λ_t takes the place of the drift process ξ_t in (8.16), so we refer to λ_t as the "adjusted" drift of $\{\sigma_t\}$. Since λ_t is arbitrary and any such D and P^E that solve (8.19) and (8.20) also solve (8.18), we see that the replicating argument fails to deliver a unique solution for the price of either derivative.[8]

Another way of seeing the problem [*c.f.* Renault and Touzi (1996, p. 282)] is that a change of measure of the form (8.17) with $\theta_{1t} = (\mu_t - r_t)/\sigma_t$ but with θ_{2t} unrestricted reduces $\{S_t/M_t\}$ to a martingale. While expectation in any such measure delivers unique prices for the underlying asset

---

[8]Scott (1987, p. 423), considering specifically the use of one call to help replicate another, sums up the problem in a very intuitive way:

> "We cannot determine the price of a call option without knowing the price of another call on the same stock, *but that is precisely the function that we are trying to determine.* (Emphasis added.)

and (trivially so) for the money fund, it does not produce a unique value for a (nonlinear) derivative that responds to volatility risk, because $\hat{E}_t[\bar{D}(S_T)M_T/M_t]$ depends on $\{\theta_{2t}\}$.

Being thus convinced that preferences will matter in pricing options and other nonlinear derivatives under stochastic volatility, the next question is "What do we do about it?" The essence of the problem is that solutions will depend on unobservable, preference-dependent parameters (or processes) like the adjusted drift $\{\lambda_t\}$ in the discussion of (8.18). For instance, $\{\lambda_t\}$ can be considered to incorporate the market's required compensation for exposure to σ risk, with D_σ representing the derivative's marginal sensitivity to such risk, just as beta in the CAPM represents a primary asset's sensitivity to market risk. One approach is to derive $\{\lambda_t\}$ from an equilibrium model that relies on strong assumptions about preferences and markets (and the processes $\{\xi_t\}$, $\{\gamma_t\}$, and $\{\rho_t\}$ that define $\{\sigma_t\}$). Alternatively, one can do the following. First,

- specify a plausible, flexible, yet reasonably (parametrically) parsimonious model for $\{\sigma_t\}$ that has a good chance of capturing the dynamics associated with the unknown equivalent martingale measure that the market actually uses; then
- infer the parameters from the structure of prices of traded derivatives.

This is in the same spirit as deducing the volatility parameter in the Black-Scholes model from the price of a single European option.[9] If the number of parameters is the same as the number of observed prices, one can ordinarily solve for the parameters uniquely, just as is done for implicit volatility. With more observations than parameters inference can be based on some robust statistical procedure, such as nonlinear least squares. For example, taking a panel of observed prices of options with various strikes and maturities, one can fit a parametric model by minimizing a measure of distance (such as the sum of squared deviations) of actual prices from those implied by the model.[10] One's confidence in the particular parameterization depends

[9]Chernov and Ghysels (1998) have considered combining information from options' prices and the time series of the underlying to infer the parameters of the equivalent martingale distribution. They find, however, that it is the options' values that contribute the most information. See also Ghysels *et al.* (1996) for further discussion of modeling issues and inference.

[10]Recall that this was the procedure used by Dumas *et al.* (1998) in estimating flexible price-dependent volatility models.

on the stability of parameter estimates from different time samples and the accuracy of out-of-sample predictions of future prices.

8.3.2 *Specific S.V. Models*

We now look at some specific models for European-style derivatives in the stochastic-volatility (s.v.) framework. For definiteness in interpreting "in the money" and "out of the money" we focus the discussion on European puts, expressing (8.20) as

$$\begin{aligned}-P^E_{T-t} &+ P^E_S r_t S_t + P^E_{SS}\sigma_t^2 S_t^2/2 + P^E_\sigma \lambda_t + P^E_{\sigma\sigma}\gamma_t^2/2 \\ &+ P^E_{S\sigma}\rho_t S_t \sigma_t \gamma_t - P^E r_t = 0. \end{aligned} \tag{8.21}$$

Determining the current value of the option, $P^E(S_t, \sigma_t, T-t)$, whether done by solving (8.21) subject to $\bar{P}^E(S_T) = (X - S_T)^+$ and boundary conditions or by exploiting the martingale property of $\{P^E(S_t, \sigma_t, T-t)/M_t\}_{0\le t\le T}$ under $\hat{\mathbb{P}}$, requires specific assumptions about the processes $\{\gamma_t\}$, $\{\rho_t\}$, and the adjusted drift parameter $\{\lambda_t\}$. The analysis is easiest when $\{\rho_t\}$, the correlation between the shocks to price and volatility, is always zero.

Independent Shocks to Price and Volatility

Regarding stochastic volatility as a mixing process that scatters probability mass into the tails of the distribution of S_T, one's intuition is that both out-of-money puts and out-of-money calls would be worth more in such an environment than under Black-Scholes dynamics. By put-call parity, it follows that deep in-the-money options would be worth more as well. These considerations suggest the presence of a smile effect in the implicit volatility curve. While the logic seems compelling, the result has been established rigorously only in the case that $\{\rho_t = 0\}$. Renault and Touzi (1996) prove two general propositions that describe the shape of the implicit volatility curve under this condition. Letting $x \equiv \ln(X/\mathsf{f}_t)$, where (assuming no dividend on the underlying) $\mathsf{f}_t = S_t/B(t,T)$ is the time-t forward price, they show that implicit volatility as a function of x is symmetric about the origin, decreasing with X when $x < 0$ (puts out of the money), reaching a minimum at $x = 0$, and increasing with X when $x > 0$ (puts in the money). This establishes firmly that a volatility smile is in fact present when $\rho_t = 0$. Indeed, the indicated symmetry of implicit volatility with

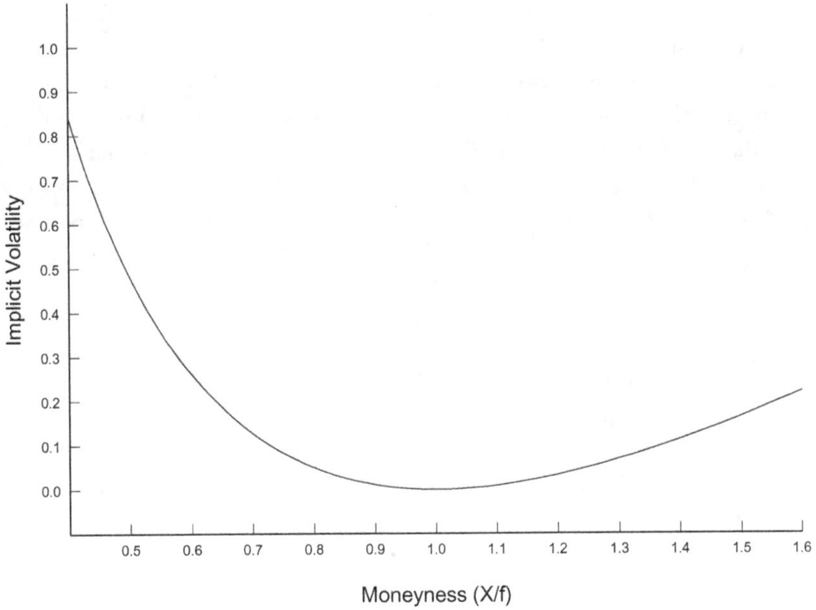

Fig. 8.3. An asymmetric smile consistent with $\rho_t = 0$.

respect to x implies an asymmetry with respect to $X/\mathsf{f}_t = e^x$ that is consistent with the smirk effect. For example, figure 8.3, which simply plots $\sigma(x) = x^2$ *vs.* e^x, illustrates an asymmetric smile that is consistent with the theoretical results of Renault and Touzi.

Besides supporting these definite qualitative results, maintaining $\{\rho_t = 0\}$ greatly simplifies the quantitative problem of pricing derivatives under stochastic volatility. This is because the independence of price and volatility shocks makes it possible to express prices of European options as mathematical expectations of Black-Scholes prices. To see this, recall our development of the Black-Scholes put and call formulas in section 6.3.2 for deterministic but time-varying volatility. Taking the current time to be $t = 0$, the put formula is

$$P^E(S_0, T; \bar{\sigma}_T) \equiv B\left\{X\Phi\left[\frac{\ln(X/\mathsf{f}_0) + \bar{\sigma}_T^2 T/2}{\bar{\sigma}_T\sqrt{T}}\right] - \mathsf{f}_0\Phi\left[\frac{\ln(X/\mathsf{f}_0) - \bar{\sigma}_T^2 T/2}{\bar{\sigma}_T\sqrt{T}}\right]\right\},$$

where $B = B(0,T)$, f_0 is the current forward price for delivery of the underlying at T, and $\bar{\sigma}_T^2 = T^{-1}\int_0^T \sigma_t^2 \cdot dt$ is the average volatility to expiration. If shocks to volatility and price are independent, then the time-T value of the underlying is lognormal conditional on average volatility; and since the validity of the put formula depends just on the lognormality of S_T, it still applies in this conditional sense. Therefore, the current value of the option under stochastic volatility is simply the mathematical expectation of $P^E(S_0, T; \bar{\sigma}_T)$, or

$$P^E(S_0, \sigma_0, T) = \hat{E}P^E(S_0, T; \bar{\sigma}_T).$$

Notice that the boundedness of the Black-Scholes price function assures the existence of the expectation.

There remains the issue of inferring the distribution for $\bar{\sigma}_T$ that is consistent with the particular preference-dependent martingale measure that actually governs $\{S_t\}_{t\geq 0}$ and $\{\sigma_t\}_{t\geq 0}$. Since in this measure $d\sigma_t = \lambda_t \cdot dt + \gamma_t \cdot d\hat{W}_{2t}$ when $\rho_t = 0$, the distribution of $\bar{\sigma}_T$ will be determined by the specification of $\{\gamma_t\}$ and adjusted drift process $\{\lambda_t\}$ in (8.21). Assuming that the appropriate risk adjustment is simply an additive constant, so that $\lambda_t = \xi_t - \lambda^*$, Scott (1987) uses Monte Carlo simulation to implement the procedure for two specific mean-reverting specifications (i) $\xi_t = \xi(\bar{\sigma}_\infty - \sigma_t)$, $\gamma_t = \gamma$; and (ii) $\xi_t = [\xi \ln(\bar{\sigma}_\infty/\sigma_t) + \gamma^2/2]\sigma_t$, $\gamma_t = \gamma\sigma_t$, where in each case $\bar{\sigma}_\infty$ is the constant value about which $\{\sigma_t\}$ fluctuates.[11] Because of the conditioning argument just related, pricing the option by Monte Carlo just involves simulating price paths of $\{\sigma_t\}_{0\leq t\leq T}$, evaluating $\bar{\sigma}_T$ for each path, and averaging $P^E(S_0, T; \bar{\sigma}_T)$ over the replications. Hull and White (1987), modeling $\{\sigma_t^2\}$ first as geometric Brownian motion with $\lambda_t = \sigma_t\lambda$ and $\gamma_t = \sigma_t\gamma$, work out an approximate formula for $\hat{E}P^E(S_0, T; \bar{\sigma}_T)$ by developing the first few moments of $\bar{\sigma}_T^2 = T^{-1}\int_0^T \sigma_t^2 \cdot dt$ and expanding $P^E(S_0, T; \bar{\sigma}_T)$ about the mean. They also use simulation to implement a mean-reverting specification, as $\lambda_t = \xi(\bar{\sigma}_\infty - \sigma_t)$. Stein and Stein (1991) adopting Scott's (Ornstein-Uhlenbeck) specification (i), exploit the conditional lognormality of S_T given $\bar{\sigma}_T$ to derive the characteristic function. Inverting by numerical integration to find the risk-neutral c.d.f. of S_T, they can estimate $P^E(S_0, \sigma_0, T)$ as $B(0,T)\int_0^X (X - s) \cdot d\hat{F}_{S_T}(s; \xi, \bar{\sigma}_\infty, \gamma, \lambda^*)$ by a second numerical integration. Unfortunately, this relatively quick analytical

---

[11]Notice that (i) is an Ornstein-Uhlenbeck process (defined in example 38 on page 106), which has the unfortunate property of allowing negative volatility.

method depends critically on the assumption that $\rho_t = 0$, whereas (as we shall see) simulation does not.

Dependent Price/Volatility Shocks

Just as one expects there to be a smile in implicit volatilities when $\{\sigma_t\}$ is stochastic, intuition suggests also that correlation between price and volatility shocks should affect the symmetry of the curve. For example, when $\rho_t > 0$ positive volatility shocks tend to be associated with positive price shocks, which therefore tend to be larger in absolute value than negative price shocks. It seems that this effect would lengthen the right tail in the distribution of $\ln S_T$ and that negative correlation, conversely, should skew the distribution of $\ln S_T$ to the left. Since extra probability mass in the right and left tail enhances the values of calls and puts, respectively—particularly for options that are out of the money—one expects skewness in the distribution of $\ln S_T$ to induce skewness in the smile curve also. While this effect has yet to be proved rigorously, the logic is well supported by numerical estimates obtained with variously specified models. Although we have seen from the work of Renault and Touzi (1996) that symmetry in the plot of implicit volatility vs. $\ln(X/\mathsf{f}_t)$ is consistent with skewness in the plot against X/f_t itself, allowing for dependence in price/volatility shocks extends the range of patterns that the model can accommodate.

The pricing problem is harder when $\rho_t \neq 0$ because the Black-Scholes formula no longer applies conditionally on $\bar{\sigma}_T$ when processes $\{S_t\}_{t\geq 0}$ and $\{\sigma_t\}_{t\geq 0}$ are dependent. Nevertheless, several attacks have been made on the problem. Hull and White (1987) simulate the joint process $\{S_t, \sigma_t\}$ and average $(X - S_T)^+$ over the replicates obtained from the resulting sample paths of $\{S_t\}_{0\leq t\leq T}$. We give details of their approach in chapter 11. Although computationally intensive, simulation does afford considerable flexibility in modeling; for example, the specification of the $\{\sigma_t\}$ process can include direct price-level effects on volatility, as in the c.e.v. model. Another approach, followed by Wiggins (1987) with Scott's (1987) specification (ii), is to solve (8.21) using finite-difference methods. However, since a three-dimensional lattice is now required, this method is much more cumbersome than when volatility depends on the price level alone. Finally, Heston (1993) has developed an analytical procedure that offers a nice balance of modeling flexibility and computational efficiency. We now give a detailed account of that method.

An Analytical Approach

Following Heston (1993), let us model the price process as in (8.11) but replace (8.12) with

$$dv_t = \xi(\bar{v}_\infty - v_t) \cdot dt + \gamma\sqrt{v_t}(\rho \cdot dW_{1t} + \bar{\rho} \cdot dW_{2t}),$$

where ρ and $\bar{\rho} \equiv \sqrt{1-\rho^2}$ are constants and $v_t \equiv \sigma_t^2$ is squared volatility. We will then assume that the risk-adjusted process under the martingale measure has the same functional form:

$$dv_t = (\alpha - \beta v_t) \cdot dt + \gamma\sqrt{v_t}(\rho \cdot d\hat{W}_{1t} + \bar{\rho} \cdot d\hat{W}_{2t}). \tag{8.22}$$

This would be consistent with a drift adjustment in $\{v_t\}$ of the form $\lambda_t = \xi(\bar{v}_\infty - v_t) - \lambda^* v_t$. Squared volatility is thus proposed to be a mean-reverting square-root process under $\hat{\mathbb{P}}$. This is the "Feller" process or "CIR" process that was introduced in expression (3.9). Like the Ornstein-Uhlenbeck process, this captures the intuitive notion that volatility should be attracted to some long-run norm (in this case α/β), but it rules out negative values of v_t. Corresponding to (8.21), the price of a European put under this specification satisfies the p.d.e.

$$\begin{aligned} 0 = &-P_{T-t}^E + P_S^E r_t S_t + P_{SS}^E v_t S_t^2/2 + P_v^E(\alpha - \beta v_t) \\ &+ P_{vv}^E v_t \gamma^2/2 + P_{Sv}^E \rho S_t v_t \gamma - P^E r_t \end{aligned} \tag{8.23}$$

subject to $\bar{P}^E(S_T) = P(S_T, \sigma_T, 0) = (X - S_T)^+$. Not surprisingly, there are no known simple tricks for obtaining a solution. The feasible approach, which we now undertake, is to develop and solve a similar p.d.e. for the conditional characteristic function of $\ln S_T$ under $\hat{\mathbb{P}}$, which can then be inverted in order to price the put via martingale methods.

In what follows we write $s_T \equiv \ln S_T$ and put $\tau \equiv T - t$. The first step is to recognize that the conditional expectations *process*

$$\{\Psi(\zeta; s_t, v_t, \tau) = \hat{E}_t e^{i\zeta s_T}\}_{0 \le t \le T}$$

is a (complex-valued) martingale adapted to $\{\mathcal{F}_t\}$. Expressing $d\Psi$ via Itô's formula and equating the drift term to zero, we obtain p.d.e.

$$0 = -\Psi_\tau + \Psi_s\left(r_t - \frac{v_t}{2}\right) + \Psi_{ss}\frac{v_t}{2} + \Psi_v(\alpha - \beta v_t) + \Psi_{vv}\gamma^2\frac{v_t}{2} + \Psi_{sv}\rho\gamma v_t, \tag{8.24}$$

which must be solved subject to initial condition

$$\Psi(\zeta; s_T, v_T, 0) = e^{i\zeta s_T} = S_T^{i\zeta}.$$

The affine structure of (8.24) makes it possible to find a solution of the form

$$\Psi(\zeta; s_t, v_t, \tau) = \exp[i\zeta s_t + g(\tau;\zeta) + h(\tau;\zeta)v_t] \tag{8.25}$$

for certain complex-valued functions g and h. Clearly, the initial condition requires

$$g(0;\zeta) = h(0;\zeta) = 0, \tag{8.26}$$

and we must have $g(\tau;0) = h(\tau;0) = 0$ as well if Ψ is to be a c.f. To verify that the solution is of this form and to identify g and h, express the derivatives of Ψ as

$$\begin{aligned}
\Psi_\tau &= \Psi \cdot (g' + h'v_t) \\
\Psi_s &= \Psi \cdot i\zeta \\
\Psi_{ss} &= -\Psi \cdot \zeta^2 \\
\Psi_v &= \Psi \cdot h \\
\Psi_{vv} &= \Psi \cdot h^2 \\
\Psi_{sv} &= \Psi \cdot i\zeta h,
\end{aligned}$$

where g' and h' are derivatives with respect to τ. Inserting these into (8.24) yields

$$0 = \Psi \cdot \left\{[-g' + i\zeta r_t + \alpha h] - \left[h' + \frac{i\zeta + \zeta^2}{2} - (i\zeta\rho\gamma - \beta)\,h - \frac{\gamma^2}{2}h^2\right] v_t\right\}.$$

While we cannot rule out that $\Psi \equiv \Psi(\zeta; s_t, v_t, \tau) = 0$ for some values of ζ and the state variables, for the above equality to hold uniformly requires that each of the two bracketed terms equal zero. This gives rise to the following ordinary differential equations:

$$\frac{dg(\tau;\zeta)}{d\tau} = i\zeta r_t + \alpha h(\tau;\zeta) \tag{8.27}$$

$$\frac{dh(\tau;\zeta)}{d\tau} = -\frac{i\zeta + \zeta^2}{2} + (i\rho\gamma\zeta - \beta)h(\tau;\zeta) + \frac{\gamma^2}{2}h(\tau;\zeta)^2. \tag{8.28}$$

Solving first for h, write (8.28) as

$$\frac{dh}{A + Bh + \gamma^2 h^2/2} = d\tau,$$

where $A = -(i\zeta + \zeta^2)/2$ and $B = i\zeta\rho\gamma - \beta$. Integrating, we have

$$\tau + K = C^{-1} \ln\left[\frac{\gamma^2 h(\tau;\zeta) + B - C}{\gamma^2 h(\tau;\zeta) + B + C}\right],$$

where we take the principal value of the logarithm. Here, K is a constant of integration and $C \equiv \sqrt{B^2 - 2A\gamma^2}$ when $\gamma\zeta \neq 0$ and $C \equiv -\beta$ otherwise. Putting $\tau = 0 = h(0;\zeta)$ forces K to satisfy $e^{CK} = (B-C)/(B+C) \equiv D$ and gives as the final solution when $\gamma \neq 0$

$$h(\tau;\zeta) = \frac{B-C}{\gamma^2}\frac{e^{C\tau}-1}{1-De^{C\tau}}. \tag{8.29}$$

Turning to (8.27), we have

$$g(\tau;\zeta) = i\zeta\int_t^{t+\tau} r_u \cdot du + \alpha\int_0^{\tau} h(u;\zeta)\cdot du.$$

Applying (8.29), the right side is

$$-i\zeta\ln B(t,T) + \frac{\alpha}{\gamma^2}(B-C)\left[\int_0^{\tau}\frac{du}{e^{-Cu}-D} - \int_0^{\tau}\frac{du}{1-De^{Cu}}\right],$$

and evaluating the integrals and simplifying lead to

$$g(\tau;\zeta) = -i\zeta\ln B(t,T) + \frac{\alpha}{\gamma^2}\left[(-B+C)\tau - 2\ln\left(\frac{1-De^{C\tau}}{1-D}\right)\right], \gamma \neq 0.$$

Note that the solutions do also satisfy $g(\tau;0) = h(\tau;0) = 0$.

For completeness, the solutions for h and g when $\gamma = 0$ are

$$h(\tau;\zeta) = \frac{A}{2\beta}(1-e^{-\beta\tau})$$

$$g(\tau;\zeta) = -i\zeta\ln B(t,T) + \alpha A\tau - \frac{\alpha A}{\beta}(1-e^{-\beta\tau}).$$

This is the case of time-varying but *deterministic* volatility and is not of particular interest.

Having found g and h, we can at last evaluate the conditional c.f. of $s_T \equiv \ln S_T$ under $\hat{\mathbb{P}}$ as

$$\Psi(\zeta) \equiv \Psi(\zeta; s_t, v_t, \tau) = \exp\{i\zeta s_t + g(\tau;\zeta) + h(\tau;\zeta)v_t\}. \tag{8.30}$$

As a simple check of the result, setting $\zeta = -i$ should give the proper expression for $\Psi(-i) = \hat{E}e^{-i^2 \ln S_T} = \hat{E}_t S_T$. With this value of ζ we have $A = 0$, $C = B$, and $D = 0$, so that $h(\tau;\zeta) = 0$ and $g(\tau;\zeta) = -\ln B(t,T) = \int_t^T r_s \cdot ds$. Thus, $\hat{E}_t S_T = \exp(\int_t^T r_s \cdot ds + s_t) = M_T S_t/M_t$, showing that the solution is consistent with the martingale property of $\{S_t/M_t\}$ under $\hat{\mathbb{P}}$, as it should be.

There are several ways to use the c.f. to value the European put at some $t \in [0, T]$. Heston's approach was to work from the risk-neutral pricing formula

$$\begin{aligned} P^E(S_t, \sigma_t, T-t) &= B(t,T) \int_{[0,\infty)} (X-s)^+ \cdot d\hat{F}_t(s) \\ &= B(t,T)X \int_{[0,X]} d\hat{F}_t(s) - B(t,T) \int_{[0,X]} s \cdot d\hat{F}_t(s) \\ &= B(t,T)X\hat{F}_t(X) - S_t\hat{G}_t(X), \end{aligned}$$

where $\hat{F}_t(X) = \hat{\mathbb{P}}_t(S_T \le X)$ and $\hat{G}_t(X) = \int_{[0,X]} \frac{s}{\hat{E}_t S_T} \cdot d\hat{F}_t(s)$. Since $d\hat{G}_t \ge 0$ and $\int_{[0,\infty)} d\hat{G}_t = 1$, the quantity $\hat{G}_t(X)$ is, like $\hat{F}_t(X)$, a conditional probability that the option expires in the money—but now in a measure different from, but equivalent to, $\hat{\mathbb{P}}$. (Assuming $\{S_t\}$ is strictly positive, it is in fact the martingale measure that applies when $\{S_t\}$ rather than $\{M_t\}$ serves as numeraire.) Since $S_T \le X$ if and only if $s_T \equiv \ln S_T \le x \equiv \ln X$, we also have

$$P^E(S_t, \sigma_t, T-t) = B(t,T)X\tilde{F}_t(x) - S_t\tilde{G}_t(x), \tag{8.31}$$

where $\tilde{F}$ and $\tilde{G}$ are the c.d.f.s of $\ln S_T$ in the respective measures. Letting $\Psi_F(\zeta) \equiv \Psi(\zeta)$ and[12] $\Psi_G(\zeta) = \Psi_F(\zeta - i)/\Psi_F(-i)$ represent the corresponding c.f.s, we can find $\tilde{F}_t(x)$ and $\tilde{G}_t(x)$ from the inversion formula for c.f.s, expression (2.47), as

$$\tilde{J}_t(x) = \frac{1}{2} - \lim_{c\to\infty} \int_{-c}^{c} \frac{e^{-i\zeta x}}{2\pi i \zeta} \Psi_J(\zeta) \cdot d\zeta,\ J \in \{F, G\}. \tag{8.32}$$

8.4 Computational Issues

As the derivation makes clear, expression (8.31) and the comparable expression for the call,

$$C^E(S_t, \sigma_t, T-t) = S_t[1 - \tilde{G}_t(x)] - B(t,T)X[1 - \tilde{F}_t(x)], \tag{8.33}$$

---

[12]Since for integrable g, $\int g(s) \cdot d\tilde{G}_t(s) = (\hat{E}_t S_T)^{-1} \int g(s)e^s \cdot d\tilde{F}_t(s)$, we have

$$\begin{aligned} \int_{-\infty}^{\infty} e^{i\zeta s} \cdot d\tilde{G}_t(s) &= (\hat{E}_t S_T)^{-1} \int_{-\infty}^{\infty} e^{(i\zeta+1)s} \cdot d\tilde{F}_t(s) \\ &= \Psi_F(-i)^{-1} \Psi_F(\zeta - i). \end{aligned}$$

are not specific to the Heston model but hold under any dynamics of $\{S_t\}_{t\geq 0}$ for which martingale methods apply. Thus, arbitrage-free values of European options can be determined if one can evaluate the conditional c.d.f.s of $\ln S_T$ in the two measures. As in the Heston model, it is often not possible to do this directly, but it can be accomplished by Fourier inversion so long as there are computationally feasible expressions for the c.f.s. Such an approach will be needed for most of the models considered in chapter 9. Here we look at some practical ways of carrying out Fourier inversion and some one-step methods that are alternatives to working from (8.31) and (8.33). We shall make us of these one-step methods repeatedly in the sequel.

8.4.1 *Inverting C.f.s*

Bohman (1975) considers several algorithms for inverting c.f.s numerically, and Waller *et al.* (1995) describe applications of the most straightforward of these, which is implemented in routine `INVERTCF` on the CD:

$$\tilde{J}_t(x) \doteq \frac{1}{2} + \frac{\eta x}{2\pi} - \sum_{\substack{j=1-N \\ j\neq 0}}^{N-1} \frac{e^{-i\eta j x}}{2\pi i j}\Psi_J(\eta j),\ J \in \{F, G\}. \tag{8.34}$$

Here $\eta > 0$ governs the fineness of the grid of ζ values on which the integral is approximated, and η and integer N jointly govern the range of integration. Values $\eta = .01$ and $N = 5000$ work well in many cases. The distribution $\tilde{J}_t$ should be put in standard form or otherwise centered and scaled so that most of the probability mass lies within a few units of the origin. Fortunately, standardizing $\ln S_T$ is easy with the information at hand. Since Ψ_J is known explicitly for $J \in \{F, G\}$, the conditional mean and variance of $\ln S_T$ can be calculated as

$$\begin{aligned} E_J s_T &= i^{-1}\Psi'_J(0) \\ V_J s_T &= -\Psi''_J(0) - (E_J s_T)^2, \end{aligned}$$

where $\Psi'_J(0)$ and $\Psi''_J(0)$ are derivatives with respect to ζ evaluated at $\zeta = 0$. Letting $\zeta^* \equiv \zeta/\sqrt{V_J s_T}$, the c.f. of the standardized variable is then

$$\Psi^*_J(\zeta^*) = \exp(-i\zeta^* E_J s_T)\Psi_J(\zeta^*),$$

and one calculates $\tilde{J}_t(x)$ as in (8.34) with Ψ^*_J replacing Ψ_J and $x^* \equiv (x - E_J s_T)/\sqrt{V_J s_T}$ replacing x. Bohman discusses several more accurate inversion algorithms that can be applied when the c.f. is in standard form.

While the application of (8.34) is straightforward, it is computationally demanding if options at many strikes are to be priced at one time, as is required in order to estimate parameters of the model implicitly from market prices of traded puts and calls. With some care it is possible to implement the fast Fourier transform (f.F.t.) to estimate c.d.f. $\tilde{J}_t(\cdot)$ on an extensive grid of strike prices all at once. Beginning with a vector of values of the complex-valued function $\Psi_J(\cdot)$ on a grid $\{\zeta_j\}_{j=1}^N$, the method returns a real-valued vector $\tilde{J}_t(\cdot)$ on a grid $\{x_j\}_{j=1}^N$. Done directly as in (8.34), this would require on the order of N^2 computations, but the f.F.t. algorithm reduces this to $O(N \log_2 N)$. Routine `INVFFT` on the CD applies the f.F.t. routine in Press *et al.* (1992) to invert a user-coded c.f. corresponding to a centered, scaled distribution.

8.4.2 *Two One-Step Approaches*

Whereas the Heston (1993) procedure based on (8.31) and (8.33) requires two separate inversions for each strike price in order to evaluate $\tilde{F}_t(x)$ and $\tilde{G}_t(x)$, Carr and Madan (1998) have developed a way to price an option with a single Fourier inversion. The method requires merely having a computable expression for the risk-neutral conditional c.f. and is not limited to the Heston model. The idea is to work with a transform of the price of the option itself, regarded as a function of the log of the strike. They show that this can be expressed as a function of the c.f. of $\tilde{F}_t$ alone, $\Psi_F \equiv \Psi$.

Here is how it works for European call options. Express the value of the call in terms of the log of the strike price as $C^E(x) = B(t,T)\int_{(x,\infty)}(e^s - e^x)\cdot d\tilde{F}_t(s)$, where $\tilde{F}_t(s) \equiv \hat{\mathbb{P}}_t(\ln S_T \le s)$ is the conditional c.d.f. of $\ln S_T$ under the risk-neutral measure. The Fourier transform of $C^E(\cdot)$, namely $\int_\Re e^{i\zeta x} C^E(x)\cdot dx$, exists if C^E itself is integrable, but this condition fails here since

$$\lim_{x\to-\infty} C^E(x) = B(t,T)\int_\Re e^s \cdot d\tilde{F}_t(s) = S_t > 0.$$

An alternative is to work with a "damped" function, $C_\theta^E(x) \equiv e^{\theta x} C^E(x)$. With $\theta > 0$ the transform of C_θ^E is then

$$\begin{aligned}(\mathcal{F}C_\theta^E)(\zeta) &= \int_\Re e^{i\zeta x} B(t,T)\int_{(x,\infty)} (e^{s+\theta x} - e^{(1+\theta)x})\cdot d\tilde{F}_t(s)\,dx \\ &= B(t,T)\int_\Re\int_{(x,\infty)} (e^{s+(\theta+i\zeta)x} - e^{(1+\theta+i\zeta)x})\cdot d\tilde{F}_t(s)\,dx\end{aligned}$$

$$
\begin{aligned}
&= B(t,T)\int_{\Re}\int_{(-\infty,s)}(e^{s+(\theta+i\zeta)x}-e^{(1+\theta+i\zeta)x})\cdot dx\,d\tilde{F}_t(s)\\
&= B(t,T)\int_{\Re}\left[\frac{e^{(1+\theta+i\zeta)s}}{\theta+i\zeta}-\frac{e^{(1+\theta+i\zeta)s}}{1+\theta+i\zeta}\right]\cdot d\tilde{F}_t(s) \qquad (8.35)\\
&= B(t,T)\frac{\Psi[\zeta-i(1+\theta)]}{(\theta+i\zeta)(1+\theta+i\zeta)}.
\end{aligned}
$$

$C^E_\theta(x)$ can now be recovered in one operation by inverting $(\mathcal{F}C^E_\theta)$, and this yields the solution for the undamped call price as

$$C^E(x)=e^{-\theta x}\mathcal{F}^{-1}(\mathcal{F}C^E_\theta)=\frac{e^{-\theta x}}{2\pi}\int_{-\infty}^{\infty}e^{-i\zeta x}(\mathcal{F}C^E_\theta)(\zeta)\cdot d\zeta. \qquad (8.36)$$

Although the method can work very well, some experimentation with θ is required for good results. Also, existence of the integrals in (8.35) requires that $\int_{\Re}e^{(1+\theta)s}\cdot d\tilde{F}_t(s)=\hat{E}_tS_T^{1+\theta}<\infty$. If not all moments of S_T exist under $\hat{\mathbb{P}}$, then this imposes an upper bound on θ

There is an alternative one-step approach that is even simpler and requires no choice of damping parameter. For this we work directly with the put, exploiting the identity

$$\hat{E}_t(X-S_T)^+=\int_0^X\hat{F}_t(s)\cdot ds,$$

which is easily established via integration by parts. Transforming S_T and X to logs and applying (8.32) give for $\hat{E}_t(X-S_T)^+$

$$
\begin{aligned}
\int_{-\infty}^{x}\tilde{F}_t(s)e^s\cdot ds &= \int_{-\infty}^{x}\left[\frac{1}{2}-\lim_{c\to\infty}\int_{-c}^{c}\frac{e^{-i\zeta s}}{2\pi i\zeta}\Psi(\zeta)\cdot d\zeta\right]e^s\cdot ds\\
&= \frac{X}{2}-\frac{1}{2}\int_{-\infty}^{x}\lim_{c\to\infty}a(c,s)e^s\cdot ds,
\end{aligned}
$$

where $a(c,s)\equiv\int_{-c}^{c}\frac{e^{-i\zeta s}}{\pi i\zeta}\Psi(\zeta)\cdot d\zeta$. Inversion formula (8.32) implies that $|\lim_{c\to\infty}a(c,s)|=|1-2\tilde{F}_t(s)|\le 1$, so for arbitrary $\varepsilon>0$ there exists c_ε such that $|\sup_s[1-2\tilde{F}_t(s)-a(c_\varepsilon,s)]|<\varepsilon$, in which case $|a(c,s)e^s|\le e^s(1+\varepsilon)$ for sufficiently large c. Since e^s is integrable on $(-\infty,\ln X)$, the dominated convergence theorem implies that

$$
\begin{aligned}
\int_{-\infty}^{x}\lim_{c\to\infty}a(c,s)e^s\cdot ds &= \lim_{c\to\infty}\int_{-\infty}^{x}a(c,s)e^s\cdot ds\\
&= \lim_{c\to\infty}\int_{-\infty}^{x}\int_{-c}^{c}\frac{e^{(1-i\zeta)s}}{\pi i\zeta}\Psi(\zeta)\cdot d\zeta\,ds.
\end{aligned}
$$

Moreover, since the limit of the double integral equals $X - 2E_t(X - S_T)^+ \in [-X, X]$, Fubini's theorem justifies reversing the order of integration. Therefore,

$$\lim_{c\to\infty} \int_{-\infty}^{x} \int_{-c}^{c} \frac{e^{(1-i\zeta)s}}{\pi i\zeta} \Psi(\zeta) \cdot d\zeta \, ds = \lim_{c\to\infty} \int_{-c}^{c} \left[\int_{-\infty}^{x} e^{(1-i\zeta)s} \cdot ds \right] \frac{\Psi(\zeta)}{\pi i\zeta} \cdot d\zeta$$
$$= \frac{X}{\pi} \lim_{c\to\infty} \int_{-c}^{c} \frac{X^{-i\zeta}}{i\zeta + \zeta^2} \Psi(\zeta) \cdot d\zeta$$

and thus, for sufficiently large c,

$$P^E(S_t, T-t) \doteq B(t,T)X \left[\frac{1}{2} - \frac{1}{2\pi} \int_{-c}^{c} X^{-i\zeta} \frac{\Psi(\zeta)}{\zeta(i+\zeta)} \cdot d\zeta \right]. \tag{8.37}$$

Of course, the singularity at $\zeta = 0$ must be avoided in the numerical integration, but even the simplest trapezoidal scheme with a central interval that straddles the origin is fast and very accurate.[13] The inequality

$$\left| e^{-i\zeta x} \frac{\Psi(\zeta)}{\zeta(i+\zeta)} \right| < \frac{1}{\zeta^2 + 1}$$

can help in setting finite integration limits. Despite the singularity it is easy to adapt f.F.t. routines to carry out the inversion.

Tables 8.1 and 8.2 compare Carr and Madan's damped estimates of prices of European calls with estimates obtained using (8.37) and put-call

Table 8.1. Damped, ICDF, and Black-Scholes estimates of call values under geometric Brownian motion: $\sigma_0 = .1$, $S_0 = 100$, $T = .5$, $r = 0$.

Strike	Carr-Madan				
X	$\theta = .1$	$\theta = .3$	$\theta = .5$	ICDF	B-S
80	15.86	19.99	20.00	20.00	20.00
85	10.88	15.02	15.02	15.02	15.02
90	6.06	10.19	10.20	10.20	10.20
95	1.80	5.93	5.94	5.94	5.94
100	−1.32	2.81	2.82	2.82	2.82
105	−3.09	1.04	1.05	1.05	1.05
110	−3.84	.30	.30	.30	.30
115	−4.07	.06	.07	.07	.07
120	−4.13	.00	.01	.01	.01

[13]The range of integration can be restricted to $(0, \infty)$ by doubling the result, but computation is actually slower because smaller step sizes are needed to maintain accuracy.

Table 8.2. Damped, ICDF, and two-step estimates of call values under s.v. dynamics: $\alpha = .02$, $\beta = 2$, $\gamma = .10$, $\sigma_0 = .1$, $\rho = -.5$, $S_0 = 100, T = .5$, $r = 0$.

Strike	Carr-Madan				
X	$\theta = .1$	$\theta = .3$	$\theta = .5$	ICDF	2-step
80	15.87	20.00	20.01	20.01	20.01
85	10.92	15.05	15.06	15.06	15.06
90	6.15	10.28	10.29	10.29	10.29
95	1.88	6.01	6.02	6.02	6.02
100	−1.36	2.78	2.78	2.78	2.78
105	−3.21	.92	.93	.93	.93
110	−3.93	.20	.21	.21	.21
115	−4.11	.02	.03	.03	.03
120	−4.14	.00	.00	.00	.00

parity. The latter estimates are labeled "ICDF" (for "integrated c.d.f."). Carr-Madan estimates are given for several values of damping parameter θ. Entries in table 8.1 are for the special case of geometric Brownian motion ($\alpha = \beta = \gamma = 0$) to allow comparison with values from Black-Scholes formulas. Entries in table 8.2 are for Heston's stochastic volatility model. Those in the "2-step" column are obtained by his two-step procedure, using the method described in Waller *et al.* (1995) to invert the c.f.s. Numerical integrations are by simple trapezoidal approximation with ζ ranging between ± 99.9 in steps of 0.2.

The estimates with $\theta = .1$ show that the choice of θ does matter in the Carr-Madan procedure. Nevertheless, damping procedures applied either to calls or directly to puts work well in this application for a wide range of θ values. Indeed, estimates even for $\theta = .1$ can be brought into line if the integration steps are made sufficiently small. However, valuing options by integrating the risk-neutral c.d.f. is simpler, and it is just as accurate as the much slower two-step procedure.

9

Discontinuous Processes

This chapter treats derivatives that depend in some way on processes with discontinuous sample paths, such as Poisson processes that count randomly spaced, discrete events in time. Such events can figure into the modeling of derivatives in two general ways. One way is to affect the timing of a derivative's potential payoffs; for example, European options with random time to expiration or loan-guarantee policies with random audit times. These models are treated in section 9.1. Another way is in the modeling of price processes for underlying assets, and the remaining sections of the chapter cover some of the many ways of letting price processes change abruptly. Section 9.2 considers mixed jump/diffusion processes whose sample paths are almost everywhere continuous but have a random, Poisson-distributed number of breaks per unit time. Such models allow explicitly for the abrupt, large changes in prices of financial assets that sometimes do occur. Section 9.3 shows that one can also allow for stochastic volatility in the diffusion part and even for volatility processes that are themselves subject to jumps. As we shall see in section 9.4, there are also useful models based on purely discontinuous—pure-jump—processes, some of which have infinitely many jumps during any finite interval of time. The final section of the chapter presents a general class of regime-switching models in which several distinct types of processes drive the price of the underlying security, alternating at random times. Not surprisingly, it will turn out that short-lived and out-of-money options can be worth considerably more when the underlying price is subject to any of these forms of abrupt change.

The special tools needed to describe and work with discontinuous processes were surveyed in section 3.5, and most of the processes encountered in this chapter have already been described. We begin with applications involving the simple Poisson process, using the differential-equation technique to price derivatives whose payoffs are triggered by Poisson events.

9.1 Derivatives with Random Payoff Times

Section 7.1.3 showed that it is relatively straightforward to develop analytical formulas for arbitrage-free prices of indefinitely lived American options when the underlying follows geometric Brownian motion. Assuming constant dividend rates and short rates of interest, values of perpetual derivatives are not time-dependent. In this case the replicating argument that leads to the Black-Scholes partial differential equation for finite-lived derivatives leads to a more tractable ordinary differential equation in the price of the underlying. Usually, this o.d.e. is much easier to solve. Interestingly, because of an elementary property of the interarrival times of Poisson events, the same simplifying circumstance is present when the timing of a derivative's payoffs is governed by a Poisson process.

Given a Poisson process $\{N_t\}_{t\geq 0}$ with intensity θ, and taking t as the current time, let T_t be the time that will elapse until the next Poisson event occurs; that is, $T_t = t^* - t$ if the next event occurs at t^* . Now for $\tau > 0$

$$\mathbb{P}(T_t \leq \tau \mid \mathcal{F}_t) = \mathbb{P}(N_{t+\tau} - N_t > 0 \mid \mathcal{F}_t) = \mathbb{P}(N_{t+\tau} - N_t > 0),$$

since the Poisson process has independent increments. The increment $N_{t+\tau} - N_t$ is distributed as Poisson with mean $\theta\tau$, so $\mathbb{P}(N_{t+\tau} - N_t = 0) = e^{-\theta\tau}$. Thus, the c.d.f. of T_t is

$$F_{T_t}(\tau) = \begin{cases} 0, & \tau \leq 0 \\ 1 - e^{-\theta\tau}, & \tau > 0. \end{cases}$$

The interarrival times of Poisson events are thus exponentially distributed with mean θ^{-1}. They are also stationary, in that the distribution of the time until the next event depends on neither the current time nor the time of the most recent event. This memoryless feature of exponential waiting times causes prices of derivatives whose payoffs are triggered by Poisson events to be time-invariant whenever the dynamics of the underlying assets under $\hat{\mathbb{P}}$ are also stationary. In such cases, the law of motion for a derivative with a single underlying asset can be described by just an ordinary differential equation, and solving this is usually easier than applying martingale methods.

For later reference notice also that the exponential model implies $F_{T_t}(\tau) \to 0$ for all τ as $\theta \to 0$ and that $F_{T_t}(\tau) \to 1$ for $\tau > 0$ as $\theta \to \infty$.

Thus, almost surely,

$$\lim_{\theta \to 0} T_t = \infty \quad \text{and} \quad \lim_{\theta \to \infty} T_t = 0. \tag{9.1}$$

We now give three examples of such derivative assets, solving ordinary differential equations to work out their arbitrage-free prices. In each case it is assumed that the risk associated with the Poisson events is diversifiable and therefore not compensated in market equilibrium. This implies that no preference parameters are present in the equilibrium solution.[1]

1. Consider a pension fund that holds a portfolio of traded assets whose value, currently S_0, follows geometric Brownian motion with volatility σ. The fund's managers now have contractual obligations to pensioners totaling L_0, but the obligation will grow over time at a fixed, continuous rate λ. The fund is currently solvent, meaning that $L_0 < S_0$. In order to guarantee that it can meet its future obligations, the fund seeks to buy insurance against insolvency. This would work in the following way. The insurer would audit the fund on a random schedule governed by a Poisson process with intensity $\theta > 0$. The value of this parameter, which equals the mean number of audits per unit time, would be agreed upon in advance. If the fund was found solvent at an audit date τ then it would continue to operate; but if $L_\tau > S_\tau$ the insurer would take over the fund's assets and make up the shortfall, $L_\tau - S_\tau$, out of its own resources. The problem is to find the value of such an insurance arrangement as a function of S_t and L_t.

The problem becomes simpler upon normalizing by L_t and valuing insurance per unit of liabilities. For this, let $s_t \equiv S_t/L_t$, so that the solvency condition becomes $s_t \geq 1$. The process s_t evolves under risk-neutral measure $\hat{\mathbb{P}}$ as

$$\begin{aligned} ds_t &= (L_t \cdot dS_t - S_t \cdot dL_t)/L_t^2 \\ &= (r - \lambda)s_t \cdot dt + \sigma s_t \cdot d\hat{W}_t, \end{aligned}$$

where r is the short rate of interest. The value of the contract per unit of insured liabilities can be written as $i(s_t; N_t) \equiv I(S_t, L_t; N_t)/L_t$, where $\{N_t\}_{t \geq 0}$ is the Poisson process that counts the audits from time 0.

[1]Shimko (1992) gives other examples that involve pricing Poisson-dependent payoff streams. Merton (1978) and Pennacchi (1987a, b) have applied techniques like those in the first two examples to analyze the fair pricing of deposit insurance at commercial banks.

Applying to $i(\cdot;\cdot)$ the extension of Itô's formula for functions of Itô and jump processes—that is, expression (3.43) on page 130 with $J_t = N_t$—gives

$$di(s_t;N_t) = i' \cdot ds_t + (i''/2) \cdot d\langle s \rangle_t + [i(s_t,N_t) - i(s_t,N_{t-})].$$

If the fund is solvent at the audit, then $i(s_t,N_t) - i(s_t,N_{t-}) = 0$ and there is no change in value. If the fund is insolvent, it receives $1 - s_t$ per unit of liabilities and the insurance terminates—along with the fund's control over assets. Thus, when $s_t < 1$

$$i(s_t;N_t) - i(s_t;N_{t-}) = 1 - s_t - i(s_t;N_{t-}).$$

The $\hat{\mathbb{P}}$-expected instantaneous change in normalized value during $[t,t+dt)$ is then $\lim_{h\downarrow 0} \hat{E}_t[i(s_{t+h},N_{t+h}) - i(s_t,N_t)]$ or

$$\hat{E}_t di(s_t;N_t) = \begin{cases} \left[(r-\lambda)s_t i' + \dfrac{1}{2}\sigma^2 s_t^2 i''\right] \cdot dt, & s_t \geq 1 \\ \left[(r-\lambda)s_t i' + \dfrac{1}{2}\sigma^2 s_t^2 i'' + (1 - s_t - i)\theta\right] \cdot dt, & 0 \leq s_t < 1. \end{cases} \tag{9.2}$$

Since $e^{-rt}I(S_t,L_t;N_t)$ is a martingale under $\hat{\mathbb{P}}$, we also have $\hat{E}dI = rI \cdot dt$, so that

$$\begin{aligned} \hat{E}di(s_t;N_t) &= L_t^{-1}(\hat{E}dI - I \cdot dL_t/L_t) \\ &= ri \cdot dt - i \cdot dL_t/L_t \\ &= (r-\lambda)i(s_t;N_t) \cdot dt. \end{aligned} \tag{9.3}$$

Equating (9.2) and (9.3) gives the ordinary differential equation that must be satisfied by the normalized value of insurance:

$$0 = \begin{cases} (r-\lambda)s_t i' + \dfrac{1}{2}\sigma^2 s_t^2 i'' - (r-\lambda)i, & s_t \geq 1 \\ (r-\lambda)s_t i' + \dfrac{1}{2}\sigma^2 s_t^2 i'' - (r-\lambda+\theta)i + (1-s_t)\theta, & 0 \leq s_t < 1. \end{cases}$$

Suppressing the Poisson counter now, the general solution to these equations has the form

$$i(s_t) = \alpha^+ s_t^{\beta^+} + \alpha^- s_t^{\beta^-} + \gamma s_t + \delta.$$

Since the value of insurance should approach zero as the asset/liability ratio goes to infinity, we have an upper boundary condition $i(+\infty) = 0$. There is also a lower boundary condition, $i(0) \leq 1$, since the value of insurance can be no greater than the current liability. The specific form of solution depends on the value of s_t. In the case $s_t \geq 1$ we have $\gamma = \delta = 0$ and the two roots $\beta^+ = 1$ and $\beta^- = -2(r-\lambda)/\sigma^2$. The upper boundary condition requires $\alpha^+ = 0$, so that β^+ becomes irrelevant when $s_t \geq 1$. If $\lambda \geq r$ the condition also requires $\alpha^- = 0$, indicating that there are no nontrivial solutions. Assuming that $\lambda < r$, solutions are restricted to $i(s_t) = \alpha^- s_t^{\beta^-}$. In the case $s_t < 1$ we have $\gamma = -1$, $\delta = \theta/(\theta + r - \lambda)$, and

$$\beta^{\pm} = 1/2 - (r-\lambda)/\sigma^2 \pm \sqrt{[(r-\lambda)/\sigma^2 + 1/2]^2 + 2\theta/\sigma^2} \tag{9.4}$$

with $\beta^+ > 1$ and $\beta^- < 0$. The lower boundary condition requires $\alpha^- = 0$ (making β^- irrelevant now) and restricts solutions to

$$i(s_t) = \alpha^+ s_t^{\beta^+} - s_t + \theta/(\theta + r - \lambda).$$

Notice that this gives $i(0) = \theta/(\theta+r-\lambda) = [1+(r-\lambda)/\theta]^{-1}$. Thus, in this worst possible case the value of insurance is L_t times a discount factor. The factor approaches unity as either the rate of growth of liabilities approaches the short rate or the expected time until the next audit (θ^{-1}) approaches zero. At these extreme values the right compensation for the insurer equals the full present value of liabilities.

To complete the problem requires finding α^+ and α^-. Requiring i and i' to be continuous at $s_t = 1$ imposes the conditions

$$\begin{aligned} \alpha^- - \alpha^+ + 1 - i(0) &= 0 \\ \alpha^-\beta^- - \alpha^+\beta^+ + 1 &= 0, \end{aligned}$$

which yield solutions

$$\alpha^+ = \frac{1 - [1 - i(0)]\beta^-}{\beta^+ - \beta^-} \tag{9.5}$$

$$\alpha^- = \frac{1 - [1 - i(0)]\beta^+}{\beta^+ - \beta^-}. \tag{9.6}$$

Merton (1990, chapter 13) shows that if a solution with continuous first derivative does exist then it must be the correct, arbitrage-free valuation. Summarizing, the solution for insurance value is

$$i(s_t) = \begin{cases} \alpha^- s_t^{\beta^-}, & s_t \geq 1 \\ \alpha^+ s_t^{\beta^+} - s_t + \theta/(\theta + r - \lambda), & s_t < 1, \end{cases}$$

with α^+ and α^- given by (9.5) and (9.6).

2. Continuing the previous example, suppose the managers of the pension fund plan to assess the beneficiaries for the cost of insuring its solvency. To do this, they will collect at periodic intervals an accumulated cash sum that accrues at the continuous rate $C_t \equiv cL_t$ so long as the insurance is maintained—that is, so long as the fund remains solvent. Payments are to be contributed by beneficiaries rather than paid out of the fund. The problem is to determine c in advance such that the value of this indefinitely lived stream equals the value of insurance. Clearly, this fair value of c will depend on the value of the fund's assets when the insurance agreement is consummated. We will value the stream in terms of an arbitrary c and then determine the c that equates the stream's value to the value of insurance.

Letting $V(S_t, L_t; N_t, c)$ represent the value of a stream at arbitrary fixed proportional rate cL_t, and again normalizing by L_t, we have $s_t \equiv S_t/L_t$ and $V(S_t, L_t; N_t, c)/L_t \equiv v(s_t; N_t)$ (with c understood). Since under $\hat{\mathbb{P}}$ the stream must yield the same expected return as a riskless bond, it follows that

$$\hat{E}dV(S_t, L_t; N_t, c) + cL_t \cdot dt = rV(S_t, L_t; N_t, c) \cdot dt.$$

Therefore,

$$\begin{aligned}
\hat{E}dv(s_t; N_t) + c \cdot dt &\equiv L_t^{-1}(\hat{E}dV - V \cdot dL_t/L_t) + c \cdot dt \\
&= L_t^{-1}(rV - cL_t - V\lambda) \cdot dt + c \cdot dt \\
&= (r - \lambda)v(s_t; N_t) \cdot dt. \qquad (9.7)
\end{aligned}$$

Expressing $\hat{E}dv$ using the extended Itô formula gives

$$\hat{E}dv(s_t; N_t) = \left\{(r - \lambda)s_t v' + \frac{1}{2}\sigma^2 s_t^2 v'' + [v(s_t; N_t) - v(s_{t-}; N_{t-})]\theta\right\} \cdot dt.$$

When the fund is solvent the audit brings no change in the value of the cash stream, so that the bracketed term is zero when $s_t \geq 1$. However, $v(s_t; N_t) - v(s_t; N_{t-}) = -v(s_t; N_{t-})$ when $s_t < 1$, since the stream terminates if the fund is found to be insolvent. Equating $\hat{E}dv + c$ with the version of (9.7) that is appropriate in each of these cases gives the following differential equation:

$$0 = \begin{cases} \frac{1}{2}\sigma^2 s_t^2 v'' + (r - \lambda)s_t v' - (r - \lambda)v + c, & s_t \geq 1 \\ \frac{1}{2}\sigma^2 s_t^2 v'' + (r - \lambda)s_t v' - (r - \lambda + \theta)v + c, & 0 \leq s_t < 1. \end{cases}$$

Still assuming that $\lambda < r$, the upper boundary condition is $v(+\infty) < \infty$, since the present value of a sure infinite stream proportional to L_t is finite if L_t grows more slowly than the interest rate. At the lower boundary we have $0 \le v(0) < \infty$. Finally, the solutions for the two cases and their derivatives should be continuous at $s_t = 1$. The general solution in each case is of the form

$$v(s_t) = \alpha^+ s_t^{\beta^+} + \alpha^- s_t^{\beta^-} + \delta.$$

The case $s_t \ge 1$ yields $\beta^+ = 1$ and $\beta^- = -2(r-\lambda)/\sigma^2 < 0$, as for the insurance value, and $\delta = c/(r-\lambda)$. The case $s_t < 1$ yields the same expressions as (9.4) for $\beta^{\pm}$, with $\delta = c/(\theta + r - \lambda)$. For $s_t \ge 1$ the upper boundary condition requires $\alpha^+ = 0$, and for $s_t < 1$ the lower boundary requires $\alpha^- = 0$. The solutions are thus

$$v(s_t) = \begin{cases} \alpha^- s_t^{\beta^-} + c/(r-\lambda), & s_t \ge 1 \\ \alpha^+ s_t^{\beta^+} + c/(\theta + r - \lambda), & 0 \le s_t < 1. \end{cases}$$

Equating the two expressions and their first derivatives at $s_t = 1$ and solving for α^+ and α^- give final expressions

$$v(s_t) = \begin{cases} c\left[-\dfrac{i(0)}{r-\lambda}\dfrac{\beta^+}{\beta^+ - \beta^-} s_t^{\beta^-} + \dfrac{1}{r-\lambda}\right], & s_t \ge 1 \\ c\left[-\dfrac{i(0)}{r-\lambda}\dfrac{\beta^-}{\beta^+ - \beta^-} s_t^{\beta^+} + \dfrac{1}{\theta + r - \lambda}\right], & 0 \le s_t < 1. \end{cases}$$

Here, $i(0) \equiv [1 + (r-\lambda)/\theta]^{-1}$ is the worst-case value of insurance. Notice that $v(s_t)$ is increasing in s_t in both the solvency and insolvency states, since $\beta^- < 0$. Notice, too, that $v \uparrow c/(r-\lambda)$ as $s_t \to \infty$ and $v \downarrow c/(r - \lambda + \theta)$ as $s_t \to 0$. The upper limit is the value of a perpetual stream of payments beginning at rate c and growing at rate λ. It is independent of θ because the rate of audit is irrelevant if the fund's solvency is not in doubt. To interpret the lower bound, notice that θ^{-1} is the expected time until the next audit, at which, in this limiting case, the stream of payments would certainly terminate.

The fair proportional rate of payment, $c(s_t) \equiv C(S_t, L_t)/L_t$, that equates $i(s_t)$ and $v(s_t)$ is given by

$$c(s_t) = \frac{\{1 - [1 - i(0)]\beta^+\} s_t^{\beta^-}}{-\beta^- + \beta^+[1 - i(0) s_t^{\beta^-}]}(r - \lambda)$$

when $s_t \geq 1$ and

$$c(s_t) = \frac{\left[\frac{1-[1-i(0)]\beta^-}{\beta^+-\beta^-} s_t^{\beta^+} - s_t + i(0)\right]}{\left[(\theta + r - \lambda)^{-1} - \frac{i(0)}{r-\lambda}\frac{\beta^-}{\beta^+-\beta^-} s_t^{\beta^+}\right]}$$

when $0 \leq s_t < 1$. Note that $c(s_t) \downarrow 0$ as $s_t \to \infty$ and that $c(s_t) \to i(0)(\theta + r - \lambda) = \theta$ as $s_t \to 0$. In the first case the interpretation is simply that the fair rate of compensation for insurance declines as the fund's enduring solvency becomes more certain. In the second case, as the fund becomes hopelessly insolvent, the required rate of compensation for insurance is greater the sooner an audit is apt to occur.

3. Consider a European-style call option with the unusual feature that the time to expiration is uncertain, being distributed as exponential with mean $\theta^{-1} > 0$. Under the risk-neutral measure the price of the underlying asset evolves as $dS_t = rS_t \cdot dt + \sigma S_t \cdot d\hat{W}_t$. Because of the dual relation between exponential waiting times and Poisson processes, we recognize that expiration will occur at the next change in a Poisson process, $\{N_t\}_{t\geq 0}$, that has intensity θ. Therefore, expressing the value of the call at t as $C^E(S_t, N_t)$ and applying (3.43), we find its law of motion under $\hat{\mathbb{P}}$ to be

$$dC^E(S_t, N_t) = C_S^E \cdot dS_t + \frac{1}{2}C_{SS}^E \cdot d\langle s \rangle_t + [C^E(S_t, N_t) - C^E(S_t, N_{t-})].$$

Normalizing $C^E(\cdot, \cdot)$ and S_t by the strike price, with $s_t \equiv S_t/X$ and $c(s_t, N_t) \equiv C^E(s_t, N_t)/X$, and noting that $c(s_t, N_t) = (s_t - 1)^+$ when $N_t - N_{t-} = 1$, the above can be written

$$\begin{aligned} dc(s_t, N_t) &= (c_s s_t r + c_{ss} s_t^2 \sigma^2/2) \cdot dt + c_s s_t \cdot d\hat{W}_t \\ &\quad + [(s_t - 1)^+ - c(s_t, N_{t-})]. \end{aligned} \tag{9.8}$$

The fair-game property of the call's discounted value under $\hat{\mathbb{P}}$ implies $\hat{E}dc(s_t, N_t)/c(s_t, N_t) = r \cdot dt$. Taking expectations in (9.8) then gives the following two-part differential equation that $c(\cdot, \cdot)$ must satisfy in the absence of arbitrage:

$$\begin{aligned} 0 &= c_s s_t r + c_{ss} s_t^2 \sigma^2/2 - c(\theta + r), \qquad & 0 \leq s_t \leq 1 \\ 0 &= c_s s_t r + c_{ss} s_t^2 \sigma^2/2 + \theta(s_t - 1) - c(\theta + r), \qquad & s_t > 1. \end{aligned} \tag{9.9}$$

The solutions of both equations are of the form $\alpha^+ s_t^{\beta^+} + \alpha^- s_t^{\beta^-} + \gamma s_t + \delta$, where

$$\beta^{\pm} = (-r/\sigma^2 + 1/2) \pm \sqrt{(r/\sigma^2 + 1/2)^2 + 2\theta/\sigma^2}.$$

Note that $\beta^+ > 1$ and $\beta^- < 0$. Lower boundary condition $c(0, N_t) = 0$ requires $\alpha^- = \gamma = \delta = 0$ in the first equation of (9.9). Assuming as an upper boundary condition that c_s is bounded requires $\alpha^+ = 0$ when $s_t > 1$. In that case we also have $\gamma = 1$ and $\delta = -\theta/(\theta+r)$. Thus, letting $t = 0$ be the current time and taking $N_0 = 0$,

$$c(s_0, 0) = \begin{cases} \alpha^+ s_0^{\beta^+}, & s_0 \le 1 \\ \alpha^- s_0^{\beta^-} + s_0 - \theta/(\theta + r), & s_0 > 1. \end{cases}$$

Finally, imposing continuity of $c(\cdot, \cdot)$ and of $c_s(\cdot, \cdot)$ at $s_0 = 1$ gives solutions for α^+ and α^-:

$$\alpha^+ = (\beta^+ - \beta^-)^{-1}[1 - \beta^- r/(\theta + r)]$$
$$\alpha^- = (\beta^+ - \beta^-)^{-1}[1 - \beta^+ r/(\theta + r)].$$

Recall from (9.1) that the exponential waiting time goes to infinity a.s. as $\theta \to 0$ and that it approaches zero a.s. as $\theta \to \infty$. Notice that our solution implies $c(s_0, 0) \to s_0$ as $\theta \to 0$, so that the value of the call approaches that of the underlying as the expiration date becomes more and more remote. This is consistent with the sharp, dynamics-free bounds that were established in section 4.3.3. Conversely, $c(s_0, 0) \to (s_0 - 1)^+$ as $\theta \to \infty$ and the time to expiration approaches zero. Of course, the Black-Scholes formula has the same limiting values as the time to expiration goes to zero or infinity.

9.2 Derivatives on Mixed Jump/Diffusions

This section treats the pricing of derivatives when underlying price process $\{S_t\}_{t \ge 0}$ is a geometric Brownian motion with added stochastic, Poisson-directed jumps. This is a mixture of an Itô process and a "J" process like that in example 51 in section 3.5.2.

9.2.1 *Jumps Plus Constant-Volatility Diffusions*

For the applications we parameterize the mixed process, as follows. With $dS_t = a_t \cdot dt + b_t \cdot dW_t + c_{t-} U \cdot dN_t$, take $EU = \upsilon$, $a_t = \mu_t S_{t-}$, $b_t = \sigma S_{t-}$, and $c_t = S_t$, giving

$$dS_t = \mu_t S_{t-} \cdot dt + \sigma S_{t-} \cdot dW_t + S_{t-} U \cdot dN_t. \qquad (9.10)$$

To see what the model implies for the conditional distribution of S_T, put $f(S_t) = \ln S_t$ and apply the extended Itô formula (3.45) as in (3.47) assuming that $\mathbb{P}(1 + U \leq 0) = 0$:

$$\begin{aligned} d\ln S_t &= (\mu_t - \sigma^2/2) \cdot dt + \sigma \cdot dW_t + (\ln S_t - \ln S_{t-}) \\ &= (\mu_t - \sigma^2/2) \cdot dt + \sigma \cdot dW_t + \ln(1 + U) \cdot dN_t. \end{aligned}$$

In integral form this is

$$\ln S_T - \ln S_t = (\mu_t - \sigma^2/2)(T - t) + \sigma(W_T - W_t) + \sum_{j=N_t+1}^{N_T} \ln(1 + U_j)$$

where the summation is taken to equal zero if $N_T - N_t = 0$. This represents the continuously compounded *ex*-dividend return on the underlying as the sum of a deterministic component, a normally distributed part arising from the Brownian motion, and a jump component. The jump part is a sum of a Poisson-distributed number $N_T - N_t$ of i.i.d. random shocks.

To complete the specification requires a model for U. A convenient assumption is that $\{1 + U_j\}_{j=1}^{\infty}$ are i.i.d. as lognormal, independent of $\{W_t\}_{t\geq 0}$ and $\{N_t\}_{t\geq 0}$, with parameters $\ln(1 + \upsilon) - \xi^2/2$ and ξ^2. With this parameterization $E(1 + U) = \exp[\ln(1 + \upsilon) - \xi^2/2 + \xi^2/2] = 1 + \upsilon$, as previously specified. Conditional on $N_T - N_t = n$, the continuously compounded return $\ln(S_T/S_t)$ is then a sum of $n+1$ independent normals with conditional mean

$$m_n \equiv (\mu_t - \sigma^2/2)\tau + n[\ln(1 + \upsilon) - \xi^2/2]$$

and conditional variance

$$v_n \equiv \sigma^2\tau + n\xi^2,$$

where $\tau \equiv T - t$. Unconditionally, the return is a Poisson mixture of normals. Press (1967) proposed such a mixture to capture the thick tails observed in empirical marginal distributions of logs of stocks' high-frequency returns. As we saw in example 58 on page 137, the process $\{\ln S_t\}_{t\geq 0}$ can also be constructed as a time change of a Brownian motion in the case $\upsilon = 0$.

The marginal distribution of $\ln S_T - \ln S_t$ can have many shapes, including bimodality when $\left|\ln(1 + \upsilon) - \xi^2/2\right|$ is sufficiently large. The moments can be found most easily from the cumulant generating function, $\mathfrak{L}_t(\zeta) \equiv$

$\ln \mathfrak{M}_t(\zeta) = \ln E \exp[\zeta \ln(S_T/S_t)]$:[2]

$$\mathfrak{L}_t(\zeta) = \tau\{\zeta[\mu_t + (\zeta - 1)\sigma^2/2] + \theta[(1+\upsilon)^\zeta e^{(\zeta^2-\zeta)\xi^2/2} - 1]\}.$$

Evaluating $\mathfrak{M}_t(\cdot)$ at $\zeta = 1$ gives the conditional expectation of the total return as

$$E_t(S_T/S_t) = e^{(\mu_t+\theta\nu)(T-t)}. \tag{9.11}$$

Differentiating the cumulant generating function three times and using expression (2.42) on page 61, one finds that coefficient of skewness α_3 is inversely proportional to $\sqrt{T-t}$ and that its sign is that of $[\ln(1+\upsilon) - \xi^2/2]$. Thus, the distribution can be right- or left-skewed or symmetric, with the asymmetry of probability mass in the tails increasing as time to expiration diminishes. Notice that the skewness of log return is negative if the mean size of jumps, υ, is less than or equal to zero. Differentiating again and applying (2.43) shows that excess kurtosis $\alpha_4 - 3$ is positive and of order $(T-t)^{-1}$ except in the degenerate cases $\theta = 0$ or $\upsilon = 0$ and $\xi = 0$. Thus, like pure Itô processes with stochastic volatility, the distribution of log return does have thicker tails than the normal. We will see later that the thick tails produce higher implicit volatilities for options far in and far out of the money, the effect being most pronounced for options with short lives.

9.2.2 *Nonuniqueness of the Martingale Measure*

As with stochastic-volatility models, derivatives pricing in this environment is complicated by the fact that random-jump dynamics do not admit a unique equivalent martingale measure. This can be seen as follows. Referring to (9.11), it is clear that we need $\mu_t + \theta\nu = r_t - \delta$ (the instantaneous short rate less the dividend rate) if normalized *cum*-dividend process $\{S_t^{c*} = S_t e^{\delta t}/M_t\}_{t\geq 0}$ is to be a martingale. Equivalently, using the extended Itô formula with $f(t, S_t) = S_t^{c*} = S_t M_0^{-1} \exp[\int_0^t (\delta - r_s) \cdot ds]$ gives

$$\begin{aligned} dS_t^{c*} &= (\delta - r_t)S_t^{c*} \cdot dt + S_{t-}^{c*}(\mu_t \cdot dt + \sigma \cdot dW_t + U \cdot dN_t) \\ &= (\delta - r_t)S_{t-}^{c*}(1 + U \cdot dN_t) \cdot dt + S_{t-}^{c*}(\mu_t \cdot dt + \sigma \cdot dW_t + U \cdot dN_t) \\ &= S_{t-}^{c*}[(\mu_t + \theta\nu + \delta - r_t) \cdot dt + \sigma \cdot dW_t + (U \cdot dN_t - \theta\nu \cdot dt)], \end{aligned}$$

[2]This may be found directly or from expression (3.49) for the log of the c.f. of the mixed process in example 51.

apart from terms of $o(dt)$. We eliminate the drift term by changing to any measure $\widetilde{\mathbb{P}}$ for which[3]

- $\{\widetilde{N}_t\}_{t\geq 0}$ is a Poisson process with rate $\tilde{\theta}$.
- Random variables $\{\ln(1+\widetilde{U}_j)\}_{j=1}^{\infty}$ are i.i.d. as $N[\ln(1+\tilde{\nu})-\tilde{\xi}^2/2, \tilde{\xi}^2]$ and independent of $\{\widetilde{N}_t\}$.
- $\{\widetilde{W}_t = W_t + \sigma^{-1}\int_0^t(r_s - \delta - \theta\nu - \mu_s)\cdot ds\}_{t\geq 0}$ is a Brownian motion independent of $\{\widetilde{N}_t\}_{t\geq 0}$ and $\{\widetilde{U}_j\}_{j=1}^{\infty}$.
- $\tilde{\theta}\tilde{\nu} = \theta\nu$.

Thus, the martingale property would be consistent with either a high mean rate ($\tilde{\theta}$) of jumps having low mean ($\tilde{\nu}$) or infrequent jumps with large expectation. Only the product $\tilde{\theta}\tilde{\nu}$ matters for the martingale property.

As we have seen, the absence of a unique martingale measure is associated with the incompleteness of markets, meaning that payoffs of derivatives cannot always be replicated with portfolios of traded assets. The inability to replicate nonlinear payoffs in the jump-diffusion setting can be seen through the following example. Take S_t to be the price of a primary underlying asset that (for simplicity) pays no dividends, and let M_t be the price of a money fund paying interest at a constant short rate r. We will try to find a self-financing portfolio of the underlying and the money fund that replicates a derivative whose value depends nonlinearly on S_t and time alone, $D(S_t, T-t)$. Setting $D(S_t, T-t) = p_t S_t + q_t M_t$ as was done in chapter 6 when S_t was a continuous Itô process, write

$$\begin{aligned} dD &= p_t \cdot dS_t + (D - p_t S_{t-})\cdot dM_t/M_t \\ &= [Dr + p_t S_{t-}(\mu_t - r)]\cdot dt + p_t S_{t-}\sigma\cdot dW_t + p_t S_{t-}U\cdot dN_t. \end{aligned}$$

Applying the extended Itô formula to $D(S_t, T-t)$ gives

$$\begin{aligned} dD = {} & [-D_{T-t} + D_S S_t \mu_t + D_{SS} S_t^2\sigma^2/2]\cdot dt + D_S S_{t-}\sigma\cdot dW_t \\ & + \{D[S_{t-}(1+U), T-t] - D(S_{t-}, T-t)\}\cdot dN_t. \end{aligned}$$

---

[3]Putting $\lambda = t^{-1}\sigma^{-1}\int_0^t(r_s - \delta - \theta\nu - \mu_s)\cdot ds$ and $Q_T = \exp(-\lambda^2 T/2 - \lambda W_T)$, the change in Brownian measure is accomplished as $\widetilde{\mathbb{P}}(A) = \int 1_A(\omega)Q_T(\omega)\cdot d\mathbb{P}(\omega)$, where A is any event in the product σ-field generated by the independent families of random variables $\{W_t\}_{0\leq t\leq T}, \{N_t\}_{0\leq t\leq T}, \{U_j\}_{j=1}^{\infty}$. Simultaneous changes in the Poisson measure are effected with the Radon-Nikodym process $\{Q'_t = (\tilde{\lambda}/\lambda)^{N_t}e^{-t(\tilde{\lambda}-\lambda)}\}$, as $\widetilde{\mathbb{P}}(A) = \int 1_A(\omega)Q_T(\omega)Q'_T(\omega)\cdot d\mathbb{P}(\omega)$, while changes of the $\{\ln(1+U_j)\}$ from $N[\ln(1+\nu)-\xi^2/2, \xi^2]$ to $N[\ln(1+\tilde{\nu})-\tilde{\xi}^2/2, \tilde{\xi}^2]$ are done with the usual Radon-Nikodym derivative.

If we maintain $p_t = D_S$, then the portfolio matches the progress of the Brownian motion, but it does not capture the jump risk unless

$$D_S = \frac{D[S_{t-}(1+U), T-t] - D(S_{t-}, T-t)}{S_{t-}(1+U) - S_{t-}} \tag{9.12}$$

for all values of random variable U. Clearly, this is possible if and only if D is linear in S, as it would be, for example, if D were a forward contract.[4]

In the absence of a unique martingale measure we are again obliged to consider how the parameters of the process $\{S_t\}_{t \geq 0}$ reflect attitudes toward risk. One simple possibility, considered by Merton (1976), is that jump risk is diversifiable. In that case, since the diffusion risk can be hedged away, only the $\{\mu_t\}$ of the diffusion process reflects attitudes toward risk. The remaining parameters of the model are the same under the risk-neutral and natural measures and can be estimated with arbitrarily high precision by sufficiently patient observation of the process. In this case, Girsanov's theorem implies a unique new measure $\hat{\mathbb{P}}$ under which $\hat{W}_t = W_t + (\int_0^t (r_s - \delta - \mu_s) \cdot ds + \theta \nu t)/\sigma$ is a Brownian motion. In effect, the change of measure is accomplished by setting $\mu_t = r_t - \delta + \theta\nu$ and $W_t = \hat{W}_t$ in (9.10), leaving all other parameters the same. However, as Jarrow and Rosenfeld (1984) argue, the occasional large breaks in market indices are highly improbable if all the jump risk is asset-specific, as it would have to be in order to be diversifiable. Bates (1996a) and Bakshi *et al.* (1997) appeal to representative-agent equilibrium models that imply a risk-adjusted version of (9.10) with the same general form, but with values of υ, ξ, and θ that register risk and time preference.[5] These implicit parameters, which we have represented as $\tilde{\upsilon}$, $\tilde{\xi}$, and $\tilde{\theta}$, can be inferred from prices of traded derivatives—for example, by nonlinear least squares. There is evidence even in pre-1987 data that these can differ considerably from the corresponding versions under $\mathbb{P}$.[6]

9.2.3 *European Options under Jump Dynamics*

A computational formula for European-style derivatives can be developed easily by means of a conditioning argument. The idea is to find the

---

[4]If the dynamics were pure-jump, having no diffusion part, and if U were *deterministic*, then the derivative could be replicated by setting p_t equal to the right side of (9.12). See Cox and Ross (1976).

[5]The model proposed by Ahn (1992) does not quite nest within the Merton framework unless $\upsilon = 0$ and there is zero correlation between the jump shocks, U and U_W, of the security and aggregate wealth.

[6]Stern (1993), estimating υ, ξ, θ, and σ by maximum likelihood from daily 1986–87 stock returns, finds that the jump model, while an improvement over Black-Scholes, still underprices short-maturity, out-of-money options.

conditional expectation of the terminal (time-T) payoff under $\hat{\mathbb{P}}$ given a (random) number of jumps, then apply the tower property. Taking $t = 0$ as the current date and continuing to assume that short rates are $\mathcal{F}_0$-measurable, the distribution of $\ln(S_T/S_0)$ under $\hat{\mathbb{P}}$ conditional on $N_T = n$ is normal with variance $\sigma^2 T + n\tilde{\xi}^2$ and mean

$$\int_0^T r_t \cdot dt - \tilde{\theta}\tilde{\upsilon}T + n\ln(1+\tilde{\upsilon}) - \delta T - (\sigma^2 T + n\tilde{\xi}^2)/2.$$

With some new notation values of options conditional on the number of jumps can now be expressed in terms of Black-Scholes formulas. Putting $B_n \equiv B(0,T)\exp(\tilde{\theta}\tilde{\upsilon}T)(1+\tilde{\upsilon})^{-n}$ and $\sigma_n^2 \equiv \sigma^2 + n\tilde{\xi}^2/T$, the conditional distribution of S_T/S_0 is then lognormal with parameters $-\ln B_n - \delta T - \sigma_n^2 T/2$ and $\sigma_n^2 T$. If B_n were actually the current price of a T-maturing bond, then conditional on $N_T = n$ the discounted expected payoffs of European call and put struck at X would be given by the Black-Scholes formulas with B_n, σ_n replacing $B(0,T), \sigma$:

$$C^E(S_0, T; B_n, \sigma_n) \equiv S_0 e^{-\delta T}\Phi[q_n^+(\mathsf{f}_n/X)] - B_n X\Phi[q_n^-(\mathsf{f}_n/X)]$$
$$P^E(S_0, T; B_n, \sigma_n) \equiv B_n X\Phi[q_n^+(X/\mathsf{f}_n)] - S_0 e^{-\delta T}\Phi[q_n^-(X/\mathsf{f}_n)],$$

where $\mathsf{f}_n = S_0 e^{-\delta T}/B_n$ and $q_n^{\pm}(s) \equiv [\ln s \pm \sigma_n^2 T/2]/(\sigma_n\sqrt{T})$.However, since the bond's price is actually $B(0,T)$ rather than B_n, the discounted expected payoffs conditional on $N_T = n$ are really $B(0,T)/B_n$ times these modified Black-Scholes prices; that is,

$$B(0,T)\hat{E}[(S_T - X)^+ \mid N_T] = C^E(S_0, T; B_{N_T}, \sigma_{N_T})B(0,T)/B_{N_T}$$
$$B(0,T)\hat{E}[(X - S_T)^+ \mid N_T] = P^E(S_0, T; B_{N_T}, \sigma_{N_T})B(0,T)/B_{N_T}.$$

The arbitrage-free prices of European calls and puts are then the expectations of these conditional values. Again, some new notation simplifies the final expressions. Representing Poisson probabilities as $p(n;\theta) = \theta^n e^{-\theta}/n!$ and letting $\tilde{\theta}_\upsilon \equiv \tilde{\theta}(1+\tilde{\upsilon})$, notice that $p(n;\tilde{\theta}T)B(0,T)/B_n = p(n;\tilde{\theta}_\upsilon T)$. With this notation, the values of the European call and put are, respectively,

$$B(0,T)\hat{E}(S_T - X)^+ = \sum_{n=0}^{\infty} p(n;\tilde{\theta}_\upsilon T)C^E(S_0, T; B_n, \sigma_n) \tag{9.13}$$

$$B(0,T)\hat{E}(X - S_T)^+ = \sum_{n=0}^{\infty} p(n;\tilde{\theta}_\upsilon T)P^E(S_0, T; B_n, \sigma_n). \tag{9.14}$$

To make the computations, one simply truncates the summations when the tail probability for the modified Poisson distribution becomes sufficiently small. Since the put's conditional values are bounded (by X), it is best to work with (9.14) and price the call by European put-call parity. One can then terminate the sum in (9.14) at a value $N(\varepsilon)$ such that $1-\sum_{n=0}^{N(\varepsilon)} p(n;\tilde{\theta}_\upsilon T) \leq \varepsilon/X$, where ε is an error tolerance. Notice that Poisson probabilities can be calculated recursively as $p(n;\theta) = p(n-1;\theta)\theta/n$ for $n \in \aleph$, with $p(0;\theta) = e^{-\theta}$. Program `JUMP` on the accompanying CD (and implemented also in the `ContinuousTime.xls` spreadsheet) values European puts and calls by these methods. Alternatively, options can be priced by Fourier inversion using one of the one-step approaches described in section 8.4.2.

9.2.4 *Properties of Jump-Dynamics Option Prices*

Figures 9.1–9.4 show smile curves from the jump-diffusion model. These are plots of implicit volatility vs. moneyness, X/S_t, when option prices are generated by (9.13) or (9.14). The figures contrast (i) times to expiration, $T-t$; (ii) Poisson intensities, $\tilde{\theta}$; (iii) volatilities of the jump shock, $\tilde{\xi}$; and (iv) means of the jump shock, $\tilde{\upsilon}$. Figure 9.1 shows that implicit volatilities at low values of X/S_t (where puts are out of the money) and at high

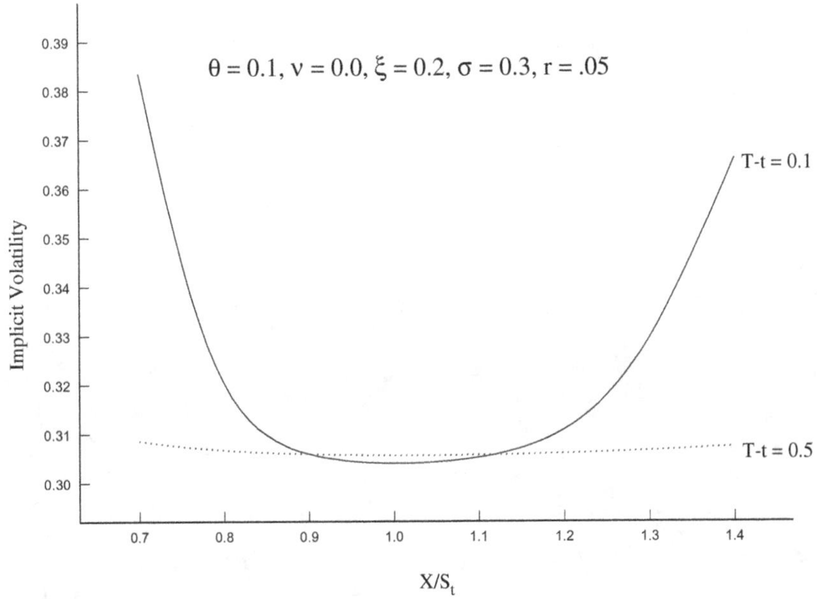

Fig. 9.1. Jump-diffusion implicit volatilites for short vs. long maturities, $T-t$.

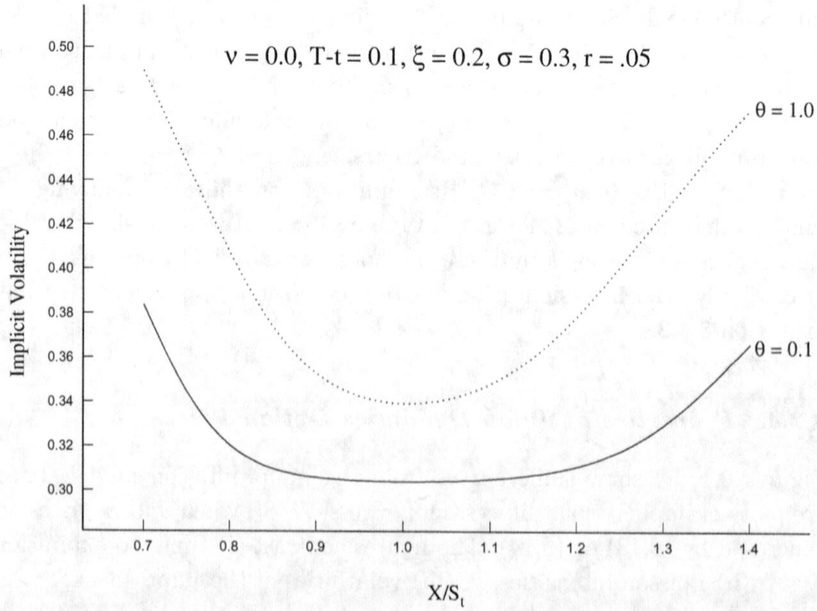

Fig. 9.2. Jump-diffusion implicit volatilities for high vs. low jump intensities, θ.

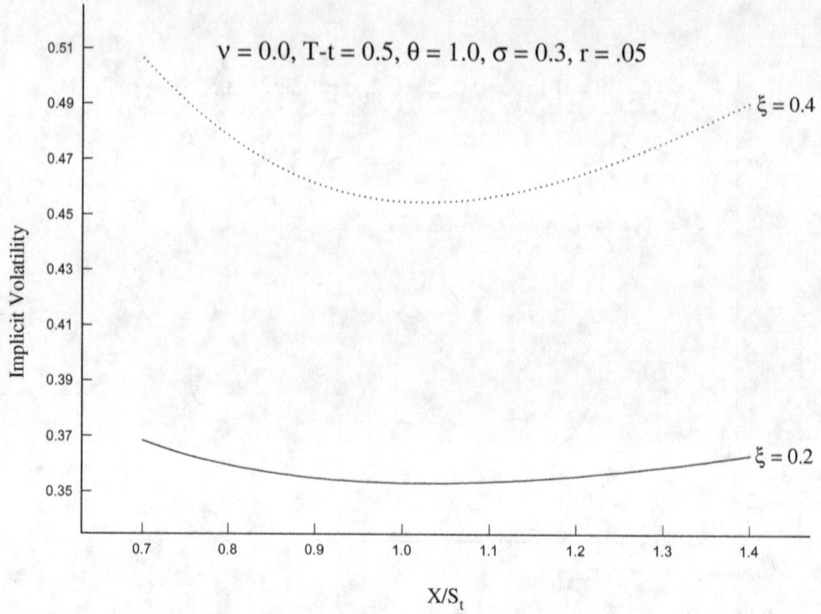

Fig. 9.3. Jump-diffusion implicit volatilities for high vs. low jump volatilities, ξ.

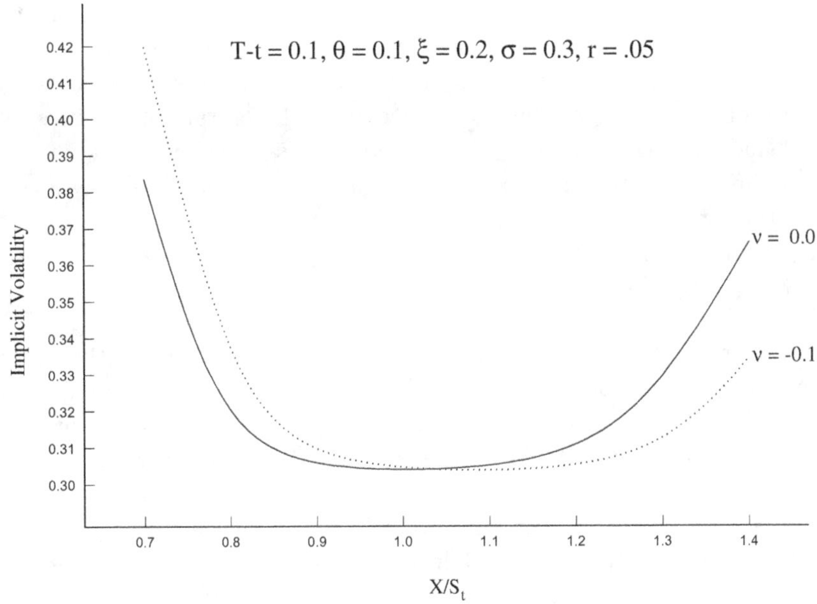

Fig. 9.4. Jump-diffusion implicit volatilities for zero vs. negative mean jump sizes, ν.

values of X/S_t (where calls are out) increase markedly as the time to expiration decreases. This results from the higher probability that short-lived, out-of-money options will hit the money when the underlying price process may be discontinuous. Figures 9.2 and 9.3 show that implicit volatilities increase with either the mean frequency or volatility of jump shocks, and more rapidly so for options away from the money. Figure 9.4 shows that the greater negative skewness in the distribution of $\ln(S_T/S_t)$ when $\tilde{\upsilon}$ is negative enhances the smirk effect in the smile curve, raising implicit volatility more for out-of-money puts than for out-of-money calls. It does make sense intuitively that the greater potential for negative shocks adds more value to puts of given degrees of moneyness than it does to calls.

9.2.5 *Options Subject to Early Exercise*

Obviously, no simple formula like (9.13) applies to American-style options on assets whose prices follow jump-diffusions. However, Amin (1993) has developed a very satisfactory method for pricing derivatives subject to early exercise by discretizing both the time and state spaces, as in the binomial method. The diffusion part of the price process is approximated using Bernoulli dynamics, with price moving either up or down by one tick during $(t, t + \Delta t]$, while jumps are modeled as multitick moves that occur with probability $\theta \cdot \Delta t$. The state space and risk-neutral probabilities can be defined in such a way that the discrete process converges weakly to the jump-diffusion as the number of time steps increases and Δt decreases. At each time step t_j (with $t_{j+1} - t_j \equiv \Delta t$) and price state s^k an American put, say, is valued recursively as

$$P^A(s^k, T - t_j) = \max\{X - s^k, e^{-r\Delta t}\hat{E}_{t_j}[P^A(S_{j+1}, T - t_{j+1}) \mid S_j = s^k]\}.$$

The method allows flexibility in modeling distributions of the jumps themselves. In the degenerate case $\mathbb{P}(U = -1) = 1$, wherein jumps correspond to bankruptcy events, Amin (1993) discovers through numerical computation that the early exercise boundary for American calls on dividend-paying stocks is higher than it is under the pure-diffusion model. In other words, even though bankruptcy would wipe out a call's value, it is optimal to hazard this event and delay exercise until the call becomes even deeper in the money than under geometric Brownian motion. The early exercise of American puts is also delayed—occurs deeper in the money—as one would expect. Things are more complicated with lognormal jumps, the exercise boundaries moving farther in the money for short-term puts and calls, but moving nearer the money for longer maturities.

9.3 Jumps Plus Stochastic Volatility

Working with 1984–1991 data for \$/DM options and estimating parameters from actual option prices via nonlinear least squares, Bates (1996a) found that a stochastic-volatility model without jumps gives better fits in sample than a mixed jump process with deterministic volatility on the Itô part. However, since the estimated volatility of the variance process is implausibly high given the time-series behavior of implicit volatilities, he concludes that combining the two models is the best way to account for the volatility smile. In this way the jump process carries some of the burden for explaining the high kurtosis in exchange-rate changes. Bakshi *et al.* (1997) find with 1988–1991 data for S&P 500 index options that such an "s.v.-jump" model that allows for jumps in price as well as stochastic volatility reduces both in-sample and out-of-sample pricing errors, particularly for short-lived options away from the money. Certainly, one expects to have greater flexibility in accounting for the smirk in implicit volatilities if the two effects are combined. In this section we look first at methods used by

Bates (1996a), Bakshi *et al.* (1997), and Scott (1997) to develop computational formulas for European options in the s.v.-jump framework.[7] There are, however, indications in both option prices and index returns that the rapid changes and sustained levels of volatility following major economic events require still more elaborate models. Eraker *et al.* (2003) argue that replicating this aspect of the data requires modeling volatility also as a discontinuous process. An alternative idea, proposed by Fang (2002), is to allow the intensity of jumps in the price process to vary stochastically with time. This feature can help to capture the abruptness and persistence of change in the tenor of the market even when price volatility is assumed to be constant; however, adding stochastic volatility does further enhance the model's flexibility. We will see that such extensions to the basic s.v.-jump model can be accommodated easily and produce computationally feasible formulas for prices of European options. Our treatment can be relatively brief since the analytical methods for treating all these models are similar to those used in the basic s.v. model of Heston (1993).

9.3.1 *The S.V.-Jump Model*

We work with the following model for the risk-neutral process for the price of the underlying:

$$
\begin{aligned}
dS_t &= S_{t-}(r_t - \theta v) \cdot dt + \sqrt{v_t} S_{t-} \cdot d\hat{W}_{1t} + S_{t-} U \cdot dN_t \\
dv_t &= (\alpha - \beta v_t) \cdot dt + \gamma \sqrt{v_t} \cdot d\hat{W}_{2t}.
\end{aligned}
$$

Here (i) $\{\hat{W}_{1t}\}$ and $\{\hat{W}_{2t}\}$ are standard Brownian motions under $\hat{\mathbb{P}}$ with $|\hat{E}W_{1t}W_{2t}| = |\rho| t < t$; (ii) $\{N_t\}$ is an independent Poisson process with (constant) intensity θ; (iii) v_t is the square of price volatility; and (iv) $\ln(1+U)$ is distributed independently (of $\{N_t\}_{t\geq 0}$ and the Brownian motions) as normal with mean $\ln(1+v) - \xi^2/2$ and variance ξ^2.[8] The notation for the jump process is the same as in the previous section, except that tildes are omitted from the risk-adjusted parameters. Likewise, the adjusted process for v_t is the same as (8.22) except that we express dv_t in terms of a single Brownian motion that is correlated with the one that

---

[7]Naik (1993) presents a related model in which volatility is a Markov process that can switch among two or more discrete levels, with accompanying nonstochastic jumps to price.

[8]The model applies to a primary asset that pays no dividend. Otherwise, we make the usual adjustments: replace r_t by $r_t - \delta_t$ if the underlying pays a continuous dividend and set the mean proportional drift rate to $-\theta v$ if S_t represents a futures price.

drives $\{S_t\}$. Applying the extended Itô formula with $ds_t \equiv d\ln S_t$ gives $ds_t = (r_t - \theta\upsilon - v_t/2) \cdot dt + \sqrt{v_t} \cdot d\hat{W}_{1t} + u \cdot dN_t$, where $u \equiv \ln(1+U)$.

To price a derivative on an underlying asset whose price follows such a process, we follow a program like that for the basic s.v. model without jumps (section 8.3.2). That is, we first exploit the martingale property of conditional expectations processes to determine $\Psi(\zeta; s_t, v_t, \tau)$, the conditional c.f. of S_T as of $t = T - \tau$. This time, discovering Ψ will require solving a partial differential integral equation (p.d.i.e.), but the method is much the same. Once Ψ has been determined, we can use one of the techniques described in section 8.4 to value a T-expiring European option via Fourier inversion.

To begin the program, apply the extended Itô formula to $\Psi(\zeta; s_t, v_t, \tau) = \hat{E}_t e^{i\zeta s_T}$ to obtain

$$\begin{aligned} d\Psi = &-\Psi_\tau \cdot dt + \Psi_s \left(r_t - \theta\nu - \frac{v_t}{2}\right) \cdot dt + \Psi_s \sqrt{v_t} \cdot dW_{1t} + \Psi_{ss} \frac{v_t}{2} \cdot dt \\ &+ \Psi_\nu(\alpha - \beta v_t) \cdot dt + \Psi_\nu \gamma \sqrt{v_t} \cdot dW_{2t} + \Psi_{\nu\nu}\gamma^2 \frac{v_t}{2} \cdot dt + \Psi_{s\nu}\rho\gamma v_t \cdot dt \\ &+ [\Psi(\zeta; s_t, v_t, \tau) - \Psi(\zeta; s_{t-}, v_t, \tau)] \cdot dN_t. \end{aligned}$$

Taking expectations conditional on $\mathcal{F}_{t-}$ and exploiting the martingale property under $\hat{\mathbb{P}}$ give the p.d.i.e.

$$\begin{aligned} 0 = &-\Psi_\tau + \Psi_s(r_t - \theta\nu - \frac{v_t}{2}) + \Psi_{ss}\frac{v_t}{2} + \Psi_\nu(\alpha - \beta v_t) + \Psi_{\nu\nu}\gamma^2 \frac{v_t}{2} \\ &+ \Psi_{s\nu}\rho\gamma v_t + \hat{E}_{t-}[\Psi(\zeta; s_t, v_t, \tau) - \Psi(\zeta; s_{t-}, v_t, \tau)]\theta, \end{aligned} \tag{9.15}$$

which must be solved subject to terminal condition $\Psi(\zeta; s_T, v_T, 0) = e^{i\zeta s_T}$.

Again, as in the basic s.v. model, the solution has a simple exponential-affine form in state variables s_t, v_t; namely,

$$\Psi(\zeta; s_t, v_t, \tau) = \exp[i\zeta s_t + g(\tau; \zeta) + h(\tau; \zeta)v_t + q(\zeta)\theta\tau] \tag{9.16}$$

for certain complex-valued functions g, h, and q satisfying

$$g(0; \zeta) = h(0; \zeta) = g(\tau; 0) = h(\tau; 0) = q(0) = 0. \tag{9.17}$$

The added term involving q in the exponent arises from the expectation term in (9.15). Given the form (9.16), that expectation term equals $\Psi(\zeta; s_t, v_t, \tau)$ times

$$\hat{E}e^{i\zeta \ln(1+U)} - 1 = (1+\nu)^{i\zeta} e^{-(i\zeta + \zeta^2)\xi^2/2} - 1.$$

Setting $q(\zeta) \equiv \hat{E}e^{i\zeta \ln(1+U)} - 1$, inserting this and the derivatives of (9.16) into (9.15), and separating the terms proportional to v_t from those that are not give the conditions

$$0 = -g' + i\zeta(r_t - \theta\nu) + \alpha h$$
$$0 = -h' - (i\zeta + \zeta^2)/2 + (i\rho\gamma\zeta - \beta)h + \gamma^2 h^2/2.$$

When $\gamma \neq 0$ solutions of these o.d.e.s for g and h subject to (9.17) are

$$h(\tau;\zeta) = \frac{B-C}{\gamma^2}\frac{e^{C\tau}-1}{1-De^{C\tau}}$$
$$g(\tau;\zeta) = -i\zeta[\ln B(t,T) + \tau\theta\nu] + \frac{\alpha}{\gamma^2}\left[(-B+C)\tau - 2\ln\left(\frac{1-De^{C\tau}}{1-D}\right)\right],$$

where $A \equiv -(i\zeta + \zeta^2)/2$, $B \equiv i\zeta\rho\gamma - \beta$,

$$C \equiv \begin{cases} \sqrt{B^2 - 2A\gamma^2}, & \zeta \neq 0 \\ -\beta, & \zeta = 0, \end{cases}$$

and $D \equiv (B-C)/(B+C)$. We thus have

$$\Psi(\zeta; s_t, v_t, \tau) = \exp\{i\zeta s_t + g(\tau;\zeta) + h(\tau;\zeta)v_t + [(1+\nu)^{i\zeta}e^{A\xi^2} - 1]\theta\tau\}. \tag{9.18}$$

With this in hand we can use the integrated c.d.f. method described in section 8.4.2 to approximate the price of a T-expiring European put as

$$P^E(S_t, v_t, T-t) = B(t,T)X\left[\frac{1}{2} - \frac{1}{2\pi}\int_{-c}^{c} X^{-i\zeta}\frac{\Psi(\zeta; s_t, v_t, \tau)}{\zeta(i+\zeta)}\cdot d\zeta\right] \tag{9.19}$$

for some large c.

Before looking at still richer models, it is instructive to see how to retrieve the simpler ones that the s.v.-jump model encompasses.

1. If $\theta = 0$ the jump component is eliminated, the last term in the exponent of (9.18) vanishes, and we are back to Heston's s.v. model. The solutions for $g(\tau;\zeta)$ and $h(\tau;\zeta)$ are the same except that the term $-i\zeta\tau\theta\nu$ disappears in the expression for g.
2. If $\gamma = 0$, $\alpha \geq 0$, $\beta > 0$, and $\theta \geq 0$ we are left with either a pure diffusion process (when $\theta = 0$) or a jump-diffusion process (when $\theta > 0$) in which volatility is time-varying but deterministic. Specifically, solving $dv_t = (\alpha - \beta v_t)\cdot dt$ subject to initial value v_0 gives $v_t = \alpha/\beta + e^{-\beta t}(v_0 - \alpha/\beta), t \geq 0$. As $t \to \infty$ this either increases or decreases toward α/β according as $v_0 < \alpha/\beta$ or $v_0 > \alpha/\beta$. The solutions for g and h now have different forms:

$$h(\tau;\zeta) = A\beta^{-1}(1 - e^{-\beta\tau})$$
$$g(\tau;\zeta) = -i\zeta\ln B(t,T) - i\zeta\theta\nu\tau - \alpha\beta^{-1}A\tau - \alpha A\beta^{-2}(1 - e^{-\beta\tau}).$$

3. If $\gamma = \alpha = \beta = 0$ but $\theta > 0$, then volatility is constant ($\sqrt{v_t} = \sqrt{v_0}$), and we are left with the simple jump-diffusion model. In this case $h(\tau;\zeta) = A\tau$ and $g(\tau;\zeta) = -i\zeta[\ln B(t,T) + \tau\theta\nu]$.
4. If $\alpha = \beta = \gamma = \theta = 0$, we are all the way back to geometric Brownian motion with volatility $\sigma = \sqrt{v_0}$. Setting $\theta = 0$ in the last expression for $g(\tau;\zeta)$, the conditional c.f. of s_T is simply

$$\Psi(\zeta; s_t, v_t, \tau) = \exp[i\zeta s_t - i\zeta \ln B(t,T) - i\zeta v_0\tau/2 - \zeta^2 v_0 \tau/2],$$

which corresponds to $N(s_t + \int_t^T r_u \cdot du - v_0\tau/2, v_0\tau)$.[9]

9.3.2 *Further Variations*

Duffie *et al.* (2000), Eraker *et al.* (2003), and Eraker (2004) propose extending the s.v.-jump model to allow for discontinuous changes in volatility as well as in price. The changes can be modeled as occurring either simultaneously or independently. We describe here just the simultaneous version, as presented by Eraker *et al.* (2003). In this model the log price and squared volatility of an underlying (no-dividend) primary asset evolve under risk-neutral measure $\hat{\mathbb{P}}$ as

$$\begin{aligned} ds_t &= (r_t - \theta\nu - v_t/2) \cdot dt + \sqrt{v_t} \cdot d\hat{W}_{1t} + u_1 \cdot dN_t \\ dv_t &= (\alpha - \beta v_t) \cdot dt + \gamma\sqrt{v_t} \cdot d\hat{W}_{2t} + u_2 \cdot dN_t. \end{aligned}$$

As in the s.v.-jump model $\{\hat{W}_{1t}\}$ and $\{\hat{W}_{2t}\}$ are standard Brownian motions with correlation ρ and $\{N_t\}$ is an independent Poisson process with intensity θ; however, the Poisson events now affect both price and volatility. Volatility shocks $\{U_{2j}\}_{j=1}^{N_t}$ during $[0,t]$ are modeled as i.i.d. exponential variates, with c.d.f. $F(u) = (1 - e^{-\varphi u})\mathbf{1}_{[0,\infty)}(u)$, while shocks $\{U_{1j}\}_{j=1}^{N_t}$ to log price are i.i.d. as $N[\ln(1+\nu) - \xi^2/2 + \kappa u_2, \xi^2]$ conditional on $\{U_{2j} = u_2\}$. Setting $\kappa < 0$ allows price shocks to move counter to volatility shocks on average, as they appear to do in most markets. With s_{t-} and v_{t-} as left-hand limits of log price and of volatility and $s_t = s_{t-} + u_1 \cdot dN_t$ and $v_t = v_{t-} + u_2 \cdot dN_t$, the same arguments that led to (9.15) show that the conditional c.f. of s_T satisfies the p.d.i.e.

$$\begin{aligned} 0 = -\Psi_\tau + \Psi_s\left(r_t - \theta\nu - \frac{v_{t-}}{2}\right) + \Psi_{ss}\frac{v_{t-}}{2} + \Psi_\nu(\alpha - \beta v_{t-}) + \Psi_{\nu\nu}\gamma^2\frac{v_{t-}}{2} \\ + \Psi_{s\nu}\rho\gamma v_{t-} + \hat{E}_{t-}[\Psi(\zeta; s_t, v_t, \tau) - \Psi(\zeta; s_{t-}, v_{t-}, \tau)]\theta \end{aligned}$$

---

[9] Applying (9.19) with this c.f. and comparing with the usual Black-Scholes computations give a good check on the program and the choice of settings in the numerical integration algorithm.

and terminal condition $\Psi(\zeta; s_T, v_T, 0) = e^{i\zeta s_T}$. The solution again has the form (9.16) with $g(\tau;\zeta)$ and $h(\tau;\zeta)$ just the same but with

$$q(\zeta) \equiv \exp[i\zeta \ln(1+\nu) + A\xi^2][1 - (i\zeta\kappa + h)\varphi]^{-1} - 1.$$

Sending $\varphi \to 0$ makes the volatility jumps vanishingly small and returns us to the s.v.-jump model.

Figures 9.5 and 9.6 present series of implicit volatility curves and p.d.f.s of $\ln S_T$ that show the incremental effects of the various features of the model. That is, we start with Black-Scholes (model BS) and progressively add jumps to price (model JD), allow for stochastic volatility without jumps in price (model SV), include both stochastic volatility and price jumps (model SVJ), and finally allow jumps in both price and stochastic volatility (model SJJ). All the illustrations except those for BS use the parameter values for the SJJ model from Eraker's (2004) Table III. These were estimated jointly from returns of the S&P 500 and options on the S&P during

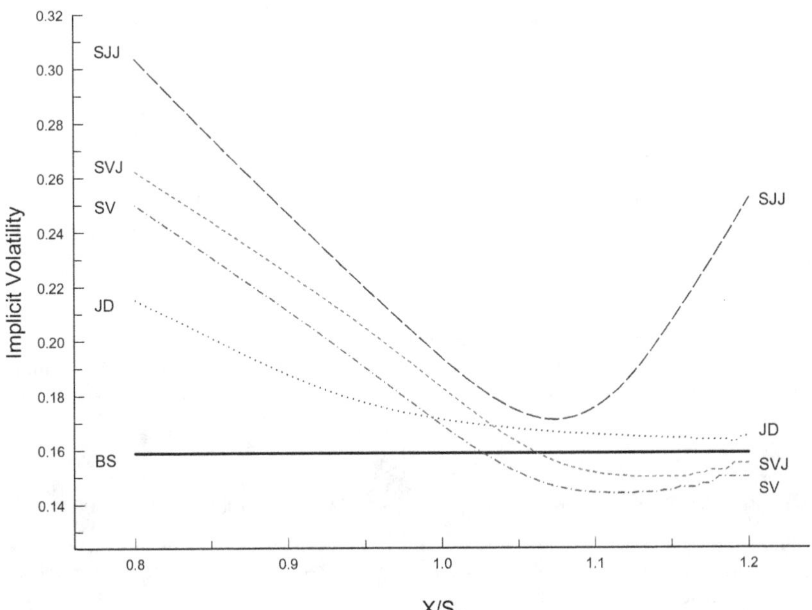

Fig. 9.5. Implicit volatility curves for jump-diffusion (JD), stochastic volatility (SV), s.v.-jump (SVJ), and s.v.-jump with jumps in volatility (SJJ).

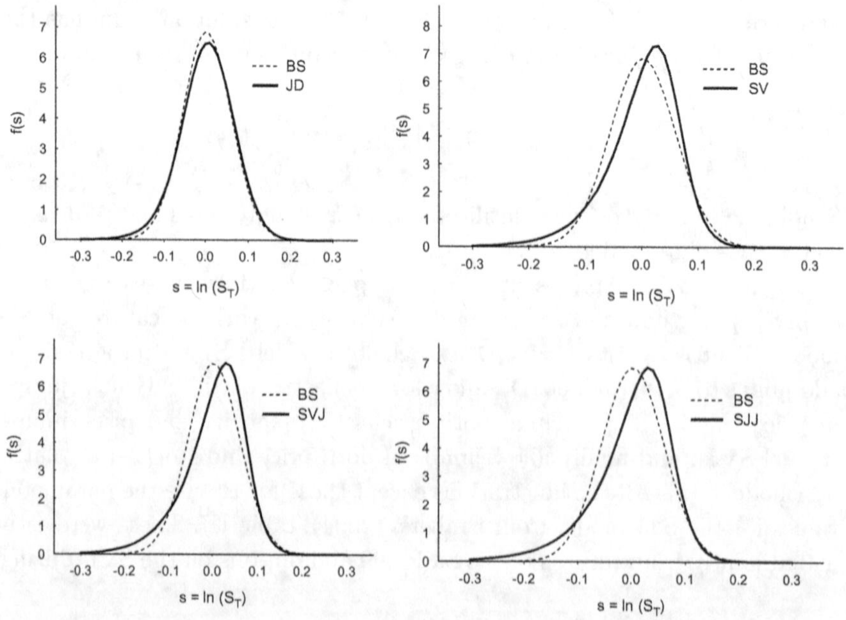

Fig. 9.6. P.d.f.s for jump-diffusion (JD), stochastic volatility (SV), s.v.-jump (SVJ), and s.v.-jump with jumps in volatility (SJJ).

1987–1990.[10] Note that since the same SJJ parameters were used in all the submodels, the volatility curves for SVJ, SV, and JD are not the best-fitting versions.[11] The plots simply show the incremental effects of enriching the model step by step. Generally, the stepwise changes progressively lengthen the left tail of the distribution of $\ln S_T$ and increase the implicit volatilities

---

[10]Eraker's estimates were rescaled to express returns in proportions rather than percentages and to record time in years rather than in days, assuming a 365-day market year. The resulting values are shown below in the text notation and in Eraker's.

Text	α	β	ρ	γ	ν	κ	ξ	φ	θ
Eraker	$\theta\kappa^Q$	κ^Q	ρ	σ_V	$e^{\sigma_y^2/2-\mu_y^Q}$-1	ρ_J	σ_y	μ_V	λ_0
Estimate	0.198	4.02	−0.582	0.595	−0.072	−0.190	0.0365	0.0598	0.730

The computations for the figures use $\sigma_0 = 0.1588$ for initial volatility, which is the average from returns data reported in Table II of Eraker *et al.* (2003). Other data for the figures: short rate $r = 0.03$, dividend rate $\delta = 0$, time to expiration $T = 0.135$ years (about the average in Eraker's sample of options).

[11]In any case Eraker's estimates from pooled time series of option and underlying prices do not minimize a loss function based on option prices alone. Thus, they are not directly comparable with the results in Bakshi *et al.* (1997) and other studies that minimize a squared-error loss.

for out-of-money puts. With these parameter values, the addition of jumps to volatility also thickens slightly the right tail of the distribution and thereby increases implicit volatility for in-the-money puts. While adding jumps to volatility clearly does add to the flexibility of the model, it is not clear from Eraker's (2004) findings that much is gained in reducing out-of-sample pricing errors.

Fang (2002) extended the s.v.-jump model in a different direction, keeping the volatility process as a diffusion but allowing stochastic variation in the intensity of the Poisson process that counts price shocks.[12] With intensity $\{\theta_t\}_{t\geq 0}$ as another mean-reverting square-root process, the full model becomes

$$\begin{aligned} ds_t &= (r_t - \theta_t \nu - v_t/2)\cdot dt + \sqrt{v_t}\cdot d\hat{W}_{1t} + u \cdot dN_t \\ dv_t &= (\alpha - \beta v_t)\cdot dt + \gamma\sqrt{v_t}\cdot d\hat{W}_{2t} \\ d\theta_t &= (\mu - \lambda\theta_t)\cdot dt + \phi\sqrt{\theta_t}\cdot d\hat{W}_{3t}. \end{aligned}$$

In this setup the random excursions in the intensity process can account for the clustering of jumps. By restricting $\{\hat{W}_{3t}\}$ to be independent of $\{\hat{W}_{1t}, \hat{W}_{2t}\}$, we get a conditional c.d.f. that is still exponentially affine in state variables s, v, θ:

$$\Psi(\zeta; s_{t-}, v_t, \tau) = \exp[i\zeta s_{t-} + g(\tau;\zeta) + h(\tau;\zeta)v_t + k(\tau;\zeta)\theta_t].$$

The standard methods yield as solutions for g, h, and k when $\gamma \neq 0$ and $\phi \neq 0$

$$\begin{aligned} g(\tau;\zeta) &= -i\zeta \ln B(t,T) + \frac{\alpha}{\gamma^2}\left[(-B+C)\tau - 2\ln\left(\frac{1-De^{C\tau}}{1-D}\right)\right] \\ &\quad + \frac{\mu}{\phi^2}\left[(\lambda + C')\tau - 2\ln\left(\frac{1-D'e^{C\tau}}{1-D'}\right)\right] \\ h(\tau;\zeta) &= \frac{B-C}{\gamma^2}\frac{e^{C\tau}-1}{1-De^{C\tau}} \\ k(\tau;\zeta) &= \frac{-\lambda - C'}{\phi^2}\frac{e^{C'\tau}-1}{1-D'e^{C'\tau}}. \end{aligned}$$

Here, A, B, C, D and $q(\zeta) = \hat{E}e^{i\zeta u} - 1$ are as in the s.v.-jump model and $A' \equiv -i\zeta\nu + q(\zeta)$, $D' \equiv (-\lambda - C')/(-\lambda + C')$, $C' \equiv \sqrt{\lambda^2 - 2A'\phi^2}$ when $\zeta \neq 0$ and $C' = -\lambda$ otherwise.

[12]Such doubly stochastic or "Cox" processes are commonly used in modeling survival times and defaults on financial obligations; c.f. Cox and Oakes (1984), Lando (1998). We describe these in more detail in section 10.5, where we look at applications to pricing defaultable bonds.

In practice, the usefulness of these models and of potential further extensions is limited by the need to estimate the large number of parameters they contain. Counting the unobservable initial volatility, the s.v.-jump model has eight parameters under the risk-neutral measure $(v_0, \alpha, \beta, \gamma, \nu, \xi, \theta, \rho)$, the volatility-jump setup adds two more (κ and φ), and the stochastic-intensity extension adds three (μ, λ, ϕ, with θ_0 replacing θ). However, Fang (2002) has found some success in pricing individual stock options using the seven-parameter submodel that restricts price volatility to be constant. This forces the excursions in jump intensity to account for all the fluctuations in price volatility that appear to be in the data, which it seems to do reasonably well.

9.4 Pure-Jump Models

This section treats four models in which the underlying price process has no continuous component at all. The first two are the well-known variance-gamma model of Madan and coauthors (1990, 1991, 1998, 1999) and the hyperbolic model of Eberlein and Keller (1995) and Eberlein *et al.* (1998). Both models are obtained by subordinating Brownian motion to a directing process with continuous sample paths. This construction yields Lévy processes with infinite Lévy measure, for which the expected number of jump discontinuities in any finite interval is infinite. To illustrate the versatility of the subordination scheme, we present a third model in which Brownian motion is directed by a pure-jump process. This again produces a Lévy process, but one having only finitely many jumps per unit time. In all three models prices move in discrete ticks, the sizes of which are randomly distributed according to the Lévy measure that governs the process. The last model of the section characterizes price as a continuous-time branching process. Here the number of jumps per unit time is again finite, but the sizes of the jumps are all integral multiples of the minimum price tick authorized for the market. Despite their differences, all four models account for skewness and thick tails in the marginal distributions of asset returns and are therefore capable of reducing the smile effect in option prices.

9.4.1 *The Variance-Gamma Model*

To motivate the continuous-time variance-gamma (v.g.) model, consider the following discrete-time model for an asset's average continuously compounded return over a unit interval of time—i.e., $\ln(S_{t+1}/S_t)$. Taking

$\{Z_t\}_{t=1}^{\infty}$ as i.i.d. standard normal and $\{U_t\}_{t=1}^{\infty}$ as i.i.d. gamma variates, independent of the $\{Z_t\}$, with shape parameter $\nu > 1$ and unit scale, let

$$R_{t,t+1} \equiv \ln S_{t+1} - \ln S_t = \mu + \sigma\sqrt{\nu - 1}Z_t/\sqrt{U_t}.$$

Algebra reduces the right side to $\mu + \sigma\sqrt{\frac{\nu-1}{\nu}}T_t$, where $T_t = Z_t\sqrt{\nu/U_t}$ has Student's distribution with 2ν degrees of freedom (not necessarily an integer). Since $ET_t = 0$, $VT_t = \nu/(\nu - 1)$, and (if $\nu > 2$) $ET_t^4/(VT_t)^2 = 3(1 + \frac{1}{\nu-2})$, this setup delivers for $R_{t,t+1}$ a symmetric distribution with mean μ and variance σ^2 but with excess kurtosis related inversely to ν. Since the change in log price is distributed as $N[\mu, \sigma^2(\nu-1)/U]$ conditional on U, the distribution can be viewed as a mixture of normals whose inverse variances are distributed as gamma. The idea of modelling centered returns of stocks as scaled Student's t was first proposed by Praetz (1972). Blattberg and Gonedes (1974) found that the model does successfully capture the thick tails observed in the empirical marginal distributions of daily returns, obtaining estimates of ν in the 3.0–6.0 range for most stocks in a sample of large US companies.

While the Student model fits the discrete-time empirical data reasonably well, it has significant shortcomings in the application to derivatives pricing. One is that the Student family is not closed under convolution, meaning that sums of independent Student variates are not in the family.[13] A more serious problem is that the model cannot account for asymmetries in the risk-neutral distribution of log price that are often implicit in option prices. Madan and Seneta (1990) proposed a related model that overcomes the first objection, and modifications by Madan and Milne (1991) and Madan, Carr , and Chang (1998) (henceforth, Madan *et al.*) subsequently overcame the second. The basic idea is to model the log price in continuous time as a time change of Brownian motion; that is, as a subordinated process.

Variance-Gamma as a Subordinated Process

Here is how the v.g. model works. Let the directing process or subordinator $\{\mathbb{T}_t\}_{t\geq 0}$ that represents operational time be a gamma process with

[13]On the other hand, the Student distribution is known to be infinitely divisible and, like the hyperbolic model described below, it therefore generates a Lévy process in continuous time. In this sense, lack of closure under convolution, while a nuisance, is not a crucial limitation.

parameters t/ν and ν and density

$$f_{\mathbb{T}_t}(\tau) = \frac{\tau^{t/\nu - 1} e^{-\tau/\nu}}{\Gamma(t/\nu)\nu^{t/\nu}}, \quad t > 0, \ \nu > 0.$$

Notice that the first two moments are $E\mathbb{T}_t = t$ and $V\mathbb{T}_t = \nu t$. Taking $\{W_t\}_{t \geq 0}$ to be an independent Brownian motion, the continuously compounded return over $[0, t]$ is modeled as

$$R_{0,t} \equiv \ln(S_t/S_0) = \mu t + \gamma \mathbb{T}_t + \sigma W_{\mathbb{T}_t}. \tag{9.20}$$

Noting that this is distributed conditionally as $N(\mu t + \gamma \mathbb{T}_t, \sigma^2 \mathbb{T}_t)$, we see that the model for return is again a mixture of normals but having stochastic mean as well as variance and with a gamma (instead of inverse gamma) variate driving the parametric change. The subordination scheme adapts the mixture model to continuous time in a mathematically elegant way. Moreover, associating operational time with information flow gives the model a natural economic interpretation.

Canonical Representation

We know from general principles that the return process $\{R_{0,t}\}$ is Lévy under model (9.20), since it comes from subordinating one Lévy process (Brownian motion) to another (the gamma process). Discovering other properties begins with a development of the c.f. Evaluating

$$\Psi_{R_{0,t}}(\zeta) = E \exp[i\zeta(\mu t + \gamma \mathbb{T}_t + \sigma W_{\mathbb{T}_t})]$$

by conditioning first on $\mathbb{T}_t$ gives

$$\Psi_{R_{0,t}}(\zeta) = e^{i\mu\zeta t}[1 - \nu(i\zeta\gamma - \zeta^2\sigma^2/2)]^{-t/\nu}. \tag{9.21}$$

Since $\Psi_{R_{0,t}}(\zeta)\Psi_{R_{t,u}}(\zeta) = \Psi_{R_{0,u}}(\zeta)$, the model is closed under convolution, and we confirm that $\{R_{0,t}\}_{t \geq 0}$ is indeed a Lévy process since (9.21) has the form $e^{t\Theta(\zeta)}$, with

$$\Theta(\zeta) = i\zeta\mu - \nu^{-1} \ln[1 - \nu(i\zeta\gamma - \zeta^2\sigma^2/2)].$$

Notice that this has a limiting value of $i(\mu + \gamma) - \sigma^2/2$ as $\nu \to 0$, indicating that the process approaches a Brownian motion as the volatility of operational time diminishes to zero. However, as we are about to see, the process is of finite variation and, indeed, is pure-jump for each $\nu > 0$.

Let us consider two independent gamma processes, $\{R^+_{0,t}\}_{t\geq 0}$ and $\{R^-_{0,t}\}_{t\geq 0}$, with c.f.s $\Psi_{R^{\pm}_{0,t}}(\zeta) = (1 - i\zeta\beta^{\pm})^{-t/\nu}$ and parameters

$$\beta^{\pm} = \left[\sqrt{\gamma^2\nu^2 + 2\sigma^2\nu} \pm \gamma\nu\right]/2.$$

Algebra shows that

$$e^{i\zeta\mu t} E \exp[i\zeta(R^+_{0,t} - R^-_{0,t})] = e^{i\zeta\mu t}\Psi_{R^+_{0,t}}(\zeta)\Psi_{R^-_{0,t}}(-\zeta) = \Psi_{R_{0,t}}(\zeta).$$

Therefore, apart from the deterministic drift, $R_{0,t}$ is the difference between these two nondecreasing processes. As such, the return process necessarily has finite variation on finite intervals. Moreover, from the canonical representations of the gamma processes as (see example 56 and (3.54))

$$\ln \Psi_{R^+_{0,t}}(\zeta) = t\nu^{-1}\int_{(0,\infty)} (e^{i\zeta x} - 1)x^{-1}e^{-x/\beta^+} \cdot dx \qquad (9.22)$$

$$\ln \Psi_{R^-_{0,t}}(-\zeta) = t\nu^{-1}\int_{(-\infty,0)} (e^{i\zeta x} - 1)|x|^{-1}e^{-|x|/\beta^-} \cdot dx \qquad (9.23)$$

it follows that $\Psi_{R_{0,t}}(\zeta)$ has canonical form[14]

$$\ln \Psi_{R_{0,t}}(\zeta) = t\left[i\zeta\mu + \int_{\Re\backslash\{0\}} (e^{i\zeta x} - 1)\Lambda(dx)\right], \qquad (9.24)$$

where

$$\Lambda(dx) = \frac{1}{\nu|x|}\exp\left\{\frac{1}{\sigma^2}\left[x\gamma - |x|\sqrt{\gamma^2 + 2\sigma^2/\nu}\right]\right\} \cdot dx. \qquad (9.25)$$

---

[14]To see this, write $\Lambda(dx)$ in (9.25) as $\lambda(x) \cdot dx$, where for $x \neq 0$

$$\begin{aligned}
\lambda(x) &\equiv \frac{1}{\nu|x|}\exp\left[-\frac{x}{2}\left(\frac{1}{\beta^+} - \frac{1}{\beta^-}\right) - \frac{|x|}{2}\left(\frac{1}{\beta^+} + \frac{1}{\beta^-}\right)\right] \\
&= \frac{1}{\nu|x|}\exp\left[-\frac{1}{2}\left(\frac{x + |x|}{\beta^+} + \frac{x - |x|}{\beta^-}\right)\right] \\
&= \begin{cases} \dfrac{1}{\nu|x|}\exp(-|x|/\beta^+), & x > 0 \\ \dfrac{1}{\nu|x|}\exp(-|x|/\beta^-), & x < 0. \end{cases}
\end{aligned}$$

Therefore, from (9.22) and (9.23) we have

$$\ln\left[\Psi_{R^+_{0,t}}(\zeta)\Psi_{R^-_{0,t}}(-\zeta)\right] = t\int_{\Re\backslash\{0\}} (e^{i\zeta x} - 1)\lambda(x) \cdot dx,$$

and hence (9.24).

The absence of a Brownian component in (9.24) indicates that the process is pure jump. Moreover, observing from (9.25) that $\Lambda(\Re\backslash\{0\}) = +\infty$, one sees that the expected number of jumps per unit time is infinite. Since the Lévy density declines monotonically from the origin in both directions, jumps of larger magnitude tend to occur less frequently than those of smaller size. Notice that positive jumps are more or less likely than negative jumps of the same size according as $\gamma > 0$ or $\gamma < 0$.

Density Function and Moments

A development of the p.d.f. of $R_{0,t}$ starts from the mixture interpretation, as

$$f_{R_{0,t}}(R) = \int_0^\infty \frac{1}{\sqrt{2\pi\sigma^2\tau}} \exp\left[-\frac{(R-\mu t-\gamma\tau)^2}{2\sigma^2\tau}\right] \frac{\tau^{t/\nu-1}e^{-\tau/\nu}}{\Gamma(t/\nu)\nu^{\tau/\nu}} \cdot d\tau.$$

Madan *et al.* (1998) develop from this the following explicit formula:

$$f_{R_{0,t}}(R) = \frac{2e^{\gamma(R-\mu t)/\sigma^2}}{\nu^{t/\nu}\sqrt{2\pi\sigma^2}\Gamma(t/\nu)} \left(\frac{|R-\mu t|}{\rho}\right)^{\frac{t}{\nu}-\frac{1}{2}} K_{\frac{t}{\nu}-\frac{1}{2}}\left(\rho\frac{|R-\mu t|}{\sigma^2}\right),$$

where

$$\rho \equiv +\sqrt{2\sigma^2/\nu+\gamma^2}$$

and $K_{t/\nu-1/2}(\cdot)$ is a modified Bessel function of the second kind of order $t/\nu - 1/2$.[15]

The m.g.f.,

$$\mathfrak{M}_{R_{0,t}}(\zeta) = e^{\zeta\mu t}[1-\nu(\zeta\gamma+\zeta^2\sigma^2/2)]^{-t/\nu},$$

exists for $\zeta\sigma^2 \in [-\gamma-\sqrt{\gamma^2+2\sigma^2/\nu}, -\gamma+\sqrt{\gamma^2+2\sigma^2/\nu}]$, indicating that all moments of $R_{0,t}$ are finite. We find the mean and variance to be $ER_{0,t} = (\mu+\gamma)t$ and $VR_{0,t} = (\gamma^2\nu+\sigma^2)t$, respectively, while the coefficients of skewness and excess kurtosis are

$$\alpha_3 = \frac{\gamma\nu(2\gamma^2\nu+3\sigma^2)}{(\gamma^2\nu+\sigma^2)^{3/2}\sqrt{t}} \propto t^{-1/2}$$

$$\alpha_4 - 3 = \frac{3\nu}{t}\left[2-\frac{\sigma^4}{(\gamma^2\nu+\sigma^2)^2}\right] \propto t^{-1}.$$

Both α_3 and $\alpha_4 - 3$ increase with ν and decrease with t. The distribution is symmetric when $\gamma = 0$. Setting $S_t = S_0e^{R_{0,t}}$, we can evaluate ES_t as

[15] Modified Bessel functions are described in section 2.1.8.

$S_0\Psi_{R_{0,t}}(-i)$. Under risk neutrality and a constant interest rate r this would equal S_0e^{rt}, requiring the total mean drift under the martingale measure to be

$$\mu + \gamma = r + \gamma + \nu^{-1}\ln[1 - \nu(\gamma + \sigma^2/2)], \tag{9.26}$$

approaching $r - \sigma^2/2$ as $\nu \to 0$.

Option Pricing under Variance-Gamma Dynamics

Regarding ν, γ, and σ as parameters of the risk-neutral distribution, and with μ satisfying (9.26), the value at $t = 0$ of a T-expiring European call struck at X would be

$$C^E(S_0, T) = e^{-rT}\hat{E}(S_T - X)^+.$$

Exploiting the lognormality of S_T conditional on $\mathbb{T}_T$, Madan and Milne (1991) express the value of the call as

$$C^E(S_0, T) = \int_0^\infty c(\tau)\frac{\tau^{t/\nu - 1}e^{-\tau/\nu}}{\Gamma(t/\nu)\nu^{\tau/\nu}} \cdot d\tau,$$

where

$$\begin{aligned} c(\tau) \equiv\, & S_0(1 - \nu\beta^2/2)^{T/\nu}e^{\beta^2\tau/2}\Phi\left(d/\sqrt{\tau} + \beta\sqrt{t}\right) \\ & - Xe^{-rT}(1 - \nu\alpha^2/2)^{T/\nu}e^{\alpha^2\tau/2}\Phi\left(d/\sqrt{\tau} + \alpha\sqrt{t}\right) \end{aligned}$$

with

$$\begin{aligned} d &\equiv (\beta - \alpha)^{-1}\left[\ln(S_0/X) + rT + \frac{T}{\nu}\ln\left(\frac{1 - \nu\beta^2/2}{1 - \nu\alpha^2/2}\right)\right] \\ \alpha &\equiv -\frac{\gamma}{\sigma\sqrt{1 + \frac{\gamma^2\nu}{2\sigma^2}}} \\ \beta &\equiv \frac{\sigma^2 - \gamma}{\sigma\sqrt{1 + \frac{\gamma^2\nu}{2\sigma^2}}}. \end{aligned}$$

Madan *et al.* (1998) develop an explicit expression in terms of modified Bessel functions. However, Carr and Madan (1998) find that applying the fast Fourier transform to a modification of the call function is faster computationally than the explicit formula. Alternatively, one can use the integrated c.d.f. approach by evaluating formula (8.37) on page 399.

Madan *et al.* (1998) apply the v.g. model to (American) futures options on the S&P 500 index, estimating the implicit risk-neutral parameters (γ, ν, and σ) weekly from the option prices by nonlinear least squares. Finding that in-sample pricing errors reveal no statistically significant relation to moneyness, they conclude that the model accounts well for smile effects in the option prices. However, the substantial variation in their parameter estimates over the 143 weeks of the sample suggests that out-of-sample predictions would be much less satisfactory.

9.4.2 *The Hyperbolic Model*

The idea of modeling the log-price process as a subordinated Brownian motion opens up many possibilities for capturing the thick tails and skewness in the risk-neutral distribution of returns. Eberlein and Keller (1995) and Eberlein *et al.* (1998) have studied another such model that uses as subordinator a process generated by the generalized inverse Gaussian distribution. As does the gamma process that clocks operational time in the v.g. model, the inverse Gaussian subordinator yields a pure-jump Lévy process having no Gaussian component and infinite Lévy measure. The implied distribution of the continuously compounded return over a unit interval is a mixture of normals known as the "hyperbolic law", which has been studied extensively by O. E. Barndorff-Nielsen and coauthors and applied in modeling the distribution of sizes of sand particles.

The Hyperbolic Distribution

To develop the model, begin with a random variable Y whose conditional distribution is normal with mean and variance proportional to a generalized inverse Gaussian variate V:

$$f_{Y|V}(y|v) = \frac{1}{\sqrt{2\pi v}} e^{-(y-\beta v)^2/(2v)}, \quad y \in \Re \tag{9.27}$$

$$f_V(v) = \frac{\sqrt{\alpha^2-\beta^2}}{2K_1\left(\sqrt{\alpha^2-\beta^2}\right)} e^{-[1/v+(\alpha^2-\beta^2)v]/2}, \quad v > 0, \tag{9.28}$$

where $\alpha > 0, 0 \leq |\beta| < \alpha$, and K_1 is a modified Bessel function of the second kind of order one.[16] Working out the integral $\int_0^\infty f_{Y|V}(y|v) f_V(v) \cdot dv$ gives

[16]To see that (9.28) is a density, set $z = \sqrt{\alpha^2-\beta^2}$ in the following integral representation for the Bessel function, which is formula 8.432.6 in Gradshteyn and Ryzhik (1980), then change variables as $t = (2v)^{-1}$:

$$K_1(z) = \int_0^\infty \frac{z \exp[-t - z^2/(4t)]}{4t^2} \cdot dt.$$

as the marginal p.d.f. of the resulting mean-variance mixture of normals the expression

$$f_Y(y;\alpha,\beta) = \frac{\sqrt{\alpha^2-\beta^2}}{2\alpha K_1\left(\sqrt{\alpha^2-\beta^2}\right)} e^{-\alpha\sqrt{1+y^2}+\beta y}, \quad y \in \Re. \tag{9.29}$$

Parameter α governs the thickness of the tails, and from the symmetry of $f_Y(y;\alpha,0)$ we see that β controls skewness, the right tail being elongated when $\beta > 0$. To see why the model is called "hyperbolic", set z equal to the exponent in (9.29), square $z - \beta y$, and collect terms to produce

$$Az^2 + Bzy + Cy^2 - \alpha^2 = 0,$$

where $A = 1$, $B = -2\beta$, and $C = \beta^2 - \alpha^2$. Since $B^2 - 4AC = 4\alpha^2 > 0$, the locus (y, z) is in fact an hyperbola.

The form of (9.29) makes it easy to develop the m.g.f. as

$$\begin{aligned}\mathfrak{M}_Y(\zeta) &= \int_\Re e^{\zeta y} f_Y(y;\alpha,\beta)\cdot dy \\ &= \frac{\sqrt{\alpha^2-\beta^2}}{\sqrt{\alpha^2-(\beta+\zeta)^2}} \frac{K_1\left(\sqrt{\alpha^2-(\beta+\zeta)^2}\right)}{K_1\left(\sqrt{\alpha^2-\beta^2}\right)} \int_\Re f_Y(y;\alpha,\beta+\zeta)\cdot dy \\ &= \frac{\sqrt{\alpha^2-\beta^2}}{\sqrt{\alpha^2-(\beta+\zeta)^2}} \frac{K_1\left(\sqrt{\alpha^2-(\beta+\zeta)^2}\right)}{K_1\left(\sqrt{\alpha^2-\beta^2}\right)}, \end{aligned} \tag{9.30}$$

for $|\beta+\zeta| < \alpha$. While the existence of the m.g.f. implies the finiteness of the moments, these unfortunately have to be expressed in terms of Bessel functions. For example, evaluating $\mathfrak{M}'_Y(\cdot)$ at $\zeta = 0$ gives[17]

$$EY = \frac{\beta}{\sqrt{\alpha^2-\beta^2}} \frac{K_2\left(\sqrt{\alpha^2-\beta^2}\right)}{K_1\left(\sqrt{\alpha^2-\beta^2}\right)}.$$

Modeling Prices and Returns

Taking Y_1 and Y_2 to be i.i.d. as Y with m.g.f. (9.30), one sees that $\mathfrak{M}_{Y_1+Y_2}(\zeta) = \mathfrak{M}_Y(\zeta)^2$ is not of the same functional form as $\mathfrak{M}_Y(\zeta)$. This

---

[17]To differentiate $\mathfrak{M}_Y(\zeta)$ use the formula (2.27), which for functions of order one gives $K_1'(z) = z^{-1}K_1(z) - K_2(z)$.

means that the hyperbolic family, like the Student family, is not closed under convolution. However, since the distribution is infinitely divisible,[18] the family does generate a class of Lévy processes $\{Y_t\}_{t\geq 0}$ from the relation

$$\Psi_{Y_t}(\zeta) = \Psi_Y(\zeta)^t, \tag{9.31}$$

where $\Psi_Y(\zeta) = \mathfrak{M}_Y(i\zeta)$ is the c.f. of Y. Eberlein and Keller (1995) exploit this fact to produce a plausible model of security prices in continuous time. Extending to allow arbitrary location and scale, the continuously compounded return over a unit time interval is defined as $R_{0,1} = \mu + \sigma Y$, so that[19]

$$\begin{aligned}\Psi_{R_{0,1}}(\zeta) &= e^{i\zeta\mu}\Psi_Y(\zeta\sigma)\\ &= e^{i\zeta\mu}\frac{\sqrt{\alpha^2-\beta^2}}{\sqrt{\alpha^2-(\beta+i\zeta\sigma)^2}}\,\frac{K_1\left(\sqrt{\alpha^2-(\beta+i\zeta\sigma)^2}\right)}{K_1\left(\sqrt{\alpha^2-\beta^2}\right)}.\end{aligned}$$

Applying (9.31), the continuously compounded return over $[0,t]$ is then a random variable having c.f.

$$\Psi_{R_{0,t}}(\zeta) = e^{i\zeta\mu t}\Psi_{Y_t}(\zeta\sigma) = e^{i\zeta\mu t}\Psi_Y(\zeta\sigma)^t. \tag{9.32}$$

With this model for the return the price at t is given by

$$S_t = S_0 e^{R_{0,t}}. \tag{9.33}$$

Under this setup the log of the price relative is, apart from added mean trend at rate μ, a time change of Brownian motion with subordinator $\mathbb{T}_t = \sigma Y_t$. Developing the c.f. in canonical form, Eberlein and Keller (1995) show that

$$t^{-1}\ln\Psi_{R_{0,t}}(\zeta) = i\zeta\mu + \int_{\Re\backslash\{0\}}(e^{i\zeta x} - 1 - i\zeta x)\Lambda(dx),$$

where the Lévy measure is such that $\Lambda(\Re\backslash\{0\}) = +\infty$. Accordingly, apart from the deterministic trend the return process is pure-jump and, as in the v.g. model, there are infinitely many jumps in each open interval of time. With S_t as in (9.33) an application of (3.46) gives

$$dS_t = S_{t-}\mu\cdot dt + S_{t-}\sigma\cdot dY_t + (S_t - S_{t-} - S_{t-}\sigma\Delta Y_t)$$

---

[18] As proved by Barndorff-Nielsen and Halgreen (1977).

[19] Here the shape parameters α and β correspond to those of Eberlein and Keller (1995) and Eberlein *et al.* (1998) multiplied by the scale (our "σ" and their "δ").

or

$$dS_t/S_{t-} = \mu \cdot dt + \sigma \cdot dY_t + e^{\sigma \Delta Y_t} - 1 - \sigma \Delta Y_t,$$

where $\Delta Y_t = Y_t - Y_{t-}$ and Y_t is the random variable defined implicitly by (9.31). One determines the distribution of S_t given S_0 from the p.d.f. of $R_{0,t}$. For $t \neq 1$ this has to be found by inversion of (9.32).

Pricing Options

As was the case for the v.g. model, payoffs of options—and nonlinear derivatives generally—cannot be replicated with portfolios of riskless bonds and the underlying alone when process $\{S_t\}_{t \geq 0}$ is hyperbolic. Accordingly, the risk-neutral measure $\hat{\mathbb{P}}$ under which assets' prices (normalized by $\{M_t\}$) are martingales in an arbitrage-free market is not unique. Taking r to be average short rate during $[0, t]$, the fair-game property under $\hat{\mathbb{P}}$ implies that $S_0/M_0 = \hat{E}(S_t/M_t)$ or $S_0 = e^{-rt} S_0 \hat{E} e^{R_{0,t}}$. This in turn implies that $r = \ln \Psi_{R_{0,1}}(-i)$ and, therefore, that the parameters satisfy

$$\mu = r - \ln \left[\frac{\sqrt{\alpha^2 - \beta^2}}{\sqrt{\alpha^2 - (\beta + \sigma)^2}} \frac{K_1\left(\sqrt{\alpha^2 - (\beta + \sigma)^2}\right)}{K_1\left(\sqrt{\alpha^2 - \beta^2}\right)} \right]. \tag{9.34}$$

If the underlying pays continuous dividends at rate δ, r is simply replaced by $r - \delta$ in the above. For simplicity, we ignore dividends from this point.

Given values of the parameters Eberlein *et al.* (1998) suggest the following direct way to price a T-expiring European-style derivative: (i) imposing (9.34), find $\hat{f}_{R_{0,T}}(\cdot)$, the risk-neutral p.d.f. of $R_{0,T}$, by Fourier inversion of the c.f. in (9.32); (ii) find $\hat{f}_{S_T}(\cdot)$, the implied density of $S_T = S_0 e^{R_{0,T}}$ from $\hat{f}_{R_{0,T}}(\cdot)$ via a change of variables; and (iii) price the derivative by numerical integration. For example, the price of a European put struck at X would be

$$P^E(S_0, T) = e^{-rT} \left[X \int_0^X \hat{f}_{S_T}(s) \cdot ds - \int_0^X s \hat{f}_{S_T}(s) \cdot ds \right].$$

While this is certainly feasible, we now know from section 8.4 that there are much faster ways to price European options from c.f.s; namely, Carr-Madan's (1998) Fourier inversion of the damped call or the integrated c.d.f. method using formula (8.37).

Restricting μ as in (9.34) and using nonlinear least squares to infer α, β, and σ from prices of options on several German equities, Eberlein *et al.* (1998) find that the model produces much smaller mean absolute error than

Black-Scholes. However, neither an in-sample comparison with competing pure-jump, mixed-jump, and s.v. models nor an assessment of predictions out of sample is yet available.

9.4.3 *A Lévy Process with Finite Lévy Measure*

To show the versatility of the subordination scheme, we outline here another pure-jump model that differs from the v.g. and hyperbolic in having only finitely many jumps per unit time. The change is accomplished by modeling the directing process as compound-Poisson rather than as a process with continuous sample paths.

Specifically, let us model operational time as $\mathbb{T}_t = \sum_{j=0}^{N_t} U_j$, where (i) $\{N_t\}_{t\geq 0}$ is a Poisson process with intensity θ, (ii) $U_0 \equiv 0$, and (iii) $\{U_j\}_{j=1}^{\infty}$ are nonnegative, i.i.d. random variables, independent of $\{N_t\}$ and with m.g.f. $\mathfrak{M}_U(\zeta)$. $\{\mathbb{T}_t\}_{t\geq 0}$ is thus a nondecreasing Lévy process that is itself pure-jump, and $\mathbb{T}_t$ has m.g.f. $\mathfrak{M}_{\mathbb{T}_t}(\zeta) = \exp\{t\theta[\mathfrak{M}_U(\zeta) - 1]\}$. Now introduce a trended Brownian motion $\{\mu t + \sigma W_t\}_{t\geq 0}$ that is independent of everything else and create the subordinated process

$$\{X_t = \mu \mathbb{T}_t + \sigma W_{\mathbb{T}_t}\}_{t\geq 0}. \tag{9.35}$$

By first conditioning on $\mathbb{T}_t$ it is easy to see that X_t has m.g.f.

$$\mathfrak{M}_{X_t}(\zeta) = \exp\{t\theta[\mathfrak{M}_U(\zeta\mu + \zeta^2\sigma^2/2) - 1]\}$$

and c.f. $\Psi_{X_t}(\zeta) = \mathfrak{M}_{X_t}(i\zeta)$ respectively. The expected value is $EX_{\mathbb{T}_t} = t\theta\mu EU_1$. Of course, $\{X_t\}$ is again a Lévy process. In the canonical form $\Psi_{X_t}(\zeta) = e^{t\Theta(\zeta)}$ we have

$$\begin{aligned}\Theta(\zeta) &= \theta[\mathfrak{M}_U(i\zeta\mu - \zeta^2\sigma^2/2) - 1] \\ &= i\zeta\theta\mu EU_1 + \int_{\Re\backslash\{0\}} (e^{i\zeta x} - 1 - i\zeta x)\Lambda(dx),\end{aligned}$$

with Lévy measure

$$\Lambda(dx) = \theta \cdot dx \int_0^{\infty} \frac{1}{\sqrt{2\pi\sigma^2 u}} \exp\left[-\frac{(x-\mu u)^2}{2\sigma^2 u}\right] \cdot dF_U(u)$$

for $x \in \Re$. Thus, $\theta^{-1}\Lambda(dx)$ is the density of a random variable distributed as $N(U\mu, U\sigma^2)$ conditional on U, where U has c.d.f. F_U. Since $\int_{\Re\backslash\{0\}} \Lambda(dx) = \theta < \infty$, the process $\{X_t\}$ has only finitely many jumps during any finite span of time; and since $\Theta(\zeta)$ contains no term in ζ^2 the process has no Brownian component.

With $\{X_t\}_{t\geq 0}$ as the corner stone, we can build a model for a price process $\{S_t\}_{t\geq 0}$ as

$$S_t = S_0 \exp\left(\int_0^t \nu_s \cdot ds + X_t\right),$$

where $\{\nu_t\}$ is a specified drift process. To model $\{S_t\}_{t\geq 0}$ under the risk-neutral measure one would take

$$\nu_t = r_t - \delta_t - \theta[\mathfrak{M}_U(\mu + \sigma^2/2) - 1],$$

where r_t, δ_t are the short rate and dividend rate, and regard $\{W_t\}$ and $\{N_t\}$ as Wiener and Poisson processes under $\hat{\mathbb{P}}$. With $M_t = M_0 e^{\int_0^t r_s \cdot ds}$ as numeraire, the normalized *cum*-dividend price $S_t^{c*} = M_t^{-1} S_t \exp(\int_0^t \delta_s \cdot ds)$ then has expected value under $\hat{\mathbb{P}}$

$$\begin{aligned}\hat{E}S_t^{c*} &= S_0^{c*} \exp\{-t\theta[\mathfrak{M}_U(\mu + \sigma^2/2) - 1]\}\hat{E}e^{X_t} \\ &= S_0^{c*} \mathfrak{M}_{X_t}(\zeta)^{-1} \mathfrak{M}_{X_t}(\zeta) \\ &= S_0^{c*},\end{aligned}$$

as required for the martingale property. $\ln S_t^{c*}$ thus follows a steady trend during any interval $(t, t+\Delta t]$ in which $N_{t+\Delta t} - N_t = 0$, while at t such that $N_t - N_{t-} = 1$ we have $\ln S_t^{c*} - \ln S_{t-}^{c*} = \mu U + \sigma\sqrt{U}Z$, where $Z \sim N(0,1)$. In words, when a Poisson event occurs $\ln S_t^{c*}$ jumps by the net change in the trended Brownian motion during the random time U.

As a specific example of such a process, take $\{U_j\}_{j=1}^{\infty}$ as i.i.d. exponential, with c.d.f. $F_U(u) = [1 - \exp(-u/\beta)]\mathbf{1}_{(0,\infty)}(u)$ and m.g.f. $M_U(\zeta) = (1 - \beta\zeta)^{-1}$ for $\zeta < \beta^{-1}$. Forcing $\theta = \beta^{-1}$ gives $E\mathbb{T}_t = t$ and

$$\mathfrak{M}_{X_t}(\zeta) = \exp\{t\beta^{-1}[1 - \beta(\zeta\mu + \zeta^2\sigma^2/2)]^{-1} - t\beta^{-1}\}.$$

By differentiating cumulant generating function $\mathcal{L}_{X_t}(\zeta) = \ln \mathfrak{M}_{X_t}(\zeta)$ one can see that $\ln(S_t/S_0)$ has coefficients of skewness and excess kurtosis given by

$$\alpha_3 = \frac{6\mu\beta(\mu^2\beta + \sigma^2)}{\sqrt{t}(2\mu^2\beta + \sigma^2)^{3/2}}$$

$$\alpha_4 - 3 = \frac{6\beta(4\mu^4\beta^2 + 6\mu^2\beta\sigma^2 + \sigma^4)}{t(2\mu^2\beta + \sigma^2)^2}.$$

The sign of α_3 is the same as that of μ, while excess kurtosis is always positive. As for any Lévy process with finite moments, both α_3 and $\alpha_4 - 3$ approach zero as $t \to \infty$.

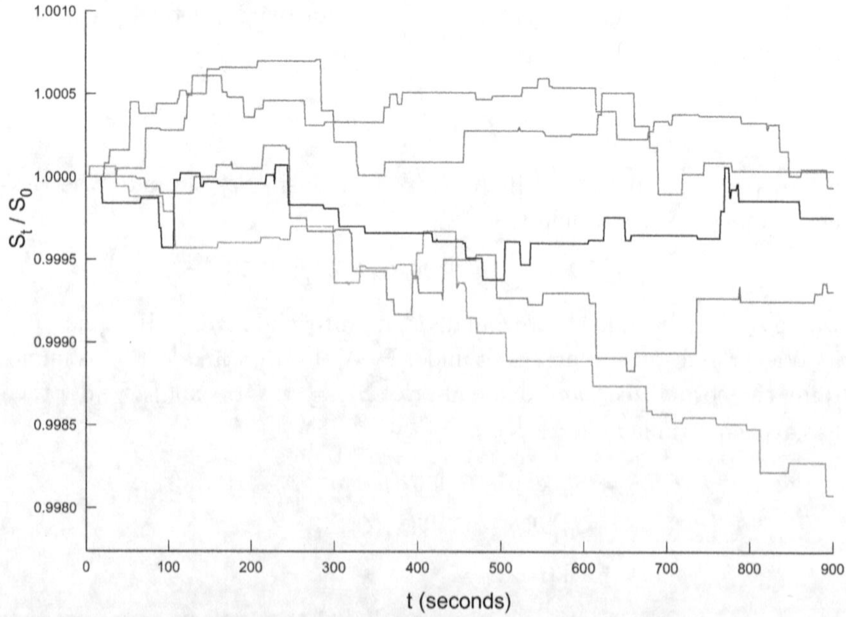

Fig. 9.7. Detrended sample paths of S&P 500 returns modeled from Brownian motion directed by a compound-Poisson process.

Figure 9.7 depicts five sample paths of $\{S_t^c/S_0 = S_t \exp(\int_0^t \delta_s \cdot ds)/S_0\}$ with parameters $\beta = 0.001$, $\mu = -0.109$, $\sigma = 0.0046$. These values approximately match the cumulants with empirical estimates from daily, dividend-adjusted closing values of the S&P 500 index during 2004–2005. The paths, which correspond to just 900 seconds (15 minutes) of market time, thus give a close-up view of the behavior of the index in the natural measure. (The data also imply an average drift rate $\bar{\nu} \doteq 0.109$, but the figures are plotted with $\bar{\nu} = 0$ to emphasize the stochastic component.)

9.4.4 *Modeling Prices as Branching Processes*

The three pure-jump models we have seen thus far have much in common. Each represents $\{\ln S_t\}_{t\geq 0}$ as a Lévy process, and—while the stochastic component of price is pure-jump—the state space for S_t is a continuum, $\Re^+$. We now describe a pure-jump model with very different properties. Returns over unit time intervals are no longer i.i.d., and price—although it evolves in continuous time—is an *integer-valued* process when denominated in units of the minimum tick size.

Taking S_0 to be a given positive integer, we begin with the construction of an integer-valued process in discrete time, $\{S_n\}_{n=0}^{\infty}$, as follows. For each n let $K_{0n} = 0$; take $\{K_{jn}\}_{j\in\aleph}$ to be i.i.d., nonnegative, integer-valued random variables; and define $S_n = \sum_{j=0}^{S_{n-1}} K_{jn}$. Thinking of S_{n-1} as the size of a population at generation $n-1$ and supposing that each member $j \in \{1, 2, \ldots, S_{n-1}\}$ of the generation dies off and leaves behind K_{jn} progeny, S_n is then the size of the population at generation n. Known as the Bienaymé-Galton-Watson (BGW) process, this discrete-time branching process has a long history in applied probability theory, with applications in demography, biology, and particle physics.[20] We extend the BGW process to continuous time as follows. Set $N_0 = 0$ and introduce a nondecreasing, integer-valued process $\{N_t\}_{t\geq 0}$, independent of the $\{K_{jn}\}$ and having stationary, independent increments. Taking $S(0) \equiv S_0$ and $\{S(t)\}_{t\geq 0} \equiv \{S_{N_t}\}_{t\geq 0}$ delivers an integer-valued process that evolves in continuous time. Dion and Epps (1999) have observed that the process, if observed at discrete intervals of time, belongs to the class of branching processes in random environments (b.p.r.e.s), introduced by Smith and Wilkinson (1969). We shall refer to the continuous-time process also as a "b.p.r.e." Epps (1996) has applied such a model to prices of common stocks measured in units of the minimum price tick. In this application $\{N_t\}_{t\geq 0}$ can be thought of as counting the number of information events up to t. The price process is clearly pure jump, since $S(t)$ does not change between information shocks; moreover, $S(t) - S(t-)$ is always an integral multiple of the minimum tick size whenever branching does occur. The model also implies a positive (and easily quantified) probability of "extinction"—the event $S(t) = 0$—which occurs if $K_{jn} = 0$ for $n = N_{t-} + 1$ and each $j \in \{1, 2, \ldots, N_{t-}\}$. Clearly, $S(t) = 0$ implies that $S(u) = 0$ for all $u > t$, so that the origin is an absorbing barrier for the process. Using this process to model the price of an underlying asset, we now look at the features that are most relevant for pricing derivatives.

Generating Function and Moments

The model's main properties are most easily inferred from the probability generating function (p.g.f.). Letting $\{p_k\}_{k=0}^{\infty} = \{\mathbb{P}(K = k)\}_{k=0}^{\infty}$ represent the probability mass function of K and $\Pi_K(\zeta) = \sum_{k=0}^{\infty} p_k \zeta^k$ the p.g.f., we have for the generating function of S_n (the price after n shocks) conditional

[20]The standard BGW process is described at length in the books by Harris (1963) and Athreya and Ney (1972).

on S_{n-1}

$$\Pi_{S_n}(\zeta \mid S_{n-1}) = E(\zeta^{S_n} \mid S_{n-1}) = E(\zeta^{\sum_{j=0}^{S_{n-1}} K_{jn}} \mid S_{n-1}) = \Pi_K(\zeta)^{S_{n-1}}.$$

Accordingly, the generating function of S_n conditional on S_{n-2} is

$$\Pi_{S_n}(\zeta \mid S_{n-2}) = \Pi_K[\Pi_K(\zeta)]^{S_{n-2}},$$

and so on, with

$$\Pi_{S_n}(\zeta \mid S_0) = \Pi_K^{[n]}(\zeta)^{S_0},$$

where "$[n]$" indicates that $\Pi_K^{[n]}(\cdot)$ is the nth "iterate" of the p.g.f. Taking $S(0) = S_0$ as given, the generating function of $S(t) \equiv S_{N_t}$ is then

$$\Pi_{S(t)}(\zeta) = E\Pi_K^{[N_t]}(\zeta)^{S_0} = \sum_{n=0}^{\infty} \Pi_K^{[n]}(\zeta)^{S_0}\mathbb{P}(N_t = n). \tag{9.36}$$

A model for the distribution of K that is both convenient and satisfactory in the application to equity prices is the two-parameter geometric,

$$p_k(\alpha, \beta) = \begin{cases} 1 - \dfrac{\alpha}{1-\beta}, & k = 0 \\ \alpha\beta^{k-1}, & k = 1, 2, \ldots, \end{cases} \tag{9.37}$$

where $0 < \beta < 1$ and $0 < \alpha \leq 1 - \beta$. This has the nice property that the iterates of the generating function can be expressed explicitly, as[21]

$$\Pi_K^{[n]}(\zeta) = \begin{cases} \dfrac{\gamma(\mu^n - 1)}{\mu^n - \gamma} + \dfrac{\mu^n \left(\frac{1-\gamma}{\mu^n-\gamma}\right)^2 \zeta}{1 - \left(\frac{\mu^n-1}{\mu^n-\gamma}\right)\zeta}, & \mu \neq 1 \\ \dfrac{n\beta - (n\beta + \beta - 1)\zeta}{1 - \beta(1 + n - n\zeta)}, & \mu = 1, 0 < \beta < 1, \end{cases}$$

where $\mu \equiv EK = \sum_{k=1}^{\infty} kp_k = \alpha/(1-\beta)^2$ and $\gamma = \beta^{-1}[1-(1-\beta)\mu]$. Besides being convenient, this gives considerable flexibility in fitting the data, since it allows μ to be very close to unity even though $\sigma^2 \equiv VK = \mu(\frac{1+\beta}{1-\beta} - \mu)$ is small. Both features are required to obtain good fits to equity prices. Epps (1996) considers four other such flexible two-parameter models, all being mixtures of a degenerate variate equal to unity for sure with a one-parameter lattice distribution—Bernoulli, Poisson, logseries, and geometric. For these the iterates must be calculated recursively as the p.g.f. is computed.

[21]See Harris (1963, p. 9).

Moments of $S(t)$ can be found from the descending factorial moments, $\mu_{(k)} = E\{S(t)[S(t)-1]\cdot\cdots\cdot[S(t)-k+1]\}$. These are obtained by evaluating the derivatives of $\Pi_{S(t)}(\cdot \mid S_0)$ at $\zeta = 1$. Unless the iterates can be expressed explicitly, differentiating $\Pi_K^{[n]}(\zeta)$ requires the recursive formula $\Pi_K^{[n]\prime}(\zeta) = \Pi_K^{[n-1]\prime}[\Pi_K(\zeta)]\Pi_K'(\zeta)$, where primes denote differentiation. Since $\Pi_K^{[n]\prime}(1) = \Pi_K'(1)^n = \mu^n$, the mean is

$$ES(t) = S_0 E(\mu^{N_t}) = S_0 \Pi_{N_t}(\mu),$$

where Π_{N_t} is the generating function of N_t. The variance is

$$VS(t) = S_0 \sigma^2 EN_t$$

when $\mu = 1$ and

$$VS(t) = [\Pi_{N_t}(\mu^2) - \Pi_{N_t}(\mu)^2]S_0^2 + \frac{\sigma^2}{\mu(\mu-1)}[\Pi_{N_t}(\mu^2) - \Pi_{N_t}(\mu)]S_0$$

when $\mu \neq 1$, where $\sigma^2 \equiv VK$. The conditional mean of the total return over $[0, t]$ is then

$$E[S(t)/S_0] = \Pi_{N_t}(\mu),$$

while the conditional variance has the form

$$V[S(t)/S_0] = \kappa_t + \eta_t S_0^{-1},$$

where $\kappa_t \geq 0$ and $\eta_t > 0$ are functions of μ, σ, and the parameters governing $\{N_t\}_{t\geq 0}$. As does the constant-elasticity-of-variance model, the b.p.r.e. thus implies that the conditional variance of total return decreases as the initial price level increases. Unlike what we see in models based on Lévy processes, returns over nonoverlapping periods are therefore not i.i.d. The skewness of the total return can be positive or negative, increasing in absolute value as holding period t increases. Excess kurtosis is always positive but declines as the holding period lengthens.

Among candidate models for $\{N_t\}_{t\geq 0}$ are the Poisson process, with

$$\Pi_{N_t}(\zeta) = e^{\theta t(\zeta - 1)}, \quad \theta > 0,$$

and the negative-binomial process, with

$$\Pi_{N_t}(\zeta) = \left[\frac{\theta}{1 - \zeta(1-\theta)}\right]^{\varphi t}, \quad 0 < \theta < 1, \ \varphi > 0.$$

In either case the number of price moves in any finite time interval is a.s. finite.

An Explicit Expression for Probabilities

While the extinction probability, $\mathbb{P}\{S(t)=0\}$, is readily expressed as

$$\Pi_{S(t)}(0)=\sum_{n=0}^{\infty}\Pi_K^{[n]}(0)^{S_0}$$

for any given distribution of the number of progeny, K, it is not generally easy to evaluate the probability mass function of $S(t)$ at other points than zero. However, Williams (2001) has shown that an explicit expression can indeed be found when the progeny distribution is given by (9.37). His results are as follows. Letting

$$A_n \equiv \frac{\mu^n-1}{\mu^n-\gamma},\quad B_n \equiv \mu^n\left(\frac{1-\gamma}{\mu^n-\gamma}\right)^2, \quad \text{and} \quad \mathfrak{p}_n \equiv \mathbb{P}(N_t=n),$$

the c.f. of $S(t)$ is

$$\Psi_{S(t)}(\zeta)=\sum_{n=0}^{\infty}\mathfrak{p}_n\left(\gamma A_n+\frac{B_n e^{i\zeta}}{1-A_n e^{i\zeta}}\right)^{S_0}. \tag{9.38}$$

When s is a positive integer, the inversion formula for c.f.s gives

$$\mathbb{P}[S(t)=s]=\frac{1}{2\pi}\sum_{n=0}^{\infty}\mathfrak{p}_n\int_{-\pi}^{\pi}e^{-i\zeta s}\left(\gamma A_n+\frac{B_n e^{i\zeta}}{1-A_n e^{i\zeta}}\right)^{S_0}\cdot d\zeta.$$

S_0 being itself an integer, a binomial expansion yields

$$\mathbb{P}[S(t)=s]=\frac{1}{2\pi}\sum_{n=0}^{\infty}\mathfrak{p}_n\sum_{j=0}^{S_0}\binom{S_0}{j}(\gamma A_n)^{S_0-j}$$
$$\times B_n^j\int_{-\pi}^{\pi}e^{i\zeta(j-s)}(1-A_n e^{i\zeta})^{-j}\cdot d\zeta.$$

The integral in this expression vanishes when $j=0$, so it is necessary to consider only the terms with $j>0$. Noting that $0<|A_n e^{i\zeta}|=A_n<1$ when $\mu>1$, we can apply to these terms the binomial series expansion, (2.5), to obtain for $\mathbb{P}[S(t)=s]$ the expression

$$\frac{1}{2\pi}\sum_{n=0}^{\infty}\mathfrak{p}_n\sum_{j=1}^{S_0}\sum_{k=0}^{\infty}\binom{S_0}{j}\binom{j-k-1}{j-1}(\gamma A_n)^{S_0-j}B_n^j A_n^k\int_{-\pi}^{\pi}e^{i\zeta(k+j-s)}\cdot d\zeta.$$

The integral being 2π when $k+j-s=0$, and zero otherwise, we wind up with

$$\mathbb{P}[S(t)=s]=\sum_{n=0}^{\infty}\mathfrak{p}_n\sum_{j=1}^{S_0\wedge s}\binom{S_0}{j}\binom{s-1}{j-1}\gamma^{S_0-j}B_n^j A_n^{S_0+s-2j}. \tag{9.39}$$

A Limiting Form

The discreteness of the price increments is a salient feature of the model, but this clearly becomes less relevant the larger the initial price, $S(0) = S_0$. Remarkably, when $\mu = 1$ it turns out that a normalization of the price at t converges in distribution as $S_0 \to \infty$ to a time change of Brownian motion. To see this, let $(VS_n)^{-1/2}(S_n - ES_n)$ be the normalized price after n shocks. When $\mu = 1$ this is proportional to $Z_n \equiv S_0^{-1/2}(S_n - S_0)$. Note that S_n can be represented as $S_n = \sum_{j=1}^{S_0} Q_{jn}$, where Q_{jn} is the number of units at stage n that descended from unit $j \in \{1, 2, \ldots, S_0\}$ at stage zero. Since for each n the $\{Q_{jn}\}$ are i.i.d. with $EQ_{jn} = 1$ and $VQ_{jn} = n\sigma^2$ when $\mu = 1$, the central limit theorem implies that Z_n converges weakly (for each fixed n) to $N(0, n\sigma^2)$ as $S_0 \to \infty$. In turn, this implies that the c.f. of Z_n (conditional on S_0) is

$$\Psi_{Z_n}(\zeta) = e^{-\zeta^2\sigma^2 n/2} + o(1),$$

where $o(1) \to 0$ as $S_0 \to \infty$. The corresponding normalization of $S(t)$ is $[S(t) - ES(t)]/\sqrt{VS(t)}$, which is proportional to

$$Z(t) \equiv S_0^{-1/2}(S_{N_t} - S_0) = Z_{N_t}$$

when $\mu = 1$. The limiting c.f. is therefore

$$\begin{aligned} \lim_{S_0 \to \infty} \Psi_{Z(t)} &= \lim_{S_0 \to \infty} [Ee^{-\zeta^2\sigma^2 N_t/2} + o(1)] \\ &= \Pi_{N_t}(e^{-\zeta^2\sigma^2/2}). \end{aligned}$$

Defining a directing process $\{\mathbb{T}_t\}_{t \geq 0}$ as $\mathbb{T}_t = \sigma N_t$ and a subordinated process $\{W(\mathbb{T}_t)\}_{t \geq 0}$, where $\{W(t)\}_{t \geq 0}$ is a Brownian motion, one sees immediately that

$$\Psi_{W(\mathbb{T}_t)}(\zeta) = \Pi_{N_t}(e^{-\zeta^2\sigma^2/2}).$$

This shows that $Z(t)$ is indeed distributed asymptotically as a time change of Brownian motion. When $\{N_t\}_{t \geq 0}$ is Poisson with intensity θ we have

$$\lim_{S_0 \to \infty} \Psi_{Z(t)} = \exp[\theta t(e^{-\zeta^2\sigma^2/2} - 1)],$$

which is the c.f. of a compound-Poisson process—a Poisson sum of normals. As explained by Epps (1996) and Dion and Epps (1999), the ability to approximate the distribution of $S(t)$ through the limiting form of $Z(t)$ is important in estimating the parameters of the process from price data.

Pricing Derivatives

As for the v.g. and hyperbolic models, the discreteness of $S(t)$ under b.p.r.e. dynamics makes it impossible to replicate nonlinear payoff structures with the underlying asset and riskless bonds alone. Accordingly, there is not a unique equivalent martingale measure within which derivatives can be priced as discounted expected values. Stated differently, the condition that $\{S(t)/M_t\}$ be a martingale under $\hat{\mathbb{P}}$ restricts the parameters of the distribution of K and N_t to satisfy

$$e^{-rt}ES(t)/S(0) = e^{-rt}\Pi_{N_t}(\mu) = 1; \tag{9.40}$$

yet there are infinitely many models in the class that meet this restriction. In the case that $\{N_t\}_{t\geq 0}$ is Poisson with intensity θ condition (9.40) requires $e^{-rt}e^{\theta t(\mu-1)} = 1$ or $\mu = 1 + r/\theta$, and when $\{N_t\}_{t\geq 0}$ is negative-binomial with parameters θ and γt it requires $\mu = (1-\theta)^{-1}(1-\theta e^{-r/\gamma})$. As for the other incomplete models we have considered, parameter values subject to the martingale restrictions can be inferred from prices of traded options of different strikes and maturities.

Williams' (2001) formula, (9.39), offers a very direct way to price the European put when (9.37) models the progeny distribution; namely, as

$$P^E(S_0, T) = B(0,T)\sum_{s=0}^{X}(X-s)\mathbb{P}[S(t) = s]. \tag{9.41}$$

More generally, one can apply the integrated c.d.f. method of section 8.4. The first step is to develop the c.f. of $S(T)$ from the p.g.f., as $\hat{\Psi}_{S(T)}(\zeta) = \hat{\Pi}_{S(T)}(e^{i\zeta})$. We can then adapt formula (8.37) by expressing $P^E(S_0, T)$ as

$$\begin{aligned} P^E(S_0, T) &= \int_0^X \hat{F}_T(s) \cdot ds \\ &= \int_0^X \left[\frac{1}{2} - \lim_{c\to\infty}\int_{-c}^{c}\frac{e^{-i\zeta x}}{2\pi i\zeta}\hat{\Psi}_{S(T)}(\zeta)\cdot d\zeta\right]\cdot ds, \end{aligned}$$

where the bracketed expression is from the usual inversion formula for c.d.f.s. Recall from theorem 3 of chapter 2 that the inversion formula applies only when argument s is a continuity point of $\hat{F}_T$. Although $\hat{F}_T$ is discontinuous in this model, it has only countably many (indeed, finitely many) jumps on $[0, X]$, so we can still use it to evaluate the Riemann integral. The result is

$$P^E(S_0, T) = \frac{X}{2} + \frac{1}{2\pi}\lim_{c\to\infty}\int_{-c}^{c}\frac{1-e^{-i\zeta X}}{\zeta^2}\hat{\Psi}_{S(T)}(\zeta)\cdot d\zeta, \tag{9.42}$$

where the integral would be evaluated numerically for some large c, avoiding the singularity at $\zeta = 0$. Neither of (9.41) nor (9.42) seems to dominate from a computational standpoint when (9.41) applies. Although (9.41) avoids the numerical integration required to evaluate $\hat{F}_T$, the multiple summations that it involves can be tedious, particularly when X (in tick units) is large. On the other hand, vector processing speeds this up considerably.

Liu (2003) has applied the b.p.r.e. model to options on a sample of individual U.S. equities. Taking K to be distributed as in (9.37) and $\{N_t\}_{t\geq 0}$ as Poisson, and inferring the parameters from transaction prices of traded options, he finds that the model typically eliminates the smile effect in sample. Results are mixed when out-of-sample predictions are compared with the *ad hoc* version of Black-Scholes with moneyness-specific implicit volatilities. He does find evidence that the predictions are better for options on low-priced stocks, where the discreteness arising from the minimum tick size would seem to be more relevant.

9.4.5 *Assessing the Pure-Jump Models*

Since a thorough empirical comparison of the Lévy and b.p.r.e. models is not available, it seems appropriate to offer a brief qualitative appraisal. All these models account in some degree for the thick-tailedness of the empirical distributions of assets' returns, and for this reason all are able to reduce the severe mispricing of short-term, out-of-money options that is now associated with the Black-Scholes theory. However, no model that represents log price as a Lévy process can account for variations over time in the moments of returns. In the branching process the variance and higher moments of total return do vary with the initial price level, but whether this picks up much of the variation that is observed empirically is not yet known. It is now widely believed that some of the thick-tailedness in the marginal distributions of returns can be attributed to temporal variation in volatility. Lévy models obviously cannot account for this, since nonoverlapping returns are i.i.d.[22] The b.p.r.e. accounts only for contemporaneous price-level effects and so does little better on this score. Thus, none of the models explains the high autocorrelation that is almost always observed in the absolute values or squares of high-frequency returns. Moreover, no Lévy model with

[22]However, Finlay and Seneta (2006) have shown that it is possible to construct a stationary process whose increments over intervals of specific length have v.g.-distributed marginals but are nevertheless dependent.

finite-variance increments to log price could successfully capture the slowly fading volatility smiles that characterize the empirical term structure of implicit volatility.[23] Of course, allowing explicitly for stochastic variation in one or more parameters, as in the s.v. diffusions, is always possible in theory; but doing so would both add to an already-extensive computational burden and fundamentally change the character of these models.

9.5 A Markov-Switching Model

In previous sections we have seen several ways of modeling discontinuous changes through the use of Lévy processes. This has been done both directly by invoking the Poisson model and indirectly by using Lévy processes to direct stochastic time changes. Here we look at an alternative approach, based on Edwards (2004, 2005), in which changes are triggered by a different type of continuous-time Markov process. Like the Poisson process, this is a continuous-time Markov chain, but instead of taking unit jumps it merely switches back and forth at random times among some finite number of positions. In the present application the switch determines which of a set of alternative processes controls the stochastic part of underlying price process $\{S_t\}_{t\geq 0}$. Such "regime-switching" models have found wide application in empirical economics since their introduction by Hamilton (1989). The modeling of economic time series as switching at random times among regimes of different growth rates and/or volatilities can be motivated in various ways: through spontaneous changes in consumers' and investors' degrees of optimism, through successive breakthroughs and plateaus in technology, changes in legal constraints, shifts in monetary and fiscal policies, natural disasters, wars and threats of war, and so forth. Sufficient motivation

[23]This is documented by Carr and Wu (2003), who circumvent this limitation of the Lévy specification by modeling increments to log price as affine functions of an assymetric-stable law with characteristic exponent $\alpha \in (1,2)$ and symmetry parameter $\beta = -1$. Stable laws are represented via the c.f., with $\ln \Psi(\zeta)$ given by

$$i\zeta\mu - |\zeta\sigma|^\alpha[1 - i\beta\mathfrak{s}(\zeta)\tan(\pi\alpha/2)],$$

where $\mathfrak{s}(\cdot)$ is the signum function, μ and σ are location and scale parameters, and $\beta \in [-1,1]$ and $\alpha \in (0,2]$ control symmetry and tail thickness. The normal and Cauchy laws are special cases with $\alpha = 2$ and $\alpha = 1, \beta = 0$ respectively. First moments exist whenever $\alpha > 1$, but the variance is finite only in the normal case. While changes in log price thus have infinite variance in the Carr-Wu setup, they do possess means. Also, with $\beta = -1$ the right tail of the distribution is thin enough that $ES_T = Ee^{\ln S_T} < \infty$, as needed for martingale pricing. Unfortunately, the inflexibility of this choice limits the model's ability to adapt to the skewness in actual returns.

for our purposes is the fact that the switching scheme offers a simple way of generating price processes that produce volatility smiles. We begin by examining some properties of the switching process itself, then see how to apply it in modeling the underlying price of a derivative security.[24] As we shall see, Lévy processes will continue to play a prominent role even though they do not actually trigger the changes in regime.

We work with a stochastic process $\{\mathbb{I}_t\}_{t\geq 0}$ having $\mathcal{S}+1$ states, associated with the set of integers $\mathbb{S} = \{0, 1, \ldots, \mathcal{S}\}$. The behavior of such processes is specified by the conditional probabilities of being in each possible state at each time $t+u$ given what is known at time t; i.e., $p_{s'}(t,u) = \mathbb{P}(\mathbb{I}_{t+u} = s' \mid \mathcal{F}_t)$. By restricting to processes that are Markovian, we can condition just on the knowledge of the present state, as $p_{ss'}(t,u) = \mathbb{P}(\mathbb{I}_{t+u} = s' \mid \mathbb{I}_t = s)$. Restricting further to time-homogeneous processes removes the implied dependence on t, so we can write the transition probabilities for the process simply as $p_{ss'}(u) = \mathbb{P}(\mathbb{I}_{t+u} = s' \mid \mathbb{I}_t = s)$. We assume that these satisfy for each $t \geq 0$, $u \geq 0$ and each state s the conditions $p_{ss'} \geq 0$ and $\sum_{s' \in \mathbb{S}} p_{ss'} = 1$, together with the Chapman-Kolmogorov equations,

$$\left\{ p_{ss'}(t+u) = \sum_{s'' \in \mathbb{S}} p_{ss''}(t) p_{s''s'}(u) \right\}_{s,s' \in \mathbb{S}}. \tag{9.43}$$

Each of these equations expresses the probability of transiting from s to s' as the probability of the union of a collection of exclusive events; namely, that there is an intermediate transition from s to each $s'' \in \mathbb{S}$ and thence from s'' to s'. Of course, such conditions must hold for all positive t and u. In addition, we set the initial transition probabilities as $p_{ss}(0) = 1$.

The usual procedure in modeling the transition probabilities is to specify their asymptotic behavior as $u \downarrow 0$. Specifically, for positive constants $\vartheta_{ss'}$ we put

$$p_{ss'}(u) = \begin{cases} \vartheta_{ss'} u + o(u), & s' \neq s \\ 1 - \vartheta_{ss} u + o(u), & s' = s, \end{cases}$$

where $o(u)/u \longrightarrow 0$ as $u \downarrow 0$. Thus, $\vartheta_{ss'}$ represents the instantaneous rate at which the system transits *from* state s *to* state s'. Of course, $\sum_{s' \in \mathbb{S}} p_{ss'} = 1$ implies that $\vartheta_{ss} = \sum_{s' \neq s} \vartheta_{ss'}$. Putting state 0 last, we can arrange these in

[24]One may consult Cox and Miller (1965) and Ross (2000) for further details about discrete-space Markov processes in continuous time.

matrix form as

$$\mathbf{\Theta} = \begin{pmatrix} -\vartheta_{11} & \vartheta_{12} & \cdots & \vartheta_{1S} & \vartheta_{10} \\ \vartheta_{21} & -\vartheta_{22} & \cdots & \vartheta_{2S} & \vartheta_{20} \\ \vdots & \vdots & \ddots & \vdots & \vdots \\ \vartheta_{S1} & \vartheta_{S2} & \cdots & -\vartheta_{SS} & \vartheta_{S0} \\ \vartheta_{01} & \vartheta_{02} & \cdots & \vartheta_{0S} & -\vartheta_{00} \end{pmatrix}.$$

Together, the asymptotic structure, the Chapman-Kolmogorov equations, and the initial conditions $p_{ss}(0) = 1$ lead to a system of differential equations that can be solved for the transition probabilities at all future times. Thus, from (9.43) we have

$$\begin{aligned} p_{ss'}(t+u) &= p_{ss'}(t)p_{s's'}(u) + \sum_{s'' \neq s'} p_{ss''}(t)p_{s''s'}(u) \\ &= p_{ss'}(t)(1 - \vartheta_{s's'}u) + \sum_{s'' \neq s'} p_{ss''}(t)\vartheta_{s''s'}u + o(u) \end{aligned}$$

and hence

$$p'_{ss'}(t) = \lim_{u \downarrow 0} u^{-1}[p_{ss'}(t+u) - p_{ss'}(t)] = -p_{ss'}(t)\vartheta_{s's'} + \sum_{s'' \neq s'} p_{ss''}(t)\vartheta_{s''s'},$$

which are Kolmogorov's "forward" equations. Represented in matrix form as $\mathbf{P}'(t) = \mathbf{P}(t)\mathbf{\Theta}$, the solution subject to $\mathbf{P}(0) = \mathbf{I}$ is $\mathbf{P}(t) = e^{\mathbf{\Theta}t}$, where the exponential matrix is defined as the convergent series

$$e^{\mathbf{\Theta}t} = \sum_{k=0}^{\infty} \mathbf{\Theta}^k t^k / k!.$$

For example, if there are just two states, $\mathbb{S} = \{0, 1\}$, the forward equations are

$$\begin{aligned} p'_{01}(t) &= -p_{01}(t)\vartheta_{11} + p_{00}(t)\vartheta_{01} = -p'_{00}(t) \\ p'_{10}(t) &= -p_{10}(t)\vartheta_{00} + p_{11}(t)\vartheta_{10} = -p'_{11}(t). \end{aligned}$$

Setting $\vartheta_{00} = \vartheta_{01} \equiv \vartheta_0$ and $\vartheta_{11} = \vartheta_{10} \equiv \vartheta_1$, the solutions have the explicit form

$$\begin{aligned} p_{01}(t) &= \frac{\vartheta_0}{\vartheta_0 + \vartheta_1}[1 - e^{-(\vartheta_0+\vartheta_1)t}] = 1 - p_{00}(t) \\ p_{10}(t) &= \frac{\vartheta_1}{\vartheta_0 + \vartheta_1}[1 - e^{-(\vartheta_0+\vartheta_1)t}] = 1 - p_{11}(t). \end{aligned} \tag{9.44}$$

What is essential for our application is to know the behavior of the *occupation times*, which are the amounts of time during $[0, t]$ that process $\{\mathbb{I}_t\}$ spends in each state. These can be represented as $T_s(t) = \int_0^t \mathbf{1}_{\{s\}}(\mathbb{I}_u) \cdot du \geq 0$ for $s \in \{1, 2, \ldots, \mathcal{S}\}$ and $T_0(t) = t - \sum_{s=1}^{\mathcal{S}} T_s(t) \geq 0$. The joint m.g.f. of $\{T_s(t)\}_{s=1}^{\mathcal{S}}$ was given by Darroch and Morris (1968, theorem 1) as

$$\mathfrak{M}_{\{T_s(t)\}}(\boldsymbol{\rho}) = \boldsymbol{\pi}' e^{t(\boldsymbol{\Theta}+\mathbf{D})} \mathbf{1},$$

where (i) $\boldsymbol{\pi}$ is the vector of initial state probabilities,[25] $\{\pi_s = \mathbb{P}(\mathbb{I}_0 = s)\}_{s=0}^{\mathcal{S}}$, (ii) $\boldsymbol{\rho} \in \mathfrak{C}_{\mathcal{S}}$ (an $\mathcal{S}$ -vector of complex numbers), (iii) $\mathbf{1}$ is an $\mathcal{S}+1$ vector of units, and (iv)

$$\mathbf{D} = \begin{pmatrix} \rho_1 & 0 & \cdots & 0 & 0 \\ 0 & \rho_2 & \cdots & 0 & 0 \\ \vdots & \vdots & \ddots & \vdots & \vdots \\ 0 & 0 & \cdots & \rho_{\mathcal{S}} & 0 \\ 0 & 0 & \cdots & 0 & 0 \end{pmatrix}.$$

We will now build on these foundations a switching model for the price process $\{S_t\}_{t\geq 0}$ of some underlying asset under the risk-neutral measure. Let $\{R_{s,t}\}_{s\in\mathbb{S}, t\geq 0}$ be a collection of $\mathcal{S}+1$ stochastic returns processes having the following properties:

- Processes $\{R_{s,t}\}_{s\in\mathbb{S}, t\geq 0}$ are mutually independent and independent of the state process $\{\mathbb{I}_t\}_{t\geq 0}$.
- Each process $\{R_{s,t}\}_{t\geq 0}$ is a nonnegative martingale under $\hat{\mathbb{P}}$, with $\hat{E}R_{s,t} = R_{s,0} = 1$.
- Each process $\{\ln R_{s,t}\}_{t\geq 0}$ is a Lévy process.

Two important features of Lévy processes will be recalled from section 2.2.8. First, Lévy processes are Markov, since they have independent increments. Second, if $\{R_{s,t}\}_{t\geq 0}$ is Lévy its c.f. under $\hat{\mathbb{P}}$ satisfies

$$\Psi_{s,t}(\zeta) = \hat{E} e^{i\zeta \ln R_{s,t}} = \Psi_{s,1}(\zeta)^t = (\hat{E} e^{i\zeta \ln R_{s,1}})^t. \quad (9.45)$$

These properties will prove critical in what follows.

We will have to represent the security's price at t in terms of the various return processes that operate during $[0, t]$. This is not hard to do, but to get it right takes a bit of notation. For this, let $N_t \in \aleph_0$ be the number of times on $(0, t)$ at which transitions from some state to another occur.

[25]These to be distinguished from the initial *transition* probabilities.

If there is at least one such transition, let $\{\tau_j\}_{j=1}^{N_t}$ be the set of (random) times themselves—these being the discontinuity points of the state process $\{\mathbb{I}_u\}_{0\le u\le t}$, and take $\tau_0 = 0$ and $\tau_{N_t+1} = t$. Next, let $s_j = \mathbb{I}_{\tau_j}$ be the state to which a transit occurs at τ_j, with $s_0 = \mathbb{I}_0$ being the initial state and $s_{N_t+1} = \mathbb{I}_t$. Now letting $\{r_t\}$ and $\{\delta_t\}$ be $\mathcal{F}_0$ -measurable short-rate and instantaneous dividend processes, we model the price of a traded asset under $\hat{\mathbb{P}}$ as

$$S_t = \begin{cases} S_0 e^{\int_0^t (r_u-\delta_u)\cdot du} \prod_{j=1}^{N_t+1} \dfrac{R_{s_{j-1},\tau_j-}}{R_{s_{j-1},\tau_{j-1}}} \cdot \dfrac{R_{s_{N_t},t}}{R_{s_{N_t},\tau_{N_t}}}, & N_t = 1,2,\ldots \\ S_0 e^{\int_0^t (r_u-\delta_u)\cdot du} \dfrac{R_{0,t}}{R_{0,0}} = S_0 e^{\int_0^t (r_u-\delta_u)\cdot du} R_{0,t}, & N_t = 0. \end{cases} \tag{9.46}$$

For example, if $N_t = 1$ then

$$S_t = S_0 e^{\int_0^t (r_u-\delta_u)\cdot du} R_{0,\tau_1-} \cdot R_{s_1,t}/R_{s_1,\tau_1}.$$

Expression (9.46) makes it appear that the distribution of S_t is extremely complicated, because S_t seems to depend on the random times of state changes, on which states occurred, and on their order. Fortunately, things are simpler than this. Because processes $\{\mathbb{I}_u\}_{0\le u\le t}$ and $\{R_{s,u}\}_{s\in\mathbb{S},0\le u\le t}$ are Markov, S_t depends just on the total *occupation time* for each state, not on how many times each state was visited and when. Thus, with $\{T_s(t)\}_{s\in\mathbb{S}}$ representing the occupation times, we have

$$S_t = S_0 e^{\int_0^t (r_u-\delta_u)\cdot du} \prod_{s=0}^{\mathcal{S}} R_{s,T_s(t)},$$

where "=" is interpreted here as "equal in distribution".

With this simplification we can now do two useful things. First, dividing by $M_t = M_0 e^{\int_0^t r_u\cdot du}$, we can verify at once that the normalized *cum*-dividend price process, $\{S_t^{c*} = S_t e^{\int_0^t \delta_u\cdot du}/M_t\}$, is a martingale under $\hat{\mathbb{P}}$:

$$\begin{aligned} \hat{E}S_t^{c*} &= S_0^*\hat{E}\prod_{s=0}^{\mathcal{S}} R_{s,T_s(t)} \\ &= S_0^*\hat{E}\left[\prod_{s=0}^{\mathcal{S}} \hat{E}(R_{s,T_s(t)} \mid \{T_s(t)\}_{s=1}^{\mathcal{S}})\right] \\ &= S_0^*\hat{E}\prod_{s=0}^{\mathcal{S}} R_{s,0} = S_0^* \equiv S_0^{c*}. \end{aligned}$$

This assures us that the model is consistent with an arbitrage-free market. Next, we can find the c.f. of $\ln S_t$ under $\hat{\mathbb{P}}$. This will open a clear path toward the goal of pricing European-style derivatives on this asset, via either of the one-step Fourier approaches described in section 8.4. Finding

$$\Psi(\zeta) = \hat{E}e^{i\zeta \ln S_t} = S_0^{i\zeta} e^{i\zeta \int_0^t (r_u - \delta_u)\cdot du} \hat{E}e^{i\zeta \sum_{s=0}^{\mathcal{S}} \ln R_{s,T_s(t)}}$$

is straightforward given what we know about Lévy processes and occupation times. Obviously, we have merely to work out the expectation in the expression above, which is itself the c.f. of the log of the normalized *cum*-dividend price relative, S_t^{c*}/S_0^*. We can write this as

$$\hat{E}e^{i\zeta \sum_{s=0}^{\mathcal{S}} \ln R_{s,T_s(t)}} = \hat{E}\exp\left(i\zeta \ln R_{0,t-T_1(t)-\cdots-T_{\mathcal{S}}(t)} + i\zeta \sum_{s=1}^{\mathcal{S}} \ln R_{s,T_s(t)}\right).$$

Conditioning on $\{T_s(t)\}_{s=1}^{\mathcal{S}}$ and using the tower property, we have

$$\begin{aligned}
\hat{E}e^{i\zeta \sum_{s=0}^{\mathcal{S}} \ln R_{s,T_s(t)}} &= \hat{E}\left[\Psi_{0,t-T_1(t)-\cdots-T_{\mathcal{S}}(t)}(\zeta) \prod_{s=1}^{\mathcal{S}} \Psi_{s,T_s(t)}(\zeta)\right] \\
&= \hat{E}\left[\Psi_{0,1}(\zeta)^{t-T_1(t)-\cdots-T_{\mathcal{S}}(t)} \prod_{s=1}^{\mathcal{S}} \Psi_{s,1}(\zeta)^{T_s(t)}\right] \\
&= \Psi_{0,1}(\zeta)^t \hat{E} \prod_{s=1}^{\mathcal{S}} \left[\frac{\Psi_{s,1}(\zeta)}{\Psi_{0,1}(\zeta)}\right]^{T_s(t)},
\end{aligned}$$

where property (9.45) is used in the second line. Note that division by $\Psi_{0,1}(\zeta)$ is always permissible, since c.f.s of Lévy processes have no zeroes. Now setting $\rho_s(\zeta) \equiv \ln[\Psi_{s,1}(\zeta)/\Psi_{0,1}(\zeta)]$ and $\boldsymbol{\rho} = (\rho_1(\zeta), \rho_2(\zeta), \ldots, \rho_{\mathcal{S}}(\zeta))'$, the c.f. of $\ln S_t$ under $\hat{\mathbb{P}}$ can be expressed in terms of the m.g.f. of the occupation times, as

$$\begin{aligned}
\Psi(\zeta) &= S_0^{i\zeta} e^{i\zeta \int_0^t (r_u - \delta_u)\cdot du} \Psi_{0,1}(\zeta)^t \mathfrak{M}_{\{T_s(t)\}}(\boldsymbol{\rho}) \\
&= S_0^{i\zeta} e^{i\zeta \int_0^t (r_u - \delta_u)\cdot du} \Psi_{0,1}(\zeta)^t \boldsymbol{\pi}' e^{t(\boldsymbol{\Theta}+\mathbf{D})} \mathbf{1}.
\end{aligned} \tag{9.47}$$

Notice that although the normalized *cum*-dividend price is a $\hat{\mathbb{P}}$ martingale, the change in log price is *not* a Lévy process, since

$$Ee^{i\zeta \ln(S_t^{c*}/S_0^*)} = \Psi_{0,1}(\zeta)^t \boldsymbol{\pi}' e^{t(\boldsymbol{\Theta}+\mathbf{D})}\mathbf{1} \neq [\Psi_{0,1}(\zeta)\boldsymbol{\pi}' e^{(\boldsymbol{\Theta}+\mathbf{D})}\mathbf{1}]^t.$$

Thus, the increments of log price are not i.i.d. As Edwards (2005) points out, the switching effect is therefore capable of slowing the convergence to normality of (the standardized) $\ln S_t^{c*}$ as $t \to \infty$. This fact holds promise for delivering models of derivatives' prices that better fit the observed term structures of implicit volatilities.

As a simple example of such a multistate switching process, take $\ln R_{s,t} = -\sigma_s^2 t/2 + \sigma_s \hat{Z}_{s,t}$, where $\{\hat{Z}_{s,t}\}_{s\in\mathbb{S}}$ are independent Brownian motions. Thus, our returns processes are simply geometric Brownian motions with state-dependent, $\mathcal{F}_0$-measurable volatilities. Then $\hat{E}R_{s,t} = 1$, as is required for the martingale property, and

$$\Psi_{s,t}(\zeta) = e^{-(i\zeta+\zeta^2)\sigma_s^2 t/2} = [e^{-(i\zeta+\zeta^2)\sigma_s^2/2}]^t = \Psi_{s,1}(\zeta)^t.$$

Thus, the c.f. of $\ln S_t$ is

$$\Psi(\zeta) = S_0^{i\zeta} e^{i\zeta \int_0^t (r_u - \delta_u)\cdot du} e^{-(i\zeta+\zeta^2)\sigma_0^2 t/2} \boldsymbol{\pi}' e^{t(\boldsymbol{\Theta}+\mathbf{D})}\mathbf{1}, \tag{9.48}$$

with

$$\mathbf{D} = \frac{i\zeta+\zeta^2}{2}\begin{pmatrix} \sigma_0^2-\sigma_1^2 & 0 & \cdots & 0 & 0 \\ 0 & \sigma_0^2-\sigma_2^2 & \cdots & 0 & 0 \\ \vdots & \vdots & \ddots & \vdots & \vdots \\ 0 & 0 & \cdots & \sigma_0^2-\sigma_{\mathcal{S}}^2 & 0 \\ 0 & 0 & \cdots & 0 & 0 \end{pmatrix}.$$

When there are just two states with instantaneous transition rates $\vartheta_0 \equiv \vartheta_{01} = \vartheta_{00}$ and $\vartheta_1 \equiv \vartheta_{10} = \vartheta_{11}$, we have

$$\boldsymbol{\Theta}+\mathbf{D} = \begin{pmatrix} -\vartheta_1 + (\frac{i\zeta+\zeta^2}{2})(\sigma_0^2-\sigma_1^2) & \vartheta_1 \\ \vartheta_0 & -\vartheta_0 \end{pmatrix}, \tag{9.49}$$

or, for general returns processes with $\rho(\zeta) \equiv \ln[\Psi_{1,1}(\zeta)/\Psi_{0,1}(\zeta)]$,

$$\boldsymbol{\Theta}+\mathbf{D} = \begin{pmatrix} -\vartheta_1 + \rho(\zeta) & \vartheta_1 \\ \vartheta_0 & -\vartheta_0 \end{pmatrix}. \tag{9.50}$$

At the cost of some flexibility, one parameter in the two-state model can be eliminated by setting $\pi_1 = \vartheta_0/(\vartheta_0+\vartheta_1)$, which corresponds to the "long-run" transition probability $p_{11}(\infty)$ in expression (9.44).

There are several ways to compute $e^{\mathbf{M}t}$ for an arbitrary square matrix $\mathbf{M}$. For large enough N any of the expressions $\sum_{j=1}^{N} \mathbf{M}^j t^j/j!$, $(\mathbf{I}+N^{-1}\mathbf{M}t)^N$, or $[(\mathbf{I}-N^{-1}\mathbf{M}t)^{-1}]^N$ can yield an acceptable approximation (assuming in the last case that $\mathbf{I}-N^{-1}\mathbf{M}t$ is invertible). However, an exact expression can be obtained with minimal computation by a procedure based on the Cayley-Hamilton theorem of matrix algebra. The theorem states that an $n\times n$ matrix $\mathbf{M}$ satisfies its own characteristic equation, which is to say that $|\mathbf{I}\lambda - \mathbf{M}| = \lambda^n + a_1\lambda^{n-1} + \cdots + a_{n-1}\lambda + a_n = 0$ implies

$\mathbf{M}^n + a_1\mathbf{M}^{n-1} + \cdots + a_{n-1}\mathbf{M} + a_n = 0$.[26] The Cayley-Hamilton method for evaluating $e^{\mathbf{M}t}$ consists of the following steps:

1. Write down the characteristic equation and solve for latent roots $\{\lambda_j\}_{j=1}^n$.
2. Construct the system of n equations

$$\begin{aligned} e^{\lambda_1 t} &= \alpha_0 + \alpha_1\lambda_1 + \cdots + \alpha_{n-1}\lambda_1^{n-1} \\ e^{\lambda_2 t} &= \alpha_0 + \alpha_1\lambda_2 + \cdots + \alpha_{n-1}\lambda_2^{n-1} \\ &\vdots \\ e^{\lambda_n t} &= \alpha_0 + \alpha_1\lambda_n + \cdots + \alpha_{n-1}\lambda_n^{n-1} \end{aligned} \tag{9.51}$$

   and solve for the coefficients $\alpha_0, \alpha_1, \ldots, \alpha_{n-1}$.[27]
3. Finally, construct $e^{\mathbf{M}t}$ as

$$e^{\mathbf{M}t} = \alpha_0\mathbf{I} + \alpha_1\mathbf{M} + \cdots + \alpha_{n-1}\mathbf{M}^{n-1}.$$

   Thus, only $n-2$ matrix multiplications are required for an $n\times n$ matrix.[28]

To illustrate, we follow these steps to evaluate $e^{(\mathbf{\Theta}+\mathbf{D})t}$ in the two-state case, where $\mathbf{\Theta} + \mathbf{D}$ is as in (9.50). First, the characteristic equation is

$$\begin{aligned} |\mathbf{I}\lambda - (\mathbf{\Theta} + \mathbf{D})| &= \lambda^2 - \lambda \cdot tr(\mathbf{\Theta} + \mathbf{D}) + |\mathbf{\Theta} + \mathbf{D}| \\ &= \lambda^2 + \lambda(\vartheta_0 + \vartheta_1 - \rho) - \vartheta_0\rho \end{aligned}$$

---

[26] Schott (1997) gives a precise statement of the theorem. For a succinct explanation of how the Cayley-Hamilton method follows from the Cayley-Hamilton theorem see Rowell (2004).

[27] If the $\{\lambda_j\}$ are not all distinct, (9.51) will comprise fewer than n independent equations and cannot be solved for the $\{\alpha_j\}$. In this case additional equations can be generated by differentiating the equation pertaining to each multiple root. For example, if root λ_1 has multiplicity k, the first k equations are generated as

$$\frac{d^{(j)}e^{\lambda_1 t}}{d\lambda_1^j} = \frac{d^{(j)}}{d\lambda_1^j}(\alpha_0 + \alpha_1\lambda_1 + \cdots + \alpha_{n-1}\lambda_1^{n-1})$$

for $j = 0, 1, \ldots, k-1$.

Nothing in the structure of the switching model rules out the possibility of equal roots for certain values of ζ, transition rates ϑ_{jk}, and returns processes. Fortunately, the conditions that lead to multiplicity are of the knife-edge sort and are not apt to be encountered in the process of numerically inverting the c.f.; however the numerical stability of the solutions near the edge must be considered.

[28] There is also a diagnonalization technique for evaluating $e^{\mathbf{M}t}$. It is based on the fact that when $\mathbf{M}$ has distinct latent roots the matrix $\mathbf{V} = (\mathbf{v}_1, \mathbf{v}_2, \ldots, \mathbf{v}_n)$ of latent vectors is nonsingular and such that $\mathbf{V}^{-1}\mathbf{M}^k\mathbf{V} = \mathbf{\Lambda}^k$, where $\mathbf{\Lambda}$ is a diagonal matrix with $\{\lambda_j\}_{j=1}^n$ on the diagonal. From these facts it is easy to deduce that $\mathbf{V}^{-1}e^{\mathbf{M}t}\mathbf{V} = \sum_{k=0}^{\infty} \mathbf{\Lambda}^k t^k/k!$ and hence that $e^{\mathbf{M}t} = \mathbf{V}\mathbf{L}_t\mathbf{V}^{-1}$, where $\mathbf{L}_t$ is a diagonal matrix with $\{e^{\lambda_j t}\}_{j=1}^n$ on the diagonal. The method can be modified to handle the case of multiple roots.

with (complex-valued) roots

$$\lambda_1 = \frac{1}{2}\left[\rho - (\vartheta_0 + \vartheta_1) + \sqrt{\rho^2 + 2\rho(\vartheta_0 - \vartheta_1) + (\vartheta_0 + \vartheta_1)^2}\right]$$
$$\lambda_2 = \frac{1}{2}\left[\rho - (\vartheta_0 + \vartheta_1) - \sqrt{\rho^2 + 2\rho(\vartheta_0 - \vartheta_1) + (\vartheta_0 + \vartheta_1)^2}\right].$$

Second, the coefficients α_0, α_1 are

$$\alpha_0 = \frac{\lambda_1 e^{\lambda_2 t} - \lambda_2 e^{\lambda_1 t}}{\lambda_1 - \lambda_2}$$
$$\alpha_1 = \frac{e^{\lambda_1 t} - e^{\lambda_2 t}}{\lambda_1 - \lambda_2},$$

and, third, $e^{(\mathbf{\Theta}+\mathbf{D})t} = \alpha_0 \mathbf{I} + \alpha_1(\mathbf{\Theta} + \mathbf{D})$ simplifies to

$$e^{(\mathbf{\Theta}+\mathbf{D})t} = \begin{bmatrix} \frac{(\lambda_1+\vartheta_1-\rho)e^{\lambda_2 t}-(\lambda_2+\vartheta_1-\rho)e^{\lambda_1 t}}{\lambda_1-\lambda_2} & \frac{\vartheta_1(e^{\lambda_1 t}-e^{\lambda_2 t})}{\lambda_1-\lambda_2} \\ \frac{\vartheta_0(e^{\lambda_1 t}-e^{\lambda_2 t})}{\lambda_1-\lambda_2} & \frac{(\lambda_1+\vartheta_0)e^{\lambda_2 t}-(\lambda_2+\vartheta_0)e^{\lambda_1 t}}{\lambda_1-\lambda_2} \end{bmatrix}.$$

Finally, with $\boldsymbol{\pi}' = (\pi_1, 1 - \pi_1)$ the expression $\boldsymbol{\pi}' e^{t(\mathbf{\Theta}+\mathbf{D})}\mathbf{1}$ in (9.47) has the simple form

$$\boldsymbol{\pi}' e^{t(\mathbf{\Theta}+\mathbf{D})}\mathbf{1} = \frac{[\lambda_1 - \pi_1\rho(\zeta)]e^{\lambda_2 t} - [\lambda_2 - \pi_1\rho(\zeta)]e^{\lambda_1 t}}{\lambda_1 - \lambda_2}.$$

Figure 9.8 illustrates the effect of Markov switching in a two-state model in which price is driven by geometric Brownian motions with volatilities $\sigma_0 = 0.1$ and $\sigma_1 = 0.6$. In this case the log of the ratio of c.f.s is $\rho = (i\zeta + \zeta^2)(\sigma_0^2 - \sigma_1^2)/2$, and $\mathbf{\Theta} + \mathbf{D}$ is as given in (9.49). The figure plots implicit volatilities *vs.* moneyness for three sets of transition rates, $(\vartheta_0, \vartheta_1) \in \{(.9, .1), (.5, .5), (1., .9)\}$, with π_1 set equal to long-run transition probability $\vartheta_0/(\vartheta_0 + \vartheta_1)$ in each case. Note that $\vartheta_0 \equiv \vartheta_{01}$ and $\vartheta_1 \equiv \vartheta_{10}$ are the instantaneous rates of transit *from* the low- and high-volatility states, respectively.[29] The top-most curve shows that implicit volatility varies little with moneyness when there is a relatively high probability of being and remaining in the high-volatility state. On the other hand, the volatility smile is much more pronounced when the low-volatility state is more sustainable but there is significant potential for volatility to increase. Figure 9.9 shows that implicit volatility curves for the case $(\vartheta_0, \vartheta_1) = (1., .9)$ flatten very slowly as time to expiration increases.

[29] We have used highly discrepant state volatilities so as to make the effects of variations in $(\vartheta_0, \vartheta_1)$ more apparent. The curves are based on prices of European options with 0.5 years to expiration, a short rate of 0.02, and a dividend rate of zero.

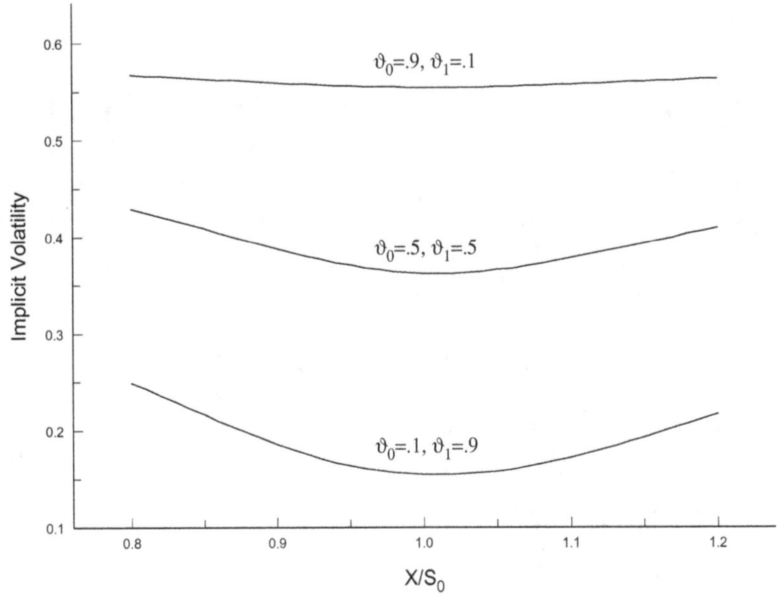

Fig. 9.8. Implicit volatility *vs.* moneyness in geometric B.m. switching model with $(\sigma_0, \sigma_1) = (0.1, 0.6)$ and three sets of transition rates $(\vartheta_0, \vartheta_1)$.

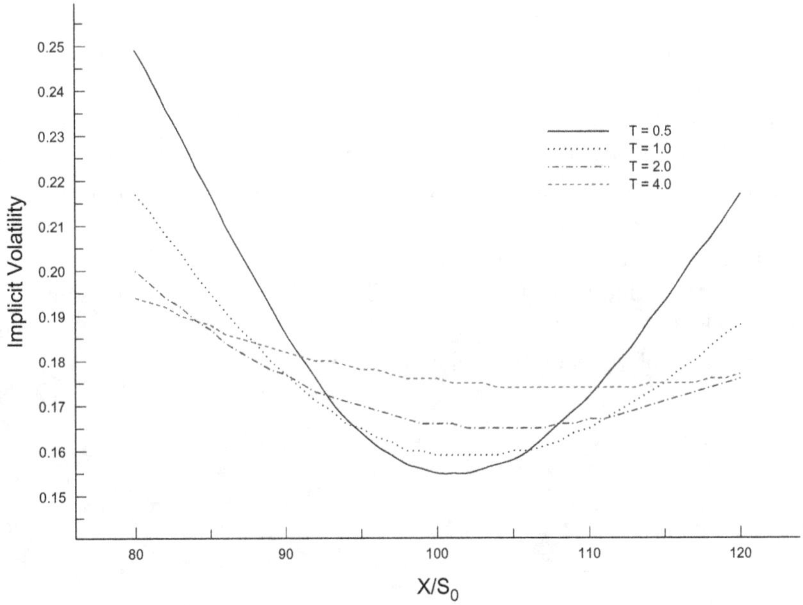

Fig. 9.9. Implicit volatility *vs.* moneyness for four times to expiration T in geometric B.m. switching model: $(\vartheta_0, \vartheta_1) = (0.1, 0.9), (\sigma_0, \sigma_1) = (0.1, 0.6)$.

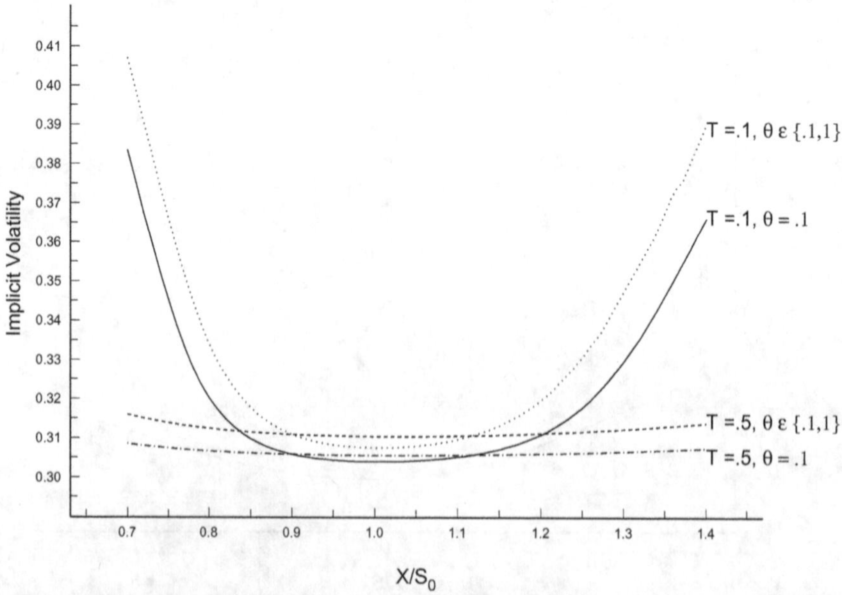

Fig. 9.10. Implicit volatility *vs.* moneyness for two times to expiration T, for jump-diffusion with intensity $\theta = .1$ and for switching model with $\theta \in \{.1, 1.0\}$.

Of course, there are many possible specifications of the switching model, since one is free to choose the number of states and the Lévy returns process that applies in each. Figure 9.10 shows implicit volatilities from a model that alternates between two constant-volatility jump-diffusion processes with radically different jump intensities, $\theta \in \{.1, 1.0\}$, but having all other features in common. The transition rates from low- and high-intensity states are $(\vartheta_0, \vartheta_1) = (.1, .9)$. For comparison the curve for each time to expiration T is paired with that for a standard jump diffusion with $\theta = .1$ (which is the same as in figure 9.1). The switching effect does somewhat improve the term-structure properties of the model by slightly accentuating the curvature of volatility smiles at later expirations.

10

Interest-Rate Dynamics

This chapter provides an introduction to the extensive literature on pricing interest-sensitive assets and derivatives, such as bonds, options on bonds, interest-rate caps and swaps, and options on caps and swaps. We have thus far worked under the assumption that future interest rates and bond prices are known in advance (condition RK of chapter 4). It is time now to recognize and begin to model their stochastic behavior. The first section provides a preliminary review of the notation and definitions introduced in chapter 4, introduces the concept of forward measure, and concludes with an overview of the models and methods considered in remaining sections.

10.1 Preliminaries

10.1.1 *A Summary of Basic Concepts*

Table 10.1 collects the notation, definitions, and fundamental properties of interest rates and bond prices that have been used in previous chapters.

Here are a few specific points that are particularly relevant for the subsequent discussion:

1. The definitions of the average and instantaneous forward rates imply that

$$r_t(t', T) = (T - t')^{-1} \int_{t'}^{T} r_t(u) \cdot du.$$

Forward prices of bonds can therefore be recovered from instantaneous forward rates as

$$B_t(t', T) = e^{-\int_{t'}^{T} r_t(u) \cdot du}.$$

Table 10.1. Notation for bond prices and interest rates.

Symbol	Interpretation	Properties/Relations
$B(t,T)$	Price at t of T-maturing	$B(t,t) = B(T,T) = 1$,
	unit discount bond	$0 < B(t,T) \leq 1$
$r(t,T)$	Average spot rate at t for	$r(t,T) = -\frac{\ln B(t,T)}{T-t}$
	borrowing from t until T	
r_t	Instantaneous spot rate,	$r_t = \frac{\partial \ln B(t,T)}{\partial t} \|_{T=t}$
	or short rate	$= -\frac{\partial \ln B(t,T)}{\partial T} \|_{T=t}$
M_t	Value of money fund that invests at the short rate	$M_t = M_0 e^{\int_0^t r_s \cdot ds}$
$B_t(t',T)$	Forward price at t to buy T-	$B_t(t',T) = \frac{B(t,T)}{B(t,t')}$,
	maturing bond at $t' \in [t,T]$	$B_t(t,T) \equiv B(t,T)$
$r_t(t',T)$	Average forward rate at t	$r_t(t',T) = -\frac{\ln B_t(t',T)}{T-t'}$
	for loans from t' until T	
$r_t(t')$	Instantaneous forward rate	$r_t(t') = -\frac{\partial \ln B(t,T)}{\partial T} \|_{T=t'}$,
	at t for borrowing at t'	$r_t(t) = r_t$

2. Likewise, since $B_t(t,T) = B(t,T)$ bonds' spot prices can be expressed in terms of forward rates as well:

$$B(t,T) = e^{-\int_t^T r_t(u) \cdot du}. \tag{10.1}$$

3. The "yield to maturity" of a T-maturing discount bond as of time t is the same as $r(t,T)$, the average spot rate for lending from t to T.

10.1.2 *Spot and Forward Measures*

Modeling interest-sensitive derivatives typically requires modeling the evolution of the entire term structure of bond prices, or, equivalently, the term structure of rates derived from them. The challenge is to derive a model that handles all of these instruments together in a way that excludes the possibility of arbitrage among them. Taking T_∞ as the date of some finite horizon at or beyond the longest term of any bond, we regard the prices of unit discount bonds of the various maturities as a collection of stochastic processes. Specifically, for each maturity date $T \in (0, T_\infty]$ the evolving price $\{B(t,T)\}_{0\le t\le T}$ of a T-maturing unit bond is an adapted process on $(\Omega, \mathcal{F}, \{\mathcal{F}_t\}_{0\le t\le T_\infty}, \mathbb{P})$ that satisfies $0 < B(t,T)$ for $t \in [0,T]$ and $B(T,T) = 1$.[1] If V_T is some $\mathcal{F}_T$-measurable contingent claim with $E|V_T| < \infty$, there is, as usual in the absence of arbitrage, a measure $\hat{\mathbb{P}}$ (not necessarily unique) with respect to which normalized process $\{V_t^* = M_t^{-1}V_t\}_{0\le t\le T}$ is a martingale. We can therefore express the claim's current arbitrage-free value as $V_t = M_t\hat{E}_t(M_T^{-1}V_T)$. In particular, with $V_t = B(t,T)$ and $V_T = B(T,T) = 1$ we have

$$B(t,T) = M_t\hat{E}_t M_T^{-1} = e^{\int_0^t r_s \cdot ds}\hat{E}_t e^{-\int_0^T r_u \cdot du} = \hat{E}_t e^{-\int_t^T r_u \cdot du}. \tag{10.2}$$

Were the entire short-rate process $\{r_u\}_{0<u\le T_\infty}$ known in advance, as was assumed in earlier chapters, arbitrages between bonds and the money fund would enforce the equality $B(t,T) = M_t M_T^{-1}$, since both the T-maturing bond and the ($\mathcal{F}_0$- and hence) $\mathcal{F}_t$-measurable number M_T^{-1} of units of the money fund would be worth one currency unit at T. In that case relation $V_t = M_t\hat{E}_t(M_T^{-1}V_T)$ would be equivalent to

$$V_t = B(t,T)\hat{E}_t V_T = B(t,T)\hat{E}_t[B(T,T)^{-1}V_T],$$

and it would make no difference whether we chose $\{M_t\}_{0\le t\le T}$ or $\{B(t,T)\}_{0\le t\le T}$ as normalizing process. That equivalence no longer holds for measure $\hat{\mathbb{P}}$ when interest rates are uncertain, but there is an alternative measure, $\mathbb{P}^T$, under which it *is* true that $V_t = B(t,T)E_t^T V_T$. Of course, both $\hat{\mathbb{P}}$ and $\mathbb{P}^T$ are martingale measures; they merely correspond to different numeraires—$\{M_t\}$ and $\{B(t,T)\}$. Both measures will be useful in pricing interest-sensitive assets, and to distinguish them we refer to $\hat{\mathbb{P}}$ and $\mathbb{P}^T$, respectively, as the "spot (martingale) measure" and the "forward (martingale) measure for date T". It is important to note that a specific forward measure corresponds to each maturity date; thus, distinct forward measures $\mathbb{P}^{T_1}$ and $\mathbb{P}^{T_2}$ correspond to numeraires $\{B(t,T_1)\}_{0\le t\le T_1}$ and $\{B(t,T_2)\}_{0\le t\le T_2}$.

The connection between $\hat{\mathbb{P}}$ and $\mathbb{P}^T$ for a particular T can be seen as follows. With T_∞ as our finite horizon and $T \le T_\infty$, the two conditions $V_0 = M_0\hat{E}(M_T^{-1}V_T) = M_0\int \frac{V_T(\omega)}{M_T(\omega)} \cdot d\hat{\mathbb{P}}(\omega)$ and $V_0 = B(0,T)E^T V_T = B(0,T)\int V_T(\omega)\cdot d\mathbb{P}^T(\omega)$ imply that $\int V_T(\omega)\cdot d\mathbb{P}^T(\omega) = \int V_T(\omega)Q_T(\omega)\cdot$

[1] It would be desirable to impose $B(t,T) \le 1$ also, in conformity with assumption RB in chapter 4; but we shall see that this is violated in some otherwise attractive models.

$d\hat{\mathbb{P}}(\omega)$, with

$$Q_T \equiv \frac{d\mathbb{P}^T}{d\hat{\mathbb{P}}} = \frac{M_0}{M_T B(0,T)} = \frac{B(T,T)/B(0,T)}{M_T/M_0}. \tag{10.3}$$

Random variable (Radon-Nikodym derivative) Q_T thus effects the change of measure from $\hat{\mathbb{P}}$ to $\mathbb{P}^T$ that is appropriate for valuation at $t = 0$, as $E^T V_T = B(0,T)\hat{E}Q_T V_T$. Now for arbitrary $t \in [0,T]$ put

$$\begin{aligned} Q_t &\equiv \hat{E}_t Q_T \\ &= \frac{M_0}{B(0,T)} \hat{E}_t \frac{B(T,T)}{M_T} \\ &= \frac{M_0}{B(0,T)M_t} (\hat{E}_t e^{-\int_t^T r_u \cdot du}) = \frac{B(t,T)/B(0,T)}{M_t/M_0}, \end{aligned}$$

noting that $Q_0 = 1$. Notice that, as a conditional expectations process, $\{Q_t\}$ is *defined* so as to be a martingale under $\hat{\mathbb{P}}$. Formula (2.51) on page 72 for conditional expectations in alternative measures (with $\mathbb{P}^T, \hat{\mathbb{P}}$ here corresponding to $\hat{\mathbb{P}}, \mathbb{P}$ there) now gives as the time-t value of an $\mathcal{F}_T$-measurable claim.

$$V_t = B(t,T) E_t^T V_T = B(t,T) \frac{\hat{E}_t Q_T V_T}{\hat{E}_t Q_T} = B(t,T)\hat{E}_t(V_T Q_T / Q_t). \tag{10.4}$$

Thus,

$$\left\{ Q_T/Q_t = \frac{M_t M_T^{-1}}{B(t,T)} = \frac{M_t M_T^{-1}}{\hat{E}_t(M_t M_T^{-1})} = \frac{e^{-\int_t^T r_u \cdot du}}{\hat{E}_t e^{-\int_t^T r_u \cdot du}} \right\}_{0 \le t \le T} \tag{10.5}$$

is the appropriate Radon-Nikodym process for valuation under the forward martingale measure for T. Later, we shall see what specific form this takes under various models for interest rates and give some examples of its application.

Regarding notation, observe that the superscript on E defines the forward measure, while (as heretofore) the subscript indicates conditioning; thus, $E_t^T(\cdot) = E^T(\cdot \mid \mathcal{F}_t) = \int (\cdot) \cdot d\mathbb{P}^T(\omega \mid \mathcal{F}_t)$.

To see why the designation "forward measure" is apt, note that if $\{V_t\}$ represents the evolving *cum*-dividend value of a financial claim, then $B(t,T)^{-1}V_t = E_t^T V_T = \hat{E}_t(V_T Q_T/Q_t)$ is the forward price that would be set at t for delivery at T. Thus, even though interest rates are recognized to be stochastic, forward prices do equal expected future spot prices in this measure. In particular, if $V_t = B(t,T)$ is the value at t of a discount bond maturing at $T \geq t$, then for $t \leq t' \leq T$ we have

$$B_t(t', T) \equiv B(t, t')^{-1} V_t = E_t^{t'} V_{t'} = E_t^{t'} B(t', T).$$

Thus, the bond's forward price as of time t is the expectation under $\mathbb{P}^{t'}$ of its spot price at t'.

10.1.3 *A Preview of Things to Come*

Comparing (10.2) with (10.1) makes it clear that the evolution of bonds' prices can be described either through the evolution of short rates, $\{r_t\}_{0 \leq t \leq T_\infty}$, or through the evolution of instantaneous forward rates at all maturities, $\{r_t(T)\}_{0 \leq t \leq T \leq T_\infty}$. Short-rate, or "spot-rate", models either specify exogenously or deduce from an equilibrium framework a model for $\{r_t\}_{0 \leq t \leq T_\infty}$, typically as a (Markovian) diffusion process. Examples are the models of Vasicek (1977),

$$dr_t = (a - br_t) \cdot dt + \sigma \cdot dW_t,$$

and Cox, Ingersoll, and Ross (1985),

$$dr_t = (a - br_t) \cdot dt + \sigma r_t^{1/2} \cdot dW_t.$$

While modeling the short rate is easier than modeling all the forward rates, the disadvantage is that spot-rate models need not fit the existing term structure of bond prices; that is, they do not in general satisfy (10.2) for each T. Needless to say, it is undesirable that a model that might be used to price derivatives on bonds cannot price the bonds themselves. Hull and White (1990) show that it is possible to fit the term structure by allowing time dependence in one or more of the parameters a, b, σ. Another approach—the one on which we focus here—is to start from an initial forward-rate curve, $\{r_0(T)\}_{0 \leq T \leq T_\infty}$, and model the evolution from there of the entire term structure of instantaneous forward rates, $\{r_t(T)\}_{0 \leq t \leq T \leq T_\infty}$. This is done by representing $\{r_t(T)\}_{0 \leq t \leq T}$ for each T as an Itô process with its own characteristic trend and volatility:

$$r_t(T) - r_0(T) = \int_0^t \mu_s(T) \cdot ds + \int_0^t \boldsymbol{\sigma}_s(T)' \cdot d\mathbf{W}_s.$$

Obviously, in view of (10.1), this does assure that bonds' initial prices are correctly fit. In this setup $\mathbf{W}$ and $\boldsymbol{\sigma}$ may be vector-valued, allowing for independent influences on rates and thereby permitting forward rates at different maturities to be imperfectly correlated. Since $r_t(t) = r_t$, the model necessarily specifies the evolution of the spot rate as well. This forward-rate approach was pioneered by Heath, Jarrow, and Morton (HJM) (1992).

Because spot-rate models are more intuitive and are still important as foundations for later developments, we work out in the next section the prices of discount, default-free bonds in the models of Vasicek and Cox *et al.* While this is usually done by solving differential equations, we show that martingale methods are not hard to apply. Section 10.3 describes the general HJM approach and its implications for bond prices. To illustrate how the model can be used, we adopt a simple one-factor version and show how to work out analytically the prices of various interest-dependent contingent claims: options on the money fund, options on discount and coupon-paying bonds, caps and floors, and options on interest-rate swaps. Using the same setup, we will also see how to take account of stochastic rates in modeling futures prices and valuing options on assets that are not highly rate-sensitive. Section 10.4 treats the more recent LIBOR-market models that deliver the simple Black-Scholes-like formulas for interest rate caps that are commonly used by traders. Because these formulas are so simple and cap markets so active, the LIBOR-model framework has been found useful in calibrating models for other interest-sensitive derivatives. Finally, the last section of the chapter treats the pricing of bonds that are subject to default and thus gives an introduction to the literature on credit risk. We shall see there that the venerable spot-rate models to which we now turn still have a role to play in the pricing of debt instruments.

10.2 Spot-Rate Models

Most continuous-time models of the spot rate can be represented as Itô processes, in the form

$$dr_t = \mu_t \cdot dt + \boldsymbol{\sigma}_t \cdot d\mathbf{W}_t,$$

where $\{\mu_t\}$ and $\{\boldsymbol{\sigma}_t\}$ are appropriate scalar- and vector-valued processes and $\{\mathbf{W}_t\}$ is a vector-valued Brownian motion on $(\Omega, \mathcal{F}, \{\mathcal{F}_t\}, \mathbb{P})$. The vector formulation allows for multiple risk factors. So long as they meet the technical conditions for Itô processes, $\{\mu_t\}$ and $\{\boldsymbol{\sigma}_t\}$ can be specified freely to depend on time, on the short rate itself, and on other observable state variables so as to represent as well as possible the observed behavior of $\{r_t\}$ under the natural measure. Of course, the more elaborate the model the harder it is to work out explicit expressions for prices of bonds and interest-sensitive derivative assets. We describe here two single-factor Markovian models in which μ_t and σ_t depend just on r_t.

10.2.1 *Bond Prices under Vasicek*

Taking $\mu_t = a - br_t$ and $\sigma_t = \sigma$ for positive constants a, b, σ yields a mean-reverting model introduced by Vasicek (1977):

$$dr_t = (a - br_t) \cdot dt + \sigma \cdot dW_t. \tag{10.6}$$

An explicit expression for r_t comes by applying Itô's formula to $r_t e^{bt}$, giving

$$\begin{aligned} r_t e^{bt} - r_0 &= a \int_0^t e^{bs} \cdot ds + \sigma \int_0^t e^{bs} \cdot dW_s \\ &= \frac{a}{b}(e^{bt} - 1) + \sigma \int_0^t e^{bs} \cdot dW_s \\ r_t &= e^{-bt} r_0 + a\beta(t) + \sigma \int_0^t e^{-b(t-s)} \cdot dW_s, \end{aligned} \tag{10.7}$$

where we set $\beta(t) \equiv b^{-1}(1 - e^{-bt})$ for $0 \leq t$. The model thus presents $\{r_t\}$ as a Gaussian process, with $Er_t = e^{-bt} r_0 + a\beta(t)$ and $Cov(r_t, r_u) = \frac{\sigma^2}{2}[\beta(u+t) - \beta(u-t)]$ for $0 < t \leq u$. Note that this has the undesirable implications that spot rates can become negative and that prices of discount bonds can exceed unity.

An expression for the arbitrage-free price of a discount bond can be found via a hedging argument similar to that used in developing the Black-Scholes formulas for options. Since $\{r_t\}$ itself is a Markov process adapted to $\{\mathcal{F}_t\}$, the price at t of a T-maturing discount bond must be a function of t, T, and r_t alone. Assuming that it has the required smoothness, we can represent $\{B(t,T)\}_{0 \leq t \leq T}$ as an Itô process with $dB(t,T) = \mu(t,T,r_t) \cdot dt + \sigma(t,T,r_t) \cdot dW_t$, say. Consider a self-financing portfolio comprising at time t one T_1-maturing unit bond and $-p_t$ unit bonds that mature at T_2. The portfolio's value at t is $P_t = B(t,T_1) - p_t B(t,T_2)$, and, because it is self-financing, $dP_t = dB(t,T_1) - p_t \cdot dB(t,T_2)$. Because of the one-factor specification the bonds' prices are (instantaneously) perfectly correlated, so the stochastic part of dP_t disappears upon setting $p_t = -\sigma(t,T_1,r_t)/\sigma(t,T_2,r_t)$. Since the resulting portfolio is riskless, we must have $dP_t = r_t P_t \cdot dt$ in order to avoid arbitrage against the money fund. Expressing dP_t and simplifying yield the following condition on the bond processes:

$$\frac{\mu(t,T_1,r_t) - r_t B(t,T_1)}{\sigma(t,T_1,r_t)} = \frac{\mu(t,T_2,r_t) - r_t B(t,T_2)}{\sigma(t,T_2,r_t)}.$$

A relation like this must hold for all pairs of bonds maturing after t. The common value of the ratio represents the added per-unit compensation for bearing the instantaneous risk associated with discount bonds—i.e., it is the "price" of risk. Assuming this price to be a constant, γ, say,[2] and using Itô's formula to express $\mu(t,T,r_t) = B_t + B_r(a - br_t) + \frac{\sigma^2}{2}B_{rr}$ and $\sigma(t,T,r_t) = \sigma B_r$, we obtain the p.d.e. $0 = B_t + B_r(a - \gamma\sigma - br_t) + \frac{\sigma^2}{2}B_{rr} - r_t B$, where $B \equiv B(t,T)$ and subscripts on B represent partial derivatives. Of course, there is the terminal condition $B(T,T) = 1$. Guessing (correctly) that the solution has the form $B(t,T) = A(T-t)\exp[-\alpha(T-t)r_t]$, one can plug in and solve the resulting ordinary differential equations for A and α subject to $\ln A(0) = \alpha(0) = 0$.

As an alternative, let us find $B(t,T)$ by martingale methods. If we apply Girsanov's theorem and change from natural measure $\mathbb{P}$ to a measure $\hat{\mathbb{P}}$ in which $\{\hat{W}_t = W_t - \gamma t\}$ is a Brownian motion, our model for $\{r_t\}$ becomes $dr_t = (a^* - br_t)\cdot dt + \sigma \cdot d\hat{W}_t$, where $a^* \equiv a + \gamma\sigma$. The p.d.e. obtained by the hedging argument is then

$$0 = B_t + B_r(a^* - br_t) + \frac{\sigma^2}{2}B_{rr} - r_t B. \tag{10.8}$$

One gets the same p.d.e. by applying Itô's formula to $B(t,T)^* \equiv B(t,T)/M_t$, replacing a by a^*, and setting drift term $\mu(t,T,r_t) - r_t B$ equal to zero. This means that $\{B(t,T)^*\}$ is a martingale under $\hat{\mathbb{P}}$, so we can now identify $\hat{\mathbb{P}}$ as the spot martingale measure under which $\{M_t\}$-normalized price processes are martingales. It follows that

$$B(t,T) = M_t\hat{E}_t[M_T^{-1}B(T,T)] = \hat{E}_t \exp\left(-\int_t^T r_u \cdot du\right), \tag{10.9}$$

so that the bond's arbitrage-free price can be found by working out the expectation.[3]

Proceeding, from (10.7) we have under the new measure

$$\int_t^T r_u \cdot du = r_0[\beta(T) - \beta(t)] + a^* \int_t^T \beta(u)\cdot du + \sigma\int_t^T \int_0^u e^{-b(u-s)}\cdot d\hat{W}_s\, du. \tag{10.10}$$

[2]That the price of risk be $\mathcal{F}_0$-measurable is enough for what follows, but allowing time dependence would complicate the formulas to little purpose.

[3]To see directly that the last expression in (10.9) satisfies the p.d.e., note that under $\hat{\mathbb{P}}$ the conditional expectations process $\{\mathcal{E}(t,r_t)\} \equiv \{\hat{E}_t \exp(-\int_0^T r_s \cdot ds)\} = \{\exp(-\int_0^t r_s \cdot ds)B(t,T)\}$ is a martingale adapted to the filtration generated by $\{\hat{W}_t\}$. By the martingale representation theorem $\{\mathcal{E}(t,r_t)\}$ can be represented as an Itô process with zero drift. Applying Itô's formula and equating the drift to zero give $0 = \mathcal{E}_t + (a^* - br_t)\mathcal{E}_r + \frac{\sigma^2}{2}\mathcal{E}_{rr} - r_t\mathcal{E}$, which simplifies to (10.8).

The last integral can be evaluated as[4]

$$\begin{aligned}
&\left(\int_0^t \int_t^T + \int_s^t \int_t^T\right) e^{-bu} e^{bs} \cdot du\, d\hat{W}_s \\
&\quad = \frac{e^{-bt} - e^{-bT}}{b} \int_0^t e^{bs} \cdot d\hat{W}_s + \frac{1}{b} \int_t^T (1 - e^{-b(T-s)}) \cdot d\hat{W}_s \\
&\quad = \frac{\beta(T-t)}{\sigma} \left[\sigma \int_0^t e^{-b(t-s)} \cdot d\hat{W}_s\right] + \int_t^T \beta(T-s) \cdot d\hat{W}_s \\
&\quad = \frac{\beta(T-t)}{\sigma} [r_t - e^{-bt} r_0 - a^* \beta(t)] + \int_t^T \beta(T-s) \cdot d\hat{W}_s.
\end{aligned}$$

Combining the three terms on the right side of (10.10) and using identities

$$\beta(T-t)e^{-bt} = \beta(T) - \beta(t) \tag{10.11}$$

$$\int_t^T \beta(T-u) \cdot du = \int_t^T \beta(u) \cdot du - \beta(T-t)\beta(t) \tag{10.12}$$

yield

$$\int_t^T r_u \cdot du = a^* \int_t^T \beta(T-u) \cdot du + \sigma \int_t^T \beta(T-u) \cdot d\hat{W}_u + \beta(T-t) r_t. \tag{10.13}$$

This being normally distributed under $\hat{\mathbb{P}}$ with time-t conditional mean

$$a^* \int_t^T \beta(T-u) \cdot du + \beta(T-t) r_t$$

and conditional variance $\sigma^2 \int_t^T \beta(T-u)^2 \cdot du$, the bond's price is

$$B(t,T) = \exp\left[-a^* \int_t^T \beta(T-u) \cdot du + \frac{\sigma^2}{2} \int_t^T \beta(T-u)^2 \cdot du - \beta(T-t) r_t\right]. \tag{10.14}$$

Thus, in view of (10.7) the value of the discount bond at time t is distributed *ex ante* as lognormal.

Notice that parameters a, b, σ can be estimated from observations of the short rate alone, whereas the effect of risk price γ shows up only in the prices of traded, interest-sensitive assets, such as the discount bonds. Once a, b, σ are determined, γ can be identified directly from (10.14) and relation $a^* = a + \gamma\sigma$.

An application of Itô's formula shows the dynamics of $\{B(t,T)\}$ to be given by $dB(t,T)/B(t,T) = r_t \cdot dt - \sigma\beta(T-t) \cdot d\hat{W}_t$ under $\hat{\mathbb{P}}$ and by

[4]To see the first step, draw a picture.

$dB(t,T)/B(t,T) = [r_t + \gamma\sigma\beta(T-t)] \cdot dt - \sigma\beta(T-t) \cdot dW_t$ under the natural measure, the extra trend component in the latter case being the required compensation for risk. Recall from (10.5) that Radon-Nikodym process $\{Q_T/Q_t = B(t,T)^{-1} \exp(-\int_t^T r_u \cdot du)\}$ effects the change of measure from $\hat{\mathbb{P}}$ to forward measure $\mathbb{P}^T$. It is easy to see from (10.13) and (10.14) that Q_T/Q_t takes here the specific form

$$Q_T/Q_t = \exp\left[-\frac{\sigma^2}{2}\int_t^T \beta(T-u)^2 \cdot du - \sigma\int_t^T \beta(T-u) \cdot d\hat{W}_u\right].$$

This corresponds to the Doléans exponential for the change of measure that converts $\{\hat{W}_t - \sigma\int_0^t \beta(T-u) \cdot du \equiv W_t^T\}$ to a Brownian motion.

$\hat{E}B(t,T)$, the time-0 expected value of $B(t,T)$ under spot measure $\hat{\mathbb{P}}$, is

$$B_0(t,T)\exp\left\{\frac{\sigma^2}{2b}[-2\beta(\tau)\beta(t) + \frac{1}{2}\beta(2\tau)\beta(2t) + \frac{1}{2}b\beta(\tau)^2\beta(2t)]\right\},$$

where $\tau \equiv T-t$. This is less than forward price $B_0(t,T)$ except at $t=0$ and $t=T$. Since the expression does not involve a^*, the same result applies under the natural measure; however, $B_0(t,T)$ itself does depend on a^* and therefore on risk price γ. Figure 10.1 depicts the ratio $\hat{E}B(t,10)/B_0(t,10)$,

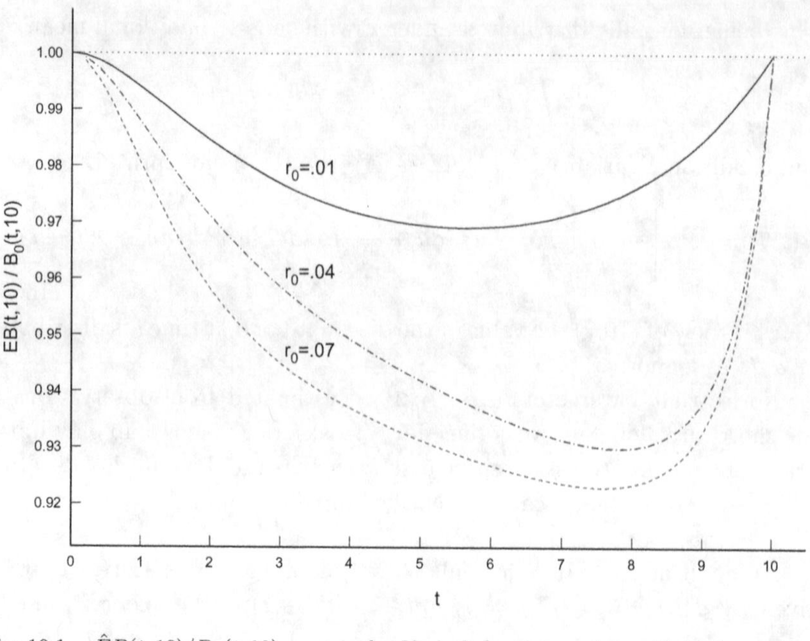

Fig. 10.1. $\hat{E}B(t,10)/B_0(t,10)$ *vs.* t under Vasicek for three values of initial spot rate, r_0. Parameters: $a^* = .025$, $b = .5$, $\sigma = .1$.

which is also independent of initial spot rate r_0. Of course, under forward measure $\mathbb{P}^T$ there is the usual result that $E^T B(t,T) = B_0(t,T)$.

Naturally, from the expressions for spot prices of bonds at all maturities one can determine all the various prices and rates derived from them: (i) average spot rates or yields to maturity, $r(t,T) = -(T-t)^{-1}\ln B(t,T)$; (ii) forward bond prices, $B_t(t',T) = B(t,T)/B(t',T)$; and (iii) average and instantaneous forward rates, $r_t(t',T) = -(T-t')\ln B_t(t',T)$ and $r_t(t') = -\partial \ln B_t(t',T)/\partial T|_{T=t'}$. For example, the instantaneous forward rate at t for loans at $T \geq t$ is

$$r_t(T) = a^*\beta(T-t) - \frac{\sigma^2}{2b}[2\beta(T-t) - \beta(2T-2t)] + r_t e^{-b(T-t)},$$

which of course reduces to r_t at $T=t$. This evolves as

$$\begin{aligned} dr_t(T) &= \sigma^2 e^{-b(T-t)}\beta(T-t)\cdot dt + \sigma e^{-b(T-t)}\cdot d\hat{W}_t \\ &= \sigma e^{-b(T-t)}[\sigma\beta(T-t) - \gamma]\cdot dt + \sigma e^{-b(T-t)}\cdot dW_t. \quad (10.15) \end{aligned}$$

Despite the model's simplicity, the mean-reverting feature of the instantaneous spot rate process allows for quite varied behavior of the term structure of yields to maturity. Specifically, as figure 10.2 shows, the yield curve (a

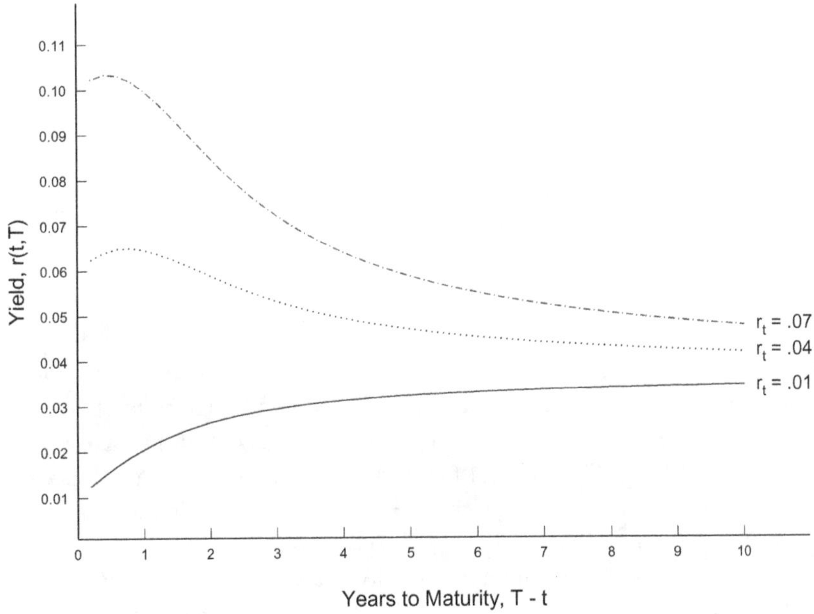

Fig. 10.2. Vasicek yields *vs.* $T-t$ for three values of initial spot rate r_t Parameters: $a^* = .025$, $b = .5$, $\sigma = .1$.

plot of $r(t,T)$ versus T) is strictly increasing when r_t is sufficiently small and hump shaped or decreasing when r_t is larger. As $T \to \infty$ the yield approaches $r(t,\infty) = \frac{a^*}{b} - \frac{\sigma^2}{2b^2}$, independently of r_t.

10.2.2 *Bond Prices under Cox, Ingersoll, Ross*

Adopting strategically simplifying assumptions about technology and preferences, Cox, Ingersoll, and Ross (1985) show that the short-rate process $\{r_t\}$ represented by $dr_t = (a - br_t)\cdot dt + \sigma r_t^{1/2}\cdot dW_t$ can result from a general equilibrium in which prices are generated by competitive, optimizing economic agents. This approach permits an interpretation of the parameters a, b, σ in terms of characteristics of individuals' tastes and of the technology that generates income. This deeper story not being relevant for our purposes, we consider the model strictly in an arbitrage framework, taking the mean-reverting square-root "CIR" process for $\{r_t\}$ as a primitive.

A hedging argument proceeds like that used for the Vasicek model but with the assumption that the price of risk is proportional to $\sqrt{r_t}$ rather than constant. Expressing $dB(t,T) = \mu(t,T)\cdot dt + \sigma(t,T)\cdot dW_t$ as before thus leads to the equality

$$\frac{\mu(t,T) - r_t B(t,T)}{\sigma(t,T)} = \gamma\sqrt{r_t}$$

for any bond maturing at $T > t$. We also get the following p.d.e. for its price: $0 = B_t + B_r(a - br_t - \sigma\gamma r_t) + \frac{\sigma^2}{2}B_{rr} - r_t B$. As in the Vasicek model, one can guess (and verify) that the solution has the form $B(t,T) = A(T-t)e^{-\alpha(T-t)r_t}$ and find the functions A and α by solving ordinary differential equations subject to $\ln A(0) = \alpha(0) = 0$. However, we will again find the solution by shifting to the spot martingale measure and evaluating $B(t,T)$ as $\hat{E}_t \exp(-\int_t^T r_u \cdot du)$. Assuming (as justified below) that $E\exp(\frac{1}{2}\int_0^{T_\infty} r_s \cdot ds) < \infty$, where T_∞ equals or exceeds the longest time to maturity of any bond, Girsanov's theorem authorizes a change of measure from $\mathbb{P}$ to $\hat{\mathbb{P}}$, in which $\{\hat{W}_t = W_t + \gamma\int_0^t \sqrt{r_s}\cdot ds\}$ is a Brownian motion on $[0, T_\infty]$. The process for $\{r_t\}$ can now be represented as $dr_t = (a - b^* r_t)\cdot dt + \sigma r_t^{1/2}\cdot d\hat{W}_t$ with $b^* = b + \gamma\sigma$, and the p.d.e. for $B(t,T)$ becomes $0 = B_t + B_r(a - b^* r_t) + \frac{\sigma^2}{2}B_{rr} - r_t B$. Again, the same p.d.e. is obtained under $\hat{\mathbb{P}}$ by applying Itô's formula to $B(t,T)^* = B(t,T)/M_t$, replacing b by b^*, and equating the drift to zero, thus confirming $\hat{\mathbb{P}}$ as the spot martingale measure.

We have encountered mean-reverting square-root processes already in sections 8.3.2 and 9.3.2 in connection with models of stochastic volatility and jump intensity, and we have seen that such processes are almost surely nonnegative. While this overcomes the salient shortcoming of the Vasicek model, the fact that $\{r_t\}$ is no longer Gaussian does make it harder to work out arbitrage-free prices of bonds and derivatives by martingale methods. Just what sort of process $\{r_t\}$ is was determined by Feller (1951), who develops the p.d.f. of r_t by solving the parabolic equation $\frac{\partial f}{\partial t} = \frac{\partial^2 (rf)}{\partial r^2}\frac{\sigma^2}{2} - \frac{\partial (a-b^* r)f}{\partial r}$ subject to an initial condition that specifies r_t at $t = 0$. The solution (here in the spot martingale measure) is[5]

$$\begin{aligned}\hat{f}(r;t,r_0) &= \theta_t \exp[-\theta_t(r + r_0 e^{-b^* t})] \left(\frac{r}{r_0 e^{-b^* t}}\right)^{\frac{a}{\sigma^2}-\frac{1}{2}} \\ &\quad \times I_{1-2a/\sigma^2}\left(2\theta_t \sqrt{r\, r_0 e^{-b^* t}}\right)\end{aligned}$$

for $r > 0$, where $\theta_t \equiv \frac{2b^*}{\sigma^2(1-e^{-b^* t})}$ and $I_q(\cdot)$ is the modified Bessel function of the first kind of order q. The p.d.f. satisfies the boundary condition $\hat{f}(0,t;r_0) < \infty$ so long as $a/\sigma^2 < 1/2$.

To gain insight into this expression, change variables as $x = 2\theta_t r$, $\lambda = 2\theta_t e^{-b^* t} r_0$, $\nu/4 = a/\sigma^2$, set $f(x;\lambda,\nu) = \hat{f}(r;t,r_0)|\frac{dr}{dx}|$, and use the fact that $I_q(\cdot) = I_{-q}(\cdot)$ to get

$$f(x;\lambda,\nu) = \frac{1}{2}e^{-(x+\lambda)/2}\left(\frac{x}{\lambda}\right)^{\nu/4-1/2} I_{\nu/2-1}\left(\sqrt{x\lambda}\right), \quad x > 0.$$

This is a common representation of the noncentral chi-squared p.d.f. with ν degrees of freedom and noncentrality parameter λ. Thus, in the original notation, one sees that r_t is $(2\theta_t)^{-1}$ times a noncentral chi-squared variate with $4a/\sigma^2$ d.f. and noncentrality $2\theta_t e^{-b^* t} r_0$.

Expressing the Bessel function in series form as in (2.24) gives another helpful expression,

$$\begin{aligned}f(x;\lambda,\nu) &= 2^{-\nu/2} e^{-(x+\lambda)/2} \sum_{j=0}^{\infty} \frac{x^{\nu/2+j-1}\lambda^j}{\Gamma(\nu/2+j)2^{2j} j!} \\ &= \sum_{j=0}^{\infty} \frac{(\lambda/2)^j}{j!} e^{-\lambda/2} f(x;\nu+2j),\end{aligned}$$

[5]This corresponds to Feller's expression (6.2) with $\sigma^2/2$ replacing a, $-b^*$ replacing b, and a replacing c. The factor $4b^2$ that appears in the first line of his expression should not be present.

where $f(\cdot;\nu+2j)$ is the central chi-squared p.d.f. with $\nu+2j$ d.f. This presents the noncentral chi-squared as a Poisson-directed mixture of central chi-squareds. Thus, $X \sim \chi^2(\nu+2Y)$ conditional on $Y \sim P(\lambda/2)$. This hierarchical representation makes it easy to calculate moments and generating functions/Laplace transforms. Specifically,

$$\begin{aligned}
EX &= E[E(X \mid Y)] = E(\nu+2Y) = \nu+\lambda \\
VX &= E[E(X^2 \mid Y)] - (\nu+\lambda)^2 = 2\nu+4\lambda \\
\mathcal{L}_X(\zeta) &= E[E(e^{-\zeta X} \mid Y)] = (1+2\zeta)^{-\nu/2}\exp\left(-\frac{\lambda\zeta}{1+2\zeta}\right), \quad \zeta \geq 0.
\end{aligned}$$

With these results we have

$$\begin{aligned}
\hat{E}r_t &= r_0 e^{-b^* t} + a\beta(t) \\
\hat{V}r_t &= \sigma^2\beta(t)[r_0 e^{-b^* t} + a\beta(t)/2] \\
\hat{\mathcal{L}}_{r_t}(\zeta) &= \left[1+\frac{\sigma^2}{2}\beta(t)\zeta\right]^{-2a/\sigma^2} \exp\left[-\frac{r_0 e^{-b^* t}\zeta}{1+\frac{\sigma^2}{2}\beta(t)\zeta}\right],
\end{aligned}$$

where $\beta(t) \equiv (1-e^{-b^* t})/b^*$.

We can now use the expression for the Laplace transform to develop a solution for $B(t,T)$. Setting $t_j = t + j(T-t)/n \equiv t + j\Delta t$, we have $\int_t^T r_s \cdot ds = \lim_{n\to\infty}\sum_{j=1}^n r_{t_j}\Delta t$ and, by dominated convergence,

$$\begin{aligned}
B(t,T) &= \lim_{n\to\infty} \hat{E}_t \exp\left(-\sum_{j=1}^{n} r_{t_j}\Delta t\right) \\
&= \lim_{n\to\infty} \hat{E}_t\left[\exp\left(-\sum_{j=1}^{n-2} r_{t_j}\Delta t\right) e^{-r_{t_{n-1}}\Delta t}\hat{E}_{t_{n-1}} e^{-r_{t_n}\Delta t}\right] \\
&= \lim_{n\to\infty}\left(1+\frac{\sigma^2}{2}\beta(\Delta t)\Delta t\right)^{-2a/\sigma^2} \\
&\quad \cdot \hat{E}_t \exp\left[-\sum_{j=1}^{n-2} r_{t_j}\Delta t - r_{t_{n-1}}\Delta t\left(1+\frac{e^{-b^*\Delta t}}{1+\frac{\sigma^2}{2}\beta(\Delta t)\Delta t}\right)\right].
\end{aligned}$$

If we now define $\alpha_{0,n} \equiv \alpha(T-t_{n-1})$ to be Δt and for $j \in \{1,2,\ldots,n\}$ set

$$\begin{aligned}
\alpha_{j,n} &\equiv \alpha(T-t_{n-j-1}) = \Delta t + \frac{e^{-b^*\Delta t}\alpha_{j-1,n}}{1+\frac{\sigma^2}{2}\beta(\Delta t)\alpha_{j-1,n}}, \\
A_{j,n} &\equiv A(T-t_{n-j}) = \prod_{i=1}^{j}\left(1+\frac{\sigma^2}{2}\beta(\Delta t)\alpha_{i-1,n}\right)^{-2a/\sigma^2},
\end{aligned}$$

then the last expression for $B(t,T)$ develops as

$$\begin{aligned}
B(t,T) &= \lim_{n\to\infty} A_{1,n}\hat{E}_t \exp\left(-\sum_{j=1}^{n-2} r_{t_j}\Delta t\right)\hat{E}_{t_{n-2}} e^{-r_{t_{n-1}}\alpha_{1,n}} \\
&= \lim_{n\to\infty} A_{2,n}\hat{E}_t \exp\left(-\sum_{j=1}^{n-3} r_{t_j}\Delta t\right)\hat{E}_{t_{n-3}} e^{-r_{t_{n-2}}\alpha_{2,n}} \\
&\ \ \vdots \\
&= \lim_{n\to\infty} A_{n,n} e^{-r_t \alpha_{n,n}}.
\end{aligned}$$

With $A(T-t) \equiv \lim_{n\to\infty} A_{n,n}$ and $\alpha(T-t) = \exp(-r_t \lim_{n\to\infty} \alpha_{n,n})$ the bond's price is then given by[6]

$$B(t,T) = A(T-t)e^{-r_t\alpha(T-t)}. \tag{10.16}$$

We solve first for $\alpha(T-t)$. Noting that $\beta(\Delta t) = \Delta t + o(\Delta t)$ as $\Delta t \to 0$, the recursion for $\alpha_{j,n}$ yields difference equation

$$\begin{aligned}
\alpha_{j,n} - \alpha_{j-1,n} &= \Delta t - \alpha_{j-1,n}(1 - e^{-b^*\Delta t}) - \frac{\sigma^2}{2}\alpha_{j-1,n}^2\Delta t + o(\Delta t) \\
&= \left[1 - b^*\alpha_{j-1,n} - \frac{\sigma^2}{2}\alpha_{j-1,n}^2\right]\Delta t + o(\Delta t)
\end{aligned}$$

and corresponding differential equation

$$\frac{d\alpha(\tau)}{d\tau} = 1 - b^*\alpha(\tau) - \frac{\sigma^2}{2}\alpha(\tau)^2.$$

The solution subject to $\alpha(0) = 0$ is

$$\alpha(\tau) = \frac{e^{2c\tau} - 1}{c(e^{2c\tau}+1) + \frac{b^*}{2}(e^{2c\tau}-1)} = \frac{\sinh(c\tau)}{c\cosh(c\tau) + \frac{b^*}{2}\sinh(c\tau)}, \tag{10.17}$$

where $\tau \equiv T - t$ and $c \equiv \frac{1}{2}\sqrt{b^{*2} + 2\sigma^2}$.

---

[6]Noting that noncentral chi-squared variates possess m.g.f.s, the Novikov condition $E\exp(\int_0^{T\infty} r_t \cdot dt) < \infty$ that justified the change to the martingale measure can now be verified by steps similar to those above.

Next, for $A(T-t)$ we have

$$\begin{aligned}\ln A(T-t) &= -\frac{2a}{\sigma^2}\lim_{n\to\infty}\sum_{j=1}^{n}\ln\left[1+\frac{\sigma^2}{2}a_{j-1,n}\Delta t + o(\Delta t)\right]\\ &= -a\lim_{n\to\infty}\sum_{j=1}^{n}[a_{j-1,n}\Delta t + o(\Delta t)]\\ &= a\int_t^T \alpha(T-u)\cdot du,\end{aligned}$$

so that $d\ln A(T-t)/dt = -d\ln A(\tau)/d\tau = -a\alpha(\tau)$. Solving this subject to $A(0)=1$ gives

$$A(\tau) = \left[\frac{c\alpha(\tau)e^{b^*\tau/2}}{\sinh(c\tau)}\right]^{2a/\sigma^2}. \tag{10.18}$$

An expression for $\hat{E}B(t,T)$ can be found by evaluating Laplace transform $\hat{\mathcal{L}}_{r_t}(\cdot)$ at $\zeta=\alpha(\tau)$, as

$$\begin{aligned}\hat{E}B(t,T) &= B_0(t,T)\frac{A(\tau)A(t)}{A(T)}\left[1+\frac{\sigma^2}{2}\beta(t)\alpha(\tau)\right]^{-2a/\sigma^2}\\ &\quad\cdot\exp\left\{-r_0\left[\frac{e^{-b^*t}\alpha(\tau)}{1+\sigma^2\beta(t)\alpha(\tau)/2}-\alpha(T)+\alpha(t)\right]\right\}.\end{aligned}$$

The expectation under $\hat{\mathbb{P}}$ is less than forward price $B_0(t,T)$ for $t\in(0,T)$. Figure 10.3 plots $\hat{E}B(t,10)/B_0(t,10)$ versus t for three values of the initial short rate. Of course, $E^T B(t,T) = B_0(t,T)$ for $t\in[0,T]$ under forward measure $\mathbb{P}^T$.

The average spot rate or yield to maturity on a $T=(t+\tau)$-maturing discount bond is

$$r(t,T) = -\frac{\ln A(\tau)}{\tau} + \frac{\alpha(\tau)}{\tau}r_t.$$

As $\tau\to\infty$ this approaches $r(t,\infty)\equiv\frac{a}{\sigma^2}[\sqrt{b^{*2}+2\sigma^2}-b^*]$ independently of r_t. As was true of the Vasicek model, the dynamics allow for yield curves of varying shapes (figure 10.4). In the same way, too, we can determine risk coefficient γ from bond prices once the remaining parameters have been estimated from observations of the short rate.

10.3 A Forward-Rate Model

Rather than base the term structure of prices and yields on a model for the spot rate alone, Heath, Jarrow, and Morton (HJM) (1992) undertake

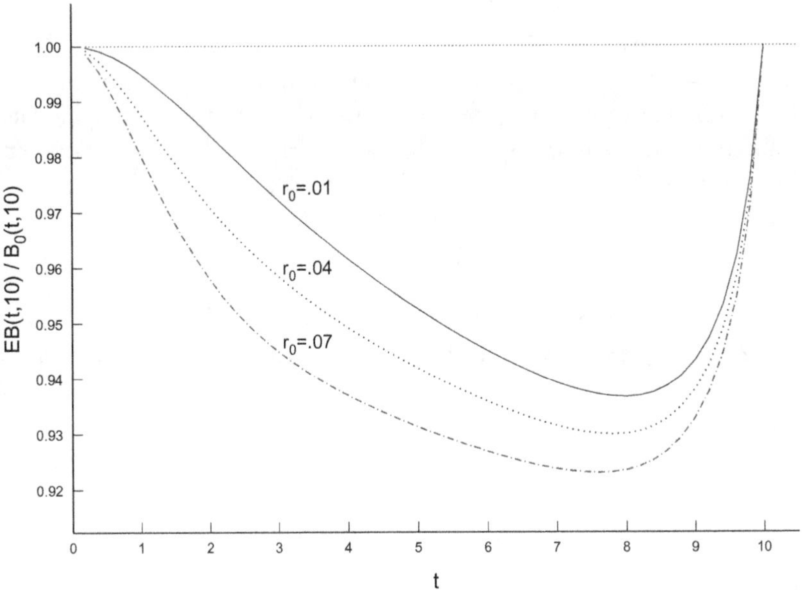

Fig. 10.3. $\hat{E}B(t,10)/B_0(t,10)$ *vs.* t under CIR model for three values of initial spot rate, r_0. Parameters: $a = .025$, $b^* = .5$, $\sigma = .8$.

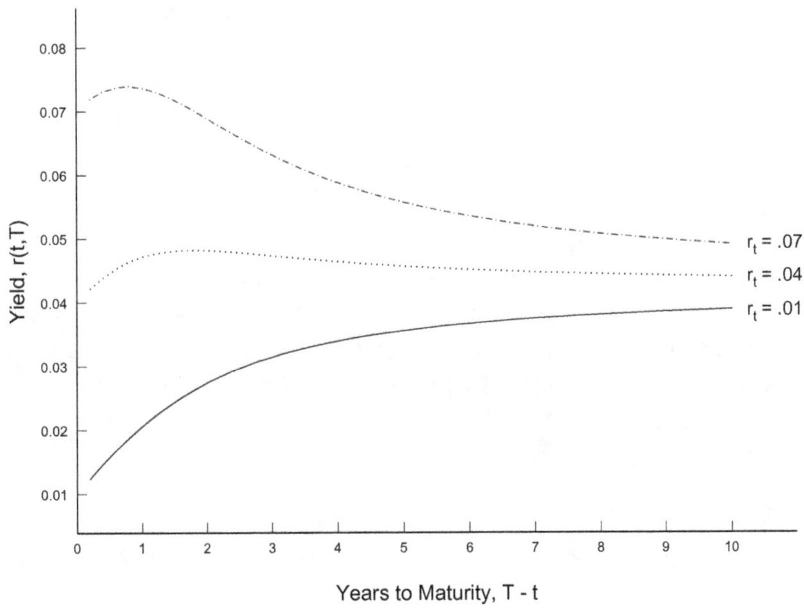

Fig. 10.4. CIR yields *vs.* $T - t$ for three values of initial spot rate, r_t Parameters: $a = .025$, $b^* = .5$, $\sigma = .8$.

to model the evolution of instantaneous forward rates across the entire spectrum of maturities. To get across the idea, we start with a one-factor version in which all forward rates respond to a single stochastic influence. After extending to multiple risk factors we return to the simple one-factor setting to illustrate how the model works to price various interest-sensitive financial instruments.

10.3.1 *The One-Factor HJM Model*

Taking as given the initial instantaneous forward rates at all maturities, $\{r_0(T)\}_{0\le T\le T_\infty}$, we begin by specifying the evolution of the various forward-rate processes, as

$$r_t(T)-r_0(T) = \int_0^t \mu_s(T)\cdot ds+\int_0^t \sigma_s(T)\cdot dW_s, 0\le t\le T\le T_\infty, \tag{10.19}$$

or, in differential form, as

$$dr_t(T) = \mu_t(T)\cdot dt + \sigma_t(T)\cdot dW_t.$$

In this one-factor version $\{W_t\}_{t\ge 0}$ is a (scalar-valued) Brownian motion on the filtered probability space $(\Omega, \mathcal{F}, \{\mathcal{F}_t\}, \mathbb{P})$. For each T the drift and diffusion are adapted processes, $\{\mu_t(T)\}_{0\le t\le T}$, $\{\sigma_t(T)\}_{0\le t\le T}$, possibly depending on present and past values of interest rates, and such that the respective integrals exist.[7] Since the time-t forward rate for lending at t is just the instantaneous spot rate—i.e., $r_t(t) = r_t$, the same framework implicitly defines the behavior of the short rate also, as $dr_t = \mu_t(t)\cdot dt+\sigma_t(t)\cdot dW_t$ or

$$r_t = r_0(t) + \int_0^t \mu_s(t)\cdot ds + \int_0^t \sigma_s(t)\cdot dW_s. \tag{10.20}$$

Notice that in this one-factor form instantaneous changes in the short rate and in the instantaneous forward rates at all future lending dates are perfectly correlated, each depending on the change in the single Brownian motion that drives the entire structure. Of course, the same is true of the one-factor versions of the spot-rate models of the previous section.

The Evolution of Bond Prices

Unlike short rates and forward rates, discount bonds themselves are traded assets. We will now see that a special structure must be imposed on the

---

[7]That is, such that $\mathbb{P}[\int_0^T |\mu_s(T)|\cdot ds < \infty] = \mathbb{P}[\int_0^T \sigma_s(T)^2\cdot ds < \infty] = 1$.

forward-rate drift processes if the bond market is to be arbitrage free. Writing $\ln B(t,T)$ in terms of the time-t instantaneous forward rates as

$$\begin{aligned}\ln B(t,T) &= -\int_t^T r_t(u)\cdot du\\ &= -\int_t^T \left[r_0(u) + \int_0^t \mu_s(u)\cdot ds + \int_0^t \sigma_s(u)\cdot dW_s\right]\cdot du\end{aligned}$$

and then reversing the order of integration give

$$\ln B(t,T) = -\int_t^T r_0(u)\cdot du - \int_0^t \mu_s(t,T)\cdot ds - \int_0^t \sigma_s(t,T)\cdot dW_s,$$

where

$$\begin{aligned}\mu_s(t,T) &\equiv \int_t^T \mu_s(u)\cdot du\\ \sigma_s(t,T) &\equiv \int_t^T \sigma_s(u)\cdot du.\end{aligned}$$

We now want to express the differential $dB(t,T)$. Note that $d\mu_s(t,T) = -\mu_s(t)\cdot dt$ and $d\sigma_s(t,T) = -\sigma_s(t)\cdot dt$, so[8]

$$\begin{aligned}d\int_0^t \mu_s(t,T)\cdot ds &= \mu_t(t,T) - \int_0^t \mu_s(t)\cdot ds\ dt\\ d\int_0^t \sigma_s(t,T)\cdot dW_s &= \sigma_t(t,T)\cdot dW_t - \int_0^t \sigma_s(t)\cdot dW_s\ dt.\end{aligned}$$

Therefore,

$$\begin{aligned}d\ln B(t,T) &= \left[r_0(t) + \int_0^t \mu_s(t)\cdot ds + \int_0^t \sigma_s(t)\cdot dW_s\right]\cdot dt\\ &\quad - \mu_t(t,T)\cdot dt - \sigma_t(t,T)\cdot dW_t\\ &= r_t\cdot dt - \mu_t(t,T)\cdot dt - \sigma_t(t,T)\cdot dW_t.\end{aligned}$$

The first term on the right supplies the customary trend in price as the bond gets closer to maturity, while the second and third terms pick up the effects of slope and shifts in the term structure, respectively. Finally, Itô's formula applied to $B(t,T) = \exp[\ln B(t,T)]$ gives

$$dB(t,T)/B(t,T) = [r_t - \mu_t(t,T) + \sigma_t(t,T)^2/2]\cdot dt - \sigma_t(t,T)\cdot dW_t. \qquad (10.21)$$

[8]The relation in the second line, the stochastic equivalent of Leibnitz' formula, follows at once from the definition of the stochastic integral.

We know that in an arbitrage-free market there is a measure $\hat{\mathbb{P}}$, equivalent to natural measure $\mathbb{P}$, under which values of traded assets relative to the per-unit price of the money fund, M_t, are martingales. Let us now see what restrictions this places on the processes $\{\mu_t(T)\}$ and $\{\sigma_t(T)\}$. Itô's formula applied to $B(t,T)^* \equiv B(t,T)/M_t$ gives

$$\begin{aligned} dB(t,T)^* &= dB(t,T)/M_t - B(t,T)^* dM_t/M_t \\ &= B(t,T)^*[dB(t,T)/B(t,T) - dM_t/M_t] \\ &= B(t,T)^*\{[-\mu_t(t,T) + \sigma_t(t,T)^2/2] \cdot dt - \sigma_t(t,T) \cdot dW_t\} \end{aligned}$$

or

$$dB(t,T)^*/B(t,T)^* = \gamma_t(t,T)\sigma_t(t,T) \cdot dt - \sigma_t(t,T) \cdot dW_t,$$

where $\gamma_s(t,T) \equiv -\mu_s(t,T)/\sigma_s(t,T) + \sigma_s(t,T)/2$. Now imagine, first, that $\gamma_s(t,T)$ were independent of t, so that we could write $\gamma_s(t,T) = \gamma_s(T)$, say; and, second, that the $\{\gamma_s(T)\}$ process were such that

$$E \exp\left(\int_0^t \gamma_s(T) \cdot dW_s - \frac{1}{2}\int_0^t \gamma_s(T)^2 \cdot ds\right) = 1. \tag{10.22}$$

Then Girsanov's theorem would give us a measure $\hat{\mathbb{P}}^T$, say (not to be confused with forward measure $\mathbb{P}^T$), such that $\hat{W}_t^T = -\int_0^t \gamma_s(T) \cdot ds + W_t$ is a Brownian motion. Under these conditions and in this new measure we would have

$$d\hat{W}_t^T = -\gamma_t(T) \cdot dt + dW_t \tag{10.23}$$

and

$$dB(t,T)^*/B(t,T)^* = -\sigma_t(t,T) \cdot d\hat{W}_t^T.$$

$B(t,T)^*$ would then be a continuous local martingale and, moreover, a martingale proper under Novikov condition $\hat{E} \exp[\frac{1}{2}\int_0^T \sigma_t(t,T)^2 \cdot dt] < \infty$.

However, the existence of process $\{\gamma_s(T)\}_{0 \le s \le T}$ and measure $\hat{\mathbb{P}}^T$ does not suffice to rule out arbitrage across the entire spectrum of available bonds. As the notation indicates, both of these are specific to the maturity date of the bond. For there to exist a single spot measure $\hat{\mathbb{P}}$ that applies to bonds of all maturities it is necessary that processes $\{\gamma_s(T)\}_{0 \le s \le T}$ be the same for all $T \in (0, T_\infty]$. This implies that $-\mu_s(t,T)/\sigma_s(t,T) + \sigma_s(t,T)/2 = \gamma_s$, say, where γ_s is independent of both t and T. This, in turn, imposes a restriction on the original drift and diffusion processes in the forward-rate model; namely, that

$$\int_t^T \mu_s(u) \cdot du = -\left[\int_t^T \sigma_s(u) \cdot du\right] \gamma_s + \frac{1}{2}\left[\int_t^T \sigma_s(u) \cdot du\right]^2$$

for $0 \leq s \leq T \leq T_\infty$ and some process $\{\gamma_s\}$ satisfying (10.22). Differentiating with respect to T yields the condition

$$\mu_s(T) = -\sigma_s(T)\gamma_s + \sigma_s(T)\sigma_s(t,T), \tag{10.24}$$

which must hold for all maturities T. Inspecting (10.21), we see that the mean drift of the bond price is $r_t + \gamma_t\sigma_t(t,T)$, giving γ_t the usual interpretation as the compensation under $\mathbb{P}$ for bearing the instantaneous risk associated with holding the discount bond. The assumption $\gamma_s(t,T) = \gamma_s$ thus amounts to requiring that the "price" of risk (although obviously not the amount of risk) be the same for bonds of all maturities. Of course, the same assumption was implicit in the development of the spot-rate models of the previous section.[9] Under this condition we then have

$$d\hat{W}_t = -\gamma_t \cdot dt + dW_t \tag{10.25}$$

for all $0 \leq t \leq T \leq T_\infty$ under the single spot measure $\hat{\mathbb{P}}$. This guarantees that there are no opportunities for arbitrage among the spectrum of bonds and the money fund.

Combining (10.24) and (10.25) with (10.19) gives for the time-t instantaneous forward rate at T under $\hat{\mathbb{P}}$

$$\begin{aligned} r_t(T) &= r_0(T) + \int_0^t [-\sigma_s(T)\gamma_s + \sigma_s(T)\sigma_s(s,T)] \cdot ds \\ &\quad + \int_0^t \sigma_s(T) \cdot (d\hat{W}_s + \gamma_s\, ds) \\ &= r_0(T) + \int_0^t \sigma_s(T)\sigma_s(s,T) \cdot ds + \int_0^t \sigma_s(T) \cdot d\hat{W}_s \end{aligned} \tag{10.26}$$

for $0 \leq t \leq T \leq T_\infty$. The representation for the short rate then follows upon setting $T = t$:

$$r_t = r_0(t) + \int_0^t \sigma_s(t)\sigma_s(s,t) \cdot ds + \int_0^t \sigma_s(t) \cdot d\hat{W}_s. \tag{10.27}$$

[9]It is easy to see that the one-factor model of Vasicek does satisfy (10.24). Referring to (10.15) we have $\sigma_t(T) = \sigma e^{-b(T-t)}$, $\sigma_t(t,T) = \sigma\beta(T-t)$, and $\mu_t(T) = \sigma_t(T)\sigma_t(t,T) - \sigma_t(T)\gamma$, so that $\mu_t(T)/\sigma_t(T) - \sigma_t(t,T) = -\gamma$.

Valuing Contingent Claims

Having at hand an expression for the short rate, we can express the value of the money fund as $M_t = M_0 \exp(\int_0^t r_s \cdot ds)$. Valuation now proceeds in the same way that it did for the spot-rate models. That is, since $\{M_t\}$-normalized values of traded assets are martingales under $\hat{\mathbb{P}}$, the value at t of a claim worth (an integrable quantity) V_T at $T \geq t$ satisfies

$$M_t^{-1} V_t = \hat{E}_t M_T^{-1} V_T,$$

so that

$$V_t = \hat{E}_t \left(e^{-\int_t^T r_u \cdot du} V_T \right). \tag{10.28}$$

Alternatively, we could work in forward measure $\mathbb{P}^T$ and find V_t as

$$V_t = B(t,T) E_t^T V_T = B(t,T) \hat{E}_t (V_T Q_T / Q_t),$$

where $Q_T/Q_t = M_t M_T^{-1} / B(t,T)$. As usual, attitudes toward risk are involved only indirectly through the change of measure from $\mathbb{P}$ to $\hat{\mathbb{P}}$ or to $\mathbb{P}^T$. Typically, evaluating the expectations must be done numerically in the HJM framework, but we will see shortly that analytical solutions are indeed possible when $\sigma_t(T)$ is simply specified.

10.3.2 *Allowing Additional Risk Sources*

In the one-factor model a single Brownian motion drives prices and yields of bonds of all maturities. This implies that yields (that is, average spot rates $r(t,T)$) on long- and short-maturity bonds always move in the same direction. In fact, yield curves do sometimes twist, with short rates moving up as long yields decline, or *vice versa*. Capturing this feature requires at least two risk sources. While this extension makes the model harder to implement and requires some additional technical conditions, we shall see that it poses no conceptual difficulty.

Taking $\boldsymbol{\sigma}_s(T)$ and $\mathbf{W}_s$ in (10.19) to be k vectors, the motion of instantaneous forward rates under $\mathbb{P}$ becomes

$$r_t(T) - r_0(T) = \int_0^t \mu_s(T) \cdot ds + \int_0^t \boldsymbol{\sigma}_s(T)' \cdot d\mathbf{W}_s$$

for $0 \leq t \leq T \leq T_\infty$. Written out, this is

$$r_t(T) - r_0(T) = \int_0^t \mu_s(T) \cdot ds + \sum_{j=1}^k \int_0^t \sigma_{js}(T) \cdot dW_{js}, \tag{10.29}$$

where the $\{W_{jt}\}$ are independent Brownian motions. The processes $\{\sigma_{jt}(T)\}$ can be specified so as to allow individual risk sources to exert different degrees of influence on rates of different term. For example, HJM (1992) propose as a two-factor model

$$r_t(T) - r_0(T) = \int_0^t \mu_s(T) \cdot ds + \sigma_1 W_{1t} + \sigma_2 e^{-\theta T} \int_0^t e^{\theta s} \cdot dW_{2s},$$

with $\theta > 0$. Here $\{W_{1t}\}$ affects the entire term structure uniformly, whereas the influence of $\{W_{2t}\}$ declines as the term of lending increases.

Paralleling the development of (10.21), the motion of the T-maturing bond is described by

$$dB(t,T)/B(t,T) = [r_t - \mu_t(t,T) + \boldsymbol{\sigma}_t(t,T)'\boldsymbol{\sigma}_t(t,T)/2] \cdot dt - \boldsymbol{\sigma}_t(t,T)' \cdot d\mathbf{W}_t$$

or

$$dB(t,T)/B(t,T) = \left[r_t - \mu_t(t,T) + \sum_{j=1}^{k} \frac{\sigma_{jt}(t,T)^2}{2} \right] \cdot dt - \sum_{j=1}^{k} \sigma_{jt}(t,T) \cdot dW_{jt},$$

where

$$\mu_s(t,T) \equiv \int_t^T \mu_s(u) \cdot du,$$

$$\sigma_{js}(t,T) \equiv \int_t^T \sigma_{js}(u) \cdot du$$

and $\boldsymbol{\sigma}_t(t,T) \equiv [\sigma_{1t}(t,T), \ldots, \sigma_{kt}(t,T)]'$. As before, we want to find conditions for the existence of a measure under which $\{B^*(t,T) = M_t^{-1}B(t,T)\}$ is a martingale. Normalizing by M_t removes r_t from the drift, but to eliminate the remaining part there must be a vector of risk prices $\boldsymbol{\gamma}_t(T) = [\gamma_{1t}(T), \ldots, \gamma_{kt}(T)]'$ that solves

$$-\mu_t(t,T) + \boldsymbol{\sigma}_t(t,T)'\boldsymbol{\sigma}_t(t,T)/2 = \boldsymbol{\sigma}_t(t,T)'\boldsymbol{\gamma}_t(T), \tag{10.30}$$

and there must exist a measure $\hat{\mathbb{P}}^T$ under which $\hat{\mathbf{W}}_t = \mathbf{W}_t - \int_0^t \boldsymbol{\gamma}_s(T) \cdot ds$ is a k-dimensional Brownian motion. Moreover, in order for the measure and the vector of risk prices in the economy to be unique, such a relation as (10.30) must hold simultaneously for bonds of all maturities. This requires the risk prices for different maturities to be the same, so that $\boldsymbol{\gamma}_t(T) = \gamma_t$ for all $T \in (t, T_\infty]$, and it restricts the drifts as

$$\mu_t(T) = -\boldsymbol{\sigma}_t(T)'\boldsymbol{\gamma}_t + \boldsymbol{\sigma}_t(T)'\boldsymbol{\sigma}_t(t,T)$$

for all $t \in [0, T]$. Further conditions on γ_t and $\boldsymbol{\sigma}_t(t,T)$ that

$$E \exp\left(\int_0^t \gamma_s' \cdot d\mathbf{W}_s - \frac{1}{2}\int_0^t \gamma_s'\gamma_s \cdot ds\right) = 1$$

and

$$E \exp\left[\frac{1}{2}\int_0^T \boldsymbol{\sigma}_t(t,T)'\boldsymbol{\sigma}_t(t,T) \cdot dt\right] < \infty$$

then assure that the unique $\hat{\mathbb{P}}$ exists under which each $\{B^*(t,T)\}$ is a martingale, with

$$dB(t,T)^*/B(t,T)^* = -\boldsymbol{\sigma}_t(t,T)' \cdot d\hat{\mathbf{W}}_t. \tag{10.31}$$

10.3.3 *Implementation and Applications*

To see how the HJM model can be put to use, let us return to the one-factor version and choose a specification that makes it relatively easy to develop some analytical results.

Setting $\sigma_t(T) = \sigma$, an $\mathcal{F}_0$-measurable constant independent of time and maturity, presents the basic forward-rate process as

$$r_t(T) - r_0(T) = \int_0^t \mu_s(T) \cdot ds + \sigma W_t, 0 \le t \le T \le T_\infty.$$

This is a continuous-time version of the discrete-time model proposed by Ho and Lee (1986). Taking $\sigma_t(t,T) \equiv \int_t^T \sigma_t(u) \cdot du = \sigma(T-t)$ in expression (10.24), the no-arbitrage condition on the drift becomes

$$\mu_t(T) = -\gamma_t\sigma + \sigma^2(T-t);$$

and, corresponding to (10.26) and (10.27), the instantaneous forward rate for time T and the short rate under the spot measure are now

$$r_t(T) = r_0(T) + \sigma^2 t(T - t/2) + \sigma\hat{W}_t \tag{10.32}$$

$$r_t = r_0(t) + \sigma^2 t^2/2 + \sigma\hat{W}_t. \tag{10.33}$$

A price paid for this simplicity is that forward rates and short rates are now normally distributed and therefore unbounded below, as in Vasicek's model. The benefit here, as there, is that we can produce explicit expressions for arbitrage-free prices of some interest-sensitive instruments. This nice tractability owes to the fact that bond prices and the value of the money fund are now *ex ante* lognormally distributed. Thus,

$$\begin{aligned}B(t,T) &= \exp\left(-\int_t^T r_t(u)\cdot du\right)\\ &= \exp\left[-\int_t^T r_0(u)\cdot du - \sigma^2 t\int_t^T (u-t/2)\cdot du - \sigma(T-t)\hat{W}_t\right]\\ &= B_0(t,T)\exp[-\sigma^2(T-t)tT/2 - \sigma(T-t)\hat{W}_t], \qquad (10.34)\end{aligned}$$

where $B_0(t,T) = \exp[-\int_t^T r_0(u)\cdot du]$ is the T-maturing bond's initial time-t forward price. The dynamics of the bond's price under $\hat{\mathbb{P}}$ are thus

$$dB(t,T)/B(t,T) = r_t\cdot dt - \sigma(T-t)\cdot d\hat{W}_t.$$

Recalling (10.25), these can be stated in terms of a Brownian motion under the natural measure as

$$B(t,T) = B_0(t,T)\exp\left[\sigma(T-t)\int_0^t \gamma_s\cdot ds - \sigma^2(T-t)tT/2 - \sigma(T-t)W_t\right].$$

and

$$dB(t,T)/B(t,T) = [r_t + \sigma(T-t)\gamma_t]\cdot dt - \sigma(T-t)\cdot dW_t.$$

Notice that in neither measure does the expectation of the spot price at t equal initial forward price $B_0(t,T)$. In particular,

$$\hat{E}B(t,T) = B_0(t,T)\exp[-\sigma^2t^2(T-t)/2] < B_0(t,T)$$

for $t\in(0,T)$.

The value of the money fund at t is

$$\begin{aligned}M_t &= M_0 e^{\int_0^t r_s\cdot ds}\\ &= M_0\exp\left\{\int_0^t [r_0(s)+\sigma^2s^2/2]\cdot ds + \sigma\int_0^t \hat{W}_s\cdot ds\right\}\\ &= M_0B(0,t)^{-1}\exp\left(\sigma^2t^3/6 + \sigma\int_0^t \hat{W}_s\cdot ds\right), \qquad (10.35)\end{aligned}$$

so that (from example 33 in chapter 3)

$$\int_0^t \hat{W}_s\cdot ds \sim N(0,t^3/3) \qquad (10.36)$$

and

$$\ln(M_t/M_0) \sim N[-\ln B(0,t) + \sigma^2t^3/6, \sigma^2t^3/3].$$

The Radon-Nikodym process $\{Q_T/Q_t\}$ that changes from the spot martingale measure to the forward martingale measure at T is given by

$$\begin{aligned}\frac{Q_T}{Q_t} &= B(t,T)^{-1}M_T^{-1}M_t \\ &= B(t,T)^{-1}\exp\left\{-\int_t^T [r_0(u)+\sigma^2u^2/2+\sigma\hat{W}_u]\cdot du\right\} \\ &= \exp\left[-\frac{\sigma^2}{6}(T-t)^3-\sigma\int_t^T \hat{W}_u\cdot du+\sigma(T-t)\hat{W}_t\right] \\ &= \exp\left[-\frac{\sigma^2}{2}\int_t^T (T-u)^2\cdot du-\sigma\int_t^T (T-u)\cdot d\hat{W}_u\right].\end{aligned}$$

This is simply the Doléans exponential that changes $\hat{W}_t-\sigma\int_0^t(T-u)\cdot du = \hat{W}_t-\sigma t(T-t/2)\equiv W_t^T$ to a Brownian motion. Referring to (10.33), we see that $r_0(T)=E^T r_t(T)$ in measure $\mathbb{P}^T$ and that $r_0(t)=E^t r_t$ in measure $\mathbb{P}^t$. Thus, if we work in the appropriate forward martingale measure, instantaneous forward rates are unbiased estimators of future spot rates.

Pricing a Sure Cash Receipt

As a warm-up in applying the model, let us verify that valuation formula (10.28) leads to a price for the bond that is consistent with (10.34). Taking $V_T=B(T,T)=1$ as the terminal value, we have $M_t^{-1}B(t,T)=\hat{E}_tM_T^{-1}$ and

$$\begin{aligned}B(t,T) &= \hat{E}_te^{-\int_t^T r_u\cdot du} \\ &= \hat{E}_t\exp\left(-\int_t^T r_0(u)\cdot du-\sigma^2\int_t^T u^2/2\cdot du-\sigma\int_t^T \hat{W}_u\cdot du\right) \\ &= B_0(t,T)\exp[-\sigma^2(T^3-t^3)/6]\,\hat{E}_t\exp\left(-\sigma\int_t^T \hat{W}_u\cdot du\right).\end{aligned}$$

To evaluate the expectation, write

$$\begin{aligned}\int_t^T \hat{W}_u\cdot du &= \int_t^T \hat{W}_t\cdot du+\int_t^T (\hat{W}_u-\hat{W}_t)\cdot du \\ &= (T-t)\hat{W}_t+\int_0^{T-t}\tilde{W}_s\cdot ds,\end{aligned}$$

where $\{\tilde{W}_t\}$ is a Brownian motion independent of $\hat{W}_t$. Since $\int_0^{T-t} \tilde{W}_s \cdot ds$ is distributed as $N[0,(T-t)^3/3]$, we have

$$\hat{E}_t \exp\left(-\sigma \int_t^T \hat{W}_u \cdot du\right) = \exp[\sigma^2(T-t)^3/6 - \sigma(T-t)\hat{W}_t]$$

and

$$B(t,T) = B_0(t,T)\exp[-\sigma^2 tT(T-t)/2 - \sigma(T-t)\hat{W}_t]$$

as in (10.34). Of course, working in the forward measure makes the result trivial, for (10.4) gives

$$V_t = B(t,T)E^T B(T,T) = B(t,T)\hat{E}_t(Q_T/Q_t) = B(t,T).$$

European Options on the Money Fund

Letting M_0 be the arbitrary initial value of one unit of the money fund, the value at $t=0$ of a European call worth $(M_T - X)^+$ at T is

$$\begin{aligned} C^E(M_0,T) &= M_0\hat{E}[M_T^{-1}(M_T - X)^+] \\ &= \hat{E}(M_0 - XM_0M_T^{-1})^+ \\ &= \hat{E}[M_0 - XB(0,T)e^{-\sigma^2T^3/6-\sigma\hat{Y}_T}]^+, \end{aligned} \tag{10.37}$$

where the last follows from (10.35) with $\int_0^T \hat{W}_s \cdot ds \equiv \hat{Y}_T$. Since $\hat{Y}_T \sim Z\sqrt{T^3/3}$ under $\hat{\mathbb{P}}$, where $Z \sim N(0,1)$, the bracketed expression is positive when

$$Z > z_X \equiv \frac{-\ln\left[\frac{M_0}{B(0,T)X}\right] - \sigma^2T^3/6}{\sigma\sqrt{T^3/3}}.$$

Thus,

$$C^E(M_0,T) = M_0\hat{E}\mathbf{1}_{(z_X,\infty)}(Z) - XB(0,T)\hat{E}e^{-\sigma^2T^3/6-\sigma\sqrt{T^3/3}\cdot Z}\mathbf{1}_{(z_X,\infty)}(Z).$$

Writing out the second expectation as an integral, completing the square, and simplifying yield the following formula for the value of the call:

$$\begin{aligned} C^E(M_0,T) = M_0\Phi&\left\{\frac{\ln\left[\frac{M_0}{B(0,T)X}\right] + \frac{\sigma^2T^3}{6}}{\sigma\sqrt{T^3/3}}\right\} \\ &- B(0,T)X\Phi\left\{\frac{\ln\left[\frac{M_0}{B(0,T)X}\right] - \frac{\sigma^2T^3}{6}}{\sigma\sqrt{T^3/3}}\right\}. \end{aligned}$$

This corresponds to the Black-Scholes formula for a T-expiring call on an underlying worth M_0 and having volatility $\sigma T/\sqrt{3}$. In the usual "q" notation this is

$$C^E(M_0, T) = M_0\Phi[q^+(M_0/BX; \sigma T/\sqrt{3})] - BX\Phi[q^-(M_0/BX; \sigma T/\sqrt{3})].$$

Working instead in the forward measure as in (10.4), we would have

$$\begin{aligned} C^E(M_0, T) &= B(0,T)E^T(M_T - X)^+ \\ &= B(0,T)\hat{E}[(M_T - X)^+ Q_T] \\ &= B(0,T)\frac{M_0\hat{E}[M_T^{-1}(M_T - X)^+]}{B(0,T)} \\ &= \hat{E}(M_0 - XM_0M_T^{-1})^+ \end{aligned}$$

as in (10.37), from which the rest follows.

Options on Discount Bonds

Let us now apply (10.28) to determine the price at time 0 of a t-expiring European call struck at X on a discount bond that matures at $T \geq t$. The initial value of the underlying is $B(0,T)$, and we can state the call's initial value as

$$C^E[B(0,T), t] = B(0,t)E^t[B(t,T) - X]^+ = M_0\hat{E}M_t^{-1}[B(t,T) - X]^+.$$

Using (10.33) and (10.34), this equals

$$B(0,t)e^{-\sigma^2 t^3/6}\hat{E}e^{-\sigma\hat{Y}_t}[B_0(t,T)e^{-\sigma^2 tT\tau/2 - \sigma\tau\hat{W}_t} - X]^+, \qquad (10.38)$$

where $\hat{Y}_t \equiv \int_0^t \hat{W}_s \cdot ds$ and $\tau \equiv T - t$. We recognize that $\hat{Y}_t \sim N(0, t^3/3)$ under $\hat{\mathbb{P}}$, but in evaluating the expectation we must take account of the dependence between $\hat{Y}_t$ and $\hat{W}_t$. The next step is to work out their joint distribution under $\hat{\mathbb{P}}$.

This is easily done by developing the joint c.f.,

$$\Psi(\zeta_1, \zeta_2) = \hat{E}\exp(i\zeta_1\hat{Y}_t + i\zeta_2\hat{W}_t).$$

The definition of the Riemann integral implies that $\hat{Y}_t = t \cdot \lim_{n\to\infty}\frac{1}{n}\sum_{j=1}^n \hat{W}_{t_j}$, where $t_j \equiv tj/n$. Applying dominated convergence, the c.f. can be expressed as

$$\Psi(\zeta_1, \zeta_2) = \lim_{n\to\infty}\hat{E}\exp\left(i\zeta_1 t \cdot n^{-1}\sum_{j=1}^n \hat{W}_{t_j} + i\zeta_2\hat{W}_t\right).$$

The terms $tn^{-1}\sum_{j=1}^{n}\hat{W}_{t_j}$ and $\hat{W}_t$ are clearly bivariate normal with zero means. Example 33 on page 98 shows that the first term has variance $t^3n^{-2}(n+1)(2n+1)/6$, while the variance of the second is, of course, t. Since the covariance under $\hat{\mathbb{P}}$ is

$$tn^{-1}\sum_{j=1}^{n}\hat{E}\hat{W}_{t_j}\hat{W}_t = tn^{-1}\sum_{j=1}^{n}t^2j/n = t^2n^{-1}(n+1)/2,$$

we conclude that

$$\begin{aligned}\Psi(\zeta_1,\zeta_2) &= \lim_{n\to\infty}\exp\left\{-\frac{1}{2}\left[\zeta_1^2\frac{t^3(n+1)(2n+1)}{6n^2}+2\zeta_1\zeta_2\frac{t^2(n+1)}{2n}+\zeta_2^2t\right]\right\}\\ &= \exp\left[-\frac{1}{2}(\zeta_1^2t^3/3+2\zeta_1\zeta_2t^2/2+\zeta_2^2t)\right]\end{aligned}$$

and therefore that $\hat{Y}_t$ and $\hat{W}_t$ are jointly normal with

$$E\hat{W}_t\hat{Y}_t \equiv E\hat{W}_t\int_0^t\hat{W}_s\cdot ds = t^2/2. \tag{10.39}$$

The variances being $t^3/3$ and t, the correlation is $\rho=\sqrt{3}/2$.

Recall from section 2.2.7 that when random variables X_1 and X_2 are bivariate normal with zero means, variances σ_1^2 and σ_2^2, and correlation ρ, then X_1 is distributed as $\rho\frac{\sigma_1}{\sigma_2}X_2+\sigma_1\sqrt{1-\rho^2}Z$, where Z is standard normal and independent of X_2. It follows in the present instance that $\hat{Y}_t$ can be decomposed as

$$\hat{Y}_t = \frac{t}{2}\hat{W}_t+\frac{t}{2\sqrt{3}}\hat{U}_t, \tag{10.40}$$

where $\{\hat{U}_t\}$ is an independent Brownian motion under $\hat{\mathbb{P}}$. Applying this decomposition to (10.38) gives

$$C^E[B(0,T),t]=B(0,t)e^{-\frac{\sigma^2t^3}{8}}\hat{E}\left[B_0(t,T)e^{-\frac{\sigma^2tT_\mathcal{T}}{2}-\sigma(T-\frac{t}{2})\hat{W}_t}-e^{-\frac{\sigma t}{2}\hat{W}_t}X\right]^+.$$

The bracketed expression is positive whenever

$$\frac{\hat{W}_t}{\sqrt{t}} < z_X \equiv \frac{\ln[B_0(t,T)/X]}{\sigma_\mathcal{T}\sqrt{t}}-\frac{\sigma T\sqrt{t}}{2}.$$

Setting $Z\equiv\hat{W}_t/\sqrt{t}\sim N(0,1)$ under $\hat{\mathbb{P}}$ gives for the expectation in the above expression

$$B_0(t,T)\hat{E}[e^{-\sigma^2tT_\mathcal{T}/2-\sigma(T-t/2)\sqrt{t}Z}\mathbf{1}_{(-\infty,z_X)}(Z)]-X\hat{E}[e^{-\sigma t^{3/2}Z/2}\mathbf{1}_{(-\infty,z_X)}(Z)].$$

Writing the expectations in integral form, completing the squares in the exponents, and recalling that the bond's forward price is $B_0 \equiv B_0(t,T) = B(0,T)/B(0,t)$, we obtain for $C^E[B(0,T),t]$ the expression

$$B(0,t)\{B_0(t,T)\Phi[q^+(B_0/X;\sigma_B)] - X\Phi[q^-(B_0/X;\sigma_B)]\}, \qquad (10.41)$$

where

$$q^\pm(B_0/X;\sigma_B) \equiv \frac{\ln(B_0/X) \pm \sigma_B^2 t/2}{\sigma_B\sqrt{t}}$$

and $\sigma_B \equiv \sigma_T \equiv \sigma(T-t)$ is the volatility of the bond's price at t.

To value the European put, we can work from put-call parity:

$$\begin{aligned} P^E[B(0,T),t] &= C^E[B(0,T),t] - M_0\hat{E}\left\{\frac{[B(t,T)-X]^+ - [X-B(t,T)]^+}{M_t}\right\} \\ &= C^E[B(0,T),t] - \hat{E}\left[\frac{B(t,T)-X}{M_t/M_0}\right] \\ &= C^E[B(0,T),t] - \hat{E}\left(e^{-\int_0^t r_s\cdot ds}\hat{E}_t e^{-\int_t^T r_s\cdot ds}\right) - \hat{E}e^{-\int_0^t r_s\cdot ds}X \\ &= C^E[B(0,T),t] - \hat{E}e^{-\int_0^T r_s\cdot ds} - B(0,t)X \\ &= C^E[B(0,T),t] - B(0,t)[B_0(t,T) - X]. \end{aligned}$$

Finally, (10.41) and the relation

$$-q^\pm(x;\sigma_B) = q^\mp(x^{-1};\sigma_B),$$

yield the following solution for $P^E[B(0,T),t]$:

$$B(0,t)\{X\Phi[q^+(X/B_0;\sigma_B)] - B_0(t,T)\Phi[q^-(X/B_0;\sigma_B)]\}. \qquad (10.42)$$

Recall now expressions (6.27) and (6.28), which are the forward-price forms of the Black-Scholes formulas for European options on an underlying whose price $\{S_t\}$ follows geometric Brownian motion. Letting σ_S denote the volatility and restating the formulas to represent the time-0 prices of calls and puts expiring at t, (6.27) and (6.28) become

$$P^E(S_0,t) = B(0,t)\{X\Phi[q^+(X/\mathsf{f};\sigma_S)] - \mathsf{f}\Phi[q^-(X/\mathsf{f};\sigma_S)]\} \qquad (10.43)$$

$$C^E(S_0,t) = B(0,t)\{\mathsf{f}\Phi[q^+(\mathsf{f}/X;\sigma_S)] - X\Phi[q^-(\mathsf{f}/X;\sigma_S)]\}, \qquad (10.44)$$

where $\mathsf{f} \equiv \mathsf{f}(0,t)$ is the forward price for delivery of the underlying at t. A comparison with (10.41) and (10.42) shows that under the Ho-Lee version of the HJM model the Black-Scholes formulas still generate the correct prices for options on discount bonds.

Options on Coupon-Paying Bonds

A coupon bond makes regular cash payments to the bearer, typically equal to a fixed proportion c of the principal value, then returns the principal value at maturity. Taking the principal value to be unity, c represents the actual periodic cash payment. Clearly, one can view the coupon bond as a portfolio of discount bonds with maturities corresponding to the payment dates. The value at $t \in [0, T_n]$ of a bond that pays cash coupon c at dates $T, T_2, \ldots, T_n$ and returns the principal value at T_n is

$$B(t, c; T_1, \ldots, T_n) = c \sum_{j=j_t}^{n} B(t, T_j) + B(t, T_n), \tag{10.45}$$

where T_{j_t} is the first payment date after t. Interestingly, in the one-factor HJM model the portfolio characterization applies not only to the coupon bond itself but also to European options written on it. In other words, a European option on a coupon bond can be valued as a portfolio of options on discount bonds that mature at the various payment dates.

The explanation for this hinges on the fact that under model (10.32) prices of discount bonds of all maturities can be expressed as functions of a single variable, the current short rate. This means that the price of the coupon bond depends just on the short rate as well. To see that these things are true, combine (10.33) and (10.34), to get

$$B(t, T_j) \mid_{r_t} = B_0(t, T_j) \exp\left\{ -\frac{\sigma^2}{2}(T_j - t)^2 t - (T_j - t)[r_t - r_0(t)] \right\}.$$

Now, since $B(t, T_j) \mid_{r_t} \uparrow +\infty$ as $r_t \to -\infty$ (negative short rates being possible in this model) and $B(t, T_j) \mid_{r_t} \downarrow 0$ as $r_t \to +\infty$, a unique value of r_t will correspond to each positive $B(t, T_j)$, and the same is true of the coupon bond as well.

Considering, then, a European call struck at X and expiring at $t \leq T_n$, there is some critical r_t—call it r_t^X—such that $B(t, c; T_1, \ldots, T_n) \mid_{r_t^X} = X$. Letting $X_j \equiv B(t, T_j) \mid_{r_t^X}$ be the corresponding value of $B(t, T_j)$, we have $X = c \sum_{j=j_t}^{n} X_j + X_n$; and the time-zero value of a t-expiring European call, $C^E[B(0, c; T_1, \ldots, T_n), t; X]$, can then be found as

$$\begin{aligned}
&M_0 \hat{E} M_T^{-1} [B(t, c; T_1, \ldots, T_n) - X]^+ \\
&\quad = \hat{E} e^{-\int_0^t r_s \cdot ds} [B(t, c; T_1, \ldots, T_n) - X] \mathbf{1}_{(-\infty, r_t^X)}(r_t) \\
&\quad = \hat{E} e^{-\int_0^t r_s \cdot ds} \left\{ c \sum_{j=j_t}^{t_n} [B(t, T_j) - X_j] + [B(t, T_n) - X_n] \right\} \mathbf{1}_{(-\infty, r_t^X)}(r_t)
\end{aligned}$$

$$= c \sum_{j=j_t}^{T_n} \hat{E} e^{-\int_0^t r_s \cdot ds}[B(t,T_j) - X_j]^+ + \hat{E} e^{-\int_0^t r_s \cdot ds}[B(t,T_n) - X_n]^+.$$

This delivers the promised portfolio characterization as

$$C^E[B(0,c;T_1,\ldots,T_n),t;X] = c\sum_{j=j_t}^{T_n} C^E[B(0,T_j),t;X_j] + C^E[B(0,T_n),t;X_n],$$

from which the arbitrage-free price of the call on the coupon bond can be found by using (10.41) to value each of the terms.

Futures/Forward Prices under Stochastic Rates

Recall that the practice of marking to market distributes the gains or losses on futures positions over time, whereas with forward contracts they are concentrated at the delivery date. Apart from default risk (which we ignore), this difference in timing is the only relevant distinction between futures and forwards from the standpoint of valuation. However, we saw in chapter 4 that even the difference in timing does not matter if there is no uncertainty about the future course of interest rates over the terms of the contracts. Under that condition, futures prices and forward prices would still be the same in arbitrage-free, frictionless markets. We are now in a position to see how things can change when interest rates and bond prices over the lives of these contracts are not predictable.

The following simple setup suffices to get across the main idea. Let $\{\mathsf{F}(t,T)\}_{0\le t\le T}$ be the futures price for a contract initiated at $t = 0$ and expiring at T, and assume that the position is to be marked to market at just one intermediate time, t^*. As usual $\{S_t\}_{t\ge 0}$ represents the price of the underlying asset or commodity, for which there is assumed to be a continuous, deterministic cost of carry at constant rate κ. For example, if S_t is the value in domestic currency of a position in a foreign certificate of deposit, then $-\kappa$ would be the guaranteed rate of interest. Upon marking to market at t^*, the futures position, which is worth nothing initially, acquires positive or negative value $\mathsf{F}(t^*,T) - \mathsf{F}(0,T)$; and there is another increment of $\mathsf{F}(T,T) - \mathsf{F}(t^*,T) = S_T - \mathsf{F}(t^*,T)$ at date T. In the absence of arbitrage the futures price at $t = t^*$ and at $t = 0$ must therefore satisfy, respectively,

$$\hat{E}_{t^*} e^{-\int_{t^*}^T r_t \cdot dt}[S_T - \mathsf{F}(t^*,T)] = 0 \qquad (10.46)$$

and

$$\hat{E} e^{-\int_0^{t^*} r_t \cdot dt}[\mathsf{F}(t^*,T) - \mathsf{F}(0,T)] + \hat{E} e^{-\int_0^T r_t \cdot dt}[S_T - \mathsf{F}(t^*,T)] = 0. \qquad (10.47)$$

Equation (10.47) is the key to the relation between $\mathsf{F}(0,T)$ and $\mathsf{f}(0,T)$, the price that would be set at $t=0$ on a T-expiring forward contract. That the second term in (10.47) is zero is seen by writing it as

$$\hat{E}\left\{e^{-\int_0^{t^*} r_t \cdot dt}\hat{E}_{t^*}e^{-\int_{t^*}^T r_t \cdot dt}[S_T - \mathsf{F}(t^*,T)]\right\}$$

and applying (10.46). The initial futures price therefore is given by

$$\mathsf{F}(0,T) = B(0,t^*)^{-1}\hat{E}e^{-\int_0^{t^*} r_t \cdot dt}\mathsf{F}(t^*,T).$$

It is clear that $\mathsf{F}(t^*,T)$ must be the same as the forward price at t^*, since from this point the two contracts have precisely the same payoff streams. Therefore, applying cost-of-carry relation (4.10),

$$\mathsf{F}(t^*,T) = \mathsf{f}(t^*,T) = B(t^*,T)^{-1}S_{t^*}e^{\kappa(T-t^*)}.$$

To progress from here and relate initial values $\mathsf{F}(0,T)$ and $\mathsf{f}(0,T)$ requires placing some structure on the underlying and on the bond price. For the latter we shall stick with the one-factor HJM model, and we shall assume that $\{S_t\}$ evolves as

$$dS_t/S_t = (r_t + \kappa)\cdot dt + \sigma_S \cdot d\tilde{W}_t.$$

Here $\{\tilde{W}_t\}_{t\geq 0}$ is a Brownian motion under $\hat{\mathbb{P}}$, distinct from but possibly correlated with the $\{\hat{W}_t\}_{t\geq 0}$ that drives interest rates. Then

$$S_{t^*} = S_0 \exp\left(\int_0^{t^*} r_t \cdot dt + \kappa t^* - \sigma_S^2 t^*/2 + \sigma_S \tilde{W}_{t^*}\right)$$

and

$$\begin{aligned}\mathsf{F}(0,T) &= B(0,t^*)^{-1}\hat{E}\left[e^{-\int_0^{t^*} r_t \cdot dt}B(t^*,T)^{-1}S_{t^*}e^{\kappa(T-t^*)}\right]\\ &= B(0,t^*)^{-1}S_0e^{\kappa T}\hat{E}\left[B(t^*,T)^{-1}e^{-\sigma_S^2 t^*/2+\sigma_S\tilde{W}_{t^*}}\right].\end{aligned}$$

From (10.34) the price of the unit bond at t^* is

$$B(t^*,T) = B(0,t^*)^{-1}B(0,T)e^{-\sigma^2(T-t^*)t^*T/2-\sigma(T-t^*)\hat{W}_{t^*}},$$

and so

$$\mathsf{F}(0,T) = [B(0,T)^{-1}S_0e^{\kappa T}]e^{\sigma^2(T-t^*)t^*T/2-\sigma_S^2 t^*/2}\hat{E}e^{\sigma(T-t^*)\hat{W}_{t^*}+\sigma_S\tilde{W}_{t^*}}.$$

The factor in brackets is the forward price, $\mathsf{f}(0,T)$. To develop the remaining part, take ρ to be the instantaneous correlation between $\{\hat{W}_t\}$ and $\{\tilde{W}_t\}$ and express $\tilde{W}_{t^*}$ as $\rho\hat{W}_{t^*} + \sqrt{1-\rho^2}\hat{U}_{t^*}$,where $\{\hat{U}_t\}_{t\geq 0}$ and $\{\hat{W}_t\}_{t\geq 0}$ are independent Brownian motions under $\hat{\mathbb{P}}$. Expression (10.33) shows that ρ

can be interpreted as the correlation between the short rate and the instantaneous return on the underlying asset. Evaluating the expectation and simplifying then give

$$\mathsf{F}(0,T) = \mathsf{f}(0,T)\exp\{\sigma t^*(T-t^*)[\rho\sigma_S + \sigma(T-t^*/2)]\}. \tag{10.48}$$

It is now possible to see what this particular model implies about the relation between forward and futures prices. Obviously, they must agree if either $t^* = 0$ or $t^* = T$, since the futures is then not marked to market between the present time and expiration, and (10.48) is clearly consistent with this. Putting $\sigma = 0$ shows that the two prices are equal also when there is no uncertainty in future bond prices, as we had already seen in chapter 4. The additional structure imposed here does provide an important new insight, however: when $t^* \in (0,T)$ and $\sigma > 0$ the futures price is unambiguously greater than the forward price so long as the return on the underlying is not negatively correlated with the short rate. The intuition for this in the case $\rho > 0$ was already provided in chapter 4. That is, when the underlying return and the short rate tend to move in concert, intermediate gains received on a long position in futures as it is marked to market can usually be reinvested at higher short rates; and, conversely, there is usually lower opportunity cost associated with losses on the futures. Since the terminal value of the futures position is then greater when $\rho > 0$ than otherwise, it does make sense that the futures price would be higher than the forward price under this condition. That $\mathsf{F}(0,T) > \mathsf{f}(0,T)$ in this model even when $\rho = 0$ is particularly interesting, since it is not so intuitive. Notice that in this case the futures price no longer depends on the volatility of the underlying, σ_S.

Equity/Index Options under Stochastic Rates

Let us now see how to price a T-expiring European option on an underlying stock or index when interest rates are stochastic. Taking S_0 as the initial price, assume that

$$dS_t/S_t = (r_t - \delta_t)\cdot dt + \sigma_S \cdot d\tilde{W}_t,$$

under $\hat{\mathbb{P}}$, where δ_t is a deterministic dividend rate. The price at date T is then

$$S_T = S_0 e^{-\int_0^T \delta_t \cdot dt} \exp\left(\int_0^T r_t \cdot dt - \sigma_S^2 T/2 + \sigma_S \tilde{W}_T\right).$$

Since the short rate is stochastic, two risk factors now generally drive the underlying price. Some work is needed to sort these out in a way that lets us price an option.

Using (10.35), the cumulative short rate is

$$\int_0^T r_t \cdot dt = -\ln B(0,T) + \sigma^2 T^3/6 + \sigma \int_0^T \hat{W}_t \cdot dt,$$

and therefore

$$S_T = \mathsf{f} \exp\left(\sigma^2 T^3/6 - \sigma_S^2 T/2 + \sigma \int_0^T \hat{W}_t \cdot dt + \sigma_S \tilde{W}_T\right), \tag{10.49}$$

where $\mathsf{f} \equiv \mathsf{f}(0,T) = B(0,T)^{-1} S_0 e^{-\int_0^T \delta_t \cdot dt}$ is the forward price for T delivery of the underlying. As in the analysis of futures and forwards let ρ be the correlation between $\{\hat{W}_t\}$ and $\{\tilde{W}_t\}$ and write

$$\int_0^T \hat{W}_t \cdot dt = \rho \int_0^T \tilde{W}_t \cdot dt + \sqrt{1-\rho^2} \int_0^T \hat{U}_t \cdot dt,$$

where $\{\hat{U}_t\}$ is a Brownian motion independent of $\{\tilde{W}_t\}$. Since $\hat{E}\tilde{W}_T \int_0^T \tilde{W}_t \cdot dt = T^2/2$ by (10.39), it follows that $\hat{E}\tilde{W}_T \int_0^T \hat{W}_t \cdot dt = \rho T^2/2$. Then, recognizing that $\int_0^T \hat{W}_t \cdot dt \sim N(0, T^3/3)$ and $\tilde{W}_T \sim N(0,T)$, we can write $\sigma \int_0^T \hat{W}_t \cdot dt + \sigma_S \tilde{W}_T$ as

$$\sigma_* \sqrt{T} Z_* \equiv \sigma_r \sqrt{T} Z_r + \sigma_S \sqrt{T} Z_S,$$

say, where (Z_r, Z_S) are bivariate standard normal with $\hat{E} Z_r Z_S \equiv \rho_{rS} = \rho\sqrt{3}/2$, $\sigma_r^2 \equiv \sigma^2 T^2/3$ and $\sigma_*^2 \equiv \sigma_r^2 + 2\rho_{rS}\sigma_r\sigma_S + \sigma_S^2$. Finally, note that

$$\begin{aligned}
\rho_{rS}\sigma_r\sigma_S &= \hat{E}(\sigma_r Z_r \cdot \sigma_S Z_S) \\
&= \hat{E}[\sigma_r Z_r(\sigma_* Z_* - \sigma_r Z_r)] \\
&= \rho_{r*}\sigma_r\sigma_* - \sigma_r^2,
\end{aligned}$$

where $\rho_{r*} \equiv \hat{E} Z_r Z_*$. With these conventions S_T can be expressed as

$$\begin{aligned}
S_T &= \mathsf{f} \exp\left[(\sigma_r^2 - \sigma_S^2)T/2 + \sigma_* \sqrt{T} Z_*\right] \\
&= \mathsf{f} \exp\left[(\sigma_r^2 + \rho_{rS}\sigma_r\sigma_S - \sigma_*^2/2)T + \sigma_* \sqrt{T} Z_*\right] \\
&= \mathsf{f} \exp\left[(\rho_{r*}\sigma_r\sigma_* - \sigma_*^2/2)T + \sigma_* \sqrt{T} Z_*\right].
\end{aligned} \tag{10.50}$$

Putting this to work to value a T-expiring European call, we have

$$\begin{aligned}
C^E(S_0,T) &= \hat{E} e^{-\int_0^T r_t \cdot dt}(S_T - X)^+ \\
&= B(0,T) e^{-\sigma_r^2 T/2} \hat{E} e^{-\sigma_r \sqrt{T} Z_r}\left[\mathsf{f} e^{(\rho_{r*}\sigma_r\sigma_* - \sigma_*^2/2)T + \sigma_* \sqrt{T} Z_*} - X\right]^+.
\end{aligned}$$

Since Z_r and Z_* are bivariate normal with correlation ρ_{r*}, Z_r can be written as

$$Z_r = \rho_{r*} Z_* + \sqrt{1-\rho_{r*}^2}\, Z_{**},$$

where Z_{**} is standard normal and independent of Z_*. Splitting Z_r in this way and setting

$$\hat{E} e^{-\sigma_r \sqrt{T}\sqrt{1-\rho_{r*}^2} Z_{**}} = e^{\sigma_r^2 (1-\rho_{r*}^2) T/2},$$

the expression for the call becomes

$$B(0,T) e^{-\rho_{r*}^2 \sigma_r^2 T/2} \hat{E} e^{-\rho_{r*}\sigma_r \sqrt{T} Z_*} \left[\mathsf{f} e^{(\rho_{r*}\sigma_r\sigma_* - \sigma_*^2/2)T + \sigma_* \sqrt{T} Z_*} - X\right]^+.$$

The bracketed expression is positive when $Z_* > -q^-(\mathsf{f}/X;\sigma_*) - \rho_{r*}\sigma_r\sqrt{T}$, where

$$q^{\pm}(x;\sigma_*) = \frac{\ln x \pm \sigma_*^2 T/2}{\sigma_* \sqrt{T}}.$$

The value of the call therefore is

$$\begin{aligned} &B(0,T)\mathsf{f} e^{-(\sigma_* - \rho_{r*}\sigma_r)^2 T/2} \hat{E} e^{(\sigma_* - \rho_{r*}\sigma_r)\sqrt{T} Z_*} \mathbf{1}_{\mathcal{A}}(Z_*) \\ &\quad - B(0,T) X e^{-\rho_{r*}^2 \sigma_r^2 T/2} \hat{E} e^{-\rho_{r*}\sigma_r \sqrt{T} Z_*} \mathbf{1}_{\mathcal{A}}(Z_*), \end{aligned} \tag{10.51}$$

where $\mathcal{A}$ is the interval $(-q^-(\mathsf{f}/X;\sigma_*) - \rho_{r*}\sigma_r\sqrt{T}, \infty)$. Writing out the expectations,

$$\begin{aligned} C^E(S_0,T) = B(0,T)\mathsf{f} &\int_{\mathcal{A}} \frac{1}{\sqrt{2\pi}} \exp\left\{-\frac{1}{2}\left[z - (\sigma_* - \rho_{r*}\sigma_r)\sqrt{T}\right]^2\right\} \cdot dz \\ -B(0,T)X &\int_{\mathcal{A}} \frac{1}{\sqrt{2\pi}} \exp\left[-\frac{1}{2}\left(z - \rho_{r*}\sigma_r\sqrt{T}\right)^2\right] \cdot dz. \end{aligned}$$

Finally, changing variables in the integrals and simplifying give as the desired result

$$C^E(S_0,T) = B(0,T)\mathsf{f}\Phi[q^+(\mathsf{f}/X;\sigma_*)] - B(0,T)X\Phi[q^-(\mathsf{f}/X;\sigma_*)]. \tag{10.52}$$

This again parallels the standard Black-Scholes formula, (10.44), in that $\sigma_*^2 T$ is the variance of $\ln(S_T/S_0)$ when interest rates are stochastic, just as $\sigma_S^2 T$ is the variance of $\ln(S_T/S_0)$ when rates are deterministic and $\{S_t\}$ has volatility σ_S. In the present instance σ_*^2 can be expressed in terms of the primitive parameters in the models for S_t and r_t as

$$\sigma_*^2 = \sigma^2 T^2/3 + \rho\sigma T\sigma_S + \sigma_S^2,$$

where σ is the standard deviation of the changes in short and forward rates, σ_S is the volatility coefficient of the underlying price, and ρ is the instantaneous correlation between movements in interest rates and in the log of the underlying price.[10]

Caps and Floors

An interest rate "cap" is an instrument that limits the interest cost to one who borrows at a floating rate. It does so by paying the holder when each interest installment is due the difference between what is owed at the floating rate and what would be owed at some prearranged, fixed rate—the "cap". An interest rate "floor" puts a lower bound on what is received by one who lends at a floating rate. It does so by paying the holder the difference between what would be due at the fixed floor rate and what is due at the floating rate. There is an extensive and very liquid over-the-counter market for interest rate caps and floors. The common arrangement is to base the floating payment on the LIBOR (London Interbank Offering Rate) as of the beginning of each payment period. Caps and floors normally extend for many payment periods, thereby comprising portfolios of one-period "caplets" and "floorlets" with staggered payment dates. Once one knows how to value individual caplets and floorlets with arbitrary fixed rates and expirations, the caps and floors themselves can be priced just by adding up.

First, we need to understand the market convention regarding contracts based on LIBOR. Let $L_{T_0}(T_0, T_1)$ be the spot LIBOR at time T_0 for loans during $[T_0, T_1]$. It is a "spot" rate since it is set at the time the loan commences. By convention, LIBOR is quoted as a simply compounded annual rate, meaning that the payment due at T_1 is the rate at T_0 times the length of the period in years, or $L_{T_0}(T_0, T_1)\tau$, where $\tau \equiv T_1 - T_0$.[11] $L_{T_0}(T_0, T_1)$ differs from $r(T_0, T_1)$, the average spot rate between T_0 and T_1, in that (by our convention) $r(T_0, T_1)$ is continuously compounded. Relating these to prices of unit discount bonds, we have $B(T_0, T_1)[1 + \tau L_{T_0}(T_0, T_1)] = B(T_0, T_1)e^{\tau r(T_0, T_1)} = 1$, since a loan of $B(T_0, T_1)$ currency units commencing at T_0 would require repayment of

[10]For generalizations of (10.52) to other volatility functions see Musiela and Rutkowski (1998, pp. 362–364).

[11]We abstract from the various "day-count" conventions that are often used in actual transactions; for example, equating $T_1 - T_0$ to the number of intervening days divided by 360 and excluding holidays that fall at either end. In what follows we also ignore the tiny default risk that is present in this interbank loan market.

one unit at T_1. Thus,

$$L_{T_0}(T_0, T_1) = \tau^{-1}[B(T_0, T_1)^{-1} - 1].$$

Forward LIBORs are related analogously to bonds' forward prices. For $t < T_0$

$$L_t(T_0, T_1) = \tau^{-1}[B_t(T_0, T_1)^{-1} - 1] = \tau^{-1}[B(t, T_0)/B(t, T_1) - 1]$$

represents forward LIBOR as of time t for a τ-year loan commencing at T_0. The dates T_0, T_1 that define the contract are known as "reset" and "settlement" dates, respectively. For interest rate caps and floors with settlement dates $\{T_j\}_{j=1}^n$ the payment at T_j is based on spot rate $L_{T_{j-1}}(T_{j-1}, T_j)$, which is thus reset at each date T_{j-1}.

In section 10.4 we introduce a specific model for forward LIBOR that yields simple, Black-Scholes-like formulas for prices of caplets and floorlets. To motivate and contrast with this later approach, let us now work out the prices implied by our simple one-factor HJM model. We begin with a caplet that extends from T_0 to T_1 with capped rate K. For each unit of principal value the holder receives at T_1 the difference between what would be owed on a τ-year loan at time-T_0 spot LIBOR and what would be owed at simply compounded rate K, provided this difference is positive. Thus, the caplet's value at T_1 is $C(T_1, K; T_0, T_1) = \tau[L_{T_0}(T_0, T_1) - K]^+$, which amounts to the payoff of τ units of T_1-expiring call options on LIBOR "in arrears"—i.e., on spot LIBOR at T_0. Since the ultimate goal is to value caps that extend for many periods, we need to price caplets that begin at future dates. We will therefore work out the caplet's arbitrage-free value as of time zero, $C(0, K; T_0, T_1)$.

Working in spot measure $\hat{\mathbb{P}}$, we have

$$\begin{aligned} C(0, K; T_0, T_1) &= M_0 \hat{E} M_{T_1}^{-1} \tau[L_{T_0}(T_0, T_1) - K]^+ \qquad (10.53) \\ &= \hat{E} e^{-\int_0^{T_1} r_s \cdot ds}[B(T_0, T_1)^{-1} - (1 + \tau K)]^+. \end{aligned}$$

Expression (10.33) gives

$$\begin{aligned} e^{-\int_0^{T_1} r_s \cdot ds} &= \exp\left\{-\int_0^{T_1} \left[r_0(s) + \frac{\sigma^2}{2}s^2 + \sigma \hat{W}_s\right] \cdot ds\right\} \\ &= B(0, T_1) \exp\left(-\sigma^2 T_1^3/6 - \sigma \hat{Y}_{T_0} - \sigma \int_{T_0}^{T_1} \hat{W}_s \cdot ds\right) \\ &= B(0, T_1) \exp\left(-\sigma^2 T_1^3/6 - \sigma \hat{Y}_{T_0} - \sigma\tau \hat{W}_{T_0} - \sigma \tilde{Y}_\tau\right), \end{aligned}$$

where $\hat{Y}_{T_0} \equiv \int_0^{T_0} \hat{W}_s \cdot ds$ and $\tilde{Y}_\tau \equiv \int_{T_0}^{T_1} (\hat{W}_s - \hat{W}_{T_0}) \cdot ds$. Note that $\tilde{Y}_\tau$ is equivalent to $\int_0^\tau \tilde{W}_s \cdot ds$, where $\{\tilde{W}_s\}_{0 \leq s \leq \tau}$ is a Brownian motion independent of events in $\mathcal{F}_{T_0}$; therefore, $\tilde{Y}_\tau$ is distributed as $N(0, \tau^3/3)$ conditional on $\mathcal{F}_{T_0}$. Also, recall from (10.40) the decomposition $\hat{Y}_{T_0} = \frac{T_0}{2}\hat{W}_{T_0} + \frac{T_0}{2\sqrt{3}}\hat{U}_{T_0}$, where $\{\hat{U}_t\}_{t \geq 0}$ is another Brownian motion independent of $\{\hat{W}_t\}_{t \geq 0}$ and therefore of $\tilde{Y}_\tau$. Thus, $\exp(-\int_0^{T_1} r_s \cdot ds)$ equals

$$B(0, T_1) \exp\left[-\sigma^2 \frac{T_1^3}{6} - \sigma \frac{T_0}{2\sqrt{3}} \hat{U}_{T_0} - \sigma \left(\tau + \frac{T_0}{2}\right) \hat{W}_{T_0} - \sigma \tilde{Y}_\tau\right].$$

Next, from (10.34) we have

$$B(T_0, T_1)^{-1} = B_0(T_0, T_1)^{-1} e^{\frac{\sigma^2}{2}\tau T_0 T_1 + \sigma\tau \hat{W}_{T_0}} = (1 + \tau L_0) e^{\frac{\sigma^2}{2}\tau T_0 T_1 + \sigma\tau \hat{W}_{T_0}},$$

where $L_0 \equiv L_0(T_0, T_1)$. Assembling the pieces and using $\hat{E}e^{-\sigma T_0 \hat{U}_{T_0}/(2\sqrt{3})} = e^{\sigma^2 T_0^3/24}$ and $\hat{E}e^{-\sigma \tilde{Y}_\tau} = e^{\sigma^2 \tau^3/6}$, we express $C(0, K; T_0, T_1)$ as $B(0, T_1) \times e^{-\frac{\sigma^2 T_0}{2}(T_1 \tau + \frac{T_0^2}{4})} \hat{E}e^{-\sigma(\frac{T_0}{2} + \tau)\hat{W}_{T_0}} [(1 + \tau L_0) e^{\frac{\sigma^2}{2}\tau T_0 T_1 + \sigma\tau \hat{W}_{T_0}} - (1 + \tau K)]^+$.

The bracketed expression is positive if and only if

$$\hat{W}_{T_0} > -\frac{\ln(1 + \tau L_0) - \ln(1 + \tau K)}{\sigma\tau} - \frac{\sigma}{2} T_1 T_0.$$

Evaluating the expectations and simplifying, one obtains as the initial per-unit-principal value of the caplet

$$C(0, K; T_0, T_1) = B(0, T_1)[(1 + \tau L_0)\Phi(q^+) - (1 + \tau K)\Phi(q^-)],$$

where

$$q^\pm \equiv \frac{\ln(1 + \tau L_0) - \ln(1 + \tau K)}{\sigma\tau\sqrt{T_0}} \pm \frac{\sigma}{2}\tau\sqrt{T_0}.$$

The floorlet, with settlement value $F(T_1, K; T_0, T_1) = \tau[K - L_{T_0}(T_0, T_1)]^+$, amounts to a portfolio comprising τ European put options on the LIBOR, so its value at $t = 0$ comes from put-call parity:

$$F(0, K; T_0, T_1) = C(0, K; T_0, T_1) - \hat{E}e^{-\int_0^{T_1} r_s \cdot ds} \tau[L_{T_0}(T_0, T_1) - K].$$

Working out the expectation, one obtains

$$\begin{aligned} F(0, K; T_0, T_1) &= C(0, K; T_0, T_1) - B(0, T_1)\tau(L_0 - K) \\ &= B(0, T_1)[(1 + \tau K)\Phi(-q^-) - (1 + \tau L_0)\Phi(-q^+)]. \end{aligned}$$

The per-unit-principal values of caps and floors that extend from $T_0 \geq 0$ to T_n with settlement dates at $T_1, T_2, \ldots, T_n$ would then be calculated as

$$C(0, K; T_0, T_1, \ldots, T_n) = \sum_{j=1}^{n} C(0, K; T_{j-1}, T_j)$$

$$F(0, K; T_0, T_1, \ldots, T_n) = \sum_{j=1}^{n} F(0, K; T_{j-1}, T_j).$$

Swaps and Swaptions

While payoff functions of caps resemble those of portfolios of options, interest-rate "swaps" are more like collections of futures or forward contracts. A simple (forward) interest-rate swap is merely an advance agreement to exchange payments at a fixed interest rate on some "notional" principal for payments made at a floating rate. The agreement becomes live at some future date T_0, and the exchanges take place at later dates $T_1, T_2, \ldots$, ending at some T_n. From the standpoint of one who pays the fixed rate, the arrangement is a "payer" swap, while to the counterparty it is a "receiver" swap. The term *notional* conveys the fact that the swap agreement involves no exchange of principal; indeed, what changes hands at intervals is merely the difference between the fixed-rate and floating-rate payments.[12] The floating-rate payment at the end of a period is commonly based on the spot LIBOR at its beginning; thus, at T_j the payment per unit notional is $(T_j - T_{j-1})L_{T_{j-1}}(T_{j-1}, T_j)$. The fixed payment for a swap initiated at $t = 0$ is at some rate $K_0 = K_0(T_0, T_1, \ldots, T_n)$, which does not change during the life of the contract. This is determined so as to give the swap zero initial value—and so depends on the initiation, commencement, and payment dates. Of course, K_0 having once been set, the swap's market value will change with time and market conditions. A prospective fixed rate K_t that would equate to zero the value of the swap at some $t \in (0, T_{n-1})$ is called the "forward swap rate". Clearly, swaps simply allow floating-rate loans to be restructured, in effect, as fixed-rate loans, or *vice versa*. "Swaptions" are options to take one position or the other in a swap arrangement at some future time.

[12] In practice, swap agreements may call for fixed and floating payments to be made at different times; e.g., annually for fixed and quarterly for floating. We abstract from this trivial complication to simplify notation.

Swaps themselves can be valued through static replicating arguments without imposing structure on the dynamics of interest rates. Consider an arrangement that calls for exchanging floating-rate payments for fixed-rate payments at dates $T_1, T_2, \ldots, T_n$. To simplify, set $T_j - T_{j-1} = \tau$, the same for each j, and take the notional to be unity. Time $T_0 = T_1 - \tau$ represents the (forward) commencement date of the swap. Now the claim on the fixed receipts is like a bond that pays fixed coupons τK_0 from T_1 to T_n but does not return the principal value. This is obviously worth $\tau K_0 \sum_{j=1}^{n} B(T_0, T_j)$ at T_0. With a little more reflection one can see also that the claim on the floating payments is worth $1 - B(T_0, T_n)$ at T_0. This is because the same stream could be purchased by selling one T_n-maturing discount bond at T_0, lending one currency unit at spot LIBOR, and reinvesting the periodic receipts at spot LIBOR at each time up to T_{n-1}. Netting fixed against floating gives the swap's value to the party receiving fixed.

While no dynamic model was needed for this, it is a good exercise to verify that our martingale pricing apparatus leads to the same result for the value of the floating leg. Recalling that $B(T_{j-1}, T_j)^{-1} = 1 + \tau L_{T_{j-1}}(T_{j-1}, T_j)$, the value of the floating leg at time T_0 is

$$
\begin{aligned}
M_{T_0} \hat{E}_{T_0} &\sum_{j=1}^{n} M_{T_j}^{-1} \tau L_{T_{j-1}}(T_{j-1}, T_j) \\
&= \sum_{j=1}^{n} \hat{E}_{T_0} e^{-\int_{T_0}^{T_j} r_t \cdot dt} [B(T_{j-1}, T_j)^{-1} - 1] \\
&= \sum_{j=1}^{n} \hat{E}_{T_0} \left\{ e^{-\int_{T_0}^{T_{j-1}} r_t \cdot dt} \hat{E}_{T_{j-1}} \left[\frac{e^{-\int_{T_{j-1}}^{T_j} r_t \cdot dt}}{B(T_{j-1}, T_j)} \right] - e^{-\int_{T_0}^{T_j} r_t \cdot dt} \right\}.
\end{aligned}
$$

The inner expectation being unity, this reduces to

$$
\sum_{j=1}^{n} [B(T_0, T_{j-1}) - B(T_0, T_j)] = 1 - B(T_0, T_n).
$$

The swap's value at T_0 to the party that receives fixed is therefore

$$
S(T_0, K_0; T_0, T_1, \ldots, T_n) = \tau K_0 \sum_{j=1}^{n} B(T_0, T_j) + B(T_0, T_n) - 1.
$$

Notice that the first two terms together amount to the value at T_0 of a unit T_n-maturing coupon bond, so that in the notation of (10.45) the value of the swap at T_0 is simply $B(T_0, K_0 \tau; T_1, \ldots, T_n) - 1$. Since one can arrange at $t \le T_0$ to possess a T_j-maturing bond at T_0 just by buying one and holding, the value of the swap at $t = 0$ is

$$\begin{aligned}S(0, K_0; T_0, T_1, \ldots, T_n) &= \tau K_0 \sum_{j=1}^{n} B(0, T_j) + B(0, T_n) - B(0, T_0) \\ &= B(0, K_0\tau; T_1, \ldots, T_n) - B(0, T_0).\end{aligned}$$

Swap rate K_0 would be set as

$$K_0(T_0, T_1, \ldots, T_n) = \frac{B(0, T_0) - B(0, T_n)}{\tau \sum_{j=1}^{n} B(0, T_j)}$$

to equate the swap's initial value to zero, while the forward swap rate at some $t \in (0, T_{n-1})$ would be

$$K_t(T_{j_t}, \ldots, T_n) = \frac{B(t, T_{j_t - 1}) - B(t, T_n)}{\tau \sum_{j=j_t}^{n} B(t, T_j)},$$

where j_t is the first payment date after t. The actual value of the swap at t can then be expressed in terms of the forward swap rate as

$$S(t, K_0; T_{j_t - 1}, T_{j_t}, \ldots, T_n) = \tau(K_t - K_0) \sum_{j=j_t}^{n} B(t, T_j).$$

At some $t \le T_0$ an option to take the fixed side of the swap at T_0 is worth

$$M_t \hat{E}_t M_{T_0}^{-1} [B(T_0, K_0\tau; T_0, T_1, \ldots, T_n) - 1]^+,$$

which is precisely the value of a call option with unit strike on a bond of unitary principal value that pays coupon $K_0\tau$.

10.4 The LIBOR Market Model

Although the HJM approach offers an elegant and comprehensive way to model interest-sensitive instruments, it has not been widely adopted by traders, who must rely on such models to make rapid pricing decisions. The fact that instantaneous forward rates are not directly observable makes it difficult to formulate good models for the volatility processes $\{\boldsymbol{\sigma}_t(T)\}_{0 \le t \le T \le T_\infty}$ that control the stochastic behavior of rates and bond prices. What *are* readily observable are dealers' quotes on caps, floors, and swaptions, for which there are very active markets. While we were able to obtain simple formulas for these instruments in the illustrative one-factor, constant-volatility setup of the previous section, things are not so easy with specifications of HJM that generate nontrivial volatility term structures and twists in yield curves. Thus, it is difficult to use these market data to calibrate the model. In fact, to price the individual caplets and floorlets that make up caps and floors, traders routinely use simple adaptations of Black's (1976b) formulas (6.33) for options on futures. As we saw in section 6.3.3, those formulas were derived by modeling the settlement value of the underlying as lognormally distributed under $\hat{\mathbb{P}}$ and regarding numeraire process $\{M_t\}$ as deterministic. How can this apply to caplets and floorlets? Treating them as options, at least, does make sense, because the settlement values of caplets and floorlets over the interval $[T_0, T_1 = T_0 + \tau]$ are the option-like functions $\tau[L_{T_0}(T_0, T_1) - K]^+$ and $\tau[K - L_{T_0}(T_0, T_1)]^+$ of spot LIBOR at T_0, $L_{T_0}(T_0, T_1) = \tau^{-1}[B(T_0, T_1)^{-1} - 1]$. The adapted Black's formulas for the initial values of these are

$$C(0, K; T_0, T_1) = B(0, T_1)\tau[L_0\Phi(l^+) - K\Phi(l^-)] \qquad (10.54\text{a})$$

$$F(0, K; T_0, T_1) = B(0, T_1)\tau[K\Phi(-l^-) - L_0\Phi(-l^+)], \qquad (10.54\text{b})$$

where

$$l^{\pm} \equiv \frac{\ln(L_0/K) \pm \bar{\lambda}^2_{T_0} T_0/2}{\bar{\lambda}_{T_0}\sqrt{T_0}},$$

$L_0 \equiv L_0(T_0, T_1)$ is forward LIBOR as of time zero, and $\bar{\lambda}_{T_0}$ is a positive "volatility" constant.

Despite the aptness of the option analogy, how to reconcile these formulas with the HJM formulation is far from apparent. In our illustrative application beginning on page 493 we valued the caplet as

$$C(0, K; T_0, T_1) = M_0 \hat{E} M_{T_1}^{-1} \tau[L_{T_0}(T_0, T_1) - K]^+,$$

using a one-factor HJM model for instantaneous forward rates with constant, maturity-invariant volatilities, $\sigma_t(T) = \sigma$. Even in that simplest possible version of HJM neither is $L_{T_0}(T_0, T_1)$ distributed as lognormal under $\hat{\mathbb{P}}$ nor—quite obviously—is M_{T_1} deterministic, so it is not surprising that (10.54a) is inconsistent with our solution:

$$C(0, K; T_0, T_1) = B(0, T_1)[(1 + \tau L_0)\Phi(q^+) - (1 + \tau K)\Phi(q^-)]$$

$$q^{\pm} \equiv \frac{\ln[(1 + \tau L_0)/(1 + \tau K)] \pm \sigma^2\tau^2 T_0/2}{\sigma\tau\sqrt{T_0}}.$$

Whether Black's formulas could be consistent with *any* specification of HJM—and therefore with an arbitrage-free market for bonds and the money fund—was for a long time in doubt, but a reconciliation was provided by Brace, Gątarek, and Musiela (1997). This section introduces their approach, now commonly referred to as the LIBOR market model.[13] We see first how to derive Black's formula in a way that is consistent with the general multidimensional form of HJM, then discuss some of the general requirements for using the LIBOR market model to price derivatives other than caps and floors.

10.4.1 *Deriving Black's Formulas*

Let us begin with a quick review of the general multidimensional version of HJM. We have a family of bond prices $\{B(t,T)\}_{0\le t\le T}$ at maturities $T \in [0, T_\infty]$, expressible in terms of instantaneous forward rates as $B(t,T) = \exp(-\int_t^T r_t(u)\cdot du)$. There is also the money market fund, whose per-unit price at t is $M_t = M_0 \exp(\int_0^t r_s \cdot ds)$, where $r_t \equiv r_t(t)$ is the instantaneous spot rate. Forward rates evolve as

$$r_t(T) - r_0(T) = \int_0^t \mu_s(T)\cdot ds + \int_0^t \boldsymbol{\sigma}_s(T)' \cdot d\mathbf{W}_s,$$

where $\{\mathbf{W}_t\}$ is an $\Re_k$-valued standard Brownian motion on $(\Omega, \mathcal{F}, \{\mathcal{F}_t\}, \mathbb{P})$. Drift processes $\{\mu_t(T)\}$, risk prices $\{\gamma_t\}$, and volatilities $\{\boldsymbol{\sigma}_t(T)\}$ satisfy the conditions (laid out in section 10.3.2) that assure that bond prices and the money fund are arbitrage-free. There are thus a spot measure $\hat{\mathbb{P}}$ and an associated Brownian motion $\{\hat{\mathbf{W}}_t\}$ such that normalized processes

$$\{B(t,T)^* = B(t,T)/M_t\}$$

are martingales, evolving as

$$dB(t,T)^*/B(t,T)^* = -\boldsymbol{\sigma}_t(t,T)' \cdot d\hat{\mathbf{W}}_t, 0 \le t \le T \le T_\infty. \qquad (10.55)$$

In addition, for each $T_0, T_1 \in (0, T_\infty]$ with $T_0 \le T_1$ there are a forward measure $\mathbb{P}^{T_1}$ and an associated $\Re_k$-valued Brownian motion $\{\mathbf{W}_t^{T_1}\}$ such that normalized processes $\{M_t/B(t,T_1)\}$ and $\{B(t,T_0)/B(t,T_1)\}_{0\le t\le T_0\le T_1}$ are martingales.

The first key to developing Black's formulas for T_1-settled caplets and floorlets is to shift from spot measure $\hat{\mathbb{P}}$ to the forward measure $\mathbb{P}^{T_1}$ that is associated with the object's particular settlement date—that is, to the

[13]Musiela and Rutkowski (2005) give a comprehensive account that encompasses the now extensive literature on models of this type. Shreve (2004) offers a brief but insightful treatment of the theory. Brigo and Mercurio (2001) and Joshi (2003) give details on implementing the model.

martingale measure that applies when $\{B(t,T_1)\}_{0\le t\le T_1}$ is numeraire. The second key is to recognize that with this numeraire the forward LIBOR process $\{L_t(T_0,T_1)\}_{0\le t\le T_0}$ is in fact the normalized price of a traded asset. That this is so can be seen directly from the relation that defines the simply compounded rate: $B(t,T_1)[1+\tau L_t(T_0,T_1)] = B(t,T_0)$. Thus,

$$B(t,T_1)L_t(T_0,T_1) = \tau^{-1}[B(t,T_0) - B(t,T_1)]$$

corresponds to the value at t of a portfolio long T_0-maturing bonds of principal value τ^{-1} and short a like number of bonds of the later maturity. In an arbitrage-free market $\{L_t(T_0,T_1)\}_{0\le t\le T_0}$ is therefore a martingale under forward measure $\mathbb{P}^{T_1}$. As such, by the martingale representation theorem there are an $\Re_k$-valued, $\{\mathcal{F}_t\}$-adapted process $\{\boldsymbol{\lambda}_{T_0}(t)\}_{0\le t\le T_0}$ and a Brownian motion $\{\mathbf{W}_t^{T_1}\}$ such that

$$dL_t(T_0,T_1) = \boldsymbol{\lambda}_{T_0}(t)' L_t(T_0,T_1)\cdot d\mathbf{W}_t^{T_1}. \tag{10.56}$$

Under the condition that $\{\boldsymbol{\lambda}_{T_0}(\cdot)\}$ is bounded and piecewise continuous and that initial forward LIBOR $L_0(T_0,T_1)$ is positive, Brace *et al.* (1997) show that (10.56) has the unique, strictly positive solution

$$\frac{L_t(T_0,T_1)}{L_0(T_0,T_1)} = \exp\left[\int_0^t \boldsymbol{\lambda}_{T_0}(s)'\cdot d\mathbf{W}_s^{T_1} - \frac{1}{2}\int_0^t \boldsymbol{\lambda}_{T_0}(s)'\boldsymbol{\lambda}_{T_0}(s)\cdot ds\right]. \tag{10.57}$$

Under the additional condition that $\{\boldsymbol{\lambda}_{T_0}(\cdot)\}$ is deterministic ($\mathcal{F}_0$-measurable), it follows that $L_t(T_0,T_1)$ is distributed as lognormal under $\mathbb{P}^{T_1}$.

That $\{\boldsymbol{\lambda}_{T_0}(\cdot)\}$ is deterministic is in fact the critical assumption, for it takes us directly to the desired simple formulas for caplets and floorlets that make it easy to calibrate the model from market prices. Indeed, we are now essentially done. Putting

$$\bar{\lambda}_{T_0}^2 \equiv T_0^{-1}\int_0^{T_0} \boldsymbol{\lambda}_{T_0}(s)'\boldsymbol{\lambda}_{T_0}(s)\cdot ds, \tag{10.58}$$

writing

$$\begin{aligned} C(0,K;T_0,T_1) &= B(0,T_1)\tau E^{T_1}[L_{T_0}(T_0,T_1) - K]^+ \\ &= B(0,T_1)\tau E\big[L_0 e^{\bar{\lambda}_{T_0}\sqrt{T_0}\ Z - \bar{\lambda}_{T_0}^2 T_0/2} - K\big]^+ \end{aligned}$$

for $Z \sim N(0,1)$, and working out the expectation yield (10.54a). Formula (10.54b) comes in a similar way from

$$F(0,K;T_0,T_1) = B(0,T_1)\tau E^{T_1}[K - L_{T_0}(T_0,T_1)]^+$$

or via the parity relation.

Of course, we have not yet reconciled with HJM, for it remains to show that processes $\{\boldsymbol{\lambda}_{T_0}(t)\}_{0\le t\le T_0}$ with the required properties correspond to some set of volatilities $\{\boldsymbol{\sigma}_t(T_0)\}_{0\le t\le T_0\le T_\infty}$ that characterize instantaneous forward-rate processes $\{r_t(T)\}$. To do this requires relating the evolution of forward LIBOR,

$$L_t(T_0,T_1)=\tau^{-1}[B(t,T_0)/B(t,T_1)-1], 0\le t\le T_0,$$

to the evolutions of discount bonds, whose dependence on forward-rate volatilities is known from the HJM theory. More particularly, we shall relate to the $\{M_t\}$-normalized processes

$$\{B(t,T_1)^*=B(t,T_1)/M_t\}$$
$$\{B(t,T_0)^*=B(t,T_0)/M_t\},$$

which we know to be martingales in spot measure $\hat{\mathbb{P}}$. Recall from (10.55) that these evolve under $\hat{\mathbb{P}}$ as $dB(t,T_1)^*/B(t,T_1)^*=-\boldsymbol{\sigma}_t(t,T_1)'\cdot d\hat{\mathbf{W}}_t$ and $dB(t,T_0)^*/B(t,T_0)^*=-\boldsymbol{\sigma}_t(t,T_0)'\cdot d\hat{\mathbf{W}}_t$. Starting with

$$dL_t(T_0,T_1)=\tau^{-1}d[B(t,T_0)/B(t,T_1)]=\tau^{-1}d[B(t,T_0)^*/B(t,T_1)^*]$$

and applying Itô's formula to the last expression lead to

$$\begin{aligned}\frac{d[B(t,T_0)^*/B(t,T_1)^*]}{B(t,T_0)^*/B(t,T_1)^*}&=\frac{dB(t,T_0)^*}{B(t,T_0)^*}-\frac{dB(t,T_1)^*}{B(t,T_1)^*}\\&\quad+\frac{d\langle B(\cdot,T_1)^*\rangle_t}{[B(t,T_1)^*]^2}-\frac{d\langle B(\cdot,T_0)^*,B(\cdot,T_1)^*\rangle_t}{B(t,T_0)^*B(t,T_1)^*}\\&=-\boldsymbol{\sigma}_t(t,T_0)'\cdot d\hat{\mathbf{W}}_t+\boldsymbol{\sigma}_t(t,T_1)'\cdot d\hat{\mathbf{W}}_t\\&\quad+\boldsymbol{\sigma}_t(t,T_1)'\boldsymbol{\sigma}_t(t,T_1)\cdot dt-\boldsymbol{\sigma}_t(t,T_0)'\boldsymbol{\sigma}_t(t,T_1)\cdot dt\\&=[\boldsymbol{\sigma}_t(t,T_1)-\boldsymbol{\sigma}_t(t,T_0)]'\cdot d\left[\hat{\mathbf{W}}_t+\int_0^t\boldsymbol{\sigma}_u(u,T_1)\cdot du\right].\end{aligned}$$

Although this shows that $\{B(t,T_0)^*/B(t,T_1)^*\}$ is not a martingale under $\hat{\mathbb{P}}$, it is in fact a martingale in forward measure $\mathbb{P}^{T_1}$, since $B(t,T_0)^*/B(t,T_1)^*=B(t,T_0)/B(t,T_1)$ is the appropriately normalized price of a traded asset. Thus, we can write the above as

$$\frac{d[B(t,T_0)^*/B(t,T_1)^*]}{B(t,T_0)^*/B(t,T_1)^*}=[\boldsymbol{\sigma}_t(t,T_1)-\boldsymbol{\sigma}_t(t,T_0)]'\cdot d\mathbf{W}_t^{T_1},$$

where $\{\mathbf{W}_t^{T_1}=\hat{\mathbf{W}}_t+\int_0^t\boldsymbol{\sigma}_u(u,T_1)\cdot du\}$ is an $\Re_k$-valued Brownian motion under $\mathbb{P}^{T_1}$. Finally, expressing the ratio of bond prices in terms of forward LIBOR gives

$$dL_t(T_0,T_1)=\boldsymbol{\lambda}_{T_0}(t)'L_t(T_0,T_1)\cdot d\mathbf{W}_t^{T_1},$$

with

$$\boldsymbol{\lambda}_{T_0}(t) \equiv \left[\frac{1+\tau L_t(T_0,T_1)}{\tau L_t(T_0,T_1)}\right][\boldsymbol{\sigma}_t(t,T_1)-\boldsymbol{\sigma}_t(t,T_0)]. \tag{10.59}$$

Expressing (10.59) for all reset/settlement pairs $\{T_{j-1},T_j\}_{j=1}^n$ that make up the market's "tenor structure" provides a recurrence relation for the $\{\boldsymbol{\sigma}_t(t,T_j)\}_{j=1}^n$. Given some specification of an initial process $\{\boldsymbol{\sigma}_t(t,T_0)\}$, this allows the bond volatilities at each maturity to be expressed in terms of prespecified LIBOR volatilities with the appropriate properties—i.e., bounded and deterministic—and of the evolving LIBORs themselves. Choosing specific LIBOR volatility functions requires calibrating the model to market data, a procedure that we discuss briefly below. While there is considerable flexibility (i.e., arbitrariness) in this, we clearly want each process $\{\boldsymbol{\lambda}_{T_{j-1}}(t)\}_{0\le t\le T_{j-1}}$ to be such that $\int_0^{T_{j-1}} \boldsymbol{\lambda}_{T_{j-1}}(s)'\boldsymbol{\lambda}_{T_{j-1}}(s)\cdot ds = T_0\bar{\lambda}^2_{T_{j-1}}$, where $\bar{\lambda}_{T_{j-1}}$ is the (scalar) volatility that equates (10.54a) and (10.54b) to market quotes for T_j-settled caplets and floorlets.

To illustrate the construction of the $\{\boldsymbol{\sigma}_t(t,T_j)\}$, let $\{T_j\}_{j=0}^{n-1}$ denote the strictly increasing sequence of extant reset dates and set $\tau_j \equiv T_j - T_{j-1}$. Then (10.59) gives

$$\boldsymbol{\sigma}_t(t,T_j) = \boldsymbol{\sigma}_t(t,T_{j-1}) + \boldsymbol{\lambda}_{T_{j-1}}(t)\left[\frac{\tau_j L_t(T_{j-1},T_j)}{1+\tau_j L_t(T_{j-1},T_j)}\right],\quad 0\le t\le T_{j-1}.$$

To start the recursion one must choose an initial process $\{\boldsymbol{\sigma}_t(t,T_0)\}_{0\le t\le T_0}$. The only logical (as opposed to empirical) requirement is that $\boldsymbol{\sigma}_{T_0}(T_0,T_0) = \mathbf{0}$, since this represents the volatility vector of a T_0-maturing, default-free bond at maturity. For example, Brace *et al.* (1997) set $\boldsymbol{\sigma}_t(t,T_0) = \mathbf{0}$ for $t\in[0,T_0]$. Now, having specified an appropriate bounded, deterministic, $\Re_k$-valued function $\{\boldsymbol{\lambda}_{T_0}(t)\}_{0\le t\le T_0}$, one finds $\boldsymbol{\sigma}_t(t,T_1)$ for $t\in[0,T_0]$ as

$$\boldsymbol{\sigma}_t(t,T_1) = \boldsymbol{\sigma}_t(t,T_0) + \boldsymbol{\lambda}_{T_0}(t)\left[\frac{\tau_1 L_t(T_0,T_1)}{1+\tau_1 L_t(T_0,T_1)}\right],$$

then extends freely for $t\in(T_0,T_1]$ subject to $\boldsymbol{\sigma}_{T_1}(T_1,T_1)=\mathbf{0}$. Continuing, $\{\boldsymbol{\sigma}_t(t,T_m)\}$ can be built up for $m\in\{2,3,\ldots,n\}$ as

$$\boldsymbol{\sigma}_t(t,T_m) = \boldsymbol{\sigma}_t(t,T_0) + \sum_{j=1}^{m}\boldsymbol{\lambda}_{T_{j-1}}(t)\left[\frac{\tau_j L_t(T_{j-1},T_j)}{1+\tau_j L_t(T_{j-1},T_j)}\right],\quad t\in(0,T_{m-1}],$$

then extended to $t\in(T_{m-1},T_m]$ in such a way that $\boldsymbol{\sigma}_{T_m}(T_m,T_m)=\mathbf{0}$. Notice that while the processes $\{\boldsymbol{\lambda}_{T_{j-1}}(t)\}$ are deterministic, the $\{\boldsymbol{\sigma}_t(t,T_j)\}$ are merely $\mathcal{F}_t$-measurable, since they depend on the evolving lognormal forward LIBORs.

The above construction shows that there do exist bond volatility processes that reconcile the market LIBOR framework with that of HJM. It also shows how, in principle, one could generate realizations of the volatilities—i.e., by simulating $\{\ln L_t(T_{j-1}, T_j)\}_{j=1}^n$ as a joint Gaussian process in some particular forward measure. As we shall see next, simulating ensembles of forward LIBORs is required in most applications of the LIBOR market model, and doing this is far from trivial.

10.4.2 *Applying the Model*

Let us now consider how the market LIBOR model can be used to price a given interest-sensitive derivative. Clearly, using it to price caps and floors is easy, but in considering other applications it helps to think precisely why this is so. Caps and floors are easy, first, because they are just portfolios of individual caplets and floorlets that can be valued one at a time and, second, because the caplets and floorlets can be valued using Black's simple formulas.[14] The second point is obvious, but there is more to the first point than may be apparent. Because each caplet or floorlet is a function of a single LIBOR, $L_{T_{j-1}}(T_{j-1}, T_j)$ say, pricing it does not require modeling the joint behavior of rates with different settlement dates. For example, the dependence between $L_{T_{j-1}}(T_{j-1}, T_j)$ and $L_{T_j}(T_j, T_{j+1})$ is of no relevance in evaluating $C(0, K; T_{j-1}, T_j)$ and $C(0, K; T_j, T_{j+1})$. Also, it is possible without inconsistency to use a different martingale measure for each caplet or floorlet—namely, the one corresponding to its particular settlement date. Thus, by using Black's formula for each $C(0, K; T_{j-1}, T_j)$ we are implicitly calculating $E^{T_j}[L_{T_{j-1}}(T_{j-1}, T_j) - K]^+$ in each LIBOR's "native" measure $\mathbb{P}^{T_j}$. Moreover, since in its native measure the LIBOR for a given reset date has no trend, there is no need to worry about drift terms in expressions for $dL_t(\cdot, \cdot)$. In general, it is easy to price any instrument whose payoff depends on the value of a single LIBOR at one point in time or, as in the case of caps and floors, can be represented as an affine function of such individual components.

In contrast, consider an option on a cap—a "caption". A T_0-expiring European-style call struck at X on a cap commencing at T_0 and with

---

[14] Indeed, the values of the caplets and floorlets would, in practice, have been deduced from market quotes for caps and floors of various durations, which in turn would have been used to calibrate the model.

settlement dates $\{T_j\}_{j=1}^n$ is worth at expiration

$$[C(T_0, K; T_0, T_1, \ldots, T_n) - X]^+ = \left[\sum_{j=1}^{n} C(T_0, K; T_{j-1}, T_j) - X\right]^+ . \tag{10.60}$$

This depends on the values of all the caplets at T_0 and therefore on the evolution of all the underlying LIBORs up to that time, and because of the $[\cdot]^+$ it is not an affine function of the caplets. There being no single measure in which all of the LIBORs are trendless, it is necessary to model the drift processes of (at least) all but one. Also, it is necessary to take into account the dependence between rates over the life of the derivative. This could be modeled by fully specifying the $\Re_k$-valued volatility processes $\{\boldsymbol{\lambda}_{T_{j-1}}(t)\}_{j=1}^n$ that govern the evolutions of the rates, but we shall see later that there is another way. Finally, to value the caption one must evaluate the expectation of its normalized value at expiration. As for most other "exotic" interest-rate derivatives, there are no simple computational formulas for these expected values, which therefore must be estimated by simulation.

In general, the key steps in applying the LIBOR market model are

- Choosing a numeraire and expressing the derivative's normalized future payoffs in terms of a set of LIBORs.
- Modeling the evolution of the relevant LIBOR processes in the martingale measure associated with the chosen numeraire.
- Simulating the LIBORs and the replicates of the derivative's payoffs in order to estimate expected values.

We first describe some specific applications that illustrate steps one and three, then take up the much more involved issue of modeling the various rates.

Some Examples

1. Consider first a T_0-expiring European put option struck at X on a discount bond maturing at T_n, which is worth $P^E[B(T_0, T_n), 0] = [X - B(T_0, T_n)]^+$ at expiration. We work with LIBOR processes

$$\{L_t(T_{j-1}, T_j = T_{j-1} + \tau_j)\}_{j=1}^n$$

and write

$$B(T_0, T_n) = \prod_{j=1}^{n} [1 + \tau_j L_{T_0}(T_{j-1}, T_j)]^{-1}.$$

The price of either a T_0-maturing bond or the underlying T_n bond itself would be sensible choices as numeraires. In the former case, with $\mathbb{P}^{T_0}$ as the relevant measure, we would value the option at $t = 0$ as

$$P^E[B(0, T_n), T_0] = B(0, T_0) E^{T_0} \left\{ X - \prod_{j=1}^{n} [1 + \tau_j L_{T_0}(T_{j-1}, T_j)]^{-1} \right\}^+ .$$

In the latter case, we would work in measure $\mathbb{P}^{T_n}$ and evaluate

$$P^E[B(0, T_n), T_0] = B(0, T_n) E^{T_n} \left\{ X \prod_{j=1}^{n} [1 + \tau_j L_{T_0}(T_{j-1}, T_j)] - 1 \right\}^+ .$$

In either case we would simulate forward LIBORs $\{L_t(T_{j-1}, T_j)\}_{j=1}^n$ out to $t = T_0$ in the pertinent measure, express the put's terminal value in terms of these, and average over replications to estimate the expectations.

2. The procedure for pricing an option on a coupon-paying, default-free bond would be much the same. The value at T_0 of a bond paying c currency units at dates $\{T_j\}_{j=1}^n$ and returning unit principal value at T_n would be

$$B(T_0, c; T_1, \ldots, T_n) = c \sum_{j=1}^{n-1} B(T_0, T_j) + (1 + c) B(T_0, T_n).$$

With $B(t, T_n)$ as numeraire the initial value of a T_0-expiring European call, $C^E[B(0, c; T_1, \ldots, T_n), T_0; X]$, would be

$$B(0, T_n) E^{T_n} \left[c \sum_{j=1}^{n-1} \frac{B(T_0, T_j)}{B(T_0, T_n)} + 1 + c - X \right]^+ ,$$

where the relative bond prices are expressed in terms of forward LIBORs as

$$\frac{B(T_0, T_j)}{B(T_0, T_n)} = \prod_{i=j+1}^{n} [1 + \tau_i L_{T_0}(T_{i-1}, T_i)].$$

Again, the LIBORs would be generated by simulation—one would need an ensemble of $n - j$ of them with reset dates out to T_{n-j}—and the expectation estimated by averaging the replications.

3. We saw in section 10.3.3 that an interest-rate receiver swap that receives payments on a unit notional principal at fixed rate K_0 and pays at floating rate $L_{T_{j-1}}(T_{j-1}, T_j)$ at equally spaced times $\{T_j = T_0 + j\tau\}_{j=1}^n$ would be worth at T_0

$$S(T_0, K_0; T_0, T_1, \ldots, T_n) = K_0\tau \sum_{j=1}^{n} B(T_0, T_j) + B(T_0, T_n) - 1.$$

Note that the first two terms amount to the value at T_0 of a bond paying fixed coupons of $K_0\tau$ at each future date and returning unit principal value at T_n. A European-style, T_0-expiring receiver "swaption" would give the holder the right to enter such a swap arrangement at T_0, at which time it would be worth $S(T_0, K_0; T_0, T_1, \ldots, T_n)^+$. This being the value of a European call struck at $X = 1$ on a T_n-maturing bond paying coupons $K_0\tau$, the swaption's value at $t = 0$, $C^E[S(0, K_0; T_0, T_1, \ldots, T_n), T_0]$, would be calculated as

$$C^E[B(0, K_0\tau; T_1, \ldots, T_n), T_0; 1],$$

or

$$B(0, T_n)K_0\tau E^{T_n}\left\{\sum_{j=1}^{n-1}\prod_{i=j+1}^{n}[1 + \tau_i L_{T_0}(T_{i-1}, T_i)] + 1\right\}^+.$$

4. Finally, consider the caption that was described earlier. This was a T_0-expiring European-style call struck at X on a cap commencing at T_0 and with settlement dates $\{T_j\}_{j=1}^n$, the caption's value at expiration being given by (10.60). Taking as numeraire the price of the T_0-maturing discount bond, we would express the caption's initial value as

$$C(0, K; T_0, T_1, \ldots, T_n) = B(0, T_0)E^{T_0}[C(T_0, K; T_0, T_1, \ldots, T_n) - X]^+. \tag{10.61}$$

How is this to be evaluated? If we knew the values at T_0 of forward LIBORs and bond prices $\{L_{T_0}(T_{j-1}, T_j), B(T_0, T_j)\}_{j=1}^n$, then each of the individual caplets could be valued using Black's formula. This would give us the T_0 value of the cap itself and thus the exercise value of the option, as

$$\left\{\sum_{j=1}^{n} B(T_0, T_j)\tau_j[L_{T_0}(T_{j-1}, T_j)\Phi(l_j^+) - K\Phi(l_j^-)] - X\right\}^+,$$

where $l_j^{\pm} \equiv \{\ln[L_{T_0}(T_{j-1},T_j)/K] \pm \frac{1}{2}\bar{\lambda}^2_{T_{j-1}}T_{j-1}\}/(\bar{\lambda}_{T_{j-1}}\sqrt{T_{j-1}})$. Expressing the prices of discount bonds at T_0 in terms of LIBORs as

$$\left\{B(T_0,T_j) = \prod_{i=1}^{j}[1+\tau_i L_{T_0}(T_{i-1},T_i)]^{-1}\right\}_{j=1}^{n},$$

we can obtain one realization of the caption's terminal value by generating the ensemble of forward LIBOR paths out to T_0. Replicating the process, averaging the realizations, and multiplying by the observable $B(0,T_0)$ give an estimate of (10.61).

Modeling the LIBORs

We have seen that each member of the collection of forward LIBOR processes

$$\{L_t(T_{j-1},T_j) = \tau_j^{-1}[B(t,T_{j-1}) - B(t,T_j)]/B(t,T_j)\}_{0\le t\le T_{j-1}, j\in\{1,2,\ldots,n\}}$$

is itself the evolving value of a portfolio of bonds, normalized by the one with later maturity. Any one such process $\{L_t(T_{m-1},T_m)\}_{0\le t\le T_{m-1}}$ is therefore a martingale in its native measure $\mathbb{P}^{T_m}$, evolving as

$$dL_t(T_{m-1},T_m) = \boldsymbol{\lambda}_{T_{m-1}}(t)' L_t(T_{m-1},T_m)\cdot d\mathbf{W}_t^{T_m}.$$

If we have chosen as numeraire the price of the T_m-maturing bond, $B(t,T_m)$, and are therefore valuing a derivative in forward measure $\mathbb{P}^{T_m}$, it follows that $\{L_t(T_{m-1},T_m)\}$ can be modeled and simulated as a trendless process. However, we must recognize that the other processes $\{L_t(T_{j-1},T_j)\}_{j\ne m}$ on which the derivative depends cannot also be martingales in this measure, for they are not representable as marketable assets normalized by $B(t,T_m)$. Instead, there will be drift terms, as in

$$dL_t(T_{j-1},T_j) = \mu_{T_{j-1},T_m}(t)L_t(T_{j-1},T_j)\cdot dt + \boldsymbol{\lambda}_{T_{j-1}}(t)' L_t(T_{j-1},T_j)\cdot d\mathbf{W}_t^{T_m}.$$

When pricing derivatives such as bond options, swaptions, and captions that depend on—but are not affine functions of—more than one LIBOR process, we must somehow model the trends of all that do not evolve in their native measures. Fortunately, we will see that the drifts can be expressed in terms of the realizations and volatility processes of the various rates. Specifying the volatilities will be the next step.

Modeling the Drifts

Given the collection

$$\{L_t(T_0,T_1)\}_{0\le t\le T_0},\{L_t(T_1,T_2)\}_{0\le t\le T_1},\ldots,\{L_t(T_{n-1},T_n)\}_{0\le t\le T_{n-1}}$$

of forward LIBOR processes, we will now see how to determine the trend of the jth member in each of the measures $\mathbb{P}^{T_n}$ and $\mathbb{P}^{T_0}$. For brevity, we shall use a simpler notation:

- B_j for $B(t,T_j)$
- L_j for $L_t(T_{j-1},T_j)$
- $dL_j = \mu_{jm}L_j \cdot dt + \boldsymbol{\lambda}_j' L_j \cdot d\mathbf{W}_t^m$ for the increment under measure $\mathbb{P}^{T_m}$, $m \in \{0,n\}$.

Also, it will be helpful to set out for reference a result from the Itô calculus: If $\{X_{1t}\},\{X_{2t}\},\ldots,\{X_{nt}\}$ are Itô processes and $P_{kt} = X_{1t}X_{2t}\cdot\cdots\cdot X_{kt}$ is the product of time-t values of the first k of these, then by Itô's formula

$$\frac{dP_{kt}}{P_{kt}} = \sum_{j=1}^{k}\frac{dX_{jt}}{X_{jt}} + \sum_{i<j}\frac{d\langle X_i,X_j\rangle_t}{X_{it}X_{jt}}. \tag{10.62}$$

Of course, when $k=2$ this is just

$$\frac{dP_{2t}}{P_{2t}} = \frac{dX_{1t}}{X_{1t}} + \frac{dX_{2t}}{X_{2t}} + \frac{d\langle X_1,X_2\rangle_t}{X_{1t}X_{2t}}. \tag{10.63}$$

Consider first measure $\mathbb{P}^{T_n}$, corresponding to numeraire $\{B_n \equiv B(t,T_n)\}$. We know that $\{L_jB_j/B_n = \tau_j^{-1}(B_{j-1}-B_j)/B_n\}$ has no trend in this measure, since it represents the normalized value of a portfolio of bonds. Likewise, $\{B_j/B_n\}$ has no trend, and this ratio can be expressed in terms of forward LIBORs as $\{B_j/B_n = \prod_{i=j+1}^{n}(1+\tau_iL_i)\}$ for $j<n$. Of course, $\{L_j\}$ itself has no trend when $j=n$, but we shall need to determine drift μ_{jn} for $j\in\{1,2,\ldots,n-1\}$. With $X_1 = L_j$ and $X_2 = B_j/B_n$ in (10.63) we have

$$\frac{d(L_jB_j/B_n)}{L_jB_j/B_n} = \frac{dL_j}{L_j} + \frac{d(B_j/B_n)}{B_j/B_n} + \frac{d\langle L_j, B_j/B_n\rangle_t}{L_jB_j/B_n}, \tag{10.64}$$

and putting $X_{jt} = (1+\tau_iL_i)$ in (10.62) gives

$$\frac{d(B_j/B_n)}{B_j/B_n} = \sum_{i=j+1}^{n}\frac{\tau_i}{1+\tau_iL_i}\cdot dL_i + \sum_{i<l}\frac{\tau_i\tau_l}{(1+\tau_iL_i)(1+\tau_lL_l)}\cdot d\langle L_i,L_l\rangle_t.$$

But since neither $\{L_j B_j/B_n\}$ nor $\{B_j/B_n\}$ has trend, that of $\{L_j\}$ must equal the negative of the last term of (10.64):

$$\mu_{jn} = -\frac{d\langle L_j, B_j/B_n\rangle_t}{L_j B_j/B_n} = -\sum_{i=j+1}^{n} \frac{\tau_i L_i \boldsymbol{\lambda}_i' \boldsymbol{\lambda}_j}{1+\tau_i L_i}.$$

Next, to get the drifts for measure $\mathbb{P}^{T_0}$ we normalize by $\{B_0 \equiv B(t, T_0)\}$. Both $\{L_j B_j/B_0\}$ and $\{B_j/B_0 = \prod_{i=1}^{j}(1+\tau_i L_i)^{-1}\}$ being trendless in this measure, it again follows that $\mu_{j0} = -d\langle L_j, B_j/B_0\rangle/(L_j B_j/B_0)$. However, since

$$dB_j/B_0 = -\sum_{i=1}^{j} \frac{\tau_i}{1+\tau_i L_i} \cdot dL_i + \sum_{i<l} \frac{\tau_i \tau_l}{(1+\tau_i L_i)(1+\tau_l L_l)} \cdot d\langle L_i, L_l\rangle_t,$$

the drifts $\{\mu_{j0}\}_{j=1}^{n}$ now come with positive signs:

$$\mu_{j0} = \sum_{i=1}^{j} \frac{\tau_i L_i \boldsymbol{\lambda}_i' \boldsymbol{\lambda}_j}{1+\tau_i L_i}.$$

Thus, forward LIBORs have positive drifts in measures in which earlier rates are martingales; and, as is consistent with this, they have negative drifts in the native measures of later rates.

To recap and express in the original notation, the drifts $\{\mu_{T_{j-1},T_m}(t)\}_{j=1}^{n}$ for LIBOR processes $\{L_t(T_{j-1}, T_j)\}_{0\le t\le T_{j-1}, j\in\{1,2,\ldots,n\}}$ in measures $\mathbb{P}^{T_0}$ and $\mathbb{P}^{T_n}$ are

$$\mu_{T_{j-1},T_0}(t) = \sum_{i=1}^{j} \frac{\tau_i L_t(T_{i-1},T_i)\boldsymbol{\lambda}_{T_{i-1}}(t)'\boldsymbol{\lambda}_{T_{j-1}}(t)}{1+\tau_i L_t(T_{i-1},T_i)}, \quad j \in \{1,2,\ldots,n\}$$

$$\mu_{T_{n-1},T_n}(t) = 0$$

$$\mu_{T_{j-1},T_n}(t) = -\sum_{i=j+1}^{n} \frac{\tau_i L_t(T_{i-1},T_i)\boldsymbol{\lambda}_{T_{i-1}}(t)'\boldsymbol{\lambda}_{T_{j-1}}(t)}{1+\tau_i L_t(T_{i-1},T_i)}, \quad j \in \{1,2,\ldots,n-1\}.$$

Notice that, unlike the volatility processes, which are by construction deterministic, the drift terms depend on past realizations of the LIBORs. This has the unfortunate implication that future LIBORs are no longer lognormally distributed in measures other than their native ones—a fact that complicates the simulation process.

Modeling the Volatilities

We have modeled the ensemble of LIBORs as driven by an $\Re_k$-valued Brownian motion $\{\mathbf{W}_t\}$ rather than by a single scalar-valued process in order to permit rates to be imperfectly correlated. Clearly, rates do tend to move together, since yield curves continually shift up and down. On the other hand, slopes of yield curves also change, indicating that the correlation in rates is not perfect. Exactly how the various rates coevolve in our model is determined by the specification of the vector-valued volatility processes, $\{\boldsymbol{\lambda}_{T_{j-1}}(t)\}_{j=1}^{n}$. However, an alternative to modeling all these $\Re_k$-valued processes is to express the rates' normal components, $\{\int_0^t \boldsymbol{\lambda}_{T_{j-1}}(s)' \cdot d\mathbf{W}_s\}_{j=1}^{n}$, in terms of n correlated, scalar-valued Brownian motions and scalar-valued volatilities. This is done as follows. Suppose we have chosen the measure $\mathbb{P}^{T_m}$ associated with numeraire $\{B(t, T_m)\}$. Then for $t \leq T_{j-1}$ we have

$$\begin{aligned}\ln L_t(T_{j-1}, T_j) &= \ln L_0(T_{j-1}, T_j) + \int_0^t \mu_{T_{j-1},T_m}(s) \cdot ds \\ &\quad - \frac{1}{2}\int_0^t \boldsymbol{\lambda}_{T_{j-1}}(s)'\boldsymbol{\lambda}_{T_{j-1}}(s) \cdot ds - \int_0^t \boldsymbol{\lambda}_{T_{j-1}}(s)' \cdot d\mathbf{W}_s^{T_m},\end{aligned}$$

where $\{\mu_{T_{j-1},T_m}(t)\}$ is the drift process under $\mathbb{P}^{T_m}$, as just described. By setting

$$\begin{aligned}\lambda_{T_{j-1}}(t)^2 &\equiv \boldsymbol{\lambda}_{T_{j-1}}(t)'\boldsymbol{\lambda}_{T_{j-1}}(t) \\ W_{jt}^{T_m} &\equiv \lambda_{T_{j-1}}(t)^{-1}\int_0^t \boldsymbol{\lambda}_{T_{j-1}}(s)' \cdot d\mathbf{W}_s^{T_m},\end{aligned} \tag{10.65}$$

we can dispense with the vector notation and write $\ln L_t(T_{j-1}, T_j)$ as

$$\ln L_0(T_{j-1}, T_j) + \int_0^t \mu_{T_{j-1},T_m}(s) \cdot ds - \frac{1}{2}\int_0^t \lambda_{T_{j-1}}(s)^2 \cdot ds - \int_0^t \lambda_{T_{j-1}}(s) \cdot dW_{js}^{T_m}.$$

Another rate $L_t(T_{l-1}, T_l)$ ($l \neq j$, $t \leq T_{l-1}$) will be driven by another scalar-valued process $\{W_{lt}^{T_m}\}$ that is also a Brownian motion under $\mathbb{P}^{T_m}$. The instantaneous correlation between the two Brownian motions at time t is

$$\rho_{jl}(t) \equiv \frac{\boldsymbol{\lambda}_{T_{j-1}}(t)'\boldsymbol{\lambda}_{T_{l-1}}(t)}{\lambda_{T_{j-1}}(t)\lambda_{T_{l-1}}(t)},$$

whereas, viewed from time zero, the correlation between $\ln L_t(T_{j-1}, T_j)$ and $\ln L_t(T_{l-1}, T_l)$ at fixed date $t \leq T_{j-1} \wedge T_{l-1}$ is

$$\rho_{jl}[0, t] \equiv \frac{\int_0^t \boldsymbol{\lambda}_{T_{j-1}}(s)'\boldsymbol{\lambda}_{T_{l-1}}(s) \cdot ds}{\int_0^t \lambda_{T_{j-1}}(s) \cdot ds \int_0^t \lambda_{T_{l-1}}(s) \cdot ds} = \frac{\int_0^t \rho_{jl}(s)\lambda_{T_{j-1}}(s)\lambda_{T_{l-1}}(s) \cdot ds}{\int_0^t \lambda_{T_{j-1}}(s) \cdot ds \int_0^t \lambda_{T_{l-1}}(s) \cdot ds}.$$

This will depend on the behavior of all three functions

$$\{\rho_{jl}(\cdot), \lambda_{T_{j-1}}(\cdot), \lambda_{T_{l-1}}(\cdot)\}$$

over $[0, t]$. Thus, to make the model accord with whatever market information or prior beliefs we have about correlations over intervals of time, we must specify all three functions appropriately.

What hard information do we have to aid in specifying the model? Not nearly enough, as it turns out. There are, of course, the implied volatilities from Black's formula and the market prices of caplets and floorlets, as inferred from prices of caps and floors. Indeed, it was the existence of liquid markets in these instruments—and traders' use of Black's formula for pricing them—that motivated the development of the LIBOR market model. Referring to (10.65) and (10.58), one sees that in our scalar version of the model the volatility $\bar{\lambda}_{T_{j-1}}$ that enters Black's formula for the initial value of a caplet settled at T_j, $C(0, K; T_{j-1}, T_j)$, is $\bar{\lambda}_{T_{j-1}} = \sqrt{T_{j-1}^{-1} \int_0^{T_{j-1}} \lambda_{T_{j-1}}(t)^2 \cdot dt}$. Observing the succession of these root-mean-squared volatilities at the (approximately) quarterly intervals at which caplets settle gives some indication of how volatilities vary with time, but these obviously do not fully specify the processes. There is also the possibility of calibrating to market quotes for swaptions, captions, etc., by adopting plausible (but, inevitably, *ad hoc*) specifications of functional forms for volatility functions and instantaneous correlations. For details of some of these schemes one may consult Rebonato (1999), Brigo and Mercurio (2001), and Joshi (2003).

10.5 Modeling Default Risk

In models treated thus far it has been assumed that parties to contracts always discharge their obligations. Thus, writers of options always supply the prescribed quantity of the underlying asset at the expiration of in-the-money calls, writers of puts always come up with the necessary funds to buy the assets, and issuers of bonds pay the coupons and face values in full and on time. While institutional arrangements do exist to reduce the risk of "nonperformance" in certain markets (e.g., clearing houses and margin requirements in option trades), and while a sovereign nation can usually be expected to pay claims denominated in the currency it controls, there are indeed financial instruments whose market prices are substantially affected by counterparty risk. As an introduction to the large literature on default risk we describe here some of the prominent models for pricing corporate debt.[15] Comprehensive treatments can be found in the books by Duffie and Singleton (2003) and Lando (2004).

There are two general approaches to the modeling of prices of defaultable bonds. What we shall call the "endogenous" approach views the credit event in terms of the relation between the value of the firm's assets and some threshold related to its liabilities. The simplest version, referred to as the Black-Scholes-Merton (BSM) model, is one in which a firm is obligated to redeem a single issue of zero-coupon bonds at some future date T. Default occurs if and only if the firm's assets at T are worth less than the face value of the bonds. An extension by Black and Cox (1976) allows default to occur at any time up to T, contingent on the value of assets falling below some time-dependent threshold. This is supposed to correspond to the requirements of protective covenants written into the bond indenture. To be operative, this setup requires modeling first-passage times of stochastic processes. As we saw in the description of barrier options in chapter 7, the theory is well developed for processes driven by Brownian motions and linear thresholds; however, extending to more realistic asset processes and barrier functions requires simulation to determine the distribution of default times and bond values. It is easier to get analytical expressions in the basic BSM framework. To illustrate, we focus here on just one fairly rich but still tractable version of that model.

A newer approach to modeling defaultable bonds' prices is to regard default events as essentially exogenous, much like the occurrences of natural disasters. The evolution of a firm's balance sheet is now unspecified, default simply being presumed to occur at the first arrival time of a Poisson process. However, to enhance flexibility and realism, the intensity of the jump process is itself modeled as varying stochastically with time, as if in response to the evolving information about the firm, its markets, and the general economic environment. One becomes more comfortable with this approach upon regarding it as yielding "reduced-form" models. That is, whatever state variables on which balance-sheet items depend—default-free interest rates, market prices of outputs and inputs, advances in technology, changes in regulatory constraints and tax laws, uninsured casualty losses, etc.—are implicitly viewed as evolving processes that drive the Poisson intensity. The obvious disadvantage of this abstract approach is that we lose a feel for what is "really" happening to the firm; nevertheless, its

[15] Our discussion is limited to defaultable bonds and does not deal with the risk of nonperformance on options, futures, or other instruments.

mathematical elegance and its complementarity with term structure modeling have brought it to prominence. In addition, since capital structure *per se* is no longer directly considered, models of exogenous risk can be applied to sovereign debt as well as to corporate liabilities.

10.5.1 *Endogenous Risk: The Black-Scholes-Merton Model*

The foundation of the BSM scheme was laid in the original papers of Black and Scholes (1973) and Merton (1973, 1978), who observed that corporate equity and liabilities have option-like features. We have already applied these ideas in section 7.2.1 to relate stock options and deposit insurance to compound calls and extendable puts. As we did there, let us consider a firm whose operations are financed by equity and debt, the latter in the form of zero-coupon bonds all maturing at future date T and having face value L. Ignoring taxes and potential claims by litigants and bankruptcy courts, ownership of the firm's assets is apportioned between stockholders and bondholders. Letting $\{A_t\}_{t\geq 0}$ represent the evolving market value of the firm's assets, the values of stockholders' and bondholders' claims at T are, respectively, $S_T = (A_T - L)^+$ and $b(T,T) = A_T - S_T$. That is, at the time of maturity stockholders (or managers acting in their interests) make the required payment of L to bondholders if and only if there is nonnegative residual value, $A_T - L$, for holders of equity. If there is not, default occurs. The available assets then pass to bondholders, and stockholders are left empty handed. Thus, in this simple framework the value of equity at any $t \in [0,T]$ is simply that of a T-expiring European call option on the firm's assets, $S_t = C^E(A_t, T-t)$, while the bonds' aggregate value is $b(t,T) = A_t - C^E(A_t, T-t)$. Alternatively, applying put-call parity, the bondholders' position can be stated as $B(t,T)L - P^E(A_t, T-t)$, where as usual $B(t,T)$ is the value of a default-free unit discount bond. In this view bondholders are short an option that gives stockholders the right to put the assets to them in the event $A_T < L$. The value of that option would correspond to the value of insuring the bondholders against default.

To see where this leads, it is helpful to normalize all market values by L, thus expressing them per unit face value of debt. Retaining the same symbols, we have S_t as the normalized value of equity and $b(t,T)$ as the price of a defaultable discount bond with unitary principal value. The bond's yield to maturity or average spot rate at t is then $r_d(t,T) = -(T-t)^{-1}\ln b(t,T)$, and the yield *spread* relative to default-free bonds is $s(t,T) = r_d(t,T) - r(t,T) = (T-t)^{-1}\ln[B(t,T)/b(t,T)]$. Our goal is to model how yield spreads vary with time to maturity, default-free interest rates, and characteristics of the firm's balance sheet and asset value process.

To show the versatility of the BSM setup, we shall use the Vasicek (1977) model for riskless rates of interest and specify $\{A_t\}$ as a jump-diffusion process. Thus, recalling the development in section 10.2.1, the short rate evolves under spot measure $\hat{\mathbb{P}}$ as $dr_t = (a^* - br_t) \cdot dt + \sigma d\hat{W}_t$, where a^* is a risk-adjusted drift parameter. The initial value of a T-maturing discount bond is

$$B(0,T) = \exp\left[-a^* \int_0^T \beta(T-u) \cdot du + \frac{\sigma^2}{2} \int_0^T \beta(T-u)^2 \cdot du - \beta(T)r_0\right],$$

where $\beta(t) \equiv b^{-1}(1-e^{-bt})$, and the value of the money-market fund at T is

$$\begin{aligned} M_T &= M_0 \exp\left[a^* \int_0^T \beta(T-u) \cdot du + \sigma \int_0^T \beta(T-u) \cdot d\hat{W}_u + \beta(T)r_0\right] \\ &= \frac{M_0}{B(0,T)} e^{\sigma \int_0^T \beta(T-u) \cdot d\hat{W}_u + \frac{\sigma^2}{2} \int_0^T \beta(T-u)^2 \cdot du}. \end{aligned} \tag{10.66}$$

Paralleling the exposition in section 9.2 (but dispensing with special symbols for risk-adjusted parameters), we model the normalized value of assets at T under $\hat{\mathbb{P}}$ as

$$A_T = A_0 \exp\left[\int_0^T r_u \cdot du - \theta\nu T - \frac{\sigma_A^2 T}{2} + \sigma_A \tilde{W}_T + \sum_{j=0}^{N_T} \ln(1+U_j)\right].$$

Here $\{N_t\}$ is a Poisson process with (constant) intensity θ; logarithms of jump shocks, $\{\ln(1+U_j)\}_{j=1}^{\infty}$, are i.i.d. as $N[\ln(1+\upsilon) - \xi^2/2, \xi^2]$; and $U_0 \equiv 0$. Brownian motions $\{\hat{W}_t, \tilde{W}_t\}$ that drive $\{r_t\}$ and $\{A_t\}$ are independent of $\{N_t\}$, and all three are independent of the jump sizes. However, we allow for correlation between interest rates and the asset process by setting $\tilde{W}_T = \rho\hat{W}_T + \bar{\rho}Y_T$, where $\bar{\rho} = \sqrt{1-\rho^2}$ and $\{Y_t\}$ is another Brownian motion independent of everything else.

The normalized (by L) value of the firm's equity at $t = 0$ is $S_0 = M_0\hat{E}M_T^{-1}(A_T - 1)^+$. As in section 9.2 we will work this out by evaluating the expectation conditional on N_T and then applying the tower property of conditional expectation. What makes this relatively easy in the Vasicek/jump-diffusion setting is that M_T is distributed as lognormal under $\hat{\mathbb{P}}$ and A_T is *conditionally* lognormal. Still, as in the option pricing example with the HJM model in section 10.3.3, it takes some effort to disentangle the stochastic interest rates from the asset value process. Here is one way to do it. Using (10.66) with $\sigma_M^2 \equiv \sigma^2 T^{-1} \int_0^T \beta(T-u)^2 \cdot du$, we can write

$$
\begin{aligned}
M_T^{-1} &= \frac{B(0,T)}{M_0} e^{-\frac{\sigma_M^2 T}{2} - [\sigma \int_0^T \beta(T-u)\cdot d\hat{W}_u]} \\
A_T &= \frac{A_0}{B(0,T)} e^{-\theta\nu T - \frac{\sigma_A^2 T}{2} + \frac{\sigma_M^2 T}{2} + [\sigma \int_0^T \beta(T-u)\cdot d\hat{W}_u + \sigma_A \tilde{W}_T + \sum_{j=1}^{N_T} \ln(1+U_j)]}.
\end{aligned}
$$

Conditioning on N_T, the bracketed term in the exponent in the expression for A_T is distributed as normal with conditional mean $N_T[\ln(1+v) - \xi^2/2] \equiv \mu_N$, and using $\tilde{W}_T = \rho \hat{W}_T + \bar{\rho} Y_T$ one sees that its conditional variance is

$$
\sigma_N^2 T \equiv \sigma_M^2 T + 2\rho\sigma_A \sigma \int_0^T \beta(T-u)\cdot du + \sigma_A^2 T + N_T \xi^2. \tag{10.67}
$$

We can therefore represent the term as $\mu_N + \sigma_N \sqrt{T} Z_N$, where $Z_N \sim N(0,1)$ conditional on N_T. Likewise, the bracketed term in the exponent in the expression for M_T^{-1} is distributed as $N(0, \sigma_M^2 T)$, so we write it as $\sigma_M \sqrt{T} Z_M$. Its covariance with $\sigma_N \sqrt{T} Z_N$ is $\sigma_{NM} T \equiv \sigma_M^2 T + \rho\sigma_A \sigma \int_0^T \beta(T-u)\cdot du$, and the correlation between Z_N and Z_M is $\rho_{NM} = \sigma_{NM}/(\sigma_N \sigma_M)$. We can now set $Z_M = \rho_{NM} Z_N + \bar{\rho}_{NM} Z_*$, say, where Z_* is conditionally independent of Z_N. Then, with $B_0 \equiv B(0,T)$,

$$
\begin{aligned}
S_0 &= \hat{E} e^{-\frac{\sigma_M^2 T}{2} - \bar{\rho}_{NM}\sigma_M \sqrt{T} Z_*} \\
&\quad \cdot \hat{E} e^{-\rho_{NM}\sigma_M \sqrt{T} Z_N} \left[A_0 e^{-\theta\nu T + \mu_N - \frac{\sigma_A^2 T}{2} + \frac{\sigma_M^2 T}{2} + \sigma_N \sqrt{T} Z_N} - B_0 \right]^+ \\
&= e^{-\frac{\rho_{NM}^2 \sigma_M^2 T}{2}} \hat{E} e^{-\rho_{NM}\sigma_M \sqrt{T} Z_N} \left[A_0 e^{-\theta\nu T + \mu_N - \frac{\sigma_A^2 T}{2} + \frac{\sigma_M^2 T}{2} + \sigma_N \sqrt{T} Z_N} - B_0 \right]^+ \\
&= A_0 e^{-\frac{\rho_{NM}^2 \sigma_M^2 T}{2} - \theta\nu T + \mu_N - \frac{\sigma_A^2 T}{2} + \frac{\sigma_M^2 T}{2}} \hat{E} e^{(\sigma_N - \rho_{NM}\sigma_M)\sqrt{T} Z_N} \mathbf{1}_{(q_N,\infty)}(Z_N) \\
&\quad - B_0 e^{-\frac{\rho_{NM}^2 \sigma_M^2 T}{2}} \hat{E} e^{-\rho_{NM}\sigma_M \sqrt{T} Z_N} \mathbf{1}_{(q_N,\infty)}(Z_N),
\end{aligned}
$$

where

$$
q_N \equiv \frac{-\ln(A_0/B_0) + \theta\nu T - \mu_N + \sigma_A^2 T/2 - \sigma_M^2 T/2}{\sigma_N \sqrt{T}}.
$$

Working out expectations conditional on N_T and setting

$$
\begin{aligned}
B_N &\equiv B_0 (1+\nu)^{-N_T} e^{\theta\nu T} \\
q_N^{\pm} &\equiv \frac{\ln(A_0/B_0) \pm \sigma_N^2 T/2}{\sigma_N \sqrt{T}}
\end{aligned}
$$

give

$$\begin{aligned} S_0 &= B_0 \hat{E} B_N^{-1}[A_0 \Phi(q_N^+) - B_N \Phi(q_N^-)] \\ &= e^{-\theta \nu T} \hat{E}(1+\nu)^{N_T} C^E(A_0, T; B_N, \sigma_N). \end{aligned}$$

Here $C^E(A_0, T; B_N, \sigma_N^2)$ represents the Black-Scholes formula for a T-expiring European call with unit strike, but with B_N as the price of a riskless discount bond and σ_N as the volatility of the underlying. As we saw in pricing options under jump dynamics in section 9.2, this can be evaluated as

$$S_0 = \sum_{n=0}^{\infty} p(n; \theta_v T) C^E(A_0, T; B_n, \sigma_n),$$

where $p(n; \theta_v T)$ represents $\hat{\mathbb{P}}(N_T = n)$ for a Poisson law with intensity $\theta_v \equiv \theta(1+\nu)$. Notice that stochastic short rates affect the result only through the pseudo-volatility parameter σ_N, as given by (10.67).

As described earlier, we can now get the initial value of the defaultable unit bond as $b(0,T) = A_0 - S_0$, and from this come the yield spreads as $s(0,T) = T^{-1} \ln[B(0,T)/b(0,T)]$. Figure 10.5 presents plots of $s(0,T)$

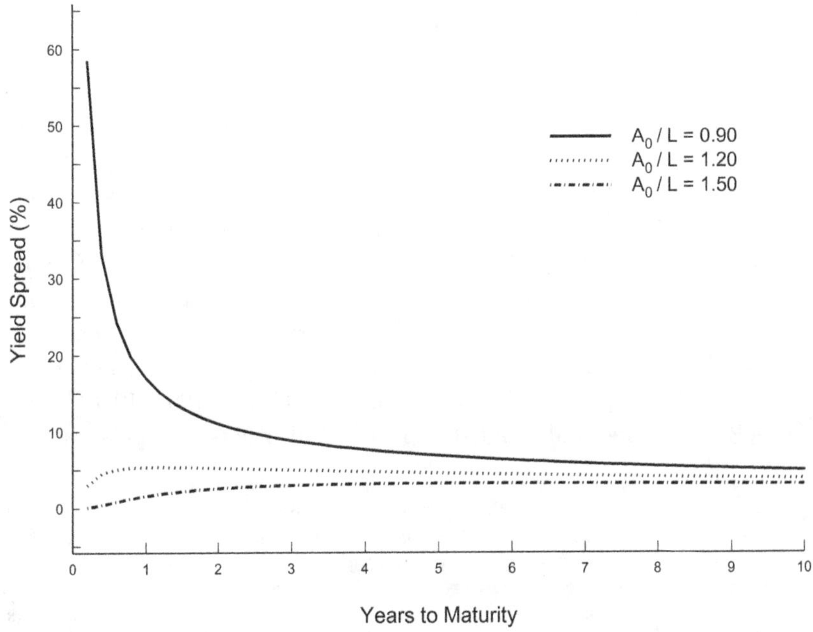

Fig. 10.5. The BSM yield spreads *vs.* time to maturity, for three initial values of assets/debt.

versus T for three values of A_0. The calculations (done with subroutine MERTON in program JUMP on the CD) are with average (default-free) spot rate $r(0,T) = 0.05$, interest rate parameters $b = 0.5$ and $\sigma = 0.1$, asset parameters $\theta = 0.1, \xi = 0.2, \sigma_A = 0.3, \nu = -0.1$, and correlation $\rho = 0.5$ between the two Brownian motions. These parameters generate enormous but rapidly declining spreads for a firm that is already insolvent, but they produce relatively flat term structures for solvent firms. Still, because of the possibility of negative jumps in the value of assets, there are positive spreads for solvent but highly leveraged firms at even the shortest maturities. All of this seems to correspond reasonably well to what we observe in markets for corporate bonds.[16]

10.5.2 *Exogenous Default Risk*

The jump-diffusion model just considered is one of many applications of Poisson processes encountered in this survey of derivatives pricing. Recalling the definition from pages 91 and 126, a Poisson process $\{N_t\}_{t\geq 0}$ with intensity θ is an adapted process on a filtered probability space $(\Omega, \mathcal{F}, \{\mathcal{F}_t\}, \mathbb{P})$ with independent increments and with

$$\mathbb{P}(N_{t+\tau} - N_t = n \mid \mathcal{F}_t) = \mathbb{P}(N_{t+\tau} - N_t = n) = (\theta\tau)^n e^{-\theta\tau}/n!, n \in \aleph_0. \tag{10.68}$$

This is often referred to as a "homogeneous" Poisson process to distinguish from versions in which θ is time-varying. While the homogeneous model has heretofore served as our main workhorse for modeling randomly occurring events, it is sometimes advantageous to extend it by allowing intensity to vary with time—either deterministically or stochastically. Processes with stochastic intensity, called "doubly stochastic processes" or "Cox processes", have long been applied to model life expectancies, failure times of industrial processes, etc.,[17] and we have already seen an application to financial derivatives in section 9.3.2. The use of Cox processes in modeling credit risk was pioneered and developed by Lando (1994, 1998), Jarrow *et al.* (1997), Duffie and Singleton (1999), and others. Before seeing how

---

[16] Of course, we have considered only discount bonds, whereas most long-term corporate issues carry coupons. It is not so easy to price coupon bonds in the BSM framework because periodic cash payments affect the balance sheet. Lando (2004) describes recursive procedures for pricing under various assumptions about how the payments are financed.

[17] For an overview of the theory and a survey of applications see Cox and Oakes (1984). Grandell (1976) presents the theory in the deeper context of point processes and random measures.

they are applied in this context, it will be helpful to learn a bit more about the theory and to trace them back to the homogeneous version in (10.68).

One will recall from page 61 that the probability generating function (p.g.f.) of a random variable X supported on $\aleph_0$ is the function $\Pi_X(\zeta) = E\zeta^X$, $\zeta \in \mathfrak{C}$, and that the probability mass function of X can be generated from Π_X by differentiation, as $\Pi_X^{(k)}(\zeta)|_{\zeta=0} = \mathbb{P}(X = k)$. In particular, $\Pi_X^{(0)}(0) \equiv \Pi_X(0) = \mathbb{P}(X = 0)$. When X is distributed as Poisson with parameter $\theta > 0$ we have $\Pi_X(\zeta) = e^{(\zeta-1)\theta}$ and $\mathbb{P}(X = 0) = e^{-\theta}$. Thus, for a homogenous Poisson process $\{N_t\}_{t\geq 0}$ with intensity θ we have $\Pi_{N_t}(\zeta) = e^{(\zeta-1)\theta t}$ and $\mathbb{P}(N_t = 0) = e^{-\theta t}$. We will construct a doubly stochastic process from this in stages. First, we obtain an inhomogeneous Poisson process by allowing $\{\theta_t\}$ to be a positive-valued, deterministic, right-continuous step function—a mapping from $\Re^+$ to a countable set of positive numbers. Putting $t_0 = 0$, introduce a strictly increasing sequence of real numbers $\{t_j\}_{j=0}^{\infty} \uparrow \infty$, a sequence of positive constants $\{\theta_j\}_{j=1}^{\infty}$, and a sequence $\{X_{jt}\}_{j=1}^{\infty}$ of independent homogeneous Poisson processes, with θ_j as the intensity of X_j. We construct a process $\{N_t\}_{t\geq 0}$ by putting $N_t = X_1(t)$ for $t \in [0, t_1)$, $N_t = X_1(t_1) + X_2(t) - X_2(t_1)$ for $t \in [t_1, t_2)$, $N_t = X_1(t_1) + X_2(t_2) - X_2(t_1) + X_3(t) - X_3(t_2)$ for $t \in [t_2, t_3)$, and so on, with $N_t = \sum_{i=1}^{j-1}[X_i(t_i) - X_i(t_{i-1})] + X_j(t) - X_j(t_{j-1})$ for $t \in [t_{j-1}, t_j)$ and $j > 1$. Increments to this process over nonoverlapping periods of time are clearly independent, and for any fixed t the p.g.f. of N_t is

$$\Pi_{N_t}(\zeta) = e^{(\zeta-1)[\sum_{j=1}^{j_t} \theta_j (t_j - t_{j-1}) + \theta_{j_t+1}(t - t_{j_t})]}, \tag{10.69}$$

where $j_t \equiv \max\{j : t_j \leq t\}$. The form of $\Pi_{N_t}(\cdot)$ shows that N_t is distributed as Poisson, and because the increments are independent, $\{N_t\}$ is indeed a Poisson process.

To generalize further, let $\theta : \Re^+ \to \Re^+$ be an $\mathcal{F}_0$-measurable (i.e., deterministic), right-continuous, Riemann-integrable function. For some $m \in \aleph$ choose the $\{t_j\}$ such that $\max_j |t_j - t_{j-1}| \leq 1/m$. Then setting $\theta_j = \theta_{t_j}$ in (10.69) and letting $m \to \infty$ give $\Pi_{N_t}(\zeta) \to e^{(\zeta-1)\int_0^t \theta_s \cdot ds}$ and $N_t \rightsquigarrow P(\int_0^t \theta_s \cdot ds)$ for each $t \geq 0$. Finally, letting $\{\theta_t\}$ be an adapted process whose sample paths $\{\theta_t(\omega)\}$ are a.s. positive, right-continuous, and integrable, a doubly stochastic adapted process $\{N_t\}$ is obtained by following the above construction for each sample path $\{\theta_t(\omega)\}$. The p.g.f. is then $\Pi_{N_t}(\zeta) = E\exp[(\zeta - 1)\int_0^t \theta_s \cdot ds]$, and the probability that there are no jumps up to t is $\mathbb{P}(N_t = 0) = Ee^{-\int_0^t \theta_s \cdot ds}$. Note that in the doubly stochastic setup the information set at t, $\mathcal{F}_t$, includes both histories, $\{\theta_s\}_{0\le s\le t}$ and $\{N_s\}_{0\le s\le t}$.

Another construction that is useful in simulating Cox processes exploits the fact that interarrival times $\{\mathfrak{T}_n - \mathfrak{T}_{n-1}\}_{n=1}^{\infty}$ of Poisson processes with deterministic intensities are distributed as exponential. The recipe is as follows. First, generate the realization of an indefinitely long sample path of intensities, $\{\theta_s(\omega) = \theta_s^*\}_{s\ge 0}$, using an appropriate model. Next, generate realizations $\{Y_j(\omega)\}_{j=1}^{\infty}$ of i.i.d. exponential random variables with expected value unity. These have distribution function $(1-e^{-y})\mathbf{1}_{[0,\infty)}(y)$ and c.f. $(1-i\zeta)^{-1}, \zeta \in \Re$. Next, setting $\mathfrak{T}_0(\omega) = 0$, determine sequentially for $n = 1, 2, \ldots$ the realizations of (stopping) times $\mathfrak{T}_n(\omega) = \inf\{t : \int_0^t \theta_s^* \cdot ds \ge \sum_{j=1}^n Y_j(\omega)\}$. Thus, $\mathfrak{T}_1(\omega)$ is the first time (strictly, the greatest lower bound of times) at which $\int_0^t \theta_s^* \cdot ds$ is at least $Y_1(\omega)$; $\mathfrak{T}_2(\omega)$ is the first time at which $\int_{\mathfrak{T}_1}^t \theta_s^* \cdot ds$ is at least $Y_2(\omega)$; and so on. Finally, for $n = 1, 2, \ldots$ set $N_t(\omega) = n-1$ for $\mathfrak{T}_{n-1}(\omega) \le t < \mathfrak{T}_n(\omega)$ and $N_{\mathfrak{T}_n}(\omega) = n$. Figure 10.6 illustrates. The resulting process $\{N_t\}$ satisfies, for the given realization $\{\theta_s^*\}$,

$$\mathbb{P}(N_t < n \mid \{\theta_s^*\}) = \mathbb{P}(\mathfrak{T}_n > t \mid \{\theta_s^*\}) = \mathbb{P}\left(\sum_{j=1}^n Y_j > \int_0^t \theta_s^* \cdot ds\right).$$

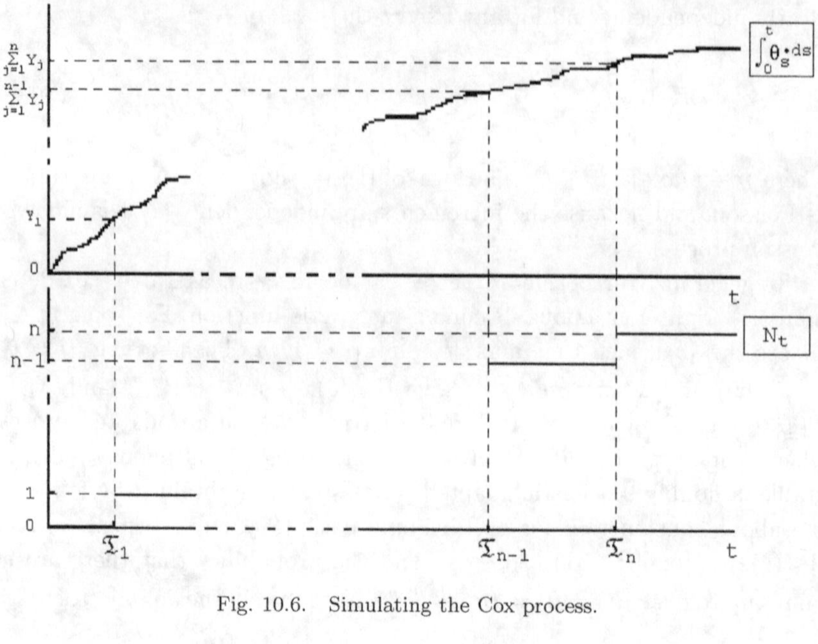

Fig. 10.6. Simulating the Cox process.

The c.f. of $\sum_{j=1}^{n} Y_j$ being $(1 - i\zeta)^{-n}$, one sees (from table 2.3) that it is distributed as $\Gamma(n, 1)$. Thus, setting $\Theta_t \equiv \int_0^t \theta_s^* \cdot ds$,

$$\mathbb{P}(N_t < n \mid \{\theta_s^*\}) = \int_{\Theta_t}^{\infty} \frac{y^{n-1}e^{-y}}{(n-1)!} \cdot dy = \sum_{j=0}^{n-1} \frac{\Theta_t^j e^{-\Theta_t}}{j!},$$

where the last equality follows upon repeated integration by parts. N_t is therefore distributed as Poisson with parameter $\int_0^t \theta_s^* \cdot ds$, conditional on $\{\theta_s\} = \{\theta_s^*\}$. In particular, $\mathbb{P}(N_t = 0 \mid \{\theta_s^*\}) = \mathbb{P}(N_t < 1 \mid \{\theta_s^*\}) = e^{-\int_0^t \theta_s^* \cdot ds}$, and averaging over realizations of $\{\theta_s\}_{s \geq 0}$ again delivers $\mathbb{P}(N_t = 0) = Ee^{-\int_0^t \theta_s \cdot ds}$.

Now consider an entity (a firm, a machine, an organism, ...) that endures and functions (in some objective sense) up to the first arrival time $\mathfrak{T}_1$ of such a doubly stochastic process, $\mathfrak{T}_1$ itself marking the precise instant of "failure". Then $\mathbb{P}(N_t = 0) = Ee^{-\int_0^t \theta_s \cdot ds} = E\mathbf{1}_{(t,\infty]}(\mathfrak{T}_1)$ represents the probability that failure will occur (if at all) *after* t, which is to say that it is the probability of survival to t.[18] Moreover, since for $T \geq t$ event $N_T = 0$ implies $N_t = 0$, we have

$$\begin{aligned} e^{-\int_t^T \theta_s \cdot ds} &= e^{-\int_0^T \theta_s \cdot ds + \int_0^t \theta_s \cdot ds} \\ &= \frac{\mathbb{P}(N_T = 0 \mid \{\theta_s\})}{\mathbb{P}(N_t = 0 \mid \{\theta_s\})} \\ &= \frac{\mathbb{P}(N_T = 0, N_t = 0 \mid \{\theta_s\})}{\mathbb{P}(N_t = 0 \mid \{\theta_s\})} \\ &= \mathbb{P}(N_T = 0 \mid N_t = 0, \{\theta_s\}). \end{aligned}$$

Therefore

$$E_t e^{-\int_t^T \theta_s \cdot ds} = \mathbb{P}(N_T = 0 \mid \mathcal{F}_t) = E_t \mathbf{1}_{(T,\infty]}(\mathfrak{T}_1)$$

is the probability of survival to T given survival to t. Notice that explicitly conditioning on $\mathfrak{T}_1 > t$ is not required since information set $\mathcal{F}_t$ includes the entity's current status. Letting $T = t + \Delta t$, we have, approximately for small Δt, $E_t e^{-\int_t^{t+\Delta t} \theta_s \cdot ds} \doteq 1 - \theta_t \Delta t$ as the probability of survival over the next small interval of time. Therefore, θ_t can be thought of as an instantaneous default rate.[19]

[18] Intensities $\{\theta_s\}$ may be such that $E\exp(-\int_0^\infty \theta_s \cdot ds) > 0$, as, for example, if $\mathbb{P}(\int_0^\infty \theta_s \cdot ds \leq m) = 1$ for some finite m. Failure might then never occur, so $\mathfrak{T}_1 : \Omega \to \Re^+ \cup \{\infty\}$ must be considered an "extended" random variable; hence, the notation $\mathbf{1}_{(t,\infty]}(\mathfrak{T}_1)$.

[19] It was to justify this interpretation that we required sample paths of $\{\theta_s\}$ to be a.s. right continuous.

Let us now apply these concepts to financial markets. Consider a firm's traded obligation to pay one currency unit at future date T conditional on the firm's remaining solvent until that time—and *nothing* otherwise. In other words, the security represents a defaultable, zero-coupon bond with zero recovery of principal upon default. With the first jump of some doubly stochastic process $\{N_t\}$ signaling default, the bond's value at T is $b_0(T,T) = \mathbf{1}_{(T,\infty]}(\mathfrak{T}_1)$, where subscript 0 signifies zero recovery. The market price of the security at $t = 0$, assuming the firm is then solvent, can be expressed as

$$b_0(0,T) = M_0 \hat{E} M_T^{-1} \mathbf{1}_{(T,\infty]}(\mathfrak{T}_1) = \hat{E} e^{-\int_0^T r_t \cdot dt} \mathbf{1}_{(T,\infty]}(\mathfrak{T}_1), \qquad (10.70)$$

where, since we are valuing a traded asset, the expectation is now in the equivalent spot measure $\hat{\mathbb{P}}$. Of course, that the firm is currently solvent is included in the market's initial information, $\mathcal{F}_0$, on which the expectation is conditioned. In measure $\hat{\mathbb{P}}$ there corresponds to $\{\theta_t\}_{t\geq 0}$ a risk-adjusted process $\{\hat{\theta}_t\}_{t\geq 0}$, in terms of which we can express the risk-neutral probability of survival:

$$\hat{\mathbb{P}}(N_T = 0) = \hat{E}\mathbf{1}_{(T,\infty]}(\mathfrak{T}_1) = \hat{E} e^{-\int_0^T \hat{\theta}_t \cdot dt}.$$

However, we would like to be able to express $b_0(0,T)$ also in terms of sample paths of $\{\hat{\theta}_t\}$, since we could then value the bond just by modeling $\{r_t\}$ and $\{\hat{\theta}_t\}$. Doing this requires a little ingenuity.

Let $\mathcal{G}_t$ be the σ-field generated by the joint history $\{\hat{\theta}_s, r_s\}_{0\leq s\leq t}$, with $\{\mathcal{G}_t\}_{t\geq 0}$ as the corresponding filtration. We insist that $\mathcal{G}_t \subseteq \mathcal{F}_t$ for each t. For example, if it were possible to express $(\hat{\theta}_t, r_t)$ as some function $\mathbf{g} : \Re_k \to \Re_2$ of t-observable state variables $\mathbf{X}_t$, we could take $\mathcal{G}_t$ to be the σ-field generated by $\{\mathbf{X}_s\}_{0\leq s\leq t}$. In any case, whether there are such state variables or not, the paths $\{\hat{\theta}_t, r_t\}_{0\leq t\leq T}$ out to maturity are $\mathcal{G}_T$-measurable, so we can write

$$\begin{aligned}\hat{E}\left[e^{-\int_0^T r_t \cdot dt} \mathbf{1}_{(T,\infty]}(\mathfrak{T}_1) \mid \mathcal{G}_T\right] &= e^{-\int_0^T r_t \cdot dt} \hat{E}[\mathbf{1}_{(T,\infty]}(\mathfrak{T}_1) \mid \mathcal{G}_T] \\ &= e^{-\int_0^T (r_t + \hat{\theta}_t) \cdot dt}.\end{aligned}$$

Now, evaluating (10.70) by conditioning first on on $\mathcal{G}_T$ and then on $\mathcal{G}_0$, we have

$$\begin{aligned} b_0(0,T) &= \hat{E}\left\{\hat{E}\left[\hat{E}\left(e^{-\int_0^T r_t \cdot dt} \mathbf{1}_{(T,\infty]}(\mathfrak{T}_1) \mid \mathcal{G}_T\right) \mid \mathcal{G}_0\right]\right\} \\ &= \hat{E}\left[\hat{E}\left(e^{-\int_0^T (r_t + \hat{\theta}_t) \cdot dt} \mid \mathcal{G}_0\right)\right] \\ &= \hat{E} e^{-\int_0^T (r_t + \hat{\theta}_t) \cdot dt}, \end{aligned}$$

where the last equality follows from $\mathcal{G}_0 \subseteq \mathcal{F}_0$. This achieves the desired result. More generally, the security's value at any $t \in [0, T]$ is

$$b_0(t,T) = \hat{E}\left[e^{-\int_t^T (r_s+\hat{\theta}_s)\cdot ds} \mid \mathcal{F}_t\right] \equiv \hat{E}_t e^{-\int_t^T (r_s+\hat{\theta}_s)\cdot ds}.$$

Modeling prices of defaultable bonds with zero recovery can thus be reduced to specifying the joint evolution of the instantaneous spot rate and the instantaneous, risk-adjusted default rate. Some ways of doing this are considered below.

Of course, the potential for partial recovery of principal does have to be considered in pricing actual corporate liabilities. This is handled automatically in the BSM framework, since the evolutions of assets' values and debt are formally modeled, but the exogenous risk approach requires a patchwork solution. To see what is involved, let us start with the following basic pricing relation for a defaultable discount bond with potential partial recovery:

$$\begin{aligned} b(t,T) &= \hat{E}_t\left[e^{-\int_t^T r_s\cdot ds}\mathbf{1}_{(T,\infty]}(\mathfrak{T}_1) + e^{-\int_t^{T^*} r_s\cdot ds} R_{T^*}\mathbf{1}_{(t,T]}(\mathfrak{T}_1)\right] \\ &= b_0(t,T) + \hat{E}_t e^{-\int_t^{T^*} r_s\cdot ds} R_{T^*}\mathbf{1}_{(t,T]}(\mathfrak{T}_1). \end{aligned}$$

Here, the first term is the bond's zero-recovery value, and the second term is the contribution from the uncertain amount R_{T^*} that is to be paid at some date T^* contingent on default. (Of course, we take for granted that the firm is solvent at t, so that $\mathbf{1}_{(t,\infty]}[\mathfrak{T}_1(\omega)] = \mathbf{1}_\Omega(\omega) = 1$.) The problem is that in any specific application there is little to guide us in modeling either the uncertain amount or the uncertain time of payment. Some of the ways proposed to do this are (i) setting $T^* = \mathfrak{T}_1$ and assuming R to be an $\mathcal{F}_t$-measurable fraction of face value (i.e., R is known in advance and paid upon default); (ii) setting $T^* = \mathfrak{T}_1$ and assuming R to be an $\mathcal{F}_{\mathfrak{T}_1}$-measurable fraction of face value (i.e., R revealed and paid upon default); (iii) setting $T^* = \mathfrak{T}_1$, with R an $\mathcal{F}_{\mathfrak{T}_1-}$-measurable fraction of the bond's value just before default, $b(\mathfrak{T}_1-, T)$.

Given the level of ignorance and the multitude of possibilities, one may feel justified in opting for an approach that has the assured virtue of simplicity. One such is to set $T^* = T$ and model R as an $\mathcal{F}_T$-measurable random variable supported on $[0, 1)$ and independent of default intensities and interest rates. This scheme yields the following very tractable result:

$$\begin{aligned}
b(t,T) &= \hat{E}_t[e^{-\int_t^T r_s \cdot ds}\mathbf{1}_{(T,\infty]}(\mathfrak{T}_1) + R_T e^{-\int_t^T r_s \cdot ds}\mathbf{1}_{(t,T]}(\mathfrak{T}_1)] \\
&= \hat{E}_t e^{-\int_t^T r_s \cdot ds}\mathbf{1}_{(T,\infty]}(\mathfrak{T}_1) + \hat{E}_t R_T \cdot \hat{E}_t e^{-\int_t^T r_s \cdot ds}[1 - \mathbf{1}_{(T,\infty]}(\mathfrak{T}_1)] \\
&= \hat{E}_t e^{-\int_t^T r_s \cdot ds} \cdot \hat{E}_t R_T + \hat{E}_t e^{-\int_t^T r_s \cdot ds}\mathbf{1}_{(T,\infty]}(\mathfrak{T}_1)(1 - \hat{E}_t R_T) \\
&= B(t,T)\hat{E}_t R_T + b_0(t,T)(1 - \hat{E}_t R_T),
\end{aligned}$$

where, as usual, $B(t,T)$ is the value of a default-free discount bond.

We can now focus on modeling $B(t,T)$ and $b_0(t,T)$, which is done by specifying the joint process $\{r_t, \hat{\theta}_t\}_{0 \le t \le T}$. Since $\hat{\theta}_t$ must be strictly positive, a bivariate version of the Vasicek (1977) model is unacceptable; but extending the mean-reverting, square-root process of Cox *et al.* (1985) is feasible, and this yields expressions that are computationally manageable. The direct approach is to model $\{r_t\}$ and $\{\hat{\theta}_t\}$ as independent CIR processes:

$$d\begin{pmatrix} r_t \\ \hat{\theta}_t \end{pmatrix} = \begin{pmatrix} a - b^* r_t \\ a_\theta - b_\theta \hat{\theta}_t \end{pmatrix} \cdot dt + \begin{pmatrix} \sigma & 0 \\ 0 & \sigma_\theta \end{pmatrix} \begin{pmatrix} r_t^{1/2} \\ \hat{\theta}_t^{1/2} \end{pmatrix} \cdot d\begin{pmatrix} \hat{W}_{1t} \\ \hat{W}_{2t} \end{pmatrix},$$

where $\{\hat{W}_{1t}\}$ and $\{\hat{W}_{2t}\}$ are independent Brownian motions under $\hat{\mathbb{P}}$, and b^* is the risk-adjusted drift parameter that we encountered in section 10.2.2. In this case $B(t,T)$ is as in (10.16), and

$$\begin{aligned}
b_0(t,T) &= B(t,T)\hat{E}_t e^{-\int_t^T \hat{\theta}_s \cdot ds} \\
&= B(t,T)A_\theta(T-t)e^{-\hat{\theta}_t \alpha_\theta (T-t)},
\end{aligned}$$

where $A_\theta(\cdot)$ and $\alpha_\theta(\cdot)$ are given by (10.18) and (10.17) with (a_θ, b_θ) in place of (a, b^*). The indirect approach, which allows for dependence between $\{r_t\}$ and $\{\hat{\theta}_t\}$, is to model each as an affine function of the same k -vector of state variables $\{\mathbf{X}_t\}$, whose components would in turn be modeled as independent CIR processes: $\{dX_{jt} = (a_j - b_j^* X_{jt}) \cdot dt + \sigma_j \sqrt{X_{jt}} \cdot d\hat{W}_{jt}\}_{j=1}^k$. This, too, yields easily computed formulas for $B(t,T)$ and $b_0(t,T)$. For details and extensions to other affine models for $\{r_t, \hat{\theta}_t\}$ one can consult Duffie and Singleton (2003) and Lando (2004).

PART III

COMPUTATIONAL METHODS

11

Simulation

Pricing derivative assets by martingale methods involves finding mathematical expectations of discounted contingent payoffs. The discounted payoffs are themselves functions (or, in the case of path-dependent derivatives, functionals) of prices of underlying primary assets, whose conditional distributions under the martingale measure are deduced from the dynamic model for the underlying. The pricing problem therefore amounts to finding expected values of various functions of random variables—that is, Lebesgue-Stieltjes integrals of functions with respect to c.d.f.s. The problem can be analytically intractable when either the integrand function is complicated or the c.d.f. is difficult to determine. In the former case numerical integration is usually the preferred approach; in the latter case, either simulation, the binomial method, or lattice solutions of p.d.e.s. When there are multiple state variables, simulation is typically the only feasible method.

Simulation is merely an application of the law of large numbers. To get the basic idea, suppose we are to evaluate $Eg(Y) = \int g(y) \cdot dF(y)$ for some (possibly vector-valued) random variable Y. We will look at more sophisticated schemes later, but the straightforward approach is simply to generate a large number of realizations $\{y_j\}_{j=1}^N$ by sampling independently from the distribution F, evaluate $g(\cdot)$ for each realization, and average to obtain the estimate, $\widehat{Eg} = N^{-1} \sum_{j=1}^N g(y_j)$. When realizations $\{y_j\}$ are generated independently, the standard error of the estimate can itself be estimated consistently as

$$\hat{\sigma}(\widehat{Eg}) = \sqrt{N^{-1} \sum_{j=1}^{N} y_j^2 - \widehat{Eg}^2}.$$

Moreover, the central limit theorem can be applied to obtain an asymptotic $100(1-\alpha)\%$ confidence interval for Eg as

$$\widehat{Eg} \pm \Phi^{-1}(1-\alpha/2)\hat{\sigma}(\widehat{Eg}),$$

where $\Phi^{-1}(\cdot)$ is the inverse of the standard normal c.d.f.[1] For example, with $\alpha = 0.05$ the approximate 95% confidence interval is $\widehat{Eg} \pm 1.96\hat{\sigma}(\widehat{Eg})$. In the application to derivatives, $g(Y)$ will represent the payoff of a derivative asset, and Y the (vector of) random influence(s) on which it depends. For example, for a vanilla European call option struck at X, random variable Y would be the price of the underlying at the expiration of the call, F would be the conditional distribution of the terminal price, and $g(Y)$ would be the discounted value of $(Y-X)^+$. For a European-style fixed-strike lookback or Asian call option Y would be, respectively, the maximum or average price over some portion of the option's life—a functional of the price path.

Simulation is well adapted to situations in which values of discounted payoffs depend on more than one stochastic state variable; as, for example, when volatility and/or interest rates are stochastic. The binomial method, which evaluates expected payoffs by discretizing the state space, requires a multidimensional lattice in such cases. Simulation is also well adapted to path-dependent derivatives, such as lookbacks and Asians, for which the entire sample path of the underlying price must be known in order to determine the payoff. As we will see in detail shortly, one approaches these problems by discretizing the time to expiration and simulating the random evolution of price one subinterval at a time. The information that determines the derivative's ultimate payoff is updated as the path progresses. Binomial pricing is cumbersome in such cases because it is not straightforward to determine the potential payoffs associated with all paths that pass through each node. However, simulation also has its limitations. Simulation proceeds by working forward in time, generating sample paths progressively from one step to the next. In pricing an American-style derivative, however, the natural approach is to compare at each step its intrinsic and continuation values—but the latter depends on the possible *subsequent* paths from that point. Of course, finding the continuation value is easy in the backwards-recursive binomial scheme. Nevertheless, despite the inherent difficulty, we will see that there are ways to use simulation to price American-style derivatives.

[1] That is, $\Phi^{-1}(1-\alpha/2)$ is the $1-\alpha/2$ quantile of the standard normal—the value z such that $\Phi(z) = 1-\alpha/2$.

At the heart of the simulation method is the task of getting realizations of random variables. While databases of random deviates are available (e.g., Marsaglia (1995)), these are ordinarily generated as part of the simulation routine. Realizations generated by digital computer are necessarily "pseudorandom", meaning they are produced by some purely deterministic, replicable process that gives the appearance of randomness in the sense of passing various statistical tests. In section 11.1 we discuss pseudorandom number generators for various distributions, giving particular attention to the uniform and normal. We then consider some ways of enhancing the efficiency of simulation estimators, focusing on methods that have the greatest relevance in financial applications. While the straightforward approach to simulation outlined above is always feasible, we will see in section 11.2 that there are usually shortcuts that produce estimators $\widehat{Eg}$ with lower variance and/or shorter execution times. Such variance-reduction schemes can be applied both in generating random deviates and in averaging the realizations of $g(Y)$.[2] Finally, in section 11.3 we will look at specific applications to the pricing of derivatives, including those that can be exercised early.

11.1 Generating Pseudorandom Deviates

Programs that generate pseudorandom variates of various forms are widely available. The International Mathematics and Statistics Library® (IMSL) contains a wide selection of executable routines in FORTRAN. The book of FORTRAN 77 routines by Press *et al.* (1992) and its counterparts in FORTRAN 90, C, Pascal, and Basic also cover many of the standard distributions and give source code with explanations of the algorithms and programming steps employed. The present discussion provides a general description and overview of the prominent methods, with emphasis on those that are of greatest relevance for financial applications.

11.1.1 *Uniform Deviates*

Standard techniques for generating random deviates from any distribution begin with realizations of uniform deviates. There are two standard options

[2] Both of these sections draw heavily from the book by Fishman (1996), which contains a wealth of additional material. Hammersley and Handscomb (1964) is still a good resource, as well. Press *et al.* (1992) gives an excellent overview of random number generators. Glasserman (2004) gives a comprehensive overview of simulation methods, with emphasis on financial applications.

for obtaining realizations of random variables distributed as uniform on the interval $(0, 1)$. One of the safest methods is to tap an existing database that is known to have passed a battery of powerful statistical tests. Marsaglia (1995) has produced such a collection, available on CDROM. The other standard option is to produce the numbers in the course of the simulation experiment using a uniform number "generator"—a body of computer code that exists either in machine language on a specific computer system or in a higher-level language, such as FORTRAN or C. Computational and programming systems like Gauss® and Matlab® usually have built-in generators. A desirable feature of such generators is that they be portable—that is, that they can be used with a variety of computers and compilers. Some understanding of generators is necessary if simulations are to produce reliable results.

An algorithm on which many number generators are based is the multiplicative congruential scheme. Beginning with a "seed" value I_0 supplied by the user, these generate new integer-valued numbers as $I_j = aI_{j-1}(\text{mod}\, m)$, where a and m are carefully chosen integers. I_j is thus the remainder after aI_{j-1} is divided by m; for example, $2(\text{mod}\, 7) = 9(\text{mod}\, 7) = 16(\text{mod}\, 7) = 2$. Since the process obviously terminates when a zero is first produced, a and m must be chosen to exclude this possibility. Since $m(\text{mod}\, m) = 0$, this limits the range of I to the integers $1, 2, \ldots, m-1$. Division by m then converts to values $u \equiv I/m$ on $(0, 1)$. As a simple example, the merely illustrative choices $a = 2$, $m = 7$, and seed $I_0 = 1$ generate the sequence $1, 2, 4, 1, 2, 4, 1 \ldots$, repeating after three digits. Clearly, this is a bad choice of a, since not even all integers to $m-1$ are obtainable. The multiplier $a = 3$ does better, generating $1, 3, 2, 6, 4, 5, 1$. The message is that $m - 1$ is the maximum period—that is, the maximum length of nonrepeating string—but that poor choices of a can leave significant gaps and produce periods much shorter than $m - 1$.

Since m limits the best possible outcome, it is obviously desirable to make it as large as possible. This is where the limits of particular machine architectures come to be felt, since m is constrained by the word size on the particular platform in use. With IBM mainframes and most PCs and workstations the limit of 31 bits (excluding the sign bit) allows integers of up to 2^{31}.[3] A standard choice for m is the prime number $2^{31} - 1 = 2147483647$,

[3]In fact, integers to $2^{32} - 1$ are managed in IBM-compatible computers by using the sign bit to record $2^{31}, \ldots, 2^{32} - 1$ as negative numbers; that is, as $-2147483648, -2147483647, \ldots, -1$.

and handling even this moderate number requires a special programming trick to assure that $a(m-1)$ does not exceed the maximum word size. Different generators come with their own special choices of a (see below). Good choices maximize the length of the nonrepeating string and eliminate obvious patterns in the output. One pattern that can remain when $a \ll m$ is that small values of I_j are apt to be followed by smaller values than average, since in such cases $aI_j \ll m$. Some routines employ a shuffling scheme to reduce this first-order dependence. These work by first creating a preliminary table of integers by the usual multiplicative scheme, then generating each I_j not directly from I_{j-1} but from a value chosen at random from the table, which is then replaced by I_j.

Notice that all sequences of numbers produced by the generator from a given seed I_0 will be the same. This is actually an attractive feature, since it enables one to replicate the random number exactly. The last value of I produced in a call to the generator serves as the seed that continues the sequence where it left off. Notice, too, that the uniformly distributed numbers $u = I/m$ produced by multiplicative congruential schemes are limited to the *open* interval $(0, 1)$.

IMSL's `RNUN` and *Numerical Recipes* (Press *et al.*, 1992) routines `RANn` $(n = 0, 1, \ldots, 3)$ are examples of portable generators. `RAN1` uses $m = 2^{31} - 1$ with $a = 16807$ and a shuffling scheme with a 32-element table. The IMSL routines allow 128-element shuffling as an option and offer a choice of multipliers $a = 16807$, 397204094, or 95070637. These progressively trade speed for increasing degrees of apparent randomness. Fishman (1996) presents a version of FORTRAN subroutine `RAND`, written by L.R. Moore of the Rand Corporation, that uses $a = 95070637$ without shuffling. FORTRAN and C++ versions of this in function subprogram format, `RANDFUN`, are on the accompanying CD. High-level languages such as Gauss and Matlab have built in generators for uniforms and other standard distributions.

Whatever is used, it is essential to run at least cursory checks on the output with one's own computer platform and software. In some cases even the supposedly portable routines are really compiler specific.[4] A test for uniformity using the Neyman (1937) N_2 statistic is included on the CD

[4]As an example, the version of `RAN1` that appeared in Press *et al.* (1986) gives nonsensical results with 32-bit FORTRAN 77 compilers by Watcom (version 9.0) and Absoft (version 5.0), yet passes cursory tests (and gives the same output) with Watfor87 and Lahey F77L (version 4.10) compilers. (With Absoft version 5.0 it outputs a single number, .86528, on every call!) However, the (1992) version of `RAN1` produces the same (satisfactory) results with all four compilers.

as `UTEST`.[5] The test, while known to have good power against broad alternatives to uniformity, is sensitive mainly to uncharacteristic behavior of the first two moments. It does not necessarily pick up problems like large gaps in the frequency distribution (indicative of a bad choice of multiplier a in multiplicative congruential generators) that a simple plot would reveal. A good supplement is simply to use a spreadsheet or statistical graphics program to construct and plot a frequency distribution of a sample of the output.

Besides the uniformity of the numbers, the lack of egregious forms of serial dependence will be important in simulating sample paths of asset prices. A simple runs test that compares the number of sequences of numbers below 0.5 is included on the CD as `RUNTEST`. Much more thorough checks can be made with tests supplied by IMSL (e.g., `RUNS`, `PAIRS`, `DSQAR`, `DCUBE`). Marsaglia (1995) includes a battery of tests for randomness with his collection of random deviates (which pass the tests).

11.1.2 *Deviates from Other Distributions*

The process of generating pseudorandom realizations from other distributions than the uniform usually begins with a uniform deviate on $(0,1)$. A very basic and often very attractive general method produces pseudorandom deviates with c.d.f. F by running uniform deviates through F^{-1}.

The Inverse Probability Integral Transform

Let F be a c.d.f., and define the functions $Y^{\pm} : (0,1) \to \Re$ as $Y^{+}(u) = \inf\{y : F(y) > u\}$ and $Y^{-}(u) = \inf\{y : F(y) \geq u\}$.

Example 68 *Exponential distribution with unit scale parameter:*

$$F(y) = \begin{cases} 0, & y \leq 0 \\ 1 - e^{-y}, & y > 0 \end{cases}$$

$$Y^{\pm}(u) = -\ln(1-u).$$

[5] For description of the procedure and discussion of its power, see D'Agostino and Stephens (1986, pp. 351–354, 357) and references cited there.

Example 69 *Bernoulli distribution with parameter θ:*

$$F(y) = \begin{cases} 0, & y < 0 \\ 1 - \theta, & 0 \le y < 1 \\ 1, & y \ge 1 \end{cases}$$

$$Y^{+}(u) = \begin{cases} 0, & 0 \le u < 1 - \theta \\ 1, & 1 - \theta \le u < 1 \end{cases}$$

$$Y^{-}(u) = \begin{cases} 0, & 0 < u \le 1 - \theta \\ 1, & 1 - \theta < u \le 1. \end{cases}$$

In general, Y^{+} and Y^{-} coincide for continuous distributions, and even in the discrete case Y^{+} and Y^{-} are indistinguishable when U is distributed as uniform on $(0,1)$, in the sense that $\mathbb{P}[Y^{+}(U) = Y^{-}(U)] = 1$. Taking an arbitrary stand, define the inverse c.d.f. as $F^{-1}(u) \equiv Y^{+}(u) = \inf\{y : F(y) > u\}$ and set $Y(u) \equiv F^{-1}(u)$. Then, if realizations of U are generated randomly from $U(0,1)$, we have

$$\mathbb{P}(Y \le y) = \mathbb{P}[F^{-1}(U) \le y] = \mathbb{P}[U \le F(y)] = F(y).$$

Thus, the random variable $Y(U) = F^{-1}(U)$ is indistinguishable from a random variable having the distribution F. The transformation that takes Y to the random variable $F(Y) \sim U(0,1)$ is often called the "probability integral transform". We require the inverse of this.

As an example, table 11.1 shows the first ten pseudorandom uniforms from `RANDFUN` with seed 1.0 and the corresponding exponential and Bernoulli deviates (with $\theta = 2/3$). Notice that since U and $1 - U$ have the same distributions, exponentials can be generated simply as $-\ln U$. Fishman (1996, table 3.1) gives expressions for inverse c.d.f.s of several prominent continuous distributions and suggestions for computational shortcuts. Routine `POIDEV` on the CD uses the inverse transform method to produce vectors of Poisson variates.

Aside from simplicity in many applications, the inverse transform method has two advantages, both relating to variance-reduction methods discussed in the next section. The first is that $Y(U)$ is a monotone-increasing function of U, which implies that $Y(U)$ and $Y(1-U)$ are negatively correlated but with identical marginal distributions. We will see that this feature often makes it possible to produce Monte Carlo estimates of $Eg(Y)$ more efficiently. The second advantage of the transform method is its ability to generate variates efficiently from modifications of F that have restricted support. For example, if Y with c.d.f. F is supported on $[a, d]$,

Table 11.1. Simulated uniforms, exponential, and Bernoulli variates.

Uniform u	Exponential $y = -\ln u$	Bernoulli $y = \mathbf{1}_{[1/3,1)}(u)$
.44271	0.58466	1
.06008	0.06196	0
.80478	1.63365	1
.17005	0.18639	0
.91588	2.47554	1
.48670	0.66689	1
.29624	0.35132	0
.74641	1.37203	1
.29842	0.35442	0
.20020	0.22339	0

then the conditional distribution given $Y \in [b, c]$, where $a \leq b < c \leq d$, is

$$F(y|Y \in [b,c]) = \begin{cases} 0, & y < b \\ \dfrac{F(y) - F(b)}{F(c) - F(b)}, & b \leq y < c \\ 1, & y \geq c. \end{cases}$$

The inverse of the conditional distribution is then

$$F^{-1}(u|Y \in [b,c]) = F^{-1}\{u[F(c) - F(b)] + F(b)\} \tag{11.1}$$

This property is relevant for generating stratified samples, which are explained below.

The main disadvantage of the inverse transform method is that it is not so simple to use when there is no convenient expression for the inverse c.d.f. Examples are the normal and gamma distributions, whose c.d.f.s themselves cannot be expressed in closed form. Satisfactory closed-form approximations for inverse c.d.f.s are sometimes available; for example, Fishman (1996, table 3.1) gives an approximation for the normal inverse. Otherwise, one can use a series expansion to represent the c.d.f., then apply an iterative root-finding approach to determine $F^{-1}(u)$. However, this takes computational and programming time, and other procedures are usually preferable.

The Rejection Method

Another general approach uses properties of conditional probability and the Radon-Nikodym derivative to generate pseudorandom variates with c.d.f. F. Let G be a c.d.f. corresponding to a measure equivalent to F; thus, a support $\mathcal{Y}$ of F is a support of G and *vice versa*. Let $R = dF/dG$ be the Radon-Nikodym derivative, which is required to be bounded—that is, G must be such that there exists $\bar{R} < \infty$ with $R(y) \le \bar{R}$ for $y \in \mathcal{Y}$. Then

$$F(y) = \int_{\mathcal{Y}\cap(-\infty,y]} R(x) \cdot dG(x).$$

Suppose now that we generate independent realizations of random variables X and U from G and $U(0,1)$, respectively. Then

$$\begin{aligned}\mathbb{P}[U \le R(X)\bar{R}^{-1}] &= \int_{\mathcal{Y}} \mathbb{P}[U \le R(x)\bar{R}^{-1}] \cdot dG(x)\\ &= \int_{\mathcal{Y}} R(x)\bar{R}^{-1} \cdot dG(x) = \bar{R}^{-1};\end{aligned}$$

and the conditional probability that $X \le y$ given that $U \le R(X)\bar{R}^{-1}$ is

$$\begin{aligned}\mathbb{P}[X \le y | U \le R(X)\bar{R}^{-1}] &= \frac{\mathbb{P}[X \le y \cap U \le R(X)\bar{R}^{-1}]}{\mathbb{P}[U \le R(X)\bar{R}^{-1}]}\\ &= \bar{R} \cdot \mathbb{P}[X \le y \cap U \le R(X)\bar{R}^{-1}]\\ &= \bar{R}\int_{\mathcal{Y}\cap(-\infty,y]} R(x)\bar{R}^{-1} \cdot dG(x)\\ &= \int_{\mathcal{Y}\cap(-\infty,y]} R(x) \cdot dG(x)\\ &= F(y).\end{aligned}$$

The equality $\mathbb{P}[X \le y|U \le R(X)\bar{R}^{-1}] = F(y)$ gives us a recipe for generating a random variable with c.d.f. F. First, find another c.d.f., G, with the same support, such that dF can be factored as the product of an almost–surely bounded function R times dG. When F is absolutely continuous, R will be a ratio of densities; otherwise, it will be a ratio of probability mass functions (p.m.f.s). Next, generate independent realizations x and u from G and $U(0,1)$. If $u \le R(x)\bar{R}^{-1}$, where $\bar{R}$ is the upper bound of R, then accept x as a realization from F; otherwise, reject and draw again. The probability of accepting is $\bar{R}^{-1}$. The process is obviously most efficient when (i) G is a distribution from which it is easy to generate realizations (as by the inverse-transform method) and (ii) $\bar{R}^{-1}$ is close to unity.

As an example, consider simulating Poisson variates, having p.m.f. $f(y) = \theta^y e^{-\theta}/y!$ for $y \in \mathcal{Y} \equiv \{0,1,2,\ldots\}$ in the special case that $\theta \in (0,1)$.

Take as the comparison distribution the geometric, $g(y) = (1-\theta)\theta^y$ for $y \in \mathcal{Y}$. The geometric c.d.f., $G(y) = 1 - \theta^{[y+1]}$, has a simple inverse, $G^{-1}(u) = [\ln u / \ln \theta - 1]$, where in both expressions $[\cdot]$ is the greatest integer function. This makes it easy to generate geometric variates by the inverse transform method. Since $R(y) = (1-\theta)^{-1}e^{-\theta}/y! \le (1-\theta)^{-1}e^{-\theta}$, we have $\bar{R}^{-1} = (1-\theta)e^{\theta}$. This declines from unity to zero as θ increases from zero to unity. The rejection method will therefore work efficiently when θ is small.

Generating Normals

While both inverse-transform and rejection methods may be used to generate pseudorandom normals, there is a simpler alternative using a transformation devised by Box and Muller (1958). Let U be distributed as uniform on $(0,1)$ and V be an independent exponential variate with unit scale. That is, $V = -\ln U'$, where U' is uniform, independent of U. The joint p.d.f. of U and V is

$$f_{UV}(u,v) = e^{-v}, \quad u \in (0,1), \quad v > 0$$

and $f_{UV}(u,v) = 0$, elsewhere. Consider the bivariate transformation $X = \sqrt{2V}\cos(2\pi U)$, $Y = \sqrt{2V}\sin(2\pi U)$. This takes $(0,1) \times \Re^+$ to $\Re_2$ and has inverse $V = (X^2 + Y^2)/2$ and $U = \tan^{-1}(X/Y)$. The Jacobian of the transformation is

$$\begin{aligned} J &\equiv \begin{vmatrix} \partial x/\partial u & \partial y/\partial u \\ \partial x/\partial v & \partial y/\partial v \end{vmatrix} \\ &= \begin{vmatrix} -2\pi\sqrt{2v}\sin(2\pi u) & 2\pi\sqrt{2v}\cos(2\pi u) \\ \dfrac{1}{\sqrt{2v}}\cos(2\pi u) & \dfrac{1}{\sqrt{2v}}\sin(2\pi u) \end{vmatrix} \\ &= -2\pi. \end{aligned}$$

Change-of-variable formula (2.17) now shows that X and Y are bivariate (spherical) normal; that is,

$$\begin{aligned} f_{XY}(x,y) &= f_{UV}(u,v) \cdot |J^{-1}| \\ &= \frac{1}{2\pi}e^{-(x^2+y^2)/2}. \end{aligned}$$

The method thus transforms independent uniform and exponential variates into a pair of independent standard normals.

While this is amply straightforward, there are some clever programming tricks that obviate the need to evaluate the sine and cosine functions. Routine `GASDEV` in Press *et al.* (1992) employs such a method. Another is given in algorithm NA in Fishman (1996, pp. 189–191). Function `GAUSFUN` on the CD is an implementation of the NA algorithm. IMSL routines `RNNOR` and `RNNOA` use inverse transformation and rejection methods, respectively, of which the second is faster. In fact, all four routines are extremely fast. Of course, speed is just one consideration. Unfortunately, none of the methods holds up especially well under powerful statistical tests for univariate normality. Routines `ADTEST` and `CFTEST` on the CD implement Anderson-Darling (1954) and integrated characteristic function (i.c.f.) tests for normality (Epps and Pulley, 1983; Baringhaus *et al.*, 1989). Applied to samples of sizes 500, 1000, and 2000 generated by the four routines, these tests reject the null hypothesis of univariate normality at frequencies much higher than nominal significance levels. `GAUSFUN` does as well as any of the routines. Since in all cases the test results are highly dependent on the initial seeds used to begin the sequence of pseudorandom uniforms, one is advised to experiment with several seeds before putting any of these routines to work. Testing multivariate normality with subsets of the output would also be desirable. For tests of this sort see Baringhaus and Henze (1988) and Mardia (1970).

11.2 Variance-Reduction Techniques

As indicated in the introduction, the straightforward application of Monte Carlo methods to estimating $Eg(Y) = \int g(y) \cdot dF(y)$ is simply to average a large number of replicates of the function evaluated at pseudorandom realizations of the random variable Y. In other words, one computes $\widehat{Eg} \equiv N^{-1}\sum_{j=1}^{N} g(y_j)$. This section explores ways to improve accuracy and/or reduce the computational time relative to this standard approach. Stratified sampling and importance sampling both have to do with how the replicates of y are produced, while the use of antithetic variates and control variates involve other aspects of the experimental design.

11.2.1 *Stratified Sampling*

As we know from elementary statistics, "random" samples are not always representative samples. The manifestation of this fact in simulation experiments is that the empirical (sample) frequency distribution, F_N, of a finite number N of replicates from a distribution F may differ considerably from F. Apart from the limitations of pseudorandom number generators, we do know that $\sup_y |F_N(y) - F(y)| \to 0$ a.s. as $N \to \infty$ (this is the Glivenko-Cantelli theorem); however, it would be nice to have some way to reduce the sizes of errors for fixed N. This is what stratified sampling is intended to accomplish. The idea is to break the support of F into strata and sample from each with the appropriate frequency.

More specifically, let F be supported on $\mathcal{Y}$, and let $\{\mathcal{Y}_s\}_{s=1}^S$ be a partition. These will be the strata from which we sample. Typically, they are contiguous, disjoint intervals. Let $p_s \equiv \mathbb{P}(Y \in \mathcal{Y}_s)$ and $E_s g(Y) \equiv E[g(Y)|Y \in \mathcal{Y}_s]$ be, respectively, the probability measure of stratum s and the conditional expectation of g given that Y takes on a value in the stratum. Taking $N_s \equiv \langle p_s N \rangle$ draws of Y from stratum s, where the angular brackets indicate "nearest integer", we can estimate $E_s g(Y)$ consistently and without bias as

$$\widehat{E_s g} = N_s^{-1} \sum_{j=1}^{N_s} g(y_{js}).$$

The stratified estimator of $Eg(Y)$, also unbiased and consistent, is then

$$\widehat{ESg} \equiv \sum_{s=1}^{S} p_s \widehat{E_s g}.$$

Considering that the straight-on estimator, $\widehat{Eg} = N^{-1} \sum_{j=1}^{N} g(y_j)$, is also unbiased and consistent, one is entitled to ask what stratification accomplishes. The answer is that the stratified estimator has lower variance for any fixed total number of replicates, N. In the process of proving this we shall discover a consideration that is involved in selecting strata. Letting

$$\gamma^2 \equiv Vg(Y) = \int_{\mathcal{Y}} g(y)^2 \cdot dF(y) - (Eg)^2, \tag{11.2}$$

where $V(\cdot)$ is the variance operator, note first that $V\widehat{Eg} = \gamma^2/N$. We can find the variance of the stratified estimator as follows:

$$V\widehat{ESg} = \sum_{s=1}^{S} p_s^2 V\widehat{E_s g} = \sum_{s=1}^{S} p_s^2 \gamma_s^2 / N_s = N^{-1} \sum_{s=1}^{S} p_s \gamma_s^2, \tag{11.3}$$

where $\gamma_s^2 \equiv V[g(Y)|Y \in \mathcal{Y}_s]$ is the conditional variance of g given that Y is from stratum s. This conditional variance can be expressed as

$$\begin{aligned}\gamma_s^2 &= \int_{\mathcal{Y}_s} [g(y) - E_s g]^2 \cdot dF(y|Y \in \mathcal{Y}_s)\\ &= \int_{\mathcal{Y}_s} \{[g(y) - Eg] + (Eg - E_s g)\}^2 \cdot dF(y|Y \in \mathcal{Y}_s)\\ &= \int_{\mathcal{Y}_s} [g(y) - Eg]^2 \cdot dF(y|Y \in \mathcal{Y}_s) - (Eg - E_s g)^2.\end{aligned}$$

Plugging into (11.3) gives

$$\begin{aligned}V\widehat{E_S g} &= \frac{1}{N}\sum_{s=1}^{S} p_s \int_{\mathcal{Y}_s} [g(y) - Eg]^2 \cdot dF(y|Y \in \mathcal{Y}_s) - \frac{1}{N}\sum_{s=1}^{S} p_s (Eg - E_s g)^2\\ &= \frac{1}{N}\sum_{s=1}^{S} \int_{\mathcal{Y}_s} [g(y) - Eg]^2 \cdot dF(y) - \frac{1}{N}\sum_{s=1}^{S} p_s (Eg - E_s g)^2\\ &= \frac{1}{N}\int_{\mathcal{Y}} [g(y) - Eg]^2 \cdot dF(y) - \frac{1}{N}\sum_{s=1}^{S} p_s^2 (Eg - E_s g)\\ &= V\widehat{Eg} - \frac{1}{N}\sum_{s=1}^{S} p_s^2 (Eg - E_s g)^2.\end{aligned}$$

The summation in the second term is simply the variance of the conditional expectations of $g(Y)$ across the strata. If there is just one stratum, or if multiple strata are chosen so that there is no variation, then the second term is zero. Otherwise, stratification does produce an estimator of lower variance than the unstratified estimator.

Getting the most out of stratification involves choosing strata that maximize the cross-stratum variation in $E_s g$. This suggests (correctly) that any refinement of the partition $\{\mathcal{Y}_s\}_{s=1}^{S}$—that is, any further stratification—can only lower, and cannot increase, the variance. However, since simulating replicates from the stratified distributions $F(\cdot|Y \in \mathcal{Y}_s)$ takes more time than simulating from F alone, there are limits to how far one should go. Given a decision as to the number of strata, however, it is clear that they should be chosen so as to maximize the variation in the conditional mean of g.

How can one generate replicates from the strata? This just involves sampling from each of the conditional distributions $F(\cdot|Y \in \mathcal{Y}_s)$. This restricted sampling is where the inverse transform method is particularly handy. Expression (11.1) gives the inverse of the conditional distribution in terms of the inverse of the unconditional in the case that the strata are intervals. Extension to more general strata, such as unions of intervals, is straightforward.

11.2.2 *Importance Sampling*

The identity

$$V\widehat{E_S g} = N^{-1}\sum_{s=1}^{S} p_s V[g(Y)|Y \in \mathcal{Y}_s]$$

that is implied by (11.3) shows that the effective application of stratified sampling involves putting more weight on regions of the support where the variation in g is low. That is, when $V[g(Y)|Y \in \mathcal{Y}_s]$ is small for some s, we want p_s to be large. This has the effect of increasing the sampling rate, $N_s = \langle p_s N\rangle$, where sampling is more informative about Eg. Importance sampling does the same thing in a more subtle manner.

Here is how it works. In estimating $Eg(Y) = \int_{\mathcal{Y}} g(y) \cdot dF(y)$, let us make a change of measure from F to a measure G with respect to which F is absolutely continuous, as

$$Eg(Y) = \int_{\mathcal{Y}} g(y)R(y) \cdot dG(y),$$

where $R \equiv dF/dG$ is the Radon-Nikodym derivative. This reminds us that not only can we estimate Eg by drawing replicates $\{y_j\}$ from F and averaging the $\{g(y_j)\}$, but also by drawing replicates $\{y_j\}$ from G and averaging $\{g(y_j)R(y_j)\}$. When would such a change of measure be advantageous?

One application is in variance reduction, as the following argument shows. Letting $\widehat{EgR} \equiv N^{-1}\sum_{j=1}^{N} g(y_j)R(y_j)$ be the modified estimator, we have

$$\begin{aligned} V\widehat{EgR} &= N^{-1}\int_{\mathcal{Y}} g(y)^2 R(y)^2 \cdot dG(y) - (Eg)^2 \\ &= N^{-1}\int_{\mathcal{Y}} g(y)^2 R(y) \cdot dF(y) - (Eg)^2, \end{aligned}$$

so that, using (11.2),

$$\begin{aligned} V\widehat{Eg} - V\widehat{EgR} &= N^{-1}\int_{\mathcal{Y}} g(y)^2 \cdot dF(y) - N^{-1}\int_{\mathcal{Y}} g(y)^2 R(y) \cdot dF(y) \\ &= N^{-1}\int_{\mathcal{Y}} g(y)^2[1 - R(y)] \cdot dF(y). \end{aligned}$$

This expression could be of either sign. Notice that since $\int_{\mathcal{Y}}[1 - R(y)] \cdot dF(y) = 0$ the integral is merely the covariance between $1 - R(Y)$ and $g(Y)^2$. The expression is positive, in which case the change of measure does effect a reduction in variance, when G is such that $1 - R(Y)$ and $g(Y)^2$ have positive covariance or, equivalently, when R and g^2 covary negatively. When this is so, more weight is placed on those regions of $\mathcal{Y}$ in which sampling is more informative about Eg, much as with stratification. In other words, the fact that G has more probability mass in those areas means that more of the given sample of replicates will be drawn from there than would be drawn under F. This explains the name "importance sampling".

Fishman (1996, section 4.1) gives several examples of the application of importance sampling in variance reduction. Another application of the method is just to swap distributions from which it is relatively hard to generate replicates for easier ones. For example, suppose $Y \sim \Gamma(\alpha, \beta)$ and that we want to estimate

$$\begin{aligned} Eg(Y) &= \int_0^\infty g(y) \frac{1}{\Gamma(\alpha)\beta^\alpha} y^{\alpha-1} e^{-y/\beta} \cdot dy \\ &= \int_0^\infty g(\beta y) \frac{1}{\Gamma(\alpha)} y^{\alpha-1} e^{-y} \cdot dy. \end{aligned}$$

Taking $dF(y) = \Gamma(\alpha)^{-1} y^{\alpha-1} e^{-y} \cdot dy$ and $dG(y) = e^{-y} \cdot dy$ (exponential), we have

$$Eg(Y) = \int_0^\infty g(\beta y) R(y) e^{-y} \cdot dy,$$

where $R(y) = \Gamma(\alpha)^{-1} y^{\alpha-1}$. Generating pseudorandom exponentials now leads to the estimate $\widehat{Eg} = \Gamma(\alpha)^{-1} N^{-1} \sum_{j=1}^N g(\beta y_j) y_j^{\alpha-1}$. Although accuracy falls off as α increases, this is much faster to calculate and will give good results for α near unity.

11.2.3 *Antithetic Variates*

If we take independent draws Y and Y' from F and calculate $g(Y)$ and $g(Y')$, then the average, $[g(Y) + g(Y')]/2$, has variance $Vg(Y)/2$. If we make N replications of this bivariate sample and form the estimator

$$(2N^{-1}) \sum_{j=1}^N [g(Y_j) + g(Y_j')],$$

then the variance is

$$\frac{1}{4N}[Vg(Y) + Vg(Y')] = \frac{1}{2N}Vg(Y).$$

Of course, this is exactly the same thing—and involves the same amount of work—as taking $2N$ replicates from F and calculating

$$\widehat{Eg} \equiv (2N)^{-1} \sum_{j=1}^{2N} g(Y_j),$$

so nothing special has been accomplished. However, suppose we could somehow generate a dependent pair Y, Y' from a joint distribution $F(y, y')$, both of whose marginals were equal to $F(y)$. Then the variance of the bivariate estimator, which we now denote $\widehat{Eg}_A$, would be

$$\begin{aligned} V\widehat{Eg}_A &= N^{-1}\{Vg(Y) + 2Cov[g(Y), g(Y')] + Vg(Y')\}/4 \\ &= (2N^{-1})Vg(Y)(1 + \rho_{gg'}), \end{aligned} \tag{11.4}$$

where $\rho_{gg'}$ is the correlation between $g(Y)$ and $g(Y')$. Clearly, relative to estimating with $2N$ independent replicates, this would achieve a reduction in variance if $\rho_{gg'} < 0$, and the gain would be greater the stronger the correlation. If Y' could be generated at least as fast as Y itself, then there would be a clear efficiency gain.

The three results below, established by Hoeffding (1940), provide guidance in putting this into practice by showing how to construct bivariate random variables having minimal correlation, subject to their having the same marginal distributions. Here X and X' are random variables with joint c.d.f., $F_{XX'}$, and with identical marginals, F, such that VX exists. For proofs of somewhat more general versions of the first two see Fishman (1996); the third is proved below.

1. $Cov(X, X') = \int_{\Re_2} F_{XX'}(x, x') \cdot dx \cdot dx' - (EX)^2$.
2. $F_{XX'}(x, x') \geq [F(x) + 1 - F(x')]^+$ for $(x, x') \in \Re_2$.
3. Construct random variables W and W' as $W = F^{-1}(U)$, $W' = F^{-1}(1-U)$, where $U \sim U(0, 1)$. Then W and W' are identically distributed as F, and

$$F_{WW'}(w, w') = [F(w) + F(w') - 1]^+.$$

Result 1, which is a product-moment extension of expression (2.39) in section 2.2.3, shows that, *ceteris paribus*, covariance increases with the integral of the joint c.d.f. Result 2 establishes a minimum for the joint c.d.f. at each point in $\Re_2$, in terms of the marginals. Finally, result 3 shows how bivariate random variables with given marginals can be constructed whose joint c.d.f. attains the lower bound. With this construction W and W' have minimal correlation subject to being identically distributed (and having finite variance). It is not hard to see (Fishman, 1996, p. 313) that this minimal correlation is in fact negative.

Applying these results to the Monte Carlo estimation of $Eg(Y)$ where Y has c.d.f. F, suppose first that g is a nondecreasing function and that Y is generated as $F^{-1}(U)$ where $U \sim U(0,1)$. Set $W = g \circ F^{-1}(U) \equiv h(U)$. Since $F^{-1}(U)$ is nondecreasing, so is h, which is the composition of two nondecreasing functions. Indeed, h^{-1} is the c.d.f. of W, since $\mathbb{P}(W \leq w) = \mathbb{P}[h(U) \leq w] = \mathbb{P}[U \leq h^{-1}(w)] = h^{-1}(w)$. Then, setting $W' = h(1-U)$, we have

$$\begin{aligned} F_{WW'}(w, w') &= \mathbb{P}[\{h(U) \leq w\} \cap \{h(1-U) \leq w'\}] \\ &= \mathbb{P}[\{U \leq h^{-1}(w)\} \cap \{1 - U \leq h^{-1}(w')\}] \\ &= \mathbb{P}[1 - h^{-1}(w') \leq U \leq h^{-1}(w)] \\ &= [h^{-1}(w) + h^{-1}(w') - 1]^+ \\ &= [F_W(w) + F_W(w') - 1]^+. \end{aligned}$$

Besides establishing Hoeffding's result 3 (take $g \equiv 1$ in the above), this in connection with results 1 and 2 shows that we induce minimal correlation in $g(Y)$ and $g(Y')$, subject to maintaining equality in the marginal distributions, by generating Y and Y' as $F^{-1}(U)$ and $F^{-1}(1-U)$, respectively. That the same result applies to nonincreasing functions g as well follows upon generating Y as $F^{-1}(1-U)$.

The dependent random variables Y and Y' that are generated in this fashion are called "antithetic variates". Since the construction guarantees that $\rho_{gg'} < 0$ when g is monotone, (11.4) shows that the use of antithetic variates necessarily reduces the variance of the estimator of Eg. If g is not monotone on the support of Y then these optimal results cannot be guaranteed; however, it is clear from (11.4) that the variance of $\widehat{Eg}_A$ constructed from N pairs of replicates Y, Y' can in no case be larger than the variance of $\widehat{Eg}$ from N realizations of Y alone. If generating Y' requires little extra work, then the antithetic procedure will usually extend the speed-accuracy frontier.

How much work, then, is involved in generating Y'? The construction presented above is limited to variates generated via the inverse-transform approach. Generating Y' in this way as $F^{-1}(1-U)$ takes twice the time as generating Y, and this can be costly when F^{-1} is not easily calculated. If Y could be generated in some faster way than by inverse transform, it seems that the antithetic procedure would not be advantageous, since one could just generate twice as many ordinary deviates. However, there is an important class of cases in which the production of Y' turns out to be trivial. Notice that since Y can be generated as $F^{-1}(U)$ and since $U - 1/2$ is distributed symmetrically about the origin, it is always possible to represent g as a function of a symmetrically distributed random variable. With such representation for g let us suppose that the distribution of Y itself is symmetric about the origin, so that $F(y) = 1 - F(-y)$. Suppose further that Y can be generated in some efficient way. For example, Y might be standard normal, which can be generated by a variant of the Box-Muller algorithm. Since Y can still be represented via the inverse-transform method, we see that $Y = F^{-1}(U)$ implies $U = F(Y) = 1 - F(-Y)$, so that $-Y = F^{-1}(1 - U)$. This shows that $\{Y, -Y\}$ constitute an antithetic pair when Y is symmetrically distributed. Therefore, expressing g as a function of a random variable Y such that Y and $-Y$ are identically distributed, the antithetic estimator of Eg is

$$\widehat{Eg}_A = (2N)^{-1} \sum_{j=1}^{N} [g(y_j) + g(-y_j)].$$

When g is monotone, this achieves maximal variance reduction; otherwise, it can do no harm. Of course, if g itself is symmetric about the origin, then this can accomplish nothing.

11.2.4 *Control Variates*

Control variates were introduced in section 5.5.2 in connection with the binomial method for pricing derivatives. In that context the idea was to use the same binomial tree to price both the derivative of interest and an analogous security whose exact value is known (such as a European option), then to adjust the estimate of interest by subtracting the known pricing error for the comparison asset. This can reduce the pricing error in the derivative if the correlation between the known and unknown pricing errors is sufficiently high, but the discrete nature of the state space and the effects of different degrees of smoothness of the two payoff functions makes a precise analysis difficult. In the simulation context the method works by using the same pseudorandom numbers to calculate payoffs of both the derivative and the comparison security. But here, since the state space is no longer highly discrete and statistical analysis is more appropriate, we will be able to adjust the scale of the pricing error of the comparison asset so as to get something close to the maximum possible reduction in mean squared error.

To see how, suppose once again that our goal is to estimate $Eg(Y) = \int g(y) \cdot dF(y)$ via simulation. Introduce a new function $h(Y)$ whose mathematical expectation Eh is actually known—h will be the payoff function of a derivative that can be priced analytically—and consider the following modified estimator of Eg:

$$\begin{aligned}\widehat{Eg}_C(\alpha) &\equiv N^{-1}\sum_{j=1}^{N}\{g(y_j) - \alpha[h(y_j) - Eh]\}\\ &= \widehat{Eg} - \alpha(\widehat{Eh} - Eh),\end{aligned}$$

where $\widehat{Eh} = N^{-1}\sum_{j=1}^{N} h(y_j)$ and where $\widehat{Eg}$ is the standard simulation estimate. The difference $\widehat{Eh} - Eh$ represents the error correction for the control variate. In choosing the weighting factor α one wants to minimize

$$V\widehat{Eg}_C(\alpha) = V\widehat{Eg} - 2\alpha Cov(\widehat{Eg}, \widehat{Eh}) + \alpha^2 V\widehat{Eh},$$

and the optimal choice is the least-squares (population) regression coefficient,

$$\alpha^* = Cov(\widehat{Eg}, \widehat{Eh})/V\widehat{Eh} = Cov[g(Y), h(Y)]/Vh(Y).$$

The same simulation that produces $\widehat{Eg}$ and $\widehat{Eh}$ generates a consistent estimate of α^*; namely,

$$\hat{\alpha}^* = \frac{\sum_{j=1}^{N} g(y_j)[h(y_j) - Eh]}{\sum_{j=1}^{N}[h(y_j) - Eh]^2}. \tag{11.5}$$

Notice that realizations of g and h do not have to be stored, because these sums can be accumulated as the simulation progresses.[6]

Since $\hat{\alpha}^*$ is also subject to sampling error, it is not necessarily the case that the estimator $\widehat{Eg}_C(\hat{\alpha}^*) \equiv \widehat{Eg} - \hat{\alpha}^*(\widehat{Eh} - Eh)$ has lower variance than $\widehat{Eg}$, but it can be shown (Fishman, 1996, section 4.2) that $V\widehat{Eg}_C(\hat{\alpha}^*) \leq V\widehat{Eg}$ if and only if $\hat{\alpha}^*$ lies between zero and $2\alpha^*$. Since the variance of $\hat{\alpha}^*$

[6]An alternative is to estimate α^* in a separate experiment with different random deviates. Ideally (abstracting from the limitations of pseudorandom number generators) this makes $\hat{\alpha}^*$ statistically independent of $\widehat{Eg}$ and $\widehat{Eh}$ and preserves the unbiasedness of the control-variate estimator.

can be estimated consistently as

$$\widehat{V\hat{\alpha}^*} = \frac{N^{-1}\sum\{g(y_j) - \hat{\alpha}^*[h(y_j) - Eh]\}^2}{\sum_{j=1}^{N}[h(y_j) - Eh]^2},$$

and since $(\widehat{V\hat{\alpha}^*})^{-1/2}(\hat{\alpha}^* - \alpha^*)$ is asymptotically standard normal, the simulation results themselves yield an estimate of the probability of the event that $V\widehat{Eg}_C(\hat{\alpha}^*) \leq V\widehat{Eg}$. This is $\mathbb{P}(|\hat{\alpha}^* - \alpha^*| \leq |\alpha^*|) \doteq 2\Phi(\hat{\alpha}^*/\sqrt{\widehat{V\hat{\alpha}^*}}) - 1$ for large n. The nice thing here is that one has complete control over the sample size on which the asymptotic approximations depend.

While one usually takes for granted the desirability of using an unbiased control variate so as to preserve (apart from sampling error in $\hat{\alpha}^*$) the unbiasedness of the final estimate, Fu *et al.* (1999) describe an application in which a biased control variate can improve the accuracy. Comparing simulation estimates of fixed-strike, continuous arithmetic-average Asian options with estimates from the analytical formula by Geman and Yor (1993), they record the performance of control variates based on both discrete and continuous geometric-average options. Exact analytical formulas for both the geometric cases are known—for example, (7.77) on page 366 for the continuous case. Since sample paths generated by the simulation actually evolve in discrete time, simulation estimates of discrete geometric options are unbiased, while those for the continuous options contain a discretization bias. Likewise, there is discretization bias in simulation estimates of the continuous arithmetic-average Asians, and the finding by Fu *et al.* (1999) is that the two biases tend to cancel when pricing errors from continuous geometrics are used as the control variate.

11.2.5 *Richardson Extrapolation*

We have explained the use of Richardson extrapolation in binomial pricing (section 5.5.2) and again in connection with the Geske-Johnson (1984) approximation for pricing American puts (section 7.1.2). In both cases the purpose was to reduce errors associated with discrete-time approximations. The method finds the same application in Monte Carlo. Discretization is required whenever it is necessary to simulate entire sample paths, as in pricing path-dependent options or when volatility is stochastic. Applying Richardson extrapolation, one generates a sequence of Monte Carlo estimates of the derivative, $\{D(h_k)\}_{k=1}^K$, where $D(h_k)$ is based on n_k subdivisions of the derivative's life into intervals each of length $h_k = T/n_k$.

Approximating $D(h)$ as a $(K-1)$th-order polynomial and solving the K equations

$$\hat{D}(h_k) = D(0) + a_1 h_k + a_2 h_k^2 + \cdots + a_{K-1} h_k^{K-1}, k = 1, 2, \ldots, K,$$

for the intercept $D(0)$ give an estimate of the limiting value of $D(h)$ as $h \to 0$. Fu *et al.* (1999) describe an application to Asian options.

11.3 Applications

This section gives several extended examples of the use of Monte Carlo methods in pricing derivatives, progressing in order of increasing complexity. We consider first how to price European options on a portfolio of assets whose log prices are correlated Brownian motions. In this simple case the price of the portfolio and the option's payoff on the expiration date can be simulated in a single step without developing entire sample paths. This offers a simple framework in which to illustrate the use of antithetic and control variates. We then turn to European options on an underlying whose price follows a process with stochastic volatility. Here, when volatility shocks and price shocks are correlated, complete price paths do have to be constructed even though the derivative's payoffs are path-independent. The added complication of path-dependent payoffs is introduced in the next example, where simulation is used to price variable-strike lookback calls in the stochastic-volatility setting. Finally, we examine in detail several schemes for pricing American/Bermudan-style options via simulation.

11.3.1 *"Basket" Options*

The value of a static portfolio of assets whose prices follow geometric Brownian motion does not itself follow this simple process. Thus, it is inconsistent to assume *both* that future prices of a collection of assets are jointly lognormal *and* that the future value of a convex combination of them is lognormally distributed. As discussed in section 6.3.3, lognormality can be a good approximation for the distribution of the value of a well-diversified portfolio, and this is the usual justification for the common practice of pricing index options by Black-Scholes. However, options on a small "basket" of underlying securities are another matter. Since analytical solutions are difficult, and since the dimensionality of the problem rules out binomial and finite-difference methods, simulation is the natural vehicle for pricing such basket options.

First, let us consider the straightforward application of Monte Carlo without the use of variance reduction. We begin with a model for the prices of the n assets in the basket. Let $\mathbf{R} = \{\rho_{jk}\}_{j,k=1}^{n}$ be the correlation matrix of the assets' log returns, and let $\{\sigma_j\}_{j=1}^{n}$ be their volatilities. Assume that the individual prices evolve as

$$dS_{jt}/S_{jt} = \mu_{jt} \cdot dt + \sigma_j \cdot dW_{jt}^*, \quad j = 1, 2, \ldots, n,$$

where the $\{W_{jt}^*\}$ are correlated standard Brownian motions, with $E_0 W_{jt}^* W_{kt}^* = \rho_{jk} t$. With this setup the individual future prices are conditionally lognormal, and prices of different assets are dependent. Assuming constant short rate and continuous dividend rates, the model becomes

$$dS_{jt}/S_{jt} = (r - \delta_j) \cdot dt + \sigma_j \cdot d\hat{W}_{jt}^*, \quad j = 1, 2, \ldots, n, \tag{11.6}$$

under risk-neutral measure $\hat{\mathbb{P}}$. The $\{\hat{W}_{jt}^*\}$ are standard Brownian motions under $\hat{\mathbb{P}}$ with the same correlations as under $\mathbb{P}$. The goal is to price a T-expiring European call on a static portfolio of these assets—the basket portfolio. Letting $\{\nu_j\}_{j=1}^{n}$ be the portfolio weights, the basket's time-t value is $K_t \equiv \sum_{j=1}^{n} \nu_j S_{jt}$. The European call is worth $(K_T - X)^+$ at expiration and

$$C^E(K_0, T) = e^{-rT} \hat{E}_0 (K_T - X)^+,$$

at $t = 0$, where $\hat{E}$ is expectation under $\hat{\mathbb{P}}$.

Using Monte Carlo to estimate the expectation requires simulating N realizations of K_T and averaging. How is this done? First, we have to see what the model implies about the terminal distribution under $\hat{\mathbb{P}}$ of the prices of the individual assets in the basket. From (11.6) each S_{jT} can be expressed as

$$S_{jT} = S_{j0} \exp[(r - \delta_j - \sigma_j^2/2)T + \sigma_j \hat{W}_{jT}^*]. \tag{11.7}$$

Representing W_{jT}^* as $Z_j^* \sqrt{T}$, where $\mathbf{Z}^* \sim N(\mathbf{0}, \mathbf{R})$, one sees that obtaining realizations of the $\{S_{jT}\}$ requires obtaining realizations of n normals with correlation matrix $\mathbf{R}$. An efficient way to do this is to construct the Cholesky decomposition of $\mathbf{R}$. This expresses $\mathbf{R}$ as the product of an $n \times n$ lower-triangular matrix $\mathbf{L}$ times its transpose; that is, as $\mathbf{R} = \mathbf{L}\mathbf{L}'$ or

$$\begin{pmatrix} 1 & \rho_{21} & \cdots & \rho_{n1} \\ \rho_{21} & 1 & \cdots & \rho_{n2} \\ \vdots & \vdots & \ddots & \vdots \\ \rho_{n1} & \rho_{n2} & \cdots & 1 \end{pmatrix} = \begin{pmatrix} 1 & 0 & \cdots & 0 \\ \lambda_{21} & \lambda_{22} & \cdots & 0 \\ \vdots & \vdots & \ddots & \vdots \\ \lambda_{n1} & \lambda_{n2} & \cdots & \lambda_{nn} \end{pmatrix} \begin{pmatrix} 1 & \lambda_{21} & \cdots & \lambda_{n1} \\ 0 & \lambda_{22} & \cdots & \lambda_{n2} \\ \vdots & \vdots & \ddots & \vdots \\ 0 & 0 & \cdots & \lambda_{nn} \end{pmatrix}.$$

Efficient routines to accomplish this are widely available; e.g., Press *et al.* (1992). Taking $\mathbf{Z} = \{Z_j\}_{j=1}^n$, a vector of pseudorandom standard normals, $\mathbf{Z}^*$ can now be constructed as $\mathbf{LZ}$. Thus,

$$\begin{pmatrix} Z_1^* \\ Z_2^* \\ \vdots \\ Z_n^* \end{pmatrix} = \begin{pmatrix} Z_1 \\ \lambda_{21}Z_1 + \lambda_{22}Z_2 \\ \vdots \\ \sum_{j=1}^n \lambda_{nj}Z_j \end{pmatrix}.$$

Multiplying each Z_j^* by $\sqrt{T}$ to get W_{jT}^*, one then constructs realizations of the S_{jT} as in (11.7) and, in turn, of $(K_T - X)^+ = (\sum_{j=1}^n \nu_j S_{jT} - X)^+$. Letting $K_T(m)$ be the mth of N realizations, the simulation estimate of the call's value is

$$\hat{C}^E(K_0, T) = N^{-1} \sum_{m=1}^N [K_T(m) - X]^+.$$

Now let us apply some variance-reduction tricks, beginning with the use of antithetic variates. Since the $\{S_{jT}\}$ are functions of the symmetrically distributed elements of $\mathbf{Z}$, antithetic versions of payoff $[K_T(m) - X]^+$ in each replication m can be had simply by constructing vectors of asset prices from both $\mathbf{Z}^*$ and $-\mathbf{Z}^*$. With $K_T(m)^+$ and $K_T(m)^-$ as the two resulting portfolio values, the estimator

$$\hat{C}_A^E(K_0, T) = (2N)^{-1} \sum_{m=1}^N \{[K_T(m)^+ - X]^+ + [K_T(m)^- - X]^+\} \quad (11.8)$$

remains unbiased and can have variance no larger than that of $\hat{C}^E(K_0, T)$.

Introducing a control variate takes a bit more work, as we must find some asset whose payoff is highly correlated with $[K_T(\cdot) - X]^+$ over the various realizations but whose initial value is known analytically. Shaw (1998) suggests as a control variate the difference between the payoff and price of a European call on an asset whose log price is normally distributed with variance matching that of the basket portfolio. In other words, one constructs an asset that mimics the basket portfolio but has a lognormal return. Realizations are obtained by using a scaled sum of the same normal deviates that were used to generate values of the basket.

Getting into the details, the model for the terminal value of the mimicking asset under $\hat{\mathbb{P}}$ is

$$K_T^* = K_0^* \exp\left[(r - \delta_K - \sigma_K^2/2)T + \sum_{j=1}^n \nu_j \sigma_j \hat{W}_{jT}^*\right],$$

where $K_0^* = K_0 = \sum_{j=1}^n \nu_j S_{j0}$, $\delta_K = \sum_{j=1}^n \nu_j \delta_j$, and $\sigma_K^2 = \sum_{j,k=1}^n \nu_j \nu_k \sigma_j \sigma_j \rho_{jk}$. Under this model the Black-Scholes formula gives the actual price of a European call as

$$C^E(K_0^*, T) = B(0,T)\{\mathsf{f}_{K^*}\Phi[q^+(\mathsf{f}_K/X)] - X\Phi[q^-(\mathsf{f}_{K^*}/X)]\},$$

where

$$q^{\pm}(x) = \frac{\ln x \pm \sigma_K^2 T/2}{\sigma_K\sqrt{T}},$$

$$\mathsf{f}_{K^*} = B(0,T)^{-1}K_0^* e^{-\delta_K T}.$$

Simulating the payoffs, one obtains antithetic versions of the call's terminal value on replication m as $[K_T^*(m)^+ - X]^+$ and $[K_T^*(m)^- - X]^+$, where $K_T^*(m)^{\pm}$ are constructed from the same realizations of the $\hat{W}_{jT}^*$ that were used to produce $K_T(m)^{\pm}$. The simulation estimate of the call on the mimicking asset, $\hat{C}_A^E(K_0^*, T)$, is then obtained as in (11.8).

Now let $\hat{C}_{AC}^E(K_0, T)$ represent the simulation estimate of the value of the basket call that applies both antithetic and control-variate techniques. This will be obtained by adjusting the antithetic estimate (11.8) by the pricing error for the call on the mimicking lognormal asset, $\hat{C}_A^E(K_0^*, T) - C^E(K_0^*, T)$, as

$$\hat{C}_{AC}^E(K_0, T) = \hat{C}_A^E(K_0, T) - \alpha[\hat{C}_A^E(K_0^*, T) - C^E(K_0^*, T)].$$

Recalling the discussion in section 11.2.4, one obtains an estimate of the optimal weighting parameter, α^*, by regressing the average of the two antithetic realizations of the payoff of the basket call on the realizations of the errors in pricing the control asset. Specifically, let

$$g_m \equiv \{[K_T(m)^+ - X]^+ + [K_T(m)^- - X]^+\}/2$$

be the payoff of the basket option on the mth realization,

$$h_m \equiv \{[K_T^*(m)^+ - X]^+ + [K_T^*(m)^- - X]^+\}/2$$

be the payoff of the control asset, and $Eh = C^E(K_0^*, T)$ be the Black-Scholes price of the control. Then, corresponding to (11.5), the estimate of α^* is

$$\hat{\alpha}^* = \frac{\sum_{j=1}^N g_m(h_m - Eh)}{\sum_{j=1}^N (h_m - Eh)^2}.$$

11.3.2 *European Options under Stochastic Volatility*

Let us now see how to use simulation to price path-independent, European-style derivatives on an underlying asset whose price is an Itô process with stochastic volatility. Specifically, consider the models discussed in section 8.3, wherein under $\hat{\mathbb{P}}$

$$\begin{aligned} dS_t &= r_t S_t \cdot dt + \sigma_t S_t \cdot d\hat{W}_{1t} \\ d\sigma_t &= \lambda_t \cdot dt + \gamma_t(\rho_t \cdot d\hat{W}_{1t} + \bar{\rho}_t \cdot d\hat{W}_{2t}). \end{aligned}$$

Here λ_t is the risk-adjusted mean drift of volatility, ρ_t is the correlation between the shocks to price and volatility (with $\bar{\rho}_t = \sqrt{1-\rho_t^2}$), and $\hat{W}_{1t}, \hat{W}_{2t}$ are independent $\hat{\mathbb{P}}$ Brownian motions.

There are two general approaches to pricing derivatives in this setup. The one to use depends on the value of the correlation between shocks. As we saw in connection with the models of Scott (1987) and Hull and White (1987), if $\rho_t = 0$ for all t the price of a European-style, path-independent derivative can be expressed in terms of the average volatility over the derivative's lifetime, $\bar{\sigma}_T^2 = T^{-1}\int_0^T \sigma_t^2 \cdot dt$. In this case, assuming that the derivative can be valued analytically when volatility is constant, only the volatility process needs to be simulated. The procedure is as follows. One realization of the path of volatility over a discrete time lattice, $\{\sigma_{t_j}\}_{j=1}^n$, is generated (in the manner described below) and the time average is calculated as $\widehat{\bar{\sigma}_T^2} = n^{-1}\sum_{j=1}^n \sigma_{t_j}^2$. Using $\widehat{\bar{\sigma}_T}$ one then evaluates analytically the expectation of $D(S_T, 0)$ (the derivative's terminal value) conditional on $\bar{\sigma}_T = \widehat{\bar{\sigma}_T}$, and discounts to the present. This produces a realization of the derivative's initial value conditional on $\bar{\sigma}_T = \widehat{\bar{\sigma}_T}$. Repeating this N times and averaging delivers the final, unconditional estimate, $\hat{D}(S_0, T)$. Thus, when shocks to price and volatility are uncorrelated, pricing a vanilla European-style put or call requires simulation only to estimate $\bar{\sigma}_T$.

Things work very differently when $\rho_t \neq 0$. In this case price and volatility shocks are not independent, and the conditioning argument that delivers an analytical expression for price given $\bar{\sigma}_T$ no longer applies. Moreover, the dependence between volatility shocks and price shocks would in any case prevent one from determining $\bar{\sigma}_T$ without modeling the evolution of price simultaneously. Dealing with the case $\rho_t \neq 0$ therefore requires simulating paths of both $\{\sigma_t\}$ and $\{S_t\}$ and estimating $D(S_0, T)$ by discounting the average of the resulting realizations of terminal value, $D(S_T, 0)$. We now describe this more general procedure in detail.

The process begins by defining a lattice of time steps, as $t_j = jT/n$. Starting at $t = t_0 = 0$ with S_0 observed and σ_0 a known parameter, and proceeding through steps $j = 1, 2, \ldots, n$, generate

$$\sigma_{t_j} = \sigma_{t_{j-1}} + \lambda_{t_{j-1}} T/n + \gamma_{t_{j-1}} \sqrt{T/n}(\rho_{t_{j-1}} Z_{1j} + \bar{\rho}_{t_{j-1}} Z_{2j}) \quad (11.9)$$

$$S_{t_j} = S_{t_{j-1}} \exp[(r_{t_{j-1}} - \sigma^2_{t_{j-1}}/2)T/n + \sigma_{t_{j-1}} \sqrt{T/n} Z_{1j}], \quad (11.10)$$

where Z_{1j} and Z_{2j} are independent pseudorandom normal deviates. The adapted processes $\{\lambda_t\}$ and $\{\gamma_t\}$ can be specified in very general ways to depend on current and past values of volatility, or price, or both. An example is Scott 's (1987) mean-reverting specification for σ_t, in which $\lambda_t = \xi(\bar{\sigma}_\infty - \sigma_t) - \lambda^*$. Price-level ("c.e.v.") effects could be accommodated by letting λ_t and/or γ_t depend on S_t. Alternatively, as in the Hobson–Rogers (1998) model (section 8.2.5), these could depend on an average of past values of S. Straight-on Monte Carlo then just requires averaging and discounting the realizations of the derivative's payoff, as

$$\hat{D}(S_0, T) = B(0, T) N^{-1} \sum_{m=1}^{N} D[S_T(m), 0].$$

Hull and White (1987) suggest the following way to employ antithetic variates in this setup. The realization of $\{Z_{1j}, Z_{2j}\}_{j=1}^n$ on replication m is used to calculate four terminal prices of the underlying and four corresponding terminal payoffs of the derivative. Terminal price $S_T(m)^{++}$ is obtained using the realizations of each pair $\{Z_{1j}, Z_{2j}\}$ that were actually generated; then prices $S_T(m)^{-+}$, $S_T(m)^{+-}$, and $S_T(m)^{--}$ are calculated by reversing, respectively, the signs of the first, the second, and both members of each pair. The resulting estimate of the current value of the derivative is then

$$\hat{D}_A(S_0, T) = (4N)^{-1} \sum_{m=1}^{N} \{D[S_T(m)^{++}, 0]^+ + D[S_T(m)^{+-}, 0]^+ \\ + D[S_T(m)^{-+}, 0]^+ + D[S_T(m)^{--}, 0]^+\}.$$

Hull and White propose as control variates the differences between payoffs of European options and Black-Scholes prices, evaluating the payoffs at terminal prices $S_T(m)^+$ and $S_T(m)^-$ generated from constant-volatility diffusions with $\sigma_t = \sigma_0$ and the realizations $\{Z_{1j}\}$ and $\{-Z_{1j}\}$, respectively. An alternative choice would be deviations of payoffs from the computational solution of Heston 's (1993) mean-reverting square-root model for volatility, (8.22). Because they do pick up variation in volatility, one expects these deviations to be more highly correlated with payoffs implied by the specific volatility model being applied. Of course, one can experiment with different control variates in search of one that is best.

Whatever variance-reduction scheme is used to reduce the required number of replications, there remains the issue of how many time steps, n, to allow. The issue here is not one of statistical efficiency, but one of bias and inconsistency. Since the choice of n will be problem-specific, the only feasible approach is to experiment by increasing n until estimates seem to approach a limit.

11.3.3 *Lookback Options under Stochastic Volatility*

Clewlow and Carverhill (1994) use an interesting adaptation of the control-variate technique to price lookback options under stochastic volatility. The method is not limited to lookbacks but is applicable whenever the entire sample path of the underlying must be simulated. The idea is to construct one or more control processes that evolve as martingales along the path and that track (as well as possible) the evolving conditional expectation of the derivative's ultimate payoff.

The essence of the technique of "martingale variance reduction" is to build control variates from innovations in adjustments to a portfolio that would approximately replicate the derivative being priced. By "innovations" is meant deviations from conditional expected values. To price a variable-strike lookback call under stochastic volatility in a model like (11.9) and (11.10), Clewlow and Carverhill develop three variates that arise from delta, gamma, and sigma hedging a lookback call under geometric Brownian motion. In the same way that delta hedging replicates a derivative's responses to the underlying price, "gamma" hedging replicates changes in the derivative's delta, and "sigma" hedging replicates its response to volatility. Since we have a specific formula—namely, (7.66)—for the value of the lookback under these Black-Scholes dynamics, hedging parameters can be developed analytically; and from these, the martingales that track the value of the lookback. Specifically, taking $d_0 = g_0 = v_0 = 0$, define processes $\{d_t\}$, $\{g_t\}$, and $\{v_t\}$ at the discrete times $\{t_j = jT/n\}_{j=1}^n$ as

$$
\begin{aligned}
d_{t_j} &= d_{t_{j-1}} + C_S^{VL}(S_{t_{j-1}}, T - t_{j-1}; \sigma_{t_{j-1}}, \underline{S}_{0,t_{j-1}})[\Delta S_{t_j} - E_{t_{j-1}} \Delta S_{t_j}] \\
g_{t_j} &= g_{t_{j-1}} + C_{SS}^{VL}(S_{t_{j-1}}, T - t_{j-1}; \sigma_{t_{j-1}}, \underline{S}_{0,t_{j-1}})[(\Delta S_{t_j})^2 - E_{t_{j-1}}(\Delta S_{t_j})^2] \\
v_{t_j} &= v_{t_{j-1}} + C_\sigma^{VL}(S_{t_{j-1}}, T - t_{j-1}; \sigma_{t_{j-1}}, \underline{S}_{0,t_{j-1}})[\Delta \sigma_{t_j} - E_{t_{j-1}} \Delta \sigma_{t_j}].
\end{aligned}
$$

Here the Δ's represent one-step changes from t_{j-1} to t_j; subscripts on C denote partial derivatives; and the argument $\underline{S}_{0,t_{j-1}}$ in C represents the minimum price up to t_{j-1}. The one-step change in variable d_t represents the innovation in the adjustment required to delta-hedge the lookback call. Likewise, changes in g_t and v_t are innovations in the adjustments for gamma hedging and sigma hedging. Having zero conditional mean, these changes are martingale differences, so that $\{d_t\}$, $\{g_t\}$, and $\{v_t\}$ are themselves discrete martingales. At the end of the sample path generated on replication m one observes $d_T(m)$, $g_T(m)$, and $v_T(m)$, as well as the call's payoff, $S_T(m) - \underline{S}_{0,T}(m)$. The control-variate estimator then takes the form

$$\hat{C}^{VL} = N^{-1} \sum_{m=1}^{N} \{S_T(m) - \underline{S}_{0,T}(m) - [\alpha_d d_T(m) + \alpha_g g_T(m) + \alpha_v v_T(m)]\},$$

where the α's are weights on the adjustment factors. Estimates of variance-minimizing values of the α's can be found by regressing the realizations of $S_T - \underline{S}_{0,T}$ on those of d_T, g_T, and v_T.

Clewlow and Carverhill (1994) report that martingale variance reduction produces a 14-fold drop in the standard error of the Monte Carlo estimates at the cost of trebling the computational time per replication. Since a 14-fold reduction in standard error would require about a 200-fold increase in replications, the method reduces by a factor of more than 60 the overall computational time to achieve a given accuracy.

A final cautionary note: Monte Carlo pricing of derivatives whose payoffs depend on extrema of the underlying price is highly sensitive to the number of time steps, n. At a minimum, n must correspond to the contractually specified sampling frequency used to identify the extremum, but larger values may be required if volatility or interest rates are modeled as stochastic.

11.3.4 *American-Style Derivatives*

Simulation was long thought to be unsuited to the task of pricing options subject to discretionary early exercise. Nevertheless, the need to value derivatives with complex payoff structures or depending on multiple state variables has motivated research that has led to significant progress. Here we summarize three very different approaches to the problem[7]: (i) flexible parametric modeling of exercise regions; (ii) flexible parametric modeling of continuation values; and (iii) a nonparametric tree-based technique that applies binomial-like backward induction along generated sample paths. While all three methods have shown promise and have found useful application, we will see that all have limitations that continue to motivate ongoing research. In explaining each method we focus on the simplest of applications—American puts under Black-Scholes dynamics—then wrap up by comparing the approaches in a more challenging setting.

Modeling Exercise Boundaries/Stopping Times

Recall from our discussion of American options in sections 5.4.3 and 7.1.2 that the initial value of a T-expiring, vanilla American put struck at X can be represented as

$$P^A(S_0, T) = \hat{E}e^{-rU}(X - S_U), \tag{11.11}$$

where $U = \inf\{t : S_t \in [0, \mathfrak{B}_t]\}$. Here $\{\mathfrak{B}_t\}_{0 \le t \le T}$ is the optimal exercise boundary and U is the optimal stopping time. The sense of (11.11) is that the put generates a receipt whose present value is $e^{-rU}(X - S_U)$ at the first time $U \le T$ at which the underlying price enters a certain time-dependent region, $\mathbb{B}_t = [0, \mathfrak{B}_t] \subset \Re$. The option's current value is simply the (risk-neutral) average of these discounted payoffs. More generally, a T-expiring derivative's value at t may depend on not just one underlying price but on some q-vector $\mathbf{Y}_t$ of state variables, as $D(\mathbf{Y}_t; T - t)$; for example, for an American put option on an underlying whose price follows a stochastic-volatility diffusion, we would have $\mathbf{Y}_t = (S_t, \sigma_t)'$ and $D(\mathbf{Y}_t; T - t) = P^A(S_t, \sigma_t; T - t)$. In this q-variate case $\mathbb{B}_t$ becomes a time-dependent region in $\Re_q$, and the optimal stopping time becomes $U = \inf\{t : \mathbf{Y}_t \in \mathbb{B}_t\}$. Thinking of a put's value in this way—i.e., as the maximum of expected discounted payoffs over possible exercise regions/stopping times—there springs to mind an obvious and natural sequence of steps for pricing by simulation.

[7]For further detail and other approaches to pricing by simulation consult Garcia (2003), Longstaff and Schwartz (2001), Broadie and Glasserman (1997), and Boyle *et al.* (1997). Glasserman (2004, chapter 8) gives an excellent overview, with emphasis on tree-based methods.

1. Adopt some flexible specification that expresses the exercise region in terms of time, the relevant state variables $\mathbf{Y}_t$, and a manageably small set of parameters, $\boldsymbol{\theta}$, as $\mathbb{B}_t(\mathbf{Y}_t;\boldsymbol{\theta})$. This implicitly defines a parametric stopping time $U(\mathbf{Y}_t;\boldsymbol{\theta}) = \inf\{t : \mathbf{Y}_t \in \mathbb{B}_t(\mathbf{Y}_t;\boldsymbol{\theta})\}$.
2. Given the maintained form of $\mathbb{B}_t(\mathbf{Y}_t;\boldsymbol{\theta})$ and a specific $\boldsymbol{\theta}$, generate N sample paths, $\{\mathbf{Y}_t(\omega)\}_{0\le t\le U(\omega;\boldsymbol{\theta}),\omega\in\{1,2,\ldots,N\}}$, under the appropriate martingale dynamics, where $U(\omega;\boldsymbol{\theta}) \equiv U(\mathbf{Y}_t(\omega);\boldsymbol{\theta})$ is the first time on path ω at which $\mathbf{Y}_t$ is within $\mathbb{B}_t(\mathbf{Y}_t;\boldsymbol{\theta})$. Letting $X(\mathbf{Y}_t)$ represent the intrinsic (exercise) value of the derivative at t, average over the paths to estimate the derivative's value (given $\boldsymbol{\theta}$) as
$$D_N(\mathbf{Y}_0,T;\boldsymbol{\theta}) = N^{-1}\sum\nolimits_{\omega=1}^{N} e^{-rU(\omega;\boldsymbol{\theta})}X(\mathbf{Y}_{U(\omega;\boldsymbol{\theta})}(\omega)).$$
3. With $\boldsymbol{\Theta}$ as the set of admissible values of θ, find (if possible) a value $\boldsymbol{\theta}_N$ such that $D_N(\mathbf{Y}_0,T;\boldsymbol{\theta}_N) = \max_{\boldsymbol{\theta}\in\boldsymbol{\Theta}} D_N(\mathbf{Y}_0,T;\boldsymbol{\theta})$.

If the specification $\mathbb{B}_t(\mathbf{Y}_t;\boldsymbol{\theta})$ is adequate, it seems that following these steps with sufficiently large N should produce an estimate of arbitrarily high precision.

While this is *almost* true, things are not quite so straightforward as they seem. The first problem—and a relatively minor one—is that $D_N(\mathbf{Y}_0,T;\boldsymbol{\theta}_N)$ is an upward-biased estimate of

$$D(\mathbf{Y}_0,T;\boldsymbol{\theta}_0) \equiv \max_{\boldsymbol{\theta}\in\boldsymbol{\Theta}} D(\mathbf{Y}_0,T;\boldsymbol{\theta}) = \max_{\boldsymbol{\theta}\in\boldsymbol{\Theta}} \hat{E}e^{-rU(\mathbf{Y}_t;\boldsymbol{\theta})}X(\mathbf{Y}_{U(\mathbf{Y}_t;\boldsymbol{\theta})})$$

(assuming that such $\boldsymbol{\theta}_0$ exists). This can be understood by recognizing that the maximum of an average can be no larger—and can be smaller—than the average of maxima. That is,

$$\begin{aligned} D(\mathbf{Y}_0,T;\boldsymbol{\theta}_0) &= \max_{\boldsymbol{\theta}\in\boldsymbol{\Theta}} ED_N(\mathbf{Y}_0,T;\boldsymbol{\theta}) \\ &\le E\max_{\boldsymbol{\theta}\in\boldsymbol{\Theta}} D_N(\mathbf{Y}_0,T;\boldsymbol{\theta}) \\ &\equiv ED_N(\mathbf{Y}_0,T;\boldsymbol{\theta}_N), \end{aligned}$$

since on the right of the inequality we average the largest values of members of an ensemble while on the left each member is evaluated at the same point. Alternatively, we can recognize that $D_N(\mathbf{Y}_0,T;\boldsymbol{\theta}_N) \ge D_N(\mathbf{Y}_0,T;\boldsymbol{\theta}_0)$ for each realization of $\boldsymbol{\theta}_N$, since $\boldsymbol{\theta}_N$ maximizes the mean of the empirical distribution of discounted payoffs rather than the actual expected value. The usual approach to confronting this problem is to undertake a fourth step, as follows:

4. With $\boldsymbol{\theta}_N$ determined from step 3, generate an additional N' sample paths and obtain a second estimate as

$$D_{N,N'}(\mathbf{Y}_0, T; \boldsymbol{\theta}_N) = N'^{-1} \sum_{\omega=1}^{N'} e^{-rU(\omega;\boldsymbol{\theta}_N)} X(\mathbf{Y}_{U(\omega;\boldsymbol{\theta}_N)}(\omega)).$$

Under the condition that $\mathbb{B}_t(\mathbf{Y}_t; \boldsymbol{\theta}_0)$ represents the true functional form of the boundary, this second estimate is now downward biased, because

$$E\{e^{-rU(\mathbf{Y}_t;\boldsymbol{\theta}_N)} X(\mathbf{Y}_{U(\mathbf{Y}_t;\boldsymbol{\theta}_N)}) \mid \boldsymbol{\theta}_N\} \leq Ee^{-rU(\mathbf{Y}_t;\boldsymbol{\theta}_0)} X(\mathbf{Y}_{U(\mathbf{Y}_t;\boldsymbol{\theta}_0)}(\omega))$$

for each path $\omega \in \{1, 2, \ldots, N'\}$. However, from the ensembles of N and N' sample paths one can estimate the standard errors of the payoffs, s_N and $s_{N'}$, say, and construct two asymptotically valid $100(1-\alpha)\%$ confidence intervals for $D(\mathbf{Y}_0, T; \boldsymbol{\theta}_0)$ as $D_N(\mathbf{Y}_0, T; \boldsymbol{\theta}_N) \pm z_{\alpha/2}\frac{s_N}{\sqrt{N}}$ and $D_{N,N'}(\mathbf{Y}_0, T; \boldsymbol{\theta}_N) \pm z_{\alpha/2}\frac{s_{N'}}{\sqrt{N'}}$, where $z_{\alpha/2}$ is the upper-$\alpha/2$ quantile of the standard normal. The meta-interval $[D_{N,N'}(\mathbf{Y}_0, T; \boldsymbol{\theta}_N) - z_{\alpha/2}\frac{s_{N'}}{\sqrt{N'}}, D_N(\mathbf{Y}_0, T; \boldsymbol{\theta}_N) + z_{\alpha/2}\frac{s_N}{\sqrt{N}}]$ then represents a conservative (asymptotic) $100(1-2\alpha)\%$ confidence interval for $D(\mathbf{Y}_0, T; \boldsymbol{\theta}_0)$.[8] In principle, the interval can be made as tight as one desires by taking N and N' sufficiently large.

While this seems a powerful result, there are several reasons not to take too much comfort in it. Continuing the list of "problems" alluded to above, one's model for the optimal exercise region is almost certain to be wrong, in the sense that there is no admissible $\boldsymbol{\theta}$ for which $\mathbb{B}_t(\mathbf{Y}_t; \boldsymbol{\theta})$ corresponds to the true optimal region. This specification error contributes a downward bias of indeterminate magnitude, which partially or wholly offsets the upward bias in $D_N(\mathbf{Y}_0, T; \boldsymbol{\theta}_N)$. Indeed, with specification error there is no basis for assuming that either $D_N(\mathbf{Y}_0, T; \boldsymbol{\theta}_N)$ or $D_{N,N'}(\mathbf{Y}_0, T; \boldsymbol{\theta}_N)$ converges in probability to $D(\mathbf{Y}_0, T)$ as $N, N' \to \infty$, and the more complicated are the dynamics of $\{\mathbf{Y}_t\}$ and the structure of the derivative the less insight one is apt to have about the proper model. Next, the fact that sample paths are generated in discrete time introduces a discretization error that limits the possible times of exercise. This is a second source of downward bias and inconsistency (with respect to N, N'). Finally, unless N is extremely large,

[8]This follows from Bonferroni's inequality: For arbitrary events A, B on $(\Omega, \mathcal{F}, \mathbb{P})$,

$$\mathbb{P}(A \cap B) \geq \mathbb{P}(A) + \mathbb{P}(B) - 1.$$

In this case A and B represent the events that the respective intervals contain $D(\mathbf{Y}_0, T; \boldsymbol{\theta}_0)$.

in even the simplest of applications one is unlikely to obtain a good image of the boundary for all values of t and $\mathbf{Y}_t$ using paths from a given initial state $\mathbf{Y}_0$. The reason is that relatively few sample paths will enter the less accessible parts of the exercise region, and so the experiment will provide little information about these areas. The implication is that new simulations will have to be performed whenever there are appreciable changes in initial conditions, even for derivatives that are far from expiration.

As an example, figure 11.1 shows three estimates of the boundary of the optimal exercise region for a vanilla put on a no-dividend stock, based on sample paths starting at $S_0 \in \{0.9, 1.0, 1.1\}$. In each case $\{S_t\}_{0 \le t \le T}$ follows Black-Scholes dynamics with $r = .05$ and $\sigma = .20$, and the put has $X = T = 1$. Since S_t is now the only state variable and the derivative is a vanilla option, the exercise region at t is determined just by specifying its boundary, as $\mathbb{B}_t = [0, \mathfrak{B}_t]$. In this experiment $\{\mathfrak{B}_t\}_{0 \le t \le T}$ was modeled as a cubic spline with knots at 21 time points, and $D_N(S_0, T; \boldsymbol{\theta}) = P_N^A(S_0, T; \boldsymbol{\theta})$ was maximized with respect to the values of $\mathfrak{B}_t$ at these 21 values of t.[9] Clearly, during the early life of the option there are large discrepancies among boundary functions estimated from different starting values, S_0. Moreover, the estimates in table 11.2 (from $N = N' = 5{,}000$ replications) display a strong downward bias relative to the corresponding binomial estimates.

We can do better than this by using some additional information. Letting $\hat{\mathfrak{B}}_t$ be the value that equates the Black-Scholes value of a European put to the intrinsic value, as $P^E(\hat{\mathfrak{B}}_t, T-t) = X - \hat{\mathfrak{B}}_t$, we can parameterize the difference between the American exercise boundary and $\hat{\mathfrak{B}}_t$ as a simple polynomial in time to expiration. This gives

$$\mathfrak{B}_t(\boldsymbol{\theta}) = \hat{\mathfrak{B}}_t - \sum_{k=1}^{K} \theta_k (T-t)^k = X - P^E(\hat{\mathfrak{B}}_t, T-t) - \sum_{k=1}^{K} \theta_k (T-t)^k \quad (11.12)$$

for the boundary and

$$\mathbb{B}_t = \left\{ S_t : X(S_t) > P^E(\hat{\mathfrak{B}}_t, T-t) + \sum_{k=1}^{K} \theta_k (T-t)^k \right\} \quad (11.13)$$

[9]Values $\hat{\mathfrak{B}}_t : P^E(\hat{\mathfrak{B}}_t, T-t) = X - \hat{\mathfrak{B}}_t$ served as an initial guess of the boundary, with $P^E(\cdot,\cdot)$ given by the Black-Scholes formula. Knots were placed at $t = 0$ and at points $\{t_j\}_{j=1}^{20}$ that divided range $[\hat{\mathfrak{B}}_0, \hat{\mathfrak{B}}_T = X] = [.895, 1.00]$ into 21 equal parts, as $\hat{\mathfrak{B}}_{t_j} = \hat{\mathfrak{B}}_0 + j(X - \hat{\mathfrak{B}}_0)/21$. The simulations used $N = N' = 5{,}000$ sample paths evaluated at 200 evenly-spaced points on $[0, 1]$. Optimization was by the simplex algorithm in Press *et al.* (1992). The table compares averages of simulated "high" and "low" estimates $P_{N,N'}^A(S_0, T; \boldsymbol{\theta}_N), P_N^A(S_0, T; \boldsymbol{\theta}_N)$ with binomial estimates from 5,000 time steps.

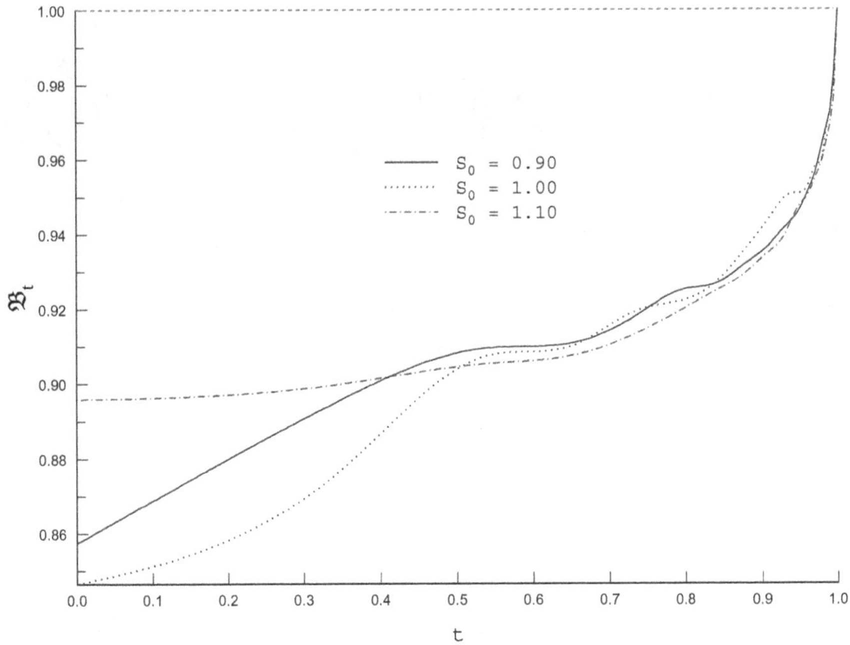

Fig. 11.1. Exercise boundaries for $P^A(S_0, T = 1; X = 1)$ by simulation and spline approximation, from initial values $S_0 \in .\{90, 1.0, 1.1\}$.

Table 11.2. Binomial *vs.* simulation estimates of $P^A(S_0, T = 1; X = 1)$ with spline-approximated exercise boundary.

S_0	Binomial	$\frac{P^A_{N,N'}+P^A_N}{2}$	$\frac{P^A_{N,N'}+P^A_N}{2} \pm z_{.025} s_{N'}/\sqrt{N'}$
0.9	.1149	.1113	[.1099, .1126]
1.0	.0609	.0598	[.0582, .0615]
1.1	.0299	.0287	[.0275, .0300]

for the corresponding exercise region, where $X(S_t) \equiv X - S_t$. Much of the structure of the optimal boundary is thus imported from that of $\{\hat{\mathfrak{B}}_t\}$, making it much simpler to approximate. Figure 11.2 shows the boundaries estimated again from three values of S_0 with a cubic approximation to $\mathfrak{B}_t(\boldsymbol{\theta})$ $(K = 3)$. The number of samples and the number of time steps are the same as before. Not only are the estimated boundaries close together, all are nicely monotonic. Table 11.3 confirms one's expectation that better estimates of values accompany the more reasonable estimates of boundaries. Note that the binomial estimates are all safely inside the confidence bins.

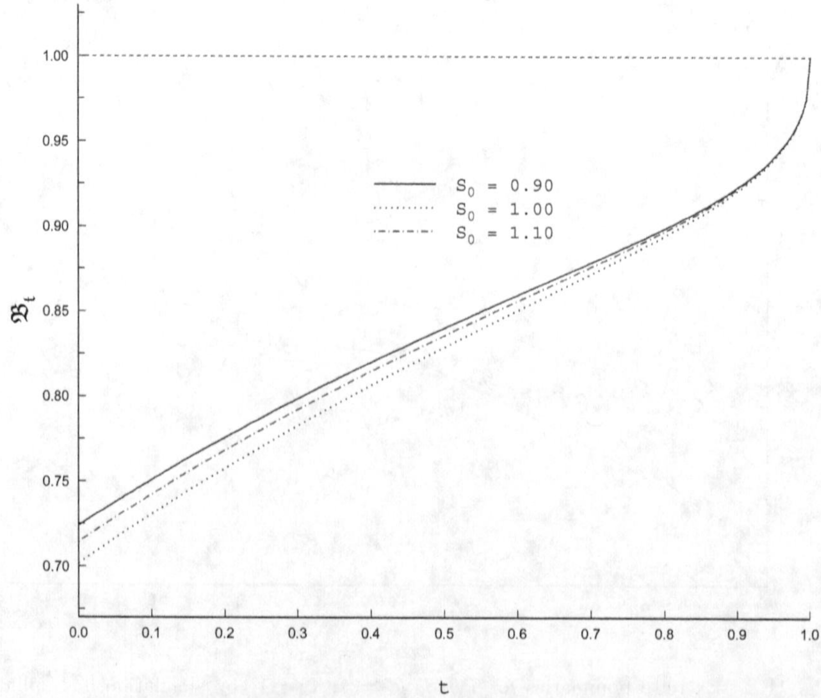

Fig. 11.2. Exercise boundaries for $P^A(S_0, T = 1; X = 1)$ by simulation and polynomial approximation, from initial values $S_0 \in .\{90, 1.0, 1.1\}$.

Table 11.3. Binomial *vs.* simulation estimates of $P^A(S_0, T = 1; X = 1)$ with polynomial-approximated exercise boundary.

S_0	Binomial	$\frac{P^A_{N,N'}+P^A_N}{2}$	$\frac{P^A_{N,N'}+P^A_N}{2} \pm z_{.025}s_{N'}/\sqrt{N'}$
0.9	.1149	.1143	[.1119, .1167]
1.0	.0609	.0607	[.0586, .0628]
1.1	.0299	.0305	[.0290, .0320]

Almost as good accuracy can be achieved by representing the boundary and corresponding exercise region as

$$\mathfrak{B}_t(\boldsymbol{\theta}) = X - P^E(S_t, T - t) - \sum_{k=1}^{K} \theta_k (T - t)^k \tag{11.14}$$

$$\mathbb{B}_t = \left\{ S_t : X(S_t) > P^E(S_t, T - t) + \sum_{k=1}^{K} \theta_k (T - t)^k \right\}. \tag{11.15}$$

This saves having to solve for $\{\hat{\mathfrak{B}}_t\}$, although at the cost of generating less-smooth estimates of the boundary.

An alternative to modeling the boundary at each $t \in [0, T]$ is to limit the feasible exercise times to some small set $\{t_j\}_{j=1}^n$, with $t_n = T$. In this case one is really valuing a Bermudan-style counterpart to an American derivative. For a vanilla option with just one state variable the parameter vector $\boldsymbol{\theta}$ would simply comprise the values $\{\mathfrak{B}_{t_j}\}_{j=0}^{n-1}$ (taking $\mathfrak{B}_T = X$). For cases involving multiple state variables $\mathbf{Y}$ the boundary of each region $\mathbb{B}_{t_j}$ would have to be specified as some flexible function of $\mathbf{Y}_{t_j}$. Unfortunately, the optimization problem quickly gets out of hand as n and the number of state variables increase.

Modeling Continuation Value

The discrete-time, discrete-state binomial approach to valuing vanilla American options relied on a different characterization than (11.11); namely (for an American put on a single underlying asset)

$$P^A(S_{t_j}, T - t_j) = (X - S_{t_j}) \vee B(t_j, t_{j+1}) \hat{E}_{t_j} P^A(S_{t_{j+1}}, T - t_{j+1}). \qquad (11.16)$$

This represents the put's value at t_j as the greater of intrinsic value, $X(S_{t_j}) \equiv X - S_{t_j}$, and continuation value—the discounted value if (optimally) held to t_{j+1}. Determining the continuation value was straightforward in the binomial scheme, and to do it by simulation will require following a similar program. Specifically, the opportunities for exercise must again be limited to a finite set of times $\{t_j\}_{j=1}^n$, and we will again have to work backwards in time, since the continuation value is known *a priori* only at T (where it is equal to zero). At each earlier time t_j the continuation value will have to be estimated from what is then known to the simulator—namely, the values of all state variables on all paths to that point. While there are a number of ways to do it, the LSM (Least Squares Monte Carlo) method proposed by Longstaff and Schwartz (2001) has proved the simplest and most popular. The main attraction is that estimation is done by ordinary least squares rather than by numerical search, which makes the optimization part very fast. We start with a brief overview and then give an example and further details.

The key insight underlying the LSM method is that a derivative's continuation value at any time is simply the (risk-neutral) conditional expectation of the discounted cash flows received from holding it until expiration. The plan is to represent the continuation value as some flexible function of the current values of the state variables $\mathbf{Y}$ on which the value of the derivative depends and then to estimate this function at each time step by regression. To make things easier, we consider first a single state variable, $Y = S$, and focus on pricing a vanilla American put. In this case the program is as follows.

We begin by generating N sample paths of S in the appropriate risk-neutral measure and under the assumed dynamics. The simulation thus supplies N realizations at each of the feasible exercise dates,

$$\{S_{t_j}(\omega)\}_{j,\omega\in\{1,2,\ldots,n\}\times\{1,2,\ldots,N\}}.$$

Next, we specify a flexible functional form for the continuation value at each t_j as

$$V(S_{t_j}, T - t_j; \boldsymbol{\theta}_j) = \sum_{k=0}^{m} \theta_{jk} b_k(S_{t_j}) = \mathbf{b}(S_{t_j})\boldsymbol{\theta}_j,$$

where $\mathbf{b}(\cdot) = (b_0(\cdot), b_1(\cdot), \ldots, b_m(\cdot))$ is a vector of prescribed "basis" functions, and $\boldsymbol{\theta}_j = (\theta_{j0}, \theta_{j1}, \ldots, \theta_{jm})'$ is a vector of parameters to be estimated. For example, we might take $b_k(S_{t_j}) = S_{t_j}^k$, making V a simple polynomial. Next, we start a backwards recursion from $t_n = T$. Since the continuation value there is $V(S_T, 0) \equiv 0$, we have for $P^A(S_T, 0)$ the option's contractually specified terminal value. The realizations of these terminal values, $\{[X - S_T(\omega)]^+\}_{\omega=1}^N$, are the actual cash flows that would be realized at T contingent on no early exercise. We store these in an N-vector $\mathbf{C}$, keeping track of the fact that they were realized at T. Next, stepping back to t_{n-1} we must estimate $V(S_{t_{n-1}}, T - t_{n-1}; \boldsymbol{\theta}_{n-1})$, recognizing this is as the approximant of the conditional expectation of the discounted future cash flows in measure $\hat{\mathbb{P}}$. To estimate it, we discount the elements of $\mathbf{C}$ back to time t_{n-1} (multiplying by the one-period bond price $B(t_{n-1}, T)$) and store the discounted values as $\mathbf{C}_{n-1}$. Letting $\mathbf{B}_{n-1}$ be the $N \times (m+1)$ matrix of realizations of the basis functions and $\mathbf{U}_n$ a vector of "errors", we have the identity $\mathbf{C}_{n-1} = \mathbf{B}_{n-1}\boldsymbol{\theta}_{n-1} + \mathbf{U}_n$. We obtain an estimate $\tilde{\boldsymbol{\theta}}_{n-1} = (\mathbf{B}_{n-1}'\mathbf{B}_{n-1})^{-1}\mathbf{B}_{n-1}'\mathbf{C}_{n-1}$ by regressing the present (date-t_{n-1}) value of the cash flows on the basis functions. Then from $\tilde{\boldsymbol{\theta}}_{n-1}$ we get an estimate of the continuation value on each path as $\tilde{V}_{n-1}(\omega) \equiv V(S_{t_{n-1}}(\omega), T - t_{n-1}; \tilde{\boldsymbol{\theta}}_{n-1}) = \mathbf{b}(S_{t_{n-1}}(\omega))\tilde{\boldsymbol{\theta}}_{n-1}$. If $\tilde{V}_{n-1}(\omega) < X - S_{t_{n-1}}(\omega)$, the option is assumed to be exercised at t_{n-1} on path ω, thereby generating a contingent cash flow. Since exercise at t_{n-1}

precludes exercise at T, we replace the old element ω in $\mathbf{C}$ with this new cash flow, again keeping track of the date at which it was realized. Once this has been done for all ω we move back to t_{n-2}, discount all elements of $\mathbf{C}$ from the dates received back to t_{n-2}, store them in $\mathbf{C}_{n-2}$, and regress $\mathbf{C}_{n-2}$ on the realizations of the basis functions, $\mathbf{B}_{n-2}$. This gives an estimate of the continuation value on each path ω at t_{n-2}, which is again compared with the intrinsic value and used to update $\mathbf{C}$ as before in preparation for time step t_{n-3}. Following these steps through $t = t_1$ yields a record $\mathbf{C}$ of the cash flows on all paths and the dates at which received. From these we get initial continuation value $V(S_0, T)$ simply by discounting the cash flows back to $t = 0$ and averaging over paths. Thus, at this step no parametric approximant is needed. At last, we obtain the estimate of the initial value of the option itself as $P^A(S_0, T) = V(S_0, T) \vee (X - S_0)$.

One further detail: Longstaff and Schwartz (2001) argue that sample paths at which the option is out of the money at a given time step are less informative about continuation value, and they recommend excluding such out-of-money realizations from the regressions. Thus, if the option were in the money on just N_j sample paths at t_j, vector $\mathbf{C}_j$ and "design" matrix $\mathbf{B}_j$ would have only N_j rows, and $\boldsymbol{\theta}_j$ would be estimated from just those N_j observations.[10]

Here is a simplified numerical example that takes us through the procedure step by step. We price a vanilla put on a no-dividend stock following Black-Scholes dynamics with $\sigma = .20$ and having initial value $S_0 = 0.9$ (see table 11.4). The option's strike is $X = 1$ and it expires at $T = 3$. We take unit time steps at $\{t_j = j\}_{j=1}^3$ and set the discount factor as $B(j, j+1) = .99$ at each step. $N = 10$ sample paths are generated as in the table, and terminal values $\{[X - S_3(\omega)]^+\}_{\omega=1}^N$ are recorded in a vector $\mathbf{C}$ (last column of table 11.4).

We take as basis functions of underlying price (the only state variable) the monomials $\{b_k(S_j) = S_j^k\}_{k=0}^2$ and so obtain a quadratic approximant to the continuation value at step $j \in \{1, 2\}$: $V(S_j, T - j; \boldsymbol{\theta}_j) \equiv V_j = \theta_{j0} + \theta_{j1} S_j + \theta_{j2} S_j^2$. Commencing at step 2, we construct the regression design matrix $\mathbf{B}_2$ and the vector $\mathbf{C}_2$ as in table 11.5. Elements of $\mathbf{C}_2$ are those of $\mathbf{C}$ in the table above multiplied by the one-step discount factor, .99. However, there are entries only for paths such that $X - S_2 > 0$, since the others

[10] Of course, to simplify programming one could simply fill the missing rows of $\mathbf{B}_j$ and $\mathbf{C}_j$ with zeroes, as this would not affect the least squares estimates.

Table 11.4. Sample paths of $\{S_t\}_{t=0}^3$ and cash flows $C = (X - S_3)^+$ at $t = 3$.

ω	S_0	S_1	S_2	S_3	**C**
1	0.9	0.905	0.737	0.882	.118
2	0.9	0.878	1.056	0.865	.135
3	0.9	1.119	0.721	0.789	.211
4	0.9	0.709	1.079	0.967	.033
5	0.9	1.100	1.508	1.636	.000
6	0.9	0.722	0.516	0.466	.534
7	0.9	1.330	1.584	1.192	.000
8	0.9	0.597	0.491	0.640	.360
9	0.9	0.887	0.825	0.830	.170
10	0.9	0.895	0.943	0.919	.081

Table 11.5. Estimated continuation values as of $t = 2$ and revised cash flows C at $t = j$.

ω		$\mathbf{B}_2$		$\mathbf{C}_2$	$\tilde{V}_2$	$(X - S_2)^+$	**C**	j
1	1.0	.737	.543	.117	.184	.263	.263	2
2	—	—	—	—	—	.000	.135	3
3	1.0	.721	.520	.209	.196	.279	.279	2
4	—	—	—	—	—	.000	.033	3
5	—	—	—	—	—	.000	.000	—
6	1.0	.516	.266	.528	.415	.484	.484	2
7	—	—	—	—	—	.000	.000	—
8	1.0	.491	.241	.357	.449	.509	.509	2
9	1.0	.825	.681	.168	.128	.175	.175	2
10	1.0	.943	.889	.080	.086	.057	.081	3

are not used in the regression. Rows of $\mathbf{B}_2$ are realizations of the basis functions on the in-the-money paths. For example, since $S_2(\omega) = 0.737 < 1.0 = X$ on path $\omega = 1$, the first row of $\mathbf{B}_2$ is $(1.0, .737, .737^2 \doteq .543)$. The least squares estimates, $\tilde{\boldsymbol{\theta}}_2 = (1.466, -2.727, 1.340)'$, yield the estimated continuation values $\tilde{V}_2(\omega) = 1.466 - 2.727 S_2(\omega) + 1.340 S_2(\omega)^2$ that appear under column head $\tilde{V}_2$. These are compared with the exercise values in the adjacent column. When the exercise value is larger, we substitute it for the old corresponding element of cash-flow vector $\mathbf{C}$ and record the time step j at which the new element of $\mathbf{C}$ was received (last column).

Now we back up to step 1 and construct new design matrix $\mathbf{B}_1$ and discounted cash-flow vector $\mathbf{C}_1$ as in table 11.6, again using only paths for which $X - S_1 > 0$. Entries in $\mathbf{C}_1$ are those in corresponding rows of $\mathbf{C}$ in the previous table multiplied by the appropriate discount factors. For

Table 11.6. Estimated continuation values as of $t = 1$ and revised cash flows C at $t = j$.

ω	$\mathbf{B}_1$			$\mathbf{C}_1$	$\tilde{V}_1$	$(X - S_1)^+$	$\mathbf{C}$	j
1	1.0	.905	.818	.261	.172	.095	.263	2
2	1.0	.878	.770	.132	.160	.122	.135	3
3	—	—	—	—	—	.000	.279	2
4	1.0	.709	.503	.033	.260	.291	.291	1
5	—	—	—	—	—	.000	.000	—
6	1.0	.722	.521	.479	.243	.278	.278	1
7	—	—	—	—	—	.000	.000	—
8	1.0	.597	.357	.504	.495	.403	.509	2
9	1.0	.887	.786	.173	.163	.113	.175	2
10	1.0	.895	.802	.080	.167	.105	.081	3

example, entries in the first, second, and fourth rows of $\mathbf{C}_1$ are (apart from rounding error) $.99(.26314) \doteq .261$, $99^2(.135) \doteq .132$, and $.99^2(.03335) \doteq .033$.

The least-squares estimates $\tilde{\boldsymbol{\theta}}_1 = (4.003, -9.060, 5.333)'$ yield entries $\tilde{V}_1(\omega) = 4.003 - 9.060S_1(\omega) + 5.333S_1(\omega)^2$ in the $\tilde{V}_1$ column. For paths 4 and 6 the exercise values exceed these continuation values, so the cash flows for those paths are updated to construct the new vectors of cash flows and exercise dates in the last two columns. The entries in the $\mathbf{C}$ and j columns now give a complete record of the cash flows after $t = 0$. From these we obtain the estimate of continuation value at $t = 0$ by averaging the discounted cash flows over the $N = 10$ paths, as

$$
\begin{aligned}
\tilde{V}(S_0, T) &= \frac{1}{10} \sum_{\omega=1}^{10} B(0, j(\omega)) C(\omega) \\
&= \frac{1}{10} [.99^2(.263) + .99^3(.135) + \cdots + .99^3(.081)] \\
&\doteq 0.197.
\end{aligned}
$$

Comparing with the initial intrinsic value, $X - S_0 = 0.1$, we value the option as $P^A(S_0, 3) = 0.197$.

Several issues must be addressed before putting this into actual practice.

Selecting basis functions. Longstaff and Schwartz (2001) find that there is considerable latitude in this choice. However, while monomials (the $\{S^k\}$ in our numerical example) are sometimes satisfactory, the resulting design matrices $\{\mathbf{B}_j\}$ are apt to be ill conditioned and to produce wild estimates of continuation values. This suggests the use of some orthogonal system, such as the Laguerre polynomials, given by[11]

$$
\begin{aligned}
L_0(s) &= 1\\
L_1(s) &= 1 - s\\
L_2(s) &= 2 - 4s + s^2\\
&\vdots\\
L_k(s) &= e^s \frac{d^k}{ds^k}(s^k e^{-s}).
\end{aligned}
$$

These are orthogonal on $[0, \infty)$ with respect to weight function e^{-s}, in the sense that $\int_0^\infty L_k(s)L_{k'}(s)e^{-s} \cdot ds = 0$ for $k \neq k'$. Thus, for a nonnegative state variable S one might take $b_0(S) = 1$ (to allow for a constant term) and $b_k(S) = e^{-S}L_{k-1}(S)$ for $k \in \{1, 2, \ldots, m\}$. Other candidates would be weighted Hermite polynomials for variables supported on $(-\infty, \infty)$ and weighted Chebychev or Legendre polynomials for variables with bounded support.

Handling multiple state variables. If the value of the derivative depends on more than one state variable (a common case in which simulation is needed), the basis functions would be products of the one-dimensional bases in the separate variables. Thus, restricting to orders no greater than m, the set of basis functions for state variables $\mathbf{Y} = (Y_1, Y_2)'$ would comprise $M_2 \equiv (m+1)(m+2)/2 \equiv (m+1)(M_1+1)/2$ terms of the form $b_i(Y_1)b_{k-i}(Y_2)$ for $i \in \{0, 1, \ldots, k\}$ and $k \in \{0, 1, \ldots, m\}$. For example, for $m = 3$ and $b_0(\cdot) = 1$ these would be

$$
\begin{matrix}
1 & b_1(Y_1) & b_2(Y_1) & b_3(Y_1)\\
b_1(Y_2) & b_1(Y_2)b_1(Y_1) & b_1(Y_2)b_2(Y_1) & \\
b_2(Y_2) & b_2(Y_2)b_1(Y_1) & & \\
b_3(Y_2) & & &
\end{matrix}.
$$

For $\mathbf{Y} = (Y_1, Y_2, Y_3)'$ there would be $M_3 = (m + 1)(M_2 + 1)/2$ terms, and so on, with $M_q = (m + 1)(M_{q-1} + 1)/2$ terms for $\mathbf{Y} = (Y_1, Y_2, \ldots, Y_q)'$. Clearly, to keep the problem manageable one must progressively curtail the maximum order as the number of state variables grows.

Performing the regressions. Design matrix $\mathbf{B}_j$ may still be ill conditioned even with weighted orthogonal polynomials as basis functions, par-

[11]Consult Abramovitz and Stegun (1970) and Davis (1963) for further description of this and other orthogonal systems. Miranda and Fackler (2002) discuss the ill-conditioned nature of design matrices built up from monomials.

ticularly when some of the state variables can assume large values and the weight functions are exponential, as for Laguerre and Hermite polynomials. For derivatives such as vanilla options that are homogenous of degree one in the underlying price and the strike, it will help to normalize by the strike and then rescale the final result. However, it is still essential to use a regression routine that can handle ill-conditioned designs.[12]

Managing memory. The LSM technique is a veritable memory hog. Programming in the most direct way requires storing in memory the realization of each state variable at each of the n time steps on each of the N sample paths. Thus, if there are q state variables, $N \cdot n \cdot q$ words of memory are needed just for the primitive data, plus another $2N$ words to record the dates and amounts of cash flows in $\mathbf{C}$, plus enough storage for the largest of design matrices and response vectors $\{\mathbf{B}_j, \mathbf{C}_j\}$, plus work space for the regression routine. Even with a single state variable, handling just 5,000 sample paths and 50 time steps is beyond the capabilities of most p.c.s. Of course, one can write the sample paths to disk and read back as needed, but that is a slow process. A better solution, at least when working in a Markov framework, is to generate the full sample paths out to T at *each* time step j (starting with the same initial seed), but to save only the realizations $\{\mathbf{Y}_{t_j}(\omega)\}_{\omega=1}^{N}$ that are needed for step j. (Realizations at $t_{j+1}, \ldots, T$ would still have to be generated—or at least the same number of calls to the generator would have to be made, since otherwise the cash flows in $\mathbf{C}$ would not have been attained on the current paths.) Perhaps the best solution attainable with straightforward programming is simply to (i) choose a value of N small enough that the full sample paths can be stored in active memory, (ii) replicate the experiment independently several times starting with new seeds, and (iii) average the final estimates. Table 11.7 compares Longstaff and Schwartz's (2001) estimates of prices of vanilla American puts from $N = 100{,}000$ paths with averages of 25 sets of estimates from $N = 4,000$ paths. By generating 2,000 pairs of paths antithetically, one can get each price in about nine seconds on a 2.8 g.h. p.c.[13]

---

[12]Longstaff and Schwartz (2001) recommend IMSL routine `LSBRR`, which uses an iterative refinement algorithm.

[13]Estimates with $N = 100,000$ and the standard errors are from Longstaff and Schwartz (2001, table 1). In each case the strike was $X = 40$, the interest rate was $r = .06$, prices followed Black-Scholes dynamics with $n = 50$ time steps, and continuation values were represented as affine functions of the first three weighted Laguerre polynomials. Computations for the $N = 25 \cdot 4000$ estimates were done in FORTRAN using IMSL routine `LBSRR` for the regressions.

Table 11.7. LSM estimates of prices of American puts from 100,000 paths vs. averages of 25 estimates from 4,000 paths.

S_0	σ	T	$N = 100,000$	s.e.	$N = 25 \cdot 4000$
36	.20	1	4.472	(.010)	4.475
36	.20	2	4.821	(.012)	4.822
36	.40	1	7.091	(.020)	7.116
36	.40	2	8.488	(.024)	8.522
40	.20	1	2.313	(.009)	2.314
40	.20	2	2.879	(.010)	2.878
40	.40	1	5.308	(.018)	5.321
40	.40	2	6.921	(.022)	6.924
44	.20	1	1.118	(.007)	1.119
44	.20	2	1.675	(.009)	1.690
44	.40	1	3.957	(.017)	3.958
44	.40	2	5.622	(.021)	5.645

Random Trees

The random tree method of Broadie and Glasserman (1997) is really just a way of adapting binomial pricing to simulation. To see how it works, it helps first to bring to mind the steps for pricing vanilla American-style derivatives in the binomial scheme. How did we do it? First, starting from initial value S_0 we generated a tree of values of underlying price in n discrete time steps using the appropriate risk-neutral Bernoulli dynamics. Specific locations in the tree were identified by specifying time step j and price level, $S_j \equiv S_{t_j}$. From S_0 and each of the subsequent time/price "nodes" prior to step n there were two branches, giving a total of $N = 2^n$ sample paths and $n+1$ distinct values of S_n for an ordinary recombining tree. Next, from each of the n nodes S_{n-1} at step $n-1$ we estimated the derivative's continuation value as the $(\hat{\pi}, 1-\hat{\pi})$-weighted average of the two discounted terminal values accessible from S_{n-1}. The derivative's value at each node S_{n-1} was then set equal to the greater of intrinsic value and continuation value, as in (11.16) for an American put. When values for all n nodes at step $n-1$ were thus determined, we moved back to step $n-2$, valued the derivative at all the $n-1$ nodes in the same way, and repeated the whole process back to step 0.

The same general steps are followed in the most straightforward version of the random tree method. First, starting from initial value S_0 we generate a tree of values of underlying price at the n discrete time steps at which exercise is to be allowed. This will now be done by simulation, using whatever measure-$\hat{\mathbb{P}}$ dynamics are appropriate for the underlying price (or other state variables) on which the derivative's intrinsic value depends. From the initial price S_0 and from each subsequent time/price node prior to step n there will now be m branches, giving a total of $N = m^n$ sample paths and (in general) the same number of distinct values of S_n. Since the price increments are now (pseudo-) random, this process will have produced a "random" tree with a continuous state space for prices. Next, from each of the m^{n-1} nodes S_{n-1} at step $n-1$ we estimate the derivative's continuation value as the sample mean of the m discounted terminal values accessible from S_{n-1}. The derivative's value at each node S_{n-1} is then set equal to the greater of intrinsic value and continuation value. When values for all m^{n-1} nodes at step $n-1$ are thus determined, we move back to step $n-2$, value the derivative at all the m^{n-2} nodes in the same way, and repeat the whole process back to step 0. Table 11.8 lays out and compares the steps for binomial and random-tree pricing of an American-style derivative worth $D(S_j) = D(S_j, n-j)$ at node S_j and having intrinsic value $X(S_j)$. There $\hat{\pi}_u$ and $\hat{\pi}_d = 1 - \hat{\pi}_u$ are the risk-neutral Bernoulli pseudo-probabilities, $S_{j,\omega}$ is the underlying price on branch ω from the price at time step $j-1 \in \{0, 1, \ldots, n-1\}$, and B represents the price of a one-step discount bond.

One can see at a glance that the straightforward random tree method is completely unworkable unless n and/or m is very small. However,

Table 11.8. Comparison of binomial and random-tree schemes.

Step	Action	Binomial	Random Tree
	From S_0 generate tree of	$n+1$ prices S_n	m^n prices S_n
$n-1$	Set $D(S_{n-1}) = X(S_{n-1}) \vee$	$B \sum_{\omega=u,d} \hat{\pi}_\omega X(S_{n,\omega})$	$\frac{B}{m} \sum_{\omega=1}^{m} X(S_{n,\omega})$
$n-2$	Set $D(S_{n-2}) = X(S_{n-2}) \vee$	$B \sum_{\omega=u,d} \hat{\pi}_\omega D(S_{n-1,\omega})$	$\frac{B}{m} \sum_{\omega=1}^{m} D(S_{n-1,\omega})$
$\vdots$	$\vdots$	$\vdots$	$\vdots$
1	Set $D(S_1) = X(S_1) \vee$	$B \sum_{\omega=u,d} \hat{\pi}_\omega D(S_{n-2,\omega})$	$\frac{B}{m} \sum_{\omega=1}^{m} D(S_{2,\omega})$
0	Set $D(S_0) = X(S_0) \vee$	$B \sum_{\omega=u,d} \hat{\pi}_\omega D(S_{n-1,\omega})$	$\frac{B}{m} \sum_{\omega=1}^{m} D(S_{1,\omega})$

since continuation values are estimated by averaging over the m replicates $\{S_{j,\omega}\}_{\omega=1}^{m}$ accessed from each node S_{j-1}, it is clear that economy must be achieved mainly by moderating the number of time steps, n. Thus, the random tree method should really be considered a means of pricing Bermudan-style derivatives rather than those that are continuously exercisable. On the other hand, there is one conspicuous advantage: no parametric approximation of either exercise boundary or continuation value is required. Rather, the continuation value at any node can be estimated with arbitrary precision merely by taking m to be sufficiently large. This means that there are virtually no limits to the number of state variables or to the complexity of their dynamics: if we can simulate the underlying variables, we can price the derivatives. In addition, there are modifications to the straightforward scheme outlined above that significantly reduce the computation time for given m and n.

Two obvious ways of increasing efficiency are the standard techniques of antithetic acceleration and, when applicable, the use of control variates. For acceleration one simply generates $m/2$ pairs of paths antithetically from each node and estimates continuation value by averaging over all m paths. Using a control variate is feasible when a corresponding European-style derivative can be valued analytically or extrinsically in some other fast, accurate way. Less obvious is a procedure known as "pruning" that saves steps at out-of-money nodes. The key insight is that the only relevant course of action at such nodes is precisely the same as for European-style derivatives—i.e., hold on. Since the proper decision is obvious, a precise estimate of continuation value is not needed. Thus, rather than generate m new paths to estimate continuation value, we can generate just *one* successor price from which to calculate $D(S_{j-1})$ as (in the notation of table 11.8) $B \cdot D(S_j)$. Doing so amounts to pruning $m-1$ branches from the tree at this point.

How to program all this is not exactly obvious. The following pseudocode shows one way to implement all three time-saving methods with $n = 3$ equally spaced time steps and an arbitrary *even* number m of branches at each in-the-money node. The specific example involves a single underlying asset worth `S(j)` at step j and following geometric Brownian motion with volatility `s` and drift `r-d` under $\hat{\mathbb{P}}$. Adapting to other dynamics and state variables requires just changing the "Generate" steps. "`D(S(j))`" and "`E(S(j))`" represent values at node `S(j)` of an American-style derivative and a European-style counterpart that serves as a control variate. "`X(S(j))`" represents the intrinsic value; for example, `X(S(j))=`$(X - S_j)^+$ for a vanilla put option. "`e(S(0))`" at the end is an extrinsic estimate of the value of the European, and "`d(S(0))`" is the final control-variate-adjusted value of the American derivative. Variables "`ip`" and "`jp`" are used to handle the antithetic acceleration, and "`jbranch`" and "`kbranch` " handle the pruning.

```
Set initial underlying price S(0), expiration date T, time
 step dt=T/3, interest rate r, one-step bond price B,
 dividend rate d, volatility s, number branches m (even)
Set emudt=exp[(r-d-s*s/2)*dt]
Initialize D(S(0))=E(S(0))=0
Loop for i=1,2,...,m/2
  Initialize D(S(1))=E(S(1))=0
  Set ip=1
1 Generate S(1,i)=S(0)*emudt*exp(ip*s*Z(i)) for Z(i)~N(0,1)
  If X(S(1))>0 set jbranch=m/2; else set jbranch=1
  Initialize D1=E1=0
  Loop for j=1,jbranch
    Initialize D(S(2))=E(S(2))=0
    Set jp=1
2   Generate S(2,j)=S(1,i)*emudt*exp(jp*s*Z(j)),Z(j)~N(0,1)
    If X(S(2))>0 set kbranch=m/2; else set kbranch=1
    Initialize D2=0
    Loop for k=1,kbranch
      Generate Sp(3,k)=S(2,j)*emudt*exp( s*Z(k))
      Generate Sn(3,k)=S(2,j)*emudt*exp(-s*Z(k))
      Accumulate D2=D2+X(Sp(3,k))+X(Sn(3,k))
    End k loop
    Accumulate E(S(2))=E(S(2))+.5*B*D2/kbranch
    Set D2 = max[X(S(2)), B*D2/kbranch]
    Accumulate D(S(2))=D(S(2))+.5*D2
    If jp=1 set jp=-1 and go to 2
    Accumulate D1=D1+D(S(2)), E1=E1+E(S(2))
  End j loop
  Accumulate E(S(1))=E(S(1))+.5*B*E1/jbranch
  Set D1=max[X(S(1)), B*D1/jbranch]
  Accumulate D(S(1))=D(S(1))+ .5*D1
  If ip=1 set ip=-1 and go to 1
  Accumulate D(S(0))=D(S(0))+D(S(1)), E(S(0))=E(S(0))+E(S(1))
```

```
End i loop
Set D(S(0)=max[X(S(0)),B*D(S(0)/m], E(S(0))=B*E(S(0))/m
Set d(S(0))=D(S(0))+[e(S(0))-E(S(0))]
End program
```

Although this is hardly the appropriate application for the method, using it to price a vanilla American put does at least give a benchmark of performance. Table 11.9 compares random-tree estimates with $m = 500$ branches and $n = 3$ time steps with binomial estimates from 5,000 time steps.[14] A European put was estimated as a control, and antithetic acceleration and pruning were used as described above. Glasserman (2004) shows that this procedure yields upward-biased estimates of true prices of Bermudan derivatives with the same exercise opportunities. He constructs matching downward-biased estimators by using different subsets of branches from a given node to decide whether to continue and to estimate continuation value, then uses the two sets of estimates and their standard deviations to construct conservative confidence intervals. Of course, the bias of the standard procedure is indeterminate when measured against the values of continuously exercisable derivatives.

A Comparison

Here we apply the three methods to a case where simulation may well be the best approach to estimation, taking both accuracy and computing time into account; namely, an American call on the maximum of two underlying stocks. This application has been considered by proponents of all three methods: Garcia (2003), Longstaff and Schwartz (2001), and Broadie and Glasserman (1997). In applying each method here we try to show it to its best advantage by exploiting what is known about this particular problem.

Table 11.9. Binomial *vs.* random tree valuations of $P^A(S_0, T = 1; X = 1)$.

S_0	Binomial	Tree without control	Tree with control
0.9	.1149	.1097	.1111
1.0	.0609	.0608	.0594
1.1	.0299	.0288	.0289

[14]The data are the same as for the earlier estimates obtained by parameterizing the exercise boundary: $X = T = 1, \sigma = .20, r = .05, \delta = 0$. Execution times for the tree estimates are about twice those of the binomial.

Specifically, we use the exact solution for the price of a European-style call on the max—formula (7.42)—to set up control variates and to model the exercise boundary and the continuation value in the two parametric approaches.

Consider a T-expiring call option on the maximum of two assets whose prices, $\{\mathbf{S}_t = (S_{1t}, S_{2t})\}_{t\geq 0}$, follow geometric Brownian motion with volatilities σ_1, σ_2 and correlation (of log price) ρ. The short rate is a constant r, and the two assets pay continuous dividends at rates δ_1, δ_2. At expiration the call is worth $C^{\max}(\mathbf{S}_T; 0) = X(\mathbf{S}_T) \equiv [X - (S_{1T} \vee S_{2T})]^+$, and the problem is to estimate initial value $C^{\max}(\mathbf{S}_0; T)$ given that the option can be exercised at any of the discrete set of times $\{t_j = jT/n\}_{j=0}^n$. To conserve notation, define $X(\mathbf{S}_j) \equiv [X - (S_{1j} \vee S_{2j})]^+$ as the intrinsic value at t_j. Letting $\{Z_{1j}, Z_{0j}\}_{j=1}^n$ be i.i.d. pairs of pseudorandom, independent standard normals, we set $Z_{2j} = \rho Z_{1j} + \sqrt{1-\rho^2} Z_{0j}$ and generate sample paths as

$$\begin{aligned} S_{1j} &= S_{1,j-1} \exp[(r - \delta_1 - \sigma_1^2/2) \cdot dt + \sigma_1 \sqrt{dt} Z_1] \\ S_{2j} &= S_{2,j-1} \exp[(r - \delta_2 - \sigma_2^2/2) \cdot dt + \sigma_2 \sqrt{dt} Z_2], \quad j \in \{1, 2, \ldots, n\}, \end{aligned}$$

where $dt \equiv T/n$ and initial values S_{10}, S_{20} are given. We follow the applications in the literature by treating the symmetric problem with

$$\begin{aligned} S_{10} &= S_{20} \in \{90, 100, 110\} \\ \sigma_1 &= \sigma_2 = .20 \\ \delta_1 &= \delta_2 = .10 \\ \rho &= .3 \\ r &= .05 \end{aligned}$$

and option parameters $T = 3$, $X = 100$.

To apply the parametric-boundary approach to this problem Garcia (2003) exploits the symmetry of the process $\{S_{1t}, S_{2t}\}$ by modeling the exercise region at t_j as

$$\mathbb{B}_j(\boldsymbol{\theta}) = \{\mathbf{S}_j : X(\mathbf{S}_j) > \theta_{1j}, |S_{1j} - S_{2j}| > \theta_{2j}\}$$

for $j \in \{0, 1, \ldots, n-1\}$ and setting $\mathbb{B}_n = \{\mathbf{S}_n : S_{1n} \vee S_{2n} > X\}$. This requires estimating the $2n$ parameters $\{\theta_{1j}, \theta_{2j}\}_{j=0}^{n-1}$. The idea is that for a given value of $S_1 \vee S_2$ the option's continuation value should diminish as the prices spread out, and the incentive for early exercise should therefore increase. Were the models for processes $\{S_{1t}\}_{t\geq 0}$ and $\{S_{2t}\}_{t\geq 0}$ not the same, some modification to $\mathbb{B}_j(\boldsymbol{\theta})$ would be needed to allow the prices to enter asymmetrically. As an alternative, one could adapt and extend the scheme that was used to price a put on a single underlying, as expressed in (11.12) and (11.13). That is, letting $C_E^{\max}(\mathbf{S}_t; T-t)$ represent the value of a European-style call on the max, given explicitly by (7.42), we could calculate at each t_j the locus of points $\mathbf{S}_j^*$ in $\Re_2$ such that $C_E^{\max}(\mathbf{S}_j^*; T-t_j) = S_{1j}^* \vee S_{2j}^* - X$. This would be done for a two-dimensional grid of values of S_{1j}, S_{2j}, with (say) linear interpolation for points between. Then we could set the exercise region as

$$\mathbb{B}_j(\boldsymbol{\theta}) = \left\{ \mathbf{S}_j : S_{1j} \vee S_{2j} - X > C_E^{\max}(\mathbf{S}_j^*; T-t_j) + \sum_{k=1}^{K} \theta_k (T-t_j)^k \right\}.$$

An alternative, corresponding to (11.15), is to skip the calculation of the locus $\mathbf{S}_j^*$ and set

$$\mathbb{B}_j(\boldsymbol{\theta}) = \left\{ \mathbf{S}_j : S_{1j} \vee S_{2j} - X > C_E^{\max}(\mathbf{S}_j; T-t_j) + \sum_{k=1}^{K} \theta_k (T-t_j)^k \right\}.$$

The corresponding exercise boundary for $S_{1j} \vee S_{2j}$ can be calculated very quickly. We use this method with $K = 3$ parameters in the comparisons given below.

To obtain their regression-based "LSM" estimates of continuation value, Longstaff and Schwartz (2001) suggest[15] using as basis functions the first few Hermite polynomials in $S_{1j} \vee S_{2j}$, the value and square of $S_{1j} \wedge S_{2j}$, and the product $(S_{1j} \vee S_{2j}) \cdot (S_{1j} \wedge S_{2j})$. They also allow an intercept in the regression functions. Again, we can simplify by making use of the formula for the European version. Specifically, we take as basis functions the price of the European call, $C_E^{\max}(\mathbf{S}_t; T-t)$, and the first two weighted Laguerre polynomials in the time to expiration, $e^{-(T-t)/2}$ and $e^{-(T-t)/2}[1-(T-t)]$. We also allow for an intercept, giving a total of four parameters to be estimated at each step in the backward recursion.

For the random-tree approach little change is needed from the steps outlined above for puts on a single underlying: merely simulate paths of the bivariate process and use the appropriate intrinsic values $X(\mathbf{S}_j)$ in the scheme laid out in table 11.8.

With all three methods we use antithetic acceleration to generate the price processes, and we estimate prices of a European put in the same simulation to use as a control. Estimates based on the parametric boundary are

[15]This is an inference from their recipe for a call on the maximum price of *five* underlying assets.

from 20,000 sample paths; those based on least-squares estimates of continuation value are averages from 25 replications of 10,000 paths each; and those from random trees are averages of 25 replications using 200 branches at each step. The multiple replications of the last two methods provide estimates of the standard deviation of the mean of the 25 values; standard errors for the parametric boundary estimates come from the cash flows realized in the 20,000 paths. Table 11.10 gives estimates of value for each of the three sets of initial underlying prices, estimates of the standard errors of the estimators, and execution times in seconds.[16] Execution times to estimate the parametric boundary are highly variable, depending on the initial guesses used in the optimization. Results for $n = 3$ time steps (one exercise opportunity per year) are given for all three methods and compared with supposedly accurate values (reported in Glasserman, 2004, p. 440), obtained by two-dimensional lattice methods. The flexibility of the two parametric schemes allows for more exercise opportunities, so we report results from these for $n = 60$ also.

Table 11.10. Parametric/nonparametric simulation estimates of price of American call on the maximum price of two underlying assets, with $n + 1$ exercise dates, three years to expiration, and strike price 100.

Method	$S_{10}, S_{20} = 90$	$S_{10}, S_{20} = 100$	$S_{10}, S_{20} = 110$
Par. Boundary			
$n = 3$	7.119	12.51	19.07
(s.e., time)	(.097, 10.1)	(.12, 16.1)	(.14, 18.3)
$n = 60$	7.962	13.39	20.43
(s.e., time)	(.087, 416)	(.10, 318)	(.12, 654)
Par. Continuation			
$n = 3$	7.239	12.39	19.02
(s.e., time)	(.018, 4.1)	(.02, 6.0)	(.03, 7.6)
$n = 60$	7.680	13.11	20.06
(s.e., time)	(.045, 25.3)	(.06, 38.4)	(.07, 49.9)
Random Tree			
$n = 3$	7.250	12.50	19.17
(s.e., time)	(.024, 20.6)	(.04, 38.9)	(.06, 57.1)
Lattice			
$n = 3$	7.234	12.41	19.06

Standard errors (s.e.) and execution times in seconds are in parentheses.

[16]Programming was done in FORTRAN using IMSL routines to compute the bivariate normal c.d.f.s in (7.42) and for the LSM regressions. Execution times are for a 2.8 g.h. single-processor p.c.

Better results for the parametric continuation method were obtained without a control variate. These better results are reported in the table.

In this particular application it is clear that the regression-based LSM technique (identified as "Par. Continuation" in the table) dominates in terms of accuracy (relative to the lattice estimates) and computational speed. The parametric boundary scheme also has the flexibility to allow arbitrarily many exercise opportunities, but it seems to offer no advantage to compensate for the long execution times. As the results here indicate, it can be easy to find good parametric models for continuation values when there is a tractable formula for a comparable European option. Still, the advantage of LSM over the nonparametric tree method depends critically on having such good models of continuation value, and the nonparametric scheme provides a natural way to check their adequacy for small n.

12

Solving P.D.E.s Numerically

This chapter gives an introduction to the numerical solution of partial differential equations by finite-difference methods.[1]

To get the basic idea of finite-difference methods, let us consider first applications in which the value of a T-expiring financial derivative depends just on S_t and t—the underlying price and time—as $D(S_t, T-t)$. The goal is to get an approximate numerical solution to a p.d.e. that involves the derivatives of D with respect to these two arguments. The procedure then involves the following steps:

1. Express the p.d.e. in terms of appropriate finite-difference approximations for the partial derivatives;
2. Set up a grid of values of t and S_t;
3. Use initial and boundary conditions to specify the derivative's values on the borders of the grid that correspond to $T - t = 0$ and to the extreme values of S_t; and
4. At each remaining point (t, S_t) in the grid use the p.d.e. to develop one or more equations from which $D(S_t, T-t)$ can be found in terms of known values of D at other points.

Figure 12.1 illustrates the basic scheme. Solid circles at nodes on the upper, lower, and right-hand borders represent values known from initial and boundary conditions, and the empty circles elsewhere are values that must

[1]Details and more advanced approaches can be found in the books by Collatz (1966), Hall and Porsching (1990), Mitchell and Griffiths (1980), and Zwillinger (1989). Wilmott *et al.* (1993) and Wilmott *et al.* (1995) give extensive treatment of the financial p.d.e. in particular.

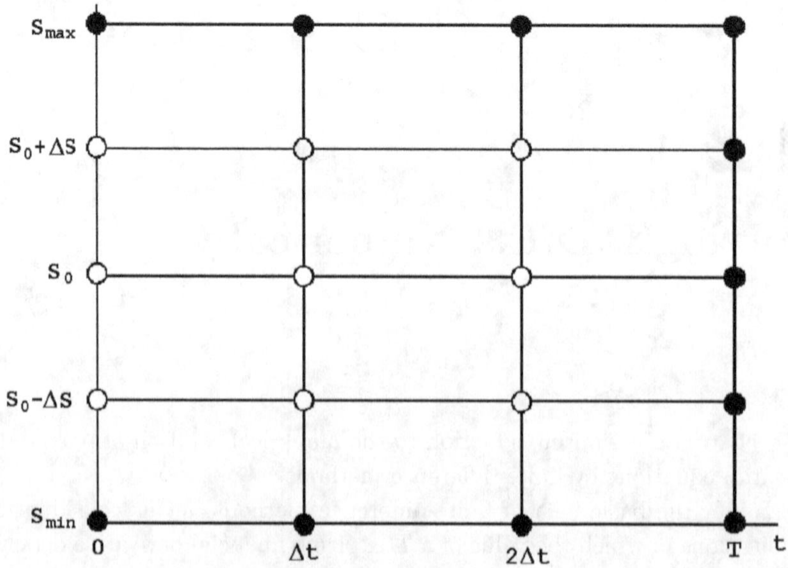

Fig. 12.1. Time-price grid for numerical solution of p.d.e.

be found. Lattices of higher dimension are required if the financial derivative depends on additional state variables.

We will take up three methods of solution in the order of increasing accuracy and complexity: an "explicit" method, a "first-order implicit" method, and a "second-order implicit" method due to Crank and Nicolson (1947). The explicit method involves solving just one equation at each node, whereas the implicit methods require a recursive procedure. Since all the techniques are somewhat involved, we will first explain how to apply them to ordinary finite-lived derivatives, such as vanilla European or American options, on assets whose prices follow geometric Brownian motion. We will also assume constant short rates and dividend rates. Under these conditions the fundamental p.d.e. takes the familiar form

$$-D_{T-t} + D_S(r-\delta)S_t + D_{SS}\sigma^2 S_t^2/2 - Dr = 0. \tag{12.1}$$

Once mastering the techniques in this setting, we can see how to apply the Crank-Nicholson approach to more complicated p.d.e.s that allow for continuous time- and asset-dependent cash receipts by the underlying and the derivative, time-varying interest rates, and more general Itô processes for S_t; for example, p.d.e.s such as

$$\kappa(S_t,t) - D_{T-t} + D_S\mu(S_t,t) + D_{SS}\sigma^2(S_t,t)/2 - Dr_t = 0$$

for various specifications of $\kappa(S_t,t)$, $\mu(S_t,t)$, $\sigma(S_t,t)$, and r_t.

The main advantage of working with the simpler p.d.e. (12.1) is that it can be transformed to make the coefficients of the partial derivatives invariant with respect to t and S_t. This is done by expressing the derivative's value in terms of $s_t \equiv \ln(S_t)$ and calendar time t instead of S_t and $T-t$. Letting $d(s_t,t) \equiv D(e^{s_t}, T-t)$, we have

$$\begin{aligned} d_s &= S_t D_S, \\ d_{ss} &= S_t^2 D_{SS} + d_s, \\ d_t &= -D_{T-t}; \end{aligned}$$

and the following new version of (12.1):

$$d_t + d_s(r - \delta - \sigma^2/2) + d_{ss}\sigma^2/2 - dr = 0. \tag{12.2}$$

12.1 Setting Up for Solution

Before looking for solutions we must carry out the first three steps on page 577.

12.1.1 *Approximating the Derivatives*

There are various ways to approximate derivatives d_t, d_s, and d_{ss} by finite differences, and the choice determines whether the method of solving the p.d.e. will be explicit or implicit. For example, d_t can be approximated as either a forward or backward difference; that is, as

$$d_t^+ \equiv \frac{d(s,t+\Delta t) - d(s,t)}{\Delta t} \tag{12.3}$$

or as

$$d_t^- \equiv \frac{d(s,t) - d(s,t-\Delta t)}{\Delta t}. \tag{12.4}$$

To see the order of accuracy of these, use Taylor's theorem to approximate d at $t+\Delta t$ and at $t-\Delta t$ near (s,t), assuming that the second time derivative exists and is continuous:[2]

$$d_t^+(s,t) = d_t(s,t) + \frac{1}{2}d_{tt}(s,t) \cdot \Delta t + O(\Delta t^2)$$
$$d_t^-(s,t) = d_t(s,t) - \frac{1}{2}d_{tt}(s,t) \cdot \Delta t + O(\Delta t^2).$$

This shows that d_t^+ and d_t^- are $O(\Delta t)$ approximations. The Δt term vanishes, however, upon averaging the forward and backward approximations. The result is a centered approximation that is accurate to $O(\Delta t^2)$:

$$d_t^0(s,t) \equiv \frac{d_t^+ + d_t^-}{2} = \frac{d(s,t+\Delta t) - d(s,t-\Delta t)}{2\Delta t} = d_t(s,t) + O(\Delta t^2). \tag{12.5}$$

Similarly, there are $O(\Delta s)$ forward and backward approximations to the s derivative,

$$d_s^\pm(s,t) = \frac{\pm d(s \pm \Delta s, t) \mp d(s,t)}{\Delta s},$$

and an $O(\Delta s^2)$ centered approximation:

$$d_s^0(s,t) \equiv \frac{d(s+\Delta s,t) - d(s-\Delta s,t)}{2\Delta s}. \tag{12.6}$$

An $O(\Delta s^2)$ approximation for d_{ss} can be obtained from the difference between forward and backward first derivatives:

$$d_{ss}^0(s,t) \equiv \frac{d_s^+(s,t) - d_s^-(s,t)}{\Delta s} = \frac{d(s+\Delta s,t) + d(s-\Delta s,t) - 2d(s,t)}{\Delta s^2}. \tag{12.7}$$

The explicit and implicit methods we describe below use various combinations of these approximations.

12.1.2 *Constructing a Discrete Time/Price Grid*

Evaluating $d(s,t)$ by finite-difference methods requires a discrete grid of s and t values. For definiteness we put t on the horizontal axis and s on the vertical. Let $t = 0$ and $t = T$ be the current date and the expiration date of the financial derivative, respectively. To construct the grid, first divide $[0, T]$ into some n_T equal intervals of length $\Delta t = T/n_T$. Larger values of n_T give more accuracy but lengthen the computation, and one has to experiment with this parameter. This gives a set of $n_T + 1$ time values,

[2]Throughout, symbols "Δt^2" and "Δs^2" represent squares of the first differences: $(\Delta t)^2$ and $(\Delta s)^2$.

$\mathcal{T} \equiv \{j\Delta t\}_{j=0}^{n_T}$. Next, specify a grid of values of s on an interval within which s_t is almost certain to remain during the lifetime of the option. This takes a bit more care. One wants points on an interval $[s_{Min}, s_{Max}]$ such that $\hat{\mathbb{P}}(s_{Min} < s_T < s_{Max}) \doteq 1$, where $\hat{\mathbb{P}}$ is the risk-neutral measure. If the value of the derivative asset is known at $S_t = 0$, then in specifying boundary conditions it is helpful to take s_{Min} such that $e^{s_{Min}} \doteq 0$. Also, since the derivative's initial value will eventually be determined just at the values of s in the grid, one also wants the log of initial price s_0 to be one of the set. Finally, it is generally helpful to choose a price step that is proportional to the square root of the time step, as $\Delta s = \sigma\sqrt{3\Delta t}$. The reason for this will be seen later.

Here is one way to accomplish all these things. Picking $\Delta s = \sigma\sqrt{3\Delta t}$, find numbers n_S^- and n_S^+ such that $\exp(s_{Min}) \equiv \exp(s_0 - n_S^-\Delta s) \doteq 0$ and $\hat{\mathbb{P}}(s_0 - n_S^-\Delta s < s_T < s_0 + n_S^+\Delta s) \doteq 1$, and let $n_S \equiv n_S^- + n_S^+$. Then with $s_{Min} = s_0 - n_S^-\Delta s$ and $s_{Max} = s_0 + n_S^+\Delta s = s_{Min} + (n_S + 1)\Delta s$ the construction gives a set of $n_S + 1 \equiv n_S^- + n_S^+ + 1$ log-price values that contains the initial value:

$$\mathcal{S} \equiv \{s_{Min} + j\Delta s\}_{j=0}^{n_S}.$$

This construction delivers an $(n_T + 1)$ by $(n_S + 1)$ array of time and log-price points, $\mathcal{T} \times \mathcal{S}$. Most of the work is done with columns in the grid, which will contain values of the financial derivative at specific t's: $\{d(s,t)\}_{s\in\mathcal{S}}$. However, specifying boundary conditions also requires some work with the top and bottom rows: $\{d(s_{Min},t)\}_{t\in\mathcal{T}}$ and $\{d(s_{Max},t)\}_{t\in\mathcal{T}}$.

12.1.3 *Specifying Boundary Conditions*

Now use the initial and boundary conditions for the financial derivative to determine the value of $d(s,t)$ at (i) each point on the right boundary of the grid, corresponding to the time column at $n_T\Delta t = T$; (ii) each point on the top boundary, corresponding to the s_{Max} row; and (iii) each point on the bottom boundary, corresponding to the row with $s = s_{Min}$. For example, suppose the derivative is a call option. Letting X be the strike, the values at the right boundary, which correspond to the initial condition at $T - t = 0$, are $\{c(s,T) = (e^s - X)^+\}_{s\in\mathcal{S}}$. At the bottom boundary, assuming $e^{s_{Min}} \doteq 0$, the values are $\{c(0,t) = 0\}_{t\in\mathcal{T}}$. If the call is American, the values at the upper boundary are $\{c(s_{Max},t) = e^{s_{Max}} - X\}_{t\in\mathcal{T}}$; if it is European, they are $\{c(s_{Max},t) = e^{s_{Max}} - e^{-r(T-t)}X\}_{t\in\mathcal{T}}$. For a put option, values at the right and top boundaries are $\{p(s,T) = (X - e^s)^+\}_{s\in\mathcal{S}}$ and

$\{p(s_{Max}, t) = 0\}_{t \in \mathcal{T}}$. Bottom values are $\{p(0,t) = X\}_{t \in \mathcal{T}}$ if the put is American and $\{p(0,t) = e^{-r(T-t)}X\}_{t \in \mathcal{T}}$ if it is European.

12.2 Obtaining a Solution

What remains is to fill in the rest of the points on the grid, the main interest usually being in the first column, $\{d(s,0)\}_{s \in \mathcal{S}}$. As in the binomial method one begins by evaluating d for each point in the next-to-last column, corresponding to $t = (n_T - 1)\Delta t$, and then works backward. Of several ways to proceed the next is the most straightforward.

12.2.1 *The Explicit Method*

This technique gets its name from the fact that the values $\{d(s,t)\}_{s \in \mathcal{S}}$ in each column will be represented explicitly in terms of just the values in the next column; that is, in terms of $\{d(s, t + \Delta t)\}_{s \in \mathcal{S}}$. Letting $p \equiv (r - \delta - \sigma^2/2)$and $q \equiv \sigma^2/2$, write the p.d.e. (12.2) as

$$d_t + d_s p + d_{ss} q - dr = 0.$$

Since $d(s,t)$ is to be calculated from values of $d(s, t + \Delta t)$, the forward approximation $d_t^+(s,t)$ will be used for d_t, and the centered approximations to d_s and d_{ss} will be evaluated at the next time step, $t + \Delta t$. These are $d_s^0(s, t + \Delta t)$ and $d_{ss}^0(s, t + \Delta t)$. This involves a second approximation, because it treats the s derivatives as constant over the interval $[t, t + \Delta t]$. With these choices the p.d.e. becomes

$$\begin{aligned} d(s,t)r &= d_t^+(s,t) + d_s^0(s, t+\Delta t)p + d_{ss}^0(s, t+\Delta t)q \\ &= \frac{d(s,t+\Delta t) - d(s,t)}{\Delta t} + \frac{d(s+\Delta s, t+\Delta t) - d(s-\Delta s, t+\Delta t)}{2\Delta s} \cdot p \\ &\quad + \frac{d(s+\Delta s, t+\Delta t) + d(s-\Delta s, t+\Delta t) - 2d(s, t+\Delta t)}{\Delta s^2} \cdot q. \end{aligned}$$

The advantage of using the forward approximation for d_t and evaluating the centered s derivatives at $t + \Delta t$ is now apparent. Since values of $d(\cdot, \cdot)$ at the next time step are known, this is an equation in a single unknown, $d(s,t)$. The solution is

$$d(s,t) = a^* d(s - \Delta s, t + \Delta t) + b^* d(s, t + \Delta t) + c^* d(s + \Delta s, t + \Delta t), \quad (12.8)$$

where

$$a^* \equiv \frac{q\Delta t/\Delta s^2 - (p/2)\Delta t/\Delta s}{1 + r\Delta t},$$

$$b^* \equiv \frac{1 - 2q\Delta t/\Delta s^2}{1 + r\Delta t},$$

$$c^* \equiv \frac{q\Delta t/\Delta s^2 + (p/2)\Delta t/\Delta s}{1 + r\Delta t}.$$

We thus have a way of expressing $d(s,t)$ in terms of three values in the $t + \Delta t$ column: one below, at $s - \Delta s$; one on the same level, at s; and one above, at $s + \Delta s$. Since values of d in the last column of the grid are known, we begin the process at the next-to-last time step, $t_{n_T-1} = (n_T - 1)\Delta t$, where $d(s, t + \Delta t) = d(s, n_T\Delta t) \equiv d(s,T)$ is known for each $s \in \mathcal{S}$. Also, the boundary conditions determine the value of $d(\cdot, t_{n_T-1})$ at the bottom of the column, $d(s_{Min}, t_{n_T-1})$, and at the top, $d(s_{Max}, t_{n_T-1})$. Values of $d(\cdot, t_{n_T-1})$ at the remaining $n_S - 1$ points in the column are found by solving (12.8) one row at a time. When this is complete, move back one time step to $t_{n_T-2} = (n_T - 2)\Delta t$ and solve for $\{d(s, t_{n_T-2})\}_{s\in\mathcal{S}}$ in the same manner. Repeating the process through column $t = 0$ ultimately produces the initial value of the financial derivative for each $s \in \mathcal{S}$, including the value at the actual current (log) price, $d(s_0, 0)$.

If the derivative is subject to early exercise, as for American options, all entries in a column are set equal to the greater of exercise value and the value calculated from entries in the next column. This has to be done one column at a time before values at the next earlier time step are calculated. For example, if valuing an American put, each entry s in the $t_{n_T-j} = (n_T - j)\Delta t$ column would be set as $\max[d(s, t_{n_T-j}), X - e^s]$, where $d(s, t_{n_T-j})$ is calculated from (12.8).

A nice feature of this setup is that the coefficients a^*, b^*, and c^* in the calculations depend on $p, q, \Delta t, \Delta s$, but not on the values of s and t themselves. This means that they are the same at each point in the grid. Had the p.d.e. been set up in levels of underlying price rather than in terms of logs, as in (12.1), the coefficients would have varied from one row to the next. This is why it is easier to work in logs when the underlying price is modeled as geometric Brownian motion. With more complicated models, as when volatility depends on S_t and/or t, then p and q will be state- or time-dependent and a^*, b^*, c^* will vary as well. We defer this complication until section 12.3.1.

As a simplified numerical example of the explicit method, consider pricing a one-year American put struck at $X = 1.0$ on a no-dividend stock whose initial price is $S_0 = 1.0$ (initial log price $s_0 = 0.0$) and has volatility $\sigma = 0.5$. The short rate is $r = 0.20$. We take $n_T = 3$ time steps, so that $\Delta t = 1/3$, and use for the step in log price $\Delta s = \sigma\sqrt{3\Delta t} = 0.5$. Taking $s_{\max} = 1.0$ and $s_{\min} = -1.0$ as the extreme values gives $n_S \equiv (s_{\max} - s_{\min})/\sigma = 4$. These values, which are used again later in examples of the other methods, are summarized in table 12.1.

From these data we have $p \equiv (r - \delta - \sigma^2/2) = 0.075$, $q \equiv \sigma^2/2 = 0.125$, and $a^* = 0.133$, $b^* = 0.625$, $c^* = 0.180$. Figure 12.2 shows along the vertical axis the $n_S + 1 = 5$ values of the log of the underlying price and, opposite the terminal nodes for time T, the terminal payoffs of the put. The zero entries below the upper-most nodes are the values determined by the upper boundary condition: $(X - e^{s_{\max}})^+ = (1 - e^1)^+ = 0.0$. Entries below the lowest nodes are given by the lower boundary condition: $(X - e^{s_{\min}})^+ = (1 - e^{-1})^+ = 0.632$. To the right and below each remaining node is a block

Table 12.1. Data for solving p.d.e.s.

X	S_0	s_0	$s_{\min}$	$s_{\max}$	n_S	Δs	T	n_T	Δt	σ	r	δ
1.0	1.0	0.0	−1.0	1.0	4	0.5	1.0	3	0.33	0.5	0.20	0.0

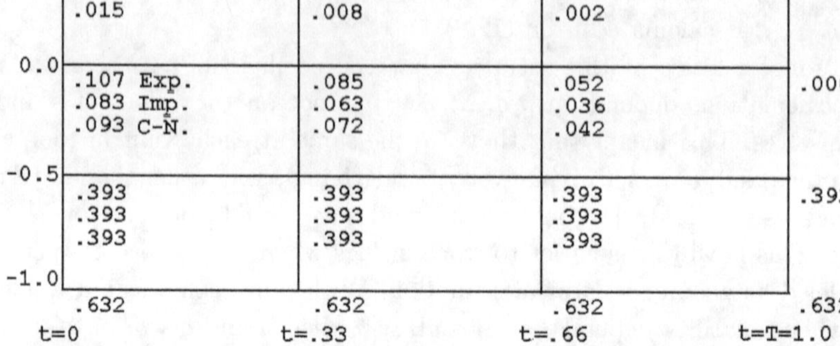

Fig. 12.2. Explicit, implicit, and Crank-Nicholson valuations for American put.

of three figures, the first of which represents the option's value given by the explicit method. (The others are described later.) Thus, the entry at the $t = 0.66, s = 0.0$ node is $a^*(.393) + b^*(.000) + c^*(.000) = 0.052$. The corresponding entry in the node below, at $t = 0.66, s = -0.5$, equals the option's intrinsic value. The American put would have been exercised at this point, since the intrinsic value of 0.393 exceeded the value if held, namely $a^*(.632) + b^*(.393) + c^*(.000) = 0.330$. Finally, the value 0.107 at the $t = 0, s = 0.0$ node is the explicit estimate corresponding to the initial value of the stock.

The explicit method is easy to apply, but it does not converge to the right solution as Δs and Δt approach zero unless Δs and Δt are chosen in special ways. An indication of whether the choice is correct is whether a^*, b^*, c^* are all positive. To see why, note that the three coefficients sum to $(1 + r\Delta t)^{-1} = B(t, t + \Delta t)$, which is the risk-free discount factor for one time step. Thus, the coefficients can be interpreted as discounted risk-neutral probabilities. For customary values of r and σ the main worry will be with b^*: specifically, we need $0 < 1 - 2q\Delta t/\Delta s^2 \equiv 1 - \sigma^2\Delta t/\Delta s^2$, or $\Delta t < \Delta s^2/\sigma^2$. Note that this is satisfied when $\Delta s = \sigma\sqrt{3\Delta t}$. Intuitively, this restriction arises because of the additional error in approximating the s derivatives at the next time step as $d_s^0(s, t + \Delta t)$ and $d_{ss}^0(s, t + \Delta t)$.

12.2.2 *A First-Order Implicit Method*

If the value of the financial derivative is less responsive to time than to price, the restriction $\Delta t < \Delta s^2/\sigma^2$ requires a smaller time step—and more computing time—than might be necessary for the desired accuracy. Here we consider a method that is known to converge regardless of the values of Δt and Δs. It is somewhat harder to program, since it does not yield an explicit representation of $d(s, t)$ in terms of the values at $t + \Delta t$. Instead, it yields an implicit representation in the form of a system of simultaneous equations for unknown d's in column t in terms of knowns in column $t + \Delta t$. With a little ingenuity, however, these equations can be solved without inverting a matrix.

We continue to work with the p.d.e. in log-price form,

$$d_t + d_s p + d_{ss} q - dr = 0,$$

and still use forward approximation $d_t^+(s, t)$ for the time derivative. However, we now evaluate d_s^0 and d_{ss}^0 at time t instead of $t + \Delta t$. This eliminates one source of error and gives approximations for the s derivatives that are correct to $O(\Delta s^2)$, but it still leaves us with the $O(\Delta t)$ approximation for d_t. This is why the method is called "first-order". With these approximations the p.d.e. becomes

$$\begin{aligned} d(s,t)r &= d_t^+(s,t) + d_s^0(s,t)p + d_{ss}^0(s,t)q \qquad (12.9) \\ &= \frac{d(s,t+\Delta t) - d(s,t)}{\Delta t} + \frac{d(s+\Delta s,t) - d(s-\Delta s,t)}{2\Delta s} \cdot p \\ &\quad + \frac{d(s+\Delta s,t) + d(s-\Delta s,t) - 2d(s,t)}{\Delta s^2} \cdot q. \end{aligned}$$

Working backwards as in the explicit method, this equation will involve one known value, $d(s, t+\Delta t)$, and three unknowns, $d(s-\Delta s, t)$, $d(s,t)$, and $d(s+\Delta s, t)$—just the reverse of the explicit method. Rearranging gives

$$d(s,t+\Delta t) = a \cdot d(s-\Delta s,t) + b \cdot d(s,t) + c \cdot d(s+\Delta s,t), \qquad (12.10)$$

where

$$\begin{aligned} a &= (p/2)\Delta t/\Delta s - q\Delta t/\Delta s^2 \\ b &= 1 + r\Delta t + 2q\Delta t/\Delta s^2 \\ c &= -(p/2)\Delta t/\Delta s - q\Delta t/\Delta s^2. \end{aligned}$$

Again, values $d(s_{Min}, t)$ at the bottom of the t column and $d(s_{Max}, t)$ at the top are known. Since each of the $n_S - 1$ intermediate points in $\mathcal{S}$ satisfies (12.10), there are $n_S - 1$ equations in as many unknowns. Fortunately, they can be solved without inverting a matrix using Gaussian elimination, since each single equation involves just three unknowns.

Simpler notation will make it easier to see how to do this. For $i \in \{0, 1, \ldots, n_S\}$ the value of s corresponding to the ith step from the bottom of a column is $s_i \equiv s_{Min} + i\Delta s$. Let $e_i = d(s_i, t+\Delta t)$ be the known value on the left side of (12.10), corresponding to the ith position of the $t + \Delta t$ column, and let $d_i = d(s_i, t)$ be the unknown value in the ith position of the t column. Expression (12.10) now comes out as

$$e_i = a \cdot d_{i-1} + b \cdot d_i + c \cdot d_{i+1}. \qquad (12.11)$$

Remember that the e's are known and the d's are not, apart from the ones at the bottom and top of the column—d_0 and d_{n_S}. Now look at the equation for $i = 1$ and move the known d_0 term to the left:

$$e_1 - a \cdot d_0 = b \cdot d_1 + c \cdot d_2. \qquad (12.12)$$

We shall see that such an equation holds for each position $i \in \{1, 2, \ldots, n_S - 1\}$, wherein a known function of the e's and d_0 is equated to a linear function of the i and $i+1$ values of d.

The steps are as follows. Use (12.12) to express d_1 in terms of d_2 and known quantities—that is, as $d_1 = b^{-1}(e_1 - a \cdot d_0 - c \cdot d_2)$—and substitute into the next equation,

$$e_2 - a \cdot d_1 = b \cdot d_2 + c \cdot d_3.$$

From this, express d_2 in terms of d_3 and known quantities, and continue in this way up to d_{n_S-1}. Each d_i will then have been expressed as a linear function of d_{i+1}. At the $n_S - 1$ step, where d_{n_S} is known, this yields an explicit solution for d_{n_S-1}. This solution is then used to find d_{n_S-2}, which is used to find d_{n_S-3}, and so on back to d_1. The trick is to keep track of the coefficients in all these linear relations that connect d_i to d_{i+1}. Fortunately, they can be generated recursively.

One more round of notation is needed to see the recursive structure. Write (12.12) as

$$k_1 = l_1 \cdot d_1 + c \cdot d_2, \tag{12.13}$$

where $k_1 = e_1 - a \cdot d_0$ and $l_1 = b$. Express d_1 in terms of d_2 as

$$d_1 = \frac{k_1 - c \cdot d_2}{l_1}$$

and then substitute in the $i = 2$ equation:

$$\begin{aligned} e_2 &= a \cdot d_1 + b \cdot d_2 + c \cdot d_3 \\ &= \frac{ak_1}{l_1} + \left(b - \frac{ac}{l_1}\right) \cdot d_2 + c \cdot d_3. \end{aligned}$$

Writing in a form like (12.13) gives

$$k_2 = l_2 \cdot d_2 + c \cdot d_3,$$

where $k_2 = e_2 - ak_1/l_1$ and $l_2 = b - ac/l_1$. The pattern of recursive relations between the k's and l's now begins to emerge. Suppose for some $i \in \{2, \ldots, n_S - 1\}$ we have

$$\begin{aligned} k_{i-1} &= l_{i-1} \cdot d_{i-1} + c \cdot d_i \\ e_i &= a \cdot d_{i-1} + b \cdot d_i + c \cdot d_{i+1}. \end{aligned}$$

Solving the first equation for d_{i-1} as

$$d_{i-1} = \frac{k_{i-1} - c \cdot d_i}{l_{i-1}} \tag{12.14}$$

and substituting in the second equation give

$$k_i = l_i \cdot d_i + c \cdot d_{i+1},$$

where

$$l_i = b - a \cdot \frac{c}{l_{i-1}} \qquad (12.15)$$
$$k_i = e_i - a \cdot \frac{k_{i-1}}{l_{i-1}}.$$

Starting with $l_1 = b$, the l's can be generated recursively up to l_{n_S-1}. Since neither b nor c depends on the time step, this is done just once. Having all the l's and taking $k_1 = e_1 - a \cdot d_0$, one can generate $k_2, \ldots, k_{n_S-1}$ for the current time step as well. This does have to be done at each time step since the e's change from one step to the next. Now, armed with the k's and l's, any d_{i-1} can be expressed in terms of d_i via (12.14). Starting at $i = n_S - 1$, where $d_{n_S} = d(s_{Max}, t)$ is known, solve for d_{n_S-1} as

$$d_{n_S-1} = \frac{k_{n_S-1} - c \cdot d_{n_S}}{l_{n_S-1}}$$

and keep applying (12.14) until reaching the solution for d_1. Having found the entire column of d's for the $n_T - 1$ time step, the e's are now available for step $n_T - 2$. Now recalculate the k's, find the d's for step $n_T - 2$, and so continue until the $t = 0$ column is complete.

Applying the first-order implicit method to the data in table 12.1 gives $a = -0.142$, $b = 1.400$, $c = -0.192$, $l_1 = b = 1.400$, $l_2 = b - a \cdot c/1.400 = 1.381$, $l_3 = b - a \cdot c/1.381 = 1.380$. Table 12.2 gives the values of k_1, k_2, k_3 for each time step. For example, the value of k_1 at $t = 0.66$ is $e_1 - a \cdot d_0 = 0.393 - a \cdot (0.632)$, where $e_1 = 0.393$ is the option's terminal value at the $t = 1.0$, $s = -0.5$ node, and $d_0 = 0.632$ is the option's lower boundary value at the $t = 0.66$ step. Applying the formulas throughout leads to the second figure in each block at the interior nodes of figure 12.2.

Table 12.2. Values of $\{k_j\}$ in first-order implicit solution of example in Fig. 12.2.

t	k_1	k_2	k_3
0.66	0.483	0.049	0.005
0.33	0.483	0.085	0.012
0.00	0.483	0.112	0.020

12.2.3 *Crank-Nicolson's Second-Order Implicit Method*

As mentioned earlier, a limitation of the first-order implicit method is that the forward approximation for $d_t(s,t)$ is correct only to $O(\Delta t)$. The Crank-Nicolson scheme is another implicit method that is second-order correct for both time and log price. One expects it to converge faster and require fewer time steps than either of the two methods previously described.

The method involves a clever "swindle" that has the effect of cutting the time step in half and yet requires no more columns in the solution array. Here is the idea. If the time step Δt were really cut in half and $d_t(s,t)$ approximated at $(s, t+\Delta t/2)$, then the centered approximation to d_t, correct to the second order, would be

$$\begin{aligned} d_t^0(s, t+\Delta t/2) &= \frac{d(s,t+\Delta t) - d(s,t)}{2(\Delta t/2)} \\ &= \frac{d(s,t+\Delta t) - d(s,t)}{\Delta t} \\ &= d_t^+(s,t). \end{aligned} \tag{12.16}$$

That is, the centered approximation at the intermediate time step is the same as the forward approximation at (s,t). If we could somehow replace the two second-order approximations for d_s and d_{ss} on the right side of (12.9) by approximations at $(s, t+\Delta t/2)$ as well, then the entire result would be second-order correct. The trick is finding how to do this without actually reducing the time step. The solution proposed by Crank and Nicolson (1947) is just to average the second-order approximations to d_s and d_{ss} at (s,t) and $(s, t+\Delta t)$. This gives for $d_s^0(s, t+\Delta t/2)$ the approximation

$$\begin{aligned} &\frac{d_s^0(s,t) + d_s^0(s,t+\Delta t)}{2} \\ &= \frac{d(s+\Delta s,t) - d(s-\Delta s,t)}{4\Delta s} + \frac{d(s+\Delta s,t+\Delta t) - d(s-\Delta s,t+\Delta t)}{4\Delta s} \end{aligned}$$

and for $d_{ss}^0(s, t+\Delta t/2)$

$$\begin{aligned} &\frac{d_{ss}^0(s,t) + d_{ss}^0(s,t+\Delta t)}{2} \\ &\quad= \frac{d(s+\Delta s,t) + d(s-\Delta s,t) - 2d(s,t)}{2\Delta s^2} \\ &\quad+ \frac{d(s+\Delta s,t+\Delta t) + d(s-\Delta s,t+\Delta t) - 2d(s,t+\Delta t)}{2\Delta s^2}. \end{aligned}$$

Substituting these into the p.d.e.

$$d_t + d_s p + d_{ss} q - dr = 0,$$

moving the knowns evaluated at $t + \Delta t$ to the left side, and using the same simplified notation as for the implicit method give an expression just like (12.11),

$$e_i = a \cdot d_{i-1} + b \cdot d_i + c \cdot d_{i+1},$$

but with different interpretations of e_i and the coefficients. Letting $s_i = s_{Min} + i\Delta s$ as before, take

$$e_i \equiv a^* d(s_{i-1}, t + \Delta t) + b^* d(s_i, t + \Delta t) + c^* d(s_{i+1}, t + \Delta t)$$

with

$$\begin{aligned} a^* &= -(p/4)\Delta t/\Delta s + (q/2)\Delta t/\Delta s^2 \\ b^* &= 1 - q\Delta t/\Delta s^2 \\ c^* &= (p/4)\Delta t/\Delta s + (q/2)\Delta t/\Delta s^2 \end{aligned}$$

and

$$\begin{aligned} a &= (p/4)\Delta t/\Delta s - (q/2)\Delta t/\Delta s^2 \\ b &= 1 + r\Delta t + q\Delta t/\Delta s^2 \\ c &= -(p/4)\Delta t/\Delta s - (q/2)\Delta t/\Delta s^2. \end{aligned}$$

Note that e_i still depends on values of the derivative at the next time step, which are always known as we work backwards in time. With these new definitions of the symbols the solution proceeds just as in the first-order version of the implicit method.

Applying the method to the data for the American put, we have $a = -0.142$, $b = 2.467$, $c = -0.192$, $a^* = 0.142$, $b^* = 1.667$, $c^* = 0.192$, $l_1 = 2.467$, $l_2 = 2.456$, $l_3 = 2.456$, and the k values shown in table 12.3. The third figure at each interior node in figure 12.2 is the Crank-Nicolson valuation.

Table 12.3. Values of $\{k_j\}$ in Crank-Nicolson solution of example in Fig. 12.2.

t	k_1	k_2	k_3
0.66	0.835	0.104	0.006
0.33	0.843	0.175	0.020
0.00	0.849	0.226	0.037

Table 12.4. Percentage pricing errors for European put: Finite-difference and binomial methods.

n_T	Explicit	1st-Order Implicit	Crank-Nicolson	Binomial
25	−0.48	−1.81	−1.14	1.15
50	−0.10	−0.75	−0.42	−0.02
100	0.03	−0.27	−0.13	0.09
200	0.06	−0.10	−0.02	0.10
400	0.04	−0.03	−0.00	0.08
800	0.01	0.00	0.00	0.04

12.2.4 *Comparison of Methods*

Program PDEgeoBm on the accompanying CD allows the user to select any of the three finite-difference methods to solve the log-price formulation of the fundamental p.d.e. with constant interest rates, dividend rate, and volatility. The user merely specifies the type of option (put or call, American or European), the specifications for underlying and option, the range of log price, and the dimensions of the time/log-price grid.

To judge the accuracies of the three finite-difference methods and to compare with the binomial approach, we apply all four methods to a problem for which Black-Scholes gives an exact answer: pricing a one-year European put with $X = 10.0$, $S_0 = 10.125$, $\sigma = 0.3$, and $r = \ln(1.05)$. Table 12.4 shows percentage deviations from the Black-Scholes price, $P^E(S_0, 1.0) = 0.8949$, for time steps ranging from $n_T = 25$ to 800. In this particular example the explicit method gives surprisingly good accuracy even for $n_T = 25$, but the faster convergence of Crank-Nicolson is apparent. The computing time is about the same for the three finite-difference methods, and all are faster than the binomial.

12.3 Extensions

This section shows how finite-difference methods can be adapted to deal with more sophisticated dynamics than geometric Brownian motion and how to handle payments by the underlying of lump-sum dividends.

12.3.1 *More General P.D.E.s*

Consider now trying to solve a financial p.d.e. of the more general form

$$\kappa(S_t, t) - D_{T-t} + D_S\mu(S_t, t) + D_{SS}\sigma^2(S_t, t)/2 - Dr_t = 0.$$

This accommodates (i) financial derivatives that themselves generate continuous time- and/or price-dependent cash receipts $\kappa(S_t, t)$; (ii) time-varying (but deterministic) short rates r_t; (iii) time/price-dependent continuous dividend payouts by the underlying; and (iv) volatility that is any function of current price and time that is bounded on $\mathcal{S} \times \mathcal{T}$. For example, (iii) could be handled by setting $\mu(S_t, t) = S_t[r_t - \delta(S_t, t)]$ for an appropriate payout function $\delta(\cdot, \cdot)$, and (iv) would accommodate a generalized c.e.v. specification such as $\sigma(S_t, t) = \sigma_0 S_t + \sigma_1 S_t^{1-\gamma}$.[3] To allow this generality it will be necessary to leave the p.d.e. in terms of absolute prices rather than transforming to logs. However, since it will still be convenient to work in forward time rather than time to expiration, we continue to work with forward-time derivative D_t and write the p.d.e. as

$$\kappa(S_t, t) + D_t + D_S \mu(S_t, t) + D_{SS} \sigma^2(S_t, t)/2 - D r_t = 0. \qquad (12.17)$$

In this general setting it may be difficult to choose Δt and ΔS so as to guarantee convergence of the explicit method. For example, the coefficient b^* that would apply in this case, $1 - \sigma^2(S_t, t)\Delta t/\Delta S$, could become negative for large values of σ. For this reason we now focus solely on the Crank-Nicolson method.

Constructing the grid in price levels requires choosing prices $S_{\max}$ and $S_{\min}$ that bound $\{S_t\}_{0 \le t \le T}$ with high probability. What bounds are suitable will depend on the model for the process. The price step ΔS should be chosen in conjunction with $S_{\min}$ so that $(S_0 - S_{\min})/\Delta S$ is an integer, thus ensuring that the initial price S_0 is in the grid $\mathcal{S}$. In constructing the time grid $\mathcal{T}$ the usual recommendation is to choose Δt to be proportional to ΔS^2. One approach is to set the number of steps n_T equal to the greatest integer in $T/\Delta S^2$, then take $\Delta t = T/n_T$. This ensures that the derivative's lifetime comprises an integral number of time steps.

Having constructed the grid, we now replace D_t, D_S, and D_{SS} in (12.17) by second-order formulas $D_t^0(S, t + \Delta t/2)$, $D_S^0(S, t + \Delta t/2)$, and $D_{SS}^0(S, t + \Delta t/2)$, corresponding to (12.16), (12.6), and (12.7) but with price and time

---

[3] Recall from section 3.2.3 that $\mu(S_t, t)$ and $\sigma(S_t, t)$ must satisfy certain conditions if there is to be a unique solution to the s.d.e. $dS_t = \mu(S_t, t) \cdot dt + \sigma(S_t, t) \cdot dW_t$ with given initial condition S_0. See Karatzas and Shreve (1991, pp. 286–291) and, in relation to the c.e.v. model in particular, Duffie (1996, pp. 291–293).

This scheme does not accommodate formulations in which μ and σ depend nontrivially on past prices, as in the Hobson and Rogers (1998) model, or on other risk sources, as in stochastic-volatility models. Since these involve additional state variables, they require lattices of dimension greater than two.

as arguments. As before, this gives an expression involving values of the financial derivative at each of three price steps and at the current and next time step, t and $t+\Delta t$. Let $S_i = S_{\min}+i\Delta S$ be the price at the ith step, $i \in \{0, 1, \ldots, n_S\}$, and let $t_j = j\Delta t$ be the time at step $j \in \{0, 1, \ldots, n_T\}$. With the time step understood to conserve notation, and setting $D_i \equiv D(S_i, t_j)$, the p.d.e. can be expressed as,

$$e_i = a_i D_{i-1} + b_i D_i + c_i D_{i+1}, \tag{12.18}$$

where

$$\begin{aligned}
e_{i+1} &\equiv \kappa(S_i, t_j)\Delta t + a_i^* D_{i-1} + b_i^* D_i + c_i^* D_{i+1} \\
a_i^* &\equiv -[\mu(S_i, t_j)/4]\Delta t/\Delta S + [\sigma(S_i, t_j)^2/4]\Delta t/\Delta S^2 \\
b_i^* &\equiv 1 - [\sigma(S_i, t_j)^2/2]\Delta t/\Delta S^2 \\
c_i^* &\equiv [\mu(S_i, t_j)/4]\Delta t/\Delta S + [\sigma(S_i, t_j)^2/4]\Delta t/\Delta S^2 \\
a_i &\equiv -a_i^* \\
b_i &\equiv 1 + r_{t_j}\Delta t + [\sigma(S_i, t_j)^2/2]\Delta t/\Delta S^2 \\
c_i &\equiv -c_i^*.
\end{aligned}$$

This is the same as (12.11) except that a, b, and c may change at each price and time step.

In solving the equations the only changes are that a, b, and c have to be calculated at each grid point and that the vector l must be reconstructed at each time step. To proceed, begin as before at next-to-last time step t_{n_T-1} and first price step $S_1 = S_{\min}+\Delta S$. Calculate and store in the first position of the $(n_S - 1)$ -vectors k and l the elements $k_1 = e_1 - a_1 D_0$ and $l_1 = b_1$. Calculate c_1 and store as the first element of vector c. Looping through the price steps, for $i = 2, 3, \ldots, n_S - 1$ fill out the c vector and generate the remaining elements of k and l recursively as

$$\begin{aligned}
k_i &= e_i - a_i \frac{k_{i-1}}{l_{i-1}} \\
l_i &= b_i - \frac{a_i c_{i-1}}{l_{i-1}}.
\end{aligned}$$

Beginning then at price step $i = n_S - 1$ where $D_{i+1} = D_{n_S}$ is known and continuing down to price step $i = 1$, solve $k_i = l_i D_i + c_i D_{i+1}$ for D_i in terms of D_{i+1}. After checking each entry to allow for the possibility of early exercise, back up to the previous time step and repeat the procedure, continuing in this way until $D_1, D_2, \ldots, D_{n_S-1}$ have been found at $t = 0$.

The program `PDEgeneral` on the CD applies the Crank-Nicolson method to the general form of the financial p.d.e., allowing flexible behavior of interest rates, continuous rates of dividend payout, and volatility.

12.3.2 *Allowing for Lump-Sum Dividends*

The first task in accommodating discrete payments by the underlying is to choose time steps Δt such that payment dates coincide approximately with elements of $\mathcal{T}$. At each *ex*-dividend date the various prices in the set $\mathcal{S}$ are reduced by the amount of the dividend. If the log-price formulation is used, dividend payments proportional to the price level can be allowed by adding $\ln(1-\delta)$ to each price, where δ is the proportional payout. If payments are fixed in currency units, one would formulate the p.d.e. in absolute prices rather than logs, subtracting the fixed dividend from each element of $\mathcal{S}$ on the *ex* date. More complicated price-dependent payout schemes require the size of the price step ΔS to vary over $\mathcal{S}$. This introduces another source of variation in the coefficients a, b, and c of the implicit methods, but it causes no special difficulty.

13

Programs

This chapter serves as a guide to the contents of the accompanying CDROM, which contains source code in FORTRAN, C++, and VBA (Visual Basic) for many of the computations described in earlier chapters.[1] The CD contains a separate folder for each language, and each such folder contains subfolders corresponding to program groups. The sections that follow correspond to the program groups. Each section contains a list of the component programs and a brief description of what each program does.

Several of the FORTRAN and C++ programs require routines in the corresponding editions of Press *et al.* (1992).[2] Others require routines included here. All the VBA routines are self-contained, with input and output handled through the spreadsheets. Routines `INVERTCF` and `INVFFT` are not implemented in VBA. `ContinuousTime.xls` executes `doB_S`, `CEV`, `COMPOUND`, `EXTEND` and `JUMP` using VBA versions of `PHI` and `GAUSSINT`.

[1]C++ programs were developed from the FORTRAN by Michael Nahas. VBA source code, visible as macros from the spreadsheets, was developed by Ben Koulibali and Ubbo Wiersema. The various versions of these programs are intended for instruction only. Although produced to the best of our ability, one cannot be certain that they are free of errors. It is nevertheless hoped that they will be found useful in guiding those who intend to conduct their own research or develop their own applications.

[2]Available on the *Numerical Recipes (2nd Ed.) Multi-Language Code CD ROM*, Cambridge University Press.

13.1 Generate and Test Random Deviates

13.1.1 *Generating Uniform Deviates*

```
function randfun(seed)
Purpose:    Generate pseudorandom uniform (0,1) deviates.
Input:      Double precision number >1, seed
Output:     One uniform deviate.
References: Fishman, G. (1996) Monte Carlo: Concepts,
            Algorithms, and Applications, Springer-Verlag:
            New York.
Notes:      Adapted to function subprogram from subroutine
            written by L.R. Moore, in Fishman (1996).
Required routines:   None
```

13.1.2 *Generating Poisson Deviates*

```
subroutine poidev(theta,n,jv,seed)
Purpose:   Generates pseudorandom Poisson variates by
           inversion of c.d.f.
Input:     Poisson parameter, theta
           Number of variates generated, n
           Double precision seed value >1, seed
Output:    n-vector of Poisson deviates, jv
Notes:     (1) A uniform u is generated.  A table of F(j) is
           constructed out to j* = {first j: F(j) > u}, and
           j* is returned as the deviate.  For new values of
           u less than umax (the max of preceding u's) the
           corresponding j's are read from the table.  When
           u > umax is generated, the table is extended.
           (2) Uses RANDFUN to generate uniform deviates.
           Replace as desired.
Required routines:  RANDFUN
```

13.1.3 *Generating Normal Deviates*

```
function gausfun(seed)
Purpose:   Generates pseudorandom standard normals by
           adaptation of algorithm  NA (see Reference) with
```

```
          exponentials generated via inverse probability
          transform.
Input:    seed, a double precision number greater than or
          equal to unity.
Output:   A single pseudorandom normal deviate.
Reference: Fishman, G.S.(1996), Monte Carlo: Concepts,
          Algorithms, and Applications, Springer: New York
Notes:    Uses RANDFUN to generate uniform deviates.
Required routines:  RANDFUN
```

13.1.4 *Testing for Randomness*

```
function runtest(u,n,nrun)
Purpose:  Perform large-sample runs test for randomness.
Input:    n-vector of observations, u
          Number of observations, n
Output:   Runs test statistic, distributed as chi-square(1)
          under null hypothesis of randomness.
          Number of runs, nrun.
Reference: Wonnacott, T. and Wonnacott, R. (1977)
          Introductory Statistics. Wiley: New York.
Notes:
              Significance level/ Critical points
   .500    .250    .100    .050    .025    .010    .005
   0.45    1.32    2.71    3.84    5.02    6.63    7.88
Required routines:  None
```

13.1.5 *Testing for Uniformity*

```
function utest(n,u)
Purpose:  Test that data are i.i.d as uniform on (0,1)
Input:    n-vector of numbers to be tested for uniformity,
          u Number of observations, n
Output:   N2 test statistic for uniformity
Reference: D'Agostino, R. and Stephens, M. (1986)
          Goodness of Fit Techniques, Dekker: New York,
          pp.351-354, 357.
Notes:    Statistic is distributed for large n
          approximately as chi-square(2).
```

Significance Level/Critical Points						
.500	.250	.100	.050	.025	.010	.005
1.386	2.773	4.605	5.991	7.378	9.210	10.597

Required routines: None

13.1.6 *Anderson-Darling Test for Normality*

```
subroutine ADtest(n,y,ybar,sdy,ad)
Purpose:    Conducts modified Anderson-Darling test of
            normality on vector y of length n.
Input:      y, vector of length n
            ybar, sdy = sample mean and variance (divisor
            n-1) If ybar,sdy = 0 on input, they will be
            computed
Output:     Anderson-Darling statistic, ad
Reference:  ''Anderson-Darling test for goodness of fit'',
            Encyclopedia of Statist. Sciences, v1, pp.81-85
            eq.(3) and a* modification for case 3, as
            specified in table 1.

                Significance level/critical points (all n)
            .100          .050          .025          .010
            .631          .752          .873         1.035
Notes:      Vector y is returned in increasing order
            ^^^^^^^^^^^^^^^^^^^^^^^^^^^^^^^^^^^^^^^^

Required routines:   PHI
From Numerical Recipes, Press et al. (1992): SORT
```

13.1.7 *ICF Test for Normality*

```
subroutine icftest(n,y,cf)
Purpose:   Test of normality, mean and variance unspecified
Input:     Number of observations, n
           Vector of observations, y
Output:    Value of CF statistic, cf
Reference: Epps, T. & Pulley, L.(1983) Biometrika 70(3),
           723-6 Baringhaus,L. et al.(1989) Commun. in
           Statist., Comp. & Simul., 18(1), 363-79
Notes:     Data vector is returned in standard form.
```

Percentage Points of CF

n\1-p	0.900	0.950	0.975	0.990
4	0.248	0.291	0.316	0.332
5	0.243	0.323	0.388	0.445
6	0.262	0.325	0.403	0.499
8	0.269	0.344	0.424	0.523
10	0.278	0.355	0.435	0.542
15	0.284	0.364	0.446	0.560
20	0.288	0.369	0.449	0.562
30	0.289	0.370	0.456	0.564
50	0.290	0.372	0.459	0.574

```
Required routines:  None
```

13.2 General Computation

13.2.1 *Standard Normal CDF*

```
function phi(x)
Purpose:    Evaluate Standard Normal CDF
Input:      x
Output:     Phi(x)= Pr(Z<x), where Z ~N(0,1)
Reference:  Moran, P.A.P. (1980), ''Calculation of the normal
            distribution function'', Biometrika 67 (3),
            675-676.
Notes:      Accuracy to 7 decimal places when |x|< 5.
Required
routines:   None
```

13.2.2 *Expectation of Function of Normal Variate*

```
function GaussInt(f,a,b,dz)
Purpose:    Integrate function f(z) between a and b with
            respect to standard normal density by rectangular
            approximation on grid spaced 'dz'.
Input:      Integration limits, a and b
            Grid spacing, dz
Output:     Integral between a and b of product of f(z) and
            standard normal p.d.f.
```

```
Notes:
  (1)  'dz' must be specified; a suggested value is .001.
  (2)  For a=-inf and/or b=+inf, set a=-1000. and/or
       b=+1000.
  (3)  'f' must appear in EXTERNAL statement in main program.
Required routines:  User-supplied function f(z)
```

13.2.3 *Standard Inversion of Characteristic Function*

```
subroutine INVERTCF(x,nx,Fx,N,h,phi)
Purpose:    Evaluate c.d.f. F(x) at nx points {x} by Fourier
            inversion of c.f.
Input:      Vector of points, x
            Length of vector, nx
            Length of grid at which c.f. is evaluated, N
            Spacing of grid for c.f., h
Output:     Vector of c.d.f. values, Fx
Reference: Waller, L et al.(1985) Amer. Statist. 49, 346-50.
Notes: (1) Random variable whose c.f. is inverted should be
            roughly centerded and with unit scale; e.g., in
            standard form.
       (2) N and h correspond to H and eta in Waller et al.
            and control the grid on which c.f. is evaluated.
            Reducing h with N fixed makes the grid denser,
            sharpening estimates of tail probabilities, but
            shrinks the range, [-Nh,Nh]. Increasing N with h
            fixed extends range and improves accuracy near
            mode.
Required routines:
          User-supplied subroutine PHI to calculate c.f. of a
          standardized random variable. Example routine
          supplied.
```

13.2.4 *Inversion of Characteristic Function by FFT*

```
subroutine invfFt(n,t,CDF)
Purpose:    Apply the fast Fourier transform to invert the
            c.f. corresponding to an absolutely continuous
            c.d.f. F, and thereby to recover values of F(t)
            on a grid of points {t}.
```

```
Input:     Number of points at which F is evaluated, n
Output:    Values of c.d.f. at points {t}, CDF
Reference: Press, W. et al (1992) Numerical Recipes in
           FORTRAN, Cambridge Press.
Required routines:
         (1) User-supplied complex-valued function PSI that
          calculates from the parameters a 'center' and a
         'scale,' then calculates 1/(i*t) times the c.f. of
         centered and scaled r.v. whose distribution is F.
         (2) From Numerical Recipes, Press et. al.(1992),
             FOUR1
Notes:   (1) n must be a power of 2 no larger than 2**14,
             depending on available memory. Larger n =>
             greater range of values where F is evaluated
             & better accuracy in the tails, but slower
             execution.
         (2) Choice of center and scale is not crucial if
             most probability mass is between -10 and
             +10 after centering and scaling.  Original
             center and scale is restored at output.
         (3) Two example psi functions, PSI1 and PSI2, are
             provided. To run one, change the name to PSI.
         (4) Set options in ***Set options*** block to
             control printing and domain on which F is
             calculated.
```

13.3 Discrete-Time Pricing

13.3.1 *Binomial Pricing*

```
program binomial
Purpose:   Prices European, American, and extendable options
           on stocks, currencies, indexes, or futures under
           Bernoulli dynamics
Inputs:    Called for on execution
Output:    Summary of input data and value of option
Required routines:  None
```

13.3.2 *Solving PDEs under Geometric Brownian Motion*

```
PROGRAM PDEgeoBm
Purpose:   Solve the financial PDE when price follows
           geometric Brownian motion by choice of EXPLICIT,
           first-order IMPLICIT, or CRANK-NICHOLSON methods.
Input:     Values in parameter statements as instructed.
Output:    Option value over range of prices around the
           strike.
Reference: Smith, G. (1985) Numerical Solution of Partial
           Differential Equations, 3d ed., Clarendon Press:
           Oxford
Required routines:  None
```

13.3.3 *Crank-Nicolson Solution of General PDE*

```
PROGRAM PDEgeneral
Purpose:   Solve financial PDE by Crank-Nicholson method,
           allowing for time-varying interest rates and
           time- and price- dependent cash flow, mean
           drift, and volatility.
Input:     Values in parameter statements as instructed in
           routine. Specify interest-rate, cash flow, drift,
           and volatility processes in function subprograms.
           Examples supplied.
Output:    Option value over range of prices around strike.
Reference: Smith, G. (1985) Numerical Solution of Partial
           Differential Equations, 3d ed., Clarendon Press:
           Oxford
Required routines:  None
```

13.4 Continuous-Time Pricing

13.4.1 *Shell for Black-Scholes with Input/Output*

```
program doB_S
Purpose:   Executes B_S subroutine with input/output
Input:     Called for at execution
Output:    Black-Scholes prices
Required routines:  B_S, PHI
```

13.4.2 *Basic Black-Scholes Routine*

```
subroutine B_S(SS,X,sigma,B,T,underlying,div,option,optval)
Purpose:   Calculates Black-Scholes price of European option
Input:     Underlying price, SS
           Strike price, X
           Volatility, sigma
           Price of unit bond maturing at expiration, B
           Expiration date, T
           Type of underlying asset, ('stock','currency',
               'index', or 'futures'), underlying
           Dividend (if underlying='stock') or dividend rate
               (if 'underlying'='currency' or 'index'), div
           Type of option (put or call), option
Output:    Value of option, optval
Required routines:  PHI
```

13.4.3 *Pricing under the C.E.V. Model*

```
program cev
Purpose:   Values European puts and calls on underlying
           asset that follows c.e.v. dynamics.
Inputs:    Called for at execution
Output:    Values of puts and calls under c.e.v. dynamics,
           Black- Scholes values at same initial volatility,
           and differences, for range of moneyness values,
           S/X
Required routines:  GAMPDF (included), B_S
From Numerical Recipes, Press et. al.(1992): GAMMLN,GAMMQ,
           GSER,GCF
```

13.4.4 *Pricing a Compound Option*

```
program compound
Purpose:   Value compound call option (call on a call)
Input:     Initial value & volatility of underlying
           Strikes of compound & primary
           Expiration times of compound & primary
Output:    Value of compound call
```

```
Reference: Geske, R. (1979), "Valuation of compound
           options," J.Financial Econ. 7, 63-91.
Required routines:  B_S, PHI, Gaussint
From Numerical Recipes, Press et. al. (1992): ZBRAC, RTBIS
```

13.4.5 *Pricing an Extendable Option*

```
program extend
Purpose:    Value a put on a no-dividend stock that must be
            exercised if in the money at t* and is otherwise
            extended to T.
Inputs:     Called for at execution
Output:     Values of European puts expiring at t* and T,
            and value of the extendable put
Required routines:  B_S, PHI, GAUSSINT
```

13.4.6 *Pricing under Jump Dynamics*

```
program jump
Purpose:    Value European put or call on underlying
            following mixed jump-diffusion dynamics.
Inputs:     Called for at execution.
Output:     Price of European option.
Required routines:  B_S, PHI
```

Bibliography

1. Abramowitz, M. and Stegun, I. (1970) *Handbook of Mathematical Functions.* New York: Dover.
2. Ahn, C.M. (1992) "Option pricing when jump risk is systematic," *Mathematical Finance* 2, 299–308.
3. Amin, K.I. (1993) "Jump diffusion option valuation in discrete time," *J. Finance* 48, 1833–1863.
4. Anderson, T.W. and Darling, D.A. (1954) "A test of goodness of fit," *J. American Statistical Association* 49, 765–769.
5. Arnold, L. (1974) *Stochastic Differential Equations: Theory and Applications.* New York: John Wiley.
6. Arrow, K.J. (1964) "The role of securities in the optimal allocation of risk-bearing," *Review of Economic Studies* 31, 91–96. Translation of "Le rôle des valeurs boursièrs pour la répartition la meilleure des risques," *Econometrie,* 41–48, Centre National de la Recherche Scientifique Paris, 1953.
7. Ash, R.B. (1972) *Real Analysis and Probability.* New York: Academic Press.
8. Athreya, K.B. and Ney, P.E. (1972) *Branching Processes.* New York: Springer-Verlag.
9. Bajeux, I. and Rochet, J.C. (1992) "Dynamic spanning: Are options an appropriate instrument?" *Mathematical Finance* 6, 1–16.
10. Bakshi, G.; Cao, C.; and Z. Chen (1997) "Empirical performance of alternative option pricing models," *J. Finance* 52, 2003–2049.
11. Baringhaus, L.; Danschke, R.; and N. Henze (1989) "Recent and classical tests for normality—a comparative study," *Communications in Statistics, Theory and Methods* 18, 363–379.
12. Baringhaus, L. and Henze, N. (1988) "A consistent test for multivariate normality based on the empirical characteristic function." *Metrika* 35, 339–348.
13. Barndorff-Nielsen, O.E. and Halgreen, O. (1977) "Infinite divisibility of the hyperbolic and generalized inverse Gaussian distributions," *Zeitschrift für Wahrscheinlichkeitstheorie und Verwandte Gebiete* 38, 309–312.
14. Barone-Adesi, G. and Whaley, R.E. (1987) "Efficient analytic approximation of American option values," *J. Finance* 42, 301–320.

15. Barone-Adesi, G. and Whaley, R.E. (1988) "On the valuation of American put options on dividend-paying stocks," *Advances in Futures and Options Research* 3, 1–13.
16. Bartoszynski, R. and Niewiadomska-Bugaj, M. (1996) *Probability and Statistical Inference.* New York: John Wiley.
17. Bates, D.S. (1996a) "Jumps and stochastic volatility: Exchange rate processes implicit in Deutschemark options," *Review of Financial Studies* 9, 69–108.
18. Bates, D.S. (1996b) "Testing option pricing models," in G.S. Maddala and C.R. Rao, eds., *Handbook of Statistics* 14, 567–611. Amsterdam: Elsevier.
19. Baxter, M. and Rennie, A. (1996) *Financial Calculus.* Cambridge (UK): Cambridge Press.
20. Beckers, S. (1980) "The constant elasticity of variance model and its implications for option pricing," *J. Finance* 35, 661–73.
21. Bergman, Y.; Grundy, B.; and Z. Wiener (1996) "General properties of option prices," *J. Finance* 51, 1573–1610.
22. Bertoin, J. (1996) *The Lévy Processes.* New York: Cambridge Press.
23. Billingsley, P. (1986) *Probability and Measure*, 2d. ed. New York: John Wiley.
24. Black, F. and Scholes, M. (1973) "The pricing of options and corporate liabilities," *J. Political Economy* 81, 637–54.
25. Black, F. (1976a) "Studies of stock price volatility changes," *Proceedings, American Statistical Assn.*, 177–181.
26. Black, F. (1976b) "The pricing of commodity options," *J. Financial Economics* 3, 167–179.
27. Black, F. and Cox, J. (1976) "Valuing corporate securities: Some effects of bond indenture provisions. *J. Finance* 31, 351–367.
28. Blattberg, R.C. and Gonedes, N.J. (1974) "A comparison of the stable and Student distributions as statistical models for stock prices," *J. Business* 47, 244–280.
29. Bohman, H. (1975) "Numerical inversion of characteristic functions," *Scandanavian Actuarial J.*, 121–124.
30. Bollerslev, T.; Chou, R.Y.; and K.F. Kroner (1992) "ARCH modeling in finance: A review of the theory and empirical evidence," *J. Econometrics* 52, 5–59.
31. Boothe, P. and Glassman, D. (1987) "The statistical distribution of exchange rates," *J. International Economics* 22, 297–319.
32. Box, G.E.P. and Muller, M.E. (1958) "A note on the generation of random normal deviates," *Annals of Mathematical Statistics* 29, 610–611.
33. Boyle, P.; Broadie, M.; and P. Glasserman (1997) "Monte Carlo methods for security pricing," *J. Economic Dynamics and Control* 21, 1267–1321.
34. Brace, A.; Gątarek, D.; and M. Musiela (1997) "The market model of interest rate dynamics," *Mathematical Finance* 7, 127–154.
35. Brennan, M.J. and Schwartz, E.S. (1977) "The valuation of American put options," *J. Finance* 32, 449–462.
36. Brigo, D. and Mercurio, F. (2001) *Interest Rate Models: Theory and Practice.* Berlin: Springer–Verlag.

37. Broadie, M. and Detemple, J. (1996) "American option valuation: New bounds, approximations, and a comparison of existing methods," *Review of Financial Studies* 9, 1211–50.
38. Broadie, M. and Glasserman, P. (1997) "Pricing American-style securities by simulation," *J. Economic Dynamics and Control* 21, 1323–1352.
39. Campbell, J.; Lo, A.; and A. MacKinlay (1997) *The Econometrics of Financial Markets.* Princeton, NJ: Princeton Press.
40. Cannon, J.R. (1984) *The One-Dimensional Heat Equation*, v. 23 in *Encyclopedia of Mathematics and Its Applications*, G. Rota, ed. Menlo Park, CA: Addison-Wesley.
41. Carr, P.; Jarrow, R.; and R. Myneni (1992) "Alternative characterizations of American put options," *Mathematical Finance* 2, 87–106.
42. Carr, P. and Madan, D. (1998) "Option valuation using the fast Fourier transform," *J. Computational Finance* 2, 61–73.
43. Carr, P. and Wu, L. (2003) "The finite moment log stable process and option pricing," *J. Finance* 58, 753–777.
44. Carverhill, A.P. and Clewlow, L.J. (1990) "Flexible convolution," *Risk* 3(4), 25–29.
45. Chernov, M. and Ghysels, E. (1998) "A study towards a unified approach to the joint estimation of objective and risk neutral measures for the purpose of options valuation," *J. Financial Economics* 56, 407–458.
46. Chou, R.Y. (1988) "Volatility persistence and stock valuations: Some empirical evidence using GARCH," *J. Applied Econometrics* 3, 279–294.
47. Christie, A.A. (1982) "The stochastic behavior of common stock variances: Value, leverage and interest rate effects," *J. Financial Economics* 10, 407–432.
48. Chung, K.L. (1974) *A Course in Probability Theory.* New York: Academic Press.
49. Chung, K.L. and Williams, R.J. (1983) *Introduction to Stochastic Integration.* Boston: Birhäuser.
50. Clark, P.K. (1973) "A subordinated stochastic process model with finite variance for speculative prices," *Econometrica* 31, 135–155.
51. Clewlow, L. and Carverhill, A. (1994) "Quicker on the curves," *Risk* 7(5), 63–64.
52. Collatz, L. (1966) *The Numerical Treatment of Differential Equations.* Berlin: Springer-Verlag.
53. Conze, A. and Viswanathan (1991) "Path dependent options: The case of lookback options," *J. Finance* 46, 1893–1907.
54. Cootner, P. (1964) *The Random Character of Stock Market Prices.* Cambridge (US): MIT Press.
55. Cox, D. and Miller, H. (1972) *The Theory of Stochastic Processes.* London: Chapman & Hall.
56. Cox, D. and Oakes, D. (1984) *Analysis of Survival Data.* London: Chapman & Hall.
57. Cox, J. (1975) "Notes on option pricing I: Constant elasticity of variance diffusions," working paper, Stanford University.

58. Cox, J.; Ingersoll, J.; and S. Ross (1985) "A theory of the term structure of interest rates," *Econometrica* 53, 385–408.
59. Cox, J. and Rubinstein, M. (1985) *Options Markets.* Englewood Cliffs, NJ: Prentice-Hall.
60. Cox, J. and Ross, S. (1976) "The valuation of options for alternative stochastic processes," *J. Financial Economics* 3, 145–66.
61. Cox, J.; Ross, S.; and M. Rubinstein (1979) "Option pricing: A simplified approach," *J. Financial Economics* 7, 229–63.
62. Crank, J. and Nicolson, P. (1947) "A practical method for numerical evaluation of solutions of partial differential equations of the heat conduction type," *Proceedings of Cambridge Philosophical Society* 43, 50–67.
63. D'Agostino, R.B. and Stephens, M.A. (1986) *Goodness-of-Fit Techniques.* New York: Marcel Dekker.
64. Darroch, J.N. and Morris, K.W. (1968) "Passage-time generating functions for continuous-time finite Markov chains," *J. Applied Probability* 5, 414–426.
65. Davis, H.F. (1963) *Fourier Series and Orthogonal Functions.* New York: Dover.
66. Debreu, G. (1959) *Theory of Value.* Cowles Foundation Monograph 17. New Haven: Yale Univ. Press.
67. Delbaen, F. and Schachermayer, W. (1994) "A general version of the fundamental theorem of asset pricing," *Mathematische Annalen* 300, 463–520.
68. Derman, E. and Kani, I. (1994) "Riding on a smile," *Risk* 7(2), 32–39.
69. Dion, J.P. and Epps, T.W. (1999) "Stock prices as branching processes in random environments: Estimation," *Communications in Statistics, Simulation and Computation* 28, 957–976.
70. Dubofsky, D.A. (1991) "Volatility increases subsequent to NYSE and AMEX stock splits," *J. Finance* 46, 421–431.
71. Duffie, D.; Pan, J.; and K. Singleton (2000) "Transform analysis and asset pricing for affine jump-diffusions," *Econometrica* 68, 1343–1376.
72. Duffie, D. and Singleton, K. (1999) " Modeling term structures of defaultable bonds," *Review of Financial Studies* 12, 687–720.
73. Duffie, D. and Singleton, K. (2003) *Credit Risk: Pricing, Measurement, and Management.* Princeton: Princeton Univ. Press.
74. Dumas, B.; Fleming, J.; and R.E. Whaley (1998) "Implied volatility functions: Empirical tests," *J. Finance* 53, 2059–2106.
75. Dupire, B. (1994) "Pricing with a smile," *Risk* 7(1), 18–20.
76. Durrett, R. (1984) *Brownian Motion and Martingales in Analysis.* Belmont, CA: Wadsworth.
77. Durrett, R. (1996) *Stochastic Calculus: A Practical Introduction.* Boca Raton, FL: CRC Press.
78. Eberlein, E. and Keller, U. (1995) "Hyperbolic distributions in finance," *Bernoulli* 1, 281–299.
79. Eberlein, E.; Keller, U.; and K. Prause (1998) "New insights into smile, mispricing, and value at risk: The hyperbolic model," *J. Business* 71, 371–407.

80. Edwards, C. (2004) *Option Pricing with Continuous-time Markov Chain Regime Switching.* Ph.D. dissertation, University of Virginia.
81. Edwards, C. (2005) "Derivative pricing models with regime switching: A general approach," *J. Derivatives* 13, 41–47.
82. Einstein, A. (1905) "On the movement of small particles suspended in a stationary liquid demanded by the molecular kinetic theory of heat," *Annals of Physics* 17.
83. Elliott, R.J. (1982) *Stochastic Calculus and Its Applications.* Berlin: Springer-Verlag.
84. Elliott, R.J. and Kopp, P.E. (1999) *Mathematics of Financial Markets.* New York: Springer-Verlag.
85. Epps, T.W. and Pulley, L.B. (1983) "A test for normality based on the empirical characteristic function," *Biometrika* 70, 723–726.
86. Epps, T.W.; Pulley, L.B.; and D.B. Humphrey (1996) "Assessing the FDIC's premium and examination policies using 'Soviet' put options," *J. Banking and Finance* 20, 699–721.
87. Epps, T.W. (1996) "Stock prices as branching processes," *Communications in Statistics, Stochastic Models* 12, 529–558.
88. Eraker, B.; Johannes, M.; and N. Polson (2003) "The impact of jumps in volatility and returns," *J. Finance* 58, 1269–1300.
89. Eraker, B. (2004) "Do stock prices and volatility jump? Reconciling evidence from spot and option prices," *J. Finance* 59, 1367–1403.
90. Fama, E. (1965) "The behavior of stock market prices," *J. Business* 38, 34–105.
91. Fang, H. (2002) *Jumps with a Stochastic Jump Rate.* Ph.D. dissertation, University of Virginia.
92. Feller, W. (1951) "Two singular diffusion problems," *Annals of Mathematics* 54, 173–182.
93. Feller, W. (1971) *An Introduction to Probabilty Theory and Its Applications* v.2, New York: John Wiley.
94. Feynman, R.J. (1948) "Space-time approach to nonrelativistic quantum mechanics," *Review of Modern Physics* 20, 367–387.
95. Finlay, R. and Seneta, E. (2006) "Stationary-increment Student and variance-gamma processes," *J. Applied Probability* 43, 1–13.
96. Fishman, G. (1996) *Monte Carlo: Concepts, Algorithms, and Applications.* New York: Springer-Verlag.
97. Fu, M.C.; Madan, D.B.; and T. Wang (1999) "Pricing continuous Asian options: A comparison of Monte Carlo and Laplace transform inversion methods," *J. Computational Finance* 2, 49–74.
98. Garcia, D. (2003) "Convergence and biases of Monte Carlo estimates of American option prices using a parametric exercise rule," *J. Economic Dynamics and Control* 27, 1855–1879.
99. Geman, H. and Eydeland, A. (1995) "Domino effect," *Risk* 8(4), 65–67.
100. Geman, H. and Yor, M. (1992) "Quelques relations entre processus de Bessel, options asiatiques, et fonctions confluentes hypergéométriques," *Compte Rendu Académie des Sciences Paris, Series I*, 471–474.

101. Geman, H. and Yor, M. (1993) "Bessel processes, Asian options and perpetutities," *Mathematical Finance* 3, 349–375.
102. Geske, R. (1979) "The valuation of compound options," *J. Financial Economics* 7, 63–81.
103. Geske, R. and H. Johnson (1984) "The American put option valued analytically," *J. Finance* 39, 1511–24.
104. Ghysels, E.; Harvey, A.C.; and E. Renault (1996) "Stochastic volatility," in G.S. Maddala and C.R. Rao, eds., *Handbook of Statistics* 14, 119–191. Amsterdam: Elsevier.
105. Glasserman, P. (2004) *Monte Carlo Methods in Financial Engineering*. New York: Springer.
106. Goldman, B.; Sosin, H.; and M. Gatto (1979) "Path dependent options: Buy at the low, sell at the high," *J. Finance* 34, 1111–1128.
107. Gradshteyn, I. and Ryzhik, I. (1980) *Table of Integrals, Series, and Products*. New York: Academic Press.
108. Grandell, J. (1976) *Doubly Stochastic Poisson Processes*. Berlin: Springer-Verlag.
109. Hall, C.A. and Porsching, T.A. (1990) *Numerical Analysis of Partial Differential Equations*. Englewood Cliffs, NJ: Prentice-Hall.
110. Hamilton, J. (1989) "A new approach to the economic analysis of nonstationary time series and the business cycle," *Econometrica* 57, 357–384.
111. Hammersley, J. and Handscomb, D. (1964) *Monte Carlo Methods*. London: Methuen.
112. Harris, T.E. (1963) *The Theory of Branching Processes*. Berlin: Springer-Verlag.
113. Harrison, J.M. and Kreps, D.M. (1979) "Martingales and arbitrage in multiperiod securities markets," *J. Economic Theory* 20, 381–408.
114. Harrison, J.M. and Pliska, S.R. (1981) "Martingales and stochastic integrals in the theory of continuous trading," *Stochastic Processes and Their Applications* 11, 215–260.
115. Heath, D.C.; Jarrow, R.A.; and A. Morton (1992) "Bond pricing and the term structure of interest rates: A new methodology for contingent claim valuation," *Econometrica* 60, 77–105.
116. Heston, S.L. (1993) "A closed-form solution for options with stochastic volatility with applications to bond and currency options," *Review of Financial Studies* 6, 327–343.
117. Ho, T.S.Y. and Lee, S.B. (1986) "Term structure movements and pricing interest rate contingent claims," *J. Finance* 41, 1011–1029.
118. Hobson, D.G. and Rogers, L.C.G. (1998) "Complete models with stochastic volatility," *Mathematical Finance* 8, 27–48.
119. Hoeffding, W. (1940) "Masstabvinvariante Korrelantionstheorie," *Schriften des Mathematischen Institut Angewandte Mathematik der Univ. Berlin* 5, 179–233.
120. Hull, J. and White, A. (1987) "The pricing of options on assets with stochastic volatilities," *J. Finance* 42, 281–300.

121. Hull, J. and White, A. (1990) "Pricing interest-rate derivative securities," *Review of Financial Studies* 3, 573–592.
122. Jackwerth, J.C. and Rubinstein, M. (1996) "Recovering probability distributions from option prices," *J. Finance* 51, 1611–1631.
123. Jarrow, R.; Lando, D.; and S. Turnbull (1997) "A Markov model for the term structure of credit risk spreads," *Review of Financial Studies* 10, 481–523.
124. Jarrow, R. and Rosenfeld, E. (1984) "Jump risks and the intertemporal capital asset pricing model," *J. Business* 57, 337–351.
125. Joshi, M. (2003) *The Concepts and Practice of Mathematical Finance.* Cambridge (UK): Cambridge Press.
126. Ju, N. (1998) "Pricing an American option by approximating its early exercise boundary as a multipiece exponential function," *Review of Financial Studies* 11, 627–646.
127. Kac, M. (1949) "On distributions of certain Wiener functionals," *Transactions of American Mathematical Society* 65, 1–13.
128. Karatzas, I. and Shreve, S. (1991) *Brownian Motion and Stochastic Calculus*, 2d. ed. Berlin: Springer.
129. Karlin, S. and Taylor, H.M. (1975) *A First Course in Stochastic Processes.* New York: Academic Press.
130. Kim, I.J. (1990) "The analytic valuation of American options," *Review of Financial Studies* 3, 547–572.
131. Kochard, L.E. (1999) *Option Pricing and Higher Order Moments of the Risk-Neutral Probability Density Function*, Ph.D. dissertation, University of Virginia.
132. Kolb, R.W. and Overdahl, J.A. (2006) *Understanding Futures Markets.* Malden, MA: Blackwell.
133. Kolmogorov, A.N. (1950) *Foundations of Probability Theory.* New York: Chelsea Pub. Co.
134. Kotz, S., Johnson, N.L., and C.B. Read, eds. (1982) *Encyclopedia of Statistical Sciences* v.1. New York: John Wiley.
135. Kullback, S. (1983) "Kullback information," in S. Kotz *et al.* (1983) *Encyclopedia of Statistical Sciences* v.4.
136. Kushner, H.J. and Dupuis, P.G. (1992) *Numerical Methods for Stochastic Control Problems in Continuous Time.* New York: Springer-Verlag.
137. Laha, R.G. and Rohatgi, V.K. (1979) *Probability Theory* . New York: John Wiley.
138. Lando, D. (1994) *Three Essays on Contingent Claims Pricing.* Ph.D. dissertation, Cornell University.
139. Lando, D. (1998) "On Cox processes and credit risky securities," *Review of Derivatives Research* 2, 99–120.
140. Lando, D. (2004) *Credit Risk Modeling.* Princeton: Princeton Univ. Press.
141. Levy, E. (1992) "Pricing European average rate currency options," *J. International Money and Finance* 11, 474–491.
142. Levy, E. and Turnbull, S. (1992) "Average intelligence," *Risk* 5(2), 53–59.

143. Lintner, J. (1965) "The valuation of risky assets and the selection of risky investments in stock portfolios and capital budgets," *Review of Economics and Statistics* 47, 13–37.
144. Liu, W. (2003) *Option pricing with pure jump models*, Ph.D. dissertation, University of Virginia.
145. Longstaff, F. (1990) "Pricing options with extendible maturities: Analysis and applications," *J. Finance* 45, 935–57.
146. Longstaff, F. and Schwartz, E. (2001) "Valuing American options by simulation: A simple least-squares approach," *Review of Financial Studies* 14, 113–147.
147. Lukacs, E. (1970) *Characteristic Functions.* London: Griffin.
148. MacMillan, L.W. (1986) "Analytic approximation for the American put option," *Advances in Futures and Options Research* 1, 119–139.
149. Madan, D.B. (1999) "Purely discontinuous asset price processes," working paper, University of Maryland.
150. Madan, D.B.; Carr, P.; and E.C. Chang (1998) "The variance gamma model and option pricing," *European Finance Review* 2, 79–105.
151. Madan, D.B. and Milne, F. (1991) " Option pricing with V.G. martingale components," *Mathematical Finance* 1, 39–55.
152. Madan, D.B. and Seneta, E. (1990) "The variance gamma (V.G.) model for share market returns," *J. Business* 63, 511–524.
153. Mandelbrot, B. (1963) "The variation of certain speculative prices," *J. Business* 36, 394–419.
154. Mandelbrot, B. and Hudson, R. (2004) *The (Mis)behavior of Markets*, New York: Basic Books.
155. Mardia, K.V. (1970) "Measures of multivariate skewness and kurtosis with applications," *Biometrika* 57, 519–530.
156. Marsaglia, G. (1995) *The Marsaglia Random Number CDROM including the Diehard Battery of Tests of Randomness*, Department of Statistics, Florida State University.
157. McCulloch, J.H. (1996) "Financial applications of stable distributions," in G.S. Maddala and C.R. Rao, eds., *Handbook of Statistics* 14, 393–425. Amsterdam: Elsevier.
158. McDonald, J.B. (1996) "Probability distributions for financial models," in G.S. Maddala and C.R. Rao, eds., *Handbook of Statistics* 14, 427–461. Amsterdam: Elsevier.
159. McLachlan, N.W. (1955) *Bessel Functions for Engineers.* Oxford: Clarendon.
160. Mee, R.W. and Owen, D.B. (1983) "A simple approximation for bivariate normal probabilities," *J. Quality Technology* 15, 72–75.
161. Merton, R.C. (1973) "The theory of rational option pricing," *Bell J. Economics and Management Science* 4, 141–183.
162. Merton, R.C. (1976) "Option pricing when underlying stock returns are discontinuous," *J. Financial Economics* 3, 125–144.
163. Merton, R.C. (1978) "On the cost of deposit insurance when there are surveillance costs," *J. Business* 51, 439–452.

164. Merton, R.C. (1990) *Continuous Time Finance.* Cambridge (US): Blackwell.
165. Milevsky, M.A. and Posner, S.E. (1998) "Asian options, the sum of lognormals, and the reciprocal gamma distribution," *J. Financial and Quantitative Analysis* 33, 409–422.
166. Miller, M.H. and Modigliani, F. (1961) "Dividend policy, growth and the valuation of shares," *J. Business* 34, 411–433.
167. Miranda, M.J. and Fackler, P.L. (2002) *Applied Computational Economics and Finance.* Cambridge (US): MIT Press.
168. Mitchell, A.R. and Griffiths, D.F. (1980) *The Finite Difference Method in Partial Differential Equations.* Chichester: John Wiley.
169. Modigliani, F. and Miller, M.H. (1958) "The cost of capital, corporation finance and the theory of investment," *American Economic Review* 48, 261–297.
170. van Moerbeke, P. (1976) "On optimal stopping and the free boundary problem," *Archives of Rational Mechanical Analysis* 60, 101–148.
171. Moskowitz, H. and Tsai, H. (1989) "An error-bounded polynomial approximation for bivariate normal probabilities," *Communications in Statistics, Simulation and Computation* 18, 1421–1437.
172. Mossin, J. (1966) "Equilibrium in a capital asset market," *Econometrica* 35, 768–783.
173. Musiela, M. and Rutkowski, M. (1998, 2005) *Martingale Methods in Financial Modelling,* 1st & 2d eds. Heidelberg: Springer-Verlag.
174. Naik, V. (1993) "Option valuation and hedging strategies with jumps in the volatility of asset returns," *J. Finance* 48, 1969–1984.
175. Naik, V. and Lee, M. (1990) "General equilibrium pricing of options on the market portfolio with discontinuous returns," *Review of Financial Studies* 3, 493–521.
176. Neyman, J. (1937) " 'Smooth' tests for goodness-of-fit," *Skandinavisk Aktuarietidskrift* 20, 149–199.
177. Officer, R.R. (1972) "The distribution of stock returns," *J. American Statistical Association* 67, 807–812.
178. Ohlson, J.A. and Penman, S.H. (1985) "Volatility increases subsequent to stock splits: An empirical aberration," *J. Financial Economics* 14, 251–266.
179. Palm, F.C. (1996) "GARCH models of volatility," in G.S. Maddala and C.R. Rao, eds., *Handbook of Statistics* 14, 209–240. Amsterdam: Elsevier.
180. Pennacchi, G.B. (1987a) "A reexamination of the over- (or under-) pricing of deposit insurance," *J. Money, Credit, and Banking* 19, 340–360.
181. Pennacchi, G.B. (1987b) "Alternative forms of deposit insurance," *J. Banking and Finance* 11, 291–312.
182. Praetz, P.D. (1972) "The distribution of share price changes," *J. Business* 45, 49–55.
183. Press, J. (1967) "A compound events model for security prices," *J. Business* 40, 317–35.
184. Press, W.H.; Flannery, B.P.; Teukolsky, S.A.; and W.T. Vettering (1986) *Numerical Recipes: The Art of Scientific Computing (FORTRAN Version).* Cambridge (UK): Cambridge Press.

185. Press, W.H.; Flannery, B.P.; Teukolsky, S.A.; and W.T. Vettering (1992a) *Numerical Recipes in FORTRAN: The Art of Scientific Computing*, 2nd. edn. Cambridge (UK): Cambridge Press.
186. Press, W.H.; Flannery, B.P.; Teukolsky, S.A.; and W.T. Vettering (1992b) *Numerical Recipes in C: The Art of Scientific Computing*, 2d ed. Cambridge (UK): Cambridge Press.
187. Protter, P. (1990) *Stochastic Integration and Differential Equations: A New Approach*. Berlin: Springer-Verlag.
188. Qiu, R. (2004) *Put Warrants: How They Are Valued and How They Affect Managerial Incentives*. Ph.D. dissertation, University of Virginia.
189. Rebonato, R. (1999) "Calibrating the BGM model," *Risk* 12, 74–79.
190. Relton, F.E. (1946) *Applied Bessel Functions*. London: Blackie & Son.
191. Renault, E. and Touzi, N. (1996) "Option hedging and implied volatilities in a stochastic volatility model," *Mathematical Finance* 6, 279–302.
192. Revuz, D. and Yor, M. (1991) *Continuous Martingales and Brownian Motion*. Berlin: Springer–Verlag.
193. Romano, M. and Touzi, N. (1997) "Contingent claims and market completeness in a stochastic-volatility model," *Mathematical Finance* 7, 399–412.
194. Ross, S.M. (2000) *Introduction to Probability Models*. San Diego, CA: Academic Press.
195. Rowell, D. (2004) "Computing the matrix exponential: The Cayley-Hamilton method," class notes for "Advanced System Dynamics and Control," Dept. of Mechanical Engineering, Massachusetts Institute of Technology: web.mit.edu/2.151/www/Handouts/CayleyHamilton.pdf.
196. Royden, H.L. (1968) *Real Analysis*. New York: Macmillan.
197. Rubinstein, M. (1983) "Displaced diffusion option pricing," *J. Finance* 38, 213–217.
198. Rubinstein, M. (1984) "A simple formula for the expected rate of return of an option over a finite holding period," *J. Finance* 39, 1503–1509.
199. Rubinstein, M. (1994) "Implied binomial trees," *J. Finance* 49, 771–818.
200. Rudin, W. (1976) *Principles of Mathematical Analysis*. New York: McGraw-Hill.
201. Schmalensee, R. and Trippi, R.R. (1978) "Common stock volatility expectations implied by option premia," *J. Finance* 33, 129–147.
202. Schott, J.R. (1997) *Matrix Analysis for Statistics*. New York: John Wiley.
203. Schroder, M. (1989) "Computing the constant elasticity of variance option pricing formula," *J. Finance* 44, 211–219.
204. Scott, L. (1987) "Option pricing when the variance changes randomly: Theory, estimation, and an application," *J. Financial and Quantitative Analysis* 22, 419–438.
205. Scott, L. (1997) "Pricing stock options in a jump-diffusion model with stochastic volatility and interest rates: Applictions of Fourier inversion methods," *Mathematical Finance* 7, 413–426.
206. Sharpe, W.F. (1964) "Capital asset prices: A theory of market equilibrium under conditions of risk," *J. Finance* 19, 425–442.

207. Shaw, W.T. (1998) *Modelling Financial Derivatives with Mathematica.* Cambridge (UK): Cambridge Press.
208. Shimko, D. (1992) *Finance in Continuous Time.* Miami: Kolb Publishing.
209. Shreve, S.E. (2004) *Stochastic Calculus for Finance II.* New York: Springer Science+Business Media.
210. Smith, W.L. and Wilkinson, W. (1969) "On branching processes in random environments". *Annals of Mathematical Statistics* 40, 814–827.
211. Sprott, J.C. (1991) *Numerical Recipes Routines and Examples in Basic.* Cambridge (UK): Cambridge Press.
212. Stein, E.M. and Stein, J.C. (1991) "Stock price distributions with stochastic volatility: An analytic approach," *Review of Financial Studies* 4, 727–752.
213. Stern, M. (1993) *Estimating and Testing the Poisson Jump-Diffusion Model for Options on Stocks.* Ph.D. dissertation, University of Virginia.
214. Stoll, H. (1969) "The relationship between put and call option prices," *J. Finance* 24, 801–824.
215. Stultz, R. (1982) "Options on the minimum or maximum of two risky assets: Analysis and applications," *J. Financial Economics* 9, 383–407.
216. Stutzer, M. (1996) "A simple nonparametric approach to derivative security valuation," *J. Finance* 51, 1633–1652.
217. Taylor, S.J. (1966) *Introduction to Measure and Integration.* Cambridge (UK): Cambridge Press.
218. Taylor, S. and Xu, X. (1994) "The term structure of volatility implied by foreign exchange options," *J. Financial and Quantitative Analysis* 29, 57–74.
219. Vasicek, O. (1977) "An equilibrium characterization of the term structure," *J. Financial Economics* 5, 177–188.
220. Waller, L.; Turnbull, B.; and J.M. Hardin (1995) "Obtaining distribution functions by numerical inversion of characteristic functions with applications," *The American Statistician* 49, 346–350.
221. Watson, G.N. (1944) *Theory of Bessel Functions.* Cambridge (UK): Cambridge Press.
222. Wiggins, J.B. (1987) "Option values under stochastic volatility," *J. Financial Economics* 19, 351–372.
223. Williams, D. (1991) *Probability with Martingales.* Cambridge (UK): Cambridge Press.
224. Williams, J.B. (1938) *The Theory of Investment Value.* Cambridge (US): Harvard U. Press.
225. Williams, T. (2001) *Option Pricing and Branching Processes.* Ph.D. dissertation, University of Virginia.
226. Wilmott, P.; Dewynne, J.; and S. Howison (1993) *Option Pricing: Mathematical Models and Computation.* Oxford: Oxford Press.
227. Wilmott, P.; Howison, S.; and J. Dewynne(1995) *The Mathematics of Financial Derivatives: A Student Introduction.* Cambridge (UK): Cambridge Press.

228. Young, M.S. and Graff, R.A. (1995) "Real estate is not normal: A fresh look at real estate return distributions," *J. Real Estate Finance and Economics* 10, 225–259.
229. Zwillinger, D. (1989) *Handbook of Differential Equations.* San Diego: Academic Press.

SUBJECT INDEX

www.ingramcontent.com/pod-product-compliance
Lightning Source LLC
LaVergne TN
LVHW010539300926
849162LV00002B/178